Imperial College London
2407598979
AF839566

Developments in Geotechnical Engineering, 85

Fundamentals of Discrete Element Methods for Rock Engineering

Developments in Geotechnical Engineering, 85

Fundamentals of Discrete Element Methods for Rock Engineering

Theory and Applications

Lanru Jing

Group of Engineering Geology and Geophysics, Department of Land and Water Resources Engineering, Royal Institute of Technology, Stockholm, Sweden

Ove Stephansson

Geo Forschungs Zentrum – Postdam,
Department of Geodynamics,
Postdam, Germany

ELSEVIER

Amsterdam • Boston • Heidelberg • London • New York • Oxford
Paris • San Diego • San Francisco • Singapore • Sydney • Tokyo

Elsevier
Radarweg 29, PO Box 211, 1000 AE Amsterdam, The Netherlands
Linacre House, Jordan Hill, Oxford OX2 8DP

First edition 2007

Copyright © 2007 Elsevier B.V. All rights reserved

No part of this publication may be reproduced, stored in a retrieval system or transmitted in any form or by any means electronic, mechanical, photocopying, recording or otherwise without the prior written permission of the publisher

Permissions may be sought directly from Elsevier's Science & Technology Rights Department in Oxford, UK: phone (+44) (0) 1865 843830; fax (+44) (0) 1865 853333; email: permissions@elsevier.com. Alternatively you can submit your request online by visiting the Elsevier web site at http://elsevier.com/locate/permissions, and selecting *Obtaining permission to use Elsevier material*

Notice
No responsibility is assumed by the publisher for any injury and/or damage to persons or property as a matter of products liability, negligence or otherwise, or from any use or operation of any methods, products, instructions or ideas contained in the material herein. Because of rapid advances in the medical sciences, in particular, independent verification of diagnoses and drug dosages should be made

Library of Congress Cataloging-in-Publication Data
A catalog record for this book is available from the Library of Congress

British Library Cataloguing in Publication Data
A catalogue record for this book is available from the British Library

ISBN: 978-0-444-82937-5
ISSN: 0165-1250

For information on all Elsevier publications
visit our website at books.elsevier.com

Transferred to digital print 2007
Printed and bound by CPI Antony Rowe, Eastbourne

07 08 09 10 11 10 9 8 7 6 5 4 3 2 1

Working together to grow
libraries in developing countries
www.elsevier.com | www.bookaid.org | www.sabre.org
ELSEVIER BOOK AID International Sabre Foundation

Further titles in this series:
Volumes 2, 3, 5–7, 9, 10, 12, 13, 15, 16A, 22 and 26 are out of print

1. G. SANGLERAT – THE PENETROMETER AND SOIL EXPLORATION
4. R. SILVESTER – COASTAL ENGINEERING. 1 AND 2
8. L.N. PERSEN – ROCK DYNAMICS AND GEOPHYSICAL EXPLORATION
   Introduction to Stress Waves in Rocks
11. H.K. GUPTA AND B.K. RASTOGI – DAMS AND EARTHQUAKES
14. B. VOIGHT (Editor) – ROCKSLIDES AND AVALANCHES. 1 and 2
17. A.P.S. SELVADURAI – ELASTIC ANALYSIS OF SOIL-FOUNDATION INTERACTION
18. J. FEDA – STRESS IN SUBSOIL AND METHODS OF FINAL SETTLEMENT CALCULATION
19. Á. KÉZDI – STABILIZED EARTH ROADS
20. E.W. BRAND AND R.P. BRENNER (Editors) – SOFT-CLAY ENGINEERING
21. A. MYSLIVE AND Z. KYSELA – THE BEARING CAPACITY OF BUILDING FOUNDATION
23. P. BRUUN – STABILITY OF TIDAL INLETS – Theory and Engineering
24. Z. BAŽANT – METHODS OF FOUNDATION ENGINEERING
25. Á. KÉZDI – SOIL PHYSICS – Selected Topics
27. D. STEPHENSON – ROCKFILL IN HYDRAULIC ENGINEERING
28. P.E. FRIVIK, N. JANBU, R. SAETERSDAL AND L.I. FINBORUD (Editors) – GROUND FREEZING 1980
29. P. PETER – CANAL AND RIVER LEVÉES
30. J. FEDA – MECHANICS OF PARTICULATE MATERIALS – The Principles
31. Q. ZÁRUBA AND V.MENCL – LANDSLIDES AND THEIR CONTROL
    Second completely revised edition
32. I.W. FARMER (Editor) – STRATA MECHANICS
33. L. HOBST AND J. ZAJÍC – ANCHORING IN ROCK AND SOIL
    Second completely revised edition
34. G. SANGLERAT, G. OLIVARI AND B. CAMBOU – PRACTICAL PROBLEMS IN SOIL MECHANICS AND FOUNDATION ENGINEERING, 1 and 2
35. L. RÉTHÁTI – GROUNDWATER IN CIVIL ENGINEERING
36. S.S. VYALOV – RHEOLOGICAL FUNDAMENTALS OF SOIL MECHANICS
37. P. BRUUN (Editor) – DESIGN AND CONSTRUCTION OF MOUNDS FOR BREAKWATER AND COASTAL PROTECTION
38. W.F. CHEN AND G.Y. BALADI – SOIL PLASTICITY - Theory and Implementation
39. E.T. HANRAHAN – THE GEOTECTONICS OF REAL MATERIALS: THE $\varepsilon_g\varepsilon_k$ METHOD
40. J. ALDORF AND K. EXNER – MINE OPENINGS - Stability and Support
41. J.E. GILLOT – CLAY IN ENGINEERING GEOLOGY
42. A.S. CAKMAK (Editor) – SOIL DYNAMICS AND LIQUEFACTION
43. A.S. CAKMAK (Editor) – SOIL-STRUCTURE INTERACTION
44. A.S. CAKMAK (Editor) – GROUND MOTION AND ENGINEERING SEISMOLOGY
45. A.S. CAKMAK (Editor) – STRUCTURES, UNDERGROUND STRUCTURES, DAMS, AND STOCHASTIC METHODS
46. L. RÉTHÁTI – PROBABILISTIC SOLUTIONS IN GEOTECTONICS
47. B.M. DAS – THEORETICAL FOUNDATION ENGINEERING
48. W. DERSKI, R. IZBICKI, I. KISIEL AND Z. MROZ – ROCK AND SOIL MECHANICS
49. T. ARIMAN, M. HAMADA, A.C. SINGHAL, M.A. HAROUN AND A.S. CAKMAK (Editors) – RECENT ADVANCES IN LIFELINE EARTHQUAKE ENGINEERING
50. B.M. DAS – EARTH ANCHORS
51. K. THIEL – ROCK MECHANICS IN HYDROENGINEERING
52. W.F. CHEN AND X.L. LIU – LIMIT ANALYSIS IN SOIL MECHANICS
53. W.F. CHEN AND E. MIZUNO – NONLINEAR ANALYSIS IN SOIL MECHANICS

54. F.H. CHEN – FOUNDATIONS ON EXPANSIVE SOILS
55. J. VERFEL – ROCK GROUTING AND DIAPHRAGM WALL CONSTRUCTION
56. B.N. WHITTAKER AND D.J. REDDISH – SUBSIDENCE – Occurrence, Prediction and Control
57. E. NONVEILLER – GROUTING, THEORY AND PRACTICE
58. V. KOLÁŘ AND I. NĚMEC – MODELLING OF SOIL-STRUCTURE INTERACTION
59A. R.S. SINHA (Editor) – UNDERGROUND STRUCTURES – Design and Instrumentation
59B. R.S. SINHA (Editor) – UNDERGROUND STRUCTURES – Design and Construction
60. R.L. HARLAN, K.E. KOLM AND E.D. GUTENTAG – WATER-WELL DESIGN AND CONSTRUCTION
61. I. KASDA – FINITE ELEMENT TECHNIQUES IN GROUNDWATER FLOW STUDIES
62. L. FIALOVSZKY (Editor) – SURVEYING INSTRUMENTS AND THEIR OPERATION PRINCIPLES
63. H. GIL – THE THEORY OF STRATA MECHANICS
64. H.K. GUPTA – RESERVOIR-INDUCED EARTHQUAKES
65. V.J. LUNARDINI – HEAT TRANSFER WITH FREEZING AND THAWING
66. T.S. NAGARAI – PRINCIPLES OF TESTING SOILS, ROCKS AND CONCRETE
67. E. JUHÁSOVÁ – SEISMIC EFFECTS ON STRUCTURES
68. J. FEDA – CREEP OF SOILS – and Related Phenomena
69. E. DULÁCSKA – SOIL SETTLEMENT EFFECTS ON BUILDINGS
70. D. MILOVIĆ – STRESSES AND DISPLACEMENTS FOR SHALLOW FOUNDATIONS
71. B.N. WHITTAKER, R.N. SINGH AND G. SUN – ROCK FRACTURE MECHANICS – Principles, Design and Applications
72. M.A. MAHTAB AND P. GRASSO – GEOMECHANICS PRINCIPLES IN THE DESIGN OF TUNNELS AND CAVERNS IN ROCK
73. R.N. YONG, A.M.O. MOHAMED AND B.P. WARKENTIN – PRINCIPLES OF CONTAMINANT TRANSPORT IN SOILS
74. H. BURGER (Editor) – OPTIONS FOR TUNNELING 1993
75. S. HANSBO – FOUNDATION ENGINEERING
76. R. PUSCH – WASTE DISPOSAL IN ROCK
77. R. PUSCH – ROCK MECHANICS ON A GEOLOGICAL BASE
78. T. SAWARAGI – COASTAL ENGINEERING - WAVES, BEACHES, WAVE-STRUCTURE INTERACTIONS
79. O. STEPHANSSON, L. JING AND CHIN-FU TSANG (Editors) – COUPLED THERMO-HYDRO- MECHANICAL PROCESSES OF FRACTURED MEDIA
80. J. HARTLÉN AND W. WOLSKI (EDITORS) – EMBANKMENTS ON ORGANIC SOILS
81. Y. KANAORI (EDITOR) – EARTHQUAKE PROOF DESIGN AND ACTIVE FAULTS
82. A.M.O. MOHAMED AND H.E. ANTIA – GEOENVIRONMENTAL ENGINEERING
83. Z. CHUHAN AND J.P. WOLF (Editors) – DYNAMIC SOIL-STRUCTURE INTERACTION – Current Research in China and Switzerland
84. Y. KANAORI, K. TANAKA AND M. CHIGIRA (Editors) – ENGINEERING GEOLOGICAL ADVANCES IN JAPAN FOR THE NEW MILLENNIUM

To my parents and grandparents with the warmest memories
and
To my beloved wife, Wenli, and daughter, Fei-Fei

Lanru Jing

To my beloved wife Almut and Samuel & Naemi

Ove Stephansson

FOREWORD

In order to coherently design structures built on and in rock masses, some form of model is required so that the future can be predicted, e.g., what will happen if a tunnel with a certain size and shape is constructed at a given orientation and depth in a particular rock mass? For the prediction to be adequate, the model must represent the rock reality to a sufficient extent: the model should contain the necessary physical variables, mechanisms and associated parameters – and be able to simulate the perturbations introduced by engineering activities. We require a model that can accommodate the discontinuous, inhomogeneous, anisotropic and non-elastic behavior of rock masses. In particular, we must be able to model fractures in the rock and the consequential systems of rock blocks which rock masses comprise. For this reason, the subject of discrete element models is crucial to the success of rock engineering structures, all of which, in their different ways, are built to advance civilization.

Accordingly it is important to understand the material in this book because of the requirement that the distinct element models supporting the engineering design must be realistic, and one can audit a model's operation and output only if one understands its content, operation, the significance of the input parameter values and hence the output. Unlike many other engineering disciplines, in rock engineering we can rarely validate the model output, and so more effort has to be expended in ensuring that the computer model is well constructed in terms of the idiosyncrasies of rock masses.

We are fortunate, therefore, that the two authors, Lanru Jing and Ove Stephansson, have found time in their busy schedules to write this authoritative book. They are experts in the subject matter and have the necessary fundamental understanding of the rock mass geometry and mechanics supporting the discrete element codes. The book systematically explains each facet of the subject from first principles. Moreover the clarity with which the book has been written ensures that, on completion of the book, the conscientious reader should be able to share the authors' complete understanding of the subject matter.

As I progressively read the draft version of the book, the sequence of the subjects, the emphasis on ensuring physical laws being satisfied, the erudition, the lucid style and the ever-present precision of the explanations pulling the reader onwards, plus the fact that the overall modeling purpose is to predict the future, reminded me of the philosophy of Aristotle, student of Plato and tutor to Alexander the Great. Aristotle considered that objects had four causes: 'material,' from which an object is made; 'formal,' the pattern in which the material is assembled; 'efficient,' the agent or force creating the object; and 'final,' the purpose of the object. In our context, the 'material' cause is the host rock mass, the 'formal' cause is the rock engineering design based on modeling, the 'efficient' cause is the rock construction process and the 'final' cause is the operational engineering function. Aristotle thought that an explanation which includes all four causes completely captures the significance and reality of the object.

Additionally Aristotle thought that the pull of the future is more important than the push of the past because, in his teleological approach, the nature of an object is inextricably linked to its goal. With distinct element models, we are not building so much on earlier more idealized elastic continuum models but being pulled forward by the need to have new types of models that incorporate fractures more realistically. He also said that 'It is the mark of an instructed mind to be satisfied with that degree of precision that the nature of the subject admits, and not to seek exactness when only an approximation of the truth is possible' – which is an implicit recurring theme in the authors' explanations in this book.

Finally Aristotle considered that happiness is achieved by the use of our intellectual faculties and by practicing intellectual virtue. Thus I hope that all readers, through combining their own enhanced understanding of discrete element models with the overall goal of realistically characterizing rock masses and simulating the physical mechanisms, will be as happy as I am to have read the book from cover to cover.

Professor John A Hudson, *FREng*
Imperial College and Rock Engineering Consultants, UK
President of the International Society for Rock Mechanics (ISRM) 2007–2011

PREFACE

This book is a summary of our collaborative teaching and research efforts over the last two decades relating to the subject of discrete element methods (DEM) and its applications to rock mechanics and rock engineering. We have not intended this book to be a thorough presentation of the most current cutting-edge research subjects in the fields of DEM and its applications in geosciences and geoengineering – because the advances are occurring too rapidly. Instead we present some fundamental concepts behind the basic theories and tools of DEM, its historical development and its wide scope of applications in geology, geophysics and rock engineering. We hope that, with this moderate ambition, it may be more helpful and useful for a larger number of practicing engineers, students and researchers alike and serve as a starting platform for more advanced and in-depth further studies.

Unlike almost all books available on the general subject of DEM, this book includes coverage of both explicit and implicit DEM approaches, namely the distinct element methods and discontinuous deformation analysis (DDA) for both rigid and deformable blocks and particle systems, and also the discrete fracture network (DFN) approach for fluid flow simulations. Actually the latter is also a discrete approach of importance for rock mechanics and rock engineering. In addition, brief introductions to some alternative approaches are also provided, such as percolation theory and Cosserat micro-mechanics equivalence to particle systems, which often appear hand in hand with the DEM in the literature.

We provide a presentation of the fundamentals of the governing equations of the discrete systems concerning motion, deformation, fluid flow and heat transfer, to an extent that is currently considered in some available DEM codes. Special attention is given to constitutive models of rock fractures and fracture system characterization methods. These two issues are the basic building blocks of DEM and also have the most significant impacts on the performance and uncertainty of the DEM models.

The DEMs were pioneered and continuously developed by Dr P. A. Cundall, the creator of the distinct element methods, and Dr Genhua Shi, the creator of DDA, block theory and numerical manifold method. The authors learned DEM basics from them and received continuous inspirations and encouragements. This book presents only some basics of the distinct element methods and discontinuous deformation analysis, only small parts of their outstanding contributions.

The first author expresses his special gratitude to Professor Xuefu Yu in China, who led the author into the fields of rock mechanics and numerical modeling in late 1970s and provided continuous guidance and encouragements.

We would like to thank friends at Itasca Consulting Group Ltd, especially Dr R. D. Hart and Dr L. Lorig, for their continued support and kindness over the past decades.

We would especially like to thank our former and current doctoral students, especially F. Lanaro, K.-B. Min, T. Koyama and A. Baghbanan, who have contributed a great deal to some of the important results presented in this book and have helped with the graphics as well.

For the presentation of the explicit DEM methods (Chapter 8), we rely heavily on the codes of the Itasca Consulting Group Ltd, i.e., UDEC, 3DEC and PFC codes, due to the fact that these codes have represented the main stream of development and application of the explicit DEM approach since the early 1970s and much of our work has been developed using this approach and these codes. Extensive use of material from the manuals of the codes has therefore been inevitable. However, we have tried to focus on the fundamental concepts behind the algorithms and coding techniques as much as possible and to avoid specific code features. Additional material concerning the basics of DEM, which come from sources different from the Itasca codes or publications, has been inserted at appropriate places. We hope that, by including these, a more balanced presentation to the subject as a whole has been achieved.

The first author would like to thank Dr J. Harrison and Mrs M. Knox in the Rock Mechanics Group at Imperial College, London, and Prof. G. Dresen and Dr T. Backers at GFZ, Potsdam, Berlin, for their hospitality and help during his two sabbatical visits to their groups when writing parts of this book. Guidance, encouragement and fruitful discussions from Dr Genhua Shi, the creator of block theory and DDA, are especially appreciated.

The second author started working on the book during his sabbatical stay at the Department of Engineering Geology, Technical University of Berlin. The kind hospitality provided by Prof. J. Tiedemann and Drs M. Alber and D. Marioni is gratefully acknowledged.

The authors would like to express their most sincere gratitude especially to Prof. John A. Hudson for his considerable efforts in checking the English, correcting errors and commenting on technical contents throughout all chapters of this book in great detail and writing the Foreword. His contribution to this book is tremendous and much appreciated.

This book would not have been completed without the devoted love, understanding and support from our families.

Lanru Jing and Ove Stephansson
December 2006, Stockholm

CONTENTS

1 INTRODUCTION

This book is about the fundamentals and some application cases of the discrete element methods (DEM). The main reason for the general difficulties in modeling rock masses, by whatever numerical method, is that rock is a natural geological material and so its physical and engineering properties cannot be established or defined through a manufacturing process as for metals or plastic materials. The rock masses are largely DIANE (discontinuous, inhomogeneous, anisotropic and non-elastic) in nature (Harrison and Hudson, 2000). Rock masses are pre-loaded, i.e. they are under stress and continuously affected by dynamic movements in the upper crust of the Earth, such as tectonic movements, earthquakes, land uplifting/subsidence, glaciation cycles and tides, in addition to gravity. A rock mass is also a fractured porous medium containing fluids in either liquid or gas phase (e.g., water, oil, natural gases and air), under complex *in situ* conditions of stresses, temperature and fluid pressures. The combination of constituents and its long history of formation make rock masses complex materials for mathematical modeling in closed forms, and numerical modeling becomes inevitable for the design and performance assessments of rock engineering projects. Knowledge of the coupled thermo-hydro-mechanical and chemical (T-H-M-C) processes about evolutions of geometrical structures and constitutive behavior of the fractured rock masses under combined static/dynamic loading conditions, fluid flow and pressure, temperature gradients and geochemical reactions becomes essential for the reliable solutions of many rock engineering problems, especially when impacts on environment are required to be understood.

To adequately capture the complex physical and/or geochemical aspects of fractured rocks and the effects of perturbations introduced by engineering, a numerical method should have the capability to represent the system geometry (especially the fracture system geometry or its effects), boundary and initial conditions, the natural and induced loading/perturbation histories, adequate constitutive laws for both the rock matrix and fractures, including scale and time effects. For projects with environmental impacts, the necessary coupled physical and chemical process models must be considered, and the problems need to be represented essentially in three-dimensional (3D) space.

Such ‘all-encompassing’ numerical models do not exist today – mainly because of our limited knowledge concerning the physical behavior of rock fractures and fractured rock masses, our limited means to represent the geometry and evolution of complex rock fracture systems and our still limited computational capacity for large or very large scale problems, even given the perpetual increase of our computing capacity nowadays. For many practical problems with complex structural conditions, numerical modeling is still largely a tool for conceptual understanding, providing guiding ideas for design and operation of rock engineering structures where there is a large degree of uncertainty involved, and generic studies which can provide more in-depth understanding of the fundamental behavior of rock masses. However, for ‘simpler’ rock engineering projects, such as tunnel design and slope stability analysis in which there is adequate information about the behavior of rock materials and fractures, numerical modeling has already become a valuable and reliable design tool. The 'model' and the 'computer' are now integral components in rock mechanics and rock engineering. The numerical methods and computing techniques have become suitable and efficient tools for formulating and testing conceptual models and mathematical theories that can integrate diverse information about geology, physics, construction technique, economy and their interactions and, last but not the least, their impacts on the environment on one compact modeling and decision-making platform. This achievement has greatly enhanced the development of modern rock mechanics from the traditional 'empirical' art of rock strength estimation and support design towards the rationalism of modern continuum mechanics, governed by and established on the three basic principles of physics: mass, momentum and energy conservation with the necessary thermodynamic constraints.

1.1 Characteristics of Fractured Rock Masses

The rock masses in the Earth's crust consist of two major components: the intact rock matrix and discontinuities. The natural discontinuities include faults, joints, dykes, fracture zones, bedding planes and other types of weakness surfaces or interfaces, and these have a significant effect on the strength, deformability and permeability of fractured rock masses as a whole. The stability and service performance of rock engineering facilities, or the permeability of oil or geothermal reservoirs in fractured rock masses, are affected by the mechanical, thermal and hydraulic properties of these discontinuities, sometimes dramatically. It is the presence of these natural discontinuities that makes a fractured rock mass a complex material that behaves quite differently from its corresponding intact matrix.

Rock masses contain discontinuities of different sizes. Large-scale geological structures, such as faults, dykes or fracture zones, usually extend tens or hundreds of meters or even kilometers, in dimension, and typically have tectonic origins (e.g., faulting). They are usually manifested in engineering problems in limited numbers (Fig. 1.1). The discontinuities at microscopic scales (e.g., grain boundaries or micro-cracks) are distributed more randomly in rock matrices and in extremely large numbers, and their effect on the

(a) (b)

Fig. 1.1 (a) A fault intersecting an open-pit quarry slope in central England with observable trace length larger than 50 m; (b) a fracture zone in Tai-Shan, China, with trace lengths larger than 100 m.

Fig. 1.2 Natural fracture systems as observed on a natural rock cliff (a) and on an excavated surface (b), with trace lengths up to about 30 m in Tai-Shan, China.

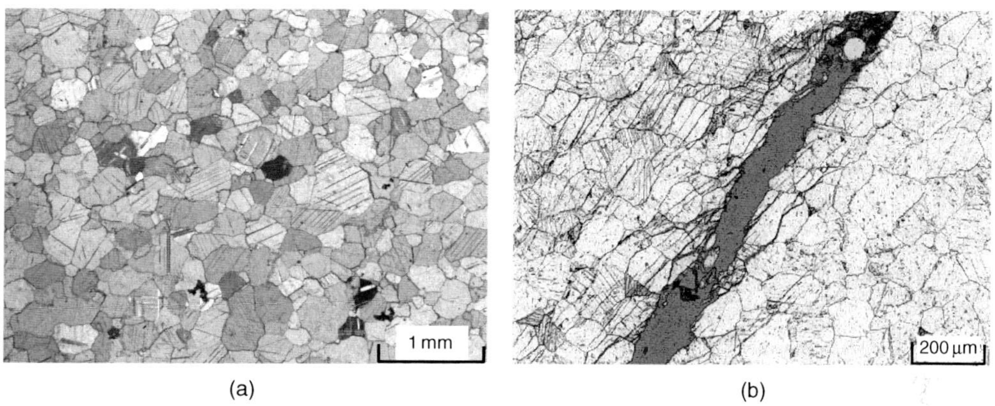

Fig. 1.3 Micrographs of a Carrara marble specimen (a) and a microcrack created during a fracture toughness test (b). Courtesy of Tobias Backer at GFZ, Germany.

behavior of intact rock is often assumed to be included in laboratory experimental results with samples of standard dimensions. The discontinuities whose dimensions lie between the above two extreme cases are usually joints, bedding planes, foliations or artificial cracks caused by engineering events such as blasting. Their linear sizes usually range from centimeters to tens of meters (Fig. 1.2). These discontinuities often appear in sets in terms of their clustered orientations. They often appear in large numbers in rocks and cut the rocks into blocks of complicated shapes. The presence of these sets of discontinuities makes the rock mass heterogeneous in configuration, highly discontinuous and non-linear in its behavior with respect to mechanical deformation and fluid flow. Due to their large numbers and complexities in geometry, their incorporation into numerical models is a challenging task. Figure 1.3 illustrates an example of the micro-structure of a marble specimen with varying grain sizes (Fig. 1.3a) and with an artificial microcrack (Fig. 1.3b).

The fractures present in rocks are not necessarily all natural features created by geological processes. They can also be created by human activities such as excavation and blasting, in both hard and soft rocks. Such fractures are the main ingredients of the so-called Excavation-induced Damage (or Disturbance) Zone (EDZ), which causes changes in the deformability and the pore structures and porosity (therefore permeability) in the rock matrix of an EDZ. Figure 1.4 below shows the excavation-induced fractures in a EDZ zone at an intersection of an old railway tunnel and an experimental drift at Tournemire, Southern

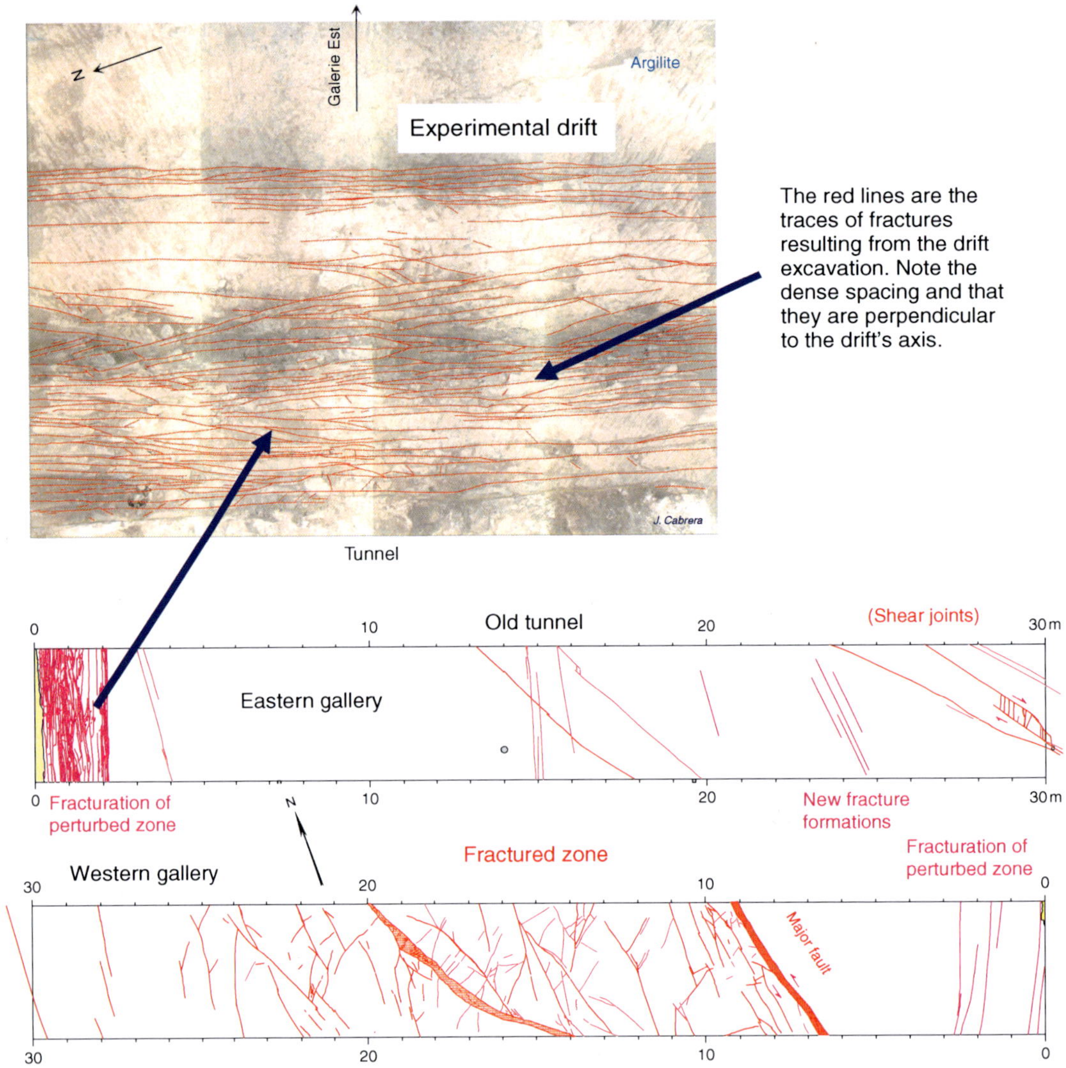

Fig. 1.4 Dense fractures created by the drift excavation at the intersection of a connection tunnel and an experimental drift at Tournemire experimental site, Southern France (Courtesy of Dr Amel Rejeb, IRSN, France).

France. Such artificially created dense fractures will have significant role in changing the coupled hydro-mechanical behavior and properties of this EDZ.

The terms 'discontinuity' and 'fracture' are used interchangeably in the rock mechanics literature for the same physical entity. The term 'fracture' is adopted throughout this book as a collective term for all types of natural or artificial discontinuities unless specifically stated otherwise. The surfaces of fractures may be simplified as nominally planar at the macroscopic level, but they may become wavy at larger scales with varying wavelengths and amplitudes. On close examination at the microscopic level, there exist numerous small-scale asperities on these surfaces. The presence of these asperities is the reason for the so-called roughness of the fracture surfaces and is the major factor causing complexities in the mechanical and hydraulic behavior of the rock fractures.

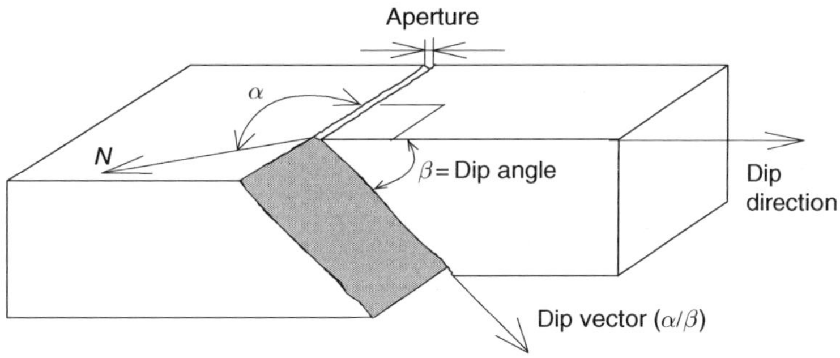

Fig. 1.5 Geometrical parameters associated with a discontinuity (ISRM, 1978). The gap between the two surfaces is exaggerated to illustrate the aperture.

A fracture in a three-dimensional space can be described with the following geometrical parameters (Fig. 1.5): dip angle, dip direction, persistence (dimension and shape) and aperture (the gap between two opposite surfaces of the discontinuity).

Most of the fractures in rock masses are pre-existing natural fractures. Although these rock fractures have occurred naturally through geological processes, their formation is governed by mechanical principles. The fractures are most often clustered in certain directions resulting from their geological modes and history of generation. A set of fractures is comprised of many fractures of same or similar orientations (dip angles and dip directions). Besides the orientation, the most important geometrical parameters for a fracture set are the trace length and spacing, i.e., the distance between two adjacent member fractures in the direction normal to the mean fracture plane. The orientation, trace length, spacing, aperture and other geometrical parameters are usually obtained by standard field measurements using scanline mapping, window mapping and/or borehole (well) logging (see Chapter 5 for more details).

Fractures may have one of the three main types of connectivity with other fractures: multiply connected (persistent), disconnected (or isolated) and singly connected, see Fig. 1.6. Multiply connected fractures are usually larger in size and have multiple (at least two) intersections with other fractures. A multiply connected fracture may be truncated at the boundaries of problem domains of interest or have dead-end segments inside rock blocks. The disconnected fractures lie completely inside one single rock block and have no intersection at all with any other fracture. The singly connected fractures have only one intersection with other fractures, and their two ends are located inside one rock block or two adjacent

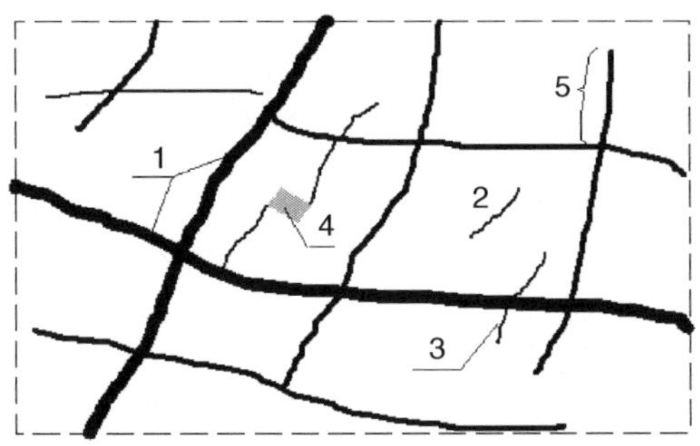

Fig. 1.6 Definition of fracture connectivity types: (1) multiply connected (persistent) fractures; (2) isolated fractures; (3) dead-end fractures; (4) rock bridges; and (5) dead-end segments of persistent fractures.

rock blocks. The portion of intact rock separating tips (or fronts) of dead-end segments of multiply or singly connected fractures is called rock bridges. These rock bridges play a significant role in the strength and deformability of rock masses, and the permeability of the rock mass is determined by the connectivity of the fracture systems, which is often called the fracture networks. Important to note here is the fact that the above definitions are constrained to a 2D image. In the reality of fractures in a 3D space and hidden in the subsurface, the connectivity and persistence patterns may be very different from that of the 2D image obtained from an observable exposure window. The modeling of the geometrical properties of fracture networks, i.e., the mathematical description of their shape, density (spacing), persistence, aperture, orientation, and connectivity has been and still is one of most important fundamental issues in rock mechanics and rock engineering and also the most uncertain in terms of unique and quantitative representations (Kulatilake, 1991; Priest, 1993; Jing, 2003).

One of the main tasks of numerical modeling in rock mechanics is to be able to characterize the fracture systems (geometry and behavior) in a computational model, either explicitly (which means that individual fractures are represented in a computational model using fracture elements, such as the thin-layer elements in FEM or contact elements in DEM) or implicitly (which means that no explicit fracture element will be included in the computational model but their influence on physical behaviour (such as deformability, strength and permeability) are considered through constitutive laws of the fractured rocks as equivalent continua, i.e., the so-called 'system and material conceptualization'). Additionally the interaction between the rock mass and the engineering structure has to be incorporated in the modeling procedure for design and performance assessment so that the consequences of the construction process can be characterized.

To adequately represent a rock mass in computational models intended for capturing the main issues such as its fracture system and its complete DIANE nature, plus the consequences of engineering, it is necessary to be able to include the following features during model conceptualization (Jing and Hudson, 2002):

- the relevant physical processes and their mathematical representations by partial differential equations, especially when coupled thermal, hydraulic and mechanical processes need to be considered simultaneously;
- the relevant mechanisms and constitutive laws of both the rock matrix and fractures, with the associated variables and parameters;
- the pre-existing state of *in situ* rock stress (the rock mass being already under stress);
- the pre-existing state of temperature and water pressure (the rock mass being porous, fractured and heated by a natural geothermal heat gradient or man-made heat sources and fully or partially saturated with groundwater or other geological fluids);
- the presence of natural fractures (the rock mass being discontinuous);
- the variations in properties at different locations (the rock mass being heterogeneous);
- the variations of properties in different directions (the rock mass being anisotropic);
- the time/rate-dependent behavior (the rock mass behavior being affected by time, such as undergoing creep or plastic deformation);
- the variations of properties at different scales (the rock mass behavior being scale-dependent);
- the effects resulting from the engineering perturbations (the problem geometry being altered according to engineering construction activities, a moving boundary problem).

The extent to which these features can actually be incorporated into a computer model will depend on the conceptualization of the physio-chemical processes involved and the modeling technique used or required; hence both modeling and any subsequent rock engineering design will contain subjective judgments.

Rock engineering projects are becoming larger and more demanding in terms of modeling requirements, due largely to the requirements of society concerning the evaluation of the impacts of rock engineering work on the environment. A truly fully coupled T-H-M-C model may be required for solving some problems of environmental importance, e.g., the performance and safety assessments of radioactive waste repositories and geothermal reservoir engineering, with more demanding knowledge of the fracture system characteristics and coupled material properties and parameters of fractured rock masses. Thus, the challenge is to know how to develop adequate models for the problems at hand. The model does not have to be complete and perfect, but it has to be adequate for the purposes of engineering.

For these reasons, rock mechanics modeling and rock engineering design are both a science and an art. They rest on a scientific foundation of continuum mechanics but require empirical judgments supported by accumulated experience through long-term practice. This is the case because the quantity and the quality of the supporting data for rock engineering design and analysis can never be complete and are very often not even adequate – even though they can be perfectly defined in models.

1.2 Mathematical Models for Discontinuous Media

In numerical modeling of engineering problems, some problems can be represented by an adequate model using a finite number of well-defined components. The behavior of such components either is well known or can be independently treated mathematically. The global behaviour of the problem can be obtained through well-defined interrelations between the individual components (elements). Such systems are termed *discrete*. One example of such a discrete system is a structure of a finite number of interconnected beams in civil engineering. The representation and solution of such systems are usually straightforward, such as the matrix solution approach for beam structures with algebraic functions representing the individual beam behavior and their interactions.

In other problems, the definition of such independent components may require an indefinite subdivision of the problem domain, and the problem can only be treated using the mathematical fiction of an *infinitesimal*, implying an infinite number of components. This usually leads to differential equations or equivalent mathematical statements to describe the system behavior at the field points. Such systems are termed *continuous* and have infinite degrees of freedom. To solve such a continuous problem by numerical methods, the problem domain is usually subdivided into a finite number of small sub-domains (elements) whose behavior can be approximated by simple mathematical descriptions with finite degrees of freedom. These sub-domains (elements) must satisfy both the governing differential equations of the problem and the continuity condition along the element boundaries. This is the so-called *discretization* of a continuum, an approximation of a continuous system with infinite degrees of freedom by a discrete system of elements with finite degrees of freedom. The continuity referred to here is a macroscopic concept. At the microscopic scale, all materials will be discrete systems. However, representing the microscopic components individually can only lead to an intractable numerical problem. Therefore approximation is needed for engineering analysis to reduce the degrees of freedom of the problems to an acceptable tolerance of the discrepancies thus generated. The finite element method (FEM) is a typical example of such an approach.

A discrete system in nature may contain a very large number of individual components whose full representations may also lead to intractable numerical difficulties. For example, a cubic kilometer of fractured rock mass may contain such a large number of rock blocks of various sizes that if every block, no matter how small it is, were to be taken into account, the memory requirement for representing them in a computational model would become too high to be practically solvable. Therefore approximation also needs to be introduced to limit the number and degrees of freedoms of individual components by either reducing the problem size (changing boundary locations) or lumping smaller components into larger ones or assuming more simple behavior of the individual blocks (such as rigid blocks) if the requirements for the problem solution permit such approximations. Special care must then be taken to consider the physical behaviour of the 'lumped' components since they are composites of individual units with interfaces between them. The continuity condition across these interfaces between the smaller units in a lumped component also needs to be imposed. The interrelations are then defined between the larger composite components. This is an important difference between a discretized continuous system and a simplified discrete system.

For fractured rock masses, the geometry and physical behavior of fractures is so different from that of the intact solid rock matrix that neglecting the fractures often leads to misinterpretation of the rock mass responses to the external loading conditions and other engineering factors, to an unknown extent. The rock mass in the Earth's upper crust, in essence, is a discrete system. Closed-form solutions rarely exist in practice for general initial-boundary value problems of rock masses concerning mechanical deformation or transport of fluids and heat, even with the assumption of an equivalent continuum. So numerical methods must be used. Due to the differences in assumptions for the system composition and material behavior for the problems concerned, different numerical methods are developed for continuous and discrete systems.

1.2.1 Continuum-Based Numerical Methods and Homogenization

The most commonly applied numerical methods for continuous systems are the finite difference method (FDM), the FEM and the boundary element method (BEM). The basic assumption adopted in these numerical methods is that the materials concerned are continuous throughout the physical processes. This assumption of continuity indicates that, at all points in a problem domain, the material cannot be torn open or broken into pieces. All material points originally in the neighborhood of a certain point in the problem domain remain in the same neighborhood throughout the simulated physical process. In the case of the presence of fractures in the materials, the continuity assumption means that the deformation along or across the fractures has the same order of magnitude as that of the solid matrix near the fractures, so that no large-scale macroscopic slip or opening of fractures should occur. Some special algorithms have been developed to deal with material fractures in continuum mechanics based methods, such as special joint elements in the FEM (Goodman, 1976) and displacement discontinuity technique in the BEM (Crouch and Starfield, 1983). However, they can only be applied with limitations:

(1) large-scale slip and opening of fracture elements are prevented in order to maintain the macroscopic material continuity;

(2) the amount of fracture elements must be kept relatively small so that the global stiffness matrix can be maintained well-posed, without causing severe numerical instabilities; and

(3) complete detachment and rotation of elements or groups of elements as a consequence of deformation are either not allowed or treated with special algorithms.

These limitations make the continuum-based methods most suitable for problems with no fractures or a small number of fractures undergoing small deformations. Although special integration algorithms or constitutive formulations have also been developed and applied to deal with problems of finite (or large)

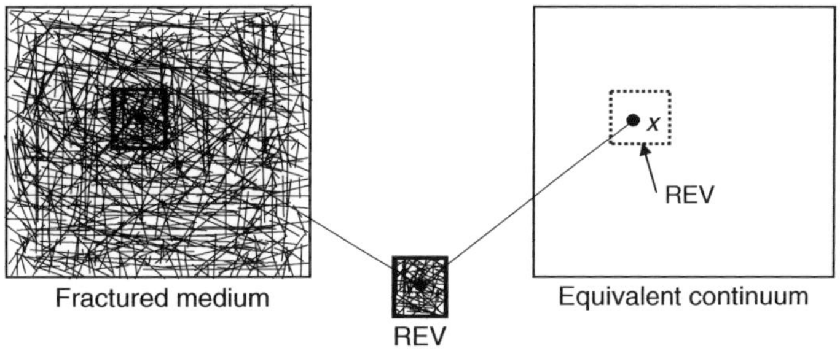

Fig. 1.7 Fractured medium, REV and equivalent continuum.

deformations and non-linear material behavior, continuum-based numerical methods are most effective for problems of small deformation (or small strain) with linear constitutive material behavior. Details of the FEM, FDM and BEM approaches can be easily found in many text books (Crouch and Starfield, 1983; Davis, 1986; Banerjee, 1993; Zienkiewitz and Taylor, 2000) and will not be repeated here.

When the continuum-based numerical methods such as FEM or BEM are applied to problems of essentially discontinuous media like fractured rocks that contain fractures at various scales, the basic assumption is that the discontinuous media behave, at the macroscopic level and in a statistical sense, as equivalent continua. It is important to formulate the constitutive models of the continua in such a way that the effects of the fractures can be properly represented in the equivalent material properties defined in the respective constitutive models of the assumed continua. This process is often called homogenization, which can be valid only in a statistical sense through an averaging operator and over a certain sampling volume of the discontinua, called the REV (representative elementary volume), see Fig. 1.7.

Let $\bar{\sigma}_{ij}$ and $\bar{\varepsilon}_{ij}$ be the macroscopic stress and strain tensors for an equivalent continuum of a fractured rock mass whose microscopic stress and strain tensors are σ_{ij} and ε_{ij}, respectively. Homogenization is the transition of constitutive relations from the microscopic to the macroscopic level, satisfying an averaging operator given by

$$\bar{\sigma}_{ij} = \frac{1}{V_{\mathrm{R}}} \int_{V_{\mathrm{R}}} \sigma_{ij}\,\mathrm{d}v = \langle \sigma_{ij} \rangle, \quad \bar{\varepsilon}_{ij} = \frac{1}{V_{\mathrm{R}}} \int_{V_{\mathrm{R}}} \varepsilon_{ij}\,\mathrm{d}v = \langle \varepsilon_{ij} \rangle \tag{1.1}$$

and

$$\bar{\sigma}_{ij} = \bar{k}_{ijkl}\bar{\varepsilon}_{kl} \tag{1.2}$$

where

$$\langle \cdot \rangle = \frac{1}{V_{\mathrm{R}}} \int_{V_{\mathrm{R}}} (\cdot)\mathrm{d}v$$

stands for the averaging operator over volume V_{R} which is equal to the REV of the fractured medium. The stiffness tensor, $\bar{k}_{klij}$, of the assumed continuum is a function of both the mechanical properties of the rock matrix and fractures and the geometrical characteristics of the fractures, contained in the REV. The validity of a homogenization, therefore, depends on the validity of the REV.

Homogenization is not a simple averaging process. It should be performed with the requirement that not only micro–macro transition rule shown by Eqn (1.1) is satisfied, but also all basic physical laws should not be violated.

Figure 1.8a illustrates the definition of the REV for a fractured porous medium (Bear and Bachmat, 1991). The domain can be decomposed into a porous medium and a fracture system. Let n_{pm} and n_{fr} be

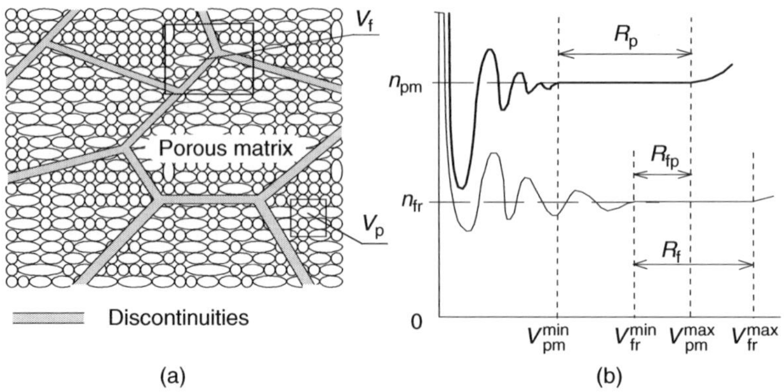

Fig. 1.8 REV for a fractured porous medium (after Bear and Bachmat, 1991 with slight modifications). (a) A fractured porous medium; (b) REV ranges expressed as constitutive properties n_{pm} and n_{fm} versus REV sizes. R_p, REV range for porous medium alone; R_f, REV range for fracture network alone; R_{fp}, REV range for the fractured porous medium.

constitutive parameters describing a physical process for the porous medium and fracture system, respectively. To find the appropriate REV for each system, a series of sampling volumes, V_p for the porous medium and V_f for the fracture system with increasing sizes of the sampling volume and at random locations inside the domain, are chosen and the values of n_{pm} and n_{fr} are evaluated by applying the averaging operators over V_p and V_f, respectively. By plotting the values of n_{pm} and n_{fr} versus sampling volume size (Fig. 1.8b), two stable regions may be found for n_{pm} and n_{fr}, defined by $R_p = [V_{pm}^{max}, V_{pm}^{min}]$ and $R_f = [V_{fr}^{max}, V_{fr}^{min}]$, respectively, over which values of n_{pm} and n_{fr} remain largely constant. Outside these two regions, the values of n_{pm} and n_{fr} may have large undulations with very small change of the sampling volumes.

The physical explanation for the undulations is that outside the stable region R_p, the sampling volume is either too small, so that the grain boundaries of the porous material may have a significant effect, or too large, so that the fractures play a significant role. Similarly explanations can also be found for n_{fr}. Outside the stable region R_f, the sampling volume is either too small, so that particles of the porous medium, rather than the fractures, dominate the process, or too large, so that fractures at much larger scales (such as faults and fracture zones) become significant contributors. Therefore, for a porous medium, any volume inside the closed interval $R_p = \left[V_{pm}^{min}, V_{pm}^{max}\right]$ can serve as its REV. Similarly for a fracture system, any volume inside the closed interval $R_f = \left[V_{fr}^{min}, V_{fr}^{max}\right]$ can be chosen as its REV. The usual practice is to choose the lower bound, V_{pm}^{min} and V_{fr}^{min}, as their respective REVs.

For the fractured porous medium, its REV may lie in an intersection of the two stable regions, $\left[V_{fr}^{max}, V_{fr}^{min}\right] \bigcap \left[V_{pm}^{max}, V_{pm}^{min}\right]$ if such an intersection exists. The lower bound of the REV should then be chosen as the maximum of the two lower bounds $\left(V_{pm}^{min}, V_{fr}^{min}\right)$. In practice, such intersection often does not exist but the REV for fractures is usually larger than the REV for a porous matrix, so the lower bounds for the fractures can be chosen as the lower bound of the fractured porous medium.

A REV is a certain volume over which a statistical equivalence can be well established between the constitutive parameters of the equivalent continuum and that of the original discontinuum in terms of the physical processes concerned (e.g., mechanical deformation, fluid flow or thermal energy transport). However, there is no guarantee that a REV will exist in all cases, especially for rocks intersected by fractures of various sizes and behavior (see last Section 3.5.1.4 for more discussion on this point). Continuum mechanics principles then cannot be applied and the discrete approach must be used to solve the problem by treating the fractured rock masses as discrete systems consisting of blocks and fractures. The individual blocks can be treated either as rigid bodies if the stresses are low and deformations are negligible or as deformable continuum bodies with appropriate constitutive behavior (e.g., linear elastic bodies). This approach provides a more straightforward representation of fractured rocks, but has a totally different methodology and algorithms, compared with that of continuum-based numerical methods.

1.2.2 Basic Features of Discrete Element Methods for Discontinua

The commonly adopted term for the numerical methods for discrete systems is discrete element methods (DEM) which includes all numerical methods treating the problem domain as an assemblage of independent units. The method is mainly applied for problems of fractured rocks, granular media and multibody systems in mechanical engineering but has also been developed for fluid mechanics problems. The individual units can be rock blocks, solid particles of granular materials, structural elements or other individual parts/members of multibody systems. For mechanical analysis, the formulation of the DEM is based on the contacts between individual members, their kinematics and their deformation mechanisms if they are deformable. For rock engineering problems, the rock blocks are defined by intersecting fractures whose locations, orientations and dimensions are required for the determination of the problem geometry. Isolated fractures or dead-end segments of fractures are generally discarded because the initiation and propagation at tips of fractures cannot be effectively simulated at the present stage of DEM development. Figure 1.9 illustrates the representation of a fractured rock mass by the FEM, BEM and DEM models. For granular materials, the solid particles can be regular circular disks (2D case) or spherical balls (3D case) or irregularly shaped solid particles, depending on the material characterization requirements. From the geometrical point of view, the DEM representation is a more natural, and therefore can be a more realistic, representation of fractured rock masses. The BEM model is the simplest for representing the problem geometry.

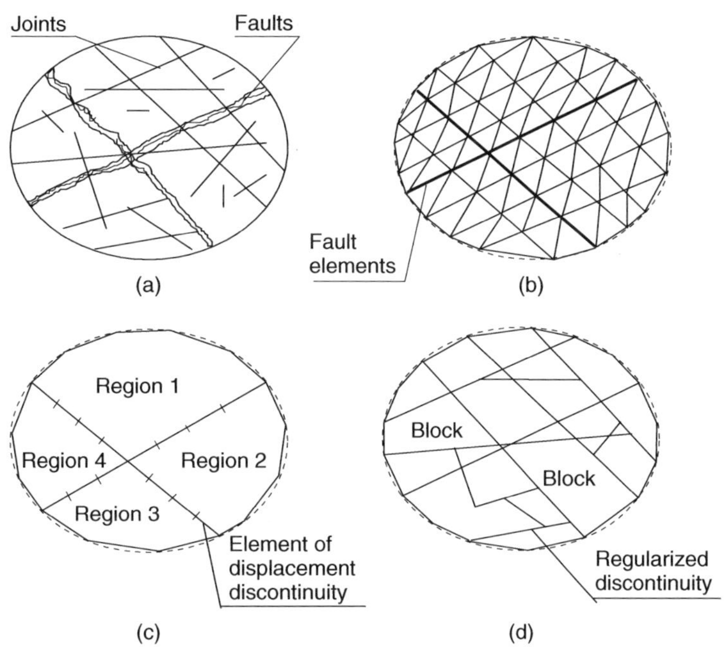

Fig. 1.9 Representation of a fractured rock mass: (a) the fractured rock mass, (b) model by FEM; (c) model by BEM and (d) model by DEM (Jing, 2003).

Both continuous and discrete numerical methods have their own advantages and shortcomings, subject to the specific practical rock engineering problem. The major drawback of DEM for a rock mechanics problem is that the locations, orientations and dimensions of the fractures, except for some major features (large faults or fracture zones, for example) that can be deterministically characterized, are largely unknown prior to the problem setup and even after the problem solutions, so that the solution of the problem contains a basically unknown uncertainty about the fracture system geometry. To obtain an impression of how the fractures are distributed inside the rock mass in question, statistical analysis for the fracture population of the rock mass is a most often used technique, as will be described in Chapter 5. The fractures are assumed to exist in sets with different orientations (dip directions and dip angles), in different statistically homogeneous regions (or sub-regions). The distributions of these set orientation parameters, together with the distributions of other parameters such as size and aperture (which is difficult to measure), are usually estimated using data from a limited number of exploration boreholes and/or mappings of fractures from a limited number of exposures using scanline or window mapping. A fracture network is then obtained from the distribution functions of these parameters by an inverse process using random number generation according to the parameter distribution functions. Such an obtained fracture network is not the real fracture network located in the considered rock mass, but a partially equivalent model in a statistical sense. This equivalent model, often called a realization, will not be unique, however, since there will exist an infinite number of such statistical models equivalent to the real fracture network, generated using the same parameter distribution functions. Hence multiple realizations need to be generated and analyzed in order to obtain a spectrum of solutions whose collective behavior may, one hopes, provide a more representative behavior of the real system. The reliability of this technique varies considerably with the quality and quantity of the fracture mapping data and the mapping and logging techniques. A topological identification process then establishes the blocks defined by the fracture network (see Chapters 6 and 7).

For a tightly packed block system, such as a fractured rock mass, the overall response of the stresses and deformations of the equivalent models may approximate that of the real fractured rock mass if the problem dimension, material properties and boundary conditions are properly defined. However, for fluid flow in fractured hard rocks, such stochastic models may not be able to provide an adequate approximation to the flow paths, especially in the near-field of an excavation, because the fluid flow in fractured hard rocks is highly path-sensitive, i.e., it depends completely on the fracture connectivity. Therefore the solution of the fluid flow in fractured rock masses is more sensitive to fracture system characterization, and therefore potentially contains even larger degree of uncertainty.

Even with this serious shortcoming, the DEM still appears as one of the most attractive methods for solving rock engineering or general geomechanical problems because of its unique advantages in dealing with fractures. Figure 1.10 illustrates how the continuum methods or discrete element methods should be used for a rock mechanics problem in different fracture situations (Brady, 1987).

In numerical modeling for fractured rock masses, using a continuous or a discrete system for a practical problem does not always depend on the problem geometry or geology alone. It also depends, to a great extent, on the problem size, dimension of the area or volume of interest (for example, an excavation), the number of discrete units (blocks), the computational capacity of the computer and the conceptualization process. It is often useful in practice to combine the use of both techniques, with the DEM representing the near-field part of a problem with discrete blocks and fractures and a continuum representation for the far-field part of the problem (see Fig. 1.11) so that the advantages of both the DEM and continuum techniques can be utilized at the appropriate scales.

Another basic difference between a continuum and a discrete approach is how to deal with the rigid body mode of motion. Rigid body motion is quite often the dominant mode of deformation for a discrete system when a large displacement situation occurs and the continuous deformations of the material

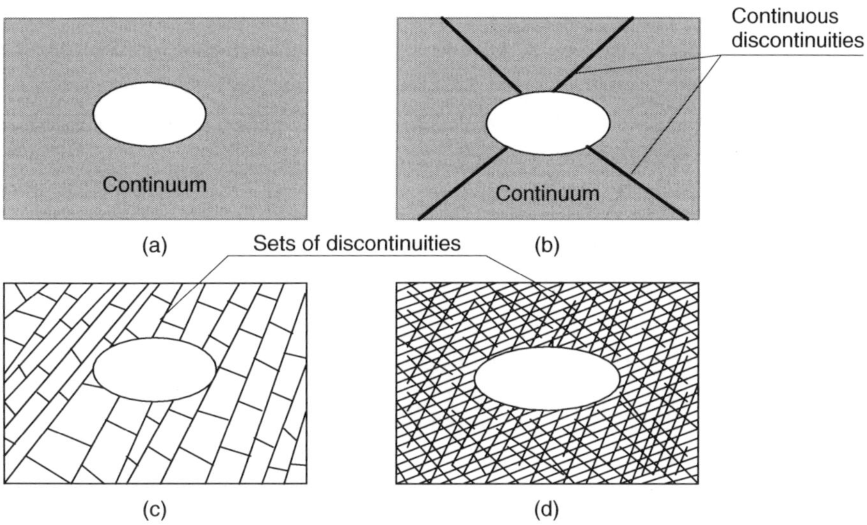

Fig. 1.10 Suitability of different numerical methods for analysis of an excavation in a fractured rock mass. (a) Continuum methods; (b) Both continuum methods and DEM; (c) DEM; (d) Continuum methods or mixed scheme (Brady, 1987).

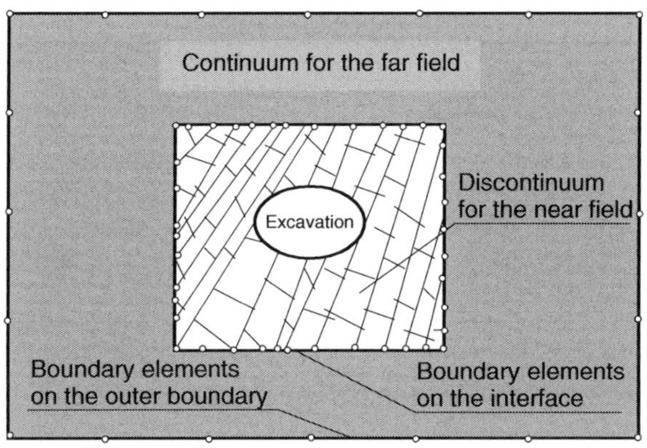

Fig. 1.11 Combined representation of a problem by DEM for the near-field area close to an excavation and BEM representation for the far-field area of the model.

blocks are generally much less in magnitude, especially for hard rocks. A typical example is rock sliding in slope problems. This is contrary to the continuum-based methods (FEM and BEM, for example) in which the rigid body motion mode of displacement is generally eliminated because it does not produce strains in the elements. This difference reflects the different focus of the physical conceptualizations concerned, between a continuum and a discrete conceptualization. In a discrete system, the individual units (blocks) are independent to move according to the force (or stress) constraints on their boundary surfaces and other external loads according to the equations of motion. The rigid body motion of a block, therefore, can be 'liberated' from other blocks. For the continuum methods, the individual units (elements) are not free to move, but are kept within the same neighborhood of other elements by the displacement compatibility conditions due to the continuity assumption. The motion of the element is not independent but constrained by other neighboring elements connected along its boundaries.

The connectivity (or contacts) of an element with other elements is fixed during the course of computation. Therefore a continuous system reflects more the 'material deformation' of the system and the discrete system reflects more the 'member (unit or component) movement' of the system. The complete decoupling of the rigid body motion mode and continuous deformation mode of individual units is usually adopted in DEM. The rigid body motion does not produce strains inside the blocks, but it does produce displacements of blocks, often on a large scale.

1.3 Historical Notes on DEM

The DEM approaches for geological media and engineering problems were developed progressively in the early 1970s–1980s. Rock mechanics and soil mechanics are the disciplines from which the ideas for motion and deformation of block/particle systems originated (Burman, 1971; Cundall, 1971, 1982, 1988; Chappel, 1972, 1974; Byrne, 1974; Cundall and Strack, 1979a,b,c, 1982; Williams et al., 1985; Williams and Mustoe, 1987; Shi, 1988; Shi and Goodman, 1988; Barbosa and Ghaboussi, 1989; Cundall and Hart, 1989), with the conceptual breakthroughs made in Cundall (1971), Cundall and Strack (1979a,b,c) and Shi (1988). The other branches from which the discrete approaches for modeling geological media and geological engineering systems were developed were the Discrete Fracture Network (DFN) approaches for flow and transport in fractured rocks (Long et al., 1982, 1985; Andersson, 1984; Endo et al., 1984; Robinson, 1984, 1986; Smith and Schwartz, 1984; Elsworth, 1986a,b; Andersson and Dverstop, 1987; Charlaix et al., 1987; Dershowitz and Einstein, 1987; Tsang and Tsang, 1987; Billaux et al., 1989; Cacas et al., 1990a,b; Stratford et al., 1990) and structural analysis (Kawai, 1977a,b, 1979; Kawai et al., 1978; Nakezawa and Kawai, 1978).

The above literature covers only some of the early original developments in this field and is not nearly complete. Extensive further developments and industrial applications have been maintained both during and after the above period in rock engineering, as presented in the following chapters, with the relevant literature referenced.

The theoretical foundation is the formulation and solution of equations of motion of rigid and deformable bodies, such as described in Wittenburg (1977) and Wang (1975), and the Navier–Stokes equation simplified for fluid flow though narrow fractures (with the mass transport equation added for the DFN approach) and the equations for heat transfer. All the above theories are based on the general principles of continuum mechanics with the basic numerical formulation techniques of the FEM (Zienkiewicz and Taylor, 2000) and the finite difference method (Wilkins, 1969) for continuum mechanics.

The DEM formulation applied to rock engineering problems has also undergone a development from the early stage of rigid body movements to motion and deformation of deformable block systems with internal finite element or finite difference discretizations for static, quasi-static or dynamic problems. The method has a broad variety of applications in rock mechanics, soil mechanics, ice mechanics, structural analysis, granular materials, material processing, fluid mechanics, multibody systems, robot simulation, computer animation, etc. It has been one of the most rapidly developing areas of computational mechanics. This is due to the three central issues of the DEM approach for these seemingly much diverse engineering branches:

(1) identification of the unit (rock blocks, material particles, mechanical parts or fracture systems) system topology;

(2) formulation and solution of equations of motion of the individual units, including or excluding deformation according to the basic conceptualization required;

(3) the detection and updating of varying contacts (or connectivity) between the units as the consequences of their motions and deformations. The basic difference between DEM and other continuum-based numerical methods is that the system topology, i.e., the contact/connectivity patterns between units of the system, is the central computational issue which may evolve with time and deformation processes, but is a fixed initial condition for continuum approaches.

The solution strategies are different for different DEM formulations. The basic difference is related to the treatment of material deformability. For rigid body analysis, it is almost universal that an explicit time-marching scheme using finite difference schemes is used to solve the dynamic equations of motion of a rigid body system or a dynamic relaxation scheme for a quasi-static problem. For deformable body systems, two schemes exist:

(1) an explicit solution with finite difference discretization of the body interiors so that only one system unknown is kept at the left-hand side) of a local equation at a time step and no matrix equations are needed in general and

(2) an implicit solution with finite element discretization of the body interior, which leads to a matrix equation representing both motion and deformation of the individual bodies.

In this book, the former is called an explicit DEM represented by the distinct element method. The later is called an implicit DEM represented by the discontinuous deformation analysis (DDA). The DFN approach, for fluid flow within connected fracture systems, can be an implicit or explicit approach when FEM or FDM is used for domain discretization and solution of fluid equations, respectively.

The most representative explicit DEM codes are the codes UDEC and 3DEC for 2D and 3D problems of block systems and the PFC 2D and PFC 3D codes for particle flow simulations for granular material problems (Itasca, 2000a,b,c).

The implicit DEM approach uses the FEM to represent deformation of the member bodies. This can be achieved by inserting a standard FEM formulation into a DEM frame of block systems. The matrix thus derived is not a purely deformation stiffness matrix, like that in FEM for continuum analysis, but a matrix of mixed contact stiffness between different bodies and deformation stiffness due to the deformations of the bodies. The matrix equation produced by this formulation will change with the ever-changing contact patterns between the individual bodies, and an efficient equation solver is needed to achieve overall computational efficiency. The equilibrium of the system at each time step is unconditional. However, due to its implicit scheme, a larger time step, compared with explicit DEM, may be used. The individual bodies can be seen to play a similar role as the 'super-elements' in a standard FEM, but without any obligation to the 'displacement compatibility condition' across body boundaries. Large displacement of the system can be treated easily because the rigid body motion mode of bodies can be fully represented and is usually decoupled from the deformation modes.

Due to the demands for modeling coupled hydro-mechanical, thermo-mechanical and thermo-hydro-mechanical processes of fractured rocks for civil engineering (e.g. dam and foundation problems), energy resources engineering (e.g. geothermal extraction and underground gas or oil storage facilities) and waste isolation and environmental protection (e.g. radioactive waste repositories), the solution of fluid flow through connected fracture networks and heat transfer through the fracture–rock systems becomes increasingly important. The function of the DEM family is expanded to consider coupled thermo-hydro-mechanical (THM) behavior of fractured rock masses. At present, the majority of DEM techniques developed assume that fluid flow is conducted in the connected fracture networks only and the rock matrix is impermeable.

In current practice of DEM, the heat conduction through the rock matrix is assumed to dominate the heat transfer process (although heat convection along fractures conducting fluid flow can be considered in UDEC), due mainly to the fact that the volume of the slow motion of the fluids in fractures is negligible compared to that of the rock, in hard crystalline rocks. This has been proven as acceptable approximation in practice. The treatment of the effect of heat on hydraulic and mechanical processes in most of the currently available DEM codes is to consider the thermally induced variation of fluid viscosity and volume expansion and thermal stresses of the rock matrix, besides temperature distribution and evolution with time. The effects of the mechanical deformation and fluid flow on heat transfer, such as the transition from dissipated mechanical work to heat, and heat convection due to flow in the rock matrix, are usually ignored. In this book, we concentrate on numerical modeling of the coupled hydro-mechanical processes using DEM.

A different type of discrete approach is the DFN method for study of fluid flow and contaminant transport through connected fracture systems. For DFN, only the void space of the connected fractures is explicitly represented and the rock blocks between the fractures are excluded. The processes of mechanical deformation and heat flow in the rock matrix therefore cannot be properly considered, in general. This method is most useful for the study of flow and transport problems in fractured hard crystalline rocks when an equivalent continuum model is difficult to establish.

Although the establishment of equations for flow and transport processes are straightforward, the characterization of the fracture systems at the field scale is a challenging task and the central issue for DFN. The DFN approach is included in this book due to the fact that fluid flow in fracture networks is also included in the general DEM approaches, even though the transport process is not included in the DEM so far. Hence for the benefit of readers of different backgrounds, we treat DFN as a member of the DEM family, despite the fact that this inclusion may not be too agreeable to those working with the DFN method in different fields of science and engineering. However, the approach is only presented briefly in a summary fashion due to the fact that excellent books are already widely circulated, such as Bear et al. (1993), Sahimi (1995), National Research Council (1996) and Adler and Thovert (1999).

Another important area studied using DEM is particle systems where deformations and stresses of the particle themselves are often not considered. This subject has been experiencing extensive research and development and enjoys extensive applications in many different areas of science and engineering, such as mineral processing, chemical engineering, soil mechanics and rock mechanics. This subject is summarized briefly in Chapter 11, but is not as extensively presented as the DDA and Distinct Element Method in this book for two reasons. One is that there are already excellent books published in this field, such as Oda and Iwashita (1999); another is that inclusion of this subject would lead to significant lengthening of this book, which we do not wish to consider.

For rock engineering applications, the basic feature of a DEM formulation is that the problem domain is subdivided into a finite number of independent blocks by introducing fractures (individually or as sets). The behavior of the block system thus formed depends on the contacts developed between the individual blocks, the deformability of the rock blocks and the fluid flow and pressure in the fracture systems. The contact behavior depends on the formulation of the constitutive models for the fractures. Therefore the inter-block contact relationship determines the overall hydro-mechanical behavior of the block system as well and controls the computational performance of the DEM codes to a very large extent. This basic feature of the DEM formulation then requires that the following problems must be solved properly in a DEM formulation for coupled hydro-mechanical problems:

(1) representation of the fracture systems according to field mapping fracture results;

(2) identification of the block–fracture system with fracture system regularization;

(3) representation of block and fracture deformation;

(4) representation of contact detection and evolution;

(5) integration of the equations of motion of blocks and fluid flow in fractures; and

(6) proper data structure in computer codes for the efficient system geometry updating.

The treatment of initial and boundary conditions is undertaken in a similar way as in other numerical methods and will not be especially treated here.

In the following chapters of this book, the above issues are addressed in necessary but different details for the different DEM approaches, with focus on the Distinct Element Method and the DDA for deformable block systems as the examples of the explicit and implicit approaches. The DFN approach is summarized briefly in Chapter 10 without details of the numerical techniques, followed by in-depth discussion concerning its applicability for solving rock engineering problems. A brief introductory presentation is given in Chapter 11 about DEM for granular materials because of the fact that they can be used to derive macroscopic mechanical properties of the rock matrix from microscopic generation.

The contents of the book are arranged in the following order:

Chapter 1 Introduction (this chapter)

Part One: Fundamentals

Chapter 2 Governing equations for the motion and deformation of block systems and heat transfer

Chapter 3 Constitutive models of rock fractures and rock masses – the basics

Chapter 4 Fluid flow equations and coupled hydro-mechanical behavior of fractures and fractured rocks

Part Two: Fracture system characterization and block model construction

Chapter 5 The basics of fracture system generation

Chapter 6 The theory of combinatorial topology for block system representation

Chapter 7 Numerical techniques for block system construction

Part Three: DEM approaches

Chapter 8 The distinct element method for deformable block systems

Chapter 9 The discontinuous deformation analysis method for deformable block systems

Chapter 10 The discrete fracture network method

Chapter 11 The discrete element method for granular materials

Part Four: Application studies

Chapter 12 Case studies of discrete element methods in geology, geophysics and rock engineering

References

Adler, P. M. and Thovert, J. F., Fractures and fracture networks. Kluwer Academic, The Netherlands, 1999.

Andersson, J., A stochastic model of a fractured rock conditioned by measured information. Water Resources Research, 1984;20(1):79–88.

Andersson, J. and Dverstop, B., Conditional simulations of fluid flow in three-dimensional networks of discrete fractures. Water Resources Research, 1987;23(10):1876–1886.

Banerjee, P. K., The boundary element methods in engineering. McGraw-Hill Book Company, London, 1993.

Barbosa, R. and Ghaboussi, J., Discrete finite element method Proc. 1st US Conf. on Discrete Element Methods, Golden, Colorado, 1989.

Bear, J. and Bachmat, Y., Introduction to modeling of transport phenomena in porous media. Kluwer Academic, The Netherlands, 1991.

Bear, J., Tsang, C.-F. and de Marsily, G. (eds), Flow and contamination transport in fractured rock. Academic Press, San Diego, CA, 1993.

Billaux, D., Chiles, J. P., Hestir, K. and Long, J. C. S., Three-dimensional statistical modeling of a fractured rock mass – an example from the Fanay-Augéres Mine. International Journal of Rock Mechanics and Mining Sciences and Geomechanics Abstracts, 1989;26(3/4):281–299.

Brady, B. H. G., Boundary element and linked methods for underground excavation design. In: Brown, E. T. (ed.), Analytical and computational methods in engineering rock mechanics. Allen & Unwin, London, pp. 164–204, 1987.

Burman, B. C., A numerical approach to the mechanics of discontinua. Ph.D. Thesis, James Cook University of North Queensland, Townsville, Australia, 1971.

Byrne, R. J., Physical and numerical model in rock and soil–slope stability. Ph.D. Thesis, James Cook University of North Queensland, Townsville, Australia, 1974.

Cacas, M. C., Ledoux, E., de Marsily, G., Tille, G., Barbreau, A., Durand, E., Feuga, B. and Peaudecerf, P., Modeling fracture flow with a stochastic discrete fracture network: Calibration and Validation 1. The flow model. Water Resources Research, 1990a;26(3):479–489.

Cacas, M. C., Ledoux, E., de Marsily, G., Tille, G., Barbreau, A., Calmel, P., Gillard, B. and Margritta, R., Modeling fracture flow with a stochastic discrete fracture network: Calibration and Validation 2. The transport model. Water Resources Research, 1990b;26(3):491–500.

Chappel, B. A., The mechanics of blocky material. Ph.D. Thesis, Australia National University, Canberra, 1972.

Chappel, B. A., Numerical and physical experiments with discontinua. Proc. 3rd Cong. ISRM, Vol. 2A, Denver, Colorado, 1974.

Charlaix, E., Guyon, E. and Roux, S., Permeability of a random array of fractures of wisely varying apertures. Transport in Porous Media, 1987;2(1):31–43.

Crouch, S. L. and Starfield, A. M., Boundary element methods in solid mechanics. Allen &, Unwin, London, 1983.

Cundall, P. A., A computer model for simulating progressive, large scale movements in blocky rock systems. Proc. Int. Symp. Rock Fracture, ISRM, Nancy, Vol. 1, PaperII-8, 1971.

Cundall, P. A., Adaptive density-scaling for time-explicit calculations Proc. 4th Int. Conf. on Numerical Methods in Geomechanics, pp. 23–26, 1982.

Cundall, P. A., Formulation of a three-dimensional distinct element model – Part I: A scheme to detect and represent contacts in a system composed of many polyhedral blocks. International Journal of Rock Mechanics and Mining Sciences and Geomechanics Abstracts, 1988;25(3):107–116.

Cundall, P. A. and Hart, R. D., Numerical modelling of discontinua. Keynote lecture. Proc. 1st US Conf. on Discrete Element Methods, Mustoe, G. G. W., Henriksen, M. and Huttelmaier, H.-P. (eds), Golden, Colorado, USA, 1989.

Cundall, P. A. and Strack, O. D. L., A discrete numerical model for granular assemblies. Géotechnique, 1979a;29(1):47–65.

Cundall, P. A. and Strack, O. D. L., The development of constitutive laws for soil using the distinct element method. In: Wittke, W. (ed.), Numerical methods in geomechanics, Aachen, pp. 289–298, 1979b.

Cundall, P. A. and Strack, O. D. L., The distinct element method as a tool for research in granular media. Report to NSF concerning grant ENG 76-20711, Part II, Dept. Civ. Engng, University of Minnesota, 1979c.

Cundall, P. A. and Strack, O. D. L., Modelling of microscopic mechanisms in granular material. Proc. US–Japan Sem. New Models Const, Rel. Mech. Gran. Mat., Ithaca, NY, 1982.

Davis, J. L., Finite difference methods in dynamics of continuous media. Macmillan, New York, 1986.

Dershowitz, W. S. and Einstein, H. H., Three-dimensional flow modelling in jointed rock masses. In: Herget, G. and Vongpaisal, S. (eds), Proc. of 6th Cong. ISRM, Montreal, Canada, Vol. 1, 87–92, 1987.

Elsworth, D., A model to evaluate the transient hydraulic response of three-dimensional sparsely fractured rock masses. Water Resources Research, 1986a;22(13):1809–1819.

Elsworth, D., A hybrid boundary-element-finite element analysis procedure for fluid flow simulation in fractured rock masses. International Journal for Numerical and Analytical Methods in Geomechanics, 1986b;10(6):569–584.

Endo, H. K., Long, J. C. S., Wilson, C. K. and Witherspoon, P. A., A model for investigating mechanical transport in fractured media. Water Resources Research, 1984;20(10):1390–1400.

Goodman, R. E., Methods of geological engineering in discontinuous rocks. West Publishing Company, San Francisco, CA, 1976.

Harrison, J. P. and Hudson, J. A., Engineering rock mechanics. Part 2: Illustrative worked examples. Pergamon, Oxford, 2000.

ISRM (International Society of Rock Mechanics), Suggested methods for the quantitative description of discontinuities in rock masses. International Journal of Rock Mechanics and Mining Sciences and Geomechanics Abstracts, 1978;15:319–368.

Itasca Consulting Group Ltd, User manual of UDEC code, Minneapolis, USA, 2000a.

Itasca Consulting Group Ltd, User manual of 3DEC code, Minneapolis, USA, 2000b.

Itasca Consulting Group Ltd, User manual of PFC codes, Minneapolis, USA, 2000c.

Jing, L., A review of techniques, advances and outstanding issues in numerical modelling for rock mechanics and rock engineering. International Journal of Rock Mechanics and Mining Sciences, 2003;40(3):283–353.

Jing, L. and Hudson, J. A., Numerical methods in rock mechanics. International Journal of Rock Mechanics and Mining Sciences, 2002;39(4):409–427. A CivilZone review paper.

Kawai, T., New discrete structural models and generalization of the method of limit analysis. Proc. Int. Conf. on Finite Elements in Nonlinear Solid and Structural Mechanics, Norway, Vol. 2, pp. G04.1–G04.20, 1977a.

Kawai, T., New element models in discrete structural analysis. Journal of the Society of Naval Architects of Japan, 1977b;141:174–180.

Kawai, T., Collapse load analysis of engineering structures by using new discrete element models. IABSE Colloquium, Copenhagen, 1979.

Kawai, T., Kawabata, K. Y., Kondou, I. and Kumagai, K., A new discrete model for analysis of solid mechanics problems. Proc. 1st Conf. Numerical Methods in Fracture Mechanics, Swansea, UK, pp. 26–27, 1978.

Kulatilake, P. H. S. W., Lecture notes on stochastic 3-D fracture network modeling including verification at the Division of Engineering Geology. Royal Institute of Technology, Stockholm, Sweden, 1991.

Long, J. C. S., Remer, J. S., Wilson, C. R. and Witherspoon, P. A., Porous media equivalents for networks of discontinuous fractures. Water Resources Research, 1982;18(3):645–658.

Long, J. C. S., Gilmour, P. and Witherspoon, P. A., A model for steady fluid flow in random three-dimensional networks of disc-shaped fractures. Water Resources Research, 1985;21(8):1105–1115.

Nakezawa, S. and Kawai, T., A rigid element spring method with applications to non-linear problems. Proc. 1st Conf. Numerical Methods in Fracture Mechanics, Swansea, UK, pp. 38–51, 1978.

National Research Council, Rock fractures and fluid flow: contemporary understanding and applications. National Academy Press, Washington, DC, 1996.

Oda, M. and Iwashita, K. (eds), Mechanics of granular materials – an introduction. Balkema, Rotterdam, 1999.

Priest, S. D., Discontinuity analysis for rock engineering. Chapman & Hall, London, 1993.

Robinson, P. C., Connectivity, flow and transport in network models of fractured media. Ph.D. Thesis, St Catherine's College, Oxford University, 1984.

Robinson, P. C., Flow modelling in three dimensional fracture networks. UK AEA Harwell, AERER 11965, 1986.

Sahimi, M., Flow and transport in porous media and fractured rock. VCH Verlagsgesellschaft GmbH, Weinheim, 1995.

Shi, G., Discontinuous deformation analysis – a new numerical model for statics and dynamics of block systems. Ph.D. Thesis, University of California, Berkeley, CA, 1988.

Shi, G. and Goodman, R. E., Discontinuous deformation analysis – a new method for computing stress, strain, and sliding of block systems. In: Cundall, P. A., Sterling, R. L. and Starfield, A. M. (eds), Key questions in mechanics, Proc. of the 29th US Symp. on Rock Mechanics, University of Minnesota, Minneapolis, June 13–15, 1988, pp. 381–393. Balkema, Rotterdam, 1988.

Smith, L. and Schwartz, F. W., An analysis of the influence of fracture geometry on mass transport in fractured media. Water Resources Research, 1984;20(9):1241–1252.

Stratford, R. G., Herbert, A. W. and Jackson, C. P., A parameter study of the influence of aperture variation on fracture flow and the consequences in a fracture network. In: Barton and Stephansson (eds), Rock joints, Balkema, Rotterdam, pp. 413–422, 1990.

Tsang, Y. W. and Tsang, C.-F., Channel model of flow through fractured media. Water Resources Research, 1987;22(3):467–479.

Wang, C. Y., Mathematical principles for continuum mechanics and magnetism – Part A. Analytical and continuum mechanics, Plenum Press, New York, 1975.

Wilkins, M. L., Calculation of elastic–plastic flow. Lawrence Radiation Laboratory, University of California, Research Report, UCRL-7322, Rev. I, 1969.

Williams, J. R. and Mustoe, G. G. W., Modal methods for the analysis of discrete systems. Computers and Geotechnics, 1987;4(1):1–19.

Williams, J. R., Hocking, G. and Mustoe, G. G. W., The theoretical basis of the discrete element method. NUMETA'85, Numerical Methods in Engineering, Theory and Application, Conf. in Swansea, January 7–11. Balkema Publishers, Rotterdam, 1985.

Wittenburg, J., Dynamics of systems of rigid bodies. B. G. Tenbner, Stuttgart, 1977.

Zienkiewitz, O. C. and Taylor, R. L., The finite element method. Volume 1 – the basics, 5th edition, Butterworth-Heinemann, Oxford, 2000.

Part One
Fundamentals

2 GOVERNING EQUATIONS FOR MOTION AND DEFORMATION OF BLOCK SYSTEMS AND HEAT TRANSFER

The DEM techniques were developed primarily for mechanical deformation/motion processes of particle or block assemblages. The governing equations are the equations of motion of systems of rigid or deformable bodies or particles. Due to the demands for modeling coupled hydro-mechanical, thermo-mechanical and THM processes of fractured rocks for civil engineering (e.g. slopes, tunnels, hydropower dams and rock foundations), energy resources engineering (e.g. geothermal reservoirs and underground gas or oil storage caverns) and environmental engineering projects (e.g. underground radioactive waste repositories), the effects of fluid and heat flows through the fractured rock masses become more and more important issues for the design, operation and performance assessments of structures in rock masses. The general conservation equations of mass, momentum and energy in continuum mechanics are the guiding principles.

A distinct feature of DEM for the coupled THM processes is that fluid flow is usually assumed to be dominated by the connected fracture networks, and the heat conduction through the rock matrix dominates the heat transfer process, due mainly to the fact that the volume of fluid is very small compared to that of rock matrix and the fluid velocity is also very low, in hard crystalline rocks. This assumption may be less appropriate for very porous rocks like sandstones, but then the DEM is mainly developed for fractured hard rocks that have large differences in fluid conductivities between the fractures and rock matrix. The matrix flow and heat convection due to fluid movement are therefore usually ignored. Accumulated experience in both field experiments and numerical modeling work has shown that this assumption is basically acceptable, especially for heat transfer processes in fractured rocks.

Going one step further in the detail, we may summarize the main governing equations as the Newton–Euler equations of motion for rigid bodies, the Cauchy equations of motion for deformable bodies, the Nervier–Stokes equation for fluid flow through fracture networks, the heat transfer equation based on Fourier's law and various constitutive equations of the rock matrix and fractures. The hydro-mechanical coupling is represented by the effect of rock deformation on the variation of fracture apertures (therefore the transmissivity) and the additional boundary stresses on blocks due to fluid pressure. The effects of pore pressure on rock (matrix) deformation or the effects of stresses on porosity/permeability of the rock matrix are not considered in this book at this stage since the rock matrix is assumed to be impermeable.

The treatment of thermal effects on hydraulic and mechanical processes in most of the currently available DEM codes is through consideration of the thermally induced variation of fluid viscosity, volume expansion and thermal stresses in the rock matrix, in addition to temperature distribution and evolution with time during transient heat conduction. The effects of mechanical deformation and fluid flow on heat transfer, such as conversion from the dissipated energy by mechanical work to heat and heat convection due to fluid flow in rock are usually ignored in practice.

The purpose of this chapter is to present a brief coverage of the governing equations most widely adopted in the current DEM approaches. We assume that the general reader is familiar with the basics of continuum mechanics, the FEM as well as tensor analysis techniques. For details on the complete definitions of the basic concepts and fundamental relations in solid and continuum mechanics, the readers are encouraged to look in the classical works by Fung (1969), Wang (1975), McDonough (1975) and Shabana (1998) for the equations of motions, Lai et al. (1993) for heat transfer and general continuum mechanics principles.

2.1 Newton's Equations of Motion for Particles

A particle is defined as a body with a constant mass, m, but its volume and shape have little effect on its dynamical behavior, according to classical mechanics. Denoting the product of mass and velocity,

$$p_i = mv_i \quad (i = 1, 2, 3) \tag{2.1}$$

as the linear momentum of a particle of a finite mass m (but negligible volume) and velocity v_i, Newton's second law of motion states that the resultant force that acts on the particle and causes its motion equals the rate of variation in its linear momentum, i.e.,

$$f_i = \dot{p}_i = \frac{\mathrm{d}p}{\mathrm{d}t} = m\dot{v}_i = \frac{\mathrm{d}(mv_i)}{\mathrm{d}t} \tag{2.2}$$

In Newtonian (non-relativistic) mechanics, the mass of a particle remains constant during motion, therefore the law can be expressed as

$$f_i = \frac{\mathrm{d}p}{\mathrm{d}t} = m\dot{v}_i = m\frac{\mathrm{d}(v_i)}{\mathrm{d}t} = m\frac{\mathrm{d}^2u_i}{\mathrm{d}t^2} = ma_i \tag{2.3}$$

where u_i and a_i are the displacement and acceleration vectors of the particle, respectively. Newton's equations of motion (2.3) consider only the translational motion of a particle since the rotation of the particle is eliminated by ignorance of its volume (therefore also the shape). This law, expressed by Eqn (2.3), is a statement of the law of linear momentum conservation.

The above equations are valid when the particle's motion is represented in a single fixed inertial frame. If, on the contrary, the particle is fixed in an arbitrarily moving (both translating and rotating) frame o–xyz relative to a fixed global inertial frame O–XYZ, with angular velocity components $(\Omega_x, \Omega_y, \Omega_z)$ of the moving frame relative to the fixed inertial frame and defined by using Euler's angles (θ, ψ, ϕ) (Fig. 2.1), the expression for the inertial-space acceleration vector of the particle is then written as (Wells, 1967)

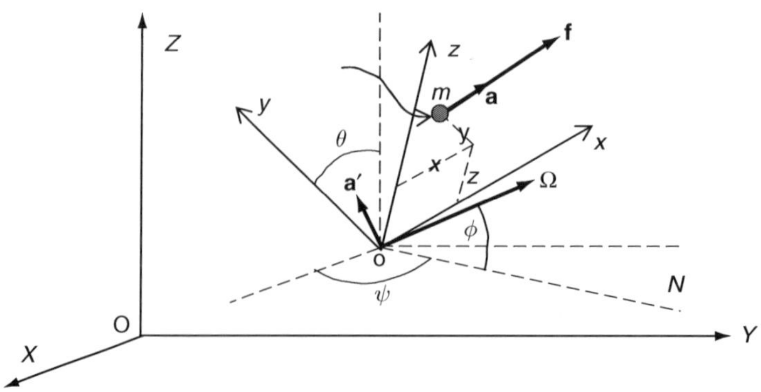

Fig. 2.1 Motion of a particle in a dual coordinate system.

$$\begin{cases} a_x = a_x' + \ddot{x} - x[(\Omega_y)^2 + (\Omega_z)^2] + y[\Omega_x\Omega_y - (\dot{\Omega}_z)] + z[\Omega_x\Omega_z + (\dot{\Omega}_y)] + 2[\dot{z}\Omega_y - \dot{y}\Omega_z] \\ a_y = a_y' + \ddot{y} + x[\Omega_x\Omega_y + (\dot{\Omega}_z)] - y[(\Omega_x)^2 + (\Omega_z)^2] + z[\Omega_y\Omega_z - (\dot{\Omega}_z)] + 2[\dot{x}\Omega_z - \dot{z}\Omega_x] \\ a_z = a_z' + \ddot{z} + x[\Omega_x\Omega_z - (\dot{\Omega}_y)] + y[\Omega_y\Omega_z + (\dot{\Omega}_x)] - z[(\Omega_x)^2 + (\Omega_y)^2] + 2[\dot{y}\Omega_x - \dot{x}\Omega_y] \end{cases} \tag{2.4}$$

where (a_x', a_y', a_z') are the translational accelerations of the origin o of the moving frame (o–*xyz*) relative to the fixed inertial frame (O–*XYZ*), and the angular velocities are given by

$$\begin{cases} \Omega_x = \dot{\psi}\sin\theta\sin\phi + \dot{\theta}\cos\phi \\ \Omega_y = \dot{\psi}\sin\theta\cos\phi - \dot{\theta}\sin\phi \\ \Omega_z = \dot{\phi} + \dot{\psi}\cos\theta \end{cases} \tag{2.5}$$

2.2 Newton–Euler Equations of Motion for Rigid Bodies

A rigid body characterized by a domain Ω of constant volume V and mass M does not deform. The distance between any two points in a rigid body remains unchanged. A rigid body is, of course, an idealization since all bodies deform, more or less, under the action of external forces. However, this idealization is acceptable in many rock engineering problems, especially large-scale block movements under low stress conditions. Rigid body dynamics is governed by Newton's law of motion and Euler's rotations of rigid bodies.

2.2.1 Moments and Products of Inertia

In rigid body dynamics, the rotation of the body must be taken into account. The volume and geometric shape of the bodies therefore become important. The most important properties of a rigid body are its moments and products of inertia. Assume dm as the mass of a differential element in the rigid body (Fig. 2.2), $\mathrm{d}m = \rho\,\mathrm{d}V = \rho\,\mathrm{d}x\,\mathrm{d}y\,\mathrm{d}z$, its position is represented by vector $\mathbf{r} = \{r_i\} = (x, y, z)$ and its perpendicular distance to an arbitrary line through the origin OB is d. The direction cosines of the line OB are then given by $(l, m, n) = (\cos\alpha, \cos\beta, \cos\gamma)$ in the adopted inertial frame O–*XYZ*. The moment of inertia of the element about axis OB is then given by

$$\mathrm{d}I_{\mathrm{OB}} = (\mathrm{d}m)d^2 = \rho[(|\mathbf{r}|)^2 - (\mathrm{OP})^2]\mathrm{d}V = \rho[(x^2 + y^2 + z^2) - (lx + my + nz)^2]\mathrm{d}V \tag{2.6}$$

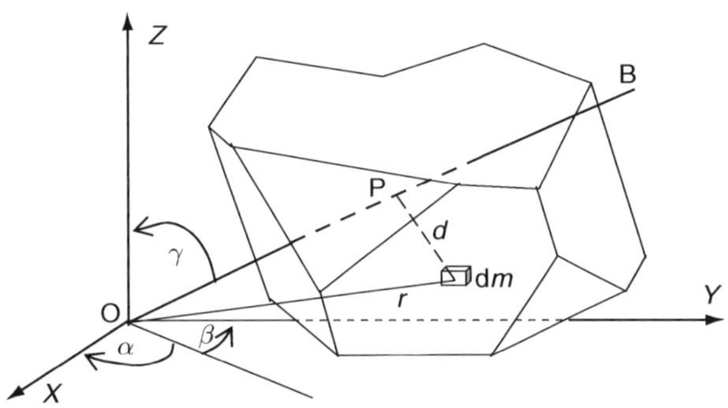

Fig. 2.2 Moments of inertia of a rigid body about an arbitrary axis OB.

Recall $l^2 + m^2 + n^2 = 1$, then the above equation can be rewritten

$$\begin{aligned} dI_{OB} &= \rho[(x^2 + y^2 + z^2)(l^2 + m^2 + n^2) - (lx + my + nz)^2]dV \\ &= \rho[l^2(y^2 + z^2) + m^2(x^2 + z^2) + n^2(x^2 + y^2) - 2lm(xy) - 2ln(xz) - 2mn(yz)]dV \end{aligned} \tag{2.7}$$

The moments of inertia of the whole body are then

$$\begin{aligned} I_{OB} &= \int_V \rho[l^2(y^2 + z^2) + m^2(x^2 + z^2) + n^2(x^2 + y^2) - 2lm(xy) - 2ln(xz) - 2mn(yz)]dV \\ &= \rho[l^2 I_{xx} + m^2 I_{yy} + n^2 I_{zz} - 2lmI_{xy} - 2lnI_{xz} - 2mnI_{yz}] \end{aligned} \tag{2.8}$$

where

$$I_{xx} = \int_V (y^2 + z^2)dV, \quad I_{yy} = \int_V (x^2 + z^2)dV, \quad I_{zz} = \int_V (x^2 + y^2)dV \tag{2.9a}$$

are called the moments of inertia about the x-axis, y-axis and z-axis, respectively, and

$$I_{xy} = \int_V xy\,dV, \quad I_{xz} = \int_V xz\,dV, \quad I_{yz} = \int_V yz\,dV \tag{2.9b}$$

are called the products of inertia of the body, respectively. The products of inertia are symmetric, i.e., $I_{xy} = I_{yx}, \ I_{xz} = I_{zx}, \ I_{yz} = I_{zy}$. The collection of moments and products of inertial of a rigid body is also often expressed by an inertia tensor, I_{ij}, given by

$$I_{ij} = \begin{bmatrix} I_{xx} & I_{xy} & I_{xz} \\ I_{yx} & I_{yy} & I_{yz} \\ I_{zx} & I_{zy} & I_{zz} \end{bmatrix} = \begin{bmatrix} I_{xx} & I_{xy} & I_{xz} \\ I_{xy} & I_{yy} & I_{yz} \\ I_{xz} & I_{yz} & I_{zz} \end{bmatrix} \tag{2.10}$$

which is a second-rank tensor. Its three principal moments of inertia, $I_x^p, \ I_y^p, \ I_z^p$, are given by three non-trivial roots of the equation

$$\begin{bmatrix} I^p - I_{xx} & I_{xy} & I_{xz} \\ I_{xy} & I^p - I_{yy} & I_{yz} \\ I_{xz} & I_{yz} & I^p - I_{zz} \end{bmatrix} \begin{Bmatrix} l \\ m \\ n \end{Bmatrix} = 0 \tag{2.11}$$

The necessary condition for the non-trivial solution of Eqn (2.11) is

$$\begin{vmatrix} I^p - I_{xx} & I_{xy} & I_{xz} \\ I_{xy} & I^p - I_{yy} & I_{yz} \\ I_{xz} & I_{yz} & I^p - I_{zz} \end{vmatrix} = 0 \tag{2.12}$$

The principal directions of inertia, $l_k, \ m_k, \ n_k, (k = x, y, z)$, can be obtained by inserting the principal moments of inertia, I_k^p $(k = x, y, z)$, into Eqn (2.11), resulting in the three orthogonal principal axes (x^p, y^p, z^p).

2.2.2 Mass, Linear and Angular Moments of Rigid Bodies

For a rigid body of mass density ρ and volume V, its mass M, linear momentum p_i and angular momentum h_i are defined by,

$$M = \iiint_V \rho dV = \rho V \tag{2.13}$$

$$p_i = \iiint\limits_V \rho v_i \, \mathrm{d}V = \rho V v_i^{\mathrm{c}} = M v_i^{\mathrm{c}} \tag{2.14}$$

$$h_i = \iiint\limits_V \rho e_{ijk} x_j v_k \, \mathrm{d}V \quad \text{or} \quad \iiint\limits_V \rho(\mathbf{r} \times \mathbf{v}) \mathrm{d}V \tag{2.15}$$

where v_i is the velocity of a point x in V with coordinates x_i, v_i^{c} the linear velocity vector at the mass center of the body, ρ the mass density and e_{ijk} is the permutation tensor. The angular momentum h_i is measured with respect to the origin of an inertial (global) frame. Vector $\mathbf{r} = (x_1, x_2, x_3)$ is the position vector and $\mathbf{v} = (v_1, v_2, v_3)$ is the velocity vector in Eqns (2.14) and (2.15).

2.3 Newton's Equations of Motion for Rigid Body Translations

Let f_i denote the resultant force vector of a set of external forces acting on a rigid body of mass M. The principle of the linear momentum conservation is expressed as

$$\frac{\mathrm{d}p_i}{\mathrm{d}t} = f_i \tag{2.16}$$

or

$$M \frac{\mathrm{d}v_i^{\mathrm{c}}}{\mathrm{d}t} = M a_i^{\mathrm{c}} = f_i \tag{2.17}$$

where a_i^{c} is the acceleration vector of the mass center of the rigid body. Besides the resultant force f_i as shown above, the set of external forces may also produce resultant torque causing rotational motions as well. However, Eqns (2.16) and (2.17) are valid regardless of the rotational motions that will be described by the equations of rotational motions as given below.

2.4 Euler's Equations of Rotational Motion – The General and Special Forms

For a representative differential element of volume dm in a rigid body with an embedded frame o–xyz with its origin located at an arbitrary point o (Fig. 2.3), its general acceleration vector can be directly deduced from Eqn (2.4), with the following simplifications: (i) x, y and z are constant and

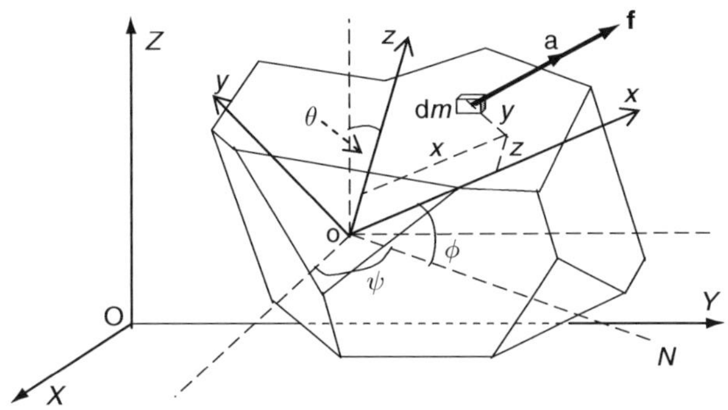

Fig. 2.3 Rotation and Euler angles of a rigid body.

(ii) $\dot{x} = \dot{y} = \dot{z} = \ddot{x} = \ddot{y} = \ddot{z} = 0$, since no relative movements of material points in the rigid body are allowed. The angular velocity $(\Omega_x, \Omega_y, \Omega_z)$ of the moving frame is now the angular velocity $(\omega_x, \omega_y, \omega_z)$ of the embedded body frame o–*xyz* relative to the global inertial frame (O–*XYZ*). The acceleration of the element is written as (Wells, 1967)

$$\begin{cases} a_x = a_x^{'} - x((\omega_y)^2 + (\omega_z)^2) + y(\omega_x\omega_y - \dot{\omega}_z) + z(\omega_x\omega_z + \dot{\omega}_y) \\ a_y = a_y^{'} + x(\omega_x\omega_y + \dot{\omega}_z) - y((\omega_x)^2 + (\omega_z)^2) + z(\omega_y\omega_z - \dot{\omega}_z) \\ a_z = a_z^{'} + x(\omega_x\omega_z - \dot{\omega}_y) + y(\omega_y\omega_z + \dot{\omega}_x) - z((\omega_x)^2 + (\omega_y)^2) \end{cases} \tag{2.18}$$

where *x*, *y* and *z* are the coordinates of the embedded (local) body frame, whose origin has the coordinates (X_o, Y_o, Z_o) in the global O–*XYZ* and acceleration $(a_x^{'}, a_y^{'}, a_z^{'})$.

For the differential element of volume d*m* in a rigid body as shown in Fig. 2.3, with acceleration $\mathbf{a} = (a_x, a_y, a_z)$ and resultant (internal) force $\mathbf{f} = (f_x, f_y, f_z)$, the 'free particle' equations of motion are written

$$(\mathrm{d}m)a_x = f_x, \quad (\mathrm{d}m)a_y = f_y, \quad (\mathrm{d}m)a_z = f_z \tag{2.19}$$

and the moments of force **f** about the body-embedded coordinate axes are

$$\begin{cases} (\mathrm{d}m)(a_z y - a_y z) = f_z y - f_y z = \mathrm{d}T_x \\ (\mathrm{d}m)(a_x z - a_z x) = f_x z - f_z x = \mathrm{d}T_y \\ (\mathrm{d}m)(a_y x - a_x y) = f_y x - f_x y = \mathrm{d}T_z \end{cases} \tag{2.20}$$

and the integration over the whole body leads to

$$\begin{cases} \int_V \rho(a_z y - a_y z)\mathrm{d}V = \int_V \rho(f_z y - f_y z)\mathrm{d}V = T_x \\ \int_V \rho(a_x z - a_z x)\mathrm{d}V = \int_V \rho(f_x z - f_z x)\mathrm{d}V = T_y \\ \int_V \rho(a_y x - a_x y)\mathrm{d}V = \int_V \rho(f_y x - f_x y)\mathrm{d}V = T_z \end{cases} \tag{2.21}$$

Equation (2.21) is the basic form of the equations of rotational motion. Substituting Eqn (2.18) into Eqn (2.21) and eliminating all acceleration components results in the general form of Euler's equations of motion for rigid body rotation

$$\begin{cases} M(a_z^{'} y_c - a_y^{'} z_c) + I_{xx}\dot{\omega}_x + (I_{zz} - I_{yy})\omega_y\omega_z + I_{xy}(\omega_x\omega_z - \dot{\omega}_y) - I_{xz}(\omega_x\omega_y + \dot{\omega}_z) + I_{yz}(\omega_z^2 - \omega_y^2) = T_x \\ M(a_x^{'} z_c - a_z^{'} x_c) + I_{yy}\dot{\omega}_y + (I_{xx} - I_{zz})\omega_x\omega_z + I_{yz}(\omega_y\omega_x - \dot{\omega}_z) - I_{xy}(\omega_y\omega_z + \dot{\omega}_x) + I_{xz}(\omega_z^2 - \omega_x^2) = T_y \\ M(a_y^{'} x_c - a_x^{'} y_c) + I_{zz}\dot{\omega}_z + (I_{yy} - I_{xx})\omega_x\omega_y + I_{xz}(\omega_y\omega_z - \dot{\omega}_x) - I_{yz}(\omega_x\omega_z + \dot{\omega}_y) + I_{xy}(\omega_y^2 - \omega_x^2) = T_z \end{cases} \tag{2.22}$$

where (x_c, y_c, z_c) are the coordinates of the center of mass of the rigid body.

Equations (2.22) and (2.17) determine completely the motion of a rigid body, with origin of the body-embedded frame o located at any point. The LHS of Eqn (2.22) is the summation of the moments of inertial forces about coordinate axes and the RHS are the summation of moments of the applied forces about the corresponding axes. The equation is therefore an expression of the 'principle of moments conservation'. The resultant moments of forces $\mathbf{T} = (T_x, T_y, T_z)$ can be calculated in the global inertial frame from the applied external forces by

$$T_x = \sum(f_Z Y - f_Y Z), \quad T_y = \sum(f_X Z - f_Z X), \quad T_z = \sum(f_Y X - f_X Y) \tag{2.23}$$

The general form of Eqn (2.22) can be simplified into special forms under certain conditions.

(a) Assuming that the origin o is located arbitrarily in the body and the local embedded coordinate axes x, y and z are oriented along the principal axes of inertia through origin o, the products of inertia $I_{xy} = I_{xz} = I_{yz} = 0$ and the rotational equations of motion become

$$\begin{cases} M(a_z' y_c - a_y' z_c) + I_{xx}\dot{\omega}_x + (I_{zz} - I_{yy})\omega_y\omega_z = T_x \\ M(a_x' z_c - a_z' x_c) + I_{yy}\dot{\omega}_y + (I_{xx} - I_{zz})\omega_x\omega_z = T_y \\ M(a_y' x_c - a_x' y_c) + I_{zz}\dot{\omega}_z + (I_{yy} - I_{xx})\omega_x\omega_y = T_z \end{cases} \tag{2.24}$$

(b) Assuming that the origin o of the embedded coordinate system (o–xyz) is located at the center of mass of the body (even though they may not be body-fixed), then $x_c = y_c = z_c = 0$. On the contrary, if the origin o is arbitrarily located but fixed in space (but can rotate around o), then $a_x' = a_y' = a_z' = 0$. In either case, Eqn (2.22) is simplified to

$$\begin{cases} I_x\dot{\omega}_x + (I_{zz} - I_{yy})\omega_y\omega_z + I_{xy}(\omega_x\omega_z - \dot{\omega}_y) - I_{xz}(\omega_x\omega_y + \dot{\omega}_z) + I_{yz}(\omega_z^2 - \omega_y^2) = T_x \\ I_y\dot{\omega}_y + (I_{xx} - I_{zz})\omega_x\omega_z + I_{yz}(\omega_y\omega_x - \dot{\omega}_z) - I_{xy}(\omega_y\omega_z + \dot{\omega}_x) + I_{xz}(\omega_x^2 - \omega_z^2) = T_y \\ I_z\dot{\omega}_z + (I_{yy} - I_{xx})\omega_x\omega_y + I_{xz}(\omega_y\omega_z - \dot{\omega}_x) - I_{yz}(\omega_x\omega_z + \dot{\omega}_y) + I_{xy}(\omega_y^2 - \omega_x^2) = T_z \end{cases} \tag{2.25}$$

(c) If the origin o of the body is either fixed in space or located at its center of mass, and the body-embedded frame is oriented along the principal inertial directions at the same time, Eqn (2.22) is then simplified to

$$\begin{cases} I_x^p\dot{\omega}_x + (I_z^p - I_y^p)\omega_y\omega_z = T_x \\ I_y^p\dot{\omega}_y + (I_x^p - I_z^p)\omega_x\omega_z = T_y \\ I_z^p\dot{\omega}_z + (I_y^p - I_x^p)\omega_x\omega_y = T_z \end{cases} \tag{2.26}$$

where (I_x^p, I_y^p, I_z^p) are the principal moments of inertia along principal axes (x^p, y^p, z^p).

The integrals $\int x^m y^n \mathrm{d}V$, $\int y^m z^n \mathrm{d}V$, $\int z^m x^n \mathrm{d}V$ (m, n = 0, 1, 2) are termed the integral properties of a rigid body and can be analytically evaluated for two-dimensional polygonal bodies using the simplex integration approach (Shi, 1993), as given in Chapter 9, when the coordinates of their vertices are known. During motion, these coordinates change continuously, so that their re-evaluation at each time step is computationally costly. This is the reason why for some DEM codes the calculations are simplified by using equivalent circular discs or spheres replacing the generally shaped polygons or polyhedra (but of identical areas and volumes) so that Eqn (2.26) is used for rotational calculations with constant inertial matrices. For rock engineering problems using tightly packed particle systems of little rotation, this simplification may be acceptable for some practical circumstances. However, it is theoretically incorrect and could not be applied for fundamental studies of the general behavior of granular media when rotation and rotational moments are critical issues that need to be considered, such as DEM simulations for equivalent Cosserat media where couple stresses due to particle rotation are the most important variable for the media's mechanical behavior, see Chapter 11 for more details. Other techniques using numerical integration based on triangularization of surfaces of polyhedra by triangle elements and constructive solid geometry (CSG) approaches using assembled regularly shaped solid parts for the representation of bodies of complex shapes can be seen in Messner and Taylor (1980) and Lee and Requicha (1982a,b).

2.5 Euler's Equations of Rotational Motion – Angular Momentum Formulation

The general form of Euler's rotational equation of motion for a rigid body is established using the concept of angular velocity, Euler angles and moments and products of inertia. A different form of the rotational equations of motion can be formulated using the concept of angular momentum. Referring to Fig. 2.4, regard the O–XYZ frame as the inertial and the o–xyz frame as body-embedded (but the origin o may or may not be attached to the body) and the two frames are parallel, with

$$X = X' + x, \quad Y = Y' + y, \quad Z = Z' + z \tag{2.27}$$

where (X', Y', Z') are the coordinates of the origin o of the body frame in the inertial frame.

Regarding $\mathbf{f} = (f_x, f_y, f_z)$ as the resultant force acting on the differential element dm, the equations of motion of the element, taking it as a free particle, can be written as

$$(\mathrm{d}m)\ddot{X} = f_x, \quad (\mathrm{d}m)\ddot{Y} = f_y, \quad (\mathrm{d}m)\ddot{Z} = f_z \tag{2.28}$$

using accelerations

$$\ddot{X} = \frac{\partial^2 x}{\partial t^2}, \quad \ddot{Y} = \frac{\partial^2 y}{\partial t^2}, \quad \ddot{Z} = \frac{\partial^2 z}{\partial t^2}$$

Similarly the moments of the body in the inertial frame can be written

$$\begin{cases} \int_V \rho(\ddot{Z}Y - \ddot{Y}Z)\mathrm{d}V = \frac{\mathrm{d}}{\mathrm{d}t}\int_V \rho(\dot{Z}Y - \dot{Y}Z)\mathrm{d}V = \int_V \rho(f_z Y - f_y Z)\mathrm{d}V = T_X \\ \int_V \rho(\ddot{X}Z - \ddot{Z}X)\mathrm{d}V = \frac{\mathrm{d}}{\mathrm{d}t}\int_V \rho(\dot{X}Z - \dot{Z}X)\mathrm{d}V = \int_V \rho(f_x Z - f_z X)\mathrm{d}V = T_Y \\ \int_V \rho(\ddot{Y}X - \ddot{X}Y)\mathrm{d}V = \frac{\mathrm{d}}{\mathrm{d}t}\int_V \rho(\dot{Y}X - \dot{X}Y)\mathrm{d}V = \int_V \rho(f_y X - f_x Y)\mathrm{d}V = T_Z \end{cases} \tag{2.29}$$

where (T_X, T_Y, T_Z) are the torques generated by the applied external forces about the X, Y and Z axes, respectively. Recall the definition of angular momentum from Eqn (2.15), and writing $\dot{X} = v_X$, $\dot{Y} = v_Y$, $\dot{Z} = v_Z$ as velocity components, Eqn (2.29) can be rewritten as

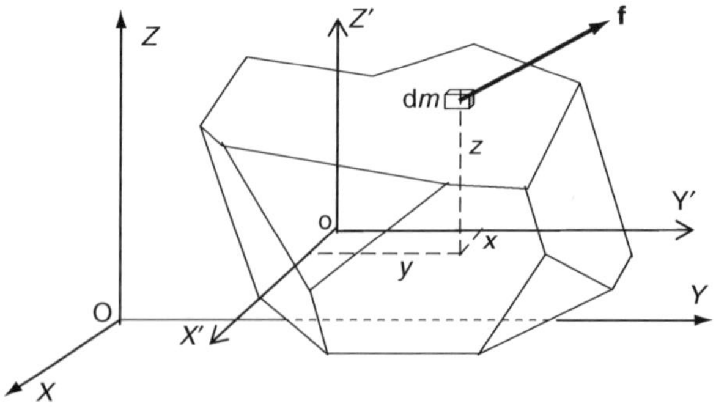

Fig. 2.4 Coordinate systems for the angular momentum formulation of Euler's rotational equations of motion of a rigid body.

$$\begin{cases} \dot{h}_X = \dfrac{\mathrm{d}}{\mathrm{d}t} \displaystyle\int_V \rho(v_Z Y - v_Y Z)\mathrm{d}V = \int_V \rho(f_z Y - f_y Z)\mathrm{d}V = T_X \\ \dot{h}_Y = \dfrac{\mathrm{d}}{\mathrm{d}t} \displaystyle\int_V \rho(v_X Z - v_Z X)\mathrm{d}V = \int_V \rho(f_x Z - f_z X)\mathrm{d}V = T_Y \\ \dot{h}_Z = \dfrac{\mathrm{d}}{\mathrm{d}t} \displaystyle\int_V \rho(v_Y X - v_X Y)\mathrm{d}V = \int_V \rho(f_y X - f_x Y)\mathrm{d}V = T_Z \end{cases} \tag{2.30}$$

This is the basic form of Euler's rotational equations of motion of a rigid body defined in the inertial frame, which uses no angular velocity, moments or products of inertia.

Substitution of relation (2.27) into Eqn (2.29) leads to

$$\begin{cases} \displaystyle\int_V \rho(\ddot{Z}Y' - \ddot{Y}Z')\mathrm{d}V + \int_V \rho(\ddot{z}y - \ddot{y}z)\mathrm{d}V + \int_V \rho(\ddot{Z}'y - \ddot{Y}'z)\mathrm{d}V = T_X \\ \displaystyle\int_V \rho(\ddot{X}Z' - \ddot{Z}X')\mathrm{d}V + \int_V \rho(\ddot{x}z' - \ddot{z}x)\mathrm{d}V + \int_V \rho(\ddot{X}'z - \ddot{Z}'x)\mathrm{d}V = T_Y \\ \displaystyle\int_V \rho(\ddot{Y}X' - \ddot{X}Y')\mathrm{d}V + \int_V \rho(\ddot{y}x - \ddot{x}y)\mathrm{d}V + \int_V \rho(\ddot{Y}'x - \ddot{X}'y)\mathrm{d}V = T_Z \end{cases} \tag{2.31}$$

Recall the relations

$$\begin{matrix} \displaystyle\int_V \rho\ddot{Z}'y\mathrm{d}V = M\ddot{Z}'y_c, & \displaystyle\int_V \rho\ddot{Y}'z\mathrm{d}V = M\ddot{Y}'z_c, & \displaystyle\int_V \rho\ddot{X}'z\mathrm{d}V = M\ddot{X}'z_c \\ \displaystyle\int_V \rho\ddot{Z}'x\mathrm{d}V = M\ddot{Z}'x_c, & \displaystyle\int_V \rho\ddot{Y}'x\mathrm{d}V = M\ddot{Y}'x_c, & \displaystyle\int_V \rho\ddot{X}'y\mathrm{d}V = M\ddot{X}'y_c \end{matrix} \tag{2.32a}$$

and

$$\begin{cases} T_X = \displaystyle\int_V \rho(f_z Y - f_y Z)\mathrm{d}V = \int_V \rho[f_z(Y' + y) - f_y(Z' + z)]\mathrm{d}V \\ T_Y = \displaystyle\int_V \rho(f_x Z - f_z X)\mathrm{d}V = \int_V \rho[f_x(Z' + z) - f_z(X' + x)]\mathrm{d}V \\ T_Z = \displaystyle\int_V \rho(f_y X - f_x Y)\mathrm{d}V = \int_V \rho[f_y(X' + x) - f_x(Y' + y)]\mathrm{d}V \end{cases} \tag{2.32b}$$

where (x_c, y_c, z_c) are the coordinates of the center of mass of the body in the moving frame, substitution of Eqns (2.28) and (2.32) into Eqn (2.31) leads to

$$\begin{cases} T_x = \displaystyle\int_V \rho(f_z y - f_y z)\mathrm{d}V = M(\ddot{Z}'y_c - \ddot{Y}'z_c) + \dfrac{\mathrm{d}}{\mathrm{d}t} \int_V \rho(\dot{z}y - \dot{y}z)\mathrm{d}V \\ T_y = \displaystyle\int_V \rho(f_x z - f_z x)\mathrm{d}V = M(\ddot{X}'z_c - \ddot{Z}'x_c) + \dfrac{\mathrm{d}}{\mathrm{d}t} \int_V \rho(\dot{x}z - z\dot{x})\mathrm{d}V \\ T_z = \displaystyle\int_V \rho(f_y x - f_x y)\mathrm{d}V = M(\ddot{Y}'x_c - \ddot{X}'y_c) + \dfrac{\mathrm{d}}{\mathrm{d}t} \int_V \rho(\dot{y}x - \dot{x}y)\mathrm{d}V \end{cases} \tag{2.33}$$

Defining rate of angular moments in the moving (local) frame by

$$\dot{h}_x = \int_V \rho(\ddot{z}y - \ddot{y}z)\mathrm{d}V, \quad \dot{h}_y = \int_V \rho(\ddot{x}z - z\ddot{x})\mathrm{d}V, \quad \dot{h}_z = \int_V \rho(\ddot{y}x - \ddot{x}y)\mathrm{d}V \tag{2.34}$$

then Eqn (2.33) can be rewritten as

$$\begin{cases} T_x = M(\ddot{Z}'y_c - \ddot{Y}'z_c) + \dot{h}_x \\ T_y = M(\ddot{X}'z_c - \ddot{Z}'x_c) + \dot{h}_y \\ T_z = M(\ddot{Y}'x_c - \ddot{X}'y_c) + \dot{h}_z \end{cases} \tag{2.35}$$

and is the usual form of Euler's rotational equations of motion of rigid blocks, based on the angular momentum concept. No moments and products of inertia or angular velocities of the block are required for numerical solutions. All quantities, except for the global coordinates (X', Y', Z') and their derivatives, are evaluated in the local body-embedded frame. If either the block is fixed at one point in space or the origin o of the moving frame is located at the mass center of the rigid body, Eqn (2.35) is simplified to

$$\begin{cases} T_x = \dot{h}_x \\ T_y = \dot{h}_y \\ T_z = \dot{h}_z \end{cases} \tag{2.36}$$

2.6 Cauchy's Equations of Motion for Deformable Bodies

Unlike a rigid body, a deformable body may translate, rotate and deform, i.e., the body changes from one configuration to another and has an infinite number of degrees of freedom.

Let a continuous body Ω have volume V, boundary surface S and undergo both translation and rotational motions under the resultant force f_i and resultant moment l_i about the origin of an inertial reference frame. As illustrated in Fig. 2.5, let (a_1, a_2, a_3) denote the coordinates of a point x in the body in the reference configuration at time $t = 0$. At a later time, the point is moved to another position (x_1, x_2, x_3) referred to in the same coordinate system, then the mapping

$$x_i = \hat{x}_i(a_1, a_2, a_3, t) \tag{2.37}$$

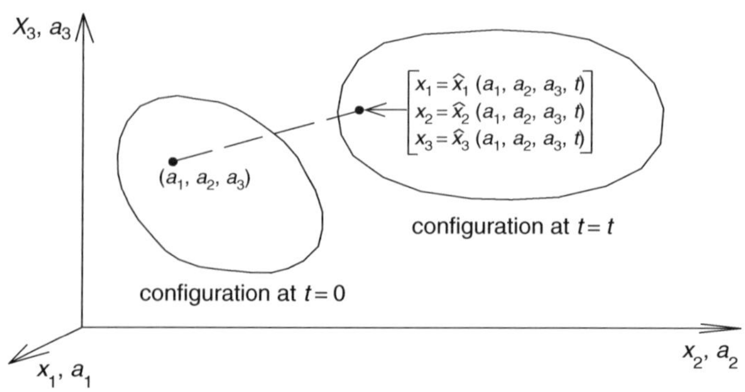

Fig. 2.5 Labeling of particles during motion.

links the instantaneous configurations of the body at different instants of time, t. The functions $\hat{x}_i(a_1, a_2, a_3, t)$is called the deformation function. When (a_1, a_2, a_3) and time, t, are considered as independent variables, the mapping by Eqn (2.37) gives the instantaneous configurations of the body at different instants of time, t. Then the description of the mechanical evolution is called a material description or Lagrangian description. If, on the contrary, the spatial location (x_1, x_2, x_3) and time, t, are taken as independent variables to describe the process, the description is called spatial description or Eulerian description. The latter is more convenient because it interprets the mechanical event that occurs at certain places, rather than following the movements of the particles.

In the material description, the velocity v_i and acceleration A_i at point (a_1, a_2, a_3) are defined as

$$v_i(a_1, a_2, a_3, t) = \frac{\partial x_i}{\partial t} \tag{2.38}$$

$$A_i(a_1, a_2, a_3, t) = \frac{\partial v_i}{\partial t} \tag{2.39}$$

while holding a_i constant. The deformation gradient of the body, f_{ij} can be expressed as

$$f_{ij} = \frac{\partial x_i}{\partial a_j} \tag{2.40}$$

and

$$\det[f_{ij}] > 0 \tag{2.41}$$

while holding time t constant.

In the spatial description, the inverse of the mapping (2.37) is used and the velocity and acceleration are, by using the chain rule, written as

$$v_i = \frac{\partial x_i}{\partial t} + \frac{\partial a_i}{\partial x_j}\frac{\partial x_j}{\partial t} = \frac{\mathrm{D}x_i}{\mathrm{D}t} \tag{2.42}$$

$$a_i = \frac{\partial v_i}{\partial t} + \frac{\partial a_i}{\partial x_j}v_j = \frac{\mathrm{D}v_i}{\mathrm{D}t} \tag{2.43}$$

The symbol $\mathrm{D}(\cdot)/\mathrm{D}t$ in Eqns (2.42) and (2.43) are the material derivatives.

The deformation gradient, the velocity and acceleration fields are the basic kinematics quantities of the motion of a continuum body. The other kinematics quantities, such as momentum and energy, can be defined from these basic quantities.

The equation of continuity is

$$\frac{\mathrm{D}M}{\mathrm{D}t} = \frac{\mathrm{D}}{\mathrm{D}t}\iiint \rho \, \mathrm{d}\Omega = \iiint \frac{\partial \rho}{\partial t}\mathrm{d}\Omega + \iint \rho v_i n_i \, \mathrm{d}S = 0 \tag{2.44}$$

where S is the surface of a representative differential volume of the continuum and n_i the unit normal vector of S. The differential form of Eqn (2.44) is then written as

$$\frac{\partial \rho}{\partial t} + \frac{\partial(\rho v_i)}{\partial x_i} = 0 \tag{2.45}$$

and is the expression for the principle of mass conservation.

The law of balance of linear and angular momentum may be written as

$$\frac{\mathrm{D}p_i}{\mathrm{D}t} = \frac{\mathrm{D}}{\mathrm{D}t}\iiint \rho v_i \,\mathrm{d}\Omega = f_i \tag{2.46}$$

$$\frac{\mathrm{D}h_i}{\mathrm{D}t} = \frac{\mathrm{D}}{\mathrm{D}t}\iiint e_{ijk}x_j v_k \,\mathrm{d}\Omega = l_i \tag{2.47}$$

The resultant force and moment of the body about the origin of the inertial frame can then be expressed as

$$f_i = \iint t_i \,\mathrm{d}S + \iiint b_i \,\mathrm{d}\Omega \tag{2.48}$$

$$l_i = \iint e_{ijk}x_j t_k \,\mathrm{d}S + \iiint e_{ijk}x_j b_k \,\mathrm{d}\Omega \tag{2.49}$$

where b_i is the body force.

Applying Gauss's theorem and Cauchy's stress formula,

$$t_i = \sigma_{ij} n_j \tag{2.50}$$

where σ_{ij} (i, j = 1, 2, 3) denote the components of the Cauchy stress tensor acting on an elementary area of the deformed body. Relations (2.48) and (2.49) then become

$$f_i = \iint t_i \,\mathrm{d}S + \iiint b_i \,\mathrm{d}\Omega = \iiint \left(\frac{\partial \sigma_{ij}}{\partial x_j} + b_i\right)\mathrm{d}\Omega \tag{2.51}$$

$$l_i = \iint e_{ijk}x_j t_k \,\mathrm{d}S + \iiint e_{ijk}x_j b_k \,\mathrm{d}\Omega = \iiint \left[e_{ijk}x_j\left(\frac{\partial \sigma_{ik}}{\partial x_i} + b_k\right)\right]\mathrm{d}\Omega \tag{2.52}$$

where S and Ω follow the deforming body. Substitution of Eqns (2.51) and (2.52) into Eqns (2.46) and (2.47) and using continuity Eqn (2.45) lead to the equations of motion for continuum bodies:

$$\frac{\mathrm{D}}{\mathrm{D}t}\iiint \rho v_i \,\mathrm{d}\Omega = \iiint \left(\frac{\partial \sigma_{ij}}{\partial x_j} + b_i\right)\mathrm{d}\Omega \tag{2.53a}$$

$$\frac{\mathrm{D}}{\mathrm{D}t}\iiint \rho e_{ijk}x_j v_k \,\mathrm{d}\Omega = \iiint \left[e_{ijk}x_j\left(\frac{\partial \sigma_{ik}}{\partial x_i} + b_k\right)\right]\mathrm{d}\Omega \tag{2.53b}$$

The differential forms of Eqns (2.53) and (2.54) may be written as

$$\rho\frac{\mathrm{D}v_i}{\mathrm{D}t} = \frac{\partial \sigma_{ij}}{\partial x_j} + b_i \tag{2.54}$$

$$e_{ijk}\sigma_{jk} = 0 \tag{2.55}$$

Equation (2.55) indicates that if the stress tensor is symmetric, i.e., $\sigma_{ij} = \sigma_{ji}$, the law of balance of angular momentum is satisfied at a point inside a continuum body. Equations (2.54) and (2.55) are the equations of motion for deformable bodies, usually called Cauchy's equations of motion.

In many applications for dynamic or quasi-static analyses of block or structural systems, damping is often used to describe the resistance effects to motions by viscous fluid, such as air. The most common formulation is to assume that damping is proportional to the velocity of the motion and the equation of motion for the damped body then becomes

$$\rho \frac{\mathrm{D}v_i}{\mathrm{D}t} + cv_i = \frac{\partial \sigma_{ij}}{\partial x_j} + b_i \tag{2.56}$$

where parameter c is called the damping coefficient that needs to be determined by experiments that could be very difficult to conduct for complex structures containing multiple deformable bodies. The damping can also serve as an artificially added force term to reach a static steady-state solution for a dynamic equation of motion. The damping term in such a case becomes a factor for a more stable numerical solution technique, rather than a physically meaningful mechanism; therefore a trial-and-error procedure may be used to reach a numerically appropriate damping coefficient value. The damping coefficient then simply plays a role as an artificial acceleration parameter for the convergence of quasi-static problems of blocks systems in DEM, see the details in Chapter 8.

2.7 Coupling of Rigid Body Motion and Deformation for Deformable Bodies

2.7.1 Complexities Caused by Rigid Body Motion and Deformation Coupling

The equations of motion for the deformable bodies, Eqns (2.55) and (2.56), are acceptable descriptions if the 'small displacement' assumption is accepted. This assumption means that, for a deformable body, the general size and shape of the body before and after the deformation process have negligible differences and that the strains caused by external and/or internal loads are small. This is an acceptable assumption for many practical problems where the total displacements are very small compared to the problem size. In other cases, rigid body movements can also be acceptable approximations for rock engineering problems if the main contributions of the deformation come from fracture displacements and the rock block deformations are rather small, such as in wedge sliding of blocks on rock slopes.

However, the small deformation or rigid body assumptions are just two extreme cases of uncoupled deformation-motion conditions, and are not necessarily universally valid. Under certain circumstances, deformable bodies may undergo large-scale displacements but have small strains that need to be taken into account (such as rock block motions during blasting, large-scale landslides or slope failures with internal rock deformation and fracturing processes, and movement and deformation/fracturing/splitting of falling rocks in highway engineering design, etc.). Under such conditions, the large-scale rigid body motion mode and the deformation mode of the bodies are coupled.

The motion and deformation of bodies undergoing large displacements have the following characteristics:

(1) The inertia of the body is no longer constant but is a function of time and displacement/deformation paths.

(2) The equations of motion become highly non-linear because of the finite rotation of the body relative to the inertial frame.

(3) The deformation of the body depends not only on the constitutive behavior of the material and the loads but also on the gross rigid body motion of the body relative to the inertial frame, i.e., the coupling between the rigid body motion mode and deformation mode.

The treatment of this coupling between rigid body motion and deformation has been an important aspect of engineering mechanics, especially in multibody system dynamics. The simplest method is to assume that the bodies are linear elastic materials following, therefore, the generalized Hooke's law. The resulting subject is called the linear theory of elastodynamics (Eringen, 1974, 1975; Shabana, 1998) which plays an important role in mechanical engineering and the aerodynamics of flight vehicles. The basic technique in solving the equations of motion for such problems is a three-step algorithm:

(1) assuming that the system consists of an assemblage of rigid bodies and solving the equations of motion to produce the inertial and interaction forces for each of the bodies and the gross rigid body translational and rotational displacements of the body as a whole;

(2) introducing these inertial and interaction forces to each of the bodies, but regarding them as elastically deformable bodies, to determine their deformation (displacement and strain) and stress fields according to analytical (if the geometry of the body permits) or numerical methods (FEM, for example);

(3) superposition of the small elastic deformation fields over the gross rigid body motion displacements.

The rigid body motion and elastic deformation are therefore de-coupled in linear elastodynamics. The rigid body motion and the static linear elasticity then can be seen as two extreme cases in general elastodynamics. The former governs the cases where the deformability and stress of the bodies are not of concern. The gross motion is the objective. Problems governed by the latter need only be concerned with the linear deformation of the body and the induced stress without gross motion.

Stepwise linearization processes can easily simulate the non-linearity of the materials without extra difficulties. Such a treatment, however, may not be suitable for problems with high motion velocities and non-linear deformations since deformation and gross rigid body mode of motion becomes significantly coupled. However, in most rock engineering problems, the cases of combinations of high velocities of motions and large, non-linear deformations are rare. Therefore using the linear elastodynamics principles can be taken as an acceptable approximation.

In the sections below, we first review the complete equations of motion of deformable bodies based on Cauchy's classical Eqns (2.54) and (2.55), and then we present the numerical treatment of the equations of motion in linear elastodynamics.

2.7.2 Extension of Equations of Motion of Deformable Bodies with Large Rotations

The complete formulation of the equations of motion, which combines the gross rigid body motion mode and deformation mode, without specifying any specific material behavior is presented in McDonough (1975). We accept that Cauchy's equations of motion, Eqns (2.55) and (2.56), are valid, but we further assume that the motion of the body (and all kinematic parameters) has also been defined relative to a non-inertial reference frame which is firmly associated with the body, but may or may not be fixed at the body, and has general translational and rotational motions relative to the inertial frame. Similar to the case of Euler's general form of rotational equations of motion for rigid bodies, the translations and rotations of the non-inertial reference frame thus represent the gross rigid body motion of the deformable body. The deformational motion of the particles inside the body, at a much smaller scale, is defined relative to the body-associated non-inertial reference frame. In the following development, we denote (X, Y, Z) as the inertial frame coordinates and $(\bar{x}, \bar{y}, \bar{z})$ are the coordinates measured in

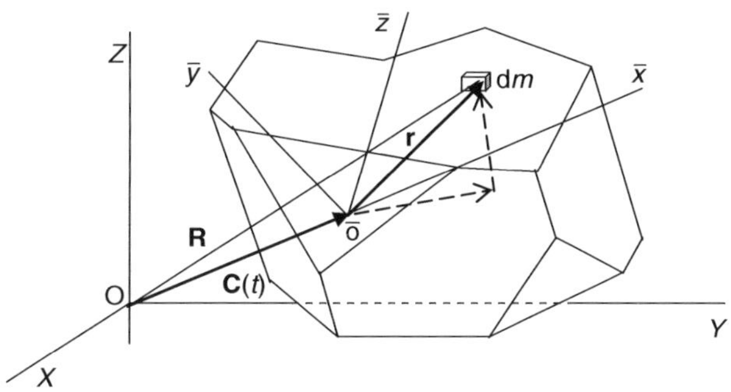

Fig. 2.6 Deformable body coordinates.

the non-inertial reference frame. The position vector in the inertial and non-inertial frames are denoted as $\mathbf{R} = (X, Y, Z)$ or $R_i = X_i$, or $\mathbf{r} = (\bar{x}, \bar{y}, \bar{z})$ or $r_i = \bar{x}_i$, (i = 1, 2, 3), respectively. The relation between the position vectors between the two frames for the same material point is given by (Fig. 2.6)

$$R_i = C_i(t) + r_i \quad \text{or} \quad X_i = C_i(t) + \bar{x}_i \tag{2.57}$$

where $\mathbf{C}(t)$ is the position vector of the origin of the non-inertial reference frame relative to the origin of the inertial frame and is a function of time, t. The material derivatives of an arbitrary vector, $\mathbf{v}$, relative to the two frames are given by

$$\dot{v}_i = \frac{\mathrm{D_R} v_i}{\mathrm{D}t} = e_{ijk}\omega_j v_k + \frac{\mathrm{D_r}\dot{v}_i}{\mathrm{D}t} \tag{2.58}$$

where $\mathrm{D}_R(\cdot)/\mathrm{D}t$ and $\mathrm{D}_r(\cdot)/\mathrm{D}t$ are the material time derivatives relative to the inertial frame and non-inertial reference frame, respectively, and $\vec{\omega} = \{\omega_X, \omega_Y, \omega_Z\}$ is the angular velocity vector of the non-inertial reference frame in the inertial frame, which is also a function of time. Insertion of relation (2.58) into Eqn (2.57) then results in

$$\frac{\mathrm{D_R R}_i}{\mathrm{D}t} = \frac{\mathrm{D_R} C_i(t)}{\mathrm{D}t} + e_{ijk}\omega_j r_k + \frac{\partial_\mathrm{r} r_i}{\partial t} \tag{2.59}$$

$$\frac{\mathrm{D_R^2 R}_i}{\mathrm{D}t^2} = \frac{\mathrm{D_R^2} C_i(t)}{\mathrm{D}t^2} + e_{ijk}\frac{\mathrm{D_R}\omega_j}{\mathrm{D}t} r_k + e_{ijk}\omega_j(\omega_k + r_k) + 2e_{ijk}\omega_j\frac{\partial_\mathrm{r} r_i}{\partial t} + \frac{\partial_\mathrm{r}^2 r_i}{\partial t^2} \tag{2.60}$$

Note that, because of the invariance of vectors and tensors transformed between the two frames, we have

$$\sigma_{ij} = \bar{\sigma}_{ij}, \quad f_i = \bar{f}_i, \quad \varepsilon_{ij} = \bar{\varepsilon}_{ij} \tag{2.61}$$

for the stress tensor, force vector and strain tensor defined in the inertial (without the short bar on the top) and non-inertial (with the short bar on the top) frames.

Recall Eqns (2.46, 2.47, 2.51 and 2.52) for the laws of linear and angular momentum conservation and apply them for a deformable body relative to a non-inertial reference frame moving relative to an inertial frame, then one has similarly

$$\mathrm{D_r}\frac{\bar{p}_i}{\mathrm{D}t} = \frac{\mathrm{D_r}}{\mathrm{D}t}\iiint \rho\bar{v}_i \, \mathrm{d}\Omega = \bar{f}_i \tag{2.62}$$

$$\mathrm{D_r}\frac{\bar{h}_i}{\mathrm{D}t} = \frac{\mathrm{D_r}}{\mathrm{D}t}\iiint e_{ijk}\bar{x}_j\bar{v}_k \, \mathrm{d}\Omega = \bar{l}_i \tag{2.63}$$

$$\bar{f}_i = \iint \bar{t}_i \, \mathrm{d}S + \iiint \bar{b}_i \, \mathrm{d}\Omega = \iiint \left(\partial \frac{\bar{\sigma}_{ij}}{\partial x_j} + \bar{b}_i \right) \mathrm{d}\Omega \tag{2.64}$$

$$\bar{l}_i = \iint e_{ijk} \bar{x}_j \bar{t}_k \, \mathrm{d}S + \iiint e_{ijk} \bar{x}_j \bar{b}_k \, \mathrm{d}\Omega = \iiint \left[e_{ijk} \bar{x}_j \left(\frac{\partial \dot{\sigma}_{ik}}{\partial x_i} + \bar{b}_k \right) \right] \mathrm{d}\Omega \tag{2.65}$$

Substitution of relations (2.57, 2.59 and 2.60) into Eqns (2.46, 2.47, 2.51 and 2.52) and recalling the invariant relation (2.61), one obtains the following relations after rearrangement of the terms

$$f_i = \bar{f}_i \tag{2.66}$$

$$l_i = e_{ijk} C_j \bar{f}_k + \bar{l}_i \tag{2.67}$$

$$p_i = M \frac{\mathrm{D_R} C_i}{\mathrm{D}t} + e_{ijk} \omega_{\mathrm{j}} G_k + \bar{p}_i \tag{2.68}$$

$$h_i = e_{ijk} C_j p_k - e_{ijk} C_j G_k + \omega_i \bar{I}_{ij} + \bar{h}_i \tag{2.69}$$

$$\frac{\mathrm{D_R} p_i}{\mathrm{D}t} = M \frac{\mathrm{D_R^2} C_i}{\mathrm{D}t^2} + e_{ijk} \omega_{\mathrm{j}} G_k + e_{ijk} \omega_{\mathrm{j}} (e_{lmn} \omega_m G_n + 2\bar{p}_i) + \frac{\mathrm{D_r} \bar{p}_i}{\mathrm{D}t} \tag{2.70}$$

$$\frac{\mathrm{D_R} h_i}{\mathrm{D}t} = e_{ijk} C_j \frac{\mathrm{D_R} p_k}{\mathrm{D}t} - e_{ijk} \frac{\mathrm{D_R^2} C_j}{\mathrm{D}t^2} G_k + \frac{\mathrm{D_R} \omega_i}{\mathrm{D}t} (\bar{I}_{ij} \omega_j) + e_{ijk} \omega_j (\omega_l \bar{I}_{lk}) + \omega_i \bar{I}_{ij} + e_{ijk} \omega_j \bar{h}_k + \frac{\mathrm{D_r} \bar{h}_i}{\mathrm{D}t} \tag{2.71}$$

where M is the mass of the body, $\bar{I}_{ij}$ the inertial tensor relative to the non-inertial reference frame and

$$G_i = \iiint_{\Omega} \rho r_i \, \mathrm{d}\Omega, = M \mathrm{r}_i^{\mathrm{c}} \tag{2.72}$$

where $r_i^{\mathrm{c}} = \bar{x}_i \mathrm{c}$ are the coordinates of the mass center in the non-inertial reference frame. The equations of motion of the deformable body defined in the inertial frame can then be written as the linear and angular momentum conservation laws:

$$f_i = \frac{\partial \sigma_{ij}}{\partial x_j} + b_i = M \frac{\mathrm{D_R^2} C_i}{\mathrm{D}t^2} + e_{ijk} \omega_j G_k + e_{ijk} \omega_j (e_{lmn} \omega_m G_n + 2\bar{p}_i) + \frac{\mathrm{D_r} \bar{p}_i}{\mathrm{D}t} \tag{2.73}$$

$$l_i = e_{ijk} x_j \left(\frac{\partial \sigma_{ik}}{\partial x_i} + b_k \right) = e_{ijk} C_j \frac{\mathrm{D_R} p_k}{\mathrm{D}t} - e_{ijk} \frac{\mathrm{D_R^2} C_j}{\mathrm{D}t^2} G_k + \frac{\mathrm{D_R} \omega_i}{\mathrm{D}t} (\bar{I}_{ij} \omega_j) + e_{ijk} \omega_j (\omega_l \bar{I}_{lk})$$

$$+ \omega_i \bar{I}_{ij} + e_{ijk} \omega_j \bar{h}_k + \frac{\mathrm{D_r} \bar{h}_i}{\mathrm{D}t} \tag{2.74}$$

Using relations (2.61), (2.66), (2.67) and (2.70), the equations of motion of a deformable body defined in the non-inertial reference frame can be derived as

$$\bar{f}_i = f_i = \frac{\partial \sigma_{ij}}{\partial x_{\mathrm{j}}} + b_i = M \frac{\mathrm{D_R^2} C_i}{\mathrm{D}t^2} + e_{ijk} \omega_j G_k + e_{ijk} \omega_j (e_{lmn} \omega_m G_n + 2\bar{p}_i) + \frac{\mathrm{D_r} \bar{p}_i}{\mathrm{D}t} \tag{2.75}$$

$$\bar{l}_i = -e_{ijk}C_jf_k + e_{ijk}x_j\left(\frac{\partial\sigma_{ik}}{\partial x_i} + b_k\right) = -e_{ijk}C_jf_k + e_{ijk}C_j\frac{\mathrm{D_R}p_k}{\mathrm{D}t} - e_{ijk}\frac{\mathrm{D_R^2}C_j}{\mathrm{D}t^2}G_k + \frac{\mathrm{D_R}\omega_i}{\mathrm{D}t}(\bar{I}_{ij}\omega_j)$$
$$+ e_{ijk}\omega_j(\omega_l\bar{I}_{lk}) + \omega_i\bar{I}_{ij} + e_{ijk}\omega_j\bar{h}_k + \frac{\mathrm{D_r}\bar{h}_i}{\mathrm{D}t} \tag{2.76}$$

The above equations can be simplified if the non-inertial reference frame is carefully selected, based on selecting $\mathbf{C}(t)$.

When $\mathbf{C}(t)$ is fixed at the mass center of the body, the coordinates of $\mathbf{C}(t)$ in the inertial frame are given by

$$C_i = X_i^{\mathrm{c}} = \frac{1}{M}\iiint_{\Omega}\rho R_i\,\mathrm{d}\Omega \tag{2.77}$$

and in the non-inertial reference frame

$$G_i = \iiint_{\Omega}\rho r_i\,\mathrm{d}\Omega, = \bar{p}_i = \frac{D_r\bar{p}}{Dt} = 0 \tag{2.78a}$$

and

$$\bar{x}^{\mathrm{c}} = \bar{y}^{\mathrm{c}} = \bar{z}^{\mathrm{c}} = 0 \tag{2.78b}$$

Eqns (2.66) and (2.67) are then reduced to

$$\frac{\partial\sigma_{ij}}{\partial x_{\mathrm{j}}} + b_i = M\frac{\mathrm{D_R^2}C_i}{\mathrm{D}t^2} \tag{2.79}$$

$$e_{ijk}\bar{x}_j\left(\frac{\partial\sigma_{ij}}{\partial xi} + b_k\right) = \frac{\mathrm{D_R}\omega_i}{\mathrm{D}t}(\bar{I}_{ij}\omega_j) + e_{ijk}\omega_j(\omega_l\bar{I}_{lk}) + \omega_i\bar{I}_{ij} + e_{ijk}\omega_j\bar{h}_k + \frac{\mathrm{D_r}\bar{h}_i}{\mathrm{D}t} \tag{2.80}$$

It is clear from the above derivation and Eqn (2.80) that the main complication in the coupling of rigid body motion and deformation concerns the coupling between rotation and deformation, since the inertial tensor is now a function of both time and deformation. This is the so-called issue of inertial coupling or co-rotation in FEM methods for problems with large rotations of deformable bodies and is especially important for problems with high velocities and slender bodies.

2.7.3 Treatment of Inertial Coupling of Motion and Deformation using FEM

The most common numerical technique for solving problems of elastodynamics of multibody systems is the FEM which is an effective technique for the treatment of inertial coupling of motion and deformation of deformable bodies. Using the FEM technique, Eqns (2.79) and (2.80) can be formulated into a matrix equation in a partitioned form (Shabana, 1998) for body i after FEM discretization

$$\begin{cases} \mathbf{M}_{\mathrm{rr}}^i\ddot{\mathbf{u}}_{\mathrm{r}}^i + \mathbf{M}_{\mathrm{rf}}^i\ddot{\mathbf{u}}_{\mathrm{f}}^i = \mathbf{F}_{\mathrm{r}}^i \\ \mathbf{M}_{\mathrm{fr}}^i\ddot{\mathbf{u}}_{\mathrm{r}}^i + \mathbf{M}_{\mathrm{rf}}^i\ddot{\mathbf{u}}_{\mathrm{f}}^i + \mathbf{K}_{\mathrm{ff}}^i\mathbf{u}_{\mathrm{f}}^i = \mathbf{F}_{\mathrm{f}}^i \end{cases} \tag{2.81}$$

where $\mathbf{M}_{\mathrm{rr}}^i$ is the mass matrix for rigid body mode, $\mathbf{M}_{\mathrm{ff}}^i$ the mass matrix for deformation mode, $\mathbf{M}_{\mathrm{rf}}^i = (\mathbf{M}_{\mathrm{fr}}^i)^{\mathrm{T}}$ are the inertial coupling matrices, $\mathbf{K}_{\mathrm{ff}}^{\mathrm{i}}$ is the stiffness matrix, $\mathbf{u}_{\mathrm{r}}^i$ and $\mathbf{u}_{\mathrm{f}}^i$ are the generalized coordinate (unknown) vectors for the partitioned rigid body motion mode and deformation mode (with the two dots indicating second-order partial differentiation with respect to time), and $\mathbf{F}_{\mathrm{r}}^i$ and $\mathbf{F}_{\mathrm{f}}^i$ are the partitioned generalized force vectors, respectively. The subscripts r and f indicate the partition of rigid body motion and deformation modes, respectively.

Assuming a linearized elastodynamic simplification, the above equation is reduced to

$$\begin{cases} \mathbf{M}^i_{\mathrm{rr}}\ddot{\mathbf{u}}^i_{\mathrm{r}} = \mathbf{F}^i_{\mathrm{r}} \\ \mathbf{M}^i_{\mathrm{rf}}\ddot{\mathbf{u}}^i_{\mathrm{f}} + \mathbf{K}^i_{\mathrm{ff}}\mathbf{u}^i_{\mathrm{f}} = \mathbf{F}^i_{\mathrm{f}} - \mathbf{M}^i_{\mathrm{fr}}\ddot{\mathbf{u}}^i_{\mathrm{r}} \end{cases} \tag{2.82}$$

since the contribution of the elastic deformation to the change of generalized coordinates is negligible. The first equation in Eqn (2.82) can then be solved by considering just rigid body motions alone, but the impact of the inertial coupling must be included for elastic deformation calculations, as indicated in the last term in the RHS of the second equation in Eqn (2.82). The key element is to ensure that the condition of zero strain generation by the rigid body motion in the body prevails – the co-rotation strain constraint.

It should be noted that the inertial coupling is most important for simulating coupled motion and deformation of slender bodies using FEM, such as beam, plate and shell structures where the linear dimensions of the bodies are much larger in one (beam) or two (shell) dimensions than that in other dimensions, where conventional FEM discretization cannot cope with large rotations since infinitesimal rotations are used as general nodal unknowns. For general elastic bodies with full FEM discretization using standard meshing and using displacements as the only nodal unknowns, the inertial coupling is automatically ensured. An example is given below in Fig. 2.7 as a demonstration using the standard 4-noded plane FEM elements, as shown in Shabana (1998).

For the four-noded rectangular FEM element as shown in Fig. 2.7 with the nodal displacement components forming the unknown vector $\mathbf{U} = (u^1_x, u^1_y, u^2_x, u^2_y, u^3_x, u^3_y, u^4_x, u^4_y)^T$, the geometry matrix of the element is given by

$$\mathbf{S} = \begin{bmatrix} N_1 & 0 & N_2 & 0 & N_3 & 0 & N_4 & 0 \\ 0 & N_1 & 0 & N_2 & 0 & N_3 & 0 & N_4 \end{bmatrix} \tag{2.83}$$

with the following shape functions

$$N_1 = \frac{1}{4bc}(b-x)(c-y), \quad N_2 = \frac{1}{4bc}(b+x)(c-y) \tag{2.84}$$

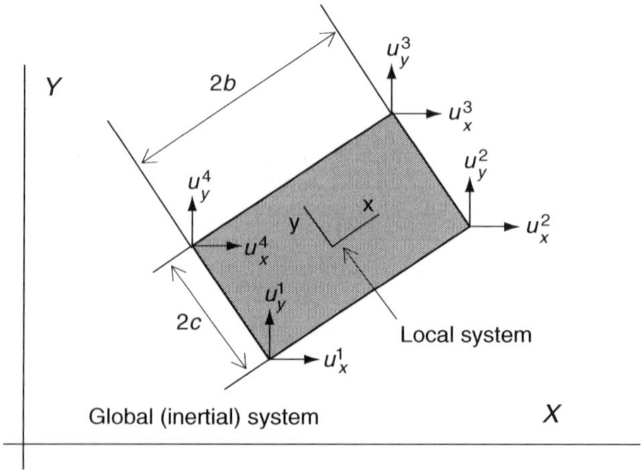

Fig. 2.7 A rectangular FEM element undergoing large rotation (Shabana, 1998).

$$N_3 = \frac{1}{4bc}(b+x)(c+y), \quad N_4 = \frac{1}{4bc}(b-x)(c+y) \tag{2.85}$$

where b and c are the length and width of the element (Fig. 2.7), and the sum of the shape functions is equal to unity. For a rigid body motion described by two translational components R_x and R_y, and a finite rotation angle θ, the nodal displacement vector **U** then becomes

$$\mathbf{U} = \begin{Bmatrix} u_x^1 \\ u_y^1 \\ u_x^2 \\ u_y^2 \\ u_x^3 \\ u_y^3 \\ u_x^4 \\ u_y^4 \end{Bmatrix} = \begin{Bmatrix} \mathrm{R}_x - b\cos\theta + c\sin\theta \\ R_y - b\sin\theta - c\cos\theta \\ R_x + b\cos\theta + c\sin\theta \\ R_y + b\sin\theta - c\cos\theta \\ R_x + b\cos\theta - c\sin\theta \\ R_y + b\sin\theta + c\cos\theta \\ R_x - b\cos\theta - c\sin\theta \\ R_y - b\sin\theta + c\cos\theta \end{Bmatrix} \tag{2.86}$$

The product of the shape function matrix **S** and displacement vector **U** leads to

$$\mathbf{SU}^{\mathrm{T}} = \begin{bmatrix} R_x + x\cos\theta - y\sin\theta \\ R_y + x\sin\theta + y\cos\theta \end{bmatrix} \tag{2.87}$$

where x and y are the coordinates of the embedded local frame in the element and the relation shows an exact rigid body motion as required. The zero strain condition is met since the rectangle elements are conforming elements of completeness and compatibility, as defined in Bathe and Wilson (1976).

2.8 Equations for Heat Transfer and Coupled Thermo-Mechanical Processes

Heat is transferred in three modes: conduction, convection and radiation. For fractured rocks, conduction and convection through fluid movement are the major modes of heat transfer. In this section, only the basic equations for conductive heat transfer and some key thermal properties (thermal conductivity and heat capacity (specific heat)) are presented. The convective heat transfer due to fluid flow through fractures will be described in Chapter 8 for DEM presentations.

2.8.1 Fourier's Law and the Heat Conduction Equation

The basic constitutive law for heat conduction in a continuum is Fourier's law. It states that the heat flux, q_i^{h}, across a cross-sectional surface of unit area in a continuum is proportional to the gradient of the temperature field, T, with a proportional coefficient κ (W/m · K), called thermal conductivity,

$$q_i^{\mathrm{h}} = -\kappa \frac{\partial T}{\partial x_{\mathrm{j}}} \tag{2.88}$$

Ignoring the conversion of the mechanical work into heat (which is usually very small for general rock engineering practice), the energy conservation equation is usually given by

$$\rho c_p \frac{\partial T}{\partial t} = -(q_i^{\mathrm{h}})_{,j} + s^{\mathrm{h}} \tag{2.89}$$

where s^{h} is the source term (W/m^3) and c_p is called the specific heat of the medium. Substitution of Eqn (2.88) into Eqn (2.89) then leads to

$$T_{,ii} = \nabla^2 T = \frac{\rho c_p}{\kappa}\frac{\partial T}{\partial t} - \frac{s^{\mathrm{h}}}{\kappa} = \frac{1}{\alpha}\frac{\partial T}{\partial t} - \frac{s^{\mathrm{h}}}{\kappa} \tag{2.90}$$

and this is called the heat conduction (or diffusion) equation. The symbol $\alpha = \kappa/\rho c_p$ is the thermal diffusivity of the medium. For steady-state problems with no source term conditions, the equation is similarly reduced to Laplace's equation

$$\nabla^2 T = 0 \tag{2.91}$$

2.8.2 Thermal Strain and the Constitutive Equation of Thermo-Elasticity

During a coupled thermo-mechanical process, the total linear strain of a material point is assumed (but well proven in practice) to be the sum of two components: the mechanical strain $\varepsilon_{ij}^{\mathrm{M}}$ caused by external forces and thermal strain $\varepsilon_{ij}^{\mathrm{T}}$ caused by the temperature gradient field

$$\varepsilon_{ij} = \varepsilon_{ij}^{\mathrm{M}} + \varepsilon_{ij}^{\mathrm{T}} \tag{2.92}$$

Assuming elastic behavior of the rock, the mechanical strain follows Hooke's law of elasticity with respect to stress, given by

$$\varepsilon_{ij}^{\mathrm{M}} = \frac{1}{2G}\left(\sigma_{ij} - \frac{\lambda}{3\lambda + 2G}\delta_{ij}\sigma_{kk}\right) \tag{2.93}$$

where λ and G are Lame's elasticity constants and δ_{ij} is the Kronecker delta.

The thermal strain is given by

$$\varepsilon_{ij}^{\mathrm{T}} = \alpha(T - T_0)\delta_{ij} \tag{2.94}$$

where α is the thermal expansion coefficient, T is the current temperature and T_0 is the initial (reference) temperature. Substitution of Eqns (2.93) and (2.94) into Eqn (2.92) lead to

$$\varepsilon_{ij} = \varepsilon_{ij}^{\mathrm{M}} + \varepsilon_{ij}^{\mathrm{T}} = \frac{1}{2G}\left(\sigma_{ij} - \frac{\lambda}{3\lambda + 2G}\delta_{ij}\sigma_{kk}\right) + \alpha(T - T_0)\delta_{ij} \tag{2.95}$$

This is the Duhamel–Neumann relation for thermoelasticity. Its reciprocal is the constitutive equation for the thermoelasticity of a continuum

$$\sigma_{ij} = \lambda\delta_{ij}\varepsilon_{kk} + 2G\varepsilon_{ij} - (3\lambda + 2G)\cdot\alpha(T - T_0)\delta_{ij} \tag{2.96}$$

The equations of motion of elastically deformable blocks of density ρ with heat transfer is the same as before, Eqn (2.79), but in a simpler and more commonly seen form

$$\frac{\partial\sigma_{ij}}{\partial x_{\mathrm{j}}} + \rho b_i = \rho\ddot{u}_i \tag{2.97a}$$

but the terms in the stress tensor given by Eqn (2.96) should be used. When the equation is written in terms of displacement instead of stress, the equation of motion becomes

$$Gu_{i,jj} + (\lambda + G)u_{j,ji} + \rho b_i + (3\lambda + 2G)\alpha(T - T_0)_{,j}\delta_{ji} = \rho\ddot{u}_i \tag{2.97b}$$

2.8.3 Heat Conduction and the Energy Conservation Equation

If the transition of internal energy into heat is to be considered, the heat transfer Eqn (2.83) needs to be modified. Introducing the heat capacity of the medium as

$$c_p = -\left[\frac{1}{\rho}\left(\frac{\partial T}{\partial t}\right)^{-1}\right]\frac{\partial q_i^{\mathrm{h}}}{\partial x_i} \tag{2.98}$$

with unit (J/kg °C) and assuming that the internal energy of the continuum is a function of both strain and temperature, then the energy balance equation representing the first law of thermodynamics can be written as

$$\kappa T_{,kk} + s^{\mathrm{h}} = \rho c_p \frac{\partial T}{\partial t} + (3\lambda + 2G)\alpha T_0 \frac{\partial \dot{\varepsilon}_{kk}}{\partial t} \tag{2.99}$$

where s^{h} is the heat source term.

The effect of mechanical deformation (represented by the volumetric strain rate as the last term in the RHS of (2.99) on the temperature field) is therefore included. If this effect is negligible, the equation is reduced to the heat conduction Eqn (2.89).

For an anisotropic media in two dimensions, the transient heat conduction equation is simply extended as

$$\frac{\partial T}{\partial t} = \frac{1}{\rho C_p}\left[\frac{\partial}{\partial x}\left(k_x \frac{\partial T}{\partial x}\right) + \frac{\partial}{\partial y}\left(k_y \frac{\partial T}{\partial y}\right)\right] \tag{2.100}$$

where k_i $(i = x, y)$ are the heat conductivities of the rock material in direction i.

References

Bathe, K.-J. and Wilson, E. L., Numerical methods in finite element analysis. Prentice-Hall, Englewood Cliff, New Jersey, 1976.

Fung, Y. C., A first course in continuum mechanics. Prentice-Hall, Englewood Cliffs, New Jersey, 1969.

Eringen, A. C., Elastodynamics. Vol. I: Finite motion. Elsevier, Rotterdam, 1974.

Eringen, A. C., Elastodynamics. Vol. II: Linear theory. Elsevier, Rotterdam, 1975.

Lai, W. M., Rubin, D. and Krempl, E., Introduction to continuum mechanics, 3rd edition, Butterworth-Heinemann, MA, 1993.

Lee, Y. T. and Requicha, A. A. G., Algorithms for computing the volume and other integral properties of solids. I. Known methods and open issues. Communication of the ACM, 1982a;25(9):635–641.

Lee, Y. T. and Requicha, A. A. G., Algorithms for computing the volume and other integral properties of solids. II. A family of algorithms based on representation convention and cellular approximation. Communication of the ACM, 1982b;25(9):642–650.

McDonough, T. B., Formulation of the global equations of motion of a deformable body. American Institute of Aeronautics and Astronautics Journal, 1975;14(5):656–660.

Messner, A. M. and Taylor, G. Q., Algorithm 550: solid polyhedron measures [Z]. ACM Transactions on Mathematical Software, 1980;6(1):121–130.

Shabana, A. A., Dynamics of multibody systems. 2nd edition, Cambridge University Press, London, 1998.

Shi, G., Block system modeling by discontinuous deformation analysis. Computational Mechanics Publications, Southampton UK, 1993.

Wang, C. Y., Mathematical principles for continuum mechanics and magnetism. Part A: Analytical and continuum mechanics, Plenum Press, New York, 1975.

Wells, D. A., Theory and problems of Lagrangian dynamics. Schaum's outline series. McGraw-Hill, NY, 1967.

3 CONSTITUTIVE MODELS OF ROCK FRACTURES AND ROCK MASSES – THE BASICS

Constitutive models of rock fractures and rock masses are key components for numerical modelling of the physical behavior of fractured rocks. The development of the models must meet two requirements:

- They must be able to capture the conceptual behavior of rocks observed in laboratory experiments and/or field observations to acceptable tolerances for the quantitative analysis of rock deformations and stresses as required by the rock engineering problems concerned.
- They must be able to simulate rock and fracture behavior under general loading conditions and the associated stress/deformation paths without violating the second law of thermodynamics.

It should be noted that constitutive models are theoretical approximations to what one observes in reality, based on certain assumptions according to the different theoretical principles and mathematical approaches adopted, and the material behavior observed in laboratory or field observations. Due to the extreme complexity of the physical behavior of the rock fractures and fractured rocks, and the limitation of currently available mathematical tools and computer methods, it is not possible to simulate every aspect of the physical behavior by mathematical models. Only the most important aspects of the overall behavior are usually considered when developing constitutive models. Different guiding principles and mathematical approaches naturally lead to different constitutive model formulations and these are associated with different sets of material properties and parameters. Together with the equations for conservation of mass, momentum and energy, and the algorithms of contact detection, the equations for the constitutive models for rock fractures and rock masses are the most important components of the discrete element methods.

There are generally two approaches to formulate a constitutive model: the empirical approach and theoretical approach. The empirical approach of model development uses empirical functions best fitting the measured results from laboratory or field experiments, and these are conducted often with simple loading paths. Curve fitting using mathematical regression techniques (e.g., least-squares regression) is most often used with the addition of artificially defined parameters determined by the regression process. Additional flexibility in the parameters or functions may also be added so that the models can cover larger ranges of data observed in different but similar laboratory or field experiments. Generally, however, there is no consideration given to the critical need for satisfying the second law of thermodynamics; therefore under special conditions that cannot be considered in experiments, the models may generate, rather than dissipate, energy. However, such models can provide agreeable results if the loading conditions and parameter ranges are suitably considered.

The theoretical approach, on the other hand, usually takes one of the solid mechanics branches, such as plasticity theory, contact mechanics, damage mechanics, etc., as the basic mathematical platforms for the formulation. Necessary assumptions used in the theories must be inherited and special considerations must be given to the choices of material parameters or properties so that the most important aspects of the fractures or rock masses may be reflected in the model, with properties/parameters that can be determined by laboratory or field experiments. The thermodynamic restrictions need to be applied so that the models will not violate the second law under general loading conditions within certain ranges of parameter values but, unfortunately, this latter practice is not always followed when developing constitutive models for rock engineering problems. The difficulty of such theoretical approaches is that special parameters are often needed to follow the guiding principles – but they may not have clear physical meanings or are difficult to determine by experiments. The theoretical models are usually 'grossly acceptable' universal models satisfying all basic physical laws and the empirical models are 'good models' under special conditions. A more rational approach is the combination of the two approaches with both a sound thermodynamics basis and the flexibility for parameterization, reflecting the most important aspects of the mechanical behavior of the subjects and processes.

In this chapter, the fundamental aspects of mathematical theories relating to the mechanical behavior of rock fractures and rock masses are presented, including the most commonly encountered criteria for shear strength and constitutive models for rock fractures, and constitutive models for rock masses based on the theory of elasticity, elastoplasticity and the crack tensor concept. The basic theory of plasticity in solid mechanics is not included here because it is extensively covered in the literature. However, the elastic model is the most often adopted constitutive behavior in DEM codes. Since the constitutive model is a very wide field with numerous publications and a subject of very active researches, only a limited number of published literature are cited here.

An important aspect of constitutive model development for rock fractures is the characterization of the roughness of the rock fracture surfaces. However, due to limits on the space and scope of this book, this aspect is not included as a systematic presentation. However, in order that readers can probe the subject in-depth, references are given when special aspects of roughness related to model formulations are described.

Many criteria for shear strength and constitutive models for rock fractures, and many more constitutive models for fractured rock masses, have been proposed over the years. We, however, cover only a limited number of them in this book with our focus on the DEM approaches – as developed in the following chapters.

3.1 Mechanical Behavior of Rock Fractures

Figure 3.1a and b illustrates the typical, much idealized, shear stress–shear displacement behavior of rock fractures under a constant normal stress or a constant normal stiffness condition during direct shear tests under laboratory test conditions, with monotonic loading. Due to the surface roughness, the extra increment of the normal stress under constant normal stiffness constraint will cause continuous increase of shear stress so that the apparent 'peak' may not appear. The fracture behavior under a normal compressive loading–unloading sequence is illustrated in Fig. 3.1c. The shifting of the stress–closure curves with repeated loading–unloading cycles with increasing normal closure indicates damage to the asperities on the fracture surfaces accumulated during the tests.

Denoting the shear and normal stresses and displacements of fractures as σ_t, σ_n, u_t and u_n, respectively, an idealized shear stress–shear displacement curve can be characterized by five parameters: shear stiffness k_t, peak shear stress (shear strength) σ_t^p, residual shear stress σ_t^r, shear displacement at peak shear stress u_t^p and shear displacement when residual shear stress starts u_t^r. The ratio $\phi_r = \sigma_t^r/\sigma_n$ is called the residual friction angle. The shear stiffness may be expressed by

$$k_t = \frac{\partial \sigma_t}{\partial u_t} \quad (0 \le u_t \le u_t^p) \tag{3.1}$$

in the range of the curve before the peak shear stress, as is often the case observed in laboratory tests.

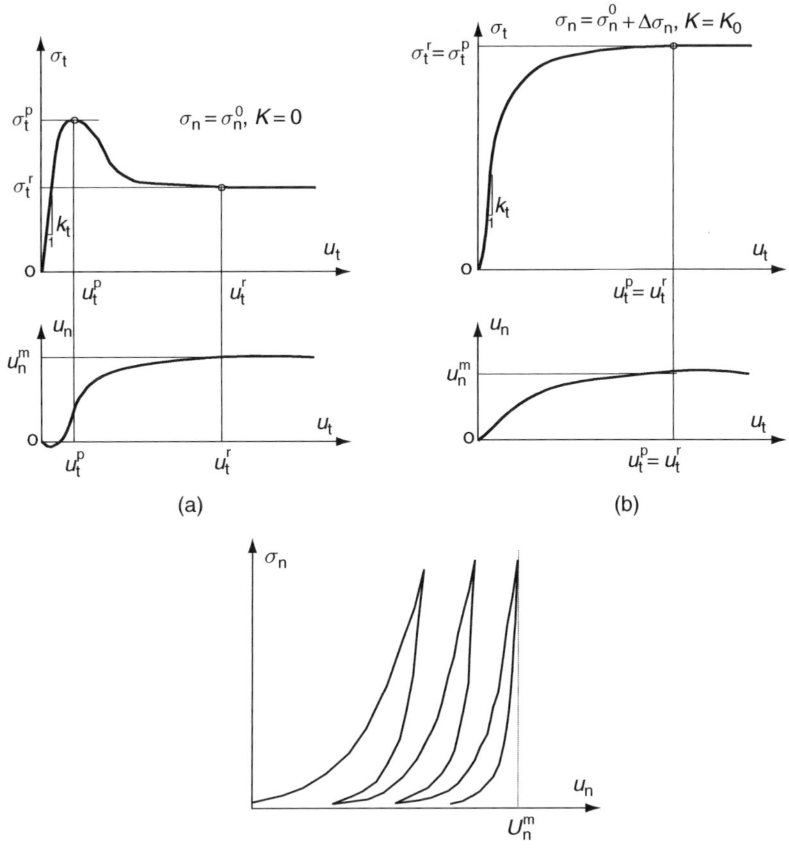

Fig. 3.1 Conceptual behavior of rock fractures during direct shear testing and normal compression testing: (a) under a constant normal stress; (b) under a constant normal stiffness; (c) during compressive loading–unloading cycles.

The peak shear stress may be predicted by various criteria (see Section 3.2) and the shear stiffness is found to depend on the magnitude of the normal stresses acting on the rock fractures. Based on a series of direct shear tests under different normal stresses, Jing (1990) proposed an empirical relation to consider this stress dependency of the shear stiffness:

$$k_t = k_{t0}\left(1 - \frac{\sigma_n}{\sigma_c}\right)^a \tag{3.2}$$

where k_{t0} is the initial shear stiffness, a a material constant, σ_n the normal stress acting on the fracture surface and σ_c the uniaxial compressive strength of the rock material.

An important phenomenon which has a critical bearing on the hydraulic conductivity and deformability of the rock fractures is the dilatancy, i.e., the increase of the fracture aperture during shearing due to over-riding of the asperities on the opposing rough surfaces of the fractures. The rate and magnitude of the dilatancy depends on many factors, such as the magnitude of the compressive normal stresses, material hardness and/or compressive strength of the rock material, presence of filling materials and fluids and the shearing velocities. However, the most important factor is the morphological characteristics of the surface roughness of the rock fractures. Various empirical models have been suggested to simulate the dilatancy phenomenon – with varying degrees of success. The simplest approach is to use a constant 'dilatancy angle' α relating the increments of the normal and tangential displacements

$$du_n = (\tan\alpha)du_t \quad (0 \le u_t \le u_t^r) \tag{3.3a}$$

$$du_n = 0 \quad (u_t^r < u_t) \tag{3.3b}$$

Although this is a crude approximation it has been accepted in many constitutive models for rock fractures just because of its simplicity. More detailed treatment of dilatancy will be given in presentations of the different constitutive models in Section 3.3.

There are two empirical models that are most often applied to represent the normal stress–normal displacement (closure) curves. One is a hyperbolic function proposed by Bandis (1980):

$$\sigma_n = \sigma_n = \frac{u_n}{a - bu_n} = k_{n0}\left(\frac{u_n}{1 - u_n/u_n^m}\right) \tag{3.4}$$

where k_{n0} is the initial normal stiffness and a and b are experimental constants with $k_{n0} = 1/a$ and $u_n^m = a/b$. Another was suggested by Goodman (1976):

$$\frac{\sigma_n - \sigma_{n0}}{\sigma_{n0}} = A\left(\frac{u_n}{u_n^m - u_n}\right)^t \tag{3.5}$$

where σ_{n0} is a reference normal stress and A and t are material constants. The parameter u_n^m is the maximum normal displacement (closure) of the fracture.

The normal stiffness of the fracture is then given by

$$k_n = \frac{\partial\sigma_n}{\partial u_n} = k_{n0}\left(1 - \frac{u_n}{u_n^m}\right)^{-2} = \left(1 - \frac{u_n}{u_n^m}\right)^{-1}\frac{\sigma_n}{u_n} \tag{3.6}$$

from Bandis' function and

$$k_n = \frac{\partial\sigma_n}{\partial u_n} = \left(1 - \frac{u_n}{u_n^m}\right)^{-1}\frac{(t-1)}{u_n}(\sigma_n - \sigma_{n0}) \tag{3.7}$$

from Goodman's function. The two functions can be managed to produce identical normal stiffness when $t = 2$ and $\sigma_{n0} = 0$ in the Goodman model. It can be observed that $k_n \to \infty$ when $u_n = u_n^m$ in the two models.

3.2 Shear Strength of Rock Fractures

3.2.1 Patton's Criterion

The effect of surface roughness on the shear strength of rock fractures was possibly recognized much earlier, but the first attempt to correlate the surface roughness with the shear strength of rock fractures was made by Patton (1966) using the assumption that the asperities on the fracture surface have identical shape and an inclination angle i. The criterion is written as (see Fig. 3.2)

$$\sigma_t^p = \sigma_n \tan(\phi_b + i) \tag{3.8}$$

where ϕ_b is the basic friction angle of smooth surfaces of the rock material and the parameter i represents the effect of irregular asperities on the fracture surface, also called the dilation angle. The value of i is determined as the statistical mean of the angles between the mean smooth reference surface and the first-order waviness along the whole rough surface concerned, thus ignoring the secondary irregularities

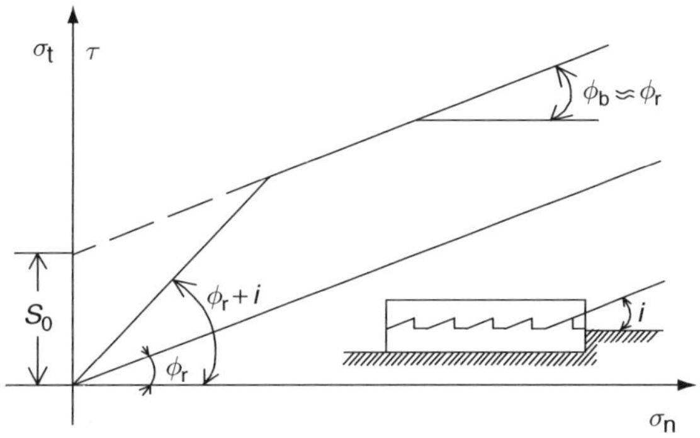

Fig. 3.2 Patton's bilinear criterion for the shear strength of rock fractures with multiple uniform inclined asperities on the fracture surface (Patton, 1966).

(asperities) superimposed over the first-order waviness. An extension to this model was made later to include a cohesive increment S_0 to the shear strength under high normal stresses. The basic friction angle was taken as equal to the residual friction angle in Patton's criterion, which, however, may be different for fractures in different rock types.

Let $\mu = \tan\,\phi_b$ and recall the identity

$$\tan\,(\phi_b + i) = \frac{\tan\,\phi_b + \tan\,i}{1 - \tan\,\phi_b \tan\,i} = \frac{\mu \cos\,i + \sin\,i}{\cos\,i - \mu \sin\,i} \tag{3.9}$$

Equation (3.8) can be transformed into an equivalent alternative form

$$(\sigma_t^p \cos\,i + \sigma_n \sin\,i) + \mu(-\sigma_t^p \sin\,i + \sigma_n \cos\,i) = 0 \tag{3.10}$$

or

$$\sigma_t^p + \sigma_n \tan\,(\phi_b + i) = 0 \tag{3.11}$$

as has often appeared in the literature. Note that different sign conventions for the normal compressive stresses are adopted in Eqn (3.8) and (3.11).

Patton's criterion represents the effect of the roughness of rock surfaces on its shear strength by a 2D simplification without regard for scale effect and roughness evolution during the deformation process. In practice, the mean first-order asperity angle may or may not become stationary within the size of samples tested in laboratory conditions. The criterion is, however, a simple function that has served as a conceptual breakthrough, stimulating the development of many shear strength criteria and constitutive models for rock fractures in subsequent years.

3.2.2 Ladanyi and Archambault's Criterion

Patton's criterion assumes that the dilation angle is a constant, independent of the magnitude of the normal stress, and ignores the interlocking effect between the asperities. The change of the surface roughness during shear due to asperity damage is also neglected. Based on this observation, Ladanyi and Archambault (1969) proposed another shear strength criterion for rock fractures given by

$$\sigma_t^p = \frac{\sigma_n(1 - a_s)[\tan(\phi_b) + \nu] + a_s[\sigma_n \tan(\phi_0) + s_0\eta]}{1 - \nu(1 - a_s)\tan(\phi_f)} \tag{3.12}$$

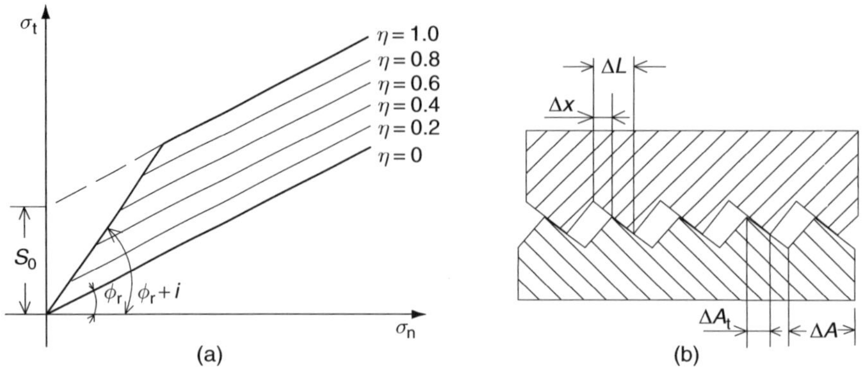

Fig. 3.3 (a) Results using Patton's bilinear criterion and (b) the definition of the interlocking coefficient, η (Ladanyi and Archambault, 1969).

$$a_s = 1 - \left[1 - \frac{\sigma_n}{\sigma_c}\right]^{k_1}, \qquad v = \left[1 - \frac{\sigma_n}{\sigma_c}\right]^{k_2} \tan(i_0) \tag{3.13}$$

where v represents the rate of dilatancy when the peak shear stress occurs, a_s the ratio of the actual contact area over the total surface area of the fracture, ϕ_0 the friction angle of smooth rock surfaces without any asperity, ϕ_f the statistical mean value of the friction angle when sliding occurs along the asperities, s_0 the cohesion between asperities, i_0 the initial dilatancy angle and k_1 and k_2 the material constants and η called the degree of interlocking defined as (see Fig. 3.3b)

$$\eta = \left|\frac{\sum \Delta A_t}{\sum \Delta A}\right| = \left|1 - \frac{\Delta x}{\Delta L}\right| \tag{3.14}$$

Under high normal stresses, $a_s = 1$ and $v = 0$, and the peak shear stress becomes equivalent to Patton's criterion, written as

$$\sigma_t^p = \sigma_n \tan(\phi_0) + s_0 \eta \tag{3.15}$$

Ladanyi and Archambault's criterion is a non-linear model for predicting the shear strength of rough rock fractures. It can provide better non-linear normal stress versus shear stress curves than Patton's criterion in some cases. However, there are also some drawbacks associated with this criterion: (i) it requires more parameters than Patton's criterion and some special tests are required to determine all of them; and (ii) parameters a_s and v represent the effect of the roughness of the fractures. Under any deformation process, the roughness of the fractures can only be a monotonically decreasing measure due to the accumulated damage incurred on the asperities. The definitions of a_s and v in Ladanyi and Archambault's criterion make them, however, reversible with respect to the normal stress. If the normal stress decreases during the process of shear deformation, then the values of a_s and v will increase rather than decrease, indicating that the roughness of the fracture will increase, which is a physical impossibility. Therefore, this criterion is more suitable to fractures undergoing small shear displacements under constant or monotonically increasing normal stresses.

3.2.3 Barton's Criterion

Barton (1971, 1973, 1974, 1976) proposed a well-known shear strength criterion for rock fractures which contains an explicit measure of roughness, called the joint roughness coefficient (JRC). The criterion is written

$$\sigma_t^p = \sigma_n \tan\left[\mathrm{JRC}\log_{10}\left(\frac{\mathrm{JCS}}{\sigma_n}\right) + \phi_b\right] \tag{3.16}$$

where σ_t^p, σ_n and ϕ_b are as defined as before and the parameter JCS is called the joint wall compressive strength, obtained using a Schmidt Hammer tests for weathered fracture surfaces or simply the compressive strength of the intact rock for fresh fracture surfaces. In high-stress situations, the equation is modified to

$$\sigma_t^p = \sigma_n \tan\left[\mathrm{JRC}\log_{10}\left(\frac{\sigma_1 - \sigma_3}{\sigma_n}\right) + \phi_b\right] \tag{3.17}$$

where σ_1 is the axial stress at failure and σ_3 the effective confining pressure. Both JRC and JCS are scale dependent. For fractures in the field with lengths L_n, the JRC and JCS values are evaluated, respectively, by

$$\mathrm{JRC}_n = \mathrm{JRC}_0\left(\frac{L_n}{L_0}\right)^{-0.02\mathrm{JRC}_0} \tag{3.18}$$

and

$$\mathrm{JCS}_n = \mathrm{JCS}_0\left(\frac{L_n}{L_0}\right)^{-0.03\mathrm{JRC}_0} \tag{3.19}$$

where JRC_0 and JCS_0 are the values of JRC and JCS determined in laboratories with samples length L_0 (having a typical value of 100 mm).

Barton's shear strength criterion, represented by Eqns (3.16) and (3.17), is essentially equivalent to Patton's, in the sense that an extra increment of friction is contributed by the roughness. The difference is that the constant roughness angle i in the Patton criterion is replaced in Barton's by a function:

$$i_B = \mathrm{JRC}\log_{10}\left(\frac{\sigma_1 - \sigma_3}{\sigma_n}\right) \quad \text{or} \quad i_B = \mathrm{JRC}\log_{10}\left(\frac{\mathrm{JCS}}{\sigma_n}\right) \tag{3.20}$$

These functions represent the effects of the normal stress and rock material strength on the mobilized friction angle of the fracture. JRC is determined by tilting tests or direct shear tests under small constant normal stresses (Barton and Choubey, 1977).

Similar to Ladanyi and Archambault's criterion, the stress variables in the logarithmic functions in Barton's criterion were intended to consider the reduction of roughness under normal stresses. However, it suffers the same shortcoming – the reversible roughness measure i_B with cyclic normal stress changes.

A shear strength criterion should be applicable without violating the second law of thermodynamics so that the wear-off of the asperities or degradation of surface roughness through energy dissipation processes will never become reversible. One way to ensure such irreversibility is to supply an additional constraint, or an evolution law, of the surfaces' roughness so that roughness will always evolve in a decreasing direction. One example of such a constraint was suggested by Plesha (1987) and later verified experimentally by Hutson (1987), through an exponential law

$$\alpha = \alpha_0 e^{-\lambda W^p} \tag{3.21}$$

where α is the roughness angle, λ an experimentally determined constant and W^p the accumulated work done by the shear stresses over the shearing displacement paths after the slip occurred. Since $W^p > 0$ and is a monotonically increasing quantity, the α value becomes irreversibly decreasing. The added difficulty for this approach is, of course, the addition of two more parameters that may not be easy to measure in laboratory experiments, besides the calculation of the work W^p.

3.2.4 A 3D Shear Strength Criterion of Rock Fractures with Anisotropic Roughness

The representation of roughness by a single constant measure of roughness angle is valid only if the fracture surface is ergodic, i.e., the geometrical properties of the surface are not only stationary and isotropic, but also can be truly represented by a line profile which is identical in all directions in the fracture plane. This is not a realistic assumption because rock surfaces are generally only conditionally stationary, generally anisotropic and not ergodic because of the strong directional effects of tectonic movements that created the fractures or the shear displacements that the fractures have undergone in their past history. Therefore a one-parameter representation of roughness has limited applicability, even in 2D cases when cyclic shearing is considered.

Based on a series of tilting and shear tests with concrete replicas of natural rock fractures, Jing (1990) investigated the anisotropy of friction of rough surfaces (see Fig. 3.4a). The shear strength is still the Patton-type, but in 3D,

$$\sigma_t^p = \sigma_n \tan\,[\phi_b + \alpha_\theta] \tag{3.22}$$

where ϕ_b is a constant (isotropic in all directions of shear), but the roughness angle α_θ, in the shear direction θ, is assumed to have an elliptical distribution on the surface of the fracture given by

$$\alpha_\theta = \sqrt{[C_1 \cos\psi - C_2 \sin\psi]^2 + [C_1 \sin\psi + C_2 \cos\psi]^2} \tag{3.23}$$

$$C_1 = \alpha_1 \cos\,(\theta - \psi), \qquad C_2 = \alpha_2 \cos\,(\theta - \psi) \tag{3.24}$$

$$\begin{cases} \alpha_1 = \alpha_1^0 e^{-d_m W^p} |\cos\,(\theta - \psi)| \\ \alpha_2 = \alpha_2^0 e^{-d_m W^p} |\sin\,(\theta - \psi)| \end{cases} \tag{3.25}$$

where α_1 and α_2 are the principal values of the roughness (asperity) angle ellipse (cf. Fig. 3.4b) and α_1^0 and α_2^0 the initial values of α_1 and α_2. The roughness degradation expressed in Eqn (3.25) is irreversible for general shearing paths so that the second law of thermodynamics is satisfied regardless of any special loading cases. The material constant $d_m > 0$, however, has to be determined by laboratory experiments, preferably over large shear displacements and wide ranges of normal stress magnitudes.

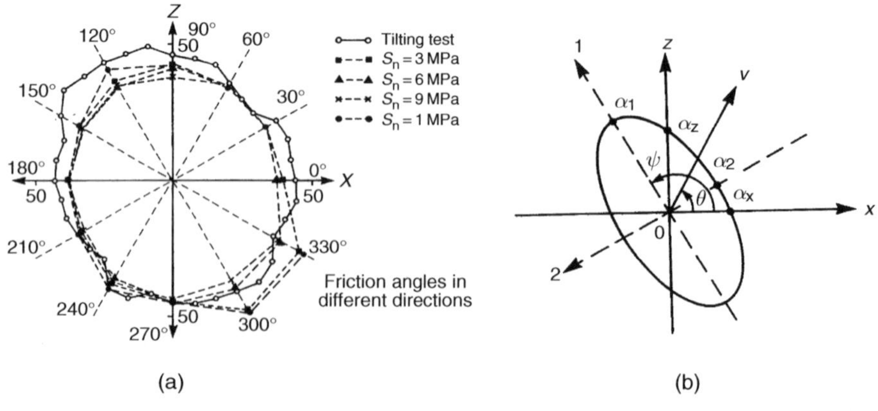

Fig. 3.4 Friction angles obtained during direct shear tests and tilting tests (Jing, 1990): (a) distributions of total friction angle (α) under different normal stresses (S_n) and (b) roughness ellipse.

3.3 Constitutive Models of Rock Fractures

Constitutive models of rock fractures have been developed and applied in rock mechanics analyses for a long time, with typical examples being the conventional Mohr–Coulomb model and Goodman's model (Goodman, 1976). Such a model determines the interactive forces/stresses between the contacting rock blocks across their interfaces (fractures), thus affecting the motion and deformation of the blocks, and therefore is one of the most important components in the DEM.

3.3.1 Goodman's Empirical Model

Goodman (1976) formulated the first comprehensive constitutive model, with an empirical approach, for 2D rock fractures. The model is formulated in total stresses and displacements, based on the shear behavior of rock fractures under constant normal stresses observed in laboratory shear tests. The normal stress–normal displacement relation (Fig. 3.5a) and normal stiffness are defined by Eqns (3.5) and (3.7), respectively, and an incremental relation may also be written as

$$\mathrm{d}\sigma_{\mathrm{n}} = k_{\mathrm{n}}\, \mathrm{d}u_{\mathrm{n}} \tag{3.26}$$

The complete curve of shear stress–shear displacement is composed of five connected parts for monotonic shear paths (Fig. 3.5b), given by

I)

$$\sigma_{\mathrm{t}} = -\sigma_{\mathrm{t}}^{\mathrm{r}}, \qquad k_{\mathrm{t}} = 0 \quad (u_{\mathrm{t}} \le u_{\mathrm{t}}^{\mathrm{r}(-)}) \tag{3.27}$$

II)

$$\sigma_{\mathrm{t}} = -\sigma_{\mathrm{t}}^{\mathrm{p}} + \frac{\sigma_{\mathrm{t}}^{\mathrm{p}} - \sigma_{\mathrm{t}}^{\mathrm{r}}}{u_{\mathrm{t}}^{\mathrm{p}} - u_{\mathrm{t}}^{\mathrm{r}}}(u_{\mathrm{t}} - u_{\mathrm{t}}^{\mathrm{p}(-)}), \qquad k_{\mathrm{t}} = \frac{\sigma_{\mathrm{t}}^{\mathrm{p}} - \sigma_{\mathrm{t}}^{\mathrm{r}}}{u_{\mathrm{t}}^{\mathrm{p}} - u_{\mathrm{t}}^{\mathrm{r}}} \quad (u_{\mathrm{t}}^{\mathrm{r}(-)} \le u_{\mathrm{t}} \le u_{\mathrm{t}}^{\mathrm{p}(-)}) \tag{3.28}$$

III)

$$\sigma_{\mathrm{t}} = k_{\mathrm{t}} u_{\mathrm{t}} + \sigma_{\mathrm{t}}^{0}, \qquad k_{\mathrm{t}} = \text{constant} \quad (u_{\mathrm{t}}^{\mathrm{p}(-)} \le u_{\mathrm{t}} \le u_{\mathrm{t}}^{\mathrm{p}(+)}) \tag{3.29}$$

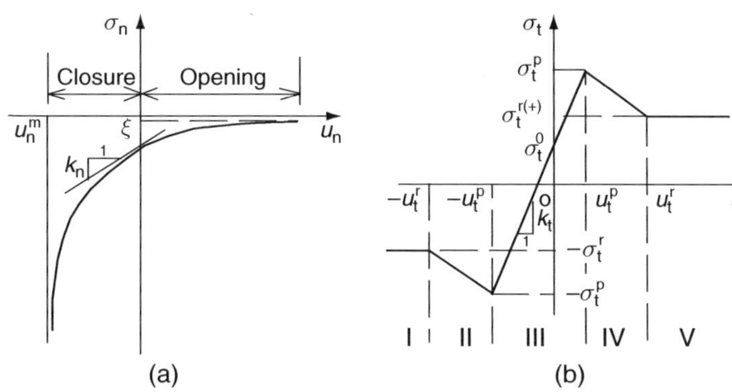

Fig. 3.5 Stress–displacement curves for Goodman's constitutive model for rock fractures (Goodman, 1976). (a) Normal stress–normal displacement curve; (b) shear stress–shear displacement curve.

IV)

$$\sigma_t = \sigma_t^p + \frac{\sigma_t^p - \sigma_t^r}{u_t^p - u_t^r}(u_t - u_t^{p(+)}), \qquad k_t = \frac{\sigma_t^p - \sigma_t^r}{u_t^p - u_t^r} \quad (u_t^{p(+)} \le u_t \le u_t^{r(+)}) \tag{3.30}$$

V)

$$\sigma_t = \sigma_t^r, \qquad k_t = 0 \quad (u_t \ge u_t^{r(+)}) \tag{3.31}$$

where σ_t^0 is the initial shear stress, $u_t^{p(+)}$ and $u_t^{p(-)}$ the shear displacements corresponding to the peak shear stresses in the positive and negative shear directions, respectively, and $u_t^{r(+)}$ and $u_t^{r(-)}$ the shear displacements corresponding to the occurrence of residual shear stresses in the positive and negative shear directions, respectively. They are given by

$$u_t^{p(+)} = u_t^p = \frac{\sigma_t^p - \sigma_t^0}{k_t}, \qquad u_t^{p(-)} = -u_t^p - \frac{2\sigma_t^0}{k_t} \tag{3.32}$$

$$u_t^{r(+)} = u_t^r = M\frac{\sigma_t^p - \sigma_t^0}{k_t}, \qquad u_t^{r(-)} = -u_t^r - \frac{2\sigma_t^0}{k_t} \tag{3.33}$$

where M is a material constant.

The increment of normal displacement due to dilatancy by roughness, in terms of differences, is given by

$$\Delta u_n = -\frac{d_p}{u_t^p}\left(|u_t| + \left|\frac{\sigma_t^0}{k_t}\right|\right) \quad (u_t^{r(-)} \le u_t \le u_t^{r(+)}) \tag{3.34}$$

$$\Delta u_n = -\frac{d_p}{u_t^p}\left(u_t^{r(+)} + \left|\frac{\sigma_t^0}{k_t}\right|\right) \quad (u_t \ge u_t^r) \tag{3.35}$$

where d_p is the dilatancy of the fracture at u_t^p.

Although Goodman's model was initially formulated in total stresses and displacements, an incremental form can also be obtained by using the derived shear stiffness in the five different parts of the curves (3.27–3.31), together with Eqn (3.26), given by

$$\begin{Bmatrix} d\sigma_n \\ d\sigma_t \end{Bmatrix} = \begin{bmatrix} k_n, & 0 \\ 0, & k_t \end{bmatrix} \begin{Bmatrix} du_n \\ du_t \end{Bmatrix} \tag{3.36}$$

which is a non-linear elastic model in the normal direction and a piecewise linear model in the shear direction.

Goodman's model represents another conceptual breakthrough in the field of constitutive models for rock fractures, because not only is it the first comprehensive constitutive model of rock fractures ever developed, but also it was formulated in a general frame of the finite element method, thereby stimulated continuous further developments and applications in the field of numerical modeling in rock mechanics and rock engineering, with implementations in many computer programs for the finite element method. However, several important factors are not considered in this model such as surface roughness degradation, cyclic shear paths, scale effects, etc., due mainly to the fact that the model was an early pioneering work based on an empirical platform without considering the restriction of the second law of ʋith general shearing paths and material damage on the

›irical Model (BB-Model)

odel) is an empirical constitutive model for rock fractures
t al., 1981; Bandis et al., 1983; Barton et al., 1985; Barton
in total stress and displacement components, rather than
er implementation. The required input parameters are the
: aperture e, mechanical aperture E and residual friction
;le ϕ_b for fresh fractures without weathering). All these
r via back analyses.

l displacement equations

.vior is specified by Eqn (3.4) and the normal stiffness is
·e proposed to determine the initial normal stiffness k_{n0}

$$\text{:S}_0 \over E_i \Big) + 1.75\,\text{JRC}_0 - 7 \tag{3.37}$$

$$\text{JRC}_0 + C_i\left(\frac{\text{JCS}_0}{E_i}\right)^{D_i} \tag{3.38}$$

:s of JRC and JCS, A_i, B_i, C_i, D_i the material constants
:iation of the ith cycle of cyclic normal compression tests.
ıe stress paths of normal compressive loading–unloading
nent is given by

$$E_0 - \sum_{k=1}^{i-1} u_{n,k}^{irr} \tag{3.39}$$

l closure at the end of the kth cycle and E_0 is the initial
:ed by using another empirical relation:

$$\left(0.2\frac{\sigma_c}{\text{JCS}_0} - 0.1\right) \tag{3.40}$$

displacement equations

the concept of 'mobilized friction'. The JRC introduced
ıear stresses (cf. Eqns (3.16–3.17)). The friction angle
given by

$$\text{peak} \log_{10}\left(\frac{\text{JCS}}{\sigma_n}\right) + \phi_r \tag{3.41}$$

RESERVER'S NAME
LEI, Q
PLEASE ISSUE TO YOUR ACCOUNT
COLLECT BY DATE
04 SEP 2014

at a shear displacement δ_{peak} which may be estimated by an empirical equation

$$\delta_{peak} = \frac{L_n}{500}\left(\frac{JRC_n}{L_n}\right)^{0.33} \tag{3.42}$$

The mobilized shear stress, σ_t^{mob}, at any given shear displacement δ is specified by

$$\sigma_t^{mob} = \sigma_n \tan\left[\left(JRC_{mob}\log\left(\frac{JCS_{mob}}{\sigma_n}\right) + \phi_r\right)\right] \tag{3.43}$$

where ϕ_r is the residual friction angle, JRC_{mob} the mobilized roughness coefficient of the fracture at the corresponding shear displacement, JCS_{mob} the mobilized uniaxial compressive strength of the rock material. Under very high normal stresses, the term JCS_{mob} is replaced by $(\sigma_1 - \sigma_3)$. This equation therefore describes the whole shear stress–shear displacement curve. The mobilized friction angle is then written as

$$\phi_{mob} = JRC_{mob}\log\left(\frac{JCS_{mob}}{\sigma_n}\right) + \phi_r \tag{3.44}$$

To facilitate the tracing of the curve during a simulation, two dimensionless coordinates,

$$\frac{JRC_{mob}}{JRC_{peak}} = \frac{\phi_{mob} - \phi_r}{\phi_{peak} - \phi_r} = \frac{\phi_{mob} - \phi_r}{i_B} \tag{3.45}$$

and a ratio of δ/δ_{peak} were proposed by Barton (1982) to represent the stage of shear, and the whole process was divided into four stages (Fig. 3.6) (Barton et al., 1985):

(1) Mobilization of the basic friction at the start of shearing with $\delta = 0$, $\phi_{mob} = 0$; therefore

$$\frac{\delta}{\delta_{peak}} = 0, \qquad \frac{JRC_{mob}}{JRC_{peak}} = -\frac{\phi_r}{i_B} \tag{3.46}$$

(2) Mobilization of roughness and dilatancy starts with the initial value

$$\Delta u_n = \frac{1}{2} JRC_{mob}\log_{10}\left(\frac{JCS_{mob}}{\sigma_n}\right) \tag{3.47}$$

(3) The peak shear stress is reached at $JRC_{mob}/JRC_{peak} = 1.0$ and $\delta/\delta_{peak} = 1.0 = 1.0$, with peak shear stress given by Eqn (3.16) at the peak shear displacement δ_{peak} and the dilatancy is given as

$$\Delta u_n = \frac{1}{2} JRC_{peak}\log_{10}\left(\frac{JCS_{peak}}{\sigma_n}\right) \tag{3.48}$$

(4) Shear stress and dilatancy decrease with increasing δ, until the residual stress is reached with $\phi_{mob} = \phi_r$, and therefore

$$\frac{JRC_{mob}}{JRC_{peak}} = 0 \quad \text{and} \quad \frac{\delta}{\delta_{peak}} = \infty \tag{3.49}$$

In practice, Barton et al. (1985) suggested that $\delta/\delta_{peak} = 100.0$ for reaching the residual stress. The dilatancy is then zero.

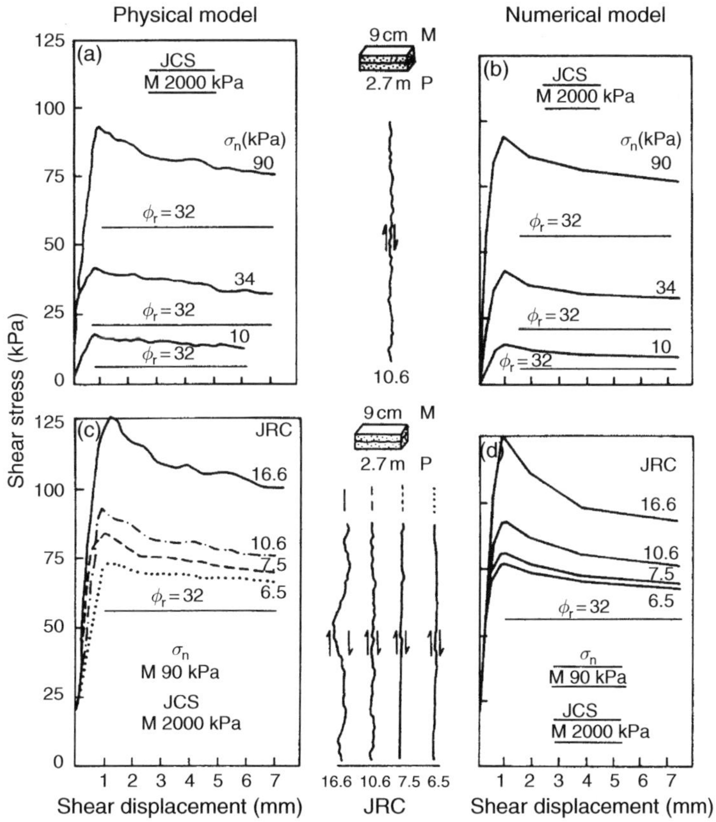

Fig. 3.6 Dimensionless curves of shear stress–shear displacement for the BB-model (Barton et al., 1985) with similar ideas also reported in Sharp (1970).

A simplified BB-model was developed by Guvanasen and Chan (1991) for simulating the coupled hydro-mechanical behavior of rock fractures. In this simplified model, the normal stress–normal displacement relation of a joint is the same as that of Eqn (3.6) and the normal stiffness is given by

$$k_{\mathrm{n}} = \frac{\partial \sigma_{\mathrm{n}}}{\partial u_{\mathrm{n}}} = k_{\mathrm{n}i}\left(\frac{u_{\mathrm{n}}^{\mathrm{m}}}{u_{\mathrm{n_n}}^{\mathrm{m}} - u_{\mathrm{n}}}\right)^2 = \frac{\sigma_{\mathrm{n}}^{\mathrm{n}} u_{\mathrm{n}}^{\mathrm{m}}}{(u_{\mathrm{n}}^{\mathrm{m}} - u_{\mathrm{n}})u_{\mathrm{n}}} \tag{3.50}$$

The shear stress–shear displacement relation is given by

$$\sigma_{\mathrm{t}} = \frac{\sigma_{\mathrm{t}}^{\mathrm{p}}}{u_{\mathrm{t}}^{\mathrm{p}}} u_{\mathrm{t}}, \qquad |u_{\mathrm{t}}| \le |u_{\mathrm{p}}| \tag{3.51}$$

$$\sigma_{\mathrm{t}} = \sigma_{\mathrm{t}}^{\mathrm{p}} + \frac{\sigma_{\mathrm{t}}^{\mathrm{r}} - \sigma_{\mathrm{t}}^{\mathrm{p}}}{u_{\mathrm{t}}^{\mathrm{r}} - u_{\mathrm{t}}^{\mathrm{p}}}(u_{\mathrm{t}} - u_{\mathrm{t}}^{\mathrm{p}}), \qquad |u_{\mathrm{t}}^{\mathrm{p}}| \le |u_{\mathrm{t}}| \le |u_{\mathrm{t}}^{\mathrm{r}}| \tag{3.52}$$

$$\sigma_{\mathrm{t}} = \sigma_{\mathrm{t}}^{\mathrm{r}}, \qquad |u_{\mathrm{t}}| \ge |u_{\mathrm{t}}^{\mathrm{r}}| \tag{3.53}$$

with

$$\sigma_{\mathrm{t}}^{\mathrm{p}} = \sigma_{\mathrm{n}} \tan\,(\phi_{\mathrm{r}} + i) = \sigma_{\mathrm{n}} \tan\left[\phi_r + \mathrm{JRC}\log_{10}\left(\frac{\mathrm{JCS}}{\sigma_{\mathrm{n}}}\right)\right] \tag{3.54}$$

$$u_t^p = A\,(\mathrm{JRC})^B \tag{3.55}$$

$$u_t^r = m(u_t^p) \tag{3.56}$$

$$\sigma_t^r = \left(\frac{0.5 + r}{1.0 + r}\right)\sigma_t^p \tag{3.57}$$

where the term i is the effective dilatancy angle of the fracture surface and term r the ratio $\frac{\phi_r}{i}$. The parameters A, B and m are empirical constants. The shear stiffness can be derived as

$$k_t = \frac{\partial \sigma_t}{\partial u_t} = \frac{\sigma_t^p}{u_p}, \qquad |u_t| \leq |u_t^p| \tag{3.58}$$

$$k_t = \frac{\partial \sigma_t}{\partial u_t} = \frac{\sigma_t^r - \sigma_t^p}{u_t^r - u_t^p}, \qquad |u_t^p| \leq |u_t| \leq |u_t^r| \tag{3.59}$$

$$k_t = 0, \qquad |u_t| \geq |u_t^r| \tag{3.60}$$

An incremental non-linear elastic model can then be implemented, written as

$$\begin{Bmatrix} d\sigma_n \\ d\sigma_t \end{Bmatrix} = \begin{bmatrix} k_n, & 0 \\ 0, & k_t \end{bmatrix} \begin{Bmatrix} du_n \\ du_t \end{Bmatrix} \tag{3.61}$$

with k_n and k_t given by Eqn (3.6) and Eqns (3.58–3.60), respectively.

The BB-model is a mathematical expression of the total stress–displacement curves and needs to be modified into an increment form for computer coding. It also exhibits a reversibility of the roughness representation term (the last term in the left-hand side of Eqn (3.54)) with changing normal stresses, which may violate the second law of thermodynamics for some special reversing stress paths. Its many material constants employed in all its empirical relations may cause difficulties in extrapolating its applicability outside its database, on which all these material constants were determined, without extensive experimental backup. The model can be applied for monotonic shear paths of increasing (or constant) normal stresses with adequate flexibility, but may have theoretical difficulties for complex loading paths due to its empirical nature.

On the other hand, the BB-model can represent the most realistic behavior of rough rock fractures observed in laboratory experiments, especially the cyclic normal stress–closure behavior, among the developed models and has been applied extensively in rock mechanics analysis due to this advantage, especially when fluid flow in fractures needs to be considered. The BB-model can provide reliable estimation of the aperture variations as a function of normal stress and shear dilation. Caution, however, needs to be taken when reversible stress/deformation paths are included in the problem conceptualization, because of its lack of constraint by the second law of thermodynamics due to its empirical framework of development.

3.3.3 Amadei–Saeb's Theoretical Model

Amadei and Saeb (1990) developed a 2D, non-linear elastic constitutive model for rock fractures with consideration of the different normal deformabilities of rock fractures with mated and unmated initial positions, and the effects of the deformability of the surrounding rock mass on the fracture behavior. The model is formulated in a more general incremental form given by

$$\begin{Bmatrix} d\sigma_n \\ d\sigma_t \end{Bmatrix} = \begin{bmatrix} k_{nn}, & k_{nt} \\ k_{tn}, & k_{tt} \end{bmatrix} \begin{Bmatrix} du_n \\ du_t \end{Bmatrix} \tag{3.62}$$

where k_{ij} (i, j = n, t) denotes the stiffness tensor of the fracture. The elements of this stiffness tensor are specified as

$$k_{nn} = \left[\frac{-vk_2}{\sigma_c} \left(1 - \frac{\sigma_n}{\sigma_c}\right)^{k_2 - 1} \tan\,(i_0) + \frac{(u_n^m)^2 k_n^0}{k_n^0 u_n^m - \sigma_n} \right]^{-1} \tag{3.63}$$

$$k_{nt} = -\left(1 - \frac{\sigma_n}{\sigma_c}\right)^{k_2} \tan(i_0) \left[\frac{-vk_2}{\sigma_c} \left(1 - \frac{\sigma_n}{\sigma_c}\right)^{k_2 - 1} \tan(i_0) + \frac{(u_n^m)^2 k_n^0}{k_n^0 u_n^m - \sigma_n} \right]^{-1} \tag{3.64}$$

$$k_{tn} = \frac{u_t}{u_t^p} k_{nn} \frac{\partial \sigma_t^p}{\partial \sigma_n} \quad \text{for } (u_t < u_t^p) \tag{3.65}$$

$$k_{tn} = \frac{k_{nn}}{u_t^p - u_t^r} \left\{ \frac{\partial \sigma_t^p}{\partial \sigma_n} (u_t - u_t^r) + (u_t^p - u_t) C \right\} \tag{3.66}$$

for $(u_t^p < u_t < u_t^r; \sigma_n < \sigma_t)$, with

$$C = \frac{\partial \sigma_t^p}{\partial \sigma_n} \left(B_0 + \frac{1 - B_0}{\sigma_c} \sigma_n \right) + \frac{\sigma_t^p}{\sigma_n} (1 - B_0) \tag{3.67}$$

$$k_{tt} = \frac{u_t}{u_t^p} k_{nt} \frac{\partial \sigma_t^p}{\partial \sigma_n} + \frac{\sigma_t^p}{u_t^p} \quad \text{for } (u_t < u_t^p) \tag{3.68}$$

$$k_{tt} = \frac{\sigma_t^p - \sigma_t^r}{u_t^p - u_t^r} + \frac{k_{nt}}{u_t^p - u_t^r} \left\{ \frac{\partial \sigma_t^p}{\partial \sigma_n} (u_t - u_t^r) + (u_t^p - u_t) C \right\} \tag{3.69}$$

for $(u_t^p \le u_t \le u_t^r;\ \sigma_n < \sigma_t)$ where ϕ_μ is the friction angle when sliding along the asperities (equivalent to ϕ_0 in the criterion by Ladanyi and Archambault (1969)). The parameter $B_0 (0 \le B_0 \le 1)$ is the ratio of residual shear stress over peak shear stress under zero normal (or very low) stresses. The expressions for σ_t^p, σ_t^r and $\partial \sigma_t^p / \partial \sigma_n$ depends on the selected criterion for the peak shear stress. Using the Ladanyi and Archambault's criterion for peak shear stress, Amadei and Saeb (1990) gave the expressions as

$$\frac{\partial \sigma_t^p}{\partial \sigma_n} = \alpha_s \tan(\varphi_0) + \frac{s_r}{\sigma_c} k_1 \left(1 - \frac{\sigma_n}{\sigma_c}\right)^{k_1 - 1} + (1 - \alpha_s) F_1 + F_2 \tag{3.70}$$

$$F_1 = \tan\,(\phi_\mu + i) \tag{3.71}$$

$$F_2 = -\frac{\sigma_n}{\sigma_t} \frac{(1 - \alpha_s) k_2}{\cos^2(\phi_\mu + i)} \frac{\tan(i_0) \left(1 - \frac{\sigma_n}{\sigma_c}\right)^{k_1 - 1}}{1 + \left(1 - \frac{\sigma_n}{\sigma_c}\right)^{2k_2} \tan^2(i_0)} - \frac{\sigma_n}{\sigma_c} k_1 \tan\,(\phi_\mu + i) \left(1 - \frac{\sigma_n}{\sigma_c}\right)^{k_1 - 1} \tag{3.72}$$

The other parameters are as defined before. When $u_t > u_t^r$, k_{tt} vanishes. The same applies when $\sigma_n \ge \sigma_c$ and $u_t \ge u_t^p = u_t^r$. When $\sigma_n \ge \sigma_c$ and $u_t < u_t^p$, $k_{tt} = \sigma_t^p / u_t^p$. The dilatancy rate can be derived from the Eqn (3.62) as

$$du_n = -\frac{k_{nt}}{k_{nn}} du_t \tag{3.73}$$

under the constant normal stress condition, $d\sigma_n = 0$.

The representation of roughness in the above model is expressed by the initial and mobilized roughness angles, i_0 and i, in the expressions for k_{nn}, k_{nt}, k_{tn} and k_{tt} (cf. Eqns 3.63–3.69). However, multiplication of $\tan(i_0)$ and $\tan(\phi_\mu + i)$ by $(1 - \sigma_n/\sigma_c)^{k_1 - 1}$ or similar expressions makes the variation of roughness reversible, like that in the BB-model. Therefore, the second law of thermodynamics may be violated for some special stress paths and the model is mostly valid only for paths with monotonic increase of the normal stress magnitude.

Souley et al. (1995) extended the Amadei–Saeb model to include cyclic shear paths through an elastic unloading with the initial shear stiffness and constant shear stress at the residual level during reversed shearing before the initial position of the fracture is recovered. Anti-symmetric shear behavior is assumed so that the overall non-linear elastic approach is maintained. The model was implemented in the DEM code UDEC.

3.3.4 Plesha's Theoretical Model and Its Extension

Plesha (1987) developed a theoretical 2D constitutive model for rock fractures with consideration of roughness degradation. The fracture surface is assumed to be macroscopically smooth and planar, but rough with uniformly shaped asperities at the microscopic scale (Fig. 3.7a). The model is based on the theory of plasticity and an assumption of uniform tooth-shaped asperities, with the stress transformation given by

$$\begin{cases} \sigma_t^c = \eta(\sigma_t \cos \alpha_k + \sigma_n \sin \alpha_k) \\ \sigma_n^c = \eta(-\sigma_t \sin \alpha_k + \sigma_n \cos \alpha_k) \end{cases} \tag{3.74}$$

where η is a constant representing the areal ratio of active asperity surface over the base surface of the representative asperity ($\eta = L_k/L_0, k = L,\ R$) and α_k is the current active asperity angle. Variables σ_t^c and σ_n^c represent the microscopic contact shear and normal stresses on the active asperity surface (Fig. 3.7b).

The theory of plasticity with a non-associated flow rule was used to derive the constitutive model, using the similarity between the elasto-plastic deformation of plastic solids and the shear stress–shear displacement behavior of rock fractures (Fig. 3.1).

Following the standard philosophy of plasticity theory, it is assumed that the displacement increment of a rock fracture, du_i, $i = t, n$, can be divided into an elastic (reversible) part du_i^e and an irreversible part du_i^p

$$du_i = du_i^e + du_i^p \tag{3.75}$$

and the elastic part is related to the increment of contact stresses by the stiffness tensor k_{ij}

$$d\sigma_i = k_{ij}\, du_j^e \tag{3.76}$$

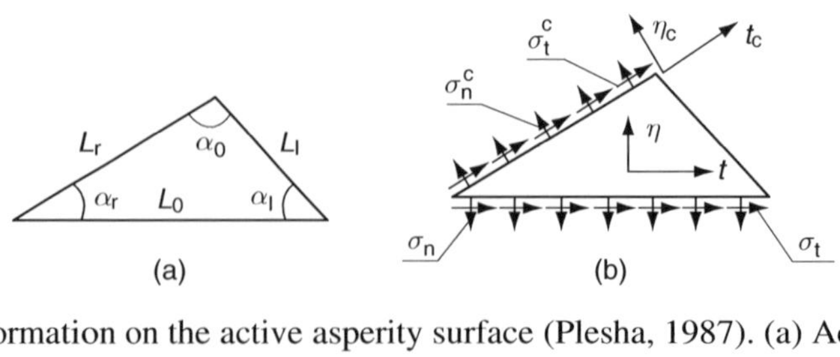

Fig. 3.7 Stress transformation on the active asperity surface (Plesha, 1987). (a) Active asperity surfaces: forwards (α_r) and backwards (α_l); (b) transformation of macroscopic (σ_t, σ_n) and microscopic (σ_t^c, σ_n^c) stress components on the active asperity surface.

with

$$[k_{ij}] = \begin{bmatrix} k_t & 0 \\ 0 & k_n \end{bmatrix} \tag{3.77}$$

The stress increment in Eqn (3.76) may then be rewritten

$$d\sigma_i = k_{ij}\, du_j^e = k_{ij}\left(du_j - du_j^p\right) \tag{3.78}$$

The irreversible displacement increment is prescribed by a non-associated flow rule in plasticity theory:

$$du_i^p = \begin{cases} 0, & F < 0 \\ \lambda \dfrac{\partial Q}{\partial \sigma_i}, & F \geq 0 \end{cases} \tag{3.79}$$

where $\lambda > 0$ is a positive scalar and Q and F are the plastic potential and yield functions, respectively, given by

$$F = |\sigma_t \cos\alpha_k + \sigma_n \sin\alpha_k| + \tan\phi_r(-\sigma_t \sin\alpha_k + \sigma_n \cos\alpha_k) - C \tag{3.80}$$

$$Q = |\sigma_t \cos\alpha_k + \sigma_n \sin\alpha_k| \tag{3.81}$$

where ϕ_r, C represent the residual friction angle and cohesion of fractures, respectively. The dissipative work done by the stresses over the displacements is given by

$$dW^p = \sigma_i\, du_i^p = \lambda\sigma_i \frac{\partial Q}{\partial \sigma_i} \tag{3.82}$$

Assuming that the loading function $f(\sigma_i, W^p)$ takes a linear form,

$$f(\sigma_i, W^p) = F(\sigma_i) - mW^p \tag{3.83}$$

where m is the modulus for work hardening or softening. The consistence condition, $df = 0$, then leads to

$$\frac{\partial F}{\partial \sigma_i} k_{ij}\left(du_j - \lambda\frac{\partial Q}{\partial \sigma_i}\right) - m\sigma_i\left(\lambda\frac{\partial Q}{\partial \sigma_i}\right) = 0 \tag{3.84}$$

Solving for λ from Eqn (3.84) yields

$$\lambda = \frac{k_{ir}\dfrac{\partial Q}{\partial \sigma_r}\dfrac{\partial F}{\partial \sigma_p}k_{pj}}{\dfrac{\partial F}{\partial \sigma_r}k_{rp}\dfrac{\partial Q}{\partial \sigma_p} + m\sigma_r\dfrac{\partial Q}{\partial \sigma_r}} \tag{3.85}$$

where $i, j, r, p = t, n$ for 2D problems. Substitution of Eqns (3.79) and (3.85) into Eqn (3.78) finally produces the incremental constitutive model for rock fractures as

$$d\sigma_i = \left(k_{ij} - \frac{k_{ir}\dfrac{\partial Q}{\partial \sigma_r}\dfrac{\partial F}{\partial \sigma_p}k_{pj}}{\dfrac{\partial F}{\partial \sigma_r}k_{rp}\dfrac{\partial Q}{\partial \sigma_p} + m\sigma_r\dfrac{\partial Q}{\partial \sigma_r}}\right) du_j \tag{3.86}$$

Plesha (1987) took $m = 0$ (perfect plastic deformation) and used the exponential roughness degradation law in Eqn (3.31).

Jing (1990) extended the original Plesha's model with different pre and post-peak shear stress behaviors as work-hardening and work-softening cases, the stress dependence of shear stiffness (cf. Eqn (3.3)), and the second law of thermodynamics were used to restrict the values of material constants and parameters. The final constitutive model is written (Jing, 1990; Jing et al., 1993) as

$$\begin{cases} d\sigma_t = \left(k_t - \dfrac{b_1 k_t^2}{b_1 k_t + b_2 k_n + mQ}\right) du_t - \dfrac{b_3 k_t k_n}{b_1 k_t + b_2 k_n + mQ} du_n \\ d\sigma_n = -\dfrac{b_4 k_t k_n}{b_1 k_t + b_2 k_n + mQ} du_t + \left(k_n - \dfrac{b_2 k_n^2}{b_1 k_t + b_2 k_n + mQ}\right) du_n \end{cases} \tag{3.87}$$

where

$$b_1 = \cos^2 \alpha_k - \tan \phi_r \operatorname{sgn}(\sigma_t^c) \sin \alpha_k \cos \alpha_k \tag{3.88a}$$

$$b_2 = \sin^2 \alpha_k + \tan \phi_r \operatorname{sgn}(\sigma_t^c) \sin \alpha_k \cos \alpha_k \tag{3.88b}$$

$$b_3 = \sin \alpha_k \cos \alpha_k + \tan \phi_r \operatorname{sgn}(\sigma_t^c) \cos^2 \alpha_k \tag{3.88c}$$

$$b_4 = \sin \alpha_k \cos \alpha_k - \tan \phi_r \operatorname{sgn}(\sigma_t^c) \sin^2 \alpha_k \tag{3.88d}$$

where $\operatorname{sgn}(\sigma_t^c)$ is the plus or minus sign of the microscopic shear stress component σ_t^c on the active asperity surface. The stiffness k_n and k_t are specified by Eqns (3.6) and (3.2), respectively, for the stress dependency of the shear stiffness and the hyperbolic normal stress–closure relation. The modulus m has two different functions, one for displacement-hardening before the peak shear stress (Fig. 3.8) given by

$$m = \frac{u_t^p - u_t}{u_t^p - u_t^0} k_t \quad (u_t^0 \le u_t \le u_t^p) \tag{3.89}$$

and one for displacement softening after the peak shear stress written as

$$m = -s_c \frac{(u_t^r - u_t)(u_t - u_t^p)}{(u_t^r - u_t^p)^2} \frac{\sin^2 \alpha}{Q} k_t \quad (u_t > u_t^p) \tag{3.90}$$

where $s_c > 0$ is a material constant.

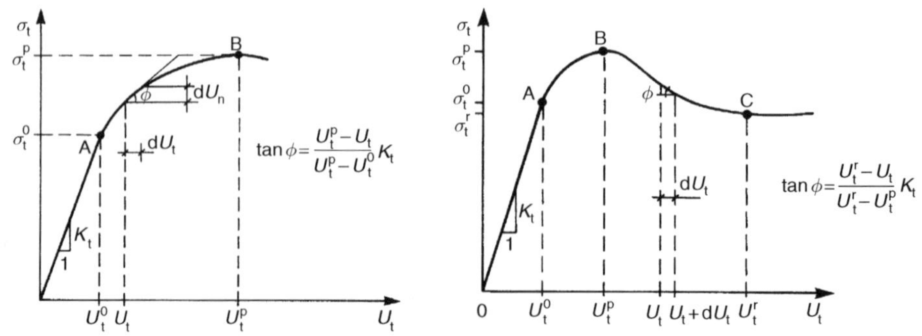

Fig. 3.8 Definitions of the (a) hardening and (b) softening moduli (Jing, 1990; Jing et al., 1993).

The second law of thermodynamics requires that the dissipative energy must be non-negative, i.e.,

$$dW^{\mathrm{p}} = \lambda \frac{\partial Q}{\partial \sigma_i} \sigma_i \geq 0 \tag{3.91}$$

This leads to the conditions

$$\left[1 + \eta \sin^2\alpha + \eta \tan\phi_{\mathrm{r}} \,\mathrm{sgn}\,(\sigma_{\mathrm{t}}^{\mathrm{c}}) \sin\alpha \cos\alpha\right] k_{\mathrm{t}} + mQ \geq 0 \tag{3.92}$$

for the displacement-hardening case and

$$\left[\cos^2\alpha + (1 - f) \sin^2\alpha + \eta \tan\phi_{\mathrm{r}} \,\mathrm{sgn}\,(\sigma_{\mathrm{t}}^{\mathrm{c}}) \sin\alpha \cos\alpha\right] k_{\mathrm{t}} \geq 0 \tag{3.93}$$

for displacement softening where

$$f = s_{\mathrm{c}}(u_{\mathrm{t}}^{\mathrm{r}} - u_{\mathrm{t}})(u_{\mathrm{t}} - u_{\mathrm{t}}^{\mathrm{p}})(u_{\mathrm{t}}^{\mathrm{r}} - u_{\mathrm{t}}^{\mathrm{p}})^{-2} \tag{3.94}$$

It was shown in Jing (1990) that the above conditions are satisfied when $-45^{\circ} < \alpha < 45^{\circ}$, $k_{\mathrm{n}} \geq k_{\mathrm{t}}$, $\phi_{\mathrm{r}} \leq 45^{\circ}$ and $0 \leq s_{\mathrm{c}} \leq 4.0$.

The model was implemented into a 2D DEM code, UDEC, and validated against laboratory experiments.

3.3.5 A 3D Constitutive Model with Anisotropic Roughness Representation

A difficulty in modeling fractured rocks is that problems can rarely be approximated by 2D simplifications, due mainly to the complex geometry of the fracture system. The intrinsic anisotropic nature of the geometric properties of the fracture surfaces also makes the 2D fracture models inappropriate for many practical problems. It is therefore imperative to have appropriate 3D constitutive models for rock fractures, especially in DEM analysis.

In a 3D space, the two rock blocks forming a fracture generally have six degrees of freedom (Fig. 3.9): two translations in the fracture plane, one translation in the normal direction and three rotations about the three coordinate axes, among which two are bending moments and one can cause frictional rotation in the

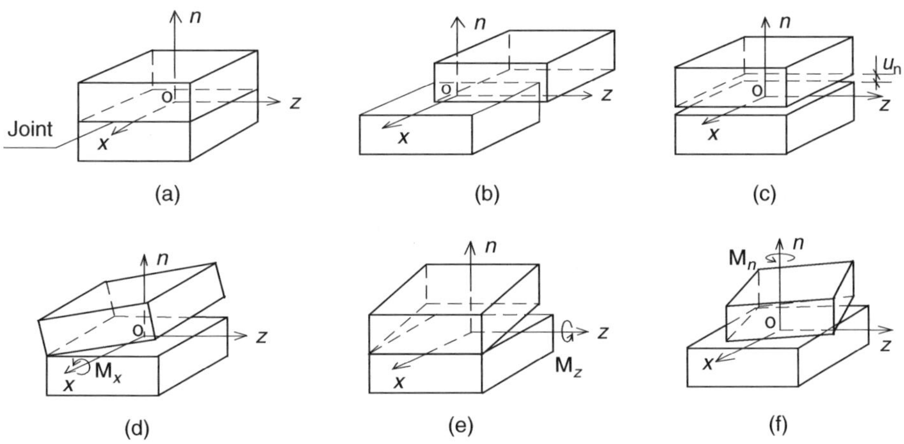

Fig. 3.9 Degrees of freedom of a rock joint in three dimensions. (a) Initial state; (b) translation displacements in the *xz*-plane; (c) translation in the normal (*n*) direction; (d) rotation about the *x*-axis (bending moments in the *xn*-plane); (e) rotation about the *z*-axis (bending moments in the *zn*-plane); and (f) rotation about the *n*-axis (frictional rotation in the *xz*-plane).

fracture plane. Since the two bending moments cause only deformations of the rock blocks, they do not affect the fracture significantly. Therefore, in DEM analyses, a 3D constitutive model for rock fractures should consider the effects of the three translational and one frictional rotation of the relative movements of the two rock blocks on the mechanical behavior of the interface between them.

A 3D constitutive model for rock fractures was developed by Jing (1990) and Jing et al. (1994) using the same general model expressed in Eqn (3.86) with *i, j, r, p* = *x, z, n* for 3D problems. The slip function F and sliding potential Q are given by

$$F = \sqrt{\left(\frac{\tau_x}{\mu_x}\right)^2 + \left(\frac{\tau_z}{\mu_z}\right)^2} + \sigma_n - C \tag{3.95}$$

$$Q = \sqrt{\left(\frac{\tau_x}{\mu_x}\right)^2 + \left(\frac{\tau_z}{\mu_z}\right)^2} + \sigma_n \sin\alpha \tag{3.96}$$

with

$$\mu_x = \tan(\phi_r + \alpha_x), \qquad \mu_z = \tan(\phi_r + \alpha_z) \tag{3.97}$$

where the roughness angle $\alpha = \alpha_\theta$, α_x and α_z are given by Eqn (3.23) with the current values of shear direction θ. The final incremental model form is written as

$$\begin{aligned} \mathrm{d}\tau_x &= \frac{1}{A}((a^2k_z + k_n \sin\alpha)k_x\, \mathrm{d}u_x - abk_xk_z\, \mathrm{d}u_z - bk_xk_n\, \mathrm{d}u_n) \\ \mathrm{d}\tau_z &= \frac{1}{A}(-abk_xk_z\, \mathrm{d}u_x(b^2k_x + k_n \sin\alpha)k_z\, \mathrm{d}u_z - ak_zk_n\, \mathrm{d}u_n) \\ \mathrm{d}\sigma_n &= \frac{1}{A}(-(bk_x\, \mathrm{d}u_x + ak_z\, \mathrm{d}u_z)k_n \sin\alpha + (b^2k_x + a^2k_z)k_n\, \mathrm{d}u_n) \end{aligned} \tag{3.98}$$

where

$$a = \frac{1}{\mu_z}\frac{\tau_z'}{\sqrt{(\tau_x')^2 + (\tau_z')^2}}, \qquad \tau_z' = \frac{\tau_z}{\mu_z} \tag{3.99a}$$

$$b = \frac{1}{\mu_z}\frac{\tau_x'}{\sqrt{(\tau_x')^2 + (\tau_z')^2}}, \qquad \tau_x' = \frac{\tau_x}{\mu_x} \tag{3.99b}$$

$$A = b^2k_x + a^2k_z + k_n \sin\alpha + mQ \tag{3.99c}$$

The anisotropy of the fracture roughness can then be approximated by an elliptical model. For validation of this model against experimental results, see Jing (1990) and Jing et al. (1994).

All the above constitutive models have only one single parameter, the asperity angle, as the representative roughness measure, which is certainly a simplistic representation. Most of them have no allowance for scale effects and some of them do not consider damage evolution of the asperities. No model considers the effects of gouge material generated during shear, or dynamic (rate) and time (creep) effects. Comprehensive reviews on the applicability of the constitutive models of rock fractures for coupled thermo-hydro-mechanical processes of fractured rocks in Stephansson and Jing (1995), Jing et al. (1996) and Jing (2003) show that the present constitutive models for rock fractures invoke still great challenges for predicting the coupled T-H-M behavior of rock fractures and fractured rock masses with acceptable accuracy, reliability, robustness and level of confidence, especially regarding fluid flow through the fracture systems. It is also recognized that understanding roughness is the key step for overcoming this difficulty.

3.4 Constitutive Models of Fractured Rock Masses as Equivalent Continua

Constitutive models are the equations that define the relations between the incremental (or total, but rarely) stress and strain components of a continuum or an equivalent (or effective) homogeneous and continuous medium (such as a fractured rock mass or an assembly of particles of varying sizes and shapes) over a certain representative size (REV). Strictly speaking, the formulation of a constitutive model for a continuum should follow the basic principles of thermodynamics of solids based on the concept of thermodynamic potential and the classical Clausius–Duhem inequality that combines both the first and second laws of thermodynamics, with carefully chosen state variables. In practice, however, constitutive laws are most often defined by intuitive understanding of the material behavior, such as that observed during laboratory experiments using small-sized samples and simple mathematics. Such models are formulated without resorting to the use of abstract state variables but compliance with the Clausius–Duhem inequality is still required. Typical examples are the laws of linear elasticity and elasto-plasticity. In the theory of elasticity, the material behavior is assumed to be conservative without generating dissipative energy; therefore, the Clausius–Duhem inequality is not needed. Plasticity models can be properly constrained by the requirement of positive plastic work increment $dW^{p} \geq 0$, which is a simplified version of the Clausius–Duhem inequality with an isothermal assumption. In this chapter, we only present some of the most popular constitutive laws for fractured rocks without considering their thermodynamic background.

3.4.1 Constitutive Laws for Elastic Continua with Small Deformation

Broadly speaking, the general form of an elastic continuum can be written in total stress and strain components as

$$\sigma_{ij} = f(\varepsilon_{ij}) \tag{3.100}$$

where $i, j = x, y, z$ for an orthogonal Cartesian xyz-coordinate system or more explicitly as

$$\begin{aligned}
\sigma_{xx} &= f_1(\varepsilon_{xx}, \varepsilon_{yy}, \varepsilon_{zz}, \varepsilon_{xy}, \varepsilon_{yz}, \varepsilon_{zx}) \\
\sigma_{yy} &= f_2(\varepsilon_{xx}, \varepsilon_{yy}, \varepsilon_{zz}, \varepsilon_{xy}, \varepsilon_{yz}, \varepsilon_{zx}) \\
\sigma_{zz} &= f_3(\varepsilon_{xx}, \varepsilon_{yy}, \varepsilon_{zz}, \varepsilon_{xy}, \varepsilon_{yz}, \varepsilon_{zx}) \\
\sigma_{xy} &= f_4(\varepsilon_{xx}, \varepsilon_{yy}, \varepsilon_{zz}, \varepsilon_{xy}, \varepsilon_{yz}, \varepsilon_{zx}) \\
\sigma_{yz} &= f_5(\varepsilon_{xx}, \varepsilon_{yy}, \varepsilon_{zz}, \varepsilon_{xy}, \varepsilon_{yz}, \varepsilon_{zx}) \\
\sigma_{zx} &= f_6(\varepsilon_{xx}, \varepsilon_{yy}, \varepsilon_{zz}, \varepsilon_{xy}, \varepsilon_{yz}, \varepsilon_{zx})
\end{aligned} \tag{3.101}$$

Assuming that only small deformation is considered, a Taylor expansion of equations (3.101) without partial derivatives equal to or higher than the second order will lead to

$$\begin{cases}
\sigma_{xx} = C_{10} + C_{11}\varepsilon_{xx} + C_{12}\varepsilon_{yy} + C_{13}\varepsilon_{zz} + C_{14}\varepsilon_{xy} + C_{15}\varepsilon_{yz} + C_{16}\varepsilon_{zx} \\
\sigma_{yy} = C_{20} + C_{21}\varepsilon_{xx} + C_{22}\varepsilon_{yy} + C_{23}\varepsilon_{zz} + C_{24}\varepsilon_{xy} + C_{25}\varepsilon_{yz} + C_{26}\varepsilon_{zx} \\
\sigma_{zz} = C_{30} + C_{31}\varepsilon_{xx} + C_{32}\varepsilon_{yy} + C_{33}\varepsilon_{zz} + C_{34}\varepsilon_{xy} + C_{35}\varepsilon_{yz} + C_{36}\varepsilon_{zx} \\
\sigma_{xy} = C_{40} + C_{41}\varepsilon_{xx} + C_{42}\varepsilon_{yy} + C_{43}\varepsilon_{zz} + C_{44}\varepsilon_{xy} + C_{45}\varepsilon_{yz} + C_{46}\varepsilon_{zx} \\
\sigma_{xx} = C_{50} + C_{51}\varepsilon_{xx} + C_{52}\varepsilon_{yy} + C_{53}\varepsilon_{zz} + C_{54}\varepsilon_{xy} + C_{55}\varepsilon_{yz} + C_{56}\varepsilon_{zx} \\
\sigma_{xx} = C_{60} + C_{61}\varepsilon_{xx} + C_{62}\varepsilon_{yy} + C_{63}\varepsilon_{zz} + C_{64}\varepsilon_{xy} + C_{65}\varepsilon_{yz} + C_{66}\varepsilon_{zx}
\end{cases} \tag{3.102}$$

where

$$C_{i0} = (f_i)_0, \quad C_{i1} = \left(\frac{\partial f_i}{\partial \varepsilon_{xx}}\right)_0, \quad C_{i2} = \left(\frac{\partial f_i}{\partial \varepsilon_{yy}}\right)_0, \quad C_{i3} = \left(\frac{\partial f_i}{\partial \varepsilon_{zz}}\right)_0 \tag{3.103a}$$

$$C_{i4} = \left(\frac{\partial f_i}{\partial \varepsilon_{xy}}\right)_0, \qquad C_{i5} = \left(\frac{\partial f_i}{\partial \varepsilon_{yz}}\right)_0, \qquad C_{i6} = \left(\frac{\partial f_i}{\partial \varepsilon_{zx}}\right)_0 \tag{3.103b}$$

for $i = 1, \ldots, 6$. The symbol $(\,)_0$ means the value of the function in the bracket at $\varepsilon_{ij} = 0$. The first term C_{i0} clearly represents the initial stress components and therefore can be discarded. The reminder is termed as the generalized Hooke's law, written as

$$\sigma_{ij} = E_{ijkl}\varepsilon_{kl} \tag{3.104}$$

$$\varepsilon_{ij} = \frac{1}{2}\left(\frac{\partial u_i}{\partial x_j} + \frac{\partial u_j}{\partial x_i}\right) \tag{3.105}$$

where u_i, $i = x, y, z$ are the displacement components and the fourth rank tensor E_{ijkl} is called the elasticity tensor. Equation (3.104) can be written in a compacted matrix form as

$$\begin{Bmatrix} \sigma_{xx} \\ \sigma_{yy} \\ \sigma_{zz} \\ \sigma_{xy} \\ \sigma_{yz} \\ \sigma_{zx} \end{Bmatrix} = \begin{bmatrix} C_{11} & C_{12} & C_{13} & C_{14} & C_{15} & C_{16} \\ C_{21} & C_{22} & C_{23} & C_{24} & C_{25} & C_{26} \\ C_{31} & C_{32} & C_{33} & C_{34} & C_{35} & C_{36} \\ C_{41} & C_{42} & C_{43} & C_{44} & C_{45} & C_{46} \\ C_{51} & C_{52} & C_{53} & C_{54} & C_{55} & C_{56} \\ C_{61} & C_{62} & C_{63} & C_{64} & C_{65} & C_{66} \end{bmatrix} \begin{Bmatrix} \varepsilon_{xx} \\ \varepsilon_{yy} \\ \varepsilon_{zz} \\ \varepsilon_{xy} \\ \varepsilon_{yz} \\ \varepsilon_{zx} \end{Bmatrix} \tag{3.106}$$

The density of the strain energy π is then given by the product

$$\pi = \frac{1}{2}\sigma_{ij}\varepsilon_{ij} = \frac{1}{2}\left(\sigma_{xx}\varepsilon_{xx} + \sigma_{yy}\varepsilon_{yy} + \sigma_{zz}\varepsilon_{zz} + \sigma_{xy}\varepsilon_{xy} + \sigma_{yz}\varepsilon_{yz} + \sigma_{zx}\varepsilon_{zx}\right) \tag{3.107}$$

The identity

$$\frac{\partial^2 \pi}{\partial \varepsilon_{ij}\partial \varepsilon_{kl}} = \frac{\partial}{\partial \varepsilon_{ij}}\left(\frac{\partial \pi}{\partial \varepsilon_{kl}}\right) = \frac{\partial}{\partial \varepsilon_{kl}}\left(\frac{\partial \pi}{\partial \varepsilon_{ij}}\right) \tag{3.108}$$

leads to symmetric coefficients in the matrix in Eqn (3.106) with 21 independent elastic coefficients

$$C_{ij} = C_{ji} \quad (\mathrm{i,j} = 1, 6) \tag{3.109}$$

which represents general anisotropy in all directions. Materials with such behavior are called extreme elastic solids.

Deriving the constitutive models of elastic solids with various degree of anisotropy is achieved with the transformation (mapping) of the stress and strain tensors in two different coordinate systems, xyz and $x'y'z'$, whose axes have the direction cosines defined as in Table 3.1.

Table 3.1 Direction cosines between two coordinate systems

	X	Y	Z
x'	$l_1 = \cos(x, x')$	$m_1 = \cos(y, x')$	$n_1 = \cos(z, x')$
y'	$l_2 = \cos(x, y')$	$m_2 = \cos(y, y')$	$n_2 = \cos(z, y')$
z'	$l_3 = \cos(x, z')$	$m_3 = \cos(y, z')$	$n_3 = \cos(z, z')$

The transformation of stresses and strains between the two coordinate systems is then given by

$$\begin{Bmatrix} \sigma_{x'x'} \\ \sigma_{y'y'} \\ \sigma_{z'z'} \\ \sigma_{y'z'} \\ \sigma_{z'x'} \\ \sigma_{x'y'} \end{Bmatrix} = \begin{bmatrix} (l_1)^2 & (m_1)^2 & (n_1)^2 & 2m_1n_1 & 2n_1l_1 & 2l_1m_1 \\ (l_2)^2 & (m_2)^2 & (n_2)^2 & 2m_2n_2 & 2n_2l_2 & 2l_2m_2 \\ (l_3)^2 & (m_3)^2 & (n_3)^2 & 2m_3n_3 & 2n_3l_3 & 2l_3m_3 \\ l_2l_3 & m_2m_3 & n_2n_3 & m_2n_3+m_3n_2 & n_2l_3+n_3l_2 & l_2m_3+l_3m_2 \\ l_3l_1 & m_3m_1 & n_3n_1 & m_3n_1+m_1n_3 & n_3l_1+n_1l_3 & l_3m_1+l_1m_3 \\ l_1l_2 & m_1m_2 & n_1n_2 & m_1n_2+m_2n_1 & n_1l_2+n_2l_1 & l_1m_1+l_1m_2 \end{bmatrix} \begin{Bmatrix} \sigma_{xx} \\ \sigma_{yy} \\ \sigma_{zz} \\ \sigma_{yz} \\ \sigma_{zx} \\ \sigma_{xy} \end{Bmatrix} \quad (3.110a)$$

$$\begin{Bmatrix} \varepsilon_{x'x'} \\ \varepsilon_{y'y'} \\ \varepsilon_{z'z'} \\ \varepsilon_{y'z'} \\ \varepsilon_{z'x'} \\ \varepsilon_{x'y'} \end{Bmatrix} = \begin{bmatrix} (l_1)^2 & (m_1)^2 & (n_1)^2 & m_1n_1 & n_1l_1 & l_1m_1 \\ (l_2)^2 & (m_2)^2 & (n_2)^2 & m_2n_2 & n_2l_2 & l_2m_2 \\ (l_3)^2 & (m_3)^2 & (n_3)^2 & m_3n_3 & n_3l_3 & l_3m_3 \\ 2l_2l_3 & 2m_2m_3 & 2n_2n_3 & m_2n_3+m_3n_2 & n_2l_3+n_3l_2 & l_2m_3+l_3m_2 \\ 2l_3l_1 & 2m_3m_1 & 2n_3n_1 & m_3n_1+m_1n_3 & n_3l_1+n_1l_3 & l_3m_1+l_1m_3 \\ 2l_1l_2 & 2m_1m_2 & 2n_1n_2 & m_1n_2+m_2n_1 & n_1l_2+n_2l_1 & l_1m_1+l_1m_2 \end{bmatrix} \begin{Bmatrix} \varepsilon_{xx} \\ \varepsilon_{yy} \\ \varepsilon_{zz} \\ \varepsilon_{yz} \\ \varepsilon_{zx} \\ \varepsilon_{xy} \end{Bmatrix} \quad (3.110b)$$

3.4.1.1 Transversely anisotropic solids

If an elastic solid has one plane of elastic symmetry at every point, then the two opposite directions symmetric about this plane have the same elastic properties. The normal direction of this plane of elastic symmetry is called the principal direction of elasticity and within this elastic symmetry plane the solid is still generally anisotropic. Such solids are called transversely anisotropic solids. Assuming without losing generality that the yz-plane is the plane of elastic symmetry in a xyz-space, the same solid in another coordinate system (Fig. 3.10a) $x'y'z'$ should have the same elastic properties as that in the original xyz system, i.e., the generalized Hooke's law by Eqn (3.106) should be identical. The directional cosines of the two coordinate systems are as in Table 3.2.

Substituting the directional cosines into Eqn (3.110) leads to the elasticity symmetry conditions for such transversal anisotropic solids

$$\sigma_{xx} = \sigma_{x'x'}, \quad \sigma_{yy} = \sigma_{y'y'}, \quad \sigma_{zz} = \sigma_{z'z'}, \quad \sigma_{xy} = -\sigma_{x'y'}, \quad \sigma_{yz} = \sigma_{y'z'}, \quad \sigma_{zx} = -\sigma_{z'x'} \quad (3.111a)$$

$$\varepsilon_{xx} = \varepsilon_{x'x'}, \quad \varepsilon_{yy} = \varepsilon_{y'y'}, \quad \varepsilon_{zz} = \varepsilon_{z'z'}, \quad \varepsilon_{xy} = -\varepsilon_{x'y'}, \quad \varepsilon_{yz} = \varepsilon_{y'z'}, \quad \varepsilon_{zx} = -\varepsilon_{z'x'} \quad (3.111b)$$

Substituting conditions (3.111) into Eqns (3.104) and (3.106) then results in

$$\begin{Bmatrix} \sigma_{x'x'} \\ \sigma_{y'y'} \\ \sigma_{z'z'} \\ -\sigma_{x'y'} \\ \sigma_{y'z'} \\ -\sigma_{z'x'} \end{Bmatrix} = \begin{bmatrix} C_{11} & C_{12} & C_{13} & -C_{14} & C_{15} & -C_{16} \\ C_{21} & C_{22} & C_{23} & -C_{24} & C_{25} & -C_{26} \\ C_{31} & C_{32} & C_{33} & -C_{34} & C_{35} & -C_{36} \\ C_{41} & C_{42} & C_{43} & -C_{44} & C_{45} & -C_{46} \\ C_{51} & C_{52} & C_{53} & -C_{54} & C_{55} & -C_{56} \\ C_{61} & C_{62} & C_{63} & -C_{64} & C_{65} & -C_{66} \end{bmatrix} \begin{Bmatrix} \varepsilon_{x'x'} \\ \varepsilon_{y'y'} \\ \varepsilon_{z'z'} \\ -\varepsilon_{x'y'} \\ \varepsilon_{y'z'} \\ -\varepsilon_{z'x'} \end{Bmatrix} \quad (3.112)$$

Since (3.112) must be identical to (3.106), i.e., the two elasticity tensors must be identical, this condition then leads to

$$C_{14} = C_{24} = C_{34} = C_{54} = C_{16} = C_{26} = C_{36} = C_{56} = 0 \quad (3.113)$$

The symmetry condition (3.109) then further yields

$$C_{41} = C_{42} = C_{43} = C_{45} = C_{61} = C_{62} = C_{63} = C_{65} = 0 \quad (3.114)$$

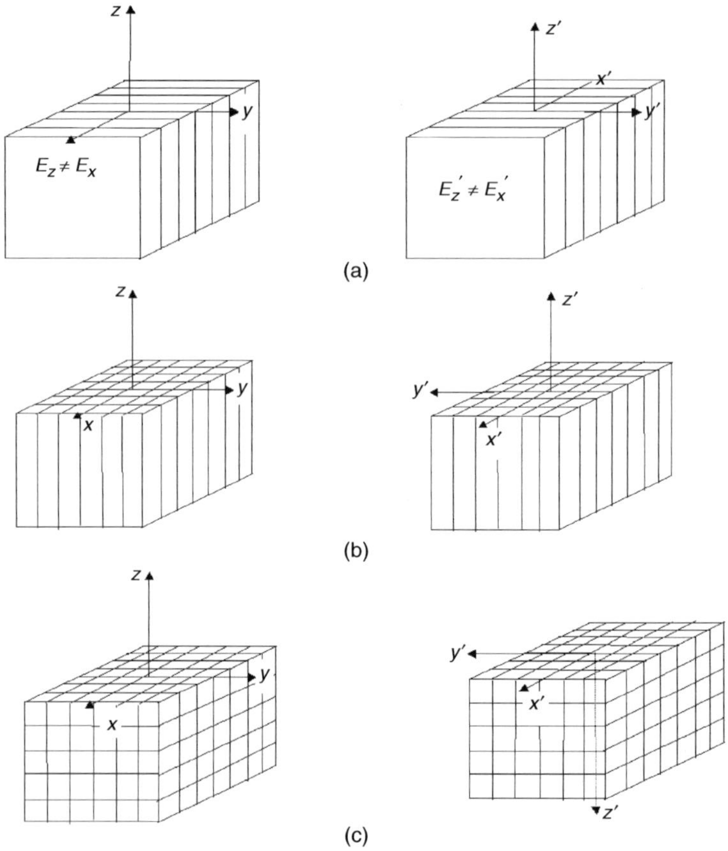

Fig. 3.10 Solids with elastic symmetry planes. (a) One plane of elastic symmetry (the *yz*-plane); (b) two planes of elastic symmetry (the *yz* and *xz*-planes); and (c) three planes of elastic symmetry (the *xy*, *yz* and *xz*-planes).

Table 3.2 Direction cosines for defining transversal anisotropic solids

	X	Y	Z
x'	$l_1 = \cos(x, x') = -1$	$m_1 = \cos(y, x') = 0$	$n_1 = \cos(z, x') = 0$
y'	$l_2 = \cos(x, y') = 0$	$m_2 = \cos(y, y') = 1$	$n_2 = \cos(z, y') = 0$
z'	$l_3 = \cos(x, z') = 0$	$m_3 = \cos(y, z') = 0$	$n_3 = \cos(z, z') = 1$

The final elasticity tensor for transversally anisotropic elastic solids (with 13 independent elasticity coefficients and $C_{ij} = C_{ji}$) becomes

$$\begin{Bmatrix} \sigma_{xx} \\ \sigma_{yy} \\ \sigma_{zz} \\ \sigma_{xy} \\ \sigma_{yz} \\ \sigma_{zx} \end{Bmatrix} = \begin{bmatrix} C_{11} & C_{12} & C_{13} & 0 & C_{15} & 0 \\ C_{21} & C_{22} & C_{23} & 0 & C_{25} & 0 \\ C_{31} & C_{32} & C_{33} & 0 & C_{35} & 0 \\ 0 & 0 & 0 & C_{44} & 0 & C_{46} \\ C_{51} & C_{52} & C_{53} & 0 & C_{55} & 0 \\ 0 & 0 & 0 & C_{64} & 0 & C_{66} \end{bmatrix} \begin{Bmatrix} \varepsilon_{xx} \\ \varepsilon_{yy} \\ \varepsilon_{zz} \\ \varepsilon_{xy} \\ \varepsilon_{yz} \\ \varepsilon_{zx} \end{Bmatrix} \tag{3.115a}$$

Similarly taking *xz*-plane or *xy*-plane as the elastic symmetry plane leads to

$$\begin{Bmatrix} \sigma_{xx} \\ \sigma_{yy} \\ \sigma_{zz} \\ \sigma_{xy} \\ \sigma_{yz} \\ \sigma_{zx} \end{Bmatrix} = \begin{bmatrix} C_{11} & C_{12} & C_{13} & C_{14} & 0 & 0 \\ C_{21} & C_{22} & C_{23} & C_{24} & 0 & 0 \\ C_{31} & C_{32} & C_{33} & C_{34} & 0 & 0 \\ C_{41} & C_{42} & C_{43} & C_{44} & 0 & 0 \\ 0 & 0 & 0 & 0 & C_{55} & C_{56} \\ 0 & 0 & 0 & 0 & C_{65} & C_{66} \end{bmatrix} \begin{Bmatrix} \varepsilon_{xx} \\ \varepsilon_{yy} \\ \varepsilon_{zz} \\ \varepsilon_{xy} \\ \varepsilon_{yz} \\ \varepsilon_{zx} \end{Bmatrix} \tag{3.115b}$$

$$\begin{Bmatrix} \sigma_{xx} \\ \sigma_{yy} \\ \sigma_{zz} \\ \sigma_{xy} \\ \sigma_{yz} \\ \sigma_{zx} \end{Bmatrix} = \begin{bmatrix} C_{11} & C_{12} & C_{13} & 0 & 0 & C_{16} \\ C_{21} & C_{22} & C_{23} & 0 & 0 & C_{26} \\ C_{31} & C_{32} & C_{33} & 0 & 0 & C_{36} \\ 0 & 0 & 0 & C_{44} & C_{54} & 0 \\ 0 & 0 & 0 & C_{54} & C_{55} & 0 \\ C_{61} & C_{62} & C_{63} & 0 & 0 & C_{66} \end{bmatrix} \begin{Bmatrix} \varepsilon_{xx} \\ \varepsilon_{yy} \\ \varepsilon_{zz} \\ \varepsilon_{xy} \\ \varepsilon_{yz} \\ \varepsilon_{zx} \end{Bmatrix} \tag{3.115c}$$

3.4.1.2 Orthogonal anisotropic solids

If, in addition to the *yz*-plane, the *xz*-plane is also an elastic symmetry plane, i.e., both the *y*- and *x*-axes are principal axes (Fig. 3.10b), then similarly one has elastic symmetry conditions, i.e.,

$$\sigma_{xx} = \sigma_{x'x'}, \quad \sigma_{yy} = \sigma_{y'y'}, \quad \sigma_{zz} = \sigma_{z'z'}, \quad \sigma_{xy} = -\sigma_{x'y'}, \quad \sigma_{yz} = -\sigma_{y'z'}, \quad \sigma_{zx} = \sigma_{z'x'} \tag{3.116a}$$

$$\varepsilon_{xx} = \varepsilon_{x'x'}, \quad \varepsilon_{yy} = \varepsilon_{y'y'}, \quad \varepsilon_{zz} = \varepsilon_{z'z'}, \quad \varepsilon_{xy} = -\varepsilon_{x'y'}, \quad \varepsilon_{yz} = -\varepsilon_{y'z'}, \quad \varepsilon_{zx} = \varepsilon_{z'x'} \tag{3.116b}$$

Substituting (3.116) into (3.115a) and making both sets of stress–strain relations identical leads to

$$C_{15} = C_{25} = C_{35} = C_{56} = C_{51} = C_{52} = C_{53} = C_{65} = 0 \tag{3.117}$$

and the stress–strain relation becomes

$$\begin{Bmatrix} \sigma_{xx} \\ \sigma_{yy} \\ \sigma_{zz} \\ \sigma_{xy} \\ \sigma_{yz} \\ \sigma_{zx} \end{Bmatrix} = \begin{bmatrix} C_{11} & C_{12} & C_{13} & 0 & 0 & 0 \\ C_{21} & C_{22} & C_{23} & 0 & 0 & 0 \\ C_{31} & C_{32} & C_{33} & 0 & 0 & 0 \\ 0 & 0 & 0 & C_{44} & 0 & 0 \\ 0 & 0 & 0 & 0 & C_{55} & 0 \\ 0 & 0 & 0 & 0 & 0 & C_{66} \end{bmatrix} \begin{Bmatrix} \varepsilon_{xx} \\ \varepsilon_{yy} \\ \varepsilon_{zz} \\ \varepsilon_{xy} \\ \varepsilon_{yz} \\ \varepsilon_{zx} \end{Bmatrix} \tag{3.118}$$

The number of the independent elasticity coefficients is then reduced to nine. The normal stresses and shear strain components, and the shear stresses in different planes, are completely de-coupled. Substituting Eqn (3.116) into Eqn (3.115b) or (3.115c) then leads to identical results.

If we further assume that the *xy*-plane also becomes an elastic symmetry plane (Fig. 3.10c), then the same results will be obtained. This implies that, for an elastic solid with two orthogonal elastic planes, the third orthogonal plane must also be an elastic plane.

3.4.1.3 Transversely isotropic elastic solids

For orthogonal anisotropic solids represented by Eqn (3.118), if we assume further that within one elastic symmetry plane the solid is isotropic, i.e., within this plane the elasticity properties of the solid are the same in all directions, then such a solid is called a transversely isotropic solid.

Assuming that the *xy*-plane in the orthogonal anisotropic solid shown in Fig. 3.10c is the isotropic plane, then the elastic properties in the positive and negative *x*-directions should be the same. Based on

Eqn (3.118), we rotate the *xyz* coordinate system 90 degrees about the *z*-axis (Fig. 3.10c) and the elastic symmetric conditions then become

$$\sigma_{x'x'} = \sigma_{xx}, \quad \sigma_{y'y'} = \sigma_{yy}, \quad \sigma_{z'z'} = \sigma_{zz}, \quad \sigma_{y'z'} = -\sigma_{xz}, \quad \sigma_{x'z'} = \sigma_{yz}, \quad \sigma_{x'y'} = -\sigma_{xy} \tag{3.119a}$$

$$\varepsilon_{x'x'} = \varepsilon_{xx}, \quad \varepsilon_{y'y'} = \varepsilon_{yy}, \quad \varepsilon_{z'z'} = \varepsilon_{zz}, \quad \varepsilon_{y'z'} = -\varepsilon_{xz}, \quad \varepsilon_{x'z'} = \varepsilon_{yz}, \quad \varepsilon_{x'y'} = -\varepsilon_{xy} \tag{3.119b}$$

Substitution of the condition given by Eqn (3.119) into Eqn (3.118) then produces

$$\begin{Bmatrix} \sigma_{x'x'} \\ \sigma_{y'y'} \\ \sigma_{z'z'} \\ -\sigma_{x'y'} \\ -\sigma_{y'z'} \\ \sigma_{z'x'} \end{Bmatrix} = \begin{bmatrix} C_{11} & C_{12} & C_{13} & 0 & 0 & 0 \\ C_{21} & C_{22} & C_{23} & 0 & 0 & 0 \\ C_{31} & C_{32} & C_{33} & 0 & 0 & 0 \\ 0 & 0 & 0 & C_{44} & 0 & 0 \\ 0 & 0 & 0 & 0 & C_{55} & 0 \\ 0 & 0 & 0 & 0 & 0 & C_{66} \end{bmatrix} \begin{Bmatrix} \varepsilon_{x'x'} \\ \varepsilon_{y'y'} \\ \varepsilon_{z'z'} \\ -\varepsilon_{x'y'} \\ -\varepsilon_{y'z'} \\ \varepsilon_{z'x'} \end{Bmatrix} \tag{3.120}$$

Making the coefficient matrices in Eqn (3.120) and Eqn (3.118) identical yields

$$C_{11} = C_{22}, \quad C_{13} = C_{23}, \quad C_{44} = C_{55} \tag{3.121}$$

The stress–strain relation, derived from Eqn (3.118), then becomes

$$\begin{Bmatrix} \sigma_{xx} \\ \sigma_{yy} \\ \sigma_{zz} \\ \sigma_{xy} \\ \sigma_{yz} \\ \sigma_{zx} \end{Bmatrix} = \begin{bmatrix} C_{11} & C_{12} & C_{13} & 0 & 0 & 0 \\ C_{12} & C_{11} & C_{13} & 0 & 0 & 0 \\ C_{13} & C_{13} & C_{33} & 0 & 0 & 0 \\ 0 & 0 & 0 & C_{44} & 0 & 0 \\ 0 & 0 & 0 & 0 & C_{44} & 0 \\ 0 & 0 & 0 & 0 & 0 & C_{66} \end{bmatrix} \begin{Bmatrix} \varepsilon_{xx} \\ \varepsilon_{yy} \\ \varepsilon_{zz} \\ \varepsilon_{xy} \\ \varepsilon_{yz} \\ \varepsilon_{zx} \end{Bmatrix} \tag{3.122}$$

with six independent elastic coefficients. However, based on Eqn (3.122), a rotation of the coordinate system about the *z*-axis for an finite angle ϕ then leads to

$$\begin{cases} \sigma_{x'y'} = \dfrac{1}{2}(\sigma_{yy} - \sigma_{xx}) \sin 2\phi + \sigma_{xy} \cos 2\phi \\ \varepsilon_{x'y'} = (\varepsilon_{yy} - \varepsilon_{xx}) \sin 2\phi + \varepsilon_{xy} \cos 2\phi \end{cases} \tag{3.123}$$

According to (3.120), $\sigma_{x'y'} = C_{44}\varepsilon_{x'y'}$, i.e.,

$$\frac{1}{2}(\sigma_{yy} - \sigma_{xx}) \sin 2\phi + \sigma_{xy} \cos 2\phi = C_{44}[(\varepsilon_{yy} - \varepsilon_{xx}) \sin 2\phi + \varepsilon_{xy} \cos 2\phi] \tag{3.124}$$

Substituting $\sigma_{xy} = C_{44}\varepsilon_{xy}$ into (3.124) then leads to

$$(\sigma_{yy} - \sigma_{xx}) = 2C_{44}(\varepsilon_{yy} - \varepsilon_{xx}) \tag{3.125}$$

From equation (3.122), the difference between the first two equations is given by

$$(\sigma_{yy} - \sigma_{xx}) = (C_{11} - C_{12})(\varepsilon_{yy} - \varepsilon_{xx}) \tag{3.126}$$

Therefore, we finally have that

$$C_{44} = \frac{1}{2}(C_{11} - C_{12}) \tag{3.127}$$

and the stress–strain relation for transversal isotropic solids becomes

$$\begin{Bmatrix} \sigma_{xx} \\ \sigma_{yy} \\ \sigma_{zz} \\ \sigma_{xy} \\ \sigma_{yz} \\ \sigma_{zx} \end{Bmatrix} = \begin{bmatrix} C_{11} & C_{12} & C_{13} & 0 & 0 & 0 \\ C_{12} & C_{11} & C_{13} & 0 & 0 & 0 \\ C_{13} & C_{13} & C_{33} & 0 & 0 & 0 \\ 0 & 0 & 0 & (C_{11}-C_{12})/2 & 0 & 0 \\ 0 & 0 & 0 & 0 & C_{44} & 0 \\ 0 & 0 & 0 & 0 & 0 & C_{66} \end{bmatrix} \begin{Bmatrix} \varepsilon_{xx} \\ \varepsilon_{yy} \\ \varepsilon_{zz} \\ \varepsilon_{xy} \\ \varepsilon_{yz} \\ \varepsilon_{zx} \end{Bmatrix} \tag{3.128}$$

with five independent elasticity coefficients. The *xy*-plane is both its isotropic plane and its elasticity symmetry plane, but the elastic properties within the *xy*-plane and in the *z*-direction are of course different.

3.4.1.4 Isotropic elastic solids

Assuming further that the solid described by Eqn (3.128) is isotropic in all directions, i.e., all elastic symmetry planes are also the isotropic planes as well, then the elastic properties are symmetric in all directions, regardless of the coordinate system chosen to express them. To derive the corresponding stress–strain relation, we define a transform of the coordinate system as shown in Fig. 3.11, with the stresses and strains having the following relations

$$\sigma_{x'x'} = \sigma_{xx}, \quad \sigma_{y'y'} = \sigma_{zz}, \quad \sigma_{z'z'} = \sigma_{yy}, \quad \sigma_{y'z'} = -\sigma_{yz}, \quad \sigma_{x'z'} = -\sigma_{xy}, \quad \sigma_{x'y'} = \sigma_{zx} \tag{3.129a}$$

$$\varepsilon_{x'x'} = \varepsilon_{xx}, \quad \varepsilon_{y'y'} = \varepsilon_{zz}, \quad \varepsilon_{z'z'} = \varepsilon_{yy}, \quad \varepsilon_{y'z'} = -\varepsilon_{yz}, \quad \varepsilon_{x'z'} = -\varepsilon_{xy}, \quad \varepsilon_{x'y'} = \varepsilon_{zx} \tag{3.129b}$$

Substituting Eqn (3.129) into Eqn (3.128) results in

$$\begin{Bmatrix} \sigma_{x'x'} \\ \sigma_{y'y'} \\ \sigma_{z'z'} \\ \sigma_{x'y'} \\ -\sigma_{y'z'} \\ -\sigma_{z'x'} \end{Bmatrix} = \begin{bmatrix} C_{11} & C_{13} & C_{12} & 0 & 0 & 0 \\ C_{13} & C_{33} & C_{13} & 0 & 0 & 0 \\ C_{12} & C_{13} & C_{11} & 0 & 0 & 0 \\ 0 & 0 & 0 & C_{44} & 0 & 0 \\ 0 & 0 & 0 & 0 & C_{44} & 0 \\ 0 & 0 & 0 & 0 & 0 & (C_{11}-C_{12})/2 \end{bmatrix} \begin{Bmatrix} \varepsilon_{x'x'} \\ \varepsilon_{y'y'} \\ \varepsilon_{z'z'} \\ \varepsilon_{x'y'} \\ -\varepsilon_{y'z'} \\ -\varepsilon_{z'x'} \end{Bmatrix} \tag{3.130}$$

The coefficient matrices in Eqns (3.130) and (3.128) should be identical and thus we have

$$C_{12} = C_{13}, \quad C_{11} = C_{33}, \quad C_{66} = (C_{11} - C_{12})/2 \tag{3.131}$$

Substituting Eqn (3.131) into Eqn (3.128) then leads to the stress–strain relation of an isotropic elastic solid

$$\begin{Bmatrix} \sigma_{xx} \\ \sigma_{yy} \\ \sigma_{zz} \\ \sigma_{xy} \\ \sigma_{yz} \\ \sigma_{zx} \end{Bmatrix} = \begin{bmatrix} C_{11} & C_{12} & C_{12} & 0 & 0 & 0 \\ C_{12} & C_{11} & C_{12} & 0 & 0 & 0 \\ C_{12} & C_{12} & C_{11} & 0 & 0 & 0 \\ 0 & 0 & 0 & (C_{11}-C_{12})/2 & 0 & 0 \\ 0 & 0 & 0 & 0 & (C_{11}-C_{12})/2 & 0 \\ 0 & 0 & 0 & 0 & 0 & (C_{11}-C_{12})/2 \end{bmatrix} \begin{Bmatrix} \varepsilon_{xx} \\ \varepsilon_{yy} \\ \varepsilon_{zz} \\ \varepsilon_{xy} \\ \varepsilon_{yz} \\ \varepsilon_{zx} \end{Bmatrix} \tag{3.132}$$

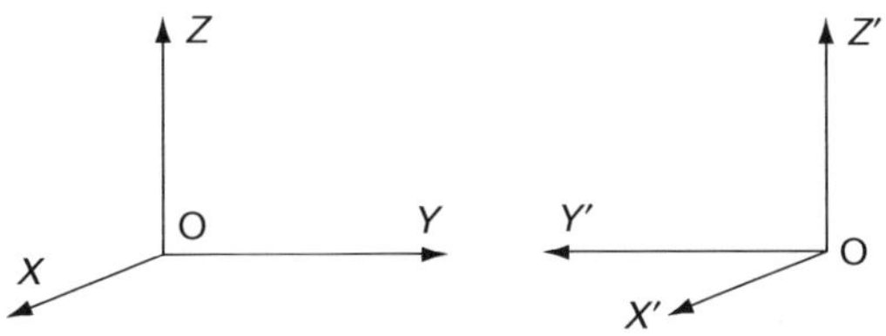

Fig. 3.11 Coordinate rotation for defining isotropic solids.

i.e., isotropic elastic solids have only two independent elasticity coefficients. Denoting

$$\lambda = C_{12}, \quad 2\mu = C_{11} - C_{12}, \quad E = \frac{\mu(3\lambda + 2\mu)}{\lambda + \mu}, \quad \nu = \frac{\lambda}{2(\lambda + \mu)} \tag{3.133}$$

where λ and μ are called the Lame's elasticity constants, E the Young's modulus, ν the Poisson's ratio and $\mu = G$ is also called the shear modulus. The stress–strain relation can be written in two more commonly encountered forms

$$\begin{Bmatrix} \sigma_{xx} \\ \sigma_{yy} \\ \sigma_{zz} \\ \sigma_{xy} \\ \sigma_{yz} \\ \sigma_{zx} \end{Bmatrix} = \frac{E}{(1+\nu)(1-2\nu)} \begin{bmatrix} 1-\nu & \nu & \nu & 0 & 0 & 0 \\ \nu & 1-\nu & \nu & 0 & 0 & 0 \\ \nu & \nu & 1-\nu & 0 & 0 & 0 \\ 0 & 0 & 0 & 1/(1-2\nu) & 0 & 0 \\ 0 & 0 & 0 & 0 & 1/(1-2\nu) & 0 \\ 0 & 0 & 0 & 0 & 0 & 1/(1-2\nu) \end{bmatrix} \begin{Bmatrix} \varepsilon_{xx} \\ \varepsilon_{yy} \\ \varepsilon_{zz} \\ \varepsilon_{xy} \\ \varepsilon_{yz} \\ \varepsilon_{zx} \end{Bmatrix} \tag{3.134}$$

or

$$\begin{Bmatrix} \sigma_{xx} \\ \sigma_{yy} \\ \sigma_{zz} \\ \sigma_{xy} \\ \sigma_{yz} \\ \sigma_{zx} \end{Bmatrix} = \begin{bmatrix} 2\mu+\lambda & \lambda & \lambda & 0 & 0 & 0 \\ \lambda & 2\mu+\lambda & \lambda & 0 & 0 & 0 \\ \lambda & \lambda & 2\mu+\lambda & 0 & 0 & 0 \\ 0 & 0 & 0 & \mu & 0 & 0 \\ 0 & 0 & 0 & 0 & \mu & 0 \\ 0 & 0 & 0 & 0 & 0 & \mu \end{bmatrix} \begin{Bmatrix} \varepsilon_{xx} \\ \varepsilon_{yy} \\ \varepsilon_{zz} \\ \varepsilon_{xy} \\ \varepsilon_{yz} \\ \varepsilon_{zx} \end{Bmatrix} \tag{3.135}$$

This is Hooke's law for isotropic elastic solids with only two independent elastic coefficients and is one of the most often adopted assumptions for material behavior in the DEM.

3.4.2 Fractured Rocks with Persistent Sets of Fractures as Equivalent Elastic Solids

The effects of fractures on the mechanical behavior of rock masses are significant, but also difficult to consider in constitutive laws. Over the years, closed-form stress–strain relations have been developed only for rocks with persistent fracture systems (Duncan and Goodman, 1968; Singh, 1973; Lekhnitskii, 1977; Huang et al., 1995) and randomly distributed fractures (Oda, 1982, 1986a,b, 1988a,b; Oda et al., 1984, 1986; Stietel et al., 1996) with different applicability in practice.

3.4.2.1 Rocks with persistent orthogonal sets of fractures

For rock mass containing three persistent, perpendicular fracture sets with a constant mean spacing, analytical relations can be derived for the elastic moduli of the fractured rock mass (Duncan and Goodman, 1968; Lekhnitskii, 1977) by using the concept of an equivalent continuum. Denoting the axes of anisotropy of the rock mass as *x*, *y* and *z* (Fig. 3.15), the elastic moduli of the equivalent rock mass are given by

$$\frac{1}{E_i} = \frac{1}{E_0} + \frac{1}{k_{\mathrm{n}}^i S_i} \tag{3.136}$$

$$\frac{1}{G_{ij}} = \frac{1}{G_0} + \frac{1}{k_{\mathrm{t}}^i S_i} + \frac{1}{K_{\mathrm{t}}^i S_j} \tag{3.137}$$

where $i, j = x, y$ and z, respectively. Symbols E_0 and G_0 are the Young's and shear modulus of the intact rock, E_i the Young's modulus of the fractured rock mass in direction i and G_{ij} the shear modulus in the ij-plane. Parameters k_{n}^i and k_{t}^i are the normal and shear stiffness of fracture sets with their unit normal

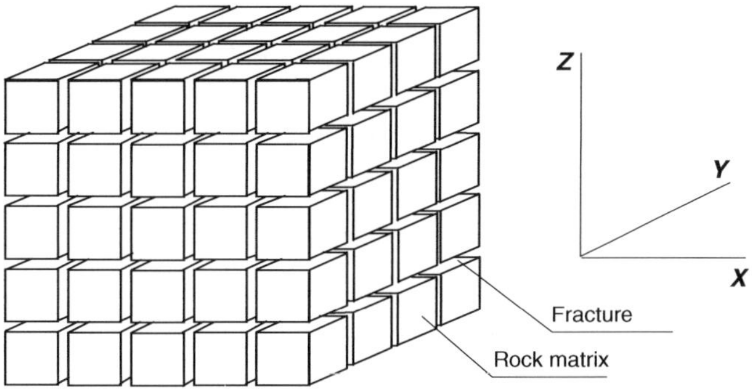

Fig. 3.12 Rock mass containing three orthogonal sets of persistent fractures (Duncan and Goodman, 1968).

vectors parallel to direction, i, and with spacing, S_i. Equations (3.136) and (3.137) are applicable for both two and three-dimensional problems by just adjusting the ranges of the indices, i and j.

Taking the fracture planes as the elasticity symmetry planes, the fractured rock is analogous to the orthogonal anisotropic solids shown in Fig. 3.12 with its stress–strain relation described by Eqn (3.118) with nine independent coefficients given by

$$\{\sigma_{ij}\} = [E_{ijkl}]\{\varepsilon_{kl}\} \tag{3.138}$$

where $\{\sigma_{ij}\} = \{\sigma_{xx}, \sigma_{yy}, \sigma_{zz}, \sigma_{xy}, \sigma_{yz}, \sigma_{zx}\}^{\mathrm{T}}$, $\{\varepsilon_{ij}\} = \{\varepsilon_{xx}, \varepsilon_{yy}, \varepsilon_{zz}, \varepsilon_{xy}, \varepsilon_{yz}, \varepsilon_{zx}\}^{\mathrm{T}}$ and E_{ijkl} is called the elastic stiffness tensor in Chen (1994) and is identical in form to Eqn (3.118). The values of the coefficients are different because of the fractures, as given by

$$C_{11} = E_x E_y [E_z - E_y (\nu_{yz})^2]/A, \quad C_{22} = (E_y)^2 [E_z - E_x (\nu_{zx})^2]/A \tag{3.139a}$$

$$C_{33} = (E_z)^2 [E_y - E_x (\nu_{xy})^2]/A, \quad C_{12} = E_x E_y [E_z \nu_{xy} + E_y \nu_{zx} \nu_{yz}]/A \tag{3.139b}$$

$$C_{13} = E_x E_y E_z [\nu_{zx} + \nu_{xy} \nu_{yz}]/A, \quad C_{23} = E_y E_z [E_x \nu_{xy} \nu_{zx} + E_y \nu_{yz}]/A \tag{3.139c}$$

$$C_{44} = G_{xy}, \quad C_{55} = G_{yz}, \quad C_{66} = G_{zx} \tag{3.139d}$$

$$A = E_y E_z - E_x E_y (\nu_{xy})^2 - E_z E_x (\nu_{zx})^2 - (E_y)^2 (\nu_{yz})^2 - 2E_x E_y \nu_{xy} \nu_{yz} \nu_{zx} \tag{3.139e}$$

where ν_{ij} are the Poisson's ratios between the directions i and j. The elastic compliance tensor is easily written as (Lekhnitskii, 1977; Wittke, 1990)

$$[C_{ijkl}] = [E_{ijkl}]^{-1} = \begin{bmatrix} \frac{1}{E_x} & -\frac{\nu_{xy}}{E_y} & -\frac{\nu_{zx}}{E_z} & 0 & 0 & 0 \\ -\frac{\nu_{xy}}{E_y} & \frac{1}{E_y} & -\frac{\nu_{yz}}{E_z} & 0 & 0 & 0 \\ -\frac{\nu_{zx}}{E_z} & -\frac{\nu_{yz}}{E_z} & \frac{1}{E_z} & 0 & 0 & 0 \\ 0 & 0 & 0 & \frac{1}{G_{xy}} & 0 & 0 \\ 0 & 0 & 0 & 0 & \frac{1}{G_{yz}} & 0 \\ 0 & 0 & 0 & 0 & 0 & \frac{1}{G_{zx}} \end{bmatrix} \tag{3.140}$$

for orthogonal anisotropic solids. The nine coefficients are the three Young's modulus, three shear moduli and three Poisson's ratios. The consistency condition $A > 0$ must be met.

For isotropic rocks with just one set of persistent fracture planes with spacing S, the constitutive relation is reduced to (Amadei and Goodman, 1981; Fossum, 1985)

$$\begin{Bmatrix} \sigma_{xx} \\ \sigma_{yy} \\ \sigma_{zz} \\ \sigma_{xy} \\ \sigma_{yz} \\ \sigma_{zx} \end{Bmatrix} = \begin{bmatrix} C_{11} & C_{12} & C_{12} & 0 & 0 & 0 \\ C_{12} & C_{22} & C_{23} & 0 & 0 & 0 \\ C_{12} & C_{23} & C_{33} & 0 & 0 & 0 \\ 0 & 0 & 0 & C_{44} & 0 & 0 \\ 0 & 0 & 0 & 0 & C_{22} - C_{23} & 0 \\ 0 & 0 & 0 & 0 & 0 & C_{44} \end{bmatrix} \begin{Bmatrix} \varepsilon_{xx} \\ \varepsilon_{yy} \\ \varepsilon_{zz} \\ \varepsilon_{xy} \\ \varepsilon_{yz} \\ \varepsilon_{zx} \end{Bmatrix} \tag{3.141}$$

where the fracture planes are in the yz-plane and the x-axis is normal to the fracture planes. The fractured rock is then characterized by the Young's modulus E, Poisson's ratio ν, fracture spacing S and fracture stiffness k_n and k_s. Recall Eqns (3.136) and (3.137), the coefficients in Eqn (3.141) are simply given by,

$$C_{11} = \frac{Sk_n E(1-\nu)}{sk_n(1+\nu)(1-2\nu) + E(1-\nu)}, \quad C_{44} = \frac{2k_s SE}{2(1+\nu)k_s S + E} \tag{3.142a}$$

$$C_{12} = \frac{Sk_n E\nu}{sk_n(1+\nu)(1-2\nu) + E(1-\nu)} \tag{3.142b}$$

$$C_{22} = \frac{E[Sk_n(1-\nu)^2 + E]}{(1+\nu)[sk_n(1+\nu)(1-2\nu) + E(1-\nu)]} \tag{3.142c}$$

$$C_{23} = \frac{E\nu[Sk_n(1+\nu) + E]}{(1+\nu)[sk_n(1+\nu)(1-2\nu) + E(1-\nu)]} \tag{3.142d}$$

3.4.2.2 *Rocks with non-orthogonal persistent sets of fractures*

Closed-form presentations of elastic stress–strain relations for non-orthogonal and persistent sets of fractures have also been developed, such as in Fossum (1985) and Yoshinaka and Yamabe (1986). In a more consistent approach, Huang et al. (1995) developed a stress–strain relation for fractured rocks with three sets of not perpendicularly intersecting fractures. The original theory was developed with incremental strains and stresses. However, since the fractured rocks were assumed as an equivalent elastic continuum, the same theory applies for total stresses and strains as described here. The basic assumption is that the total strain of the fractured rock, ε_{ij}, can be divided into two additive components, one from the intact rock matrix, $\varepsilon_{ij}^{\mathrm{I}}$, and one from the fractures, $\varepsilon_{ij}^{\mathrm{f}}$,

$$\varepsilon_{ij} = \varepsilon_{ij}^{\mathrm{I}} + \varepsilon_{ij}^{\mathrm{f}} \tag{3.143}$$

The intact rock matrix is assumed to be elastic and satisfy Hooke's law

$$\varepsilon_{ij}^{\mathrm{I}} = C_{ijkl}^{\mathrm{I}} \sigma_{ij} \tag{3.144}$$

where C_{ijkl}^{I} is the compliance tensor. The strain of the fractures is similarly given by

$$\varepsilon_{ij}^{\mathrm{f}} = C_{ijkl}^{\mathrm{f}} \sigma_{ij} \tag{3.145}$$

where C^{f}_{ijkl} is the compliance tensor of fractures. An elastic constitutive equation relating the normal and shear components of the fracture stresses (or more correctly, tractions on fracture surfaces) and displacements is written as

$$\begin{Bmatrix} u_n \\ u_s \\ u_t \end{Bmatrix} = \begin{bmatrix} c_{nn} & c_{ns} & c_{nt} \\ c_{sn} & c_{ss} & c_{st} \\ c_{tn} & c_{ts} & c_{tt} \end{bmatrix} \begin{Bmatrix} \sigma_n \\ \sigma_s \\ \sigma_t \end{Bmatrix} \quad \text{or} \quad u^{\mathrm{f}}_i = C^{\mathrm{f}}_{ij}\sigma^{\mathrm{f}}_j \tag{3.146}$$

where $i, j = n, s, t$ are the normal (n) and two orthogonal shear axes defined for the fractures. The relation between contacting stress components (as tractions on the fracture planes) and the overall stress components in the rock is given by Cauchy's formula

$$\sigma^{\mathrm{f}}_i = \sigma_{ij} n_j \tag{3.147}$$

where n_j is the unit outward normal vector of the fracture plane. In a representative volume of rock, V, with M fractures of area A_k, the work done by the tractions (contact stresses) over the fracture deformation is given as

$$W^{\mathrm{f}} = \sigma_{ij}\varepsilon^{\mathrm{f}}_{ij} = \frac{1}{V}\sum_{k=1}^{M} (\sigma^{\mathrm{f}}_i)_k (u^{\mathrm{f}}_j)_k A_k \tag{3.148}$$

Substituting Eqn (3.147) into Eqn (3.148) leads to

$$\varepsilon^{\mathrm{f}}_{ij} = \sum_{k=1}^{M} (n_i)_k (u^{\mathrm{f}}_j)_k \frac{1}{S_k} \tag{3.149}$$

where $S_k = V/A_k$ is the mean fracture spacing. Substituting Eqn (3.146) into Eqn (3.149) then yields

$$\varepsilon^{\mathrm{f}}_{ij} = \sum_{k=1}^{M} (n_i)_k (u^{\mathrm{f}}_j)_k \frac{1}{S_k} = \left[\sum_{k=1}^{M} (n_i)_k (C^{\mathrm{f}}_{ij})_k (n_j)_k \frac{1}{S_k} \right] \sigma_{ij} \tag{3.150}$$

Denoting the coordinate transformation matrix as T_{ij} between the local system on the fracture planes and the global *xyz*-system, the tensor C^{f}_{ij} can be replaced by

$$C^{\mathrm{f}}_{ijkl} = T_{ij} C^{\mathrm{f}}_{jk} T^{\mathrm{T}}_{kl} \tag{3.151}$$

The compliance tensor in Eqn (3.145) can then finally be described by

$$C^{\mathrm{f}}_{ijkl} = \sum_{k=1}^{M} (n_i)_k (T_{ij} C^{\mathrm{f}}_{jk} T^{\mathrm{T}}_{kl})_k (n_l)_k \frac{1}{S_k} \tag{3.152}$$

The total compliance tensor of the fractured rock is then given as the sum

$$C_{ijkl} = C^{\mathrm{I}}_{ijkl} + C^{\mathrm{f}}_{ijkl} \tag{3.153}$$

Assuming isotropic shear properties of fractures, and ignoring couplings between the shear components, the compliance matrix of the fractures can be simplified to

$$C_{ij} = \begin{bmatrix} c_{\mathrm{n}} & 0 & 0 \\ 0 & c_{\mathrm{t}} & 0 \\ 0 & 0 & c_{\mathrm{t}} \end{bmatrix} = \begin{bmatrix} 1/k_{\mathrm{n}} & 0 & 0 \\ 0 & 1/k_{\mathrm{t}} & 0 \\ 0 & 0 & 1/k_{\mathrm{t}} \end{bmatrix} = (k_{ij})^{-1} \tag{3.154}$$

where c_{n} represents the normal compliance, c_{t} the shear compliance and the dilatancy is ignored in order to hold the elastic symmetry (otherwise the off-diagonal terms will not be zero). For a representative

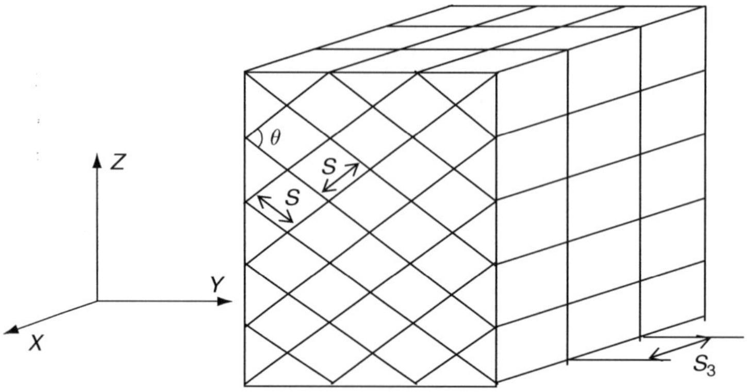

Fig. 3.13 A rock mass with intersecting sets of fractures (Huang et al., 1995).

volume of rock with three fracture sets, with an angle θ between the first two sets with spacing S and the third set having a spacing S_3 and oriented perpendicular to the first two sets (Fig. 3.13), the total compliance matrix is given by (Huang et al., 1995)

$$[C_{ijkl}] = \begin{bmatrix} \frac{1}{E_0}+\frac{1}{E_x} & -\frac{\nu_0}{E_0}-\frac{\nu_{yx}}{E_y} & -\frac{\nu_0}{E_0}-\frac{\nu_{zx}}{E_z} & 0 & 0 & 0 \\ -\frac{\nu_0}{E_0}-\frac{\nu_{xy}}{E_y} & \frac{1}{E_0}+\frac{1}{E_y} & -\frac{\nu_0}{E_0}-\frac{\nu_{yz}}{E_y} & 0 & 0 & 0 \\ -\frac{\nu_0}{E_0}-\frac{\nu_{xz}}{E_y} & -\frac{\nu_0}{E_0}-\frac{\nu_{zy}}{E_y} & \frac{1}{E_0}+\frac{1}{E_z} & 0 & 0 & 0 \\ 0 & 0 & 0 & \frac{1}{G_0}+\frac{1}{G_{xy}} & 0 & 0 \\ 0 & 0 & 0 & 0 & \frac{1}{G_0}+\frac{1}{G_{yz}} & 0 \\ 0 & 0 & 0 & 0 & 0 & \frac{1}{G_0}+\frac{1}{G_{zx}} \end{bmatrix} \tag{3.155}$$

$$\frac{1}{E_x} = \frac{1}{(k_n)_3 S_3} \tag{3.156a}$$

$$\frac{1}{E_y} = 2\sin^2\theta\left[\frac{k_n \cos\theta + k_s \sin^2\theta}{k_n k_s S}\right] \tag{3.156b}$$

$$\frac{1}{E_z} = 2\cos^2\theta\left[\frac{k_n \cos\theta + k_s \sin^2\theta}{k_n k_s S}\right] \tag{3.156c}$$

$$\frac{1}{G_{xy}} = \frac{1}{(k_s)_3 S_3} + \frac{2\sin 2\theta \cos^2\theta}{k_s S} \tag{3.156d}$$

$$\frac{1}{G_{yz}} = \frac{2(k_n + k_s)\sin^2\theta}{k_s k_n S} \tag{3.156e}$$

$$\frac{\nu_{yz}}{E_z} = \frac{\nu_{zy}}{E_y} = \frac{k_n - k_s}{2k_n k_s}\sin^2\theta \tag{3.156f}$$

$$\frac{\nu_{xy}}{E_x} = \frac{\nu_{yx}}{E_y} = 0 \tag{3.156g}$$

$$\frac{\nu_{zx}}{E_z} = \frac{\nu_{xz}}{E_x} = 0 \tag{3.156h}$$

where E_0, G_0 and ν_0 are the Young's modulus, shear modulus and Poisson's ratio of the intact rock matrix, respectively. Paramcters k_n and k_s are the normal and shear stiffness of the two intersecting fracture sets with angle θ and spacing S and $(k_n)_3$ and $(k_s)_3$ are the stiffness of the fracture set perpendicular to the two intersecting non-orthogonal sets with spacing S_3. When $\theta = 90$ degrees the relation (3.155) is reduced to the model proposed by Amadei and Goodman (1981). No dilatancy is considered in (3.152), however, because the dilatancy (with non-zero off-diagonal terms in (3.152)) would destroy the elastic symmetry condition in relations (3.151) and (3.152).

3.4.2.3 Singh's elastic solids with non-persistent fracture sets

For cases with non-persistent fractures, a two-dimensional example is provided by Singh (1973) for a transversely fractured rock mass containing one persistent set and another staggered set of fractures (see Fig. 3.14). The moduli of elasticity, E_n and E_s, the shear modulus G_{ns} and the Poisson's ratio ν_{ns} are given as

$$\frac{1}{E_n} = \frac{1}{E_0} + \frac{b_{nn}}{k_{nn}S_n} \tag{3.157a}$$

$$\frac{1}{E_s} = \frac{1}{E_0} + \frac{1}{k_{ns}S_s} \tag{3.157b}$$

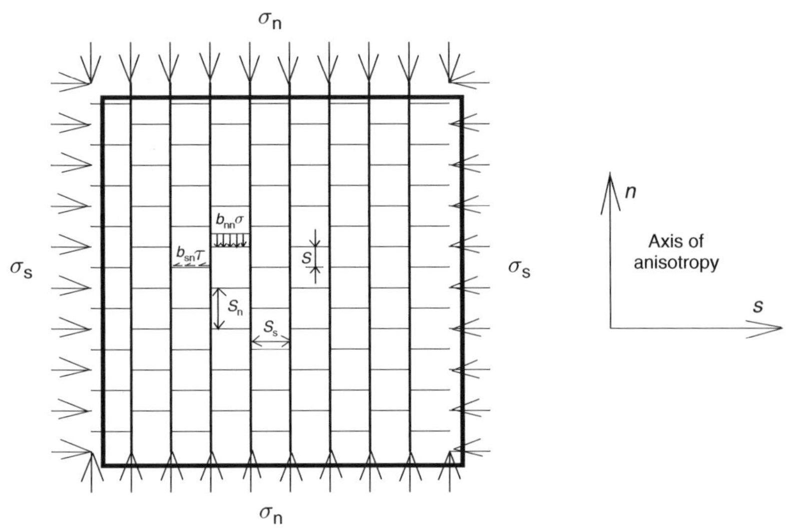

Fig. 3.14 Rock mass containing two orthogonal fracture sets: one persistent and one staggered (Singh, 1973).

$$G_{\rm ns} = \frac{G_0 S_{\rm n} S_{\rm s} k_{\rm sn} k_{\rm ss}}{k_{\rm sn} k_{\rm ss} S_{\rm n} S_{\rm s} + G_0 b_{\rm sn} k_{\rm ss} S_{\rm s} + G_0 k_{\rm sn} S_{\rm n}} \tag{3.157c}$$

$$\nu_{\rm ns} = \frac{\nu_0 k_{\rm nn} S_{\rm n}}{k_{\rm nn} S_{\rm n} + b_{\rm nn} E_0} \tag{3.157d}$$

where $E_{\rm n}$ is the modulus of elasticity in the n-direction, $E_{\rm s}$ is the modulus of elasticity in the s-direction, $G_{\rm ns}$ is the shear modulus, $\nu_{\rm ns}$ is the Poisson's ratio when loading in the n-direction, E_0, G_0 and ν_0 are the elastic modulus, shear modulus and Poisson's ratio of the intact rock, $k_{\rm ns}$, $k_{\rm nn}$ are the normal stiffness of the sub-vertical and the sub-horizontal fractures, respectively, $k_{\rm ss}$, $k_{\rm sn}$ are the shear stiffness of the sub-vertical and the sub-horizontal fractures, respectively, $S_{\rm s}$ is the mean spacing of the sub-vertical fractures in the s-direction, $S_{\rm n}$ is the mean spacing of the sub-horizontal fractures in the n-direction and

$$b_{nn} = \left[1 + \frac{k_{\rm ss}}{k_{\rm nn}}\left(1 - \frac{s}{S_{\rm n}}\right)\frac{s}{S_{\rm s}}\right]^{-1} \tag{3.158}$$

$$b_{\rm sn} = \left[1 + \frac{k_{\rm ns}}{k_{\rm sn}}\left(1 - \frac{s}{S_{\rm n}}\right)\frac{s}{S_{\rm s}}\right]^{-1} \tag{3.159}$$

which are stress concentration factors due to the staggered fractures. The symbol s is the fracture offset of the staggered sub-horizontal fractures. These two factors are defined by Singh as the ratio of the mean normal and shear stresses along the fracture to the corresponding overall stresses on a plane parallel to that fracture within the rock mass. Equations (3.158) and (3.159) are derived for rigid blocks. The complete stress–strain relations are characterized by the 2D elastic stiffness tensor

$$\begin{Bmatrix} \sigma_{xx} \\ \sigma_{yy} \\ \sigma_{xy} \end{Bmatrix} = \begin{bmatrix} \dfrac{E_x E_y}{E_y - E_x(\nu_{xy})^2} & \dfrac{E_x E_y \nu_{xy}}{E_y - E_x(\nu_{xy})^2} & 0 \\ \dfrac{E_x E_y \nu_{xy}}{E_y - E_x(\nu_{xy})^2} & \dfrac{(E_y)^2}{E_y - E_x(\nu_{xy})^2} & 0 \\ 0 & 0 & G_{xy} \end{bmatrix} \begin{Bmatrix} \varepsilon_{xx} \\ \varepsilon_{yy} \\ \varepsilon_{xy} \end{Bmatrix} \tag{3.160}$$

with the consistency condition $E_y > E_x(\nu_{xy})^2$.

The equivalent elastic continuum models are valid as long as the deformation is small and occurs without shear failures and openings in the fractures. Large deformation and shear failure/openings of fractures will destroy the geometrical configuration of the model geometry to such an extent that the basic assumptions of the formulations may no longer hold.

It should also be noted that the deformation moduli of the fractures depend on the stiffness of the fractures (cf. Eqns (3.136), (3.137) and (3.157)) which, in turn, depends on the normal stresses (or closure) as indicated by Bandis' and Goodman's hyperbolic models (cf. Eqns (3.6) and (3.7)). This implies that the deformability of the fractured rock masses is stress dependent and path dependent. The equivalent elastic models presented above can only serve for estimating the initial elastic properties of rock masses with persistent fractures. For fractured rocks with generally more randomly distributed fracture systems, and under changing stress paths with large shear displacements, numerical modeling using DEM techniques must be applied (see Chapter 12 for examples).

3.4.3 Fractured Rocks with Randomly Distributed Fractures of Finite Sizes

Although fractures appear often in rock masses in sets, their sizes vary and they are often not at all persistent. A considerable portion of the total fracture population is distributed rather randomly and does not belong to any particular set. The models defined for rock masses with regular sets of fractures cannot, therefore, be applied to such more general situations.

3.4.3.1 Oda's crack tensor model

Oda (1982) proposed the concept of a 'fabric' or 'crack' tensor to describe the geometry of distributed fractures in a rock mass. Assuming that the fractures are planar, circular features of radius r (or for non-circular fractures with equivalent radius $r = \sqrt{A/\pi}$ where A is the area of the fracture), the dimensions of the fractures are described by a probability density function $f(r)$ with condition

$$\int_0^\infty f(r)\mathrm{d}r = 1 \tag{3.161}$$

The nth moment of $f(r)$ is defined as

$$\langle r^n \rangle = \int_0^\infty r^n f(r)\mathrm{d}r \tag{3.162}$$

The orientation of the fractures is described by another probability density function $E(\mathbf{n} \cdot r)$ following a similar condition

$$\iiint\limits_{0\ \Omega}^{\infty} E(\mathbf{n}, r)\mathrm{d}\Omega\,\mathrm{d}r = 1 \tag{3.163}$$

where the vector $\mathbf{n}$ is the outward unit normal of the fracture surface and $\mathrm{d}\Omega$ is an increment of solid angle in a 3D space (Fig. 3.15) and can be simply written as

$$\mathrm{d}\Omega = \sin\beta\,\mathrm{d}\alpha\,\mathrm{d}\beta \tag{3.164}$$

The product $E(\mathbf{n}, r)\mathrm{d}\Omega\mathrm{d}r$ then yields a fraction of fracture surface whose unit normal $\mathbf{n}$ is contained within the solid angle $\mathrm{d}\Omega$ and whose radius is located in the interval $(r, r + \mathrm{d}r)$. For mutually independent $\mathbf{n}$ and r, one has

$$E(\mathbf{n}, r) = E(\mathbf{n})f(r) \tag{3.165}$$

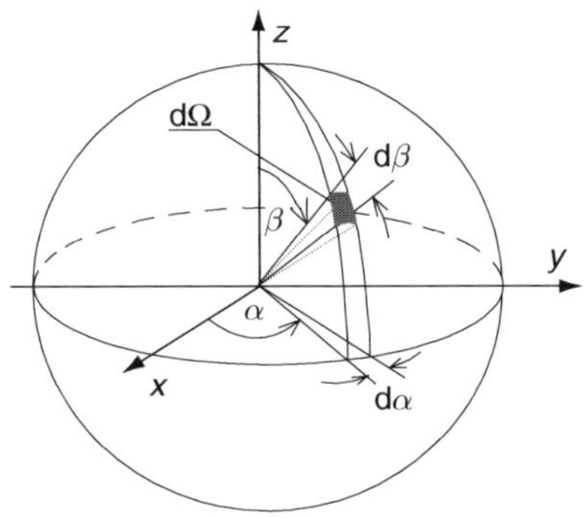

Fig. 3.15 Definition of a fabric tensor and the unit sphere for defining the solid angle $\mathrm{d}\Omega$ (Oda, 1982).

and

$$E(\mathbf{n}) = E(-\mathbf{n}) \tag{3.166}$$

If the orientation **n** of the fractures is distributed isotropically, then $E(\mathbf{n}) = 1/(4\pi)$. Denoting ρ as the density of fractures in a representative volume V, $\rho = M_V/V$, where M_V is the number of fractures in V, the density of fractures intersecting a straight scanline is then given by

$$\eta = 2\pi r^2 n_i E(\mathbf{n}, r)\mathrm{d}\Omega\,\mathrm{d}r \tag{3.167}$$

where n_i is the projection of the unit normal **n** on the unit direction vector **i** of a scanline. Denoting a new vector $\mathbf{m} = 2r\mathbf{n}$ with the same direction as the unit normal of the fracture plane, but with the magnitude of the fracture diameter, a vector sum of all fractures intersecting a unit length of a scanline in the ***i*** direction is given by

$$\eta \cdot \mathbf{m} = \left[4\pi r^3 n_i E(\mathbf{n}, r)\mathrm{d}\Omega\,\mathrm{d}r\right]\mathbf{n} \tag{3.168}$$

The projection of this vector to an orthogonal direction ***j*** then leads to a new density function

$$f_{ij} = 4\pi r^3 n_i n_j E(\mathbf{n}, r)\mathrm{d}\Omega\,\mathrm{d}r \tag{3.169}$$

The fabric tensor is then defined as the integration of the density function over the upper hemisphere ($\Omega/2$) and $0 < r < \infty$.

$$F_{ij} = 4\rho\pi \int_0^\infty \iint_{\Omega/2} r^3 n_i n_j E(\mathbf{n}, r)\mathrm{d}\Omega\,\mathrm{d}r \tag{3.170}$$

where *i*, *j=1, 2, 3* or *x*, *y*, *z* form an orthogonal coordinate system. If we assume independent **n** and *r*, and isotropic **n**, then

$$E(\mathbf{n}, r) = E(\mathbf{n})f(r) = \frac{f(r)}{4\pi} \tag{3.171}$$

and the fabric tensor is simplified to

$$F_{ij} = 4\rho\pi \int_0^\infty \int_0^{2\pi} \int_0^\pi r^3 n_i n_j E(\mathbf{n})f(r)\sin\beta\,\mathrm{d}\alpha\,\mathrm{d}\beta\;\mathrm{d}r = 2\rho\pi n_i n_j \int_0^\infty r^3 f(r)\mathrm{d}r = 2\rho\pi n_i n_j \langle r^3\rangle \tag{3.172}$$

where $\langle r^3\rangle$ is the third moment of the probability density function *f(r)*. For computer implementations, the fabric tensor can be simply written

$$F_{ij} = \frac{2\pi}{V}\sum_{k=1}^{M_V} (r^3 n_i n_j)_k \tag{3.173}$$

where $(r^3 n_i n_j)_k$ stands for the product of $(r^3 n_i n_j)$ of the *k*th fracture in the volume *V*.

The fabric tensor therefore characterizes the sizes and orientations of fractures randomly distributed in a volume *V*.

Oda (1986a,b) extended the original fabric tensor to a crack tensor concept especially for fractured rocks, with the inclusion of the fracture aperture, *t*, in the probability density function, which now becomes $E(\mathbf{n}, r \cdot t)$ instead of $E(\mathbf{n}, r)$ with the condition

$$\int_0^{r_m}\int_0^{t_m}\iint_{\Omega} E(\mathbf{n},r,t)\mathrm{d}\Omega\,\mathrm{d}r\,\mathrm{d}t = 2\int_0^{r_m}\int_0^{t_m}\iint_{\Omega/2} E(\mathbf{n},r,t)\mathrm{d}\Omega\,\mathrm{d}r\,\mathrm{d}t = 1 \tag{3.174}$$

where $\Omega/2$ is the upper hemisphere and r_m, t_m are the maximum fracture radius and aperture, respectively. Similarly $E(\mathbf{n}, r\cdot t) = E(-\mathbf{n}, r\cdot t)$ and for independent r and t,

$$E(\mathbf{n},r,t) = E(\mathbf{n})f(r,t) = \frac{f(r,t)}{4\pi} \tag{3.175}$$

The fractures are assumed to be composed of two parallel planar surfaces of aperture t, connected by a normal spring of stiffness $\bar{k}_n$, and a shear spring of stiffness $\bar{k}_s$. Using Bandis' hyperbolic function for normal stress–normal displacement (cf. Eqn (3.4)) and Cauchy's stress formula (Eqn (3.147)), the secant normal stiffness, $\bar{k}_n$, of an individual fracture is given by

$$\bar{k}_n = \frac{\sigma_n}{u_n} = \frac{1+b\sigma_n}{a} = \frac{1}{a} + \frac{b\sigma_n}{a} = k_{n0} + \frac{1}{t_0}\sigma_n = k_{n0} + \frac{1}{t_0}\sigma_{ij}n_in_j \tag{3.176}$$

where σ_{ij} is the stress tensor in the global system and the normal stress σ_n is defined in the local system on the fracture surface. Introducing an aspect ratio $c = r/t_0$, the secant normal stiffness is then written

$$\bar{k}_n = \frac{1}{r}\left[rk_{n0} + c\sigma_{ij}n_in_j\right] = \frac{1}{r}\left[h + c\sigma_{ij}n_in_j\right] \tag{3.177}$$

which takes the scale effect into account. The mean normal stiffness over the entire solid angle is then written

$$\hat{k}_n = \int_{\Omega}\bar{k}_n E(\mathbf{n})\mathrm{d}\Omega = \int_{\Omega}\frac{h}{r}E(n)\mathrm{d}\Omega + \int_{\Omega}\frac{c}{r}\sigma_{ij}n_in_jE(\mathbf{n})\mathrm{d}\Omega = \frac{1}{r}\left(h + c\sigma_{ij}N_{ij}\right) = \frac{\bar{h}}{r} \tag{3.178}$$

where $E(\mathbf{n})$ is the probability density function of normal orientations of the fractures only and

$$\int_{\Omega} E(n)\mathrm{d}\Omega = 1 \tag{3.179}$$

$$N_{ij} = \int_{\Omega} n_in_jE(\mathbf{n})\,\mathrm{d}\Omega \tag{3.180}$$

$$\bar{h} = h + c\sigma_{ij}N_{ij} \tag{3.181}$$

The shear stiffness is given in a much simplified form

$$\bar{k}_t = \frac{g}{r}\sigma_n = \frac{g}{r}\sigma_{ij}n_in_j \tag{3.182}$$

where g is a constant independent of the normal stress and fracture size. Similarly the mean shear stiffness over the entire solid angle is given by

$$\hat{k}_t = \int_{\Omega}\bar{k}_tE(\mathbf{n})\mathrm{d}\Omega = \frac{g}{r}\sigma_{ij}N_{ij} = \frac{\bar{g}}{r} \tag{3.183}$$

where

$$\bar{g} = g\sigma_{ij}N_{ij} \tag{3.184}$$

Similarly, assuming now that the total deformation of a fractured rock can be divided into the sum of the deformation of the intact rock matrix and the deformation of the rock fractures, namely

$$\varepsilon_{ij} = \varepsilon_{ij}^{\mathrm{I}} + \varepsilon_{ij}^{\mathrm{f}} \tag{3.185}$$

where ε_{ij} is the total strain, $\varepsilon_{ij}^{\mathrm{I}}$ the strain of the rock matrix and $\varepsilon_{ij}^{\mathrm{f}}$ the strain of the fracture. The rock matrix is assumed to be homogeneous, isotropic, linearly elastic and described by Hooke's law for isotropic elasticity

$$\varepsilon_{ij}^{\mathrm{I}} = C_{ijkl}^{\mathrm{I}}\sigma_{kl} \tag{3.186}$$

where C_{ijkl}^{I} is the elastic compliance tensor of the rock matrix and given by the inverse of the elastic stiffness matrix, written as

$$C_{ijkl}^{\mathrm{I}} = \frac{1}{E_0}\left[(1+\nu_0)\delta_{ik}\delta_{jl} - \nu_0\delta_{ij}\delta_{kl}\right] \tag{3.187}$$

Let $\mathbf{u}_{\mathrm{n}}^{\mathrm{f}}$ be the normal displacement vector and $\mathbf{u}_{\mathrm{t}}^{\mathrm{f}}$ be the maximum shear displacement vector, respectively, in the global system. Since $\mathbf{u}_{\mathrm{n}}^{\mathrm{f}}$ is parallel with the unit normal $\mathbf{n}$ of the fracture, its components can be written

$$(u_{\mathrm{n}}^{\mathrm{f}})_i = \frac{1}{\hat{k}_{\mathrm{n}}}\sigma_{jk}n_i n_j n_k \quad (i = x, y, z \text{ or } 1, 2, 3) \tag{3.188}$$

The shear displacement vector $\mathbf{u}_{\mathrm{t}}$ is parallel with the direction of the maximum shear stress component and its magnitude is given by

$$(u_{\mathrm{t}}^{\mathrm{f}})_i = \frac{1}{\ddot{k}_{\mathrm{t}}}(\sigma_{ij}n_j - \sigma_{jk}n_i n_j n_k) \tag{3.189}$$

To define the crack tensor, it is necessary to sum up the fracture displacement along a scanline of length L_i which points in the $\mathbf{i}$-direction and is long enough to ensure that the volume V is representative. The density of the fractures in the volume as the number of fractures whose unit normal vectors intersect the scanline, and are contained in the incremental solid angle dΩ, is given by

$$\bar{N} = \frac{\pi}{4}L_i\rho r^2 n_i(2E(\mathbf{n}, r, t)\,\mathrm{d}\Omega\,\mathrm{d}r\,\mathrm{d}t) \tag{3.190}$$

The total displacement component of fractures, u_i^{f}, is the sum of the products of the fracture displacements by Eqns (3.188) and (3.189) and $\bar{N}$, written as

$$u_i^{\mathrm{f}} = \sum_{k=1}^{N}[(u_{\mathrm{n}}^{\mathrm{f}})_i + (u_{\mathrm{t}}^{\mathrm{f}})_i]_k = \frac{\pi}{2}L_i\rho\left[\left(\frac{1}{\bar{h}} - \frac{1}{\bar{g}}\right)n_i n_j n_k n_l + \frac{1}{\bar{g}}n_i n_l \delta_{jk}\right]r^3 E(\mathbf{n}, r, t)\mathrm{d}\Omega\,\mathrm{d}r\,\mathrm{d}t \tag{3.191}$$

Integration over the upper hemisphere $\Omega/2$, $0 \le r \le r_{\mathrm{m}}$, and $0 \le t \le t_{\mathrm{m}}$ results in

$$\frac{1}{L_i}(u_{\mathrm{i}}^{\mathrm{f}}) = \frac{1}{L}\sum_{k=1}^{M_V}\left[(u_{\mathrm{n}}^{\mathrm{f}})_i + (u_{\mathrm{t}}^{\mathrm{f}})_i\right]_k = \left[\left(\frac{1}{\bar{h}} - \frac{1}{\bar{g}}\right)F_{ijkl} + \frac{1}{\bar{g}}\delta_{jk}F_{il}\right]\sigma_{kl} \tag{3.192}$$

where M_V is the total number of fractures in the volume V and

$$F_{ij......k} = \frac{\pi\rho}{4}\int_0^{r_\mathrm{m}}\int_0^{t_\mathrm{m}}\iint_\Omega r^3 n_i n_j .....n_k E(\mathbf{n}, r, t)\mathrm{d}\Omega\,\mathrm{d}r\,\mathrm{d}t \tag{3.193}$$

is a non-dimensional, and positively definite, geometry tensor of the fracture system called the crack tensor, with

$$F_0 = \frac{\pi\rho}{4}\int_0^{r_\mathrm{m}}\int_0^{t_\mathrm{m}} r^3 f(r, t)\mathrm{d}r \tag{3.194}$$

$$F_{ij} = \frac{\pi\rho}{4}\int_0^{r_\mathrm{m}}\int_0^{t_\mathrm{m}}\iint_\Omega r^3 n_i n_j E(\mathbf{n}, r, t)\mathrm{d}\Omega\,\mathrm{d}r\,\mathrm{d}t \tag{3.195}$$

$$F_{ijkl} = \frac{\pi\rho}{4}\int_0^{r_\mathrm{m}}\int_0^{t_\mathrm{m}}\iint_\Omega r^3 n_i n_j n_k n_l E(\mathbf{n}, r, t)\mathrm{d}\Omega\,\mathrm{d}r\,\mathrm{d}t \tag{3.196}$$

Defining the fracture strain as

$$\varepsilon_{ij}^\mathrm{f} = \frac{1}{2}\left[\frac{u_i^\mathrm{f}}{L_j} + \frac{u_j^\mathrm{f}}{L_i}\right] \tag{3.197}$$

where $i, j = x, y, z$ or 1, 2, 3 defining a orthogonal coordinate system, the fracture strain can be finally defined by substituting Eqns (3.191) and (3.192) into Eqn (3.197)

$$\varepsilon_{ij}^\mathrm{f} = \left[\left(\frac{1}{\bar{h}} - \frac{1}{\bar{g}}\right)F_{ijkl} + \frac{1}{4\bar{g}}\left(\delta_{ik}F_{ji} + \delta_{jk}F_{il} + \delta_{il}F_{jk} + \delta_{jl}F_{ik}\right)\right]\sigma_{kl} = C_{ijkl}^\mathrm{f}\sigma_{kl} \tag{3.198}$$

where the tensor

$$C_{ijkl}^\mathrm{f} = \left[\left(\frac{1}{\bar{h}} - \frac{1}{\bar{g}}\right)F_{ijkl} + \frac{1}{4\bar{g}}\left(\delta_{ik}F_{ji} + \delta_{jk}F_{il} + \delta_{il}F_{jk} + \delta_{jl}F_{ik}\right)\right] \tag{3.199}$$

is the compliance tensor of the fracture system. Substitution of Eqns (3.186), (3.197), (3.198) and (3.199) into Eqn (3.185) leads to the total compliance tensor of the fractured rock, written as

$$C_{ijkl} = C_{ijkl}^\mathrm{I} + C_{ijkl}^\mathrm{f} \tag{3.200}$$

The crack tensor is a unique measure combining five significant aspects of the mechanical behavior of fractured rocks: volume, fracture size, fracture orientation, fracture stiffness and fracture aperture. The latter is crucial for coupled stress-flow analysis. Its applicability to problems of fractured rocks has been demonstrated in a number of experimental investigations (Oda, 1988a,b; Oda et al. 1984, 1986).

Oda's crack tensor model is attractive since it considers randomly oriented fractures with finite sizes in a closed-form fashion. The theory has enjoyed wide applications in rock engineering. However, like every theory, certain limitations also exist, as given below:

(1) Since the distributions of orientations, fracture size and aperture are involved in the definition of the crack tensor, a lack of uniqueness may be involved, since the same distribution may give different realizations of fracture distribution patterns, which may lead to different crack tensors. A Monte Carlo simulation with a large number of realizations is needed to reduce the uncertainty caused by this lack of uniqueness based on a REV size support.

(2) Interaction between fractures was not considered in the definition of the crack tensor (see also comments by Horii and Sahasakmontri, 1988). The deformation of one fracture, therefore, does not affect its neighboring fractures. This implies that the theory is best suited to rocks containing isolated fractures that are located far away from each other so that their interactions, either mechanical or hydraulic, will not occur. For intersecting fractures, the deformation of one fracture will affect the deformation of its intersecting partner fracture or fractures, especially considering shearing, dilatancy and aperture variation. The validity of simple superposition of fracture effects by many fractures in this theory may not apply.

(3) All fractures are assumed to be closed, but with non-zero shear stiffness $\bar{g} \neq 0$ (otherwise the compliance tensor of fractures tends to infinity). However, the fractures are not allowed to slip since no frictional stresses are permitted to be mobilized. This implies that the theory may be applied to problems involving only small elastic deformations without causing slip of fractures.

Using Oda's theory for 2D problems, Stietel et al. (1996) presented a formulation of the crack tensor for equivalent poroelastic properties of fractured rocks. Denoting θ_i ($i = x$, y, or 1, 2) as the inclination angle of a fracture segment with respect to the corresponding coordinate axes and replacing $\bar{h}$ and $\bar{g}$ by $k_{\rm n}$ and $k_{\rm t}$, the compliance tensor of the fracture system is written

$$C^{\rm f}_{ijkl} = \frac{1}{A}\sum_{f=1}^{M_{\rm V}}\left[\left(\frac{1}{k_{\rm n}} - \frac{1}{k_{\rm t}}\right)_f L_f\,(\cos\theta_i\cos\theta_j\cos\theta_k\cos\theta_l)_f \right. \\ \left. + \frac{L_f}{(k_{\rm t})_f}(\cos\theta_i\cos\theta_j\delta_{kl} + \cos\theta_j\cos\theta_l\delta_{ik} + \cos\theta_j\cos\theta_k\delta_{il} + \cos\theta_i\cos\theta_l\delta_{jk})_f\right] \quad (3.201)$$

where i, j, k, $l = x$, y, or 1, 2 and subscript f indicates the fth fracture.

3.4.3.2 *Kaneko and Shiba model for fractured rocks as equivalent elastic continua*

Based on the elastic strain contributions of a single crack, open or closed, in a 2D domain of area A and by applying a linear fracture mechanics theory (Fig. 13.16a), Kaneko and Shiba (1990) proposed a different method to formulate the equivalent elastic compliance tensor of a fractured rock mass, which treats open and closed fractures separately. The interaction between fractures is also ignored and so the fractures are assumed not to intersect (Fig. 3.16b).

For an elastic body containing a single fracture of length $2a$ and inclination angle θ (Fig. 3.16a), the strain for an open fracture is given by

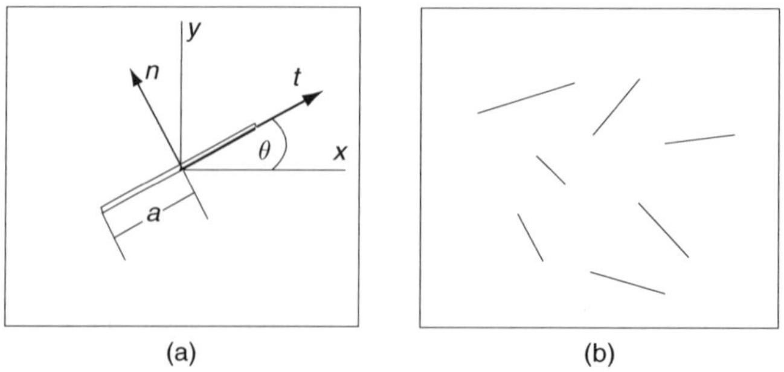

Fig. 3.16 (a) An open fracture of length $2a$ and inclination angle θ in a elastic continuum and (b) an elastic continuum containing multiple isolated fractures.

$$\varepsilon_{ij}^{\mathrm{f}} = (cwn_i n_k \delta_{jl})\sigma_{kl} \tag{3.202}$$

where $c = 1/E_0$ for plane stress problems and $c = [1 - (\nu_0)^2]/E_0$ for plane strain problems. The symbol δ_{ij} stands for the Kronecker delta and $w = 2a^2h$, where h is the thickness of the plate for plane stress problems. For a closed fracture undergoing frictional slip, the strain is written

$$\varepsilon_{ij}^{\mathrm{f}} = (cwn_i n_k g_{jl})\sigma_{kl} \tag{3.203}$$

where

$$g_{ij} = \begin{bmatrix} n_y n_y + f\mu n_x n_y & -n_x n_y + f\mu n_x n_y \\ -n_x n_y + f\mu n_x n_x & n_x n_x - f\mu n_x n_y \end{bmatrix} \tag{3.204a}$$

$$f = \mathrm{sign}(\sigma_{xy}) \quad \text{(the sign of the shear stress)} \tag{3.204b}$$

$$\mu = \tan\phi \quad (\phi \text{ is the friction angle of the fracture surface}) \tag{3.204c}$$

$$\sigma_{xy} = N_{ki} N_{lj} \sigma_{kl} \tag{3.205a}$$

$$N_{ij} = \begin{pmatrix} n_y & n_x \\ -n_x & n_y \end{pmatrix} \tag{3.205b}$$

The total elastic compliance tensor for an elastic body containing N open fractures and M closed fractures is written through simple superposition with the assumption that fractures are isolated and will not interfere with each other's behavior. Recalling Eqn (3.200) the elastic compliance of the fractured rock is given as

$$\begin{aligned} C_{ijkl} &= C_{ijkl}^{\mathrm{I}} + C_{ijkl}^{\mathrm{f}} \\ &= \frac{1}{E_0}\left[(1+\nu_0)\delta_{ik}\delta_{jl} - \nu_0\delta_{ij}\delta_{kl}\right] + \sum_{f=1}^{N}(cwn_i n_k \delta_{jl})_f + \sum_{f=1}^{M}(cwn_i n_k g_{jl})_f \end{aligned} \tag{3.206}$$

This approach has similarities with the crack tensor theory and also shares similar attractiveness and limitations, but is more simple and straightforward than the crack tensor theory in the derivation of the final elastic compliance tensor, without requiring probabilistic density functions of the fracture geometry parameters as the crack tensor model does, which, however, may or may not be an advantage. Similarly as the crack tensor model, considering fracture interactions will need numerical simulations and the attractiveness of the closed-form expressions will be lost. In addition, a common limitation shared with the crack tensor theory is the difficulty in validating the model directly with laboratory experiments – due to the difficulties of reproducing the fracture systems and their characterization.

3.4.4 Fractured Rocks as Elasto-plastic Continua

A different technique to represent the constitutive behavior of fractured rocks is the use of non-linear material models, most commonly based on the theory of plasticity. The deformation contributed by fractures is then assumed as plastic deformation of the equivalent continuum and the hardening–softening rules of plasticity are adopted to simulate similar behavior of fractured rocks. It is beyond the scope of this book to present a lengthy review of the theory of plasticity and its applications to rock mechanics problems. In this section the basic mathematical tools required to derive an incremental stress–strain relation for elasto-perfectly plastic deformation processes with and without hardening–softening considerations are presented briefly, using Mohr–Coulomb and Hoek–Brown criteria as the yielding functions, since these criteria are used widely in many DEM models for rock block behavior. It is assumed that the readers have a basic knowledge of the plasticity theory of solid mechanics so that common definitions of concepts will not be repeated here.

3.4.4.1 The elasto-perfectly plastic model: general formulation

For a plastic body, the total strain increment $d\varepsilon_{ij}$ is assumed to be the simple sum of a reversible (elastic) component, $d\varepsilon_{ij}^{e}$ and an irreversible (plastic) component, $d\varepsilon_{ij}^{p}$

$$d\varepsilon_{ij} = d\varepsilon_{ij}^{e} + d\varepsilon_{ij}^{p} \tag{3.207}$$

where Hooke's law for isotropic elasticity hold for elastic strain through

$$d\sigma_{ij} = D_{ijkl}\, d\varepsilon_{kl}^{e} \tag{3.208}$$

or

$$d\varepsilon_{ij}^{e} = C_{ijkl}\, d\sigma_{kl} = (D_{ijkl})^{-1} d\sigma_{kl} \tag{3.209}$$

where D_{ijkl} is called the stiffness tensor and its inverse C_{ijkl} the compliance tensor.

Denoting the yielding function as $F(\sigma_{ij}, m)$ and the plastic potential function as $Q(\sigma_{ij})$, respectively, where m is a scalar function representing work-hardening or work-softening behavior of the body, the flow rule to determine the plastic strain is given by

$$d\varepsilon_{ij}^{p} = \begin{cases} 0, & (F < 0) \\ \lambda \dfrac{\partial Q}{\partial \sigma_{ij}}, & (F \geq 0) \end{cases} \tag{3.210}$$

where $\lambda > 0$ is a scalar. When $F(\sigma_{ij}) \neq Q(\sigma_{ij})$, it is called a non-associated flow rule; otherwise, it is an associated flow rule. The plastic work, representing the dissipated energy during the plastic deformation process, is given by

$$dW^{p} = \sigma_{ij}\, d\varepsilon_{ij}^{p} = \lambda \frac{\partial Q}{\partial \sigma_{ij}} \sigma_{ij} \tag{3.211}$$

Therefore, during plastic deformation the stress–strain relation becomes

$$d\sigma_{ij} = D_{ijkl}\, d\varepsilon_{kl}^{e} = D_{ijkl}(d\varepsilon_{ij} - d\varepsilon_{ij}^{p}) = D_{ijkl}\left(d\varepsilon_{ij} - \lambda \frac{\partial Q}{\partial \sigma_{ij}}\right) \tag{3.212}$$

The consistency condition in plasticity, which ensures that the stress points stay on the yield surface during work-hardening/softening processes, requires that

$$F(\sigma_{ij}) = 0 \tag{3.213a}$$

or

$$F(\sigma_{ij} + d\sigma_{ij}, W^{p} + dW^{p}) = F(\sigma_{ij}) + dF(\sigma_{ij}, W^{p}) = 0 \tag{3.213b}$$

or in incremental form

$$dF(\sigma_{ij}, W^{p}) = \frac{\partial F}{\partial \sigma_{ij}} d\sigma_{ij} + \frac{\partial F}{\partial W^{p}} dW^{p} = 0 \tag{3.214}$$

Denoting

$$-m = \frac{\partial F}{\partial W^{p}} \tag{3.215}$$

and substitution of Eqns (3.211) and (3.212) into the consistency condition given by Eqn (3.214) then leads to

$$\frac{\partial F}{\partial \sigma_{ij}} D_{ijkl} \left(d\varepsilon_{ij} - \lambda \frac{\partial Q}{\partial \sigma_{ij}} \right) + m\lambda \frac{\partial Q}{\partial \sigma_{ij}} \sigma_{ij} = 0 \tag{3.216}$$

which in turn yields the expression for the scalar λ

$$\lambda = \frac{\dfrac{\partial F}{\partial \sigma_{ij}} D_{ijkl}\, d\varepsilon_{kl}}{\dfrac{\partial F}{\partial \sigma_{ij}} D_{ijkl} \dfrac{\partial Q}{\partial \sigma_{kl}} + m \dfrac{\partial Q}{\partial \sigma_{ij}} \sigma_{ij}} \tag{3.217}$$

Substitution of λ by Eqn (3.217) into Eqn (3.212) finally results in

$$d\sigma_{ij} = \left(D_{ijkl} - \frac{D_{ijmn} \dfrac{\partial F}{\partial \sigma_{mn}} \dfrac{\partial Q}{\partial \sigma_{rs}} D_{rskl}}{\dfrac{\partial F}{\partial \sigma_{mn}} D_{mnrs} \dfrac{\partial Q}{\partial \sigma_{rs}} + m \dfrac{\partial Q}{\partial \sigma_{rs}} \sigma_{rs}} \right) d\varepsilon_{kl} = D^{ep}_{ijkl}\, d\varepsilon_{kl} \tag{3.218}$$

The scalar m is an experimentally determined constant or function of the plastic work for specific work (or kinematic) hardening or softening. For elastic-perfectly-plastic behavior, no hardening/softening is considered and $m = 0$. For detailed hardening/softening rules, see Chen (1994). For fractured rocks, both hardening and softening phenomena can be observed during laboratory experiments.

Denoting

$$f_1 = \frac{\partial F}{\partial \sigma_{xx}}, \quad f_2 = \frac{\partial F}{\partial \sigma_{yy}}, \quad f_3 = \frac{\partial F}{\partial \sigma_{zz}}, \quad f_4 = \frac{\partial F}{\partial \sigma_{xy}}, \quad f_5 = \frac{\partial F}{\partial \sigma_{yz}}, \quad f_6 = \frac{\partial F}{\partial \sigma_{zx}} \tag{3.219a}$$

$$q_1 = \frac{\partial Q}{\partial \sigma_{xx}}, \quad q_2 = \frac{\partial Q}{\partial \sigma_{yy}}, \quad q_3 = \frac{\partial Q}{\partial \sigma_{zz}}, \quad q_4 = \frac{\partial Q}{\partial \sigma_{xy}}, \quad q_5 = \frac{\partial Q}{\partial \sigma_{yz}}, \quad q_6 = \frac{\partial Q}{\partial \sigma_{zx}} \tag{3.219b}$$

and recalling the elastic stiffness tensor for isotropic elastic continuum (cf. Eqn (3.135)),

$$D_{ijkl} = \begin{bmatrix} D_{11} & D_{12} & D_{13} & D_{14} & D_{15} & D_{16} \\ D_{21} & D_{22} & D_{23} & D_{24} & D_{25} & D_{26} \\ D_{31} & D_{32} & D_{33} & D_{34} & D_{35} & D_{36} \\ D_{41} & D_{42} & D_{43} & D_{44} & D_{45} & D_{46} \\ D_{51} & D_{52} & D_{53} & D_{54} & D_{55} & D_{56} \\ D_{61} & D_{62} & D_{63} & D_{64} & D_{65} & D_{66} \end{bmatrix} = \begin{bmatrix} 2\mu+\lambda & \lambda & \lambda & 0 & 0 & 0 \\ \lambda & 2\mu+\lambda & \lambda & 0 & 0 & 0 \\ \lambda & \lambda & 2\mu+\lambda & 0 & 0 & 0 \\ 0 & 0 & 0 & \mu & 0 & 0 \\ 0 & 0 & 0 & 0 & \mu & 0 \\ 0 & 0 & 0 & 0 & 0 & \mu \end{bmatrix} \tag{3.219c}$$

the elasto-perfectly plastic constitutive relation in tensor form in Eqn (3.218) can be written in matrix form for easier computer implementation as

$$\begin{Bmatrix} d\sigma_{xx} \\ d\sigma_{yy} \\ d\sigma_{zz} \\ d\sigma_{xy} \\ d\sigma_{yz} \\ d\sigma_{zx} \end{Bmatrix} = \begin{bmatrix} D_{11}^{ep} & D_{12}^{ep} & D_{13}^{ep} & D_{14}^{ep} & D_{15}^{ep} & D_{16}^{ep} \\ D_{21}^{ep} & D_{22}^{ep} & D_{23}^{ep} & D_{24}^{ep} & D_{25}^{ep} & D_{26}^{ep} \\ D_{31}^{ep} & D_{32}^{ep} & D_{33}^{ep} & D_{34}^{ep} & D_{35}^{ep} & D_{36}^{ep} \\ D_{41}^{ep} & D_{42}^{ep} & D_{43}^{ep} & D_{44}^{ep} & D_{45}^{ep} & D_{46}^{ep} \\ D_{51}^{ep} & D_{52}^{ep} & D_{53}^{ep} & D_{54}^{ep} & D_{55}^{ep} & D_{56}^{ep} \\ D_{61}^{ep} & D_{62}^{ep} & D_{63}^{ep} & D_{64}^{ep} & D_{65}^{ep} & D_{66}^{ep} \end{bmatrix} \begin{Bmatrix} d\varepsilon_{xx} \\ d\varepsilon_{yy} \\ d\varepsilon_{zz} \\ d\varepsilon_{xy} \\ d\varepsilon_{yz} \\ d\varepsilon_{zx} \end{Bmatrix} \tag{3.220}$$

where

$$D_{ij}^{ep} = D_{ij} - \frac{1}{H}\left(\sum_{k=1}^{6} f_k D_{ik}\right)\left(\sum_{k=1}^{6} q_k D_{kj}\right) \tag{3.221}$$

$$H = \sum_{k=1}^{6}\left[q_k \sum_{l=1}^{6} f_l D_{lk}\right] + m(q_1\sigma_{xx} + q_2\sigma_{yy} + q_3\sigma_{zz} + q_4\sigma_{xy} + q_5\sigma_{yz} + q_6\sigma_{zx}) \tag{3.222}$$

3.4.4.2 *Elasto-perfect-plastic models based on the Mohr–Coulomb yielding criterion*

An example is given below for a non-associated flow rule using the Mohr–Coulomb criterion as both the yielding surface and the plastic potential, given by

$$\sigma_t = C - \sigma_n \tan\phi \tag{3.223}$$

where σ_t and σ_n are the maximum shear stress and normal stress acting at a point, ϕ is the internal friction angle and C the cohesion. A more convenient form for model formulation is written

$$\begin{aligned} F &= A\sigma_1 - \sigma_3 - B \\ &= A(\sigma_{xx} + \sigma_{yy} + \sigma_{zz}) - \left[(\sigma_{xx}\sigma_{yy}\sigma_{zz}) + 2\sigma_{xy}\sigma_{yz}\sigma_{zx} - \sigma_{xx}(\sigma_{yz})^2 - \sigma_{yy}(\sigma_{zx})^2 - \sigma_{zz}(\sigma_{xy})^2\right] - B = 0 \end{aligned} \tag{3.224}$$

where σ_1 and σ_3 are the maximum and minimum principal stresses $(\sigma_1 > \sigma_2 > \sigma_3)$ and

$$A = \frac{1+\sin\phi}{1-\sin\phi} \quad \text{and} \quad B = \frac{2C\cos\phi}{1-\sin\phi} \tag{3.225}$$

The criterion is therefore characterized by two parameters: internal friction angle ϕ and internal cohesion C. Figure 3.17 illustrates the Mohr–Coulomb criterion in the principal stress space. The choice for the plastic potential can be many and different and it depends on the different problems and materials. One common use is to assume an associated flow rule so that $F = Q$, which usually results in some

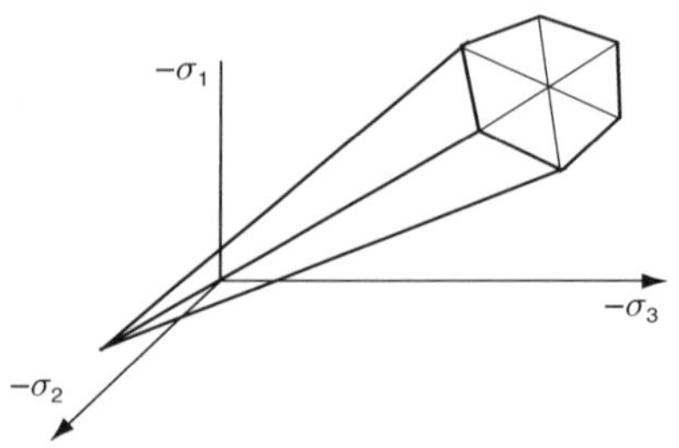

Fig. 3.17 Graphical representation of the Mohr–Coulomb criterion in principal stress space.

excessive volume expansion. A non-associated flow rule can be simply taken as the maximum shear stress, or its function, for the plastic potential, such as

$$\begin{aligned} Q &= b(\sigma_1 - \sigma_3) \\ &= b((\sigma_{xx} + \sigma_{yy} + \sigma_{zz}) - ((\sigma_{xx}\sigma_{yy}\sigma_{zz}) + 2\sigma_{xy}\sigma_{yz}\sigma_{zx} - \sigma_{xx}(\sigma_{yz})^2 - \sigma_{yy}(\sigma_{zx})^2 - \sigma_{zz}(\sigma_{xy})^2)) \end{aligned} \tag{3.226}$$

where $b > 0$ is a constant.

For two-dimensional problems, the incremental stress–strain relations are reduced to

$$\begin{Bmatrix} \mathrm{d}\sigma_{xx} \\ \mathrm{d}\sigma_{yy} \\ \mathrm{d}\sigma_{xy} \end{Bmatrix} = \begin{bmatrix} D_{11}^{\mathrm{ep}} & D_{12}^{\mathrm{ep}} & D_{13}^{\mathrm{ep}} \\ D_{21}^{\mathrm{ep}} & D_{22}^{\mathrm{ep}} & D_{23}^{\mathrm{ep}} \\ D_{31}^{\mathrm{ep}} & D_{32}^{\mathrm{ep}} & D_{33}^{\mathrm{ep}} \end{bmatrix} \begin{Bmatrix} \mathrm{d}\varepsilon_{xx} \\ \mathrm{d}\varepsilon_{yy} \\ \mathrm{d}\varepsilon_{xy} \end{Bmatrix} \tag{3.227}$$

$$D_{ij}^{\mathrm{ep}} = D_{ij} - \frac{1}{H}\left(\sum_{k=1}^{3} f_k D_{ik}\right)\left(\sum_{k=1}^{3} q_k D_{kj}\right) \tag{3.228}$$

with corresponding properties

$$H = \sum_{k=1}^{3}\left[q_k \sum_{l=1}^{3} f_l D_{lk}\right] + m(q_1\sigma_{xx} + q_2\sigma_{yy} + q_3\sigma_{xy}) \tag{3.229}$$

Below is an example of a non-associative Mohr–Coulomb model with the yield function and plastic potential given by (cf. Fig. 3.18) (Desai and Siriwardane, 1984).

$$\begin{cases} F = \dfrac{\sigma_{xx} + \sigma_{yy}}{2}\sin\phi - \left[\left(\dfrac{\sigma_{xx} - \sigma_{yy}}{2}\right)^2 + (\sigma_{xy})^2\right]^{\frac{1}{2}} + c \cdot \cos\phi \\ Q = \dfrac{\sigma_{xx} + \sigma_{yy}}{2}\sin\psi - \left[\left(\dfrac{\sigma_{xx} - \sigma_{yy}}{2}\right)^2 + (\sigma_{xy})^2\right]^{\frac{1}{2}} + c \cdot \cos\psi \end{cases} \tag{3.230}$$

where ψ is called the dilation angle representing the volume expansion effects of rock-like materials during shear. An associated flow rule results when $\phi = \psi$, but usually $\psi < \phi$.

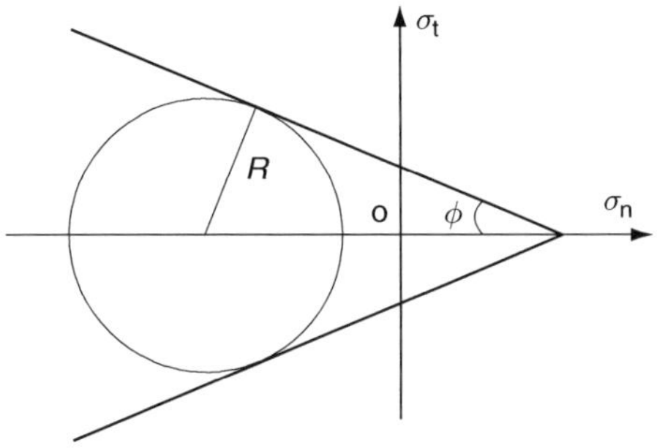

Fig. 3.18 Mohr–Coulomb criterion for plane problems.

The derivatives of the yield function and plastic potential are given by defining

$$G_1 = \left[\left(\frac{\sigma_{xx} - \sigma_{yy}}{2}\right)^2 + \left(\sigma_{xy}\right)^2\right]^{-\frac{1}{2}}, \qquad G_2 = \frac{\sigma_{xx} - \sigma_{yy}}{4} \tag{3.231a}$$

$$f_1 = \frac{\partial F}{\partial \sigma_{xx}} = \frac{\sin\phi}{2} - G_2 G_1, \quad f_2 = \frac{\partial F}{\partial \sigma_{yy}} = \frac{\sin\phi}{2} + G_2 G_1, \quad f_3 = \frac{\partial F}{\partial \sigma_{xy}} = -\sigma_{xy} G_1 \tag{3.231b}$$

$$q_1 = \frac{\partial Q}{\partial \sigma_{xx}} = \frac{\sin\psi}{2} - G_2 G_1, \quad q_2 = \frac{\partial Q}{\partial \sigma_{yy}} = \frac{\sin\psi}{2} + G_2 G_1, \quad q_3 = \frac{\partial Q}{\partial \sigma_{xy}} = f_3 \tag{3.231c}$$

For elasto-perfectly-plastic behavior, $m = 0$,

$$D_{ijkl} = \begin{bmatrix} 2\mu + \lambda & \lambda & 0 \\ \lambda & 2\mu + \lambda & 0 \\ 0 & 0 & \lambda \end{bmatrix}, \qquad H = 2(3\mu + \lambda) f_1 q_1 + \mu(f_3 q_3) \tag{3.232}$$

The coefficient matrix in the incremental stress–strain relation (3.227) is then given by

$$\left[D_{ij}^{\mathrm{ep}}\right] = \begin{bmatrix} 2\mu + \lambda - \frac{f_1 q_1}{H}(3\mu + \lambda) & \mu - \frac{f_1 q_1}{H}(3\mu + \lambda) & \frac{-f_1 q_3}{H}\mu(3\mu + \lambda) \\ \mu - \frac{f_1 q_1}{H}(3\mu + \lambda) & 2\mu + \lambda - \frac{f_1 q_1}{H}(3\mu + \lambda)(2\mu + \lambda) & -\frac{f_1 q_3}{H}\mu(2\mu + \lambda) \\ \frac{-f_1 q_3}{H}\mu(3\mu + \lambda) & -\frac{f_1 q_3}{H}\mu(2\mu + \lambda) & -\frac{f_3 q_3}{H}\mu^2 \end{bmatrix} \tag{3.233}$$

3.4.4.3 *Elasto-perfectly-plastic models based on the Hoek–Brown yielding criterion*

For fractured rocks, another commonly used failure criterion was proposed by Hoek and Brown (1980, 1988, 1997) and Hoek (1983, 1994), and written as

$$\sigma_1 = \sigma_3 + \sigma_c \left(\frac{m_i}{\sigma_c}\sigma_3 + 1\right)^{\frac{1}{2}} \tag{3.234}$$

where m_i is a material constant for the intact rock matrix and σ_c is the uniaxial compressive strength. This criterion was modified by Hoek et al. (1992) as

$$\sigma_1 = \sigma_3 + \sigma_c \left(\frac{m_b}{\sigma_c}\sigma_3\right)^a \tag{3.235}$$

where m_b and a are material constants depending on the composition, structure and surface conditions of the rock mass, including fractures. Tables 3.3 and 3.4 lists the recommended values for m_i, m_b and a (Hoek et al., 1992).

Together with the Mohr–Coulomb criterion, the Hoek–Brown criterion enjoys wide applications, especially for fractured hard rocks, for both numerical modeling and rock classification. The two criteria can be transformed from one to another (Hoek and Brown, 1997).

Table 3.3 Values of the constant m_i for intact rock by rock group (Hoek et al., 1992)

Grain size		Coarse	Medium	Fine	Very fine
Sedimentary	Carbonate	Dolomite 10.1	Chalk 7.2	Limestone 8.4	
	Detrital	Conglomerate (20)	Sandstone 18.8	Siltstone 9.6	Claystone 3.4
	Chemical		Chert 19.3	Gypstone 15.5	Anhydrite 13.2
Metamorphic	Carbonate	Marble 9.3			
	Silicate	Gneiss 29.2	Amphibolite 31.2	Quartzite 23.7	Slate 11.4
Igneous	Felsic	Granite 32.7		Rhyolite (20)	
	Mafic	Gabbro 25.8	Dolerite 15.2	Andesite 18.9	
	Mafic	Norite 25.8		Basalt (17)	

The modified 2D Hoek–Brown criterion can be taken as the yielding surface given as

$$F=\sigma_1-\sigma_3-\sigma_c\left(\frac{m_b}{\sigma_c}\sigma_3\right)^a=\sigma_1-\sigma_3-\sigma_c\left(\frac{m_b}{\sigma_c}\right)^a(\sigma_3)^a$$
$$=2\left[\left(\frac{\sigma_{xx}-\sigma_{yy}}{2}\right)^2+(\sigma_{xy})^2\right]^{\frac{1}{2}}-\sigma_c\left(\frac{m_b}{\sigma_c}\right)^a\left\{\left(\frac{\sigma_{xx}+\sigma_{yy}}{2}\right)-\left[\left(\frac{\sigma_{xx}-\sigma_{yy}}{2}\right)^2+(\sigma_{xy})^2\right]^{\frac{1}{2}}\right\}^a \tag{3.236}$$

$$f_1=\frac{\partial F}{\partial\sigma_{xx}}=\frac{\sigma_{xx}-\sigma_{yy}}{\sigma_1-\sigma_3}-\frac{a\sigma_c}{2}\left(\frac{m_b}{\sigma_c}\right)^a\left[1-\frac{\sigma_{xx}-\sigma_{yy}}{\sigma_1-\sigma_3}\right](\sigma_3)^{a-1}=q_1 \tag{3.237a}$$

$$f_2=\frac{\partial F}{\partial\sigma_{yy}}=-\frac{\sigma_{xx}-\sigma_{yy}}{\sigma_1-\sigma_3}-\frac{a\sigma_c}{2}\left(\frac{m_b}{\sigma_c}\right)^a\left[1+\frac{\sigma_{xx}-\sigma_{yy}}{\sigma_1-\sigma_3}\right](\sigma_3)^{a-1}=q_2 \tag{3.237b}$$

$$f_2=\frac{\partial F}{\partial\sigma_{xy}}=\frac{\sigma_{xy}}{\sigma_1-\sigma_3}\left[8+2a\sigma_c\left(\frac{m_b}{\sigma_c}\right)^a(\sigma_3)^{a-1}\right]=q_3 \tag{3.237c}$$

Assuming an associated flow rule by $F = Q$ and taking the derivatives of the yielding function, substitution of Eqn (3.237) into Eqns (3.228) and (3.229), with $m = 0$, leads to

$$\left[D_{ij}^{ep}\right]=\begin{bmatrix} 2\mu+\lambda-\frac{(f_1)^2}{H}(3\mu+\lambda) & \mu-\frac{(f_1)^2}{H}(3\mu+\lambda) & \frac{-f_1f_3}{H}\mu(3\mu+\lambda) \\ \mu-\frac{(f_1)^2}{H}(3\mu+\lambda) & 2\mu+\lambda-\frac{(f_1)^2}{H}(3\mu+\lambda)(2\mu+\lambda) & -\frac{f_1f_3}{H}\mu(2\mu+\lambda) \\ \frac{-f_1f_3}{H}\mu(3\mu+\lambda) & -\frac{f_1f_3}{H}\mu(2\mu+\lambda) & -\frac{(f_3)^2}{H}\mu^2 \end{bmatrix} \tag{3.238}$$

where

$$H=2(3\mu+\lambda)(f_1)^2+\mu(f_3)^2 \tag{3.239}$$

Table 3.4 Estimation of $r = m_b/m_i$ and based on rock structure and surface condition (Hoek et al., 1992).

Structure		Surface condition				
		Very Good Unweathered, discontinuous, very tight aperture, very rough surface, no infilling	Good Slightly weathered, continuous, tight aperture, rough surface, iron staining to no infilling	Fair Moderately weathered, continuous, very narrow, polished/ slickensided surfaces, hard infilling	Poor Highly weathered, continuous, very narrow, polished/ slickensided surfaces, hard infilling	Very poor Highly weathered, continuous, narrow polished/ slickensided surfaces, soft infilling
Blocky (1)	r	0.7	0.5	0.3	0.1	
	a	0.3	0.35	0.4	0.45	
Very (2)	r	0.3	0.2	0.1	0.04	
blocky	a	0.4	0.45	0.5	0.5	
Blocky or	r		0.08	0.04	0.01	0.004
seamy (3)	a		0.5	0.5	0.55	0.6
Crushed (4)	r		0.03	0.015	0.003	0.001
	a		0.5	0.55	0.6	0.65

Note:
(1) Well interlocked, undisturbed rock mass; large to very large block size.
(2) Interlocked, partially disturbed rock mass, medium block size.
(3) Folded and faulted, many intersecting joints, small blocks.
(4) Poorly interlocked, highly broken rock mass, very small blocks.

3.5 Summary Remarks

What we have presented above in this chapter are the 'classical' models widely applied in rock mechanics research and rock engineering applications and are the typical models used in DEM. They provide the 'backbone' materials for learning the subjects, but do not necessarily represent the cutting edge developments in this very active field of research since the main purpose of this chapter is to provide the fundamentals. However, a summary on the current state of understanding is helpful for readers aiming to probe deeper into the different, and also often more advanced, approaches and their relations. A similar summary of this kind was published in a review paper (Jing, 2003) with more references listed. Here we only include the fundamental statements and most necessary and representative literature sources with focus on developments not covered in the above sections in this chapter.

The constitutive equations of rocks determine the stresses and strains of the intact or equivalently intact rock blocks in the DEM models. The movements and the final geometry of the block systems in the DEM models are determined by their equations of motion, the contact detection algorithm and the constitutive equations for rock fractures and rock blocks. Since the number of the rock fractures is often too large for them to be completely included explicitly in DEM models, the majority of the fracture population of small sizes is usually not considered. Their effects, therefore, can only be reflected through their impacts on the equivalent properties in the constitutive equations of the blocks containing fractures of smaller sizes.

There are three difficult problems associated with the development of the constitutive equations of fractured rocks:

- The first is the scale effect that reflects the variation of mechanical properties with the size of the rock volumes.
- The second is the stress dependency, or more generally, path dependency. This means that the deformability and permeability of fractured rocks depend on the magnitude and evolution paths of the stresses. Both difficulties are related to the scale effects and stress dependency of the fractures, which, in turn, depend on their surface roughness and its damage evolution.
- The third is the possible transition from small-scale ductile deformation with strain localization to large-scale structural disintegration and motion of the rock, as demonstrated by 'shear-bands' of rock samples under compression. Typical examples of this bifurcation are landslides at large scales and creep rupture at small scales. Consideration of the transition from microscopic to macroscopic is therefore also needed.

All these complexities require specially developed constitutive equations for fractured rocks. The plasticity and damage mechanics theories can supply the basic platforms for their development (Krajcinovic, 1989; Chen, 1994; Shao and Rudnicki, 2000; Chen et al., 2004), but cannot be directly applied to overcome the above difficulties without further development. In the subsequent sections, we present some general remarks on the current states of understanding of some of the issues related to the above difficult problems in establishing appropriate constitutive models of rock masses and fractures.

3.5.1 Classical Constitutive Models of Rock Materials and Rock Masses

3.5.1.1 Classical models

The classical constitutive models are the models based on the theories of elasticity and plasticity but with special considerations for fracture effects. The model of linear elasticity based on the generalized Hooke's law is still by far the most widely adopted assumption in DEM for the mechanical behavior of rock blocks, especially for hard rocks. The reason is its simplicity. When the CHILE (continuous, homogeneous, isotropic, linear elastic) assumption is adopted, the constitutive law is simply characterized by two independent material properties, either the most commonly known Young's modulus (E) and Poisson's ratio (ν), or Lame's two parameters G and λ, which can be routinely determined by simple laboratory tests. The slightly more sophisticated models of anisotropic elasticity with alternative elastic symmetry conditions for intact rocks (such as transversely isotropic elasticity) or equivalent or effective continuum elastic rocks intersected by orthogonal sets of infinitely large or finite fractures are theoretically straightforward, although at different levels. However, they require additional laboratory tests to obtain additional elasticity and/or fracture properties (mainly fracture stiffness), good knowledge of *in situ* fracture patterns and principal directions of the elastic symmetry, which demands considerable increase in laboratory and field works for the required parameterization. The crack tensor approach needs more extensive fracture system characterization work, but may also introduce an additional advantage by providing a stochastic treatment of the material property ranges.

Plasticity and elasto-plastic models have been developed and widely applied to fractured rocks since the 1970s, with the Mohr–Coulomb and Hoek–Brown failure criteria as the most widely adopted yield functions and plastic potentials. Additional material properties of the internal friction angle and cohesion in the Mohr–Coulomb criterion, or the m and s parameters in the Hoek–Brown criterion, can be determined either through triaxial tests in the laboratory for homogeneous rock materials or estimated using Rock Classification schemes for large-scale fractured rocks as reported in Hoek and Brown (1997). The standard formulation of plasticity models can be seen in Owen and Hinton (1980) and Desai and Siriwardane (1984), for example.

Strain hardening and strain softening are the two main features of the plastic behavior of rocks with different models using different parameters. The main difficulty in using these models is not the theoretical complexities but the laboratory determination of the characteristic parameters for the hardening and softening properties. This will be more difficult if they are scale dependent, such as is probably the case for fractured rocks. This difficulty is often the main reason why such comprehensive models are less applied in practice.

The failure criteria of rocks are important components of constitutive relations and are usually used as yield surfaces and/or plastic potential functions in plasticity models. Besides the most well known and perhaps also the most widely used Mohr–Coulomb and Hoek–Brown criteria, different failure (or strength) criteria have also been proposed for rock masses over the years. Sheorey (1997), Mostyn and Douglas (2000) and Parry (2000) provided comprehensive reviews of the subject. It is important to note, however, that most of these criteria were developed based on experimental results or field observations with simple or limited loading and boundary conditions considered, and some of them are reversible in stress paths that may then violate the second law of thermodynamics. Care should be taken, therefore, to apply them according to their model assumptions in order to avoid fundamental mistakes.

3.5.1.2 Models based on damage mechanics approaches

Constitutive models of rocks have also been developed using continuum damage mechanics principles proposed first by Kachanov (1958). Damage can be defined in scalar, vector or tensor forms for void formation, micro-cracking or embedded fracture growth phenomena in rocks under static or dynamic loading. The theory is closely related to both continuum mechanics and fracture mechanics and can be seen as a bridge connecting the two. It has a certain parallelism in formulation with the plasticity models, such as using damage evolution laws in place of flow rules in the theory of plasticity, and the parallel damage/plastic potential functions. Damage mechanics theory also has a certain advantage in simulating the strain localization phenomenon using continuum approaches and for the study of brittle–ductile deformation mode transitions observed during testing rock samples. Comprehensive reviews on its development, characteristics, trends and weaknesses are given in Krajcinovic (2000) and de Borst (2002).

The damage mechanics approach has been applied to study strength degradation and strain localization phenomena in rocks, such as the excavation damage zone (EDZ), and to formulate damage-related constitutive models of rock and rock-like materials, with some of the most relevant contemporary literature included in Jing (2003).

3.5.1.3 Time effects and viscosity

Time effect is one of the most important and also perhaps one of the least understood aspects of the physical behavior of rock masses. There are two main aspects: the effects caused by the rock (and fracture) viscosity and the effects caused by the dynamic loading conditions. The former, concerns mainly the time-history of rock properties over long or extremely long terms, such as geological time scales; and the latter is just the opposite – dynamic and even violent behavior over short durations, such as earthquake effects. In the case of the latter, the magnitudes, directions, durations and the rates of change in loading parameters are important. The time and rate effects (also called dynamic effects) are therefore often discussed in combination. The measurements of the properties of rocks are loading rate-dependent, as demonstrated in the significant differences between the dynamic and static elastic moduli, measured using elastic wave transmission and laboratory quasi-static testing, respectively.

The effect of viscosity, as a time-dependent effect, is not very commonly included in numerical modeling for problems with hard rocks (because the hard rocks' viscosity is too small to induce significant strains or displacements in human time, say decades or hundreds of years commensurate with the design life of engineering structures), but is significant in rock salts, clays and other weak rocks of sedimentary origins. It is characterized by two mechanisms: creep and relaxation, and is very much temperature dependent. The former is the behavior of increasing deformation (strain) under constant loading (stress); and the latter is the decreasing of the loading (stress) while the deformation (strain) state is kept constant, often tested under isothermal conditions for long-term experiments for engineering analysis purposes. In fact, the *in situ* case may be somewhere between the two when unloading occurs. The fundamentals of the physics, experimental basis and the constitutive models for these two main mechanisms of rock viscosity have been described in detail in Jaeger and Cook (1969) and Cristesu and Hunsche (1998), with applications in tunnel/cavern failure and borehole closure in mining and petroleum engineering.

The effect of viscosity has been considered in constitutive modeling in combination with other basic deformation mechanisms, such as elasticity and elasto-plasticity or plasticity (leading to constitutive models of so-called viscoelasticity, and its plasticity counterparts, visco-elasto-plasticity and visco-plasticity). Basic descriptions of the constitutive models using viscoplasticity have been given in Valanis (1976) and Owen and Hinton (1980).

The consideration of rate (dynamic) effect in constitutive models and properties of rocks was studied in the rock mechanics field from early times, concerning mostly the fatigue and dynamic loading effects (Burdine, 1963; Haimson and Kim, 1972; Brown and Hudson, 1974). The subject is especially important for issues such as damage and fracturing/rupturing of rock materials caused by blasting or seismic/tectonic movements, and has significant impacts on the EDZ and rock burst phenomena. Therefore, the development of constitutive models concerning dynamic effects is mostly concentrated on dynamic damage initiation and evolution, with the micro-mechanics approach as the most commonly adopted platform, as reported in Taylor et al. (1986) and Chen (1995). Numerous publications have appeared on static continuum damage approaches, as cited in Jing (2003). An important recent development was reported in Li et al. (2001) where extensive laboratory experiments were conducted for investigating dynamic deformation properties and strengths of fractured rock samples at different loading levels and frequencies, leading to a rock fatigue model.

The dynamic problems in rock mechanics are often treated as transient (time-dependent) problems using static material properties, typically in DEM. This simplification is acceptable for some rock engineering problems when the material properties can be assumed to be loading-rate-independent. However, when investigating dynamic damaging or wave propagation problems, such as seismic events and explosion/blasting problems, dynamic properties with loading rate effects under investigations may need to be considered. The problem is accommodating the extra experimental work needed to measure the site-specific rate-dependent dynamic properties within the frequency ranges.

3.5.1.4 Scale effects and the equivalent continuum approach

Scale effect is a special feature of fractured rocks, mainly due to two factors caused by the existence of fractures of various sizes in rock masses.

The first is the fact that the fracture systems divides a fractured rock mass into a large number of sub-domains or blocks, whose sizes and interactions dominate the overall behavior of the rock masses.

The second is the fact that the physical behavior of the fractures themselves is dependent on their sizes, due to the scale dependence of surface roughness of the rock fractures whose stationarity thresholds is a scaling parameter (Lanaro et al., 1998; Lanaro, 2000; Fardin et al., 2001; Jing and Hudson, 2004). The earlier state of the art of this subject was presented in two edited volumes by Da Cunha (1990, 1993) and a recent comprehensive survey on the size effects in the strength and behavior of structures, including geotechnical structures, given by Bažant (2000).

Since most of the measured rock properties are obtained through small-scale laboratory tests, the measured values are at best valid only at these small scales, representing only the behavior of the intact rock matrix and not the *in situ* fractured rock masses at field scales. For large-scale problems, rock masses are often assumed to be equivalent continua and the equivalent properties therefore need to be evaluated mathematically using the REV or effective media approaches through a combined homogenization and upscaling process or using empirical rock classification/characterization schemes such as RMR, GSI or others.

Homogenization is the averaging process of properties inside a region of certain size, and upscaling is the process of homogenization with increasing size of the region until stationary limits are reached for the averaged properties under study, usually in a statistical sense. Therefore the two terms are often used together and mixed. The properties thus established are called the effective or equivalent properties, with the former more used in material sciences and solid mechanics and the latter more in rock mechanics.

The REV concept presents a dilemma for the analyses of fractured rocks using equivalent continuum approaches. In theory, the REV size should be large enough (or the number of fractures included in this volume should be large enough) so that a statistical equivalence between the original fractured medium and the equivalent continuum can be established. On the other hand, discretization of the solution domain into continuum elements needs to be small enough so that a continuous gradient of deformation or fluid flow can be obtained within the solution domain (Long et al., 1982). These two requirements may contradict each other. In addition, a REV may not exist, especially for hard rocks containing fractures of very different sizes. In such cases, it may be difficult to justify the use of continuum-based numerical methods. This issue was analyzed by Pariseau (1993, 1995, 1999) with a non-representative volume element (NERV) approach proposed to replace the original discrete materials, based on a numerical evaluation of equivalent properties of local continuum elements.

The focus of debate is whether such REVs can exist physically considering the presence of hierarchical structures of fracture sizes and widths (or apertures) from small joints at the centimeter scale to large-sized faults and fracture zones at the kilometer scale, and their vastly different physical behavior and properties as often observed in the field. On the other hand, the existence of the REV is relative to fracture density, connectivity and model size. Large features like faults and fault zones at or near the top of the hierarchical pyramid of the fracture systems are usually small in numbers and can be treated as deterministic in distribution. By separating these large features, the REVs may be established for the more numerous fractures population that are moderate to small in size and more suitable for stochastic homogenization and upscaling for defining the REV. The question is to what level the fracture size, density and model sizes and the lower/upper cut-offs of the fracture sizes in field mapping are needed to be included in the upscaling process (see more details in Chapter 5).

The DEM approach is especially suitable in the fields of upscaling–homogenizing the hydro-mechanical properties of fractured rocks with generally irregular fracture systems using deterministic or stochastic approaches, as demonstrated in Min and Jing (2003, 2004) and Min et al. (2004a,b) as presented in Section 12.8 of Chapter 12, due to the fact that the explicit presentation of relatively large numbers of fractures is needed in the computational models for homogenization and upscaling, which poses numerical difficulties for continuum approaches like the FEM, BEM or DEM.

3.5.2 Constitutive Models for Rock Fractures

Constitutive models for rock fractures play an important role in almost every aspect of rock mechanics and rock engineering, especially in the fields of numerical modeling in general and in DEM in particular. It is therefore not surprising that the mechanics and models of rock fractures have become one of the main themes in almost every international or national conference on rock mechanics, rock physics and rock engineering, with explosively increasing publications. The most directly relevant works are the volumes edited by Stephansson (1985), Barton and Stephansson (1990), Myer et al. (1995) and Rossmanith (1990, 1995, 1998). The subject has also become an inevitable part of many text and reference books, such as Chernyshev and Dearman (1991), Lee and Farmer (1993), Selvadurai and Boulon (1995), Hudson (1993), Indrarantna and Haque (2000), Indrarantna and Ranjith (2001) and Aliabadi (1999). Some early reviews on the experimental aspects and formulations of constitutive models and strength envelopes of rock fractures are given in Stephansson and Jing (1995) and Ohnishi et al. (1996), respectively.

The constitutive models of rock fractures are formulated mainly with two approaches: empirical and theoretical as introduced in Sections 3.1–3.3. The primary variables are contact tractions and relative displacements (instead of stresses and strains in continuum models), from which hydraulic apertures may be derived for flow calculations. Implementation of constitutive models into continuum-based numerical methods such as FEM often leads to so-called joint or interface elements, which may also cause numerical instabilities when a zero-thickness of interface elements is employed without necessary modification, such as discussed in Kaliakin and Li (1995) and Day and Potts (1994). Implementation into discrete element methods is generally a more straightforward use of contact mechanics principles, but the prevention of inter-penetration of solid blocks must be applied, using methods like a penalty function, Lagrangian multipliers or augmented Lagrangian multiplier techniques.

The approach of using the theory of plasticity as the basic mathematical platform for formulating constitutive models of rock fractures seems to comprise, at present, the majority of the constitutive modeling approaches used in practice, especially with the Mohr–Coulomb friction model. There are, however, many other constitutive models for rock fractures, being both empirical and theoretical in approach, such as are summarized in Jing (2003). A special class of models used the principles of contact mechanics of rough surfaces, based mainly on principles established in Greenwood and Williamson (1966) and Greenwood and Tripp (1971) simulating the contacts, friction and wear of rough surfaces of solids. The approach needs a comprehensive representation of surface roughness with statistical and probabilistic measures. The practical applicability of such models depends largely on the unique quantification of surface roughness and its impact on fracture behavior, which still remain as a challenging topic today, especially when a non-stationary roughness nature of rock fractures is present.

Besides JRC, many other measures for fracture roughness have been proposed using basically random field theory, geostatistics and fractal models. Some of the most recent research results in this direction are reported in Lanaro et al. (1998) and Fardin et al. (2001). Use of fractals to represent the roughness of rock fractures has become an important subject, as indicated in the above publications, but still remains a controversial topic in debate, as also in other fields (Whitehouse, 2001).

The constitutive models as presented above have been applied in many practical rock engineering and rock mechanics problems with varying degrees of success. It is often that more successful applications are about generic studies on fractured rock behavior rather than site-specific engineering applications because of the fact that uncertainty due to *in situ* fracture system geometry has at least equal, if not more, significant impacts on the final results than the constitutive models of individual fractures. Despite the shortcomings, the developed constitutive models have served their purposes in rock engineering design and analysis. They also serve as good starting points for further development of more reliable and robust models.

Besides the fractures, there are other interfaces in rock engineering practice, such as the interfaces between different materials and components (rock and soil, rock and buffer or backfilling materials, rock and reinforcement elements, e.g. bolts, grouts, cables, etc., and rock and other construction materials such as concrete, etc.), which are also important subjects in DEM modeling for practical problems. In-depth research on the constitutive models for such interfaces are scarce in the literature, but the subject is important for design and performance assessments of rock structures.

3.5.3 Rock Fracture Testing and Outstanding Issues

Although the main focus of this book is DEM modeling, it is essential that a few words being said about rock fracture testing since the subject is essential for understanding and application of the method.

The major progress in laboratory experiments on rock fractures was through normal loading–unloading testing, direct shear testing (under both constant normal stress and normal stiffness) and coupled normal stress–fluid flow testing. The majority of these tests have been performed under normal stress constraint. Progress in the direct shear tests under constant normal stiffness provides a better understanding of the mechanical behavior of rock fractures under confined situations, such as underground rock engineering. Shear strengthening is the major feature of rock fractures under this loading condition, contrary to the shear weakening under constant normal stresses corresponding to near free surface conditions. The hyperbolic relations can be taken as a sound description between the normal stress and normal deformation of rock fractures, although different loading and unloading paths need to be considered in building mathematical models. The shear strength of the fracture may be temperature dependent and the heat transfer between the rock matrix and fluid flow in the fractures depends on the fluid velocity that may, in turn, depend on the mechanical deformation process. Based on the experimental data collected so far, it can be said with confidence, at least qualitatively, that these general conclusions are valid.

On the other hand, a number of outstanding issues of importance still remain relating to rock fractures.

3.5.3.1 Roughness

All experimental findings clearly indicate that the surface roughness of rock fractures is the decisive factor for almost all aspects of the mechanical, hydraulic and coupled hydro-mechanical behavior of rock fractures. The characterization of the roughness remains a lasting challenge. The current measures of JRC and fractal dimension still have limitations in characterizing the fracture roughness uniquely. The strong correlation between roughness and other aspects of fracture behavior (e.g. scale effects, anisotropy, stress (path)-dependency and conductivity) demonstrates the strong requirement to represent, quantitatively and uniquely, the roughness of rock fractures in both two and three dimensions, and its evolution with time and deformation paths, in order to develop more reliable constitutive models.

3.5.3.2 Scale effect

Scale effect certainly exists in fracture behavior and is a manifestation of the scale dependency of the roughness. Most of the experiments on scale effect have been performed in direct shear tests under constant normal stress in the past. Investigations concerning scale effects under other test conditions (normal compression tests, direct shear tests under constant normal stiffness and other coupled stress–flow tests) may also be needed to generate ample data so that proper linkage between the laboratory tests, mathematical model development and field scale applications can be established with confidence.

It is common practice that rock fracture properties are obtained from laboratory tests on samples of limited size, say 100–400 mm. This size may or may not be large enough to reach the stationarity thresholds of the fracture samples, depending on the roughness characteristics of the sample surfaces. There is an acute lack of understanding of the hydro-mechanical behavior of large fractures (such as faults or fracture zones) with trace lengths from tens of meters to kilometers in scale and large widths (say, 10 mm–50 m). This type of rock fractures has been associated with problems in many engineering projects and they are the most important geologic feature for the design and safety of hydropower projects, radioactive waste repositories, geothermal and hydrocarbon reservoirs and other large-scale facilities in rocks with potentially significant impacts on rock structure performance and the environment. Such large features are often treated as fracture elements in numerical models with similar constitutive laws developed for smaller features like joints, as approximations. Scaling of the model properties, such as stiffness and friction angles, is sometimes adopted based on the widths and material properties of the faults/fracture zones, but uncertainties due to such simplification are difficult to estimate. The difficulty lies in the fact that such large features cannot be tested in well-controlled testing conditions, such as one expects in laboratory test conditions, and field monitoring results may be the only means for model verifications using back-analysis techniques. Even if the field monitoring data are available, they still contain the uncertainties introduced by the largely unknown *in situ* boundary conditions and the validity of the assumed constitutive models.

3.5.3.3 Gouge materials

The production of gouge material during fracture sliding affects the fracture behavior, especially the friction, shear strength and fluid transmissivity, due to particle-induced lubrication and aperture changes by particle settlement. Progress in research on this subject has not, however, been impressive for rock fractures. The difficulties lie in measuring the rate of gouge production, its distribution on the fracture surface and the distribution of the actual contact areas during tests, and quantifying its hydro-mechanical effects. More experimental data are needed for more specific and quantitative conclusions regarding the effect of gouge materials (mechanisms, production rates, consequences). The subject is important, however, in understanding the damage evolution on the fracture surfaces, which is another challenging aspect for developing constitutive models for rock fractures.

3.5.3.4 Three-dimensional effects

Most of experimental studies performed so far are one-dimensional (normal loading–unloading) or two-dimensional (direct shear) tests. The orientation of the fracture surface in space, its finite dimension and the general stress state *in situ*, however, hardly justify such simplifications. The three-dimensional models for rock fractures with anisotropic frictional behavior (Jing, 1990; Jing et al., 1994) require extra experimental work for determination of input parameters. Recent work by Grasselli et al. (2002) and Grasselli and Egger (2003) reported new 3D constitutive models for rough rock fractures based on experimentally determined relations between the contact areas under normal loads and asperity inclination angles.

The complex effects of the anisotropy of fracture roughness call for more systematic true three-dimensional experiments performed under combined shear and normal loading, under constant normal stiffness and normal stresses and other environmental conditions (e.g., temperature and fluid presence). Such experiments require 3D shear experimental devices with flexible control on the normal constraint conditions (Boulon, 1995; Armand et al., 1998).

3.5.3.5 *Dynamic and time effects*

Most of the reported tests on rock fractures are quasi-static tests, i.e., the loading rate is maintained low, steady and constant to avoid unwanted dynamic effects. Dynamic shear tests under a normal stress constraint have been conducted in the past (Barla et al., 1990) and the data accumulated so far is still too limited to draw definite conclusions. The importance of the subject, however, cannot be neglected due to demands for more reliable numerical modeling of earthquake events, industrial blasting operations and protection of underground military or civilian shelters against surface or deep-penetration explosions, etc. The experimental results have demonstrated the different fracture behavior with respect to the shearing velocity, varying from velocity strengthening, velocity weakening, to velocity independence.

Modeling of dynamic rock fracture sliding is usually performed using the state variable friction models, as developed in Rice et al. (2002), Rice and Ruina (1983), Ruina (1983) and Gu et al. (1984), treating the shear stresses as functions of both the sliding history and velocity, representing the evolution of the path-dependence of the frictional properties and rate effects. Application of this theory was reported in Lorig and Hobbs (1990) for simulating rock slope instability processes. Recent work on landslide problems (Qin et al., 2001) investigated the dynamic frictional processes of the fracture surfaces using theories of dynamic chaos and catastrophe for an integrated treatment of the interactions between the fracture surfaces regarding fracture stiffness, friction and elastic materials for the rocks. Creeping and dynamic sliding along surfaces (that are often the contact zones or fault zones) are often treated together, especially for landslide and slope stability problems since these two factors dominate the process, such as reported in Chau (1995, 1999). In applications, the extra demand for reliable parameterization is also an important issue due to the challenging work required in laboratory and field measurements.

3.5.3.6 *Summary remarks on outstanding issues*

Coupling processes in rock fractures becomes an important outstanding issue and will be addressed at the end of Chapter 4 after the flow problems in rock fractures are presented.

In summary, although tremendous efforts have been made to develop constitutive models for rock fractures, the currently available models still have significant limitations in predicting fracture behavior with an adequate level of confidence. The major difficulty is the lack of unique and quantitative representation of fracture surface roughness, confident prediction of surface damage evolution during a general deformation process and its impact on the thermo-hydro-mechanical coupling processes and properties of rock fractures. Other difficulties include the models for large-scale features, such as faults or fracture zones with large widths, time-scale dependence and hydro-mechanical coupling effects. New subjects of interest are the effects of chemical coupling and transport properties of fractures (such as mineral precipitation and dissolution), as affected by flow path tortuosity, estimation of the initial contact area and its evolution, and the fracture–rock interaction in terms of flow and transport during the deformation and chemical processes of rock fractures.

All the above add to the ever-increasing complexity in modeling the physico-chemical behavior and properties of rock fractures, and introduce increasing difficulties for formulating realistic and applicable constitutive models. On the other hand, it should also be noted that constitutive models are developed for understanding the overall behavior and helping to solve practical problems. Therefore, proper and prudent simplifications and idealization will always be needed to derive models simplified enough for the problems at hand, while still retaining the necessary levels of scientific sophistication so that the basic laws of physics and chemistry will not be violated. A further step that needs to be strengthened is to address the uncertainties introduced by such simplifications at the different levels of the processes, properties and parameters, so that their sources, changes and propagation paths and their impacts on the final results can be estimated with reasonable confidence.

References

Aliabadi, M. H. (ed.), Fractures of rock. WIT Press/Computational Mechanics Publications, Boston/Southampton, 1999.

Amadei, B. and Goodman, R. E., A 3-D constitutive relation for fractured rock masses. Proc. of Int. Symp. on Mechanical Behaviour of Structured Media, Ottawa, Canada, pp. 249–268, 1981.

Amadei, B. and Saeb, S., Constitutive models of rock joints. In: Barton, N. and Stephansson, O. (eds), Rock joints (Proc. of Int. Symp. on Rock Joints, Loen, Norway, June 4–6, 1990), pp. 585–604. Balkema, Rotterdam, 1990.

Armand, G., Boulon, M., Papadopoulos, C., Basanou, M. E. and Vardoulakis, P. P., Mechanical behaviour of Dionysos marble smooth joints: I. Experiments. In: Rossmanith, H.-P. (ed.), Mechanics of jointed and faulted rocks (Proc. of MJFR-3 Conf., Vienna, Austria, 1998), pp. 159–164. Balkema, Rotterdam, 1998.

Bandis, S. C., Experimental studies of scale effects on shear strength and deformation of rock joints. Ph.D. Thesis, Department of Earth Sciences, University of Leeds, UK, 1980.

Bandis, S. C., Lunsden, A. C. and Barton. N. R., Experimental studies of scale effects on the shear behaviour of rock joints. International Journal of Rock Mechanics and Mining Sciences and Geomechanics Abstracts, 1981;18(1):1–21.

Bandis, S. C., Lunsden, A. C. and Barton. N. R., Fundamentals of rock joint deformation. International Journal of Rock Mechanics and Mining Sciences and Geomechanics Abstracts, 1983;20(6):249–268.

Barla, G., Barbero, M., Scavia, C. and Zaninetti, A., Direct shear testing of single joints under dynamic loading. In: Barton, N. and Stephansson, O. (eds), Rock joints (Proc. of Int. Symp. on Rock Joints, Loen, Norway, June 4–6, 1990), pp. 447–454. Balkema, Rotterdam, 1990.

Barton, N., A relationship between joint roughness and joint shear strength. Proc. Int. Symp. Rock Mech. Nancy, Rock Fracture, Paper **I**-8, 1971.

Barton, N., Review of a new shear strength criterion for rock joints. Engineering Geology, 1973;7(4):287–332.

Barton, N., Estimating the shear strength of rock joint. Advances in Rock Mechanics, 1974; **II A**: 219–220.

Barton, N., The shear strength of rock and rock joints. International Journal of Rock Mechanics and Mining Sciences and Geomechanics Abstracts, 1976;13(9):255–279.

Barton, N., Modelling rock joint behaviour from in situ block tests: Implications for nuclear waste repository design. Office of Waste Isolation, Columbus, OH, ONWI-308, 1982.

Barton, N. and Bandis, S. C., Rock joint model for analyses of geological discontinua. In: Desai, C. S. (ed.), Proc. of the 2nd Int. Conf. and Short Courses on Constitutive Laws for Engineering Materials: Theory and Applications, January 5–10, 1987, The Univ. of Arizona, Tucson, AZ, USA, pp. 993–1002. Elsevier, Amsterdam, 1987.

Barton, N. and Choubey, V. The shear strength of rock joints in theory and practice. Rock Mechanics, 1977;10(1–2):1–54.

Barton, N. and Stephansson, O. (eds), Rock joints (Proc. of Int. Symp. on Rock Joints, Loen, Norway, June 4–6, 1990). Balkema, Rotterdam, 1990.

Barton, N., Bandis, S. and Bakhtar, K. Strength, deformation and conductivity coupling of rock fractures. International Journal of Rock Mechanics and Mining Sciences and Geomechanics Abstracts, 1985;22(3):121–140.

Bažant, Z., Size effect. International Journal of Solids and Structures, 2000;37(1–2):69–80.

Boulon, M., A 3D direct shear device for testing thermomechanical behaviour and hydraulic conductivity of rock joints. In: Rossmanith, H.-P. (ed.), Mechanics of jointed and faulted rocks (Proc. of MJFR-2 Conf., Vienna, Austria, 1995), pp. 407–413. Balkema, 1995.

Brown, E. T. and Hudson, J. A., Fatigue failure characteristics of some models of jointed rock. Earthquake Engineering & Structural Dynamics, 1974;2(4):379–386.

Burdine, N. T., Rock failure under dynamic loading conditions. Society of Petroleum Engineers Journal, 1963;3(1):1–8.

Chau, K. T., Landslide modeled as bifurcations of creeping slopes with non-linear friction law. International Journal of Solids and Structures, 1995;32(23):3451–3464.

Chau, K. T., Onset of natural terrain landslides modeled by linear stability analysis of creeping slopes with a two-state variable friction law. International Journal for Numerical and Analytical Methods in Geomechanics, 1999;23(15):1835–1855.

Chen, W.-F., Constitutive equations for engineering materials. Vol. 2: Plasticity and Modeling. Elsevier, Amsterdam, 1994.

Chen, E. P., Dynamic brittle material response based on a continuum damage model. In: Barta, R. C., Mal, A. K. and MacSithigh, G. P. (eds), Impact waves and fracture, pp. 21–34. ASME, New York, 1995.

Chen, W. Z., Zhu, W. S. and Shao, J. F., Damage coupled time-dependent model of a jointed rock mass and application to large underground cavern excavation. International Journal of Rock Mechanics and Mining Sciences, 2004;41(4):669–677.

Chernyshev, S. N., and Dearman, W. R., Rock fractures. Butterworth-Heinemann, London, 1991.

Cristescu, N. D. and Hunsche U., The time effects in rock mechanics. Wiley, Chichster, UK, 1998.

Da Cunha, A. P., Scale effects in rock masses. Balkema, Rotterdam, 1990.

Da Cunha, A. P., Scale effects in rock masses. Balkema, Rotterdam, 1993.

Day, R. A. and Potts, D. M., Zero thickness interface elements-numerical stability and applications. International Journal for Numerical and Analytical Methods in Geomechanics, 1994;18(10):689–708.

de Borst, R. Fracture in quasi-brittle materials: A review of continuum damage-based approaches. Engineering Fracture Mechanics, 2002;69(2):95–112.

Desai, C. S. and Siriwardane, H. J., Constitutive laws for engineering materials with emphasis on geological materials. Prentice-Hall, Englewood Cliffs, NJ, 1984.

Duncan, J. M. and Goodman, R. E., Finite element analysis of slopes in jointed rock. Final report to US Army Corps of Engineers, Vicksburg, Mississippi, Report S-68-3, 1968.

Fardin, N., Stephansson, O. and Jing, L., The scale dependence of rock joint surface roughness. International Journal of Rock Mechanics and Mining Sciences, 2001;38(5):659–669.

Fossum, A. F., Effective elastic properties for a randomly jointed rock mass. International Journal of Rock Mechanics and Mining Sciences and Geomechanics Abstracts, 1985;22(6):467–470.

Goodman, R. E., Methods of geological engineering in discontinuous rocks. West Publishing Company, San Francisco, CA, 1976.

Grasselli, G. and Egger, P., Constitutive law for the shear strength of rock joints based on three-dimensional surface parameters. International Journal of Rock Mechanics and Mining Sciences, 2003;40(1):25–40.

Grasselli, G., Wirth, J. and Egger, P., Quantitative three-dimensional description of a rough surface and parameter evolution with shearing. International Journal of Rock Mechanics and Mining Sciences, 2002;39(6):789–800.

Greenwood, J. A. and Williamson, J. B. P., Contact of nominally flat rough surfaces. Proceedings of the Royal Society of London, series A 1966;295(1442):300–319.

Greenwood, J. A. and Tripp, J. H. The contact of two nominally flat rough surfaces. Proceedings of the Institution of Mechanical Engineers, Part C, 1971;185(48):625–633.

Gu, J.-C., Rice, J. R., Ruina, A. L. and Tse, S. T., Slip motion and stability of a single degree of freedom elastic system with rate and state dependent friction. Journal of the Mechanics and Physics of Solids, 1984;32(3):167–196.

Guvanasen, V. and Chan, T., A three-dimensional finite element solution for heat and fluid transport in deformable rock masses with discrete fractures. Proc. of Int. Conf. of the Int. Assoc. Computer Methods and Advances in Geomechanics, May 6–10, Cairns, Australia, pp. 1547–1552. Balkema, Rotterdam, 1991.

Haimson, B. C. and Kim, C. M., Mechanical behavior of rock under cyclic fatigue. In: Cording, E. J. (ed.), Stability of rock slopes. Proc. 13th US Symp. on Rock Mechanics, pp. 845–863. ASCE, New York, 1972.

Hoek, E., Strength of jointed rock masses. Rankine Lecture. Geotechnique, 1983;33(3):187–223.

Hoek, E., Strength of rock and rock masses. ISRM News Journal, 1994;2(2):4–16.

Hoek, E. and Brown, E. T., Underground excavations in rock. Institution of Mining and Metallurgy, London, 1980.

Hoek, E. and Brown, E. T., The Hoek–Brown failure criterion – a 1988 update. In: Rock Engineering for Underground Excavations. Proc. 15th Canadian Rock Mech. Symp., Curran, J. C. (ed.), University of Toronto, Toronto, Canada.

Hoek, E. and Brown, E. T., Practical esitimates of rock mass strength. International Journal of Rock Mechanics and Mining Sciences and Geomechanics Abstracts, 1997;34(8):1165–1186.

Hoek, E., Wood, D. and Shah, S., A modified Hoek–Brown failure criterion for jointed rock masses. Proc. of EUROCK'92, pp. 209–214. Thomas Telford, London, 1992.

Horii, H. and Sahasakmontri, K., Is the fabric tensor sufficient? In: Satake, M. and Jenkins, J. T. (eds), Micromechanics of granular materials, pp. 91–94. Elsevier, Amsterdam, 1988.

Huang, T. H., Chang, C. S. and Yang, Z. Y., Elastic moduli for fractured rock mass. Rock Mechanics and Rock Engineering, 1995;28(3):135–144.

Hudson, J. A. (ed. in-chief), Comprehensive rock engineering, Vols 1–5, 4407 pages. Pergamon Press, Elsevier Science, Oxford, 1993.

Hutson, R. W., Preparation of duplicate rock joints and their changing dilatancy under cyclic shear. Ph.D. Thesis, Northwestern University, Evanston, IL, 1987.

Indrarantna, B. and Haque, A., Shear behaviour of rock joints. Balkema, Rotterdam, 2000.

Indrarantna, B. and Ranjith, P. G., Hydromechanical aspects and unsaturated flow in jointed rock. Balkema, Rotterdam, 2001.

Jaeger J. C., Cook, N. G. W., Fundamentals of rock mechanics. Methuen, London, UK, 1969.

Jing, L., Numerical modelling of fractured rock masses by distinct element method for two and three dimensional problems. Ph.D. Thesis, Luleå University of Technology, Luleå, Sweden, 1990.

Jing, L., A review of techniques, advances and outstanding issues in numerical modelling for rock mechanics and rock engineering. International Journal of Rock Mechanics and Mining Sciences, 2003;40(3):283–353.

Jing, L. and Hudson, J. A., Fundamentals of hydro-mechanical behaviour of rock fractures: Roughness characterization and experimental aspects. International Journal of Rock Mechanics and Mining Sciences, 2004;41(3):383. SINOROCK2004 Paper 1A 26.

Jing, L., Stephanson, O. and Nordlund, E. Study of rock joints under cyclic loading conditions. Rock Mechanics and Rock Engineering, 1993;26(3):215–232.

Jing, L., Nordlund, E. and Stephansson, O., A 3-D constitutive model for rock joints with anisotropic friction and stress dependency in shear stiffness. International Journal of Rock Mechanics and Mining Sciences and Geomechanics Abstracts, 1994;31(2):173–178.

Jing, L., Tsang, C.-F., Stephansson, O. and Kautsky, F., Validation of mathematical models against experiments for radioactive waste repositories – DECOVALEX experience. In: Stephansson, O., Jing, L. and Tsang, C.-F., Mathematical models for coupled thermo-hydro-mechanical processes in fractured media, pp. 25–56. Elsevier, Rotterdam, 1996.

Kachanov, L. M., Time of the rupture process under creep conditions. Izv. Akad. Nauk. SSR Otd., Teckh. Nauk, 1958; Vol. 8, p. 26–31.

Kaliakin, V. N. and Li, J., Insight into deficiencies associated with commonly used zero-thickness interface elements. Computers and Geotechnics, 1995;17(2):225–252.

Kaneko, K. and Shiba, T., Equivalent volume defect model for estimation of deformation behaviour of jointed rock. In: Rossmanith, H.-P. (ed.), Proc. of Int. symp. on Mechanics of Jointed and Faulted Rock (MJFR-1), pp. 277–284. Balkema, Rotterdam, 1990.

Krajcinovic, D., Damage mechanics. Mechanics of Materials, 1989;8(2–3):117–197.

Krajcinovic, D., Damage mechanics: Accomplishments, trends and needs. International Journal of Solids and Structures, 2000;37(1–2):267–277.

Ladanyi, B. and Archambault, G., Simulation of shear behaviour of a fractured rock mass. Chapter 7 in Rock mechanics – theory and practice (Proc. 11th US Symp. of Rock Mech., Berkeley, June 1969), pp. 105–125, 1969.

Lanaro, F., A random field model for surface roughness and aperture of rock fractures. International Journal of Rock Mechanics and Mining Sciences, 2000;37(8):1195–1210.

Lanaro, F., Jing, L. and Stephansson, O., 3-D-laser measurements and representation of roughness of rock fractures. In: Rossmanith, H.-P. (ed.), Proc. of the Int. Conf. on Mechanics, Jointed and Faulted Rock, MJFR-3, 1998, Vienna, Austria, pp. 185–189. Balkema, Rotterdam, 1998.

Lee, C.-H. and Farmer, I., Fluid flow in discontinuous rocks, 164 pages. Chapman and Hall, London, 1993.

Lekhnitskii, S. G., Theory of elasticity of an anisotropic body. Mir Publishers, Moscow, 1977.

Li, N., Chen, W., Zhang, P. and Swoboda, G., The mechanical properties and a fatigue-damage model for jointed rock masses subjected to dynamic cyclical loading. International Journal of Rock Mechanics and Mining Sciences, 2001;38(7):1071–1079.

Long, J. C. S., Remer, J. S., Wilson, C. R. and Witherspoon, P. A., Porous media equivalents for networks of discontinuous fractures. Water Resources Research, 1982;18(3):645–658.

Lorig, L. J. and Hobbs, B. E., Numerical modelling of slip instability using the distinct element method with state variable frictional laws. International Journal of Rock Mechanics and Mining Sciences and Geomechanics Abstracts, 1990;27(6):525–534.

Min, K.-B. and Jing, L., Numerical determination of the equivalent elastic compliance tensor for fractured rock masses using the distinct element method. International Journal of Rock Mechanics and Mining Sciences, 2003;40(6):795–816.

Min, K.-B and Jing, L., Stress-dependent mechanical properties and bounds of Poisson's ratio for fractured rock masses investigated by a DFN-DEM technique. International Journal of Rock Mechanics and Mining Sciences, 2004;41(3):431–432.

Min, K.-B., Stephansson, O. and Jing, L., Determining the equivalent permeability tensor for fractured rock masses using a stochastic REV approach: Method and application to the field data from Sellafield, UK. Hydrogeology Journal, 2004a;12(5):497–510.

Min, K.-B., Rutqvist, J., Tsang, C.-F. and Jing, L., Stress-dependent permeability of fractured rock masses: A numerical study. International Journal of Rock Mechanics and Mining Sciences, 2004b;41(7):1191–1210.

Mostyn, G. and Douglas, K., Strength of intact rock and rock masses. Proc. of GeoEng2000 (Int. Conf. on Geotechnical and Geological Engineering, November 19–24, 2000, Melbourne, Australia), Vol. 1, pp. 1389–1421. Technomic Publishing, Lancaster, 2000.

Myer, L. R., Cook, N. G.W., Goodman, R. E. and Tsang, C.-F. (eds.), Fractured and jointed rock masses (Proc. of Conf. on Fractured and Jointed Rock Masses, Lake Tahoe, USA, June 3–5, 1992). Balkema, Rotterdam, 1995.

Oda, M., Fabric tensor for discontinuous geological materials. Soils and Foundations, Japanese Society of Soil Mechanics and Foundation Engineering, 1982;22(4):96–108.

Oda, M., An equivalent continuum model for coupled stress and fluid flow analysis in fractured rock masses. Water Resources Research, 1986a;22(13):1845–1856.

Oda, M. A theory for coupled stress and fluid flow analysis in jointed rock masses. In: Gittus, J., Zarka, J. and Nemat-Nasser, S. (eds), Large deformation of solids – physical basis and mathematical modelling, pp. 349–373, 1986b.

Oda, M., An experimental study of the elasticity of mylonite rock with random cracks. International Journal of Rock Mechanics and Mining Sciences and Geomechanics Abstracts, 1988a;25(2):59–69.

Oda, M., A method for evaluating the representative elementary volume based on joint survey of rock masses. Canadian Geotechnical Journal, 1988b;25(3):440–447.

Oda, M., Suzuki, K. and Maeshibu, T. Elastic compliance for rock-like materials with random cracks. Journal of Soils and Foundations, Japanese Society for Soils and Foundations, 1984;24(3): 27–40.

Oda, M., Yamabe, T. and Kamemura, K., A crack tensor and its relation to wave velocity anisotropy in jointed rock masses. International Journal of Rock Mechanics and Mining Sciences and Geomechanics Abstracts, 1986;23(6):387–397.

Ohnishi, Y., Chan, T. and Jing, L., Constitutive models of rock joints. In: Stephansson, O., Jing, L. and Tsang, C.-F. (eds), Coupled thermo-hydro-mechanical processes of fractured media, Developments in Goetechnical Engineering, Vol. 79, pp. 57–92. Elsevier Science B. V., Amsterdam, 1996.

Owen, D. R. J. and Hinton, E., Finite elements in plasticity: Theory and applications. Pineridge Press, Swansea, UK, 1980.

Pariseau, W. G., Equivalent properties of a jointed Biot material. International Journal of Rock Mechanics and Mining Sciences and Geomechanics Abstracts, 1993;30(7):1151–1157.

Pariseau, W. G., Non-representative volume element modelling of equivalent jointed rock mass properties. In: Rossmanith, H.-P. (ed.), Proc. Int. Symp. on Mechanics of Jointed and Faulted Rock (MJFR-2), pp. 563–568. Balkema, Rotterdam, 1995.

Pariseau, W. G., An equivalent plasticity theory for jointed rock masses. International Journal of Rock Mechanics and Mining Sciences, 1999;36(7):907–918.

Parry, R. H. G., Shear strength of geomaterials – a brief historical perspective. Proc. of GeoEng2000 (Int. Conf. on Geotechnical and Geological Engineering, November 19–24, 2000, Melbourne, Australia, Vol. 1, pp. 1592–1617. Technomic Publishing, Lancaster, 2000.

Patton, F. D., Multiple modes of shear failure in rock. Proc. 1st Cong. ISRM, Lisbon, Vol. 1, pp. 509–513, 1966.

Plesha, M. E., Constitutive models for rock fractures with dilatancy and surface degradation. International Journal for Numerical and Analytical Methods in Geomechanics, 1987;11(4):345–362.

Qin, S., Jiao, J.-J., Wang, S. and Long, H., A non-linear catastrophe model of instability of planar-slip slope and chaotic dynamical mechanisms of its evolution process. International Journal of Solids and Structures, 2001;38(44–45):8093–8109.

Rice, J. R. and Ruina, A. L., Stability of steady frictional slipping. Journal of Applied Mechanics, 1983;50(2):343–349.

Rice, J. R., Lapusta, N. and Ranjith, K., Rate and state dependent friction and the stability of sliding between elastically deformable solids. Journal of the Mechanics and Physics of Solids. 2002;49(9):1865–1898.

Rossmanith, H.-P. (ed.), Mechanics of jointed and faulted rock (Proc. of the 1st Int. Conf. on the Mechanics of Jointed and Faulted Rock-MJFR-1 Vienna, April 18–20, 1990). Balkema, Rotterdam, 1990.

Rossmanith, H.-P. (ed.), Mechanics of jointed and faulted rock (Proc. of the 2nd Int. Conf. on the Mechanics of Jointed and Faulted Rock-MJFR-2, Vienna, April 10–14, 1995). Balkema, Rotterdam, 1995.

Rossmanith, H.-P. (ed.), Mechanics of jointed and faulted rock (Proc. of the 3rd Int. Conf. on the Mechanics of Jointed and Faulted Rock-MJFR-3, Vienna, April 6–9, 1998). Balkema, Rotterdam, 1998.

Ruina, A. L., Slip instability and state variable friction laws. Journal of Geophysical Research, 1983;88(B12):10359–10370.

Selvadurai, A. P. S. and Boulon, M. J. (eds). Mechanics of geomaterial interfaces. Elsevier, Amsterdam, 1995.

Shao, J. F., and Rudnicki, J. W. A microcrack-based continuous damage model for brittle geomaterials. Mechanics of Materials, 2000;32(10):609–619.

Sharp, J. C., Fluid flow through fissured media. Ph.D. Thesis, University of London, 1970.

Sheorey, P. R., Empirical rock failure criteria. Balkema, Rotterdam, 1997.

Singh, B., Continuum characterization of jointed rock masses, Part I – The constitutive equations. International Journal of Rock Mechanics and Mining Sciences and Geomechanics Abstracts, 1973(10):337–345.

Souley, M., Homand, F. and Amadei, B., An extension to the Saeb and Amadei constitutive model for rock joints to include cyclic loading paths. International Journal of Rock Mechanics and Mining Sciences and Geomechanics Abstracts, 1995;32(2):101–109.

Stephansson, O. (ed.), Fundamentals of rock joints (Proc. of Int. Symp. on Fundamentals of Rock Joints, Björkliden, Sweden, September 15–20, 1985). CENTIK Publishers, Luleå, Sweden, 1985.

Stephansson, O. and Jing, L., Mechanics of rock joints: Experimental aspects. In: Selvadurai, A.P.S. and Boulon, M. J. (eds), Mechanics of geomaterial interfaces, pp. 317–342. Elsevier, Amsterdam, 1995.

Stietel, A., Millard, A., Treille, E., Vuillod, E., Thoraval, A. and Ababou, R., Continuum representation of coupled hydromechanic processes of fractured media: Homogenization and parameter identification. In: Stephansson, O., Jing, L. and Tsang, C.-F. (eds), Coupled thermo-hydro-mechanical processes of fractured media – Mathematical and experimental studies: Recent developments of DECOVALEX Project for radioactive waste repositories, pp. 135–164. Elsevier, Amsterdam, 1996.

Taylor, L. M., Chen, E. P. and Kuszmaul, J. S., Microcrack-induced damage accumulation in brittle rock under dynamic loading. Journal of Computer Methods in Applied Mechanics and Engineering, 1986;55(3):301–329.

Valanis, K. C. (ed.), Constitutive equations in viscoplasticity: Phenomenological and physical aspects (Proc. of the Winter Annual Meeting of the ASME, NY, December 5–10, 1976). ASME, NY, 1976.

Whitehouse, D. J., Fractal or fiction. Wear, 2001;249(5–6):345–353.

Wittke, W., Rock mechanics – Theory and applications with case histories, Springer-Verlag, Berlin, 1990.

Yoshinaka, R. and Yamabe, T., Joint stiffness and the deformation behaviour of discontinuous rock. International Journal of Rock Mechanics and Mining Sciences and Geomechanics Abstracts, 1986;23(1):19–28.

4 FLUID FLOW AND COUPLED HYDRO-MECHANICAL BEHAVIOR OF ROCK FRACTURES

Mechanical processes, analyzed mainly through the evolution of stresses and deformations, are only one aspect of the physical behavior of rock masses, concerning basically the strength and stability of the fractures and fractured rock masses. In rock engineering, the fluid flow (groundwater, oil, gas) and thermal transport (for geothermal energy extraction, gas storage, nuclear waste repositories, etc.) in fractured rocks are equally important. The major concerns are the hydraulic and thermal conductivity, distributions of pressure, velocity and flow rate, temperature and thermal gradients. These mechanical, hydraulic and thermal processes occurring in fractured rocks are not independent but are affecting and being affected by each other. This interdependence is termed coupled THM processes in the fractured rocks.

The coupling mechanisms for the hydro-mechanical processes are the interdependencies between the fracture aperture, rock porosity/permeability, fluid pressure and rock stress (Fig. 4.1). The coupling link for thermo-mechanical processes is the volume expansion and thermal stress increment of the rock matrix due to the thermal gradient (and thus indirectly the fracture aperture change), and the conversion of dissipated mechanical work into thermal energy which is usually negligible in rock engineering practice but may be significant in geophysical and tectonic problems, such as the frictional heating induced by fault movements. The coupling links for thermo-hydraulic processes are more complex, involving variations in volume (density), viscosity and phase changes (evaporation, condensation, etc.) of the fluid phases by the thermal gradient, and conductive–convective thermal transfer through fractured rocks by motion of fluids and gases. In this context, the discipline of rock mechanics, therefore, is much broader than its traditional field concerning the stress, deformation, strength and stability of rocks and rock engineering facilities. It must now be extended to concern the more general physical aspects of the fractured rock masses driven by mechanical, hydraulic and thermal loading mechanisms.

The coupling mechanisms in fractured rocks are actually much wider than just THM, with typical examples of effects from electromagnetic, geochemical or even biochemical processes on the THM processes. Although these interactions are generally excluded in rock mechanics analyses, they are often of concern in rock engineering practice. With the ever stronger demands for environment protection and sustainable development by modern society, the geochemical aspects of fractured rocks become more and more important in design, construction, operation and performance/safety assessments of rock engineering projects, especially concerning their impact on the environment. A typical example is the closer links between the THM and geochemical processes, especially the coupling with reactive or non-reactive transport of contaminants or other harmful elements through groundwater flow in fractured rocks, which is one of the central issues for geological disposal of radioactive wastes and geothermal reservoir engineering.

Fluid flow phenomena in fractured rock masses have been studied extensively since the 1960s, but many uncertainties still remain. The basic knowledge regarding flow in fractured rocks comes from laboratory experiments on fluid flow, or coupled stress-flow of rock fractures under various normal stresses with or without shear, and determination of the permeability of fractured rock masses through

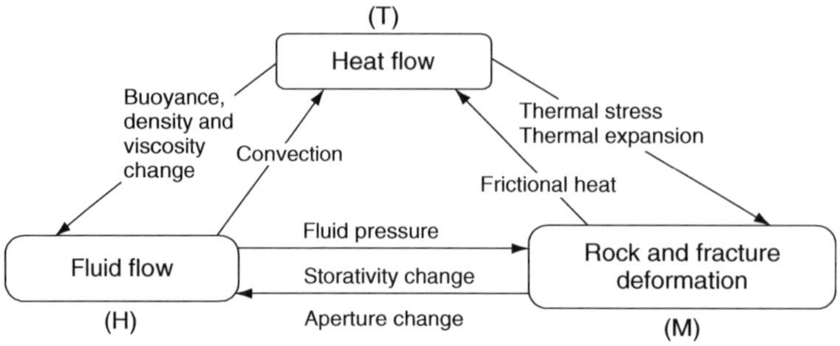

Fig. 4.1 Coupled THM processes of rock fractures.

pumping and tracer tests at field scales. Some relatively recent experiments can be seen in Esaki et al. (1999), Olsson and Barton (2001) and Jiang et al. (2004). The basic modeling approach is the solution of the continuity equation and equations of motion with the rocks represented as either discrete fracture systems or deformable porous media.

There are basically two types of fluid flow through fractured rock masses: flow through connected fracture networks and seepage through the solid rock matrix (blocks), which is usually assumed as either impermeable (such as hard rocks of low permeability, such as granites) or porous media (such as sandstones). For many practical problems, especially in hard rocks, flow through fracture networks often dominates the overall flow patterns.

The fluid conductivity of a fracture is mainly determined by its effective aperture – and that can change during the deformation process. The permeability of the rock matrix, on the contrary, depends on the porosity of the rock matrix, which also changes under stress. The overall permeability of a fractured rock mass, therefore, depends on the topology (geometry and connectivity) of the fracture networks, the deformability and aperture of the fractures, and the deformability and porosity of the rock matrix.

In this book we limit ourselves to the hydro-mechanical coupling because it is the most important link for the majority of important rock engineering problems. Thermal flow in rocks is dominated by conduction, except in geothermal reservoir engineering, and heat transfer can often be treated as a one-way coupling concerning only thermal stresses and volume expansion for the mechanical processes, and fluid viscosity/buoyancy change for the fluid flow portion. The solution of the heat transfer equation can often be treated independently and therefore will not be treated in detail in this book.

In practice, the problems of fluid flow through fractured porous media are often treated by using continuum mechanics models with dual porosity, dual permeability or multiple-continua approaches, using mainly FEM or FDM approaches. The problems of fluid flow through fractured hard rocks are most often treated by discrete approaches such as the DFN or combinations of the DEM and DFN methods. This book concerns the latter with Newtonian fluids. However, in order to be more complete in terms of the basic principles and for those who require more in-depth understanding of the physics of fluid flow in fractured media, we include a brief outline on the guiding principles of Newtonian fluid flow through porous media.

4.1 Governing Equations for Fluid Flow in Porous Continua

The governing equations for fluid flow in a porous continuum are the continuity equation, equations of motion and energy equation, based on the conservation laws of mass, momentum and energy. Thermal transport by fluid flow is not considered in this book. In this chapter, only the continuity equation (derived using the mass conservation law) and equations of motion (derived using the momentum conservation law) are presented.

A basic experimental law for viscous fluid flow in porous media is Darcy's law, written in terms of fluid flux components as

$$v_x = -k_x \frac{\partial h}{\partial x}, \quad v_y = -k_y \frac{\partial h}{\partial y}, \quad v_z = -k_z \frac{\partial h}{\partial z} \tag{4.1}$$

where v_i and k_i ($i = x, y, z$) are the components of flow velocity (volume discharge of fluid per unit time across a unit area of cross section) and permeability of the porous medium in direction i. Symbol h is called the piezometric head, defined as

$$h = Z + \frac{p}{\rho_f g} \tag{4.2}$$

where ρ_f is the density of the fluid and g the gravitational acceleration, Z a vertical elevation relative to an arbitrary datum and p the fluid pressure, sometimes also called pressure head. Both the total head and pressure are functions of position (co-ordinates x, y, z). The negative sign in Eqn (4.1) indicates flow direction toward head loss. Darcy's law through Eqn (4.1) summarizes much of the basic physics of groundwater flow in porous rocks. It can be written in a more general form as

$$\begin{Bmatrix} v_x \\ v_y \\ v_z \end{Bmatrix} = - \begin{bmatrix} k_{xx} & k_{xy} & k_{xz} \\ k_{yx} & k_{yy} & k_{yz} \\ k_{zx} & k_{zy} & k_{zz} \end{bmatrix} \begin{Bmatrix} \partial h/\partial x \\ \partial h/\partial y \\ \partial h/\partial z \end{Bmatrix} \tag{4.3}$$

for general anisotropic media or in vector form

$$\mathbf{v} = -\mathbf{K}\mathrm{grad}(h) \tag{4.4}$$

where $\mathbf{K}$ is a second-rank tensor called the permeability tensor of the medium.

4.1.1 Continuity Equation for Fluid Flow in Porous Media

From the law of mass conservation, for a unit volume in space (Fig. 4.2), the net outflow volume rate must be equal to the net inflow volume rate, plus the storage release rate of the medium per unit time and the volume rate of fluid contributed by sources or extracted by sinks within the domain of interest. We denote the fluid velocity components in a Cartesian co-ordinate system as v_x, v_y, v_z, and the side lengths of the unit element are Δx, Δy, Δz. Considering the mass conservation of the unit element volume in Fig. 4.2, the mass of fluid within the unit element due to transient flow should be equal to the mass flowing into the element, minus the mass flowing out of the element, plus the contribution from any source or sink contained inside the element, together with the fluid released from the storage of the element due to head change.

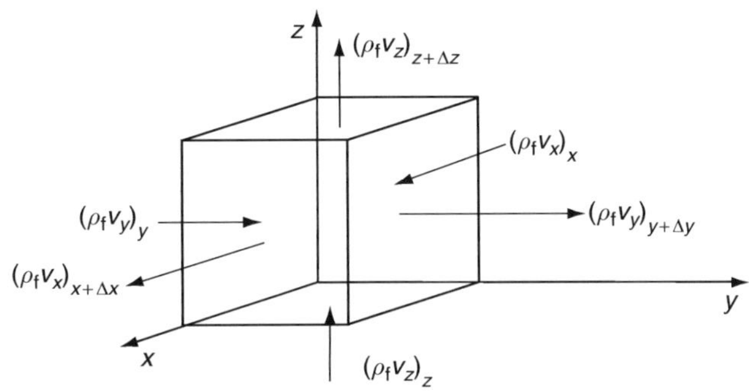

Fig. 4.2 A unit element for derivation of the continuity equation.

The accumulation of fluid per unit time is given by

$$\frac{\partial(\bar{\rho}_f)}{\partial t}\Delta x\Delta y\Delta z \tag{4.5a}$$

where $\bar{\rho}_f$ is the mean fluid density within the unit element, and the respective fluxes at the six faces of the element are

$$(\rho_f v_x)_x\Delta y\Delta z + (\rho_f v_y)_y\Delta x\Delta z + (\rho_f v_z)_z\Delta x\Delta y \tag{4.5b}$$

for in-flux and

$$(\rho_f v_x)_{x+\Delta x}\Delta y\Delta z + (\rho_f v_y)_{y+\Delta y}\Delta x\Delta z + (\rho_f v_z)_{z+\Delta z}\Delta x\Delta y \tag{4.5c}$$

for out-flux.

Denoting S as the storage coefficient of the rock, representing the mass of fluid released from the unit volume of rock matrix per unit decrease of head h, i.e.,

$$S = -\frac{\bar{\rho}_f\Delta(V_f)}{(\Delta x\Delta y\Delta z)\Delta h} \tag{4.6a}$$

where $\Delta(V_f)$ represents the volume of fluid released from the volume of the unit element. The presence of the negative sign indicates the opposite sign between $\Delta(V_f)$ and head change $\Delta h(\Delta(V_f) > 0$ when fluid is released from the matrix and $\Delta h > 0$; $\Delta(V_f) > 0$ when fluid is taken up into the matrix and $\Delta h > 0)$. The rate of fluid release is $\Delta(V_f)/\Delta t$ and, from Eqn (4.6a), can be written as

$$\frac{\Delta(V_f)}{\Delta t} = -S\left(\frac{\Delta h}{\Delta t}\right)(\Delta x\Delta y\Delta z) \tag{4.6b}$$

Assuming again that the source/sink term contribution per unit volume of rock matrix is $r(x, y, z, t)$, then the mass conservation law requires that

$$\begin{aligned}\frac{\partial(\bar{\rho}_f)}{\partial t}\Delta x\Delta y\Delta z = &\Big[(\rho_f v_x)_x\Delta y\Delta z + (\rho_f v_y)_y\Delta x\Delta z + (\rho_f v_z)_z\Delta x\Delta y\\ &- (\rho_f v_x)_{x+\Delta x}\Delta y\Delta z - (\rho_f v_y)_{y+\Delta y}\Delta x\Delta z - (\rho_f v_z)_{z+\Delta z}\Delta x\Delta y\Big]\\ &- S\left(\frac{\Delta h}{\Delta t}\right)(\Delta x\Delta y\Delta z) + r(x, y, z, t)\Delta x\Delta y\Delta z\end{aligned} \tag{4.7}$$

Dividing by the element volume $\Delta x\Delta y\Delta z$ and letting $\Delta x \to 0,\ \Delta y \to 0,\ \Delta z \to 0$ and $\Delta t \to 0$ leads to the continuity equation

$$\frac{\partial(\rho_f)}{\partial t} + \frac{\partial(\rho_f v_x)}{\partial x} + \frac{\partial(\rho_f v_y)}{\partial y} + \frac{\partial(\rho_f v_z)}{\partial z} = -S\frac{\partial h}{\partial t} + r(x, y, z, t) \tag{4.8}$$

For incompressible fluid flow, ρ_f = constant and $\partial(\rho_f)/\partial t = 0$, the continuity equation becomes

$$\frac{\partial(v_x)}{\partial x} + \frac{\partial(v_y)}{\partial y} + \frac{\partial(v_z)}{\partial z} = \frac{1}{\rho_f}\left[-S\frac{\partial h}{\partial t} + r(x, y, z, t)\right] \tag{4.9}$$

Substituting the Darcy's law into Eqn (4.8) leads to

$$\frac{\partial^2 h}{\partial x^2} + \frac{\partial^2 h}{\partial y^2} + \frac{\partial^2 h}{\partial z^2} = \frac{1}{\rho_f K}\left[S\frac{\partial h}{\partial t} - r(x, y, z, t)\right] \tag{4.10}$$

for a homogeneous medium of constant permeability **K** or

$$\frac{\partial}{\partial x}\left(K_x\frac{\partial h}{\partial x}\right)+\frac{\partial}{\partial y}\left(K_y\frac{\partial h}{\partial y}\right)+\frac{\partial}{\partial z}\left(K_z\frac{\partial h}{\partial z}\right)=\frac{1}{\rho_{\mathrm{f}}}\left[S\frac{\partial h}{\partial t}-r(x,y,z,t)\right] \tag{4.11}$$

for an orthotropic medium where the x, y, and z directions are the principal directions of the permeability tensor. For a general anisotropic medium, the general form of the continuity equation represented by Eqn (4.9) should be used.

For steady-state problems, $\partial h/\partial t=0$, the continuity equation is reduced to a Poisson equation for incompressible flow

$$\frac{\partial}{\partial x}\left(K_x\frac{\partial h}{\partial x}\right)+\frac{\partial}{\partial y}\left(K_y\frac{\partial h}{\partial y}\right)+\frac{\partial}{\partial z}\left(K_z\frac{\partial h}{\partial z}\right)=-\frac{r(x,y,z,t)}{\rho_{\mathrm{f}}} \tag{4.12}$$

and when the source term is also zero, the continuity equation becomes a Laplace equation

$$\frac{\partial}{\partial x}\left(K_x\frac{\partial h}{\partial x}\right)+\frac{\partial}{\partial y}\left(K_y\frac{\partial h}{\partial y}\right)+\frac{\partial}{\partial z}\left(K_z\frac{\partial h}{\partial z}\right)=0 \tag{4.13}$$

4.1.2 Equations of Motion for Fluid Flows

The equations of motion for a Newtonian fluid flow in a porous media are derived by using the momentum conservation law (Newton's second law of motion). Denote V as the volume occupied by a Newtonian fluid of body force **b**, and with its boundary surface S of unit outward normal vector **n**. On the surface S acts the stress σ_{ij} that yields a surface traction $T_i=\sigma_{ij}n_j$ by the Cauchy stress formula. Denoting also **v** as the velocity vector of the fluid, the momentum conservation law states that

$$\frac{\mathrm{D}}{\mathrm{D}t}\iiint_V \rho_{\mathrm{f}}v_i\,\mathrm{d}V=\iint_S \sigma_{ij}n_j\,\mathrm{d}S+\iiint_V \rho_{\mathrm{f}}b_i\,\mathrm{d}V\quad (i=x,y,z) \tag{4.14}$$

Since

$$\frac{\mathrm{D}}{\mathrm{D}t}\iiint_V \rho_{\mathrm{f}}v_i\,\mathrm{d}V=\iiint_V \rho_{\mathrm{f}}\frac{\mathrm{D}v_i}{\mathrm{D}t}\mathrm{d}V \tag{4.15a}$$

and from Gauss theorem

$$\iint_S \sigma_{ij}n_j\,\mathrm{d}S=\iiint_V \frac{\partial\sigma_{ij}}{\partial x_j}\mathrm{d}V \tag{4.15b}$$

Eqn (4.14) can be rewritten

$$\iiint_V\left[\rho_{\mathrm{f}}\frac{\mathrm{D}v_i}{\mathrm{D}t}-\rho_{\mathrm{f}}b_i-\frac{\partial\sigma_{ij}}{\partial x_j}\right]\mathrm{d}V=0 \tag{4.16}$$

or in its differential form

$$\rho_{\mathrm{f}}\frac{\mathrm{D}v_i}{\mathrm{D}t}=\rho_{\mathrm{f}}b_i+\frac{\partial\sigma_{ij}}{\partial x_j} \tag{4.17}$$

The deformation gradient tensor of a continuum is written as

$$e_{ij} = \frac{1}{2}\left(\frac{\partial v_i}{\partial x_j} + \frac{\partial v_j}{\partial x_i}\right) \quad (i,j,x_i = x,y,z) \tag{4.18}$$

For Newtonian fluids with dynamic viscosity μ, the constitutive relation between the stress and deformation rate tensors is written as

$$\sigma_{ij} = \left[-p + \left(\mu' - \frac{2}{3}\mu\right)\left(\frac{\partial v_x}{\partial x} + \frac{\partial v_y}{\partial y} + \frac{\partial v_z}{\partial z}\right)\right]\delta_{ij} + 2\mu e_{ij} \tag{4.19}$$

where μ' is called the secondary or expansion viscosity of the fluid, p the fluid pressure and δ_{ij} Kroneker delta. Substituting Eqns (4.19) and (4.18) into Eqn (4.17) leads to

$$\rho_f \frac{\mathrm{D}v_i}{\mathrm{D}t} = \rho_f b_i - \frac{\partial p}{\partial x_i} + \frac{\partial}{\partial x_i}\left[\left(\mu' - \frac{2}{3}\mu\right)\theta\right] + \frac{\partial}{\partial x_i}\left[\mu\left(\frac{\partial v_i}{\partial x_j} + \frac{\partial v_j}{\partial x_i}\right)\right] \tag{4.20a}$$

where

$$\theta = \left(\frac{\partial v_x}{\partial x} + \frac{\partial v_y}{\partial y} + \frac{\partial v_z}{\partial z}\right) \tag{4.20b}$$

representing the volume change rate. This equation is the equation of motion, or momentum equation, of Newtonian fluids in a porous continuum, often is called the Navier–Stokes (N–S) equation for Newtonian fluids.

With the Stokes condition, $\mu' = 0$, which indicates that the pressure p is defined as the average of normal stresses for a compressible fluid at rest, the N–S equation becomes

$$\frac{\mathrm{D}v_i}{\mathrm{D}t} = b_i - \frac{1}{\rho_f}\frac{\partial p}{\partial x_i} - \nu\frac{2}{3}\frac{\partial}{\partial x_i}\left(\frac{\partial v_i}{\partial x_i}\right) + \nu\frac{\partial}{\partial x_i}\left(\frac{\partial v_i}{\partial x_j} + \frac{\partial v_j}{\partial x_i}\right) \tag{4.21a}$$

where $\nu = \mu/\rho_f$ is called the kinematic viscosity of the fluids. Recall Eqn (4.2) and dropping the arbitrary Z, the N–S equation for compressible Newtonian fluids of constant viscosity is often written in the following more familiar form

$$\begin{cases} \dfrac{\mathrm{D}v_x}{\mathrm{D}t} = b_x - g\dfrac{\partial h}{\partial x} + \nu\left(\dfrac{\partial^2 v_x}{\partial x^2} + \dfrac{\partial^2 v_x}{\partial y^2} + \dfrac{\partial^2 v_x}{\partial z^2}\right) \\ \dfrac{\mathrm{D}v_y}{\mathrm{D}t} = b_y - g\dfrac{\partial h}{\partial y} + \nu\left(\dfrac{\partial^2 v_y}{\partial x^2} + \dfrac{\partial^2 v_y}{\partial y^2} + \dfrac{\partial^2 v_y}{\partial z^2}\right) \\ \dfrac{\mathrm{D}v_z}{\mathrm{D}t} = b_z - g\dfrac{\partial h}{\partial z} + \nu\left(\dfrac{\partial^2 v_z}{\partial x^2} + \dfrac{\partial^2 v_z}{\partial y^2} + \dfrac{\partial^2 v_z}{\partial z^2}\right) \end{cases} \tag{4.21b}$$

The viscosity of a fluid is generally a function of temperature but, since the thermal process is not considered in this book, we assume here that it is a constant. The equations of state describing relations between pressure (stress), temperature and density are therefore also not required, and the continuity Equation (4.9) and the N–S (momentum) Equation (4.21) form a closed equation system for complete solution of problems of flow in porous media, with properly specified boundary and initial conditions.

4.2 Equation of Fluid Flow Through Smooth Fractures

4.2.1 Flow Equation for Smooth Parallel Fractures

The most commonly applied conceptual model for flow through a single fracture is derived from a much simplified N–S equation of viscous fluid flow through a pair of smooth parallel surfaces of narrow width (aperture), e, often called the 'parallel plate model' or the Cubic Law in rock mechanics literature. We adopted a fracture-oriented local co-ordinate system (x, y, z) such that the x–y plane is the middle plane of the fracture and axis z is normal to the fracture surface.

The N–S equation (4.21b) is obtained by considering the equilibrium of the forces (inertia forces, pressure forces, gravity and frictional forces) acting on an infinitesimal differential volume. We assume that the flow is non-turbulent and irrotational, therefore steady and laminar, between the two parallel and smooth plates, with the velocity component in the direction normal to the fracture, v_z, equal to zero. We assume further that the inertial effect and the second derivatives of velocity components are negligible (i.e., isotropy of flow in the plane of the fracture). Then the N–S equation (4.21b) becomes

$$\begin{cases} \nu \dfrac{\partial^2 v_x}{\partial z^2} = -b_x + g\dfrac{\partial h}{\partial x} \\ \nu \dfrac{\partial^2 v_y}{\partial z^2} = -b_y + g\dfrac{\partial h}{\partial y} \\ -b_z + g\dfrac{\partial h}{\partial z} = 0 \end{cases} \tag{4.22a}$$

Integrating the first two equations of the Eqn (4.22a) twice leads to

$$\begin{cases} v_x = \dfrac{g}{2\nu}\dfrac{\partial(h - b_x x)}{\partial x} z^2 + c_1 z + c_2 \\ v_y = \dfrac{g}{2\nu}\dfrac{\partial(h - b_y y)}{\partial y} z^2 + c_3 z + c_4 \end{cases} \tag{4.22b}$$

Using boundary conditions $v_x = v_y = 0|_{z=\pm e/2}$, the integration constants c_i (i = 1, 2, 3, 4) can be determined, which leads to a parabolic distributions of velocity components v_x and v_y (see Fig. 4.3),

$$\begin{cases} v_x = \dfrac{g}{2\nu}\dfrac{\partial(h - b_x x)}{\partial x}\left[z^2 - \left(\dfrac{e}{2}\right)^2\right] \\ v_y = \dfrac{g}{2\nu}\dfrac{\partial(h - b_y y)}{\partial y}\left[z^2 - \left(\dfrac{e}{2}\right)^2\right] \end{cases} \tag{4.23a}$$

Only when the xy plane is horizontal, does Eqn (4.23a) becomes the more familiar form

$$\begin{cases} v_x = \dfrac{g}{2\nu}\dfrac{\partial h}{\partial x}\left[z^2 - \left(\dfrac{e}{2}\right)^2\right] \\ v_y = \dfrac{g}{2\nu}\dfrac{\partial h}{\partial y}\left[z^2 - \left(\dfrac{e}{2}\right)^2\right] \end{cases} \tag{4.23b}$$

Equations (4.23a) and (4.23b) are called the Reynolds equation for laminar flow between two parallel plates separated by a distance e, representing the hydraulic aperture of the fracture. The mean velocity

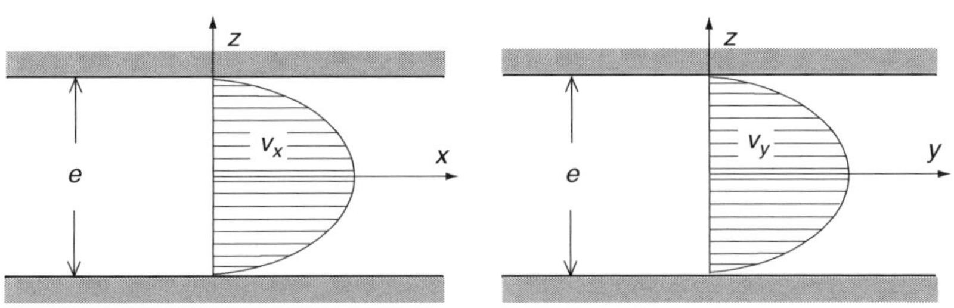

Fig. 4.3 Parabolic distribution of flow velocity in a single fracture.

components $\bar{v}_x$ and $\bar{v}_y$ of the flow between the two parallel plates are calculated by integrating Eqn (4.23a) in the z direction over the distance (aperture) e, obtained as

$$\begin{cases} \bar{v}_x = \dfrac{1}{e}\displaystyle\int_{-e/2}^{e/2} \mathrm{v}_x \, \mathrm{d}z = -\dfrac{ge^2}{12\nu}\dfrac{\partial(h - b_x x)}{\partial x} \\ \bar{v}_y = \dfrac{1}{e}\displaystyle\int_{-e/2}^{e/2} \mathrm{v}_y \, \mathrm{d}z = -\dfrac{ge^2}{12\nu}\dfrac{\partial(p - b_y y)}{\partial y} \end{cases} \tag{4.24}$$

The product of the mean velocity and the aperture is the flux component in the respective directions, i.e.,

$$\begin{cases} q_x = e\bar{v}_x = \left(-\dfrac{ge^3}{12\nu}\right)\dfrac{\partial(h - b_x x)}{\partial x} \\ q_y = e\bar{v}_y = \left(-\dfrac{ge^3}{12\nu}\right)\dfrac{\partial(h - b_y y)}{\partial y} \end{cases} \tag{4.25}$$

The term

$$T = \frac{ge^3}{12\nu} = \frac{\rho_f g e^3}{12\mu} \tag{4.26}$$

is called the transmissivity of the fracture. Dividing T by aperture e leads to another parameter called the hydraulic conductivity

$$k = \frac{ge^2}{12\nu} = \frac{\rho_f g e^2}{12\mu} \tag{4.27}$$

The above relations and parameters are derived by assuming that the flow properties in the fracture plane are isotropic; therefore T and k are constant and uniform. For anisotropic cases, the transmissivity may be different from x to y directions and the flow rate components can be written as

$$\begin{cases} q_x = -T_x \dfrac{\partial h}{\partial x} \\ q_y = -T_y \dfrac{\partial h}{\partial y} \end{cases} \tag{4.28}$$

where T_x and T_y must be determined by experiments. The continuity equation then becomes

$$\frac{\partial}{\partial x}\left(T_x \frac{\partial h}{\partial x}\right) + \frac{\partial}{\partial y}\left(T_y \frac{\partial h}{\partial y}\right) = \frac{1}{\rho_f}\left[S_f \frac{\partial h}{\partial t} - r(x, y, t)\right] \tag{4.29}$$

where S_f is called the storativity of the fracture, which may be estimated as

$$S_f = \rho_f g\left(\frac{1}{k_n} + eC_f\right) \tag{4.30}$$

where k_n is the normal stiffness of the fracture and C_f is the compressibility of the fluid (Doe and Osnes, 1985). Another definition of S_f is given by Lesnic et al. (1997) as

$$S_f = \rho_f g(\phi C_f) \tag{4.31}$$

for porous rock where ϕ is the porosity of the rock.

Substituting Eqn (4.28) into the continuity equation (4.29) then leads to

$$T_x \frac{\partial^2 h}{\partial x^2} + T_y \frac{\partial^2 h}{\partial y^2} = \frac{1}{\rho_f}\left[S_f \frac{\partial h}{\partial t} - r(x,y,t)\right] \tag{4.32}$$

for orthotropic hydraulic fractures with constant but different transmissivities in two orthogonal directions x and y.

When there is no source term and for steady-state problems, i.e., $(\partial h/\partial t = 0)$, $r(x,y,t) = 0$, the continuity equation becomes a Laplace equation

$$T_x \frac{\partial^2 h}{\partial x^2} + T_y \frac{\partial^2 h}{\partial y^2} = 0 \tag{4.33}$$

which combines both the N–S equation (for using the Cubic Law for the transmissivity) and the continuity equation, and therefore is closed for the flow problem.

4.2.2 Transmissivities of Smooth Non-Parallel Fractures

Fractures with parallel surfaces are only crude approximations. In reality, fracture surfaces most likely are not parallel, or initially parallel fracture surfaces may become wedge-shaped after a deformation process. Based on the same logic for deriving the Cubic Law of parallel fractures, Iwai (1976a,b) derived flow equations for non-parallel 'wedge-shaped' fractures (Fig. 4.4a), given by

$$\frac{q_i^{\mathrm{n}}}{q_i^{\mathrm{m}}} = 16\frac{r^2}{(1+r)^4} \tag{4.34}$$

with

$$r = \frac{e_a}{e_b} \quad \left(\text{or } r = \frac{e_b}{e_a}\right) \tag{4.35}$$

where e_a and e_b are the hydraulic apertures at the two ends of the wedge-shaped fracture of the flow rate given by q_i^{n}, and q_i^{m} is the flow rate with a mean hydraulic aperture

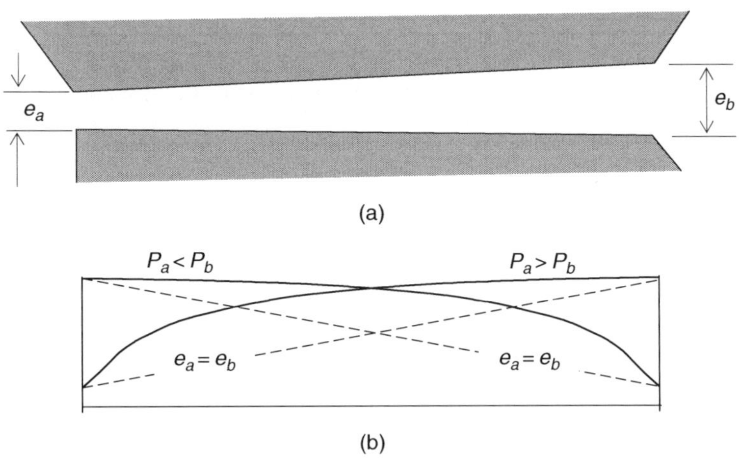

Fig. 4.4 Flow through wedge-shaped fractures. (a) A wedge-shaped fracture; (b) Pressure distribution (Iwai, 1976b).

$$e_{\mathrm{m}} = \frac{1}{2}(e_a + e_b) \tag{4.36}$$

according to the parallel plate model (Cubic Law). In other words, the hydraulic aperture of a wedge-shaped, non-parallel fracture can be taken as equivalent to a parallel fracture of an equivalent aperture e_{n} with a modification factor $F = (1 + r)^4/16r^2$ to the mean hydraulic aperture e_{m}

$$e_{\mathrm{n}} = \frac{e_{\mathrm{m}}}{F} = e_{\mathrm{m}}\left[\frac{16r^2}{(1 + r)^4}\right]^{1/3} \tag{4.37}$$

The transmissivity of the non-parallel fracture can be written as

$$T_{\mathrm{n}} = \frac{g(e_{\mathrm{n}})^3}{12\nu} = \frac{\rho_{\mathrm{f}} g(e_{\mathrm{n}})^3}{12\mu} \tag{4.38}$$

Figure 4.4b shows graphically the variation of pressure and the ratio of flow rates as a function of the aperture ratio r. The broken lines are pressures with a constant aperture.

4.3 Empirical Models for Fluid Flow Through Rough Fractures

4.3.1 Flow Models Based on the Validity of the Cubic Law

The Cubic Law was derived for solid interfaces of smooth walls, established basically from theoretical analysis. It was found to be inappropriate for some laboratory experimental results for rough rock fractures. A number of attempts were made to modify it so that the surface roughness could be considered – while maintaining the fundamental concept that the conductivity is related to the square of the aperture. The empirical relations thus proposed can be said to be a modified Cubic Law.

4.3.1.1 Lomitze, Louis and de Quadros models

Lomitze (1951) used macroscopically smooth glass plates to study the effect of microscopic asperities on the fluid conductivity and found that

$$k = \frac{ge^2}{12\nu}\left[1 + 17\left(\frac{R_t}{2e}\right)^{1.5}\right] \tag{4.39}$$

where $R_t = z_{\mathrm{max}} - z_{\mathrm{min}}$ is a measure of plate roughness with z_{max} and z_{min} the maximum and minimum heights of the asperities measured from a central-line-average datum. The model is a modification of the Cubic Law with an extra roughness coefficient. A similar approach was also taken by Louis (1969) with expression

$$k = \frac{ge^2}{12\nu}\left[1 + 8.8\left(\frac{R_t}{2e}\right)^{1.5}\right] \tag{4.40}$$

and in Barton and de Quadros (1997) with expression

$$k = \frac{ge^2}{12\nu}\left[1 + 20.5\left(\frac{R_t}{2e}\right)^{1.5}\right] \tag{4.41}$$

The difference in the constants 17, 8.8 and 20.5 reflects the different samples of fractures used for deriving these empirical equations: smooth glass surfaces used by Lomitze; and different rock fractures by Louis and de Quadros.

4.3.1.2 Barton's Aperture model using JRC

Barton et al. (1985) proposed that a rock fracture has a mechanical aperture E and a hydraulic aperture e related through an empirical relation

$$e = \mathrm{JRC}^{2.5}(E/e)^{-2} \tag{4.42}$$

in the unit of μm, and the hydraulic conductivity is then given by the Cubic Law with the hydraulic aperture e. Based on the values of JRC and JCS recorded by Bandis (1980), Barton and Bakhtar (1983) suggested that the initial mechanical aperture E_0 was given by another empirical relation

$$E_0 = \frac{\mathrm{JRC}_0}{5}\left(0.2\frac{\sigma_c}{\mathrm{JCS}} - 0.1\right) \tag{4.43}$$

or alternatively, the initial *in situ* hydraulic aperture can be back-calculated from borehole hydraulic test data. The current mechanical aperture is

$$E = E_0 - u_n \tag{4.44}$$

where u_n is the accumulated normal closure of the rock fracture, JRC_0 is the initial value of JRC and σ_c is the uniaxial compressive strength of the rock material. This empirical model was also used in Barton and de Quadros (1997). For a more comprehensive review of the approach, see Bandis (1993).

The unit dimensions of the two sides of Eqns (4.42) and (4.43) do not match (the right sides are dimensionless but the left sides are with unit μm and may present physical incompatibilities). The constants (values 2.5, 5, 0.2 and 0.1) in these equations come from experimental data, and caution is needed when these constants are directly used for other site-specific applications.

Very limited experimental data existed with which all the required parameters could be measured or estimated. Barton et al. (1985) and Barton and Bakhtar (1983) showed that the normal stress – flow and shear dilation – flow responses calculated by their empirical models were consistent with the laboratory experiments by Maini and Hocking (1977), with *in situ* flat-jack loaded block tests in welded tuff in the G-tunnel in the Nevada Test Site, USA, and with the heated block test conducted by Hardin et al. (1982).

Gale et al. (1993) found that the hydraulic apertures under normal stresses given by the model by Barton et al. (1985), as well as the derived fluid velocities, did not agree with their flow experiments. The reasons, he suggested, might be that either the friction factor and the roughness were not properly characterized by the measurement of the index parameters or they were not properly represented in the model.

4.3.1.3 Tsang and Witherspoon model

Tsang and Witherspoon (1981) developed a dual-concept model which encompassed a void deformation model for stress-normal closure behavior and a distributed aperture (distributed asperity height) model for the stress-flow relation. In this model, a rock fracture is represented by a collection of flat voids between contacting asperities, and the normal closure under stresses is due to deformation of the voids. Following an earlier formulation of Walsh (1965), an expression was obtained for the ratio between the effective modulus of the rock with voids and the intrinsic modulus of the intact rock in terms of an average length *1* of the void, d. The effective/intrinsic moduli ratio also varies in proportional to *1/d*. Thus, the experimental stress-displacement curve for the rock samples containing a fracture can be used to calculate the values of d and the number of contacts, N_c, and hence also the asperity height distribution $n(h)$ by numerical differentiation. Once these roughness characteristics are determined, the statistical average aperture $\langle e \rangle$ can be calculated, as a function of normal deformation, u_n, and effective normal stress, σ_n', without profiling the surface and without fitting parameters to stress-flow data. The relation is given by

$$\langle e^3(u_n, \sigma_n') \rangle = \frac{\int_0^{e_0 - \Delta u_n} (e_0 - u_n - h)^3 n(h) \mathrm{d}h}{\int_0^{e_0} n(h) \mathrm{d}h} \tag{4.45}$$

where e_0 is the maximum aperture of the fracture at zero normal stress and can be obtained from an estimate of the fracture contact area on the fracture walls at a specified stress. The transmissivity of the fracture is given by the Cubic Law where the statistical average $\langle e^3 \rangle$ is used in place of the smoothed-wall aperture for the usual parallel plate model. The model showed good agreement with the experimental data of Iwai (1976a) for tensile fractures in granite and basalt with contact area ratios between 10 and 20%. Gale et al. (1993), however, found a poor fit of this model to their laboratory data for the URL samples. They concluded that, since this model treats both the loading and unloading parts of the stress-flow curve of a fracture, further efforts may be needed to investigate the proper approach to extracting the key model parameters from the experimental data.

4.3.2 Flow Models without Assuming the Validity of the Cubic Law

It is reasoned that the roughness of the rock fracture surfaces is the main cause for deviations from the Cubic Law, especially when the normal stresses are high and apertures are small. Different empirical models have been proposed over the years to capture the essence of the flow through rock fractures without obeying the Cubic Law, with different representations of roughness and with the presence of the normal stresses.

4.3.2.1 Model by Gangi (1978)

Gangi (1978) developed a model for the variation of the fluid conductivity of a single rock fracture using the Hertzian approach. The rough surface of a fracture was approximated by vertical rods of uniform cross-sectional areas placed on a mean planar datum surface, thus representing the distribution of the asperities, and can be generated with different statistical distribution functions. The conductivity is written as

$$k = k_0 \left[1 - \left(\frac{\sigma_n'}{E} \frac{A}{A_c} \right)^{1/N(x)} \right]^3 \tag{4.46}$$

where σ_n' is the effective normal stress, E the Young's modulus of the rock, A the nominal area of the fracture, A_c the contact area (area covered by asperities-rods) of the fracture, k_0 the initial conductivity (under zero normal stress) and $N(x)$ a cumulative probability distribution function of the (rod) asperity shortness (defined as the difference between the maximum asperity (rod) height and the height of the asperity (rod) in question), $(z_{max} - z)$, given by a power law,

$$N(x) = m \left(\frac{x}{z_{max} - z} \right)^{n-1} \quad (1 \leq n < \infty) \tag{4.47}$$

where n is an experimentally determined constant. The function $N(x)$ represents the number of rods with shortness less than or equal to x and the parameter m the total number of rod asperities.

Gangi found that his model provided good fitting to the laboratory test data with artificial fractures in sandstone by Nelson and Hardin (1977) and in carbonate rocks by Jones (1975). Other workers, such as

Gale et al. (1993), Barr and Stesky (1980), Tsang and Witherspoon (1981) and Elliot et al. (1985), were unable to achieve a reasonable fit between experimental data and Gangi's model.

4.3.2.2 Model by Walsh (1981)

Walsh and Grosenbaugh (1979) combined the normal stiffness of rock fractures with the tribological model of Greenwood and Williamson (1966) (G–W model), based on the assumption of a Gaussian height distribution of asperities, for the elastic deformation of contacting rough surfaces to derive a normal stress – closure relation for a rock fracture. Their model relates changes in both the aperture and contact area to changes in the normal stress. Assuming the tips of the asperities to be spheres with the same radius of curvature and assuming an exponential distribution function for asperity height, they showed that the normal stiffness of a rock fracture should vary linearly with the effective normal stress with the standard deviation of asperity height as the constant of proportionality, which is equivalent to the hyperbolic relations between the normal stress and closure found by Bandis (1980), Goodman (1976) and others. They found that *in situ* normal stiffness of rock fractures measured by Pratt et al. (1977) in a large-block experiment did vary linearly with applied normal stress, as predicted by their model.

Using the above closure model and taking the advantage of the analogy between heat flow in a sheet containing non-conducting cylindrical inclusions and fluid flow in a planar interface with contacting asperities to show how contact area affects the hydraulic conductivity of a rough fracture, Walsh (1981) deduced the following relation between the fracture hydraulic conductivity, k, and the effective normal stress

$$k = k_0\left[1 - 2\sqrt{2}\frac{R_q}{e_0}\ \ln\left(\frac{\sigma_n'}{\sigma_0'}\right)\right]^3 \tag{4.48}$$

where k_0 and e_0 are the hydraulic conductivity and aperture at some reference effective stress, σ_0', and R_q is the standard deviation of the asperity height distribution.

Walsh found good agreement between Eqn (4.48) and experimental results for artificial fractures by Jones (1975), Kranz et al. (1979) and Barr and Stesky (1980). Gale et al. (1993) fitted the Walsh model by Eqn (4.48) to their experimental results (normal stress versus normalized flow rate data) and found moderately strong to weak correlations between pairs of best-fitting parameters, but a lack of correlation between any of the model parameters and the flow rate, due perhaps to the use of the asperity height distribution measured by 2D profilometer rather than 3D scanner of rough surfaces. The results did not agree with the best-fit values of R_q and e_0. The model may also have difficulties when shear deformation must also be considered when the G–W model no longer applies.

4.3.2.3 Model by Gale

Gale (1975, 1982, 1987) introduced a power law relation between the fracture's hydraulic conductivity, k, and effective normal stress, σ_n', given by

$$k = s\left(\sigma_n'\right)^n \tag{4.49}$$

where s and n are experimental constants. This model can be used to interpret many stress-flow tests without considering possible effects of shear dilatancy.

4.3.2.4 Model by Swan (1983)

The model by Swan (1980, 1983) is basically the same as that of Walsh and Grosenbaugh (1979), but assumes an exponential distribution of asperity heights and peak-to-peak contacts for the asperities. The contact area and the normal stiffness are then functions of R_q/e_0 and are directly proportional to the normal stress. This was confirmed with results of topographic profiling measurements on several slate cleavage plane surfaces and normal loading tests for different normal stress magnitudes less than 30 MPa. Combining his roughness and closure models with the assumed validity of the Cubic Law, the hydraulic conductivity of a rough rock fracture was derived as follows

$$k = k_0 \left[\left(1 - \frac{a}{e_0}\right) - \left(\frac{R_q}{e_0}\right) \ln\left(\sigma_n'\right) \right]^2 \tag{4.50}$$

where k_0 is the hydraulic conductivity at zero stress or self-weight, a is a constant, and e_0, R_t and σ_n' were defined above.

Swan (1980, 1983) did not measure the contact areas or flow rates in his experiments but used a numerical model to simulate the closure of rough surfaces in contact. He found that the model results agreed qualitatively well with Iwai's (1976a) measurements of contact area as a function of normal loading. His model indicated that the contact area of the fracture increased linearly with normal stress, decreased with increasing initial aperture and was less than 5% of the initial value at a normal stress of 20 MPa. The small contact area implies peak-to-peak contacts at a few higher asperities only, which is probably more valid at lower stresses. Some of the profiling data by Gale et al. (1993) generally confirmed this, while a small percentage of their data indicated mated surfaces. On the contrary, experimental results obtained by Pyrak-Nolte et al. (1987) and Gentier (1989) suggested that contact area varies non-linearly with increasing stress.

Gale et al. (1993) compared their normal stress-normalized flow rate experimental results with the Swan model for different loading cycles and found that, as with the Gangi model, the best-fit model parameters varied from cycle to cycle and, furthermore, there was hardly any correlation between any of the best-fitting parameters and the normalized flow rate.

4.3.2.5 Model by Cook (1988)

Cook (1988) and his colleagues proposed an empirical model for fluid flow through a single, rough, natural rock fracture, based on the observed discrepancy between the results from laboratory experiments and predictions by using the Cubic Law. Let the initial mean aperture of the fracture be e_0 and the mean closure be d, under a normal stress, the increment of the fracture's mean effective aperture can be written as $\Delta(e_0 - d) = \Delta e_c$ and is distributed over the whole nominal area, A, of the fracture. This increment of mean aperture must be accompanied by a change in the mean thickness of the void space adjacent to the contact area, Δe_a, between two rough surfaces of the fracture. Δe_a is, however, distributed over an area $(A - a)$ where a is the actual contact area. Under a uniform displacement in the direction normal to the mean plane of the fracture, the following relation must hold

$$\Delta e_a = \frac{\Delta e_c}{1 - a/A} \tag{4.51}$$

Equation (4.51) is the relation between the mechanical closure and the hydraulic aperture of a fracture under normal stress. Assuming that the fraction of the fracture area in contact at any stress can also be approximately given by the ratio of the mechanical closure under that stress over the mean initial aperture under zero stress, i.e., d/e_0, then

$$\bar{a} = \frac{a}{A} = \frac{d}{e_0} = 1 - \frac{e_c}{e_0} \tag{4.52}$$

Substituting Eqn (4.52) into Eqn (4.51) leads to

$$\Delta e_a = e_0 \frac{\Delta e_c}{e_c} \tag{4.53a}$$

or in a differential form

$$de_a = e_0 \frac{de_c}{e_c} \tag{4.53b}$$

where

$$e_a = e_0 \left[1 + \ln\left(\frac{e_c}{e_0} \right) \right] \tag{4.54}$$

since $e_a = e_c = e_0$ at zero stress.

Defining the out-of-plane tortuosity, ς, at any stress, as the ratio of the mean fracture aperture at zero stress to the mean fracture aperture at the current stress, i.e., $\varsigma = e_0/e_c$, the empirical flow law is then given in the form

$$\bar{q}_x = -L_y \frac{g}{12\nu} \{ e_0 [1 - \ln(\varsigma)\varsigma] \}^3 \left(\frac{\varsigma}{2\varsigma - 1} \right) + q_r \tag{4.55a}$$

or

$$\ln\left[\frac{-12(\bar{q}_x - q_r)}{L_y (e_0)^3 dh/dx} \right] = \ln\left\{ [1 - \ln(\varsigma)\varsigma]^3 \left(\frac{\varsigma}{2\varsigma - 1} \right) \right\} \tag{4.55b}$$

which is easier to plot in a logarithm axis system and where $\bar{q}_x$ is the mean flow rate in the x-direction, q_r the independent residual flow rate, L_y the dimension of fractures perpendicular to the flow direction (x-direction), i.e., the width of the flow, and h the hydraulic head. The slope of the line relating the logarithms of the specific flow $q/(\Delta h) = \bar{q}_x/(dh/dx)$ and the aperture by this relation is larger than six, not three as given by the Cubic Law. This empirical flow law fits well with many features of the experimental data, including both the higher exponents at low stresses, the aperture and stress-independent flow rate, but also requires more material parameters to be determined by laboratory tests or extrapolations.

The fluid flow through rock fractures is a complex phenomenon because of roughness of the fracture surfaces. The uneven distribution of the contact areas on the fracture surface creates an open space between the two surfaces of fractures with a complex pattern of pathways (Fig. 4.5a). The flow is therefore not

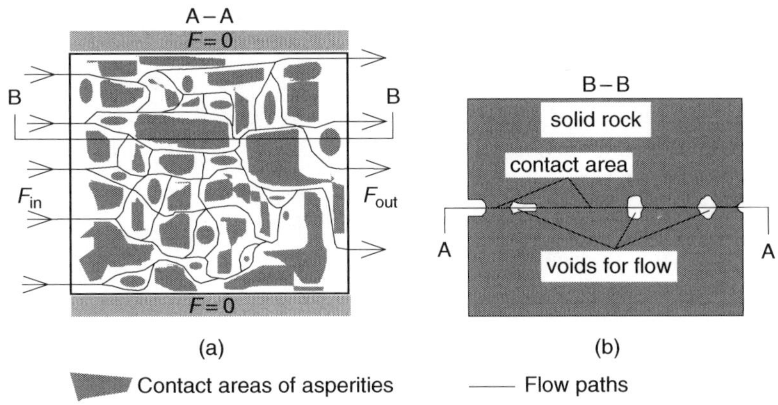

Fig. 4.5 Channel-like flow pathways in rough rock fractures, (a) plan and (b) section.

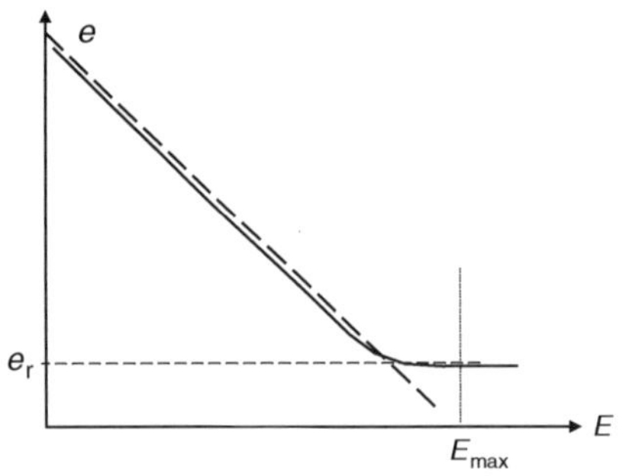

Fig. 4.6 A linear relationship between mechanical aperture, E, and hydraulic aperture, e, of rock fractures proposed by Zhao and Brown (1992), with a residual hydraulic aperture.

uniformly spread out in the fracture plane as assumed by the Cubic Law but rather through channel-like connected void spaces between contacting asperities, so-called 'channel flow' (Tsang and Tsang, 1987). This has been demonstrated in experimental work in Hakami (1988, 1995). Parts of these open spaces will never be closed even under extremely high normal stresses, therefore leading to a residual hydraulic aperture. This is the reason for the small but undiminished residual flow observed under high normal stresses as discussed above. The hydraulic aperture is, therefore, different from the mechanical aperture for the same fracture. The mechanical aperture can be estimated by normal compression tests or using Bandis' or Goodman's normal stress-normal closure models (cf. Eqns (3.4) and (3.5)). The hydraulic aperture, however, is more difficult to determine experimentally without using any flow laws. Figure 4.6 shows the experimental relation between the mechanical and hydraulic apertures of rock fractures by Zhao and Brown (1992). Similar results were also reported by Niemi et al. (1997). It should be pointed out that because of the observed residual flow under very high normal stresses in experiments, as presented above, the curve of hydraulic-mechanic apertures in Fig. 4.6 should level out at large mechanical closures, indicating the residual hydraulic aperture e_r.

4.4 Flow Equations of Connected Fracture Systems

The flow analysis of a fracture network is based on the elements of fracture segments, intersections and cycles. Intersections are the locations where two or more fractures meet and are the most important geometrical properties of a network for conducting flow. The part of a fracture between two adjacent intersections is called a segment and the set of fracture segments which form a complete block is called a fracture cycle. These elements are shown in Fig. 4.7 for an idealized 2D fracture network (after removing all dead-end segments and isolated fractures). Similar definitions can be extended to 3D networks.

Assume that there are n_i fracture segments connected at intersection i, where there exists also an external resultant recharge (or discharge) rate q_i^s. From the principle of fluid mass conservation, the sum of total inflow rate plus outflow rate should equal to the recharge (positive) or discharge (negative) rate, i.e.,

$$\sum_{j=1}^{n_i} \frac{\rho_f g (e_{ij})^3}{12\mu} \frac{h_i - h_j}{L_{ij}} = q_i^s \quad \text{or} \quad \sum_{j=1}^{n_i} (e_{ij})^3 \frac{h_i - h_j}{L_{ij}} = \frac{12\mu_f}{g\rho_f} q_i^s \tag{4.56}$$

where h_i and h_j are the piezometric heads at intersection i and j ($j = 1, 2, \ldots, n_i$), ρ_f is the mass density of fluid (= 1.0 for water), e_{ij} and L_{ij} are the equivalent hydraulic aperture and length of the fracture segment between intersection i and j ($j = 1, 2. \ldots, n_i$), μ is the dynamic viscosity of the fluid and g the gravitational acceleration. Note that, for wedge-shaped fracture segments, the equivalent aperture according to Eqn (4.37) should be

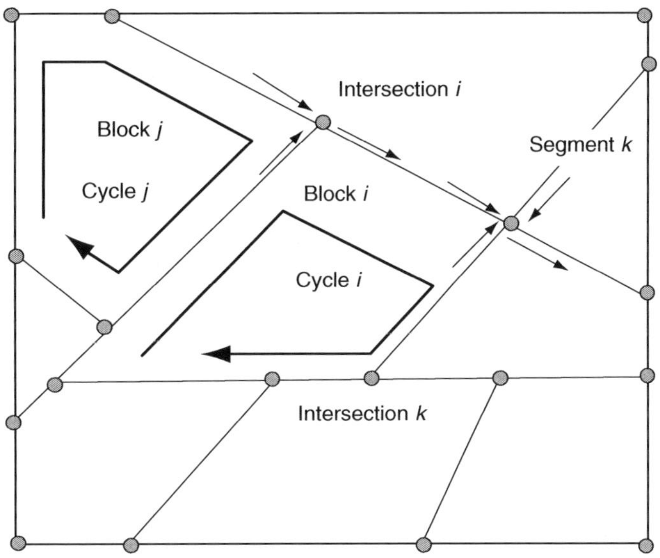

Fig. 4.7 An idealized fracture network in 2D for flow analysis.

used. From the collection of all similar equations at all intersections (including the ones at boundaries with known values of piezometric heads), one can obtain a simultaneous set of algebraic equations

$$[T_{ij}]\{h_j\} = \{\hat{q}_j\} \quad \text{or} \quad \mathbf{TH} = \mathbf{Q} \tag{4.57}$$

after moving the terms with known piezometric heads into the RHS of the equation. The matrix $[T_{ij}]$ is called the global transmissivity matrix of the fracture system, with

$$T_{ii} = \sum_{j=1}^{n_i} \frac{(e_{ij})^3}{L_{ij}} \tag{4.58}$$

$$\begin{cases} T_{ij} = -\dfrac{(e_{ij})^3}{L_{ij}}, \; i \neq j, \text{ but } i \text{ and } j \text{ are adjacent intersections} \\ T_{ij} = 0, \text{ and } i \text{ and } j \text{ are not adjacent intersections} \end{cases} \tag{4.59}$$

The RHS vector $\{\hat{q}_j\}$ is given by

$$\hat{q}_j = \frac{12\mu}{\rho_f g} q_j^s + \sum_{k=1}^{n_i} \frac{(e_{ik})^3}{L_{ik}} \hat{h}_k \tag{4.60}$$

where $\hat{h}_k$ $(k = 1, 2, \ldots, n_i)$ is the known piezometric head at intersections k adjacent to intersection i. Solution of Eqn (4.57) will lead to values of piezometric heads at all intersections, and the rest of the unknowns (pressure and flow velocity, etc.) can be obtained readily with the head information. For unconfined flow field, an iterative procedure with initial head (or pressure) should be used to determine the final geometric locations of the free piezometric surface (water table) as a set of additional discrete intersection points on fracture segments where the pressure is equal to the atmospheric pressure. The criterion used to check the iteration process is that, after each iteration, the vertical co-ordinate values of these additional intersection points determining the water table should be compared with the currently available piezometric heads at the corresponding points. If the value of the vertical ordinate is greater than the current piezometric head, the point is above the water table (phreatic surface), and should be changed to the current value. If two intersections of a fracture segment are both located above the phreatic surface, the transmissivity of this segment should be set to zero (or in practice a very small value) to ensure that no fluid flow will enter that segment at the next iteration. Equation (4.57) will then be reformed by modifying both the transmissivity matrix **T** and RHS vector **Q** and resolved to produce a new set of values of phreatic surface intersection co-ordinates. The process continues until the two sets of values reach a preset tolerance margin.

The solution technique presented above is shown in 2D, but the principle applies also for 3D problems with, however, much more complexity in handling fracture system geometry and connectivity.

4.5 Coupling of Fluid Flow and Deformation of Fractures

The technique for modeling the fluid flow and deformation of rock fractures described in this section assumes that rock blocks are impermeable, so that there is no fluid interaction between the fracture and its parent rock matrix. The fluid flow and block motion/fracture deformation are coupled through a two-way interaction:

(1) change of fluid pressures on the boundary surfaces of blocks affects the motion and deformation of blocks and, in turn, the deformation of fractures (therefore the aperture changes); and

(2) the change of hydraulic apertures of fractures affects its transmissivity, flow rate and fluid pressure distribution along the fracture surfaces, i.e., the boundary surfaces tractions of blocks, which, in turn, affects the block motion and deformation.

4.5.1 Coupling of Fluid Pressure and Block Motion/Deformation

The fluid pressures on the boundary surfaces of rock blocks are just the extra boundary tractions for the motion/deformation of the rock blocks. For the rigid blocks case, contributions from the fluid pressure are calculated as extra net distributional force and torque increments. Assuming that the representative block has N boundary edges (surfaces) and N intersections points (nodes) with piezometric heads h_i ($i = 1, 2, \ldots, N$). Denoting that the two intersections of the ith edge have the co-ordinates (x_1, y_1) and (x_2, y_2), with heads h_1 and h_2, respectively (Fig. 4.8a), the magnitude of the net force and the co-ordinates of its acting point e on the edge i are then given by

$$\Delta F = \frac{\rho_f g}{2}(h_1 + h_2)\sqrt{(x_2 - x_1)^2 + (y_2 - y_1)^2} \tag{4.61a}$$

$$\begin{cases} x_e = \dfrac{(h_2 - h_1)(x_2 - x_1)}{6(h_1 + h_2)} + \dfrac{(x_1 + x_2)}{2} \\ y_e = \dfrac{(h_2 - h_1)(y_2 - y_1)}{6(h_1 + h_2)} + \dfrac{(y_1 + y_2)}{2} \end{cases} \tag{4.61b}$$

The direction cosines (l_i, m_i) of the edge i can be calculated by

$$l_i = \frac{x_2 - x_1}{\sqrt{(x_2 - x_1)^2 + (y_2 - y_1)^2}}, \quad \mathrm{m}_i = \frac{y_2 - y_1}{\sqrt{(x_2 - x_1)^2 + (y_2 - y_1)^2}} \tag{4.62}$$

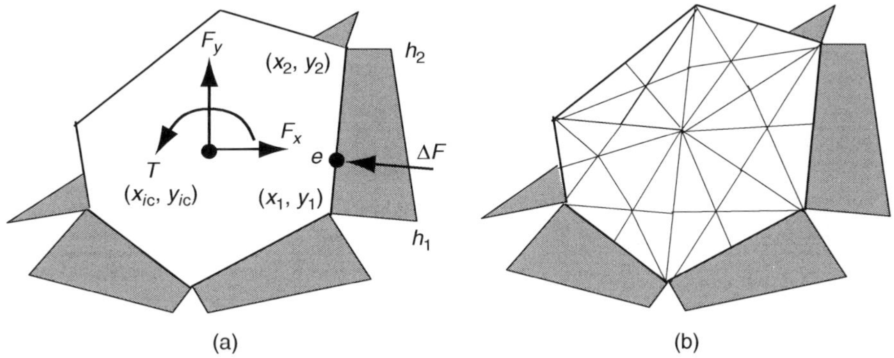

Fig. 4.8 Hydraulic pressure on the block boundaries – 2D case: (a) rigid block and (b) deformable block with FEM mesh (Jing et al., 2001).

and the direction cosines (l_F, m_F) of the resultant net force ΔF are given by the solution of equations

$$\begin{cases} l_F l_i + m_F m_i = 0 \\ (l_F)^2 + (m_F)^2 = 1 \end{cases} \tag{4.63}$$

since ΔF is perpendicular to the edge i. Care should be taken to ensure that the direction of ΔF points toward the edge. The resultant net increments of force components and torque of the block due to fluid pressure are then calculated by

$$F_x = \sum_{i=1}^{N} \Delta F_x = \sum_{i=1}^{N} (l_F \Delta F)_i \tag{4.64a}$$

$$F_y = \sum_{i=1}^{N} \Delta F_y = \sum_{i=1}^{N} (m_F \Delta F)_i \tag{4.64b}$$

$$T = \sum_{i=1}^{N} [\bar{x}\Delta F_y - \bar{y}\Delta F_x] = \sum_{i=1}^{N} [(x_e - x_{ic})(m_F \Delta F) - (y_e - y_{ic})(l_F \Delta F)] \tag{4.64c}$$

where (x_{ic}, y_{ic}) are the co-ordinates of the mass center of the block, where the resultant force components and torque are applied for block motion calculations in DEM.

For wedge-shaped fractures, the pressure distribution is not linear (cf. Fig. 4.4b) and Eqn (4.64) cannot be applied directly. An approximation can be made by dividing each fracture segment into a small number of sub-segments and by then inserting a small number of auxiliary nodes (additional artificial intersection points); the pressure values at these additional points can be obtained analytically from Eqns (4.34–4.37), corresponding to different ratio of r $(0 < r < 1)$. Then linear approximations of pressure distribution can be obtained between two adjacent auxiliary nodes and Eqn (4.64) can be applied directly to solve for the resultant force components and torque.

For deformable blocks, the interiors of the blocks are discretized into a number of finite difference zones or finite elements (Fig. 4.8b), introducing additional auxiliary nodes between the two natural adjacent intersections, thus defining a number of sub-segments (depending, of course on the mesh density). The same techniques as for the wedge-shaped fractures apply for this case also, with the difference being that the equivalent nodal forces due to pressure should be solved using the element shape functions. Since pressure has the same unit as stress, the resulting additional pressure increments from fluid pressure become just an additional set of boundary tractions for the block. The principle can be easily extended to 3D.

4.5.2 Coupling of Fluid Pressure and Fracture Deformation

The effect of fluid pressure on the deformation and change of hydraulic aperture of fractures is represented by a similar 'effective stress' concept and calculated through the constitutive laws of the fractures or point contacts in DEM.

For rigid blocks, the effect of the fluid pressure is an additional increment of boundary traction on block surfaces, as mentioned in Section 4.5.1, resulting in the change of relative positions of the two parent blocks forming the fracture, therefore changing the fracture aperture. The normal force component at point contacts may be changed directly by fluid pressure.

For deformable blocks, the normal stress component σ_n is modified to an effective normal stress $\sigma'_n = \sigma_n - p$ by the pressure p, and the effective stress rather than the total stress is applied in the constitutive laws for fractures. The final aperture is also similarly determined by the motion and deformation of the parent blocks forming the fracture, the constitutive laws assigned to the fracture concerned and how the contacts are dealt with in the algorithms in a DEM programs.

The modified fracture apertures are then applied in the Cubic Law of the flow equations (or other empirical laws as presented in Section 4.4) to produce modified pressure distributions over the fractures. The whole coupling process is therefore interactive and must be solved through iterative processes to obtain final solutions. To illustrate the different techniques for considering the coupling between fracture stress/deformation and fluid pressure, some typical models are summarized below.

4.5.2.1 Model in Pine and Cundall (1985)

Pine and Cundall (1985) proposed a hydro-mechanical coupling model for rock fractures, applied for geothermal energy extraction from underground hot dry/wet rocks with injection. The assumption was that the fractures cut the rock masses into a system of rectangular blocks for a 2D problem, and each block was further divided into four finite difference elements. The hydraulic aperture was assumed to be the sum of four components

$$e_h = e_r + re_e + e_d + e_j \tag{4.65}$$

where e_r is the residual aperture (Fig. 4.9), e_e the contribution from elastic dilation of the rock matrix when a confining effective stress $\sigma'_n = 0$ is applied, e_d the additional aperture contributed by shearing dilation of the fracture (which is determined by the law of shear dilation in the constitutive models of the fracture) and e_j jacking dilation due to gross motion of the blocks forming the fractures.

The coefficient r considers the effect of the confining pressure via the following values

$$r = \begin{cases} 0, & (\sigma'_n > \sigma'_{n0}) \\ 1, & (\sigma'_n < 0) \\ 1 - \sigma'_n/\sigma'_{nr}, & \text{(otherwise)} \end{cases} \tag{4.66}$$

where σ'_{nr} is the effective normal stress, corresponding to the residual aperture e_r. Both σ'_n and σ'_{nr} are calculated according to the pressure and contact forces at intersections of fractures (block corners). The Cubic Law is then invoked to calculate the fluid flow. The interaction between the fluid pressure and normal stress is shown in Fig. 4.9. The initial pressure and the normal stress are set to be equal (step 1 in Fig. 4.9b) with flow rate q. At step 2, the fracture opens up by an increment of normal displacement (aperture) Δu_n, given by

$$\Delta u_n = \dot{u}_n \Delta t = \frac{q}{L}\Delta t \tag{4.67}$$

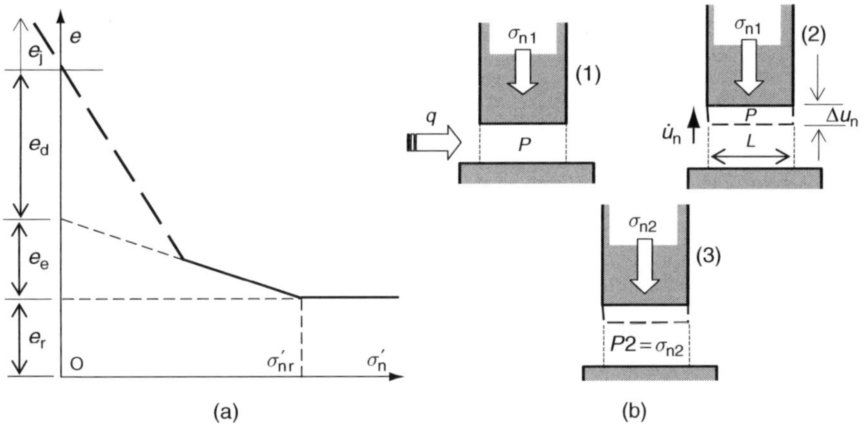

Fig. 4.9 Model of pressure-aperture-normal stress coupling by Pine and Cundall (1985). (a) Components of hydraulic aperture; (b) normal stress–fluid pressure interaction.

where L is the length of the fracture and Δt the time step. At step 3, the final pressure is set equal to the updated normal stress due to the normal deformation.

4.5.2.2 *Model in Kafritsas and Einstein (1987)*

Kafritsas and Einstein (1987) proposed another model to simulate the fracture stress-flow coupling, based on a conversion algorithm from block deformation to hydraulic aperture. For the two blocks in a vertex-to-edge contact (Fig. 4.10a), and assuming that the penetration (fracture closure) displacement is positive and separation (fracture opening) is negative, the mean fracture aperture e_m is given by

$$e_m = (e_1 + e_2)/2 \tag{4.68}$$

where, by convention, $e_1 > 0$ is the overlap (penetration) at contact 1 and $e_2 < 0$ the separation (opening) at contact 2 of the fracture of length L.

The conversion of overlap and separation into equivalent hydraulic aperture is given by

$$e = e_0 - f_\delta e_m = e_0 - \left(\frac{E/L}{K_N}\left(1 + \frac{E/k_n}{L}\right)^{-1}\right)e_m \tag{4.69}$$

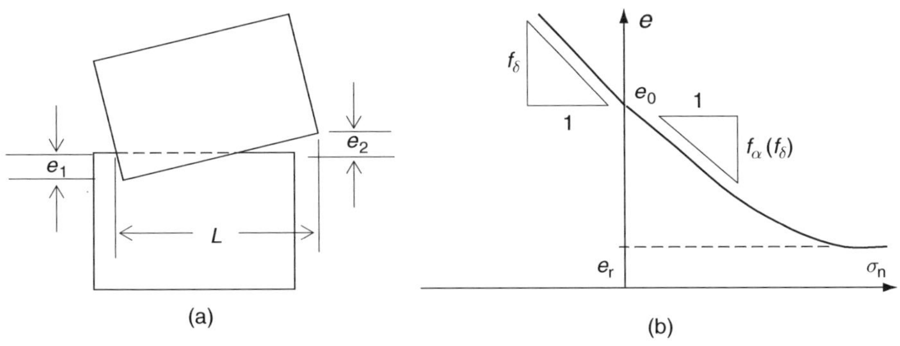

Fig. 4.10 Conversion of block overlap to fracture aperture for coupled flow-stress analysis of rock fractures by Kafritsas and Einstein (1987). (a) Definitions of overlap (penetration) and separation (opening) by block contact positions; (b) stress-aperture curve.

for $e_m < 0$, where K_N is the normal point contact stiffness, k_n the normal stiffness of the fracture (measured), E the Young's modulus of the rock and e_0 the initial aperture. The coefficient f_δ represents the composite contact normal stiffness of the fracture and rock matrix (Fig. 4.10b). For the case of $e_m > 0$, the hydraulic aperture is converted by

$$e = e_r + (e_0 - e_r)\exp\left(-\frac{f_\alpha f_\delta e_m}{e_0 - e_r}\right) \tag{4.70}$$

$$f_\alpha = \left(\frac{Lk_n}{E} + 1\right)^{-1} \tag{4.71}$$

and e_r is the residual aperture. The flow is then governed by the Cubic Law using the equivalent hydraulic aperture e converted by the above relations. Aperture variation due to shear dilation is not included in the relations but can be readily added.

4.5.2.3 *Model in Harper and Last (1989)*

Harper and Last (1989) proposed that the hydraulic aperture e of a fracture is the sum of three components (Fig. 4.11):

$$e = e_r + r\frac{\sigma_n - p}{k_n} + e_d \tag{4.72}$$

where e_r is the residual aperture when the magnitude of normal stress equals to the compressive strength of the rock matrix σ_c, k_n the normal stiffness of the fracture, σ_n the normal stress, p the fluid pressure and e_d the additional aperture component contributed by shearing dilation of the fracture, determined by the law of shear dilation. The coefficient $r = 1$ if $0 < |\sigma_n| < \sigma_c$ and $r = 0$ otherwise. The flow rate of the fracture is then calculated by using the Cubic Law. At each time step Δt, the pressure gradient across a fracture is given by

$$\Delta P = \left(\frac{Q}{V} + \dot{V}\right)C_s\Delta t \tag{4.73}$$

where Q is the net flow rate along the fracture (difference between the flow rates at two ends of the fracture along the flow direction), V the fracture volume (or area for 2D cases) and $\dot{V}$ the rate of fracture volume variation (a measure of volumetric strain rate of the fracture) and coefficient C_s the bulk stiffness of the fluid (or the reciprocal of fluid compressibility).

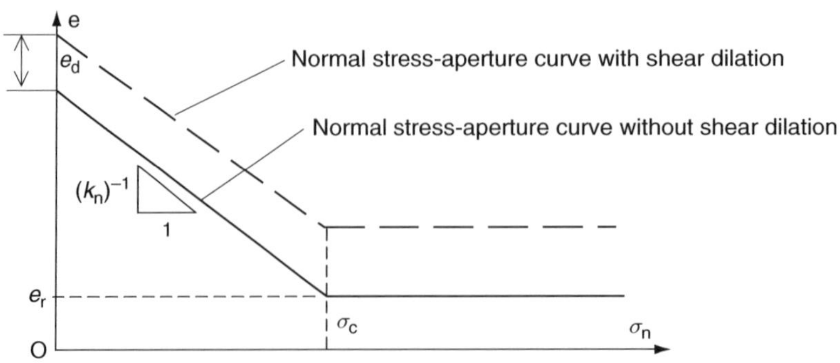

Fig. 4.11 Hydraulic aperture – normal stress model by Harper and Last (1989).

4.5.2.4 *Model in Wei (1992)*

Using also the concept of imaginary 'overlap' between two contacting blocks and the resulting hydraulic aperture, and based on models proposed by Pine and Cundall (1985), Kafritsas and Einstein (1987) and Harper and Last (1989), Wei (1992) proposed a simplified relation between the 'overlap' and hydraulic aperture. Adopting the same convention that 'overlapping' is positive and separation is negative, as used in Kafritsas and Einstein (1987), the hydraulic apertures at the two contacts of a fracture 1 and 2 (Fig. 4.12a) are given by

$$e_i = e_0 - d_i \quad (\text{for } d_i \leq 0, \text{separation}) \tag{4.74}$$

and

$$e_i = e_\mathrm{r} + (e_0 - e_\mathrm{r})\exp\left(-\frac{\alpha}{e_0 - e_\mathrm{r}} d_i\right) + e_{\mathrm{d},i} \quad (\text{for } d_i \geq 0, \text{penetration}) \tag{4.75}$$

where subscript $i = 1$ and 2, d_i denotes the normal contact deformation at the two ends of the fracture and coefficient α represents the stress-dependency of the fracture deformation, related to a critical normal stress σ_{nc}. Beyond this critical normal stress, the flow behavior becomes practically stress independent and reaches the residual behavior, as proposed by Pyrak-Nolte et al. (1987) based on experiments. $e_{\mathrm{d},i}$ denotes the hydraulic aperture increment at end i of the fracture by shear dilation. The other symbols represent the same physical terms as defined before in this section. Let d_c be the overlap or penetration

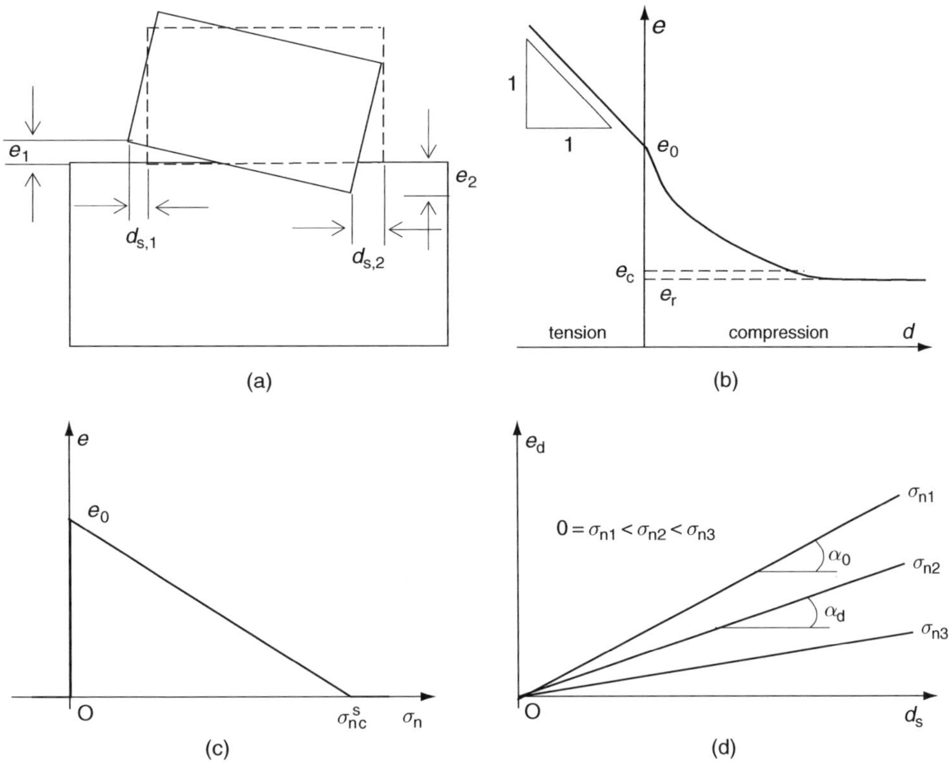

Fig. 4.12 Model of overlap-aperture conversion and shear dilation by Wei (1992). (a) Definition of contact deformation; (b) conversion between overlap and hydraulic aperture; (c) linear variation of dilation angle with normal stress and (d) hydraulic aperture variation with shear dilation under different normal stress levels.

corresponding to the critical normal stress σ_{nc} and a critical hydraulic aperture e_c which can be taken as a small fraction of maximum aperture $(e_0 - e_r)$, given by

$$e_c = r(e_0 - e_r) + e_r \tag{4.76}$$

where $r << 1.0$ is a very small percentage value (e.g., 0.01). Substitution of Eqn (4.76) into Eqn (4.75) (with $e_{d,i} = 0$ since no shearing is involved when calculating α) then leads to

$$\alpha = -\ln(r)\frac{e_0 - e_r}{d_c} = |\ln(r)|\frac{e_0 - e_r}{d_c} = |\ln(r)|\frac{(e_0 - e_r)k_n}{\sigma_{nc}} \tag{4.77}$$

assuming that the normal stiffness of the fracture k_n is a constant. Substitution of Eqn (4.77) into Eqn (4.75) then results in

$$e_i = e_r + (e_0 - e_r)\exp\left(-\frac{|\ln(r)|k_n}{\sigma_{nc}}d_i\right) + e_{d,i} \quad (\text{for } d_i \geq 0 \text{ penetration}, \ i = 1, 2) \tag{4.78}$$

For the threshold of $r = 0.01$ as used by Wei (1992), Eqn (4.78) becomes

$$e_i = e_r + (e_0 - e_r)\exp\left(-\frac{4.6052k_n}{\sigma_{nc}}d_i\right) + e_{d,i} \tag{4.79}$$

The shear-dilation-induced contribution to the hydraulic fracture, $e_{d,i}$, is different with different constitutive laws of the rock fracture. In the model proposed by Wei (1992), it was represented by a simple relation

$$e_{d,i} = d_{s,i}\tan(\alpha_d) = d_{s,i}\tan\left(\alpha_{d0}\frac{\sigma_{nc}^s - \sigma_n}{\sigma_{nc}^s}\right) \tag{4.80}$$

where $d_{s,i}$ is the shear displacement at end i (Fig. 4.12a), α_d the dilation angle of the initial value α_{d0} and σ_{nc}^s another critical normal stress when the dilation angle approaches zero. Relation (4.80) holds for $\sigma_n > 0$ where σ_n is the magnitude of compressive stress. For tensile stresses $\alpha_d = 0$. Figure 4.12b shows the normal stress-hydraulic aperture behavior for the model of Wei (1992). The linear relation between the dilation angle α_d and normal stress is shown in Fig. 4.12c, and Fig. 4.12d shows the contributions of shear dilation to the hydraulic aperture under different normal stress levels.

4.6 Remarks on Outstanding Issues

Some of the outstanding issues discussed in Section 3.6 concerning the mechanical behavior of rock fractures equally apply to their hydraulic behavior, such as the dominating effects of roughness, size and time effects and the influence of gouge materials, and will not be repeated here. Some representative ideas and results can be seen in Brown and Scholz (1985, 1986), Brown (1987, 1989) for roughness characterization, closure and transmissivity of rock fractures based on the GM-model. In-depth reviews are given in Cheng et al. (1993), Detournay and Cheng (1993), Rutqvist and Stephansson (2003) and Zimmerman and Main (2004). A few issues of importance are discussed here, focusing on flow and coupled stress-flow behavior.

4.6.1 Tests and Models of Fluid Flow in Rock Fractures

4.6.1.1 Three-dimensional effects

An emphasis is given here that for fluid flow in rock fractures, interaction between stress and fluid transmissivity depend almost entirely on the roughness and aperture evolutions during deformation paths and the problem is essentially 3D in nature. Studies by Yeo et al. (1998) and Koyama et al. (2004, 2006) show that the mode of relative displacements of the two opposite surfaces of a rock fracture has a significant impact on the flow behavior. The induced flow rate increase may become significant in the third direction perpendicular to the shearing direction, thus increasing anisotropy of the flow field in the fracture plane. However, slight relative rotations (say 5°) between the two surfaces could eliminate the flow anisotropy and increase the flow rate to a not-so-small extent. This demonstrates that fluid flow in a rock fracture should be represented in true 3D space, even though this means significant increase in computing power demands and code developments.

4.6.1.2 Effects of fracture intersections

Intersections of fractures play important roles in both the hydraulic and mechanical behavior of fractured rocks and their interactions. These intersections often form significant pathways for fluid flow and are the focal points attracting stress concentration. Retaining mass conservation at intersections is the usual condition to derive flow equations for fracture networks, but locally concentrated flow at the intersections, especially during deformation processes, are not well understood and not properly represented in numerical models yet.

In theory, DEM is a more natural approach for representing fracture intersection effects since they are naturally and explicitly represented in the DEM models. However, numerical difficulties exist in simulating interactions of blocks at these intersections due to the fact that their sizes are so small compared with the FEM or FDM elements of the blocks forming the fractures; thus, maintaining contact displacement compatibility (no overlapping or interpenetration) at these intersections demands special numerical treatments.

4.6.1.3 Definition and determination of fracture aperture

Although aperture is the dominating property for fluid flow in fractures, there is no universally agreed definition of aperture among different disciplines. From the literature there exist basically three definitions for a rock fracture (see Adler and Thovert, 1999): geometric aperture, mechanical aperture and hydraulic aperture. Assuming that a fracture is composed of two nominally planar but rough parallel surfaces, the geometric aperture is the (nominally) normal distance between the two opposite surfaces of the fracture.

Assuming that two rough surfaces have their mean surfaces parallel, then the mechanical aperture is the distance between these two mean surfaces. The hydraulic aperture is the void space, which actually conducts fluid flow, divided by the apparent area of the fracture surface. Conceptually, the orders of magnitude of these three apertures are: geometric, mechanical and hydraulic; and they all are functions of stresses and deformation paths of the fracture. These definitions are commonly adopted in hydrogeology. In rock mechanics, geometric aperture is not usually used. Due to the randomness of the surface roughness and scale effects, apertures are not constant, but functions of location. Therefore mean aperture and local aperture are used at different scales and applications. As also presented in this chapter, apertures are stress and deformation-path dependent.

The difficulty lies in the fact that it is not so easy and straightforward to determine apertures by simple experiments, especially regarding their initial values.

The geometric aperture may be estimated using accurate laser surface scanning, but an accurate relocation technique is needed, as discussed in Lanaro et al. (1998). Since the aperture values are often expressed in micrometers, demand for low error margins of the relocation is required and in practice is difficult to achieve. However, geometric aperture is needed, especially in laboratory experiments, to estimate the initial mechanical aperture.

In rock mechanics, the mechanical aperture is usually determined by normal compression tests, as the maximum closure value at the end of a cyclic loading–unloading sequence. It is also sometimes used as the initial hydraulic aperture in practice.

The hydraulic aperture is mostly determined by back analysis from laboratory compression-flow tests or field-scale pumping tests with isolated individual fractures, by assuming most commonly the validity of the Cubic Law; it is therefore subject to the validity of the flow law assumption. The Cubic Law is most commonly applied and proven to be applicable in many cases. Exceptions exist, of course, as demonstrated by Cook (1988) and many others. The question is that either aperture or the flow law exponent must be assumed, so that the other can be determined from the results of the measured flow rate which is the only known value in the experiment. Such experiments, especially laboratory experiments, are subject also to size effects. In theory, representative aperture value of a fracture can only be determined with a fracture size equal to or larger than its stationarity threshold of surface roughness (Fardin, 2003). In practice, the extra experimental work needed to determine the thresholds with large enough sample sizes makes such tests often not practical.

It should be pointed out that the surface roughness of rock fractures is the main reason for all the difficulties and complexities in characterizing and establishing models for fluid flow and the mechanics of the rock fractures. A quantitative and universally accepted mathematical representation of the roughness has yet to be developed.

4.6.2 Coupled THM Processes

The couplings between the processes of heat transfer, fluid flow, solute transport and stress/deformation in fractured rocks have become an increasingly important subject in rock mechanics research and rock engineering applications since the early 1980s, mainly due to the modeling requirements for the design and performance assessment of underground radioactive waste repositories and other engineering fields in which heat transfer and fluid flows play important roles, such as gas/oil reservoir engineering, geothermal energy extraction, large-scale hydropower facilities, land subsidence, landslides, contaminant transport analysis and environment impact evaluation in general. Although this book is not focused on coupled THM processes, a few words are needed to alert readers to their impact on DEM and DFN applications in rock engineering, i.e., within a broader perspective.

The subject of coupled modeling has attracted active researches because of its wide-reaching impacts on environmental issues. Extensive publications have been generated. The fundamentals are systematically presented in many written or edited volumes such as in Whitaker (1977), Domenico and Schwartz (1990), Charlez (1991), Charlez and Keramsi (1995), Coussy (1995), Sahimi (1995), Selvadurai (1996), Lewis and Schrefler (1987, 1998), and Bai and Elsworth (2000) with focus on multiphase flow and transport in porous media, targeted more for reservoir and environmental engineering applications. The volumes by Tsang (1987) and Stephansson et al. (1996, 2004) are more focused on coupled stress/deformation, fluid flow and heat transfer in fractured rocks, targeted especially for nuclear waste disposal applications. A comprehensive review on coupled processes in rock fractures is given in Tsang (1991).

Due to the complexity encountered in physics and the importance of time effects, mathematical models and associated computational methods are often the only quantitative means for scientists and engineers to gain understanding of complex coupled systems, through using multiple stochastic system realizations and parameter sensitivity analysis to account for the interactions among so many processes, properties and parameters, plus the uncertainty of parameter values. The reason is that mathematical models are the only means that can integrate so many and such complex interactions into a compact platform for sensitivity-parameter-scenario analyses for the long-terms required (e.g., 10 000–100 000 years for nuclear waste repositories) which cannot be reproduced in laboratory conditions.

The complexity of the coupled problems is increased by the presence of rock fractures of various dimensions, whose physical behavior under thermal, hydraulic and mechanical loadings is far from clearly understood, due mainly to the mostly unpredictable geometrical complexities of their surfaces.

Research on coupled THM process in geological media appears mainly in work for the mechanics of porous media, which is generally applicable to fractured rocks. The first theory may be traced back to Terzaghi's 1D consolidation theory of soils (von Terzaghi, 1923), but the fundamentals are established by Biot's theory of isothermal consolidation of elastic porous media, a phenomenological approach to poroelasticity (Biot, 1941, 1955, 1956). Another approach is the mixture theory by Morland (1972), Bowen (1982) and others. Non-isothermal consolidation of deformable porous media is the basis of modern coupled THM models using either an averaging approach as proposed first by Hassanizadeh and Gray (1979a,b, 1980, 1990) and Achanta et al. (1994), or an extension to Biot's phenomenological approach with a thermal component (de Boer, 1998). The former is more suitable for understanding the thermodynamic behavior of porous media at the microscopic level, and the latter is better suited for macroscopic description and computer modeling.

The THM coupling models have been developed according to two basic 'partial' coupling mechanisms which are well established within the principles of continuum mechanics: thermo-elasticity of solids (T–M) (interaction between the stress/strain and temperature fields through thermal stress and expansion) and poroelasticity theory (H–M) (interaction between the deformability and permeability fields of porous media). They are, in turn, based on Hooke's law of elasticity, Darcy's law of flow in porous media and Fourier's law of heat conduction. The effects of the THM coupling are formulated as three interrelated partial differential equations expressing the conservation of mass, energy and momentum, for describing interactions among fluid flow, heat transfer and solid deformation processes.

The solution of the coupled sets of conservation equations can use either continuum or discrete approaches. For the continuum approach, the FEM and finite volume method (FVM) are the most commonly applied methods (Pruess, 1991; Noorishad et al., 1992; Millard, 1996; Noorishad and Tsang, 1996; Ohnishi and Kobayashi et al., 1996; Börgesson et al., 2001; Nguyen et al., 2001; Rutqvist et al., 2001a,b). A general framework of the equations and the FEM formulation for porous media is given in Schrefler (2001). The continuum solution approach is based on the established equivalent properties of the fractured porous media. It is not computationally efficient when a large number of fractures are explicitly represented and derivation of the equivalent properties, especially given their scale-dependency, often requires discrete numerical methods.

Comprehensive studies, using both continuum and discrete approaches, have been conducted in the international DECOVALEX projects for coupled THM processes in fractured rocks and buffer materials for underground radioactive waste disposal since 1992. Results have been summarized in Stephansson et al. (1996, 2004) and three special issues of the International Journal of Rock Mechanics and Mining Sciences (1995:32(5), 2001:38(1) and 2005:42(5–6)). The projects contributed greatly to the understanding of coupled processes and the mathematical models for fractured rocks.

The numerical methods for THM processes using discrete approaches have not reached the same degree of development compared with their continuum counterparts, mainly because fluid flow is most often limited to fractures. The flow in the rock matrix, therefore also the fracture–matrix interaction, is not considered. The most representative example of the discrete numerical method for coupled THM processes in fractured rocks is the UDEC/3DEC DEM code group. Heat convection can be considered (Abdaliah et al., 1995), but partial saturation and fluid phase change have not been incorporated yet, because no fluid is assumed in the rock matrix. However, the experience from the DECOVALEX project demonstrated that the DEM approach is especially important for studying the near-field THM behavior of fractured rocks, especially the stress–fluid flow interactions (Jing et al., 1996). The DEM has also proved to be an efficient tool in the establishment of REVs and deriving equivalent hydro-mechanical properties of fractured rocks, as demonstrated in Stietel et al. (1996), Min and Jing (2003, 2004), Min et al. (2004a,b) involving thousands of fractures forming irregular stochastic networks, which cannot be simulated by continuum approaches. The work proved that, for fundamental studies on the constitutive behavior of fractured rocks, the discrete approach is a useful and often the most straightforward tool.

One of the important lacks of progress in the field of modeling coupled THM processes in fractured rocks is the lack of laboratory experiments on coupled THM processes of rock fractures. The early works by Zhao and Brown (1992) and later works by Pöllä et al. (1996) remain to be the only experiments considering complete coupling between heating, fluid flow and stress of rock fractures tested in triaxial chambers. However, these tests were performed without shearing and could not reveal shear-induced anisotropy in fluid flow and heat convection. This lack of progress in experiments plays an important role in limitations on development and validations of more advanced and reliable constitutive models of rock fractures and fractured rocks, for coupled THM processes.

References

Abdaliah, G., Thoraval, A., Sfeir, A. and Piguet, J. P., Thermal convection of fluid in fractured media. International Journal of Rock Mechanics and Mining Sciences and Geomechanics Abstracts 1995;32(5):481–490.

Achanta, S., Cushman, J. H. and Okos, M. R., On multicomponent, multiphase thermomechanics with interfaces. International Journal of Engineering Science, 1994;32(11):1717–1738.

Adler, P. M. and Thovert, J. F., Fractures and fracture networks. Kluwer Academic, Dordrecht, 1999.

Bai, M. and Elsworth, D., Coupled processes in subsurface deformation, flow and transport. ASCE Press, Reston, VA, 2000.

Bandis, S. C. Experimental studies of scale effects on shear strength and deformation of rock joints. Ph.D. Thesis, University of Leads, Department of Earth Sciences, Leads, UK, 1980.

Bandis, S. C., Engineering properties and characterization of rock discontinuities. In: Hudson, J. A. (ed.-in-chief), Comprehensive Rock Engineering – Principles, Practice & Projects, Chapter 6, Vol. I, pp. 155–183. Pergamon Press, Oxford, 1993.

Barr, N. S. and Stesky, R. M., Permeability of intact and jointed rock. EOS, 1980;61(46):361.

Barton, N. and Bakhtar, K., Rock joint description and modeling for the hydrothermo-mechanical design of nuclear waste repositories. TerraTek Engineering, Report 83-10, November, 1983.

Barton, N. and de Quadros, E. F., Joint aperture and roughness in the prediction of flow and groutability of rock masses. International Journal of Rock Mechanics and Mining Sciences and Geomechanics Abstracts, 1997;34(3–4):paper no. 252.

Barton, N., Bandis, S. C. and Bakhtar, K., Strength, deformation and conductivity coupling of rock joints. International Journal of Rock Mechanics and Mining Sciences and Geomechanics Abstracts, 1985;22(3):121–140.

Biot, M. A., General theory of three-dimensional consolidation. Journal Applied Physics, 1941;12(2):155–164.

Biot, M. A., Theory of elasticity and consolidation for a porous anisotropic solid. Journal Applied Physics, 1955;26(2):182–185.

Biot, M. A., General solutions of the equations of elasticity and consolidation for a porous material. Journal of Applied Mechanics and Transactions. ASME, 1956;78:91–96.

Bowen, R. M., Compressible porous media models by use of theories of mixtures. International Journal of Engineering Sciences, 1982;20(6):697–735.

Brown, S. R., Fluid flow through rock joints: the effect of surface roughness. Journal Geophysical Research, 1987;92(B2):1337–1347.

Brown, S. R., Transport of fluid and electric current through a single fracture. Journal Geophysical Research, 1989;94(B7):9429–9438.

Brown, S. R. and Scholz, C. H., Closure of random elastic surfaces in contact. Journal Geophysical Research, 1985;90(B7):5531–5545.

Brown, S. R. and Scholz, C. H,. Closure of rock joints. Journal Geophysical Research, 1986;91(B5):4939–4948.

Börgesson, L., Chijimatsu, M., Fujita, T., Nguyen, T. S., Rutqvist, J., Jing, L., Thermo-hydro-mechanical characterization of a bentonite-based buffer material by laboratory tests and numerical back analysis. International Journal of Rock Mechanics and Mining Sciences, 2001;38(1):95–104.

Charlez, P. A., Rock mechanics. Vol. 1, Theoretical Fundamentals, Editions Technip, Paris, 1991.

Charlez, P. and Keramsi, D. (eds), Mechanics of porous media (Lecture notes of the Mechanics of Porous Media summer school, June 1994). Balkema, Rotterdam, 1995.

Cheng, A. H.-D., Abousleiman, Y. and Roegiers, J.-C., Review of some poroelastic effects in rock mechanics. International Journal of Rock Mechanics and Mining Sciences and Geomechanics Abstracts, 1993;30(7):1119–1126.

Cook, N. G. W., Natural joints in rock: mechanical, hydraulic and seismic behaviour and properties under normal stress. First Jaeger memorial lecture, 29th US Symp. on Rock Mechanics, Minneapolis, MN, 1988.

Coussy O., Mechanics of porous media, Wiley, Chichester, 1995.

de Boer R., The thermodynamic structure and constitutive equations for fluid-saturated compressible and incompressible elastic porous solids. International Journal of Solids & Structures, 1998;35(34–35):4557–4573.

Detournay, E. and Cheng, A. H. D., Fundamentals of poroelasticity. In: Hudson, J. A. (ed.-in-chief), Comprehensive Rock Engineering, Vol. 2, pp. 113–171. Pergamon Press, Oxford, 1993.

Doe, T. W. and Osnes, J. D., Interpretation of fracture geometry from well test. In: Stephansson, O. (ed.), Rock joints, Proc. of Int. Symp. on Rock Joints, Björkliden, 1985, pp. 281–292. CENTIK Publishers, Luleå, 1985.

Domenico, P. A. and Schwartz, F. W., Physical and chemical hydrogeology, Wiley, New York, 1990.

Elliot, G. M., Brown, E. T., Boodt, P. I., and Hudson, J. A., Hydromechanical behaviour of joints in the Carnmenellis granite, S. W. England. In: Stephansson, O.(ed.), Proc. of Int. Symp. on Fundamentals of Rock Joints. Björkliden, Sweden, pp. 249–258. CENTEK Publishers, 1985.

Esaki, T., Du, S., Mitani, Y., Ikusada, K. And Jing, L., Development of a shear-flow test apparatus and determination of coupled properties for a single rock joint. International Journal of Rock Mechanics and Mining Sciences, 1999;36(5):641–650.

Fardin, N., The effect of scale on the morphology, mechanics and transmissivity of single rock fractures. Ph.D. Thesis, Royal Institute of Technology, Stockholm, Sweden, 2003.

Gale, J. E., A numerical, field and laboratory study of flow in rocks with deformable fractures. Ph.D. Thesis, University of California, Berkeley, CA, 1975.

Gale, J. E., The effects of fracture type (induced versus natural) on the stress-fracture closure-fracture permeability relationships. Proc. 23rd US Symp. on Rock Mech., p.290. Berkeley, California, 1982.

Gale, J. E., Comparison of coupled fracture deformation and fluid flow models with direct measurements of fracture pore structure and stress-flow properties. Proc. 28th US Symp. on Rock Mech., Tucson, June 29–July 1, 1987, pp. 1213–1223, 1987.

Gale, J. E., McLeod, R., Gutierrez, M., Dacker. L. and Makurat, A., Integration and analysis of coupled stress-flow: laboratory tests data on natural fractures. MUN and NGI Tests. Report submitted to Atomic Energy Limited of Canada Limited by FracFlow Consultsnts, Inc. and Norwegian Geotechnical Institute, 1993.

Gangi, A. F., Variation of whole and fractured porous rock permeability with confining pressure. International Journal of Rock Mechanics and Mining Sciences and Geomechanics Abstracts, 1978;15(5):249–257.

Gentier, S., Morphological analysis of a natural fracture. In: Simpson and Sharp (eds), Selected Papers on Hydrogeology from the 28th Int. Geological Cong., Washington D.C., USA, International Association of Hydrogeologists, (1989);1:315–326.

Goodman, R. E., Methods of geological engineering in discontinuous rocks. West Publishing Company, San Francisco, 1976.

Greenwood, J. A., and Williamson, J. B. P., Contact of nominally flat surfaces. Proceedings of the Royal Society of London, 1966;**A295**(1442):300–319.

Hakami, E., Water flow in single rock joints. Licentiate Thesis, Luleå University of Technology, Luelå, Sweden, 1988.

Hakami, E., Aperture distribution of rock fractures. Ph.D. Thesis, Royal Institute of Technology, Stockholm, Sweden, 1995.

Hardin, E. L., Barton, N., Lingle, R., Board, P. M. and Voegele, M. D., A heated flatjack test series to measure the thermo-mechanical and transport properties of in situ rock masses. Office of Nuclear Waste Isolation, Columbus, Ohio, ONWI-260, p. 193, 1982.

Harper, T. R. and Last, N. C., Interpretation by numerical modelling of changes of fracture system hydraulic conductivity induced by fluid injection. Geotechnique, 1989;39(1):1–11.

Hassanizadeh, M. and Gray, W. G., General conservation equations for multiphase systems: 1. Averaging procedures. Advances in Water Resources, 1979a;2:131–144.

Hassanizadeh, M. and Gray, W. G., General conservation equations for multiphase systems: 2. Mass moments, energy and entropy equations. Advances in Water Resources, 1979b;2:191–203.

Hassanizadeh, M. and Gray, W. G., General conservation equations for multiphase systems: 3. Constitutive theory for porous media flow. Advances in Water Resources, 1980;3(1):25–40.

Hassanizadeh, M. and Gray, W. G., Mechanics and thermodynamics of multiphase flow in porous media including interphase boundaries. Advances in Water Resources, 1990;13(4):169–186.

Iwai, K., Fluid flow in simulated fractures. American Institute of Chemical Engineering Journal, 1976a;2:259–263.

Iwai, K., Fundamental studies of fluid flow through a single fracture. Ph.D. Thesis, University of California, Berkeley, CA, 1976b.

Jiang, Y., Tanabashi, T., Xiao, J. and Nagaie, K., An improved shear-flow test apparatus and its application to deep underground construction. International Journal of Rock Mechanics and Mining Sciences, 2004;41(3):385–386.

Jing, L., Stephansson, O., Tsang, C.-F. and Kautsky, F., Validation of mathematical models against experiments for radioactive waste repositories – DECOVALEX experience. In: Stephansson, O.,

Jing, L. and Tsang, C.-F. (eds), Mathematical models for coupled thermo-hydro-mechanical processes in fractured media, pp. 25–56. Elsevier, Rotterdam, 1996.

Jing, L., Ma, Y. and Fang, Z., Modeling of fluid flow and solid deformation for fractured rocks with discontinuous deformation analysis (DDA) method. International Journal of Rock Mechanics and Mining Sciences, 2001;38(3),343–356.

Jones, F. O., A laboratory study of the effects of confining pressure on fracture flow and storage capacity in carbonate rocks. Journal of Petroleum Technology, 1975;January:21–27.

Kafritsas, J. C. and Einstein, H. H., Coupled flow/deformation analysis of a dam foundation with the distinct element method. Proc. 28th US Symp. on Rock Mechanics, Tucson. pp. 481–489, 1987.

Koyama, T. Fardin, N. and Jing, L., Shear-induced anisotropy and heterogeneity of fluid flow in a single rock fracture with translational and rotary shear displacements – a numerical study. Int. J. Rock Mech. Min. Sci. (2004);41:p.426, SINOROCK2004 Paper 2A 08

Koyama, T., Fardin, N., Jing, L. and Stephansson, O., Numerical simulation of shear-induced flow anisotropy and scale-dependent aperture and transmissivity evolution of rock fracture replicas. International Journal of Rock Mechanics and Mining Sciences, 2006;43(1):89–106.

Kranz, R. L., Frankel, A. D., Engelder, T. E. and Scholz, C. H., The permeability of whole and jointed Barry Granite. International Journal of Rock Mechanics and Mining Sciences and Geomechanics Abstracts, 1979;16(4):225–234.

Lanaro, F., Jing, L. and Stephansson, O., 3-D-laser measurements and representation of roughness of rock fractures. In: Rossmanith, H.-P. (ed.), Proc. of the Int. Conf. on Mechanics, Jointed and Faulted Rock, MJFR-3, Vienna, Austria, (pp. 185–189. Balkema, Rotterdam, 1998.)

Lesnic, D., Elliott, L., Ingham, D. B., Clennell, B. and Knipe, R. J., A mathematical model and numerical investigation for determining the hydraulic conductivity of rocks. International Journal of Rock Mechanics and Mining Sciences and Geomechanics Abstracts, 1997;34(5):741–759.

Lewis, R.W. and Schrefler, B.A., The finite element method in the deformation and consolidation of porous media. Wiley, Chichester, 1987.

Lewis, R. W. and Schrefler, B. A., The finite element method in the static and dynamic deformation and consolidation of porous media, 2nd edition, Wiley, Chichester, 1998.

Lomitze, G. Fluid flow in fissured formation, 1951 (In Russian). Cited from Louis (1969).

Louis, C., A study of groundwater flow in fractured rock and its influence on the stability of rock masses. Rock Mechanics Research Report No. 10, Imperial College, University of London, September, 1969.

Maini, T. and Hocking, G., An examination of the feasibility of hydrologic isolation of a high level repository in crystalline rock. Proc. of Geologic Disposal of High-level Radioactive Waste Session, Annual Meeting of the Geological Society of America, Seattle, Washington, pp. 535–540. Balkema, Rotterdam, 1977.

Millard, A., Short description of CASTEM 2000 and TRIO-EF. In: Stephansson, O., Jing, L. and Tsang, C.-F. (eds), Coupled Thermo-hydro-mechanical Processes of Fractured Media, pp. 559–564. Elsevier, Rotterdam, 1996.

Min, K.-B. and Jing, L., Numerical determination of the equivalent elastic compliance tensor for fractured rock masses using the distinct element method. International Journal of Rock Mechanics and Mining Sciences, 2003;40(6):795–816.

Min, K.-B. and Jing, L., Stress-dependent mechanical properties and bounds of Poisson's ratio for fractured rock masses investigated by a DFN–DEM technique. International Journal of Rock Mechanics and Mining Sciences, 2004;41(Supplement 1):390–395.

Min, K.-B., Stephansson, O. and Jing, L., Determining the equivalent permeability tensor for fractured rock masses using a stochastic rev approach: Method and application to the field data from Sellafield, UK. Hydrogeology Journal, 2004a;12(5):497–510.

Min, K.-B., Rutqvist, J., Tsang, C.-F. and Jing, L., Stress-dependent permeability of fractured rock masses: a numerical study. International Journal of Rock Mechanics and Mining Sciences, 2004b;41(7):1191–1210.

Morland, L. W., A simple constitutive theory for fluid saturated porous solids. Journal of Geophysical Research, 1972;77:890–900.

Nelson, R. A. and Hardin, J. W., Experimental study of fracture permeability in porous rock. American Association of Petroleum Geology Bulletin, 1977;61(2): 227–236.

Nguyen, T. S., Börgesson, L., Chijimatsu, M., Rutqvist, J., Fujita, T., Hernelind, J., Kobayashi, A., Ohnishi, Y., Tanaka, M. and Jing, L., Hydro-mechanical response of a fractured granite rock mass to excavation of a test pit – the Kamaishi Mine experiments in Japan. International Journal of Rock Mechanics and Mining Sciences, 2001;38(1):79–94.

Niemi, A. P., Vaittinen, T. A., Vuopio, J. A. and Pöllä, J. P., Simulation of heterogeneous flow in a natural fracture under varying normal stress. International Journal of Rock Mechanics and Mining Sciences and Geomechanics Abstracts, 1997;34(3/4):Paper No. 227.

Noorishad, J., Tsang, C. F. and Witherspoon, P. A., Theoretical and field studies of coupled hydromechanical behaviour of fractured rocks – 1. Development and verification of a numerical simulator. International Journal of Rock Mechanics and Mining Sciences and Geomechanics Abstracts, 1992;29(4):401–409.

Noorishad, J. and Tsang, C. F., Coupled thermohydroelasticity phenomena in variably saturated fractured porous rocks – formulation and numerical solution. In: Stephansson, O., Jing, L. and Tsang, C.-F. (eds), Coupled Thermo-hydro-mechanical Processes of Fractured Media, pp. 93–134. Elsevier, Rotterdam, 1996.

Ohnishi, Y. and Kobayashi, A., THAMES. In: Stephansson, O., Jing, L. and Tsang, C.-F. (eds), Coupled Thermo-hydro-mechanical Processes of Fractured Media, pp. 545–549. Elsevier, Rotterdam, 1996.

Olsson, R. and Barton, N., An improved model for hydromechanical coupling during shearing of rock joints. International Journal of Rock Mechanics and Mining Sciences 2001;38: 317–329.

Pine, R. J. and Cundall, P. A., Applications of the flow-rock interaction program (FRIP) to the modelling of hot dry rock geothermal energy systems. Proc. of Int. Symp. on Fundamentals of Rock Joints. Björkliden, Sweden, pp. 292–302. CENTEK Publishers, 1985.

Pöllä, J., Kuusela-Lahtinen, A. and Kajanen, J., Experimental study on the coupled T-H-M processes of single rock joint with a triaxial chamber. In: Stephansson, O., Jing, L. and Tsang, C.-F. (eds), Coupled Thermo-hydro-mechanical Processes of Fractured Media, pp. 449–465. Elsevier, Rotterdam, 1996.

Pratt, H. R., Awolfs, H. S., Brace, W. F., Black, A. D. and Hardin, J. W., Elastic and transport properties of in situ jointed granite. International Journal of Rock Mechanics and Mining Sciences and Geomechanics Abstracts, 1977;14:35–45.

Pruess K., TOUGH2 – A general purpose numerical simulator for multiphase fluid and heat flow. Lawrence Berkeley Laboratory Report LBL-29400, Berkeley, CA, 1991.

Pyrak-Nolte, L. T., Myer, L. R., Cook, N. G. W. and Witherspoon, P. A., Hydraulic and mechanical properties of natural fractures in low permeability rock. Proc. 6th ISRM Cong. Montreal, Vol. 1, pp. 225–231, 1987.

Rutqvist, J., Börgesson, L., Chijimatsu, M., Kobayashi, A., Jing, L., Nguyen, T. S., Noorishad, J. and Tsang, C. F., Thermohydro-mechanics of partially saturated geological media: Governing equations and formulation of four finite element models. International Journal of Rock Mechanics and Mining Sciences, 2001a;38(1);105–127.

Rutqvist, J., Börgsson, L., Chijimatsu, M., Nguyen, T. S., Jing, L., Noorishad, J., Tsang, C. F., Coupled thermo-hydro-mechanical analysis of a heater test in fractured rock and bentonite at Kamaishi Mine – comparison of field results to predictions of four finite element codes. International Journal of Rock Mechanics and Mining Sciences, 2001b;38(1):129–142.

Rutqvist, J. and Stephansson, O., The role of hydromechanical coupling in fractured rock engineering. Hydrogeology Journal, 2003;11(1):7–40.

Sahimi, M., Flow and transport in porous media and fractured rock. VCH Verlagsgesellschaft GmbH, Weinheim, 1995.

Schrefler, B. A., Computer modelling in environmental geomechanics. Computers and Structures, 2001;79:2209–2223.

Selvadurai, A. P. S. (ed.), Mechanics of poroelastic media. Kluwer Academic, Dordrecht, 1996.

Stephansson, O., Jing, L. and Tsang, C.-F. (eds), Coupled thermo-hydro-mechanical processes of fractured media. Developments in Geotechnical Engineering 79, Elsevier, Rotterdam, 1996.

Stephansson, O., Jing, L. and Hudson, J. A. (eds), Coupled T-H-M-C processes in geo-systems: Fundamentals, modeling, experiments and applications. Elsevier Geo-Engineering Book Series, Vol. 2, Elsevier, Rotterdam, 2004.

Stietel, A., Millard, A., Treillem E., Vuillodm E., Thoraval, A. and Ababou, A., Continuum representation of coupled hydromechanical processes of fractured media: Homogenisation and parameter identification, In: Stephansson, O., Jing L. and Tsang C.-F. (eds), Coupled Thermo-hydro-mechanical Processes of Fractured Media, pp. 135–163. Elsevier, Amsterdam, 1996.

Swan, G., Stiffness and associated joint properties of rock. Proc. Conf. on Applications of Rock Mechanics to Cut-and-Fill Mining, University of Luleå, Sweden, June 1–3, 1980, pp. 169–178. Institution of Mining and Metallurgy, London, 1980.

Swan, G., Determination of stiffness and other joint properties from roughness measurements. Rock Mechanics and Rock Engineering, 1983;16:16–38.

Tsang, C.-F (ed.), Coupled processes associated with nuclear waste repositories. Academic Press, New York, 1987.

Tsang, C.-F., Coupled thermomechanical and hydrochemical processes in rock fractures. Review of Geophysics, 1991;29:537–548.

Tsang, Y. W. and Tsang, C. F., Channel model of flow through fractured media. Water Resources Research, 1987;23(3):467–479.

Tsang, Y. W. and Witherspoon, P. A., Hydromechanical behavior of a deformable rock fracture subject to normal stress, Journal of Geophysical Research, 1981;86(B10):9287–9298.

von Terzaghi, K., Die berechnug der Durchlässigkeitsziffer des Tones aus dem Verlauf der hydrodynamischen Spannungserscheinungen. Sitzungsber. Akad. Wiss., Math.-Naturwiss, Section IIa, 1923;132(3/4):125–138.

Walsh, J. B., The effect of cracks on the uniaxial elastic compression of rocks. Journal of Geophysical Research 1965;70:300–411.

Walsh, J. B., Effects of pore pressure and confining pressure on fracture permeability. International Journal of Rock Mechanics and Mining Sciences and Geomechanics Abstracts, 1981;18:429–435.

Walsh, J. B. and Grosenbaugh, M. A., A new model for analyzing the effect of fractures on compressibility. Journal of Geophysical Research, 1979;84(B7):3532–3536.

Wei, L., Numerical studies of the hydromechanical behaviour of jointed rocks. Ph.D. Thesis, Imperial College of Science and Technology, University of London, London, 1992.

Whitaker, S., Simultaneous heat, mass and momentum transfer in porous media: A theory of drying. Academic Press, New York, 1977.

Yeo, I. W., De Freitas, M. H. and Zimmerman, R. W., Effect of shear displacement on the aperture and permeability of a rock fracture. International Journal of Rock Mechanics and Mining Sciences, 1998:35(8):1051–1070.

Zhao, J. and Brown, E. T., Hydro-thermo-mechanical properties of fractures in the Carnmenellis granite. Quarterly Journal of Engineering Geology, 1992;25:279–290.

Zimmerman, R. and Main, I., Hydromechanical behaviour of fractured rocks. In: Guéguen, Y. and Boutéca, M. (eds), Mechanics of Fluid-saturated Rocks, pp. 363–421. Elsevier Academic Press, Chapter 7, 2004.

Part Two
Fracture System Characterization and Block Model Construction

5 THE BASICS OF FRACTURE SYSTEM CHARACTERIZATION – FIELD MAPPING AND STOCHASTIC SIMULATIONS

5.1 Introduction

The fracture system geometry of rock masses is one of the most important aspects of model characterization in rock engineering practice. For a fractured rock mass, there are three types of basic structural elements: the fractures, the rock matrix and fluids (water, oil, gases, etc.). Stresses and temperature are the loading or environmental factors. One of the major issues for numerical modeling in rock mechanics and rock engineering is to represent the fracture systems of rock masses as geometrical models. This requires quantitative descriptions of the locations, orientations, sizes, shapes and apertures of all the fractures, their connectivity and the resultant geometry of the rock block assemblage defined by these fractures, using a properly organized data structure for numerical analysis.

Theoretically if the number of fractures involved is relatively small for a particular problem, the task of fracture system characterization is straightforward. An explicit representation of all individual fractures (including their locations, orientations, shapes, sizes and apertures) and the rock blocks formed by them can be directly obtained to form a deterministic geometrical model for analysis or design. This is, however, usually an impractical or inadequate idealization: rarely, if ever, is such a characterization suitable for practical problems which usually require treatment of a large number of fractures of various sizes with large degrees of uncertainty in the parameters, especially the shapes, sizes and spatial distribution. The common practice is to treat only the large-scale fractures in limited numbers as individual and deterministic entities (Fig. 5.1a), and the effects of grain boundaries, flaws and micro-cracks are considered to be properties of the intact rock. The fractures of intermediate scales are most difficult to model, since the effects of their large numbers and not-so-small sizes on the overall behavior of the rock masses of interest cannot usually be neglected.

This leads to the need for systematic characterization of the sets of intermediate-scale fractures (Fig. 5.1b), usually using stochastic methods. By this approach, each fracture set has its own representation of the statistical distributions for the orientation, spacing, size and apertures of the fracture population, based on the results of field mapping on exposures of limited areas or borehole logging data. These statistical models are then extrapolated to the entire problem domain or its statistically homogeneous sub-regions as the representative models of fracture parameters. The quality of the models, i.e., how representative they are for the fracture system *in situ*, depends on the sampling techniques, and the quantity and quality of the data collected. By the law of large numbers in probability theory, the population of fractures in the mapping area or boreholes should be large enough to reduce the bias of sampling within a limit of tolerance, providing that due allowance is made for sampling orientation in, for example, scanline/borehole and areal mapping.

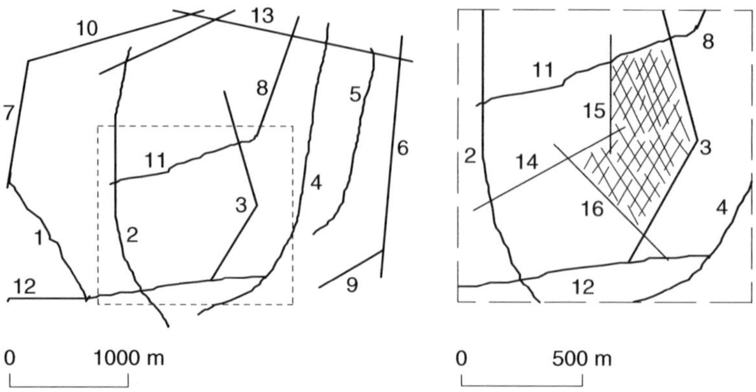

Fig. 5.1 Treatment of fractures at different scales. (a) Deterministic representation of large-scale fractures (lineaments, faults, fracture zones, etc.); (b) stochastic representation of sets of intermediate-scale fractures within statistically homogeneous sub-regions.

Certain kinds of analysis can be performed directly with the statistical information, such as the estimation of rock quality designation (RQD) (Deere, 1964) and block size distributions. However, more rigorous analyses, such as numerical analysis of rock mass behavior under external loading and environmental conditions, requires a geometrical model of the fracture system, not just some parameter distributions. To this end, stochastic realizations of fracture systems can be obtained using the reverse process of random fracture generation based on the statistical distribution functions of the fracture parameters with assumptions for the location and shapes of the fractures. Each of these resultant realizations of the target fracture system is only a partial representation of the real fracture system in the field, assuming that they share the same statistical behavior in terms of fracture parameter distributions. The collection of a large number of the system realizations using the same distribution functions for the fracture parameters is then a better representation of the fracture system *in situ*. This technique is the so-called Monte Carlo simulation process.

Although the fracture parameters, especially their locations, shapes, orientations and dimensions are typically three-dimensional parameters, almost all the information and data collected about these parameters are obtained through one or two-dimensional measurements. The major techniques are borehole logging of fracture orientation, and scanline and window mappings on surface outcrops. The information thus obtained is basically one or two-dimensional in nature and the true three-dimensional properties have to be extrapolated from them, based on some assumptions.

This book will not deal with the methods of fracture mapping techniques and fracture system characterization at site scales in detail. We present only the basics of the mapping techniques most often applied in practice, the most basic mathematical theories of stochastic simulation using Monte Carlo techniques and the basic concept of the integrated fracture system characterization at site scales – since this knowledge is fundamental for developing and applying the DEM approach for rock engineering.

5.2 Field Mapping and Geometric Properties of Fractures

5.2.1 Geometric Parameters and Field Mapping

For DEM and DFN analyses, the properties defining the geometry of a single rock fracture include (cf. Fig. 1.5):

- *Orientation*: dip direction angle α (with respect to north) and dip angle β (measured from a horizontal plane). Used for separation of different sets of fractures using stereographic projection techniques.
- *Density*: the number of fractures per unit volume of the rock mass. The value is closely related to frequency and/or spacing, i.e., the distance between two adjacent fractures of the same set following the same distribution function for their orientations.
- *Spacing*: Spacing is the distance between two adjacent fractures of the same set following the same distribution function for their orientations. Used for determination of the volume density of the fracture population.
- *Shape*: basically unknown, often assumed to be circular, rectangular or planar polygonal.
- *Size*: largely unknown and cannot be determined directly from surface mapping or core logging. It is usually estimated from the trace length distribution and shape assumption. In practice, it is often assumed that the size (e.g., the diameter of the circular fracture discs) has the same distribution as the trace length, with some bias corrections.
- *Trace length*: used to estimate size distribution.
- *Aperture*: difficult to measure in the field directly. 2D photogrammetric measurement of aperture can be applied only for ergodic fracture surfaces, i.e., when the fracture surface is representative of the population.
- *Location (coordinates) of its geometric center*: largely unknown and cannot be determined directly from surface mapping or core logging except for small numbers appearing in the mapping windows or in borehole logging records.

The above variables are the parameter sets to be measured in field mapping using borehole logging or outcrop mapping. It is clear that a large degree of uncertainty is associated with the geometric properties of fractures, especially their locations, shapes and sizes. These uncertainties are the very reasons for using stochastic distributions as their mathematical representations, rather than explicit and deterministic representations.

Scanline mapping (or survey) is one of the most common techniques used to sample fracture trace lengths in the field. Upon a section of an exposure of rock surface, a tape with length marks is fixed onto the exposure in a chosen direction (Fig. 5.2). The distance to the fracture's intersection with the tape (measured from the starting end of the tape), the orientation angle α (on the face) and the trace length L (or half-trace length) of the fractures intersected by the measuring tape are measured and recorded in sequential order. The true orientation angles of the fractures in the mapping windows should be calculated by considering the orientation of the exposed face of the outcrop. The total fracture population is then divided into an appropriate number of sets according to their true orientations. Within each set, the spacings between adjacent fractures are calculated from their distances to the tape origin. The distributions for the spacing and trace length can then be obtained by statistical analysis of the corresponding data populations. The fractures not intersecting the tape are not recorded. More details on the scanline mapping approach is given in Priest (1993).

The scanline mapping technique can lead to an underestimation of the fracture densities because fractures not intersected by the tape are not considered. This limit naturally leads to the window mapping technique in which a square, rectangular or circular area on an exposed rock face is selected and all fracture traces (or part of traces) inside this window are measured (Fig. 5.3). This technique provides better estimation of the fracture trace lengths and densities and therefore improves the reliability of the

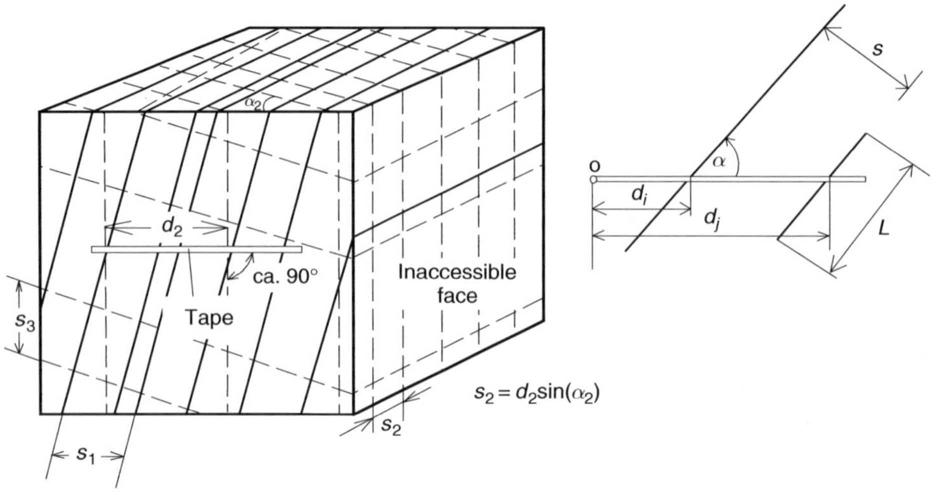

Fig. 5.2 Sets of fractures and measurement of spacing by the scanline method (ISRM, 1978). α, the angle between a fracture trace and the tape; d, the distance between the intersection of a fracture with the tape and the origin of the tape; L, the trace length of a fracture.

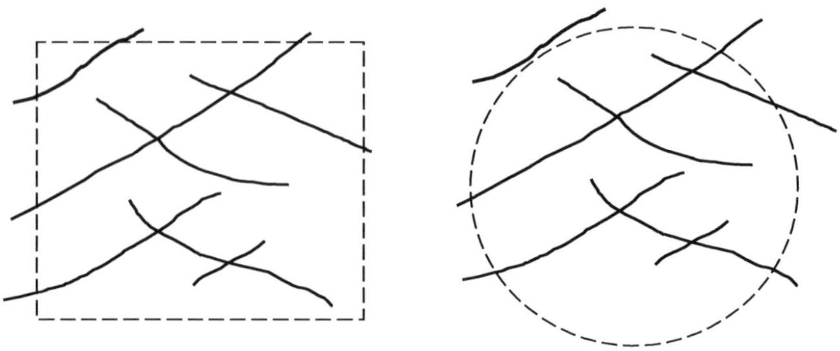

Fig. 5.3 Mapping of fracture trace lengths with square or circular windows.

sampling data. In addition, and as demonstrated in Mauldon et al. (2001), the circular window mapping provides a better performance for eliminating sampling biases due to the relative orientation of the windows with respect to fracture orientations and also in correcting errors due to length bias and censoring (cut-off at the window boundary) that are inherent in scanline and rectangular window mapping techniques.

Figure 5.4a shows examples of window mapping of fracture traces at a surface outcrop in Äspö, Southern Sweden (Bossart et al., 2001). Figure 5.4b shows a map of integrated fracture trace mapping results from a tunnel in the Äspö Hard Rock Laboratory of the Swedish Nuclear Fuel and Waste Management Co. (SKB), including the roof and two walls of the tunnel. Figure 5.4c shows the fracture trace map from one of the tunnel wall.

The measurement of fracture orientations and spacing using borehole data can also be made on images of borehole walls using digitalized photographs of the borehole walls. The traces of fractures with surfaces perpendicular to the borehole axis will be horizontal lines, but the traces of the inclined fractures, with respect to the borehole axis, will be manifested as curves that can be best fitted to sine (or cosine) curves for calculation of the orientation parameters (Fig. 5.5a). The end results of the borehole wall image

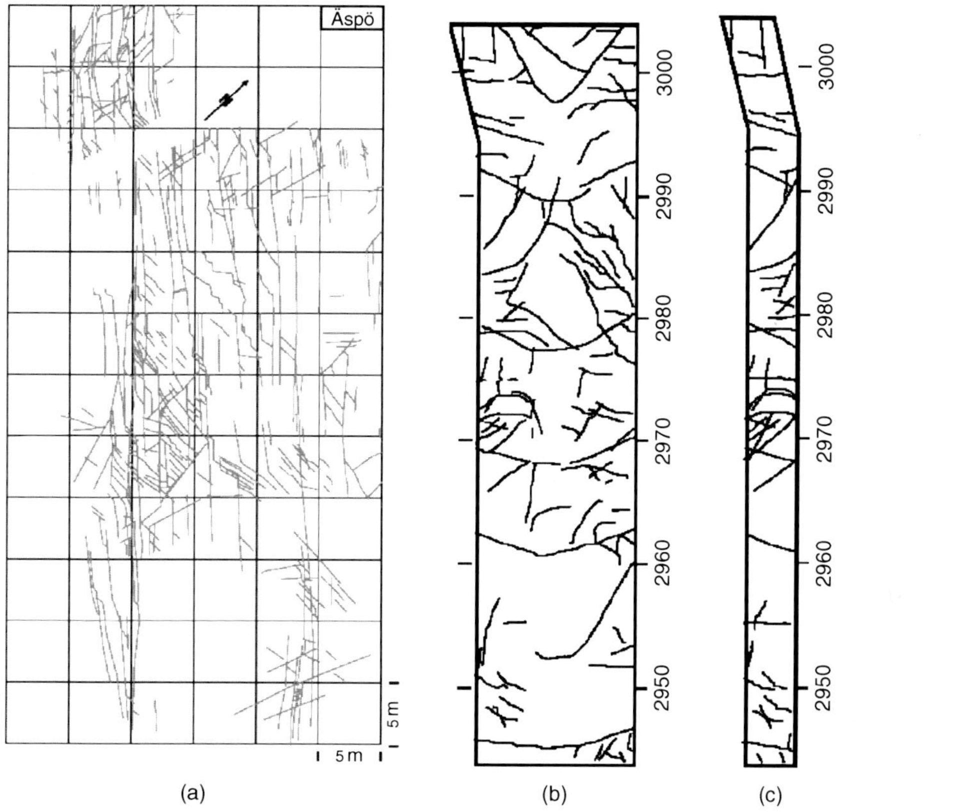

Fig. 5.4 Fracture trace maps using rectangular windows near to and in Äspö Hard Rock Laboratory, SKB, Äspö, Sweden (Bossart et al., 2001). (a) Fracture trace map from a surface outcrop at Äspö, southern Sweden; (b) an integrated fracture trace map of a tunnel including its roof and two vertical walls and (c) fracture trace map from one of the tunnel walls.

analysis will be a systematic representation of the locations, orientations, aperture, filling minerals and lithology of the core column of the borehole, as shown in Fig. 5.5b as an example (Bossart et al., 2001).

The spacing and orientation (including dip angle and dip direction) of fractures intersecting the oriented boreholes can be measured and calculated directly from the cores or borehole wall images (Fig. 5.6), with the spacing being calculated between the adjacent fractures belonging to the same set given the set delineation (see below).

The borehole (core) logging can only provide spacing and orientation parameters, not the fracture size – since the usually very small diameters of borehole make extrapolation to fracture size impossible. However, it is the only means that can provide spacing and orientational data for fractures at great depths in the rock masses concerned, thus providing information over large volumes of rock masses. The main uncertainty in this mapping approach is that any change of fracture orientation beyond the borehole cannot be estimated.

Direct mapping of fractures is not possible when the exposure cannot be accessed. In such a case, photogrammetry and remote laser point recording of spatial coordinates are techniques that can be used. The 'total station' technique is one of them (Feng et al., 2001), as shown in Fig. 5.7. The system does not require reflectors on the exposure and the high-resolution laser beam ensures that the coordinates recorded for fracture surfaces or traces are accurate enough for back-calculating the fracture orientations, densities and spacing.

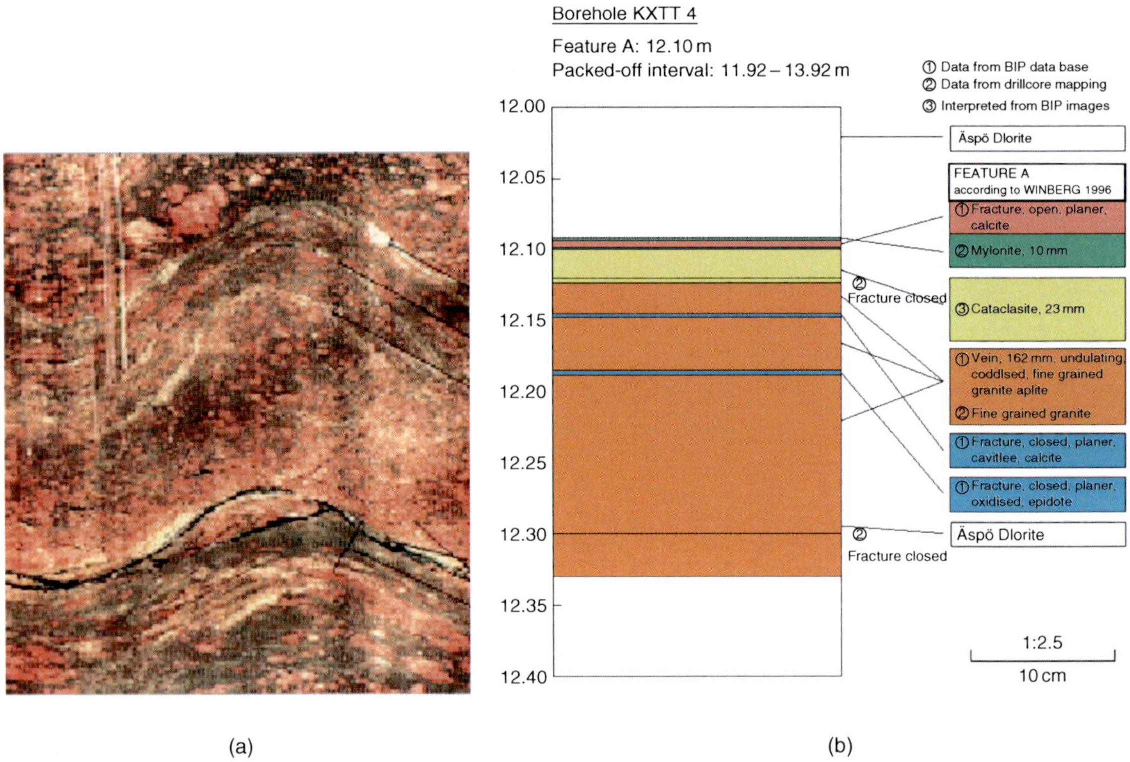

Fig. 5.5 (a) Curved traces of inclined fractures shown on a borehole wall televiewer image and (b) idealized core logging from a borehole wall image analysis (Bossart et al., 2001).

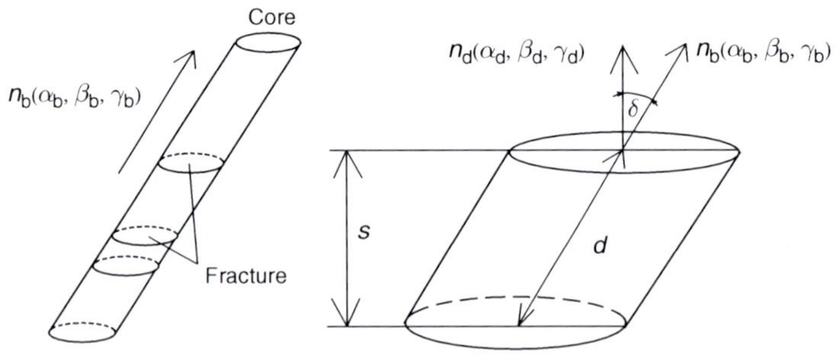

Fig. 5.6 Measurement of spacing (s) and orientation (n) of fractures from the core of an oriented borehole, where d is the intersection distance of two adjacent fractures belonging to the same orientation set and δ the angle between the borehole axis and the normal to the fracture.

5.2.2 Data Processing for Parameter Identification of Fracture Systems

5.2.2.1 Orientation

The dip direction angle (α) and the dip angle (β) of a fracture can be represented by a single point (pole or normal point) on a horizontal circular projection plan of the lower (or upper) hemisphere of radius R (see Fig. 5.8 and Table 5.1). The coordinates of the points are determined by choosing one of the two methods: equal angle or equal area projections. Sets of fractures are identified according to the clusters of their poles.

Fig. 5.7 Non-reflector total station field fracture mapping (Feng et al., 2001): (a) the total station system; (b) the imaginary scanline on exposure; (c) measurements of 3D fracture traces and (d) point-wise records of coordinates at fracture surfaces.

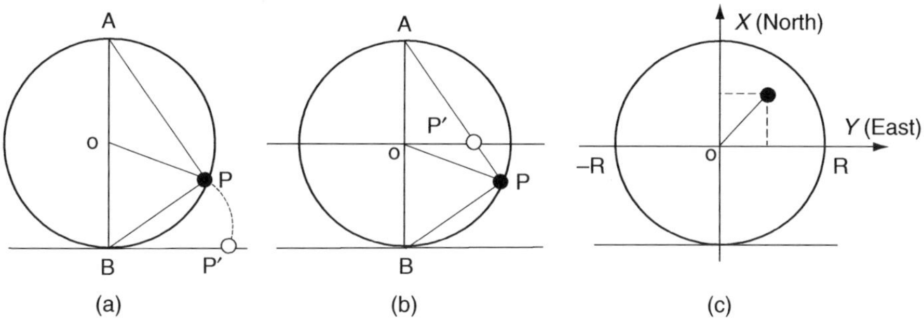

Fig. 5.8 Stereo-projection coordinates of the normal of a fracture on a projection plane (Priest, 1993): (a) equal angle projection; (b) equal area projection; (c) coordinates on the projection plane.

At each fracture pole on a hemispherical projection, a user-defined finite solid cone angle ψ around the pole is selected. The number of all the other fracture poles falling in this selected angle are counted and identified with the selected pole. Repetition of this process for all fracture poles in turn leads to the assignment of an integer representing the number of fractures closely associated with the each pole. The contours of the numbers, or the percentages with respect to the total number of poles, then produce clusters around some major concentration poles (peaks with closed contours) which form the major fracture sets. A number of different ψ values need to be tried for the best cluster analysis. Figure 5.9 shows an example of the identification of four sets of fractures from borehole logging data (Park et al., 2002).

Table 5.1 Calculations for the coordinates in a projection plane (Priest, 1993)

	X-coordinate (North 000°)	Y-coordinate (East 090°)
Equal angle	$R\cos\alpha\tan(\pi/4-\beta/2)$	$R\sin\alpha\tan(\pi/4-\beta/2)$
Equal area	$R\sqrt{2}\cos\alpha\cos(\pi/4+\beta/2)$	$R\sqrt{2}\sin\alpha\cos(\pi/4+\beta/2)$

If the orientation data were obtained by scanline mapping, a sampling bias exists due to the angle between the sampling line and the normal of the fracture planes, written δ. Assuming that the dip direction angles and dip angles of the sampling line and its intersected fracture plane are $\alpha_s, \beta_s, \alpha_n, \beta_n$, respectively, the magnitude of the acute angle δ can be calculated with the formula below

$$\cos\delta = \cos(\alpha_n - \alpha_s)\cos\beta_n\cos\beta_s + \sin\beta_n\sin\beta_s \tag{5.1}$$

A correcting factor $w = 1/\cos\delta$ should then be applied to all the orientational data.

After identification of the sets, the fractures should be regrouped into their corresponding sets for further statistical analysis to determine the mean values for dip direction and dip angle and their probabilistic density functions, using standard one-variable statistical methods. One of the common types of distribution functions for orientation is the Fisher distribution in which the variable x (θ in Priest, 1993) represents the solid angle between the mean (representative) pole (normal vector) and one of the member fractures of the set. A constant called Fisher's constant (or coefficient, denoted as K in Eqn (5.36)) is a measure of the degree of the clustering or divergence of the set: a larger value of K indicates a more clustered set (Priest, 1993).

With the identification of the sets complete, the rest of the parameters, i.e., the density, frequency/spacing, trace length/size and aperture are analyzed to produce the respective probability density functions (PDFs) for each set. The generation of the realizations of the total fracture system is achieved by combined realizations from all sets.

Besides the above standard technique for establishing fracture sets, cluster analysis methods, such as the fuzzy K-means algorithm (Hammah and Curran, 1999), have also been reported. Special problems for quantifying the standard deviation of strike angle data using conventional field measurement techniques (geological compass, for example) were discussed in Herda (1999).

5.2.2.2 *Frequency and spacing*

The frequency of fractures, λ, is often quoted as the mean number of fractures per unit length (1D), per unit area (2D) or per unit volume (3D). The 1D and 2D frequencies depend on the orientation of the measuring line and plane, but the volumetric frequency does not. The reciprocal of the linear frequency, s, is the spacing: $s = 1/\lambda$.

For linear frequency, if the angle δ between the sampling line and the normal of the fracture plane is not zero, then the frequency λ^* and spacing should be corrected by

$$\lambda^* = (N/L)\cos\delta \tag{5.2}$$

$$s = 1/\lambda^* \tag{5.3}$$

For a rock mass with N sets of fractures with frequencies $\lambda_1, \lambda_2, \ldots, \lambda_n$, Hudson and Priest (1983) suggest that the overall linear frequency λ of the rock mass is additive, given by

$$\lambda = \sum_{i=1}^{N}(\lambda_i\cos\delta_i) \tag{5.4}$$

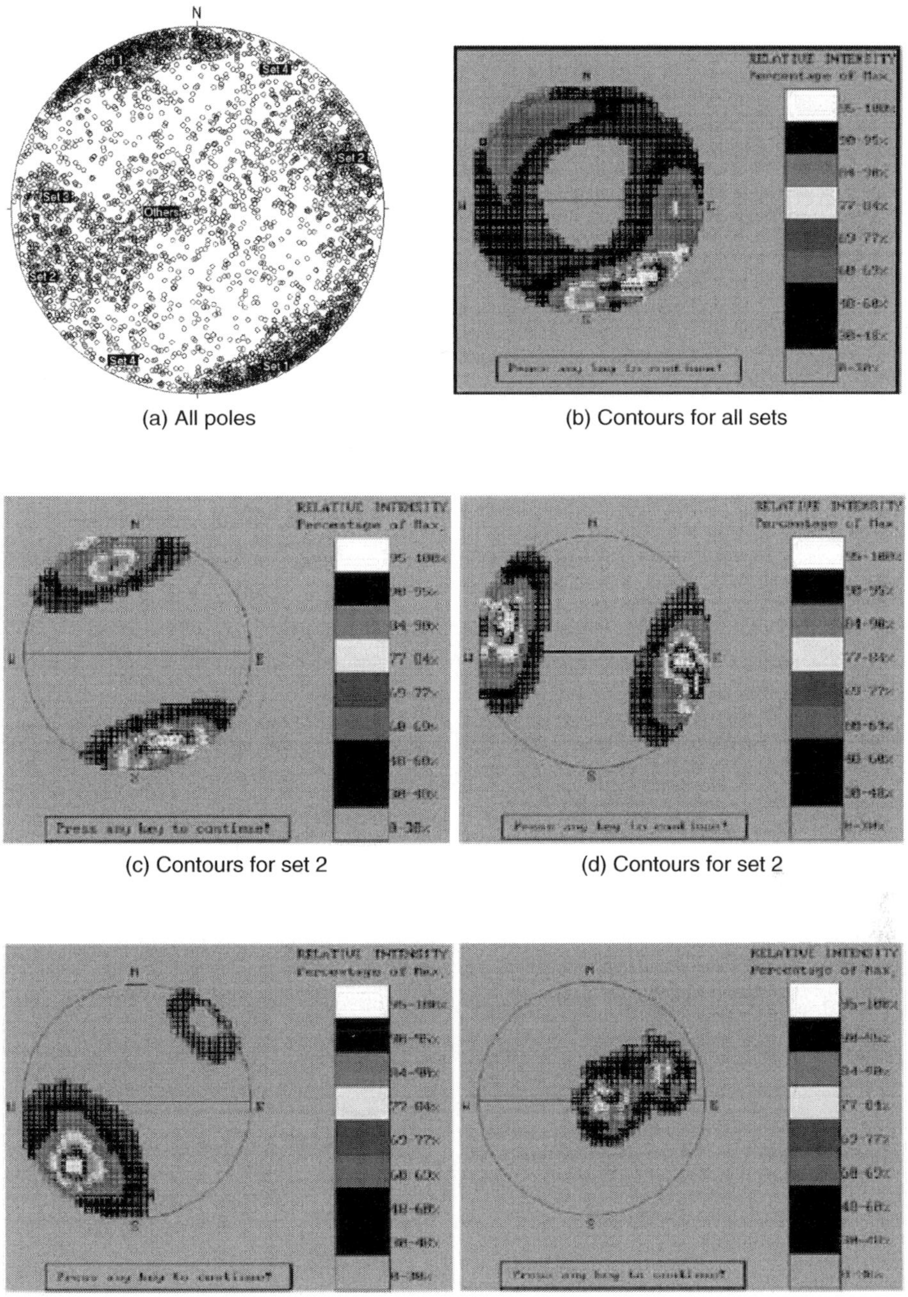

(a) All poles (b) Contours for all sets

(c) Contours for set 2 (d) Contours for set 2

(e) Contours for set 3 (f) Contours for set 4

Fig. 5.9 Identification of fracture sets from borehole logging data (Park et al., 2002): (a) the stereographic projection of poles and (b–f) density contours of the poles divided into four sets. In total, 4609 poles were observed by logging in 10 boreholes.

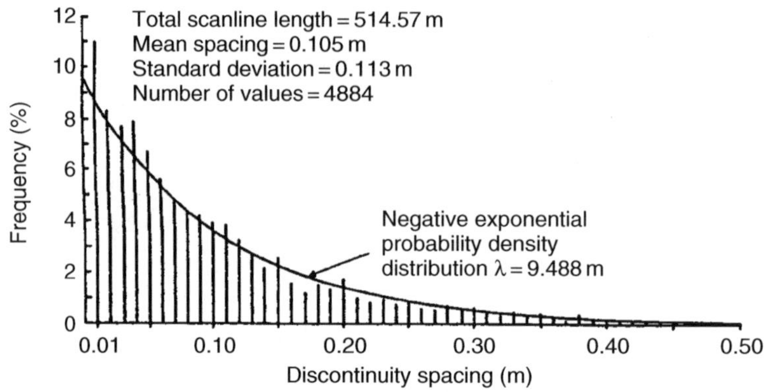

Fig. 5.10 An example of a fracture spacing histogram (Priest and Hudson, 1976).

which is often used to estimate the theoretical RQD index of fractured rock masses

$$RQD = 100e^{-\lambda t}(\lambda t + 1) \tag{5.5}$$

where t is the RQD cut-off threshold for the spacing and a negative exponential distribution for the spacing values is assumed. The standard value for t is 0.1 m.

The frequency or spacing is generally not a constant for a set of fractures, but is also a distribution. The probabilistic functions can be found through standard statistic analysis via histogram and density function approaches as introduced earlier. Figure 5.10 gives an example of a fracture spacing histogram obtained from a scanline survey for a chalk tunnel, Chinnor, UK (Priest and Hudson, 1976).

For multiple sets of fractures, this type of analysis should be carried out for each set separately, so that the different means, standard deviations and distribution functions can be determined. The most commonly adopted distribution functions for the spacing are the negative exponential, normal and log-normal distribution functions. However, other types are also possible and the properties are always site specific. Figure 5.11 illustrates an example of the anisotropy of frequency with three sets. Note that, given the frequencies and geometries of each fracture set, it is then possible to establish the frequency and hence estimated RQD values along a line at any orientation through a rock mass.

5.2.2.3 *Density*

The density of fractures of a given fracture set is defined as the mean number of fractures per unit rock volume, written as D_3, which is dependent on the sampling techniques for its value. With borehole logging data, the volumetric fracture density is given by (Chilés and de Marsily, 1993)

$$D_3 = \frac{1}{L}\sum_{i=1}^{N}\frac{1}{\sin\theta_i} \tag{5.6}$$

where N is the number of fractures of the specific set intersecting the borehole of length L and θ the inclination angle of the borehole (with respect to the horizontal plane). The overall fracture density is then the sum of the densities of all fractures sets within a given sampled region.

Density can also be defined using scanline mapping which yields a linear (1D) density D_1 and window mapping which produces an areal (2D) density D_2, as the mean numbers of fractures intersecting the scanline or contained within the mapping window, respectively. Both D_1 and D_2 are dependent on the orientations of the scanline and mapping window surface, and are therefore termed apparent fracture densities.

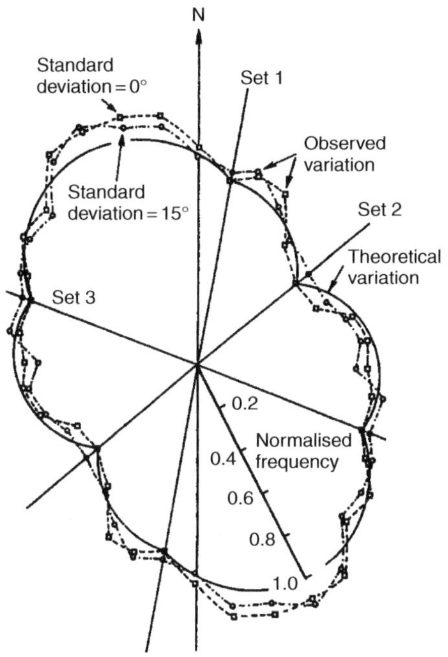

Fig. 5.11 An example of the observed and theoretical variation of the normalized frequency of three sets of fractures (Priest and Hudson, 1983).

If all fractures are parallel, i.e., they belong to the same set and θ (borehole inclination angle) is a constant, the D_1 and D_3 are related simply by

$$D_3 = \frac{N}{L \sin \theta} = \frac{D_1}{\sin \theta} \tag{5.7}$$

For isotropically distributed random fractures, the mean value of $\sin \theta$ between a fracture plane and a 1D scanline is 0.5, and is $\pi/4$ between a fracture plane and a 2D window mapping plane. Therefore we have

$$D_1 = \frac{D_3}{2}, \quad D_2 = \frac{\pi}{4} D_3 \tag{5.8}$$

for isotropically distributed random fracture systems.

Fracture density is also not a constant, but varies from boreholes, scanlines or mapping windows and their locations. It is therefore likely that density itself is also a probability density distribution.

5.2.2.4 Shape

Shape and size are the two most difficult fracture geometry parameters to establish because there is no reliable direct method for measuring them, even with very strong assumptions. Evidence from laboratory experiments and field observations has demonstrated that for a tensile fracture in an isotropic and homogeneous rock the fracture shape, or at least its initial shape, is indeed circular. However, this circularity can be quickly modified by successive tectonic movements and deformation processes (folding, faulting and jointing), and thus causing induced changes in fracture shapes.

The fracture shape is in reality too complex to be confidently idealized into regular plane shapes, such as circular, elliptic, square or rectangular. General polygons may be a more acceptable conceptual model, but such shapes are difficult for computational idealization, with their varying numbers of vertices and edges, for both convex and concave shapes. In practice, a common solution is to assume that the fractures are either circular as in the FracMan code (Golder Associates, 1993), elliptic, square or rectangular as in the NAPSAC code (Wilcock, 1996) for computational simplicity. However, if a very large number of fractures are involved, for example in a flow analysis, the significance of the fracture shape decreases with an increase in the fracture population size.

5.2.2.5 *Size and trace length*

The size of fractures is, however, a far more significant parameter since it directly affects fracture connectivity that determines the percolation threshold and permeability of fractured rock masses, and the formation of blocks and the associated block size distributions that determine the deformability and stability of the block system. The sole source of fracture size information is the trace length distributions observed at exposures of often very limited mapping window areas. Trace length distributions obtained from scanline or window mappings are themselves biased by sampling errors, such as truncation (removal of small fractures), inclusion of non-natural fractures (such as blast-induced cracks) and hidden fractures larger than the sampling windows. The size can be studied through an iterative trial-and-error process (Chilés and de Marsily, 1993).

Due to the 1D nature of the trace length, the size distribution cannot be estimated by using the trace length distribution alone and an assumption concerning fracture shape must also be used so that some analytical relation can be established between the 1D trace length and the 2D fracture size distributions.

Assuming that all fractures are circular disks in shape, with a distribution of the diameter r following a 3D PDF $P(r)$, volumetric disk center density T_3 and areal disk center density T_2 and a 2D trace length distribution $P(l)$, then according to (Chilés and de Marsily, 1993)

$$T_2 = T_3 \int_0^{\infty} rP(r)\mathrm{d}r \tag{5.9}$$

$$P(l) = \frac{1}{r} \int_l^{\infty} \frac{P(r)}{\sqrt{r^2 - l^2}} \mathrm{d}r \tag{5.10}$$

Then the relation between the volumetric fracture density D_3 and the 3D circular disc center density T_3 is given by

$$D_3 = \frac{\pi}{4} \left[\left(\int_0^{\infty} rP(r)\mathrm{d}r \right)^2 + (\sigma_r)^2 \right] T_3 \tag{5.11}$$

where $(\sigma_r)^2$ is the standard deviation of the 3D disc center distribution.

Direct inversion of Eqn (5.11) for obtaining the function $P(r)$ is theoretically possible, but may not be generally practical (Chilés and de Marsily, 1993). Experience in practice also shows that the trace length and 3D disc diameter R tend to follow the same PDFs, either negative exponential or log-normal. Therefore one may simply assume that $P(r) = P(l)$ in practice.

Table 5.2 lists the distribution forms found for fracture trace lengths as reported in the literature.

As mentioned, the trace length and the (assumed) shape are the two parameters that determine the size of the fractures. For reasons of simplicity, fractures are often assumed as circular, elliptical or rectangular (Robertson, 1970; Einstein and Baecher, 1983; Long and Witherspoon, 1985; Rasmussen et al., 1985) so that their size (area) can be estimated through use of their trace length observed on

Table 5.2 Distribution functions of spacing and trace length of rock fractures (Kulatilake, 1991), with the addition of the power law distributions used by the authors

Spacing		Trace Length	
Distribution form	Sources	Distribution form	Sources
Log-normal	Steffen (1975), Bridges (1975), Barton (1977), Einstein et al. (1979), Sen and Kazi (1984)	Log-Normal	McMahon (1971), Bridges (1975), Barton (1977), Baecher et al. (1977), Einstein et al. (1979)
Negative exponential	Call et al. (1976), Priest and Hudson (1976), Baecher et al. (1977), Einstein et al. (1979), Wallis and King (1980)	Negative Exponential	Robertson (1970), Call et al. (1976), Cruden (1977), Priest and Hudson (1983)
Power law (fractal)			Babadagli (2002), La Pointe et al. (1999)

exposures. Robertson (1970) estimated that the real size, A, of a fracture with an observed trace length L on an exposure is proportional to the size, A', of a circular disc of diameter L via the following relation

$$A = \left(\frac{4}{\pi}\right)^2 A' \tag{5.12a}$$

So that the radius of the fracture, r, is estimated as

$$r = \sqrt{A/\pi} = \frac{4}{\pi}\sqrt{A'/\pi} = \frac{2L}{\pi} \tag{5.12b}$$

The trace lengths obtained from field mapping contain different kinds of bias due to the relative orientations of the sampling window and fractures, relative sizes of fracture trace lengths and window (scanline) sizes, upper and lower cut-off limits (censoring and truncation) and the termination conditions of fractures inside and outside the mapping windows, such as reported in Attewell and Farmer (1976), Einstein and Baecher (1983), Kulatilake (1988) and Mauldon (1998). For a circular window and considering different ending conditions of fractures, Zhang and Einstein (1998) estimated the mean fracture trace length as

$$\mu_l = \frac{\pi}{2}\frac{(2N_0 + N_1)}{(N_1 + 2N_2)} R_{\mathrm{w}} \tag{5.13}$$

where R_{w} is the radius of the circular sampling window, N_0 the number of traces with both ends outside the mapping window limit, N_1 the number of traces with one end inside and one end outside the window limit and N_2 the number of traces with both ends inside the window limit. The same expression was also reported in Mauldon (1998). The assumptions made for deriving Eqn (5.13) are that fractures are planar, the distribution of the central points of the traces in the window follows a uniform distribution and the trace lengths and orientations of the fractures inside the windows follow independent distributions.

Warburton (1980) provided a theoretical estimation of the trace length and fracture size relation as

$$f(l) = \frac{1}{\mu_D} \int_l^{\infty} \frac{g(D)\mathrm{d}D}{\sqrt{D^2 - l^2}} \tag{5.14}$$

where D is the diameter of the fracture, l its trace length on an exposure, $g(D)$ the PDF of D, $f(l)$ the PDF of trace length l and μ_D the mean value of D. The mean trace length μ_l is then given by (also see Zhang and Einstein, 1998)

$$\mu_l = \int_0^{\infty} lf(l)\mathrm{d}l \tag{5.15}$$

If $g(D)$ is log-normally distributed, then

$$\mu_l = \frac{\pi}{4}\left[1 + \left(\frac{\sigma_D}{\mu_D}\right)^2\right]\mu_D \tag{5.16}$$

where σ_D is the standard deviation of the fracture diameters. If $g(D)$ has a negative exponential distribution, then

$$\mu_l = \frac{\pi}{2}\mu_D \tag{5.17}$$

The above results indicate that the mean trace lengths of fractures (on exposures) are generally larger than the mean fracture diameters (assuming they are circular in shape). Therefore the direct use of trace length for representing fracture sizes may lead to an overestimation of fracture size.

5.2.2.6 *Aperture*

Besides shape and size, aperture is the next most difficult fracture property to establish, mainly due to practical limitations in sample size and accessibility, the difference between *in situ* and laboratory test conditions and the definition of aperture itself.

Aperture can be defined in different ways: geometric aperture, mechanical aperture and hydraulic (or conducting) aperture among them. The geometric aperture is the void space width between the two rough surfaces of a rock fracture. For a fracture sample, it can be calculated using laser scanner data of both surfaces of the fracture with a precise relocation process (see Lanaro et al., 1998; Fardin et al., 2002). The mechanical aperture often means the maximum normal closure obtained during cyclic tests of normal compression. It represents the normal displacement with no shear involved. The hydraulic aperture cannot be directly measured, but is often inferred using the Cubic Law, via a fluid flow test through fracture samples and/or *in situ* flow tests with known fracture geometry and connectivity.

All apertures depend on applied stresses, fracture surface roughness, deformation paths, infilling materials, initial states and even fracture size. It is one of the most important property affecting both the permeability and deformability of the fractured rocks, but also the least understood or known.

With the sets of fractures established and the orientation, density, trace length and aperture values of fractures in each set known, the next step in the data processing is to establish the statistical distribution functions of these parameters in order that Monte Carlo simulations can be performed to generate multiple realizations of the fracture systems. Such realizations represent, to a certain extent, a collective and statistically equivalent representation of the *in situ* fracture system, as partially represented by the mapping data.

5.3 Statistical Distributions of the Fracture Geometry Parameters

Monte Carlo simulation is a stochastic process which addresses the 'randomness' of the fracture network geometry, by representing the fracture properties of location, size, orientation and aperture as random variables following their own specific PDFs, after assuming the shapes of fractures and calculating the densities of all sets of fractures. The aim of the simulation is to generate a large number of realizations of a fracture system, each of which corresponds to a particular set of individual random variables for the locations (of fracture geometric centers), orientations, size and apertures, generated according to their specific PDFs. Each realization, although statistically equivalent to the exposed part of the real but hidden fracture system, has a low probability of representing the actual total *in situ* fracture system geometry, but the collection of a large number of such realizations, all of them are statistically equivalent to but geometrically different from each other, will have a greater probability of representing the statistical properties of the actual fracture system. The approach requires therefore a repeated simulation process. The quality of the representation depends on the qualities of the PDFs, which depend, in turn, on the quality and quantity of the source data collected during the sampling processes.

These realizations can then be used as geometric models for numerical modeling of fractured rock masses for particular physical processes, such as deformation or fluid flow. The results from the numerical models, such as stresses, displacements, flow rates or velocities, can then be treated as statistical values instead of fixed deterministic values at specific locations, as in conventional deterministic models. The mean values and distributions of these results should then provide a better foundation for design and performance assessment of the engineering works.

Monte Carlo simulation is attractive in concept because it provides a means of reducing the uncertainties caused by the largely unknown fracture system geometry and improving the quantification of the variability of the properties. It is especially useful for underground rock engineering in fractured rocks. However, it also requires much more time and resource consumptions in view of the computational requirements given that a large number of realizations must be generated and used as the geometric models supporting the numerical modeling.

5.3.1 Statistical Principles

The mathematical basis for Monte Carlo simulation is the Large Number Law and Central Limit Theorem in probability theory. Based on these principles, Monte Carlo simulations generate random numbers obeying different distributions for fracture parameters so that a numerical realization can be obtained.

5.3.1.1 The law of large numbers

This law includes three statements:

(1) Bernoulli's Theorem: for a number n of independent experiments, the occurrence frequency of a random event, ν/n, converges to the probability of the event when the number of experiments increases, i.e., for any $\varepsilon > 0$

$$\lim_{n\to\infty} P\left(\left|\frac{\nu}{n} - p\right| < \varepsilon\right) = 1 \tag{5.18}$$

whereν is the number of the event concerned that occurs in n number of experiments.

(2) If there exists a mean value μ and a standard deviation σ for a set of independent random variables $\xi_1, \xi_2, \ldots$ or the random variables $\xi_1, \xi_2, \ldots$ have the same distribution with a finite mean value μ, then the mean of n random variables from the same set, $\xi_1, \xi_2, \ldots \xi_n$, converges to μ when n increases, i.e., for any $\varepsilon > 0$

$$\operatorname*{Lim}_{n\to\infty} P\left(\left|\frac{1}{n}\sum_{i=1}^{n}\xi_i - \mu\right| < \varepsilon\right) = 1 \tag{5.19}$$

(3) If independent random variables $\xi_1, \xi_2, \ldots$ have the same distribution and their mean value μ and deviation σ also exist, then the deviation of n random variables from the same set, $\xi_1, \xi_2, \ldots \xi_n$, converges to σ when n increases, i.e., for any $\varepsilon > 0$

$$\operatorname*{Lim}_{n\to\infty} P\left[\left|\frac{1}{n}\sum_{i=1}^{n}\left(\xi_i - \frac{1}{n}\sum_{k=1}^{n}\xi_k\right)^2 - \sigma^2\right| < \varepsilon\right] = 1 \tag{5.20}$$

The importance of the Large Number Law for rock fracture analysis is that, when dealing with fractures by statistical means, the fracture population must be large enough so that the bias of the statistical models representing the *in situ* system, through the sampled data and their distributions, can be reduced to an acceptable tolerance.

5.3.1.2 Central Limit Theorem

If $\xi_1, \xi_2, \ldots \xi_n$ are a set of independent random variables identically distributed with mean value μ and deviation σ, then a new random variable R representing the mean of these n variables, normalized by μ and σ, follows a standard normal distribution $N(0,1)$, i.e.,

$$\operatorname*{Lim}_{n\to\infty} P(R \le x) = \operatorname*{Lim}_{n\to\infty} P\left(\frac{\frac{1}{n}\sum_{i=1}^{n}\xi_i - \mu}{\frac{1}{\sqrt{n}}} \le x\right) = \operatorname*{Lim}_{n\to\infty} P\left(\frac{\frac{1}{n}\sum_{i=1}^{n}\xi_i - \mu}{\frac{1}{\sqrt{n}}\sqrt{\frac{1}{n}\sum_{i=1}^{n}\left(\xi_i - \frac{1}{n}\sum_{k=1}^{n}\xi_k\right)^2}}\right) \tag{5.21}$$

$$= \int_{-\infty}^{x} \frac{1}{\sqrt{2\pi}} e^{-\frac{t^2}{2}} \, dt$$

The importance of the Central Limit Theorem is that, if we have distributions other than a normal distribution (e.g., log-normal, exponential, etc., as often encountered in the statistical analysis of fractures), this theorem enables us to convert random numbers following a $N(0,1)$ normal distribution into random numbers obeying the non-normal distributions of the concerned parameters via simple transformations. The number of generated values for a parameter must be large enough so that, by the Central Limit Theorem, the frequency of the generated values will produce the same, or a very close, shape as the parent continuous distribution.

The establishment of statistical distributions of the geometric variables of the fractures follows the standard statistical techniques with one random variable. Below are some of the basic statistical definitions useful for this purpose.

5.3.2 Statistical Techniques for Stochastic Fracture System Models

5.3.2.1 Definitions of statistical properties of a random data set

For a set of univariate data of size *n*, $(x_1, x_2, \ldots, x_n)$ with minimum and maximum values $x_{\min}$ and $x_{\max}$, the interval $[x_{\min}, x_{\max}]$ can be divided into *m* sub-intervals of equal length $\Delta x = (x_{\max} - x_{\min})/m$, arranged along the *x*-axis of a *xy*-coordinate system

$$x_{\min}, x_{\min} + \Delta x, x_{\min} + 2\Delta x, \ldots, x_{\max} \tag{5.22}$$

Assuming that the number of data points in the set $x \in \{x_1, x_2, \ldots x_n\}$ falling in each of the above sub-intervals is written $k_i (i = 1, 2, \ldots m)$, the frequency of the set based on the chosen spectrum of *m* sub-intervals, *F(m),* is defined as

$$F(m) = \{k_1/m, k_2/m, \ldots k_m/m\} \tag{5.23}$$

The frequency values can also be given as percentages through multiplication by 100 of each of the above frequency values. The special property of the frequency spectrum is

$$\sum_{i=1}^{m} k_i/m = 1 \tag{5.24}$$

These discrete frequency values can be plotted as the *y*-values of the coordinate system, called histograms. They are diagrams showing the distribution of the data values in the whole frequency spectrum (Fig. 5.12a). The area of the histogram is always equal to one.

There are some statistical properties concerning the mean values, dispersions and shapes. The measures for mean values are usually defined by:

Arithmetic mean:

$$\mu = (x_1 + x_2 + \cdots x_m)/m \tag{5.25}$$

Geometric mean:

$$\mu = (x_1 x_2 \ldots x_m)^{\frac{1}{m}} \tag{5.26}$$

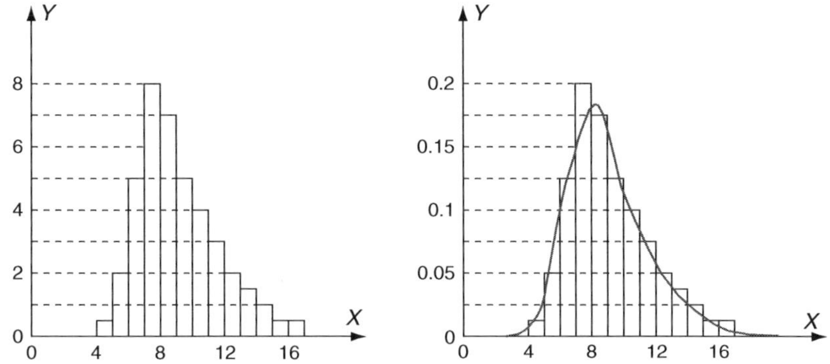

(a) Count numbers (b) Frequency histogram

Fig. 5.12 Histogram and frequency distribution of a random variable.

Median: the value of x that divides the sample population by half

Mode: the value of x corresponding to the highest frequency

The arithmetic mean is used often for symmetric distributions and the geometric mean for skewed distributions. For a skewed distribution, the median lies between the mode and mean.

The measures for dispersion are defined mainly by the standard deviation σ and variance σ^2 given by

$$\sigma^2 = \frac{1}{N-1}\sum_{i=1}^{N}(x_i - \mu)^2 \tag{5.27}$$

together with the range $[x_{min}, x_{max}]$ and the interquartile deviation defined as the modified range by removing the top 25% and bottom 25% of the random variable (x) values.

The skewness of the distribution shape is defined by two parameters: the Pearson's measure of skewness

$$S(x) = (\mu - mode)/\sigma \tag{5.28}$$

and the Fisher's measure of skewness

$$S(x) = \frac{1}{\sigma^3}\frac{1}{N-1}\sum_{i=1}^{N}(x_i - \mu)^3 \tag{5.29}$$

For a symmetric distribution, $S = 0$. $S > 0$ represents a positive skew (long tail of high values) and $S < 0$ for a negative skew (long head of high values). Skewness indicates the 'peakedness' of the distribution, which is less often needed in practice.

5.3.2.2 *Probabilistic density functions (PDF)*

The outlines of histograms are often approximated by smooth, continuous curves for analytical expressions, called frequency density functions, written here as $f(x)$, see Fig. 5.12b. They are often used as idealized PDFs as the hypothetical, population distributions of the random variables. Similar to condition (5.24), a property of a PDF is

$$\int p(x)\mathrm{d}x = 1 \tag{5.30}$$

A frequency distribution has four basic properties that completely describe the distribution: the mean value and its location, the dispersion (the extent to which the frequency is spread out along the scale), how much the values vary from the mean (or central) value(s) and the shape (the symmetry and pattern of the distribution).

Listed below are some typical continuous PDFs often found in describing fracture parameter distributions. All but the Poisson distribution are continuous distributions.

(1) *Uniform distribution:*

$$p(x) = 1/(x_{\max} - x_{\min}), \quad (x \in [x_{\min}, x_{\max}]) \tag{5.31}$$

(2) *Negative exponential function:*

$$p(x) = \lambda e^{-\lambda x}, \quad (x > 0) \tag{5.32}$$

(3) *Normal distribution:*

$$p(x) = \frac{1}{\sigma\sqrt{2\pi}} e^{-\left(\frac{x-\mu}{\sigma\sqrt{2}}\right)^2}, \quad (\sigma > 0; \quad x \in (-\infty, \infty)) \tag{5.33}$$

(4) *Log-normal distribution:*

$$p(x) = \frac{1}{\sigma x\sqrt{2\pi}} e^{-\left(\frac{\ln(x)-\mu}{\sigma\sqrt{2}}\right)^2}, \quad (x \in (0, \infty)) \tag{5.34}$$

(5) *Weibull distribution:*

$$p(x) = \frac{m}{a} x^{m-1} e^{-\frac{x^m}{a}}, \quad (a > 0; m > 0; x \in [0, \infty)) \tag{5.35}$$

(6) *Fisher distribution:*

$$p(x) = \frac{K \sin(x e^{K \cos x})}{e^K - e^{-K}}, \quad (x \in [0, \infty)) \tag{5.36}$$

(7) *Poisson distribution (for discrete events):*

$$p(x) = \frac{\lambda^x}{x!} e^{-\lambda}, \quad (x = 0, 1, 2, \ldots) \tag{5.37}$$

(8) *Power law distribution:*

$$p(x) = Ax^{-D}, \quad (x \in [0, \infty)) \tag{5.38}$$

5.3.2.3 Random number generation from known PDFs

Establishment of the fracture system realizations depends on the generation of random numbers from the PDFs of the fracture properties (location, orientation, size and aperture) after determination of the fracture set density and assumption of fracture shape.

Generation of random variables of a prescribed PDF involves two steps:

- generation of random numbers following a uniform distribution of a standard unit interval [0,1];
- transformation of the uniformly distributed random numbers into random numbers following the prescribed PDF by analytical or numerical methods.

(1) *Generation of uniformly distributed random numbers on [0,1]*

For a uniformly distributed random number on interval [a, b] (see Eqn (5.31)), the cumulative distribution function (CDF) is given by the integration

$$F(x) = \int_a^x p(t)\mathrm{d}t = \int_a^x \frac{1}{b-a}\mathrm{d}t = \frac{x-a}{b-a} \tag{5.39}$$

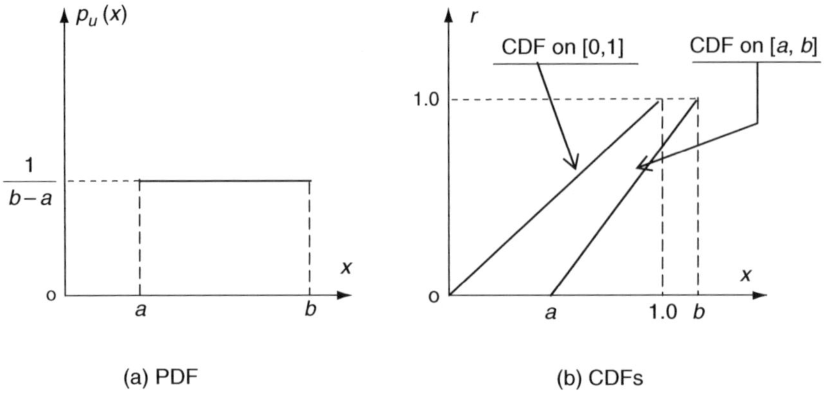

Fig. 5.13 The PDF and CDFs of the uniformly distributed random numbers. (a) PDF on [a,b], (b) CDFs on [a,b] and [0,1].

Denote $r = F(x)$, the uniform random variable x on $[a, b]$ is given by (Fig. 5.13)

$$x = r(b - a) + a \tag{5.40}$$

When $a=0$ and $b=1$, $r=x$. This means that the PDF and CDF are the same for a uniformly distributed random number on [0,1], written as $R_u[0, 1]$. Note that the number 1-r is also a standard uniformly distributed random number on [0,1]. The desired random number x_p following a specific PDF $P(x)$ can then be generated by using its inverse CDF (Fig. 5.14).

For random numbers from a negative exponential distribution (cf. Eqn (5.32)), the CDF is given by

$$F_e(x) = \int_{-\infty}^{x} p_e(t)\mathrm{d}t = 1 - e^{-\lambda x} \tag{5.41}$$

Let $r = F_e(x)$, and recall $(1 - r) \in R_u[0, 1]$, then the negatively distributed random number x_e is given by the inverse transformation

$$x_e = -\frac{1}{\lambda} \ln r, r \in R_u[0, 1] \tag{5.42}$$

For random numbers following a normal distribution $N(\mu, \sigma)$ (cf. Eqn (5.33)), its PDF cannot be integrated in closed form. Numerical approximation must be accepted, such as the standard Newton or

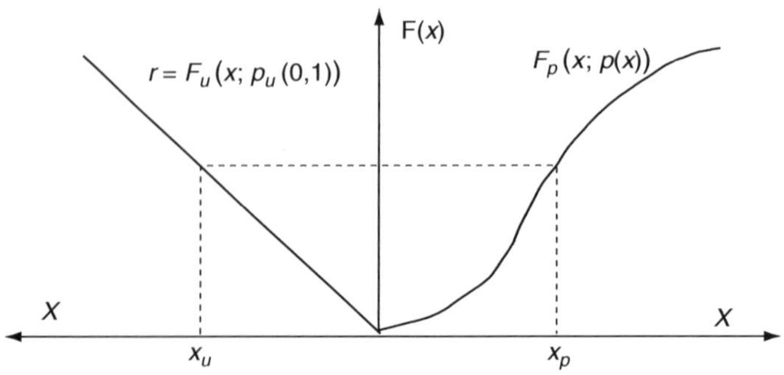

Fig. 5.14 Inverse generation of a random number x_p following PDF $p(x)$ from a standard uniform random number x_u on the [0,1] interval.

Runge-Kutta numerical integration methods. Another technique is to apply the Central Limit Theorem in probability theory and the random numbers following a normal distribution, $x_N \in N[\mu, \sigma]$, given by

$$x_N \approx \mu + \sigma\sqrt{\frac{12}{\pi}}\left(\sum_{\iota=1}^{M} \rho_\iota - \frac{M}{2}\right), \rho \in P_\upsilon[0, 1] \tag{5.43}$$

Usually $M = 10$–12 can provide reasonably good results. Random numbers for a log-normal distribution can be obtained by the simple mappings

$$x' = \ln x, \quad \mu' = \ln \mu - \frac{\sigma^2}{2}, \quad \sigma' = \left\{\ln\left[\left(\frac{\sigma}{\mu}\right)^2 + 1\right]\right\}^{\frac{1}{2}} \tag{5.44}$$

Since normally distributed random numbers have many applications in natural sciences and engineering, they are included in many computer codes and compilers and so can be readily used directly. Similar techniques can be used to generate random numbers for other PDFs.

(2) *Generation of fracture network realizations*

Generation of a realization of a fracture network system can be achieved by different methods. The one described below is one of the simple cases. It involves the following multiple steps:

(a) choosing a generation domain in 2D or 3D space within which a realization of a fracture network system is to be generated;
(b) starting from fracture set 1, calculating the number, N, of fractures to be generated according to the density values ($D2$ or $D3$) and spacing;
(c) generating the locations of N fracture centers as a Poisson process;
(d) at each fracture center, generate two random numbers representing its values of dip direction and dip angle according to its Fisher PDF;
(e) generating a random number representing its size, according to its trace length PDF, and generating a string of vertex coordinates defining its outer boundaries according to the shape assumptions (circular disks or rectangles, etc.);
(f) generating a random number representing its aperture according to the respective PDF;
(g) repeat steps (a) to (f) for all sets of fractures;
(h) determination of intersections of fractures;
(i) for flow analysis, regularizing the fracture network system by removing all sub-sets of the fractures or clusters of fractures which have no connection to the boundary of generation domain and other fracture clusters, thus implying a fracture cluster analysis according to intersection information; and
(j) for 2D flow or stress analysis, all dead-end fracture segments should also be removed.

The above steps can establish one fracture network realization. They must be repeated to generate the desired number of fracture network realizations for a proper stochastic simulation, usually by using different seed numbers (used for initializing the random number generation process).

Fracture network generation routines have been developed continuously over the years and some complicated techniques have been developed to overcome the original simplicity and to meet *in situ* conditions. The most typical of such developments is represented in the code FracMan (Golder Associates, 1993) in which additional PDFs, besides those introduced above, were used to generate more realistic or constrained fracture systems according *to in situ* information.

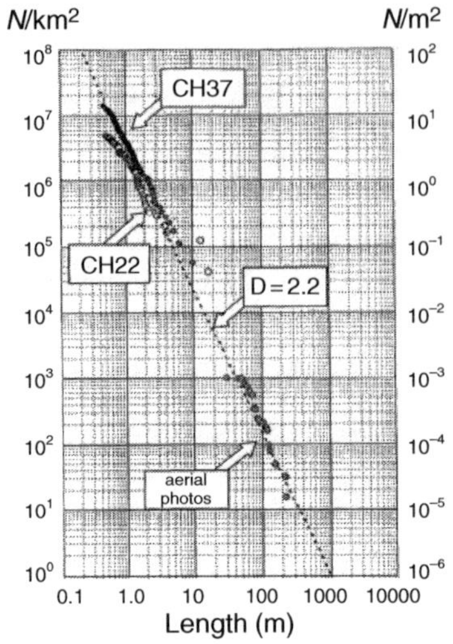

(b)

Set	Dip/dip direction	Fisher constant (K)
1	8/145	5.9
2	88/148	9.0
3	76/21	10.0
4	69/87	10.0

(a)

Fig. 5.15 Site data for the fractures at Sellafield, UK. (a) Orientations and Fisher constants of the fractures sets; (b) the power law for the trace length distribution (Min et al., 2004).

(3) *An example*

This section presents an example of generating stochastic fracture system realizations aimed at deriving constitutive properties of fractured rocks (Min and Jing, 2003; Min et al., 2004) in two dimensions. The geometric property data came from a site investigation program (Nirex, 1997), which identified four sets of fractures in one of the rock formations in the area of Sellafield, Cumbria, UK (Fig. 5.15a). A power law was used for fitting the trace lengths against numbers of fractures per square kilometer for surface mapping sites CH22 and CH37 (Fig. 5.15b) and for long fracture samples from the photo-lineament. Data fall on a line establishing a power-law scaling relation for trace length in the range of 0.5–250 m. The power law is written as

$$N = 4 \times L^{-D} \tag{5.45}$$

where L is the trace length, $D = 2.2$ is the fitted fractal dimension chosen from a range of 2.2 ± 0.2 and N is the number of fractures of trace length not less than L. The fracture density (= 4.6 in this case for all sets) is defined as the number of fractures per area for each set and is derived from the cumulative number of fractures calculated from Eqn (5.45). The curves in Fig. 5.16 are the inverse cumulative PDFs of the trace length using different D values.

Figure 5.17 presents the generated locations of the fractures using a Poisson process, as commonly adopted in practice. Ten realizations of fracture networks (Fig. 5.18) were generated for Monte Carlo simulations of the rock behavior for hydraulic, mechanical and coupled hydro-mechanical processes, using an independently developed code whose flow chart is shown in Fig. 5.19. The details of the study with the models are presented in Chapter 12 of this book.

The geometry of the fracture networks generated look similar but are different and follow the same distributions for the orientation (Fisher distribution), trace length (power law or fractal distribution) and location (Poisson distribution). Therefore, they are statistically equivalent.

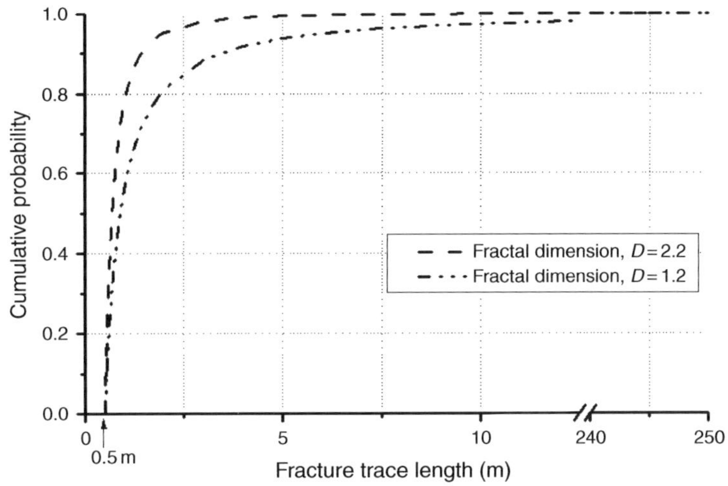

Fig. 5.16 The cumulative PDF of the trace length (Min et al., 2004).

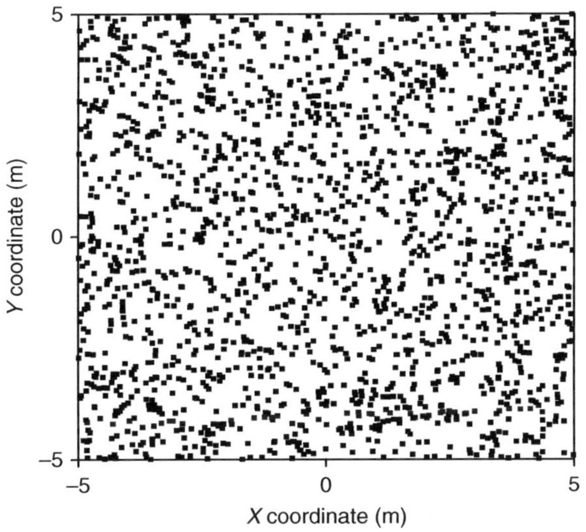

Fig. 5.17 Locations of fracture centers generated by using a Poisson process (Min et al., 2004).

The above example was aimed for generic studies of behavior of the fractured rocks, not for a large-scale field application. Therefore, site specific features such as large-scale faults, different rock units and different fluid conducting characteristics of the fractures and faults were neglected for good reasons. The example is presented here to demonstrate the process of fracture system generation using the Monte Carlo simulation technique. However, for field applications, especially large-scale field applications in which fracture system generations are needed, these site-specific features must be considered in addition to the Monte Carlo simulation, using different approaches at different scales. This is the so-called 'integrated fracture system characterization' at site scales. The generation of fracture system realizations also needs to be conditioned using the measured fracture data in boreholes and/or outcrops through the process of integration.

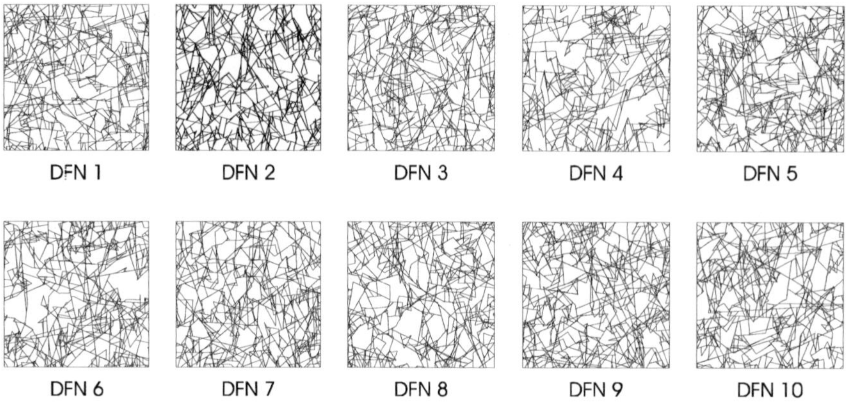

Fig. 5.18 Ten realizations generated for stochastic analyses of the hydraulic REV (Min and Jing, 2003; Min et al., 2004), with size 5 m by 5 m, after fracture regularization.

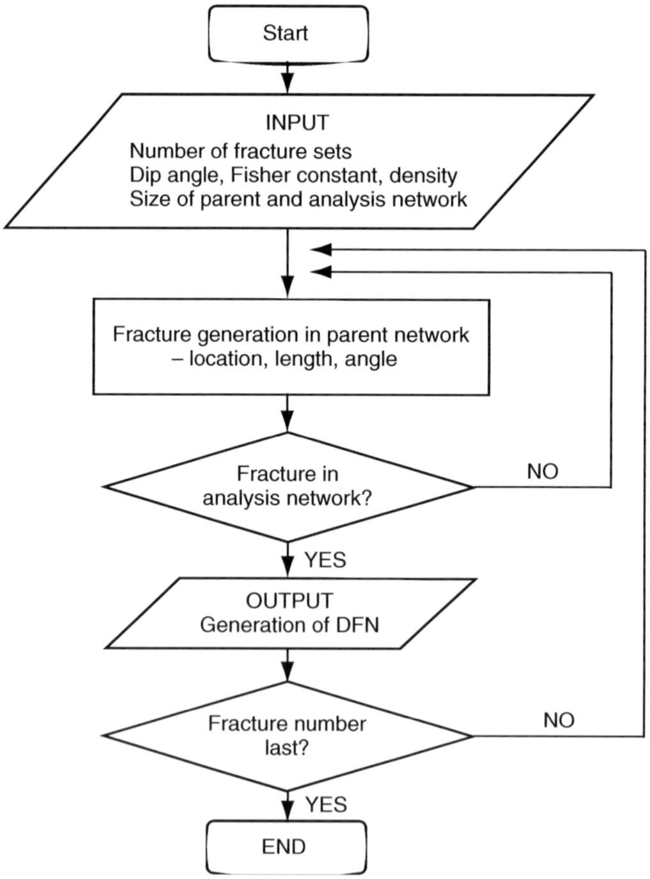

Fig. 5.19 A flow chart of the fracture system generator. The step 'Fracture in analysis network?' decides whether the generated fractures in the parent network reaches an analysis network (Min et al., 2004).

5.4 Integrated Fracture System Characterization Under Site-Specific Conditions

The fracture system characterization at field scales can be performed in different ways. One of them is to use a stepped approach.

(1) Divide the site into different domains according to large-scale structural features (such as fracture zones) and/or rock lithology so that fracture systems within each unit can be approximately regarded as homogeneous in a statistical sense, such as Domain 1-1, 2-2, ... etc., in Fig. 5.20. These large-scale features are usually small in number but large in size and usually have global impacts on the physio-chemical processes of the rocks at the site. They can be regarded as deterministic features because of their smaller number and larger size and are usually the main targets of site investigations using the different survey methods of geophysics, structural geology or hydrogeology with borehole logging and surface mapping. Further division of domains into sub-domains can also be used if necessary to ensure that statistical homogeneity of the fracture system properties (such as orientation) can be approximated at an acceptable tolerance for the sub-domains according to the objectives of the applications and site investigation conditions.

2) Perform Monte Carlo simulations for each domain (sub-domain) so that stochastic models of the fracture geometry can be established in a statistical sense.

3) Superpose the domain-scale fracture system models on the deterministically determined large-scale features so that an integrated fracture system of the whole site can be achieved. Such an integrated model is a composite model in the sense that it is deterministic on the large scale and stochastic on the smaller (domain) scale.

The schematic view of the unit division at site scale is an idealization. In reality, the fractures, large or small, have complex shapes. Figure 5.21 presents an example of the deterministic determination of the orientations, thickness and sizes of large-scale fractures zones in the Ävrö area, Sweden, using geophysical survey methods, together with borehole information.

Figure 5.22 illustrates such an integration process using the initial data from boreholes drilled in horizontal drifts within the same domain at the Äspö Hard Rock Laboratory (HRL) in southern Sweden (Bossart et al., 2001). At the start, the fracture locations and orientations are recorded along the borehole

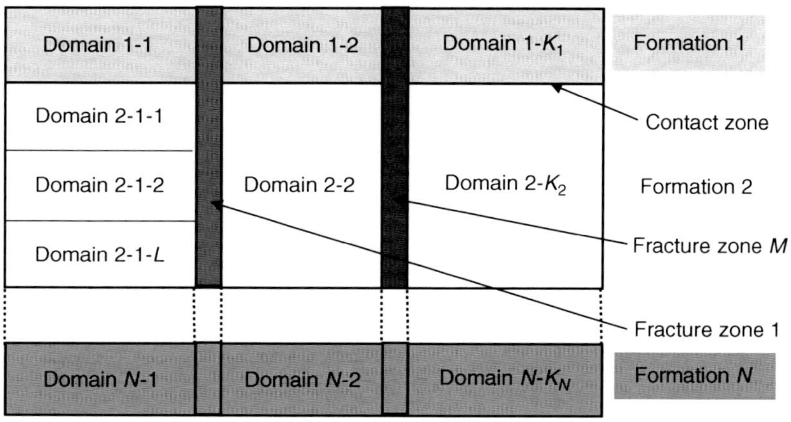

Fig. 5.20 An idealized schematic site division for fracture system characterization.

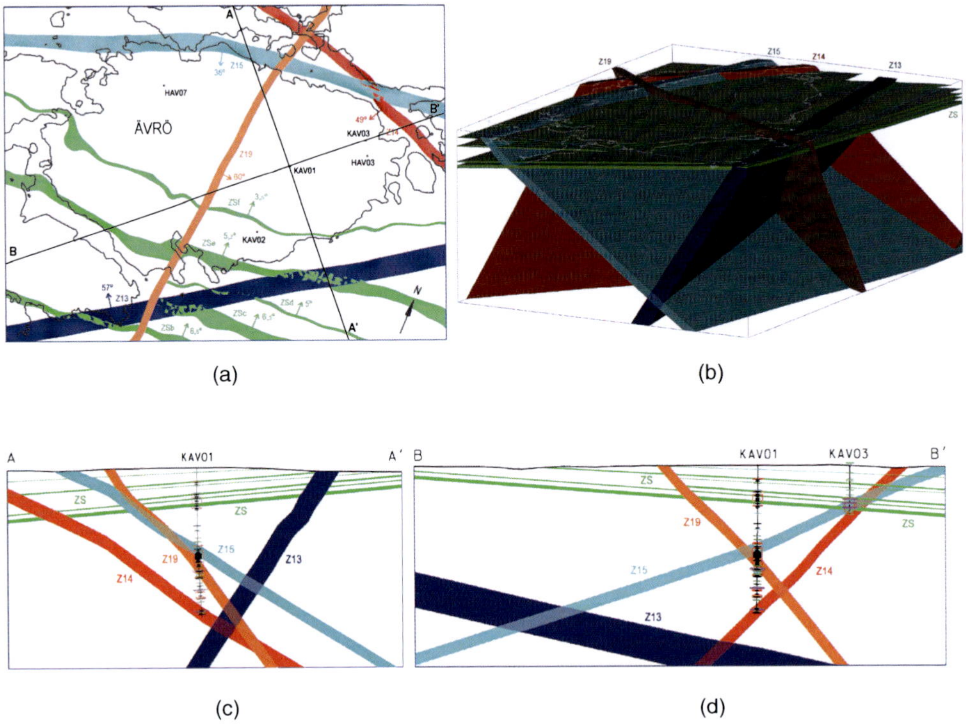

Fig. 5.21 Deterministic large-scale fracture zones surveyed using geophysical methods in the Ävrö area, Sweden (Courtesy of SKB, Sweden): (a) the plan view of the fracture zones; (b) the idealized fracture zone system; (c) a vertical cross-section A–A' with borehole KVA01; (d) a vertical cross-section B–B' with borehole KVA01 and KVA03 (Courtesy of SKB, Sweden).

length (Fig. 5.22a). Then the fractures between the nearby boreholes with a good likelihood of being the same features are connected (Fig. 5.22b). This cross-hole connection can only be implemented for relatively larger scales.

This operation needs both judgment based on the geological conditions provided by the borehole logging data regarding rock types and possible features related to fracture sizes (such as width/apertures, mineral filling characteristics, water bearing conditions, etc.) and the guidance of the unit division so that unit boundaries are usually not crossed except possible large faults. Such artificially connected cross-hole fractures are still deterministic features. Finally superposition of the deterministically identified cross-hole fractures (relatively large in size) and the stochastically simulated fracture networks of smaller fractures is performed (as shown in the background in Fig. 5.22c) with conditioning, using borehole logging data of the fractures as much as possible. A number of such superposition may be performed due to the stochastic nature of the network models.

Site characterization for integrated fracture systems is an important but challenging work – due to the uncertainty involved in the scarce raw data that can be measured *in situ* using boreholes and window mappings compared with the rock volume involved. For fluid flow analysis purposes, an additional difficulty lies in the fact that usually only a small portion of the fracture population conducts water and the geometrical distribution of these conducting features is even more difficult to estimate. This chapter only aims to arm the readers with the most necessary knowledge about Monte Carlo simulations of statistically homogeneous fracture systems and provides a brief description of the field mapping techniques. More details of both can be found in Geological Society of London (1970) for borehole logging, ISRM (1978) for scanline mapping and Pahl (1981), Mauldon

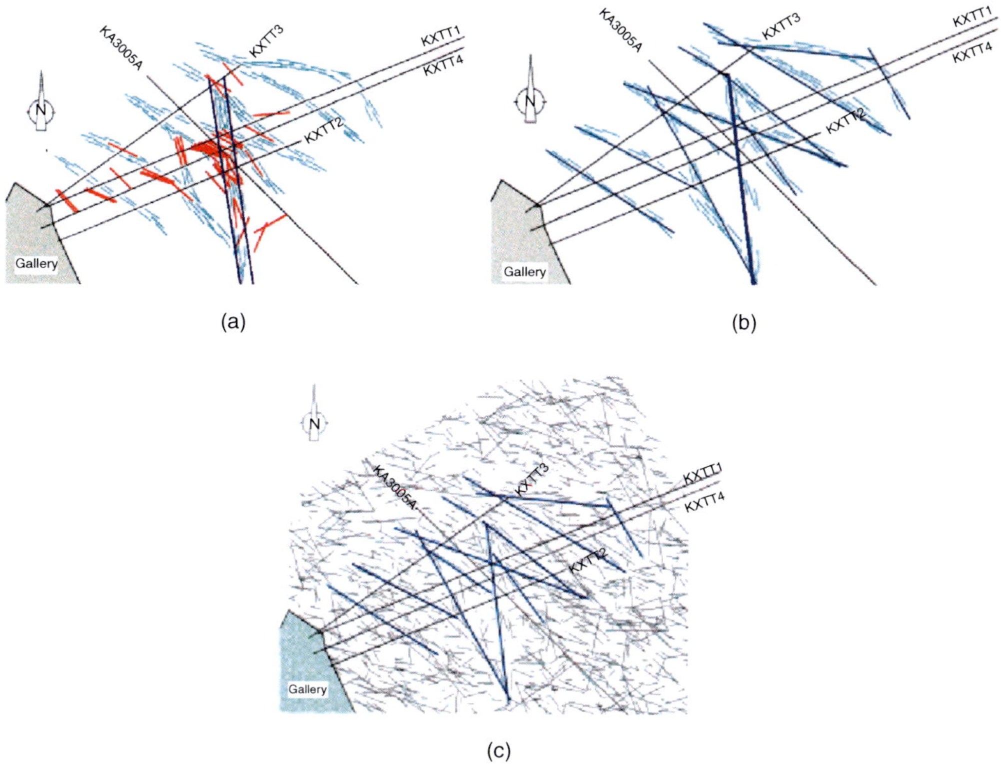

Fig. 5.22 Integrated fracture network realization steps at Äspö Hard Rock Laboratory (HRL), Southern Sweden (Bossart et al., 2001): (a) record of fractures in boreholes; (b) correlating fractures most likely to be connected into larger fractures intersecting multiple boreholes; and (c) superposition of the correlated larger (deterministic) fractures on the randomly generated realizations.

(1995, 1998) and Mauldon et al. (2001) for window mapping. More detailed information can also be found in related literature and textbooks (e.g., McClay, 1987; Priest, 1993). Additional information can also be gained from Terzaghi (1965), Mathab et al. (1972), Priest and Hudson (1976), Cruden (1977), Hudson and Priest (1979, 1983), Baecher (1980), Baecher et al. (1977), Barton (1978), La Pointe (1980), Long et al. (1987), Dershowitz (1992), Dershowitz and Einstein (1988), Kulatilake and Wu (1984), Kulatilake (1985, 1986, 1988, 1991), Grossmann (1990), Wang and Chen (1992) and Pan and Jing (1987).

Fracture networks have special geometric and physical properties, such as scaling and flow clustering or compartmentalization. Berkowitz (2002) gives a more detailed and systematic survey of the current state-of-the-arts and outstanding issues for DFN approaches for flow and transport applications. The special features in theory, methods and applications are summarized in Chapter 10.

References

Attewell, P. B. and Farmer, I. W., Principles of engineering geology. Chapman and Hall, London, 1976.

Babadagli, T., Scan-line method to determine the fractal nature of 2-d fracture networks. Mathematical Geology, 2002;34(6):647–670.

Baecher, G. B., Progressively censored sampling of rock joint traces. Mathematical Geology, 1980;12:33–40.

Baecher, G. B., Lanney, N. A. and Einstein, H. H., Statistical description of rock properties and sampling. Proc. of 18th US Symp. on Rock Mechanics, pp. 5C-1–5C1-8, 1977.

Barton, C. M., Geotechnical analysis of rock structure and fabric in C.S.A. Mine, Cobar, New South Wales. Applied Geomechanics Technical Paper 24, CSIRO, Australia, 1977.

Barton, C. M., Analysis of joint traces. Proc. of 17th US Symp. on Rock Mechanics, Vol. 1, pp. 38–41, 1978.

Berkowitz, B., Characterizing flow and transport in fractured geological media: A review. Advances in Water Resources, 2002;25:861–884.

Bossart, P., Hermanson, J. and Mazurek, M., Äspö Hard Rock Laboratory: Analysis of fracture networks based on the integration of structural and hydrogeological observations on different scales. Technical Report, TR-01-21, Swedish Nuclear Fuel and Waste Management Co. (SKB), Stockholm, Sweden, 2001.

Bridges, M. C., Presentation of fracture data for rock mechanics. Proc. 2nd Australia–New Zealand Conf. on Geomechanics, Brisbane, Australia, pp. 144–148, 1975.

Call, R. B., Savely, J. and Nicholas, D. E., Estimation of joint set characteristics from surface mapping data. Proc. 17th US Symp. on Rock Mechanics, pp. 2B-1–2B2-9, 1976.

Chilés, J.-P. and de Marsily, G., Models of fracture systems. In: Bear, J., Tsang, C.-F. and de Marsily, G. (eds), Flow and contamination transport in fractured rock. Academic Press, San Diego, pp. 169–236, 1993.

Cruden, D. M., Describing the size of discontinuities. International Journal of Rock Mechanics and Mining Sciences and Geomechanics Abstracts, 1977;14:133–137.

Deere, D. U., Technical description of rock cores for engineering purposes. Rock Mechanics and Rock Engineering, 1964;1:17–22.

Dershowitz, W. S., Interpretation and synthesis of discrete fracture orientation, size, shape, spatial structure and hydrologic data by forward modeling. Proc. ISRM Regional Conf. on Fractured and Fractured Rock Masses (Preprints), pp. 680–687, 1992.

Dershowitz, W. S. and Einstein, H. H., Characterizing rock joint geometry with joint system models. Rock Mechanics and Rock Engineering, 1988;21:21–51.

Einstein, H. H., Baecher, G. B. and Veneziano, D., Risk analysis for rock slopes in open pit mines – Part I and IV. Technical Report to US Bureau of Mines, Contract JO2575015, MIT, Massachusetts, 1979.

Einstein, H. H. and Baecher, G. B., Probabilistic and statistical methods in engineering geology – specific methods and examples. Rock Mechanics and Rock Engineering, 1983;16:39–72.

Fardin, N., Stephansson, O. and Jing, L., The scale dependence of rock joint surface roughness. International Journal of Rock Mechanics and Mining Sciences, 2002;38:659–669.

Feng, Q., Sjögren, P., Stephansson, O. and Jing, L., Measuring fracture orientation at exposed rock faces by using a non-reflector total station. Engineering Geology, 2001;59:133–146.

Geological Society of London, The logging of rock cores for Engineering purposes. Geological Society Engineering Group Working Party Report. Quarterly Journal of Engineering Geology, 1970;3:1–24.

Golder Associate Ltd., The Manual of FracMan code, 1993.

Grossmann, N. G., Joint Statistics – state-of-the-art and practical applications. Int. Workshop on Survey and Testing Method for Discontinuous Rock Masses, Tokyo, Japan, 1990.

Hammah, R. E. and Curran, J. H., On distance measures for the fuzzy K-means algorithm for joint data. Rock Mechanics and Rock Engineering, 1999;32(1):1–27.

Herda, H. H., Strike standard deviation for shallow-dipping rock fracture sets. Rock Mechanics and Rock Engineering, 1999;32(4):241–255.

Hudson, J. A. and Priest, S. D., Discontinuities and rock mass geometry. International Journal of Rock Mechanics and Mining Sciences and Geomechanics Abstracts, 1979;16:339–362.

Hudson, J. A. and Priest, S. D., Discontinuity frequency in rock masses. International Journal of Rock Mechanics and Mining Sciences and Geomechanics Abstracts, 1983;20:73–89.

ISRM (International Society of Rock Mechanics), Suggested methods for the quantitative description of discontinuities in rock masses. International Journal of Rock Mechanics and Mining Sciences and Geomechanics Abstracts, 1978;15):319–368.

Kulatilake, P. H. S. W., Fitting Fisher distributions to discontinuity orientation data. Journal of Geological Education, 1985;33(5):266–269.

Kulatilake, P. H. S. W., Bivariate normal distribution fitting on discontinuity orientation clusters. Mathematical Geology, 1986;18(2):181–195.

Kulatilake, P. H. S. W., A correction for sampling bias on joint orientation for finite size joints intersecting finite size exposures. Proc. 6th Int. Conf. Numerical Methods in Geomechanics, Innsbruck, Austria, pp. 871–876, 1988.

Kulatilake, P. H. S. W., Lecture notes on stochastic 3-D fracture network modeling including verification at the Division of Engineering Geology. Royal Institute of Technology, Stockholm, Sweden, 1991.

Kulatilake, P. H. S. W. and Wu, T. H., Sampling bias on orientation of discontinuities. Rock Mechanics and Rock Engineering, 1984;17:243–253.

Lanaro, F., Jing, L. and Stephansson, O., 3-D-laser measurements and representation of roughness of rock fractures. In: Rossmanith, H.-P. (ed.), Proc. of the Int. Conf. on Mechanics, Jointed and Faulted Rock, MJFR-3, Vienna, Austria, pp. 185–189. Balkema, Rotterdam, 1998

La Pointe, P. R., Analysis of spatial variation in rock mass properties through geostatistics. Proc. of US Symp. on Rock Mechanics, Missouri-Rolla, pp. 570–580, 1980.

La Pointe, P., Cladouhos, T. and Follin, S., Calculation of displacements on fractures intersecting canisters induced by earthquakes: Aberg, Beberg and Ceberg examples. Technical Report, TR-99-03, Swedish Nuclear Fuel and Waste Management Company, Stockholm, 1999.

Long, J. C. S. and Witherspoon, P. A., The relationship of the degree of interconnection to permeability of fractured networks. Journal of Geophysical Research, 1985;90(B4):3087–3098.

Long, J. C. S., Billaux, D., Hestir, K. and Chiles, J. F., Some geostatistical tools for incorporating spatial structures in fracture network modeling. Proc. 6th ISRM Cong., Montreal, pp. 171–176, 1987.

Mathab, M. A., Bolstad, D. D., Allredge, J. R. and Shanley, R. J., Analysis of fracture orientations for the input to structural models of discontinuous rocks. Report of investigations 7669, US Dept. of the Interior-Bureau of Mines, Washington, DC, 1972.

Mauldon, M., Keyblock probabilities and size distributions: A first model for impersistent 2-d fractures. International Journal of Rock Mechanics and Mining Sciences and Geomechanics Abstracts, 1995;32(6):575–583.

Mauldon, M., Estimating mean fracture trace length and density from observations in convex windows. Rock Mechanics and Rock Engineering, 1998;31(4):201–216.

Mauldon, M., Dunne, W. M. and Rohrbaugh, M. B., Jr, Circular scanlines and circular windows: New tools for characterizing the geometry of fracture traces. Journal of Structural Geology, 2001;23:247–258.

McClay, K., The mapping of geological structures, Geological Society of London Handbook, The Geological Field Guide Series. Wiley, Chichester, 1987.

McMahon, B., A statistical method for the design of rock slopes. Proc. 1st Australia–New Zealand Conf. on Geomechanics, pp. 314–321, 1971.

Min, K.-B. and Jing, L., Numerical determination of the equivalent elastic compliance tensor for fractured rock masses using the distinct element method. International Journal of Rock Mechanics and Mining Sciences, 2003;40(6):795–816.

Min, K.-B., Stephansson, O. and Jing, L., Fracture system characterization and evaluation of the equivalent permeability tensor of fractured rock masses using a stochastic REV approach. Hydrogeology Journal, 2004;12(5):497–510.

Nirex UK Ltd., Evaluation of heterogeneity and scaling of fractures in the Borrowdale Volcanic Group in the Sellafield Area, Nirex Report SA/97/028, 1997.

Pahl, P. J., Estimating the mean length of discontinuity traces. International Journal of Rock Mechanics and Mining Sciences and Geomechanics Abstracts, 1981;18(3):221–228.

Pan, B. and Jing, L., Computer simulation methods and applications of statistical models of rockmass structure. In: New development in rock mechanics, pp. 55–80. Chinese Society for Rock Mechanics and Engineering, Northeast University Press, 1987.

Park, B. Y., Kim, K. S., Kwon, S., Kim, C., Bae, D. S., Hartley, L. J. and Lee, H. K., Determination of the hydraulic conductivity components using a three-dimensional fracture network model in volcanic rock. Engineering Geology, 2002;66(1–2):127–141.

Priest, S. D., Discontinuity analysis for rock engineering. Chapman and Hall, London, 1993.

Priest, S. D. and Hudson, J. A., Discontinuity spacings in rock. International Journal of Rock Mechanics and Mining Sciences and Geomechanics Abstracts, 1976;13(5):135–148.

Priest, S. D. and Hudson, J. A., Estimation of discontinuity spacing and trace length using scanline surveys. International Journal of Rock Mechanics and Mining Sciences and Geomechanics Abstracts, 1983;18(3):183–197.

Rasmussen, T. C., Huang, C. H. and Evans, D. D., Numerical experiments on artificially generated three-dimensional fracture networks: An examination of scale and aggregation effects. International Association of Hydrogeologists, Memoirs, 17, Hydrogeology of Rocks of Low Permeability, pp. 676–680, 1985.

Robertson, A., The interpretation of geologic factors for use in slope theory. In: van Rensburg, P. W. J. (ed.), Proc. Symp. Theoretical Background to the Planning of Open Pit Mines, Johannesburg, South Africa, pp. 55–71. Balkema, Cape Town, 1970.

Sen, Z. and Kazi, A., Discontinuity spacing and RQD estimates from finite length scanlines. International Journal of Rock Mechanics and Mining Sciences and Geomechanics Abstracts, 1984;21(4):203–212.

Steffen, O., Recent developments in the interpretation of data from joint surveys in rock masses. Proc. 6th Regional Conf. for Africa on Soil Mech. and Found., Vol. II, pp. 7–26, 1975.

Terzaghi, R. D., Sources of error in joint surveys. Geotechnique, 1965;15:287–304.

Wallis, P. F. and King, M. S., Discontinuity spacings in a crystalline rock. International Journal of Rock Mechanics and Mining Sciences and Geomechanics Abstracts, 1980;17(1):63–66.

Wang, X. and Chen, Z., Statistical survey of rock fracture systems and computer simulations. Research Report, Institute of Water Conservancy and Hydroelectric Power researches, Beijing, China, 1992 (in Chinese).

Warburton, P. M., A stereological interpretation of joint trace data. International Journal of Rock Mechanics and Mining Sciences and Geomechanics Abstracts, 1980;17(4):181–190.

Wilcock, P., The NAPSAC fracture network code. In: Stephansson, O., Jing, L. and Tsang, C.-F. (eds), Coupled thermo-hydro-mechanical processes of fractured media – mathematical and experimental studies. Development in Geotechnical Engineering, 79, pp. 529–538. Elsevier, Amsterdam, 1996.

Zhang, L. and Einstein, H. H., Estimating the mean trace length of rock discontinuities. Rock Mechanics and Rock Engineering, 1998;31(4):217–235.

6 THE BASICS OF COMBINATORIAL TOPOLOGY FOR BLOCK SYSTEM REPRESENTATION

The mathematical representation of a block system formed by a network of fractures is the first task for a DEM method. It requires definition of the geometry of blocks, both individually and collectively, and they can be quite complex for three-dimensional problems. From the DEM and solid geometry modeling literature, there exist basically three methods for block system generation: constructive solid geometry (CSG), successive space division (SSD) and boundary representation (BR).

In the CSG method, the primary solid elements of simple geometry (cubes, balls, etc.) are used as the basic components that are combined by topological transformation and identification processes to form more complex solid forms (Mäntylä, 1988). It is widely used in computer graphics, solid geometry modeling and robotics research. The creation of particle assemblages for the simulation of granular materials is achieved mostly by the CSG method, using circular, elliptical, spherical or ellipsoidal forms of various sizes (Cundall and Strack, 1979).

The SSD approach, successive space division, depends on a successive division process of pre-defined master blocks by newly introduced fractures of infinitely large sizes. The master blocks, usually regular in form, serve also as the domains of interest (Fig. 6.1a–c). The method is used in the three-dimensional DEM code 3DEC that is widely used for rock engineering problems (Itasca, 2000). The fractures are assumed also to be smooth and planar, without thickness. The input parameters required to introduce a new fracture into the model are its orientation (dip direction and dip angle) and the coordinates of a reference point so that the fracture can be placed at specific locations in the model. The shape and size data are not required. To prevent some existing blocks from being cut by a new fracture, these blocks can be hidden away and their intersections with the new fracture are discarded.

The SSD approach is conceptually simple, straightforward to program and probably more efficient in computation. On the other hand, due to the infinitely large fractures, all blocks generated are convex. The artificial fractures that need to be introduced to defined engineering boundaries (such as tunnels, caverns and slopes) are also infinitely large and thus produce unnecessary parts of artificial fractures and blocks, which may make the block geometry and mechanical behavior of the near-field rocks more complex than reality. All the above shortcomings are due to one factor: the assumption of infinitely large fractures, both natural and artificial. The more significant shortcoming for this method is the fact that it has great difficulties in representing natural fracture systems, deterministic or stochastic, formed by fractures of finite size and general shapes which is important for fluid flow analysis in the fracture systems.

The BR approach, boundary representation, uses principles of closed surfaces and polyhedra in combinatorial topology to represent boundaries of blocks. It was first reported in Lin et al. (1987) for the representation of block system geometry for rock mechanics problems, with extensive use of the terminology and operations of set theory – which is not so easily understood by practicing engineers without a thorough knowledge of set algebra and topological geometry. On the other hand, the method is more natural for representing the geometry of blocks of general shapes, convex or concave, and the

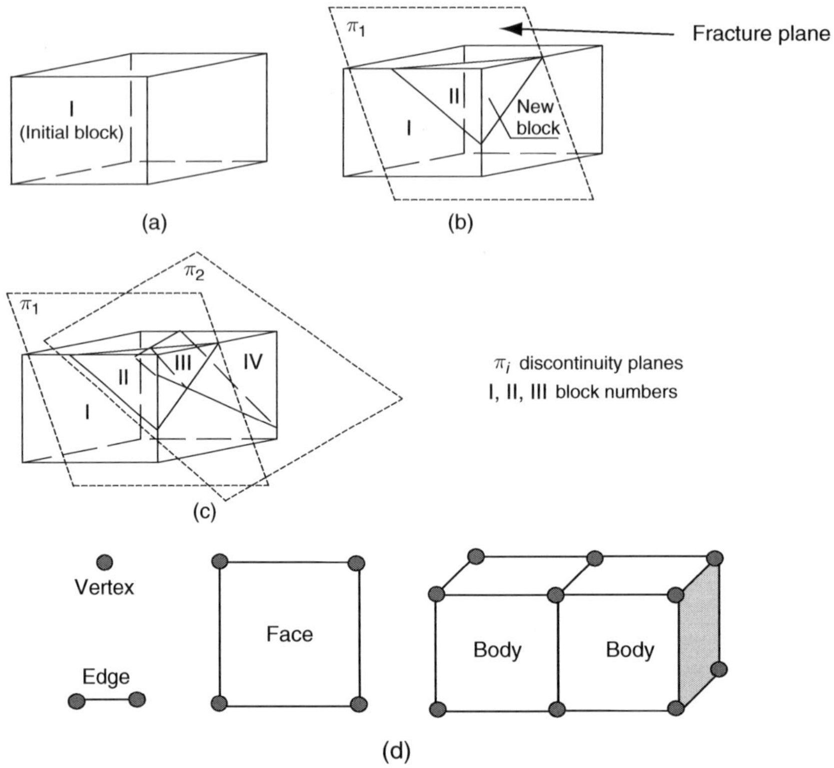

Fig. 6.1 Successive space division approach with infinitely large fractures. (a) The master block defining the domain of interest; (b) splitting of the master block by the first fracture; (c) successive splitting of existing blocks by new fractures; (d) basic topological elements for boundary representation (BR) method of block system generation in DEM: vertex, edge, face and body.

connectivity with neighboring blocks, and is computationally more efficient with skillful programming. The definitions of the basic topological elements of vertex, edge, face and body are shown in Fig. 6.1d.

The BR method uses the basic topological elements of vertices, edges, faces and bodies to define fractures of finite size, shape, orientation and location, bodies of general shapes and sizes and location, and the block systems through the connectivity relations between the bodies. A vertex is a point element representing an intersection point between two or more edges. An edge is a linear element representing intersection line segments between two or more faces. A face is a surface element defined by at least three intersecting edges and represents an intersection surface between two bodies (Fig. 6.2). An edge connects only two vertices, but can connect multiple faces. A face connects only two bodies, but can connect multiple edges and vertices. In topology, edges do not need to be straight and so can be curves. Faces do not need to be planar and smooth and so can be curved rough surfaces. However, in DEM or any other numerical

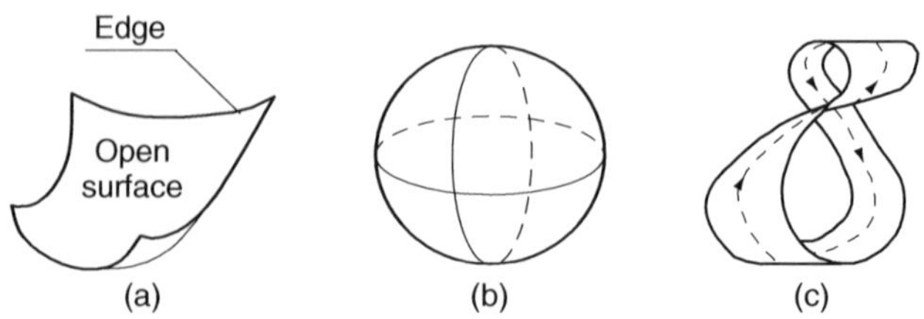

Fig. 6.2 (a) An open surface with edge; (b) a closed spherical surface; (c) a Möbius band.

method, we treat the edges as straight linear segments, faces as planar polygons and bodies as generally shaped polyhedra that can be convex, concave or multiply connected (i.e., having holes).

In the BR method, the fractures are assumed to be either regular (rectangle, circular or elliptical) or generally irregularly shaped planar polygons of finite size, without thickness. The data for fracture orientation (dip angle and dip direction), side lengths (for rectangle fractures), radius (for circular fractures) or lengths of the two semi-principal axes (for elliptical fractures) and the coordinates of a reference point (mass center, for example) are used as input data for regular fracture definition. The complete set of coordinates of boundary vertices are used to define irregular fractures whose dip direction and dip angle can be calculated from the coordinates of the boundary vertices. All fractures are introduced at once, not successively as in SSD method, and the intersections are calculated to define the sets of vertices, edges and faces (general polygons formed by single closed loops of edges) characterizing the individual blocks. The individual blocks are then identified one by one from the sets of vertices, edges and faces by using a boundary operator, and the Euler–Poincáre formula for a polyhedron is invoked to ensure the correctness of the block identification (termed tracing) process. The advantage of this algorithm is that blocks generated are generally shaped (i.e. convex, concave, singly or multiply connected with holes), and that the resultant block system provides a more realistic representation of the fracture connectivity and block system formation. This is because of the finite sizes of the fractures that are used. A minimum of artificial fractures is needed with precise shapes and dimensions to the defined engineering boundaries, and hence without introducing extra unnecessary fractures and blocks.

In this chapter, the basic concepts and principles of combinatorial topology of polyhedra are presented; these are used to develop the boundary representation algorithm. Although a complete coverage of combinatorial topology of polyhedra is beyond the scope of this book (see Henle, 1974), some most relevant concepts, based more on intuitive understanding of solid geometry and algebra, are presented to make the algorithm as readily understandable as possible to both practicing engineers and academic researchers alike.

6.1 Surfaces and Homeomorphism

A surface in combinatorial topology can have different topological properties of being open or closed and oriented or non-oriented (uniorientable). An open surface is a surface with boundaries (edges) (Fig. 6.2a). A moving point on one side of an open surface cannot move continuously onto another side without crossing its edge, but it can do so on a uniorientable surface.

A closed surface is a surface without boundaries (edges). A moving point on a closed surface cannot move continuously from one side to another, unless a hole is made in it. A surface enclosing a solid ball or a generally shaped solid polyhedron is a closed surface (Fig. 2b). A typical (also the most famous) uniorientable (or one-sided) surface is the Möbius band on which a moving point can move continuously from one side of the band onto another without crossing its edge (Fig. 6.2c).

The topology of surfaces is a branch of mathematics the subject of which is to study the properties of surfaces under topological transformations, i.e., continuous transformation of surfaces under continuous deformation, including stretching, shrinking, folding, crumbling without breaking, tearing, punching or overlapping. The property of a surface that is invariant under all topological transformations is called a topological property of the surface. If two surfaces can be transformed topologically from one to another, we say that the two surfaces are topologically equivalent, or more formally homeomorphic, or one surface is a homeomorphism of another. For example, a circular disk can be transformed by continuous deformation (without tearing, breaking and overlapping) into an ellipse, a curvilinear or straight-edged polygon and a cap-shaped surface (Fig. 6.3a). All these figures are homeomorphic to each other and we say that they are all topologically equivalent to a circular disk. Similarly, a spherical surface is homeomorphic to an ellipsoid and a straight-edged polyhedron without any through-going holes (Fig. 6.3b).

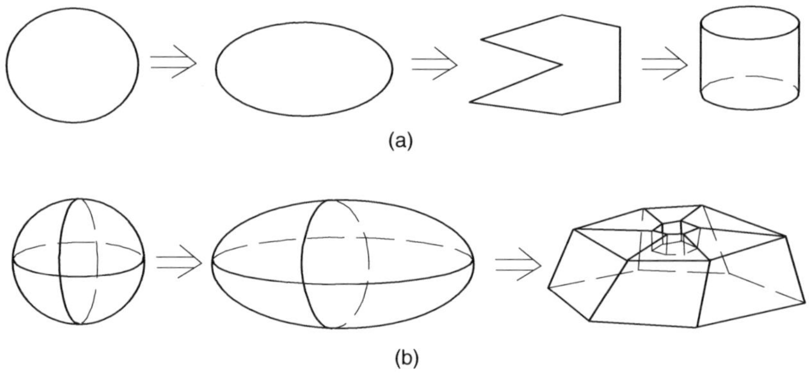

Fig. 6.3 Homeomorphic figures by topological transformation.

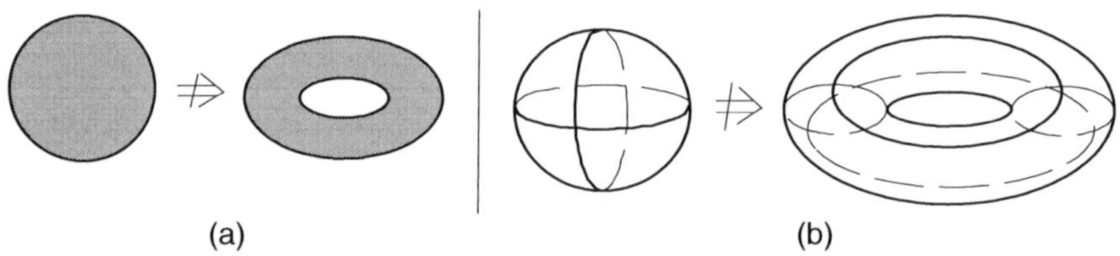

Fig. 6.4 Some non-homeomorphic figures: (a) a disk and an annulus; (b) a sphere and a torus.

However, a circular disk is not homeomorphic to an annulus and a spherical surface is not homeomorphic to a torus (Fig. 6.4). There are two categories of oriented, closed surfaces in topology that are most relevant to the study in this chapter: Those homeomorphic to a sphere and those homeomorphic to a torus with one hole or more complex closed surfaces of multiple holes.

6.2 The Polyhedron and Its Characteristics

An oriented, closed surface in a three-dimensional Euclidean space R^3 is called a polyhedron. The term polyhedron has two meanings. One refers to the solid block with its volume and mass and another to the polyhedral surface enclosing the solid mass. The meaning of a polyhedron in this book is the latter, unless stated otherwise. A polyhedron is homeomorphic to either a sphere, a torus or the closed surface of a generally shaped block with multiple holes.

Let us divide a sphere (or a torus) into a finite number of curvilinear polygons (termed a polygonal division) in such a way that two and only two of its curvilinear polygons share each curvilinear edge. By a homeomorphism, the sphere (or the torus) can be transformed into a polyhedron of planar polygonal faces and straight edges, corresponding to the same numbers of curvilinear edges and polygons on the sphere or torus (Fig. 6.5).

A polyhedron must satisfy the following conditions:

(1) Any of its two polygons have no common interior point.

(2) The edges of these polygons coincide in pairs (indicating that the total number of edges of the polygon set is even).

(3) The set of polygons cannot be separated into two disjoint subsets (implying that the polyhedron must be of one piece).

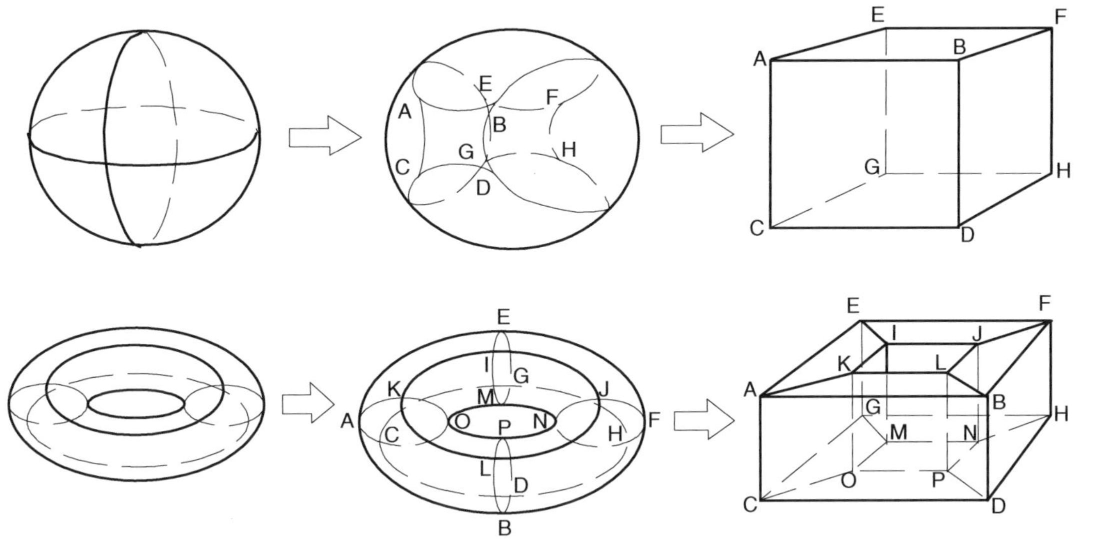

Fig. 6.5 Polygonal divisions of a sphere and a torus into two polyhedra with curvilinear and then planar faces (polygons) of straight edges.

(4) Around every vertex (intersection of edges) of the polyhedron, the polygons and edges sharing this vertex can be arranged in an alternative cyclic order of faces and edges (f_1, e_1 f_2, e_2 $\ldots$ f_n, e_n) such that f_i and f_{i+1} ($1 \le i \le n$, where $f_{n+1} = f_1$ and $e_n = e_1$) share a common edge e_i connected with this vertex (Fig. 6.6).

The topological property of a polyhedron is invariant under all homeomorphisms. That means the topological property of a polyhedron does not change for its different polygonal divisions meeting the above-mentioned conditions. This property is represented by a formula

$$N_\mathrm{v} + N_\mathrm{f} - N_\mathrm{e} = 2(N_\mathrm{b} - N_\mathrm{h}) \tag{6.1}$$

where $N_\mathrm{v}, N_\mathrm{f}, N_\mathrm{e}, N_\mathrm{b}, N_\mathrm{h}$ are the numbers of the vertices, faces, edges, bodies (polyhedra) and holes. This formula is called the Euler–Poincáre formula. For one polyhedron without a hole, the above formula becomes

$$N_\mathrm{v} + N_\mathrm{f} - N_\mathrm{e} = 2 \tag{6.2}$$

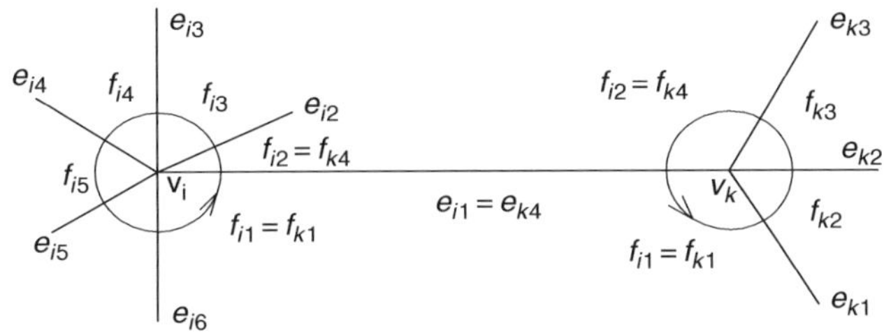

Fig. 6.6 Cyclic orders of edges and faces around a vertex of a polyhedron.

which is usually called Euler's or Euler-Poincáre polyhedron formula. For a torus, the formula becomes

$$N_v + N_f - N_e = 0 \tag{6.3}$$

The numbers 2 and 0 are said to be the characteristics of the sphere and torus, which are invariant under all homeomorphisms. Relations (6.1–6.3) are the governing relations for reconstruction of the block system of rocks using fracture data.

6.3 Simplex and Complex

The vertices, edges and faces (polygons) are all called simplexes in combinatorial topology. Figure 6.7 illustrates some simple simplexes in three dimensions. They are called 0-simplex (one single vertex), 1-simplex (an edge with two vertices), 2-simplex (a triangle with three edges and three vertices) and 3-simplex (a tetrahedron with four vertices, six edges and four faces). The simplexes formed by vertices located at the origin and on the axes with unit coordinates are termed unit simplexes (Fig. 6.7a–d). Their homeomorphisms are simplexes with general coordinates of vertices in the three-dimensional Euclidean space. As a rule, a simple n-simplex has always $n + 1$ vertices. The complete subsets of simplexes for the four cases in Fig. 6.7 are:

(a) and (e): 0-simplex (a_0): (a_0) – a single vertex;

(b) and (f): 1-simplex (a_0, a_1): (a_0), (a_1), (a_0, a_1) – two vertices and an edge;

(c) and (g): 2-simplex (a_0, a_1, a_2): (a_0), (a_1), (a_2), (a_0, a_1), (a_0, a_2), (a_1, a_2), (a_0, a_1, a_2) – three vertices, three edges and one polygon (triangle);

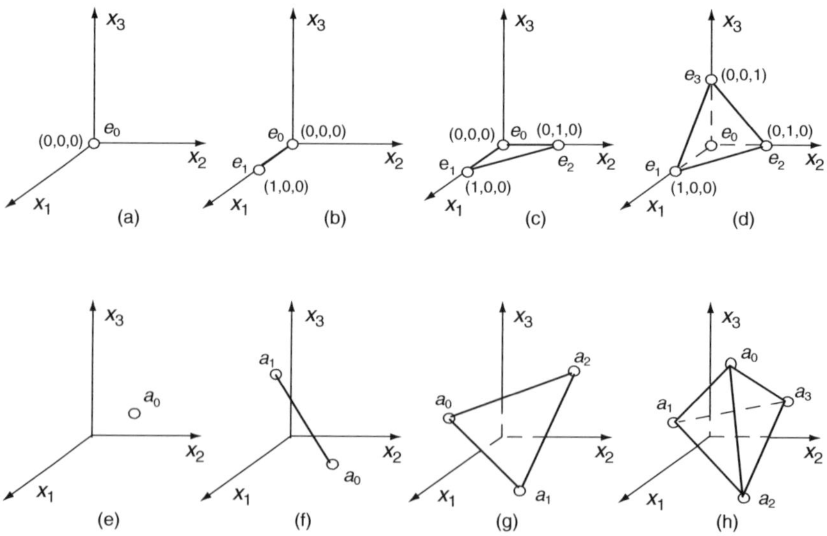

Fig. 6.7 The 0, 1, 2 and 3-simplexes in three dimensions. (a), (b), (c) and (d) are unit simplexes, (e), (f), (g) and (h) are their respective homeomorphisms.

(d) and (h): 3-simplex (a_0, a_1, a_2, a_3): $(a_0), (a_1), (a_2), (a_3), (a_0, a_1), (a_0, a_2), (a_0, a_3), (a_1, a_2), (a_1, a_3), (a_2, a_3), (a_0, a_1, a_2), (a_0, a_2, a_3), (a_0, a_3, a_1), (a_1, a_3, a_2), (a_0, a_1, a_2, a_3)$ – four vertices, six edges, four faces and one polyhedron.

Let $S = (a_0, a_1, a_2, \ldots, a_n)$ represent an n-simplex, in which $(a_0, a_1, a_2, \ldots, a_n)$ is the ordered list of the $(n + 1)$ vertices of the simplex and $(x_{1i}, x_{2i}, \ldots, x_{ni})$ be the coordinates of vertex a_i. Let $S^e = (a_0, a_1, a_2, \ldots, a_n, 1)$ be an extension of S with the addition of a vertex (1, 1, ... 1), the $(n + 1)$ vector $(x_{1i}, x_{2i}, \ldots, x_{ni}, 1)$, $(i = 0, 1, \ldots, n)$, forms a linearly independent $(n + 1)$-dimensional linear space. The orientation of the n-simplex $S = (a_0, a_1, a_2, \ldots, a_n)$ is then defined by the determinant of this linear space $D(a_0, a_1, a_2, \ldots, a_n, 1)$, given by the product

$$(-1)^n D(a_0, a_1, a_2, \ldots, a_n) = (-1)^n \begin{vmatrix} x_{10} & x_{20} & \cdots & x_{n0} & 1 \\ x_{11} & x_{21} & \cdots & x_{n1} & 1 \\ \cdots & \cdots & \cdots & \cdots & \cdots \\ x_{1i} & x_{2i} & \cdots & x_{ni} & 1 \\ \cdots & \cdots & \cdots & \cdots & \cdots \\ x_{1n} & x_{2n} & \cdots & x_{nn} & 1 \end{vmatrix} \neq 0 \tag{6.4}$$

The n-simplex is positively oriented if the product $(-1)^n D(a_0, a_1, a_2, \ldots, a_n) > 0$ and negatively oriented if the product $(-1)^n D(a_0, a_1, a_2, \ldots, a_n) < 0$. This implies that, if a n-simplex has the same orientation as the unit n-simplex, $S^0 = (e_0, e_1, e_2, \ldots, e_n)$ in which $\mathbf{e}_0 = \{0, 0, \ldots, 0\}$ is the origin, $\mathbf{e}_1 = \{1, 0, \ldots, 0\}$, $\mathbf{e}_2 = \{0, 1, \ldots, 0\}$, $\ldots \mathbf{e}_n = \{0, 0, \ldots, 0, 1\}$ are the vertices on the axes of the unit coordinates, and

$$D(\mathbf{e}_1, \mathbf{e}_2, \ldots, \mathbf{e}_n, 1) = \begin{vmatrix} 0 & 0 & \cdots & 0 & 1 \\ 1 & 0 & \cdots & 0 & 1 \\ \cdots & \cdots & \cdots & \cdots & \cdots \\ 0 & 0 & \cdots & 1 & 1 \end{vmatrix} = (-1)^n \tag{6.5}$$

A 0-simplex is only one vertex and has no orientation; a 1-simplex is an edge with two vertices, (x_{10}) and (x_{11}), and its orientation is given by

$$(-1^1)D(a_0, a_1) = -\begin{vmatrix} x_{10} & 1 \\ x_{11} & 1 \end{vmatrix} = (x_{11} - x_{10}) \tag{6.6}$$

Therefore, a 1-simplex (edge (a_0, a_1)) is positively oriented if $x_{11} > x_{10}$ and negatively oriented if $x_{11} < x_{10}$. This means that there are two classes of ordered pairs of vertices corresponding to the positive and negative orientation of a 1-simplex: (a_0, a_1) and (a_1, a_0), respectively, with $(a_0, a_1) = -(a_1, a_0)$.

The orientation of a 2-simplex is given by

$$(-1)^2 D(a_0, a_1, a_2) = \begin{vmatrix} x_{10} & x_{20} & 1 \\ x_{11} & x_{21} & 1 \\ x_{12} & x_{22} & 1 \end{vmatrix} \tag{6.7}$$

It has a positive orientation if the determinant is larger than zero, i.e., if vertices $\{a_0, a_1, a_2\}$ are ordered counter-clockwise. The even permutations $\{a_1, a_2, a_0\}$ and $\{a_2, a_0, a_1\}$ will also provide a positive orientation of the triangle simplex. The odd permutations of vertices order $\{a_1, a_0, a_2\}$, $\{a_0, a_2, a_1\}$ and $\{a_2, a_1, a_0\}$ represent the negative orientation of the 2-simplex.

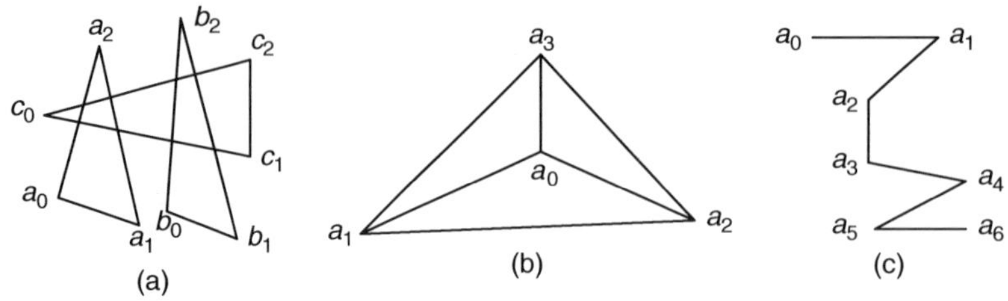

Fig. 6.8 Three sets of simplexes, (a) is not a complex.

A 3-simplex is positively oriented if

$$(-1)^3 D(a_0, a_1, a_2, a_3) = -\begin{vmatrix} x_{10} & x_{20} & x_{30} & 1 \\ x_{11} & x_{21} & x_{31} & 1 \\ x_{12} & x_{22} & x_{32} & 1 \\ x_{13} & x_{23} & x_{33} & 1 \end{vmatrix} > 0 \tag{6.8}$$

Meeting this condition, it can be directly deduced that all its 2-simplexes are positively oriented. Even and odd permutations of the vertex order will make the positive and negative classes of the simplex orientation, respectively.

A finite set K of simplexes in an n-dimensional Euclidean space R^n is called a finite Euclidean simplicial complex (simply called a complex hereafter) if and only if it has the following properties: (i) each edge (or face) of a simplex in K is also a simplex in K; (ii) the intersection of every two simplex in K is either a empty set or a common edge (or face) of these two simplexes. Simply speaking, the above definition means that all the simplexes cover exactly the exterior surface of the complex without overlapping, and each edge is shared by two and only two simplexes. Figure 6.8 illustrates three sets of simplexes in R^2. Sets (b) and (c) are complexes and set (a) is not.

A complex K is said to be oriented if and only if a definite orientation (which may be chosen arbitrarily) is assigned to each of the simplexes in K. On the other hand, if a n-dimensional complex K has all its n-simplexes having the same orientation in R^n, then the complex K is called a simplicial subdivision. The orientation of K is then the same as its n-simplexes and K is called an n-dimensional simplicial complex. Figure 6.9 illustrates a two-dimensional simplicial complex of a polygon with seven vertices (called a 7-gon) with orientations of the simplexes and the complex indicated by arrows. An oriented n-dimensional simplicial complex K can be said to be a generalized n-simplex that can be used to construct $(n + 1)$-dimensional simplicial complexes of more complex form.

For a n-dimensional simplicial complex K, denote the m-dimensional ($m \leq n$), oriented simplexes as s_i, (i = 1, 2, ... k), in K, then the formal sum expression $\sum_{i=1}^{k} c_i s_i$ is called a m-chain of simplexes in K

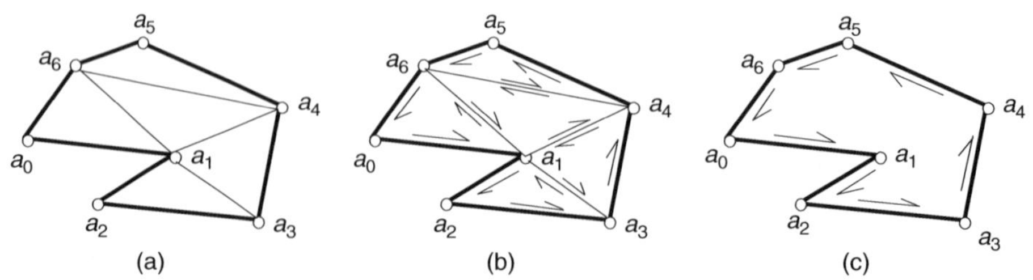

Fig. 6.9 Simplicial subdivision of a complex (a 7-gon): (a) the subdivision of K by simplexes; (b) assigning the orientation of simplexes; (c) orientation of the complex.

(written as $C_m(K)$) and c_i are real numbers called the coefficients of the m-chain. Two chains are equal if and only if they have the same dimension and same coefficients. The sum is a formal sum without any operations between the simplexes in the m-chain. However, operations of addition and multiplication are defined on a set of chains (of the same dimension) under the following rules (a, b, c, 1 and 0 are real numbers):

$$1 \cdot s = s \tag{6.9a}$$

$$(-1)s = -s \tag{6.9b}$$

$$0 \cdot s = 0 \tag{6.9c}$$

$$a(b)s = (ab)s \tag{6.9d}$$

$$(a+b)s = as + bs \tag{6.9e}$$

$$\sum_{i=1}^{k} a_i s_i + \sum_{i=1}^{k} b_i s_i = \sum_{i=1}^{k} b_i s_i + \sum_{i=1}^{k} a_i s_i \tag{6.9f}$$

$$c\left(\sum_{i=1}^{k} a_i s_i + \sum_{i=1}^{k} b_i s_i +\right) = c\sum_{i=1}^{k} a_i s_i + c\sum_{i=1}^{k} a_i s_i \tag{6.9g}$$

These operations define an Abelian group on chains (Henle, 1974).

The concept of chains is necessary to define and calculate the boundary of the simplexes and complexes. If $S = (a_0, a_1, \ldots, a_n)$ is an n-dimensional simplex in R^n and $s_r = (a_0, a_1, \ldots, \hat{a}_r, \ldots a_n) = (a_0, a_1, \ldots, a_{r-1}, a_{r+1}, \ldots a_n)$, $(r = 0, 1, 2, \ldots, n)$ is its $(n-1)$-dimensional side without vertex a_r (see Table 6.1), then the topological boundary of this simplex is defined by the set

$$\bigcup_{i=0}^{n} (a_0, a_1, \ldots, \hat{a}_r, \ldots, a_n) \tag{6.10}$$

i.e., the union of all its $(n-1)$-dimensional sides, written as $\partial(a_0, a_1, \ldots, a_n)$. The symbol ∂ is called the boundary operator. Table 6.1 lists the topological boundaries of i-simplexes in R^i.

The definition of topological boundary by (6.10) suggests that the boundary of an n-simplex, ∂S, is a chain of its n-dimensional sides, but its coefficient has not yet been determined. These linearly independent sides form an n-dimensional linear space with its determinant $D(a_0, a_1, \ldots, \hat{a}_r, \ldots, a_n)$

Table 6.1 Topological boundaries of i-simplexes in R^i

n-simplex	$(n$-$1)$-dimensional sides
(a_0, a_1)	$(a_0), (a_1)$
(a_0, a_1, a_2)	$(a_0, a_1), (a_0, a_2), (a_1, a_2)$
(a_0, a_1, a_2, a_3)	$(a_0, a_1, a_2), (a_0, a_1, a_3), (a_0, a_2, a_3), (a_1, a_2, a_3)$
....	....
$(a_0, a_1, \ldots, a_n)$	$\{(a_0, a_1, \ldots, \hat{a}_r, \ldots a_n),\ r = 0, 1, 2, \ldots, n\}$

given by

$$D(a_0, a_1, \ldots, \hat{a}_r, \ldots, a_n) = \begin{vmatrix} x_{10} & x_{20} & \cdots & x_{n0} \\ \cdots & \cdots & \cdots & \cdots \\ x_{1,r-1} & x_{2,r-1} & \cdots & x_{n,r-1} \\ x_{1,r+1} & x_{2,r+1} & \cdots & x_{n,r+1} \\ \cdots & \cdots & \cdots & \cdots \\ x_{1n} & x_{2n} & \cdots & x_{nn} \end{vmatrix} \tag{6.11}$$

For an oriented n-simplex, S, its orientation is determined by the product expressed in Eqn (6.4). Expanding the $(n + 1)$-dimensional determinant by the minors of the elements in the column of 1s, the product for the orientation becomes

$$(-1)^n D(a_0, a_1, a_2, \ldots, a_n, 1) = (-1)^{2n+2} \sum_{r=0}^{n} (-1)^r D(a_0, a_1, \ldots \hat{a}_r, \ldots, a_n) \tag{6.12}$$

The term $(-1)^r, (r = 0, 1, \ldots, n)$, then determines the missing coefficients (always $+1$ or -1) in the $(n - 1)$-chain $C_{n-1}(K)$, the topological boundary of the n-simplex $S = (a_0, a_1, \ldots, a_n)$. The resultant $(n - 1)$-chain is called the algebraic boundary of the oriented n-simplex (see Table 6.2). Therefore, the precise definition of the algebraic boundary ∂S of an oriented n-simplex $S = (a_0, a_1, \ldots, a_n)$ in an oriented simplicial complex K is given by the formal sum of its $(n - 1)$-chains $C_{n-1}(K)$, written

$$\partial S = \sum_{i=0}^{n} (-1)^i (a_0, a_1, \ldots, \hat{a}_i, \ldots, a_n) = \sum_{i=0}^{n} (-1)^i (a_0, a_1, \ldots, a_{i-1}, a_{r+1}, \ldots, a_n) \tag{6.13}$$

Table 6.2 lists the algebraic boundaries of i-simplexes in R^i, corresponding to the topological boundaries in Table 6.1.

For an oriented simplicial complex K in R^n with M oriented n-simplexes, S_i, (i = 1, 2, ... M), the algebraic boundary of K, ∂K, is defined to be the union of boundaries of all n-simplexes, given by

Table 6.2 Algebraic boundaries of i-simplexes R^i.

Determinant expression for orientation	Algebraic boundary ($C_{n-1}(K)$)
$(-1)^1 D(a_0, a_1, 1) = D(a_1) - D(a_0)$	$\partial(a_0, a_1) = (a_1) - (a_0)$
$(-1)^2 D(a_0, a_1, a_2, 1) = D(a_1, a_2) - D(a_0, a_2) + D(a_0, a_1)$	$\partial(a_0, a_1, a_2) = (a_1, a_2) - (a_0, a_2) + (a_0, a_1)$
$(-1)^3 D(a_0, a_1, a_2, a_3, 1) = D(a_1, a_2, a_3) - D(a_0, a_2, a_3) + D(a_0, a_1, a_3) - D(a_0, a_1, a_2)$	$\partial(a_0, a_1, a_2, a_3) = (a_1, a_2, a_3) - (a_0, a_2, a_3) + (a_0, a_1, a_3) - (a_0, a_1, a_2)$
....	....
$(-1)^n D(a_0, a_1, a_2, \ldots, a_n, 1) = \sum_{r=0}^{n} (-1)^r D(a_0, a_1, \ldots \hat{a}_r, \ldots, a_n)$	$\partial(a_0, a_1, a_2, \ldots, a_n) = \sum_{r=0}^{n} (-1)^r (a_0, a_1, \ldots \hat{a}_r, \ldots, a_n)$

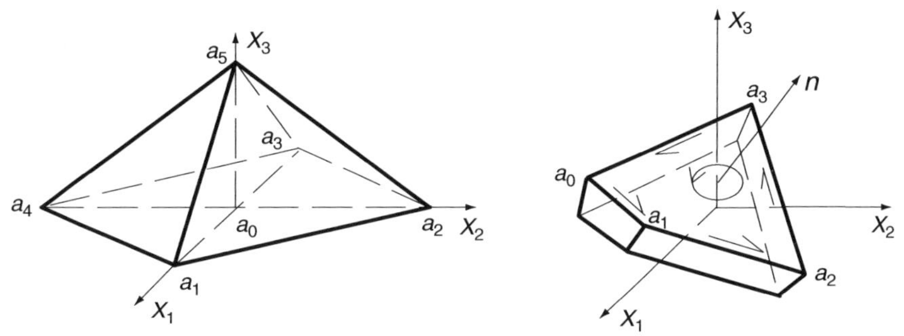

Fig. 6.10 (a) A complex in R^3 of four simplexes and (b) the right-hand rule for positive orientation of a face of a polyhedron.

$$\partial K = \partial \sum_{i=1}^{M} S_i = \sum_{i=1}^{M} \partial(S_i) = \sum_{i=1}^{M} \left(\sum_{r=0}^{n} (-1)^r (a_0, a_1, \ldots, \hat{a}_r, \ldots, a_n) \right) \tag{6.14}$$

As an example, see the oriented simplicial complex K in R^2 in Fig. 6.9. The complex K consists of five positively oriented 2-simplexes: $S_1 = (a_0, a_1, a_6)$, $S_2 = (a_1, a_2, a_3)$, $S_3 = (a_1, a_3, a_4)$, $S_4 = (a_1, a_4, a_6)$ and $S_5 = (a_4, a_5, a_6)$. The algebraic boundary of the complex K is then given by

$$\begin{aligned}
\partial K &= \partial(S_1 + S_2 + S_3 + S_4 + S_5) = \partial S_1 + \partial S_2 + \partial S_3 + \partial S_4 + \partial S_5 \\
&= [(a_1, a_6) - (a_0, a_6) + (a_0, a_1)] + [(a_2, a_3) - (a_1, a_3) \\
&\quad + (a_1, a_2)] + [(a_3, a_4) - (a_1, a_4) + (a_1, a_3)] \\
&\quad + [(a_4, a_6) - (a_1, a_6) + (a_1, a_4)] + [(a_5, a_6) - (a_4, a_6) + (a_4, a_5)] \\
&= (a_0, a_1) + (a_1, a_2) + (a_2, a_3) + (a_3, a_4) + (a_4, a_5) + (a_5, a_6) - (a_0, a_6)
\end{aligned} \tag{6.15}$$

as shown in Fig. 6.9c. It is a chain of 1-simplexes, called a 1-chain.

Figure 6.10a shows a complex K in R^3 of four simplexes. Applying the relation (6.4) for assignment of simplex orientations, the ordered lists of the positively oriented simplexes are: $S_1 = (a_0, a_1, a_2, a_5)$, $S_2 = (a_0, a_2, a_3, a_5)$, $S_3 = (a_0, a_3, a_4, a_5)$ and $S_4 = (a_0, a_4, a_1, a_5)$. The algebraic boundary of the complex K is then calculated as

$$\begin{aligned}
\partial K &= \partial(S_1 + S_2 + S_3 + S_4) = \partial S_1 + \partial S_2 + \partial S_3 + \partial S_4 \\
&= [(a_1, a_2, a_5) - (a_0, a_2, a_5) + (a_0, a_1, a_5) - (a_0, a_1, a_2)] \\
&\quad + [(a_2, a_3, a_5) - (a_0, a_3, a_5) + (a_0, a_2, a_5) - (a_0, a_2, a_3)] \\
&\quad + [(a_3, a_4, a_5) - (a_0, a_4, a_5) + (a_0, a_3, a_5) - (a_0, a_3, a_4)] \\
&\quad + [(a_4, a_1, a_5) - (a_0, a_1, a_5) + (a_0, a_4, a_5) - (a_0, a_4, a_1)] \\
&= (a_1, a_2, a_5) - (a_0, a_1, a_2) + (a_2, a_3, a_5) - (a_0, a_2, a_3) \\
&\quad + (a_3, a_4, a_5) - (a_0, a_3, a_4) + (a_4, a_1, a_5) - (a_0, a_4, a_1)
\end{aligned} \tag{6.16}$$

which is a chain of 2-simplexes, a 2-chain. The relation (6.4) also implies that, for a oriented polyhedron in R^3, if the vertices of a face (polygon) are ordered counter-clockwise (positive orientation), then the outward normal direction of the face points away from the interior of the polygon, i.e., the right-hand rule applies (Fig. 6.10b).

If K is an oriented, simplicial complex in R^n consisting of M n-simplexes $S_r = (a_0, a_1, \ldots, \hat{a}_r, \ldots, a_n)$, $(r = 0, 1, 2, \ldots, n)$, then the boundary of the simplexes is a 0-chain, i.e.,

$$\partial(\partial S_r) = \partial\partial S_r = 0 \tag{6.17}$$

Therefore, the boundary of the $(n - 1)$-chains of the complex K is also a 0-chain, i.e.,

$$\partial(\partial K) = \partial\partial K = \partial\left(\partial \sum_{i=0}^{M} S_i\right) = \partial\left(\sum_{i=0}^{M} \partial S_i\right) = \sum_{i=0}^{M} \partial(\partial S_i) = 0 \tag{6.18}$$

For the two examples shown in Figs 6.9 and 6.10a, application of relations (6.17) and (6.18) leads to

(1) for the complex K in Fig. 6.9:

$$\begin{aligned}\partial(\partial K) &= \partial(\partial S_1 + \partial S_2 + \partial S_3 + \partial S_4 + \partial S_5) = \partial(\partial S_1) + \partial(\partial S_2)\\ &\quad + \partial(\partial S_3) + \partial(\partial S_4) + \partial(\partial S_5)\\ &= \partial(a_0, a_1) + \partial(a_1, a_2) + \partial(a_2, a_3) + \partial(a_3, a_4) + \partial(a_4, a_5)\\ &\quad + \partial(a_5, a_6) - \partial(a_0, a_6)\\ &= [(a_1) - (a_0)] + [(a_2) - (a_1)] + [(a_3) - (a_2)] + [(a_4) - (a_3)] + [(a_5) - (a_4)]\\ &\quad + [(a_6) - (a_5)] - [(a_6) - (a_0)] = 0\end{aligned} \tag{6.19}$$

(2) for the complex K in Fig. 6.10a:

$$\begin{aligned}\partial(\partial K) &= \partial(\partial S_1 + \partial S_2 + \partial S_3 + \partial S_4) = \partial(\partial S_1) + \partial(\partial S_2) + \partial(\partial S_3) + \partial(\partial S_4)\\ &= \partial(a_1, a_2, a_5) - \partial(a_0, a_1, a_2) + \partial(a_2, a_3, a_5) - \partial(a_0, a_2, a_3)\\ &\quad + \partial(a_3, a_4, a_5) - \partial(a_0, a_3, a_4) + \partial(a_4, a_1, a_5) - \partial(a_0, a_4, a_1)\\ &= [(a_2, a_5) - (a_1, a_5) + (a_1, a_2)] - [(a_1, a_2) - (a_0, a_2) + (a_0, a_1)]\\ &\quad + [(a_3, a_5) - (a_2, a_5) + (a_2, a_3)] - [(a_2, a_3) - (a_0, a_3) + (a_0, a_2)]\\ &\quad + [(a_4, a_5) - (a_3, a_5) + (a_3, a_4)] - [(a_3, a_4) - (a_0, a_4) + (a_0, a_3)]\\ &\quad + [(a_1, a_5) - (a_4, a_5) + (a_4, a_1)] - [(a_4, a_1) - (a_0, a_1) + (a_0, a_4)] = 0\end{aligned} \tag{6.20}$$

Relation (6.18) is a manifestation of the topological property of a closed surface, i.e., the sum of oriented edges of an oriented simplicial complex in R^3 representing a polyhedron is empty. Otherwise the complex represents an open surface.

6.4. Planar Schema of Polyhedra

A polyhedron (curvilinear or planar) can be represented as a planar graph in two-dimensions, called a planar schema of the polyhedron. The making of a planar schema of a polyhedron homeomorphic to a sphere can be done by removing one of the faces and flattening and stretching of the rest of the surface, see Fig. 6.11 for the making of a planar graph of a cube. A planar graph representation of a polyhedron is a boundary representation.

The boundary representation of a polyhedron homeomorphic to a sphere by a planar graph is straightforward, see Fig. 6.12 for more examples. The orientations of all faces are positive (the vertices are ordered counter-clockwise and the right-hand rule is applied to determine the outward normal directions of the faces) and all faces are simple (defined by one single closed chain of edges). The

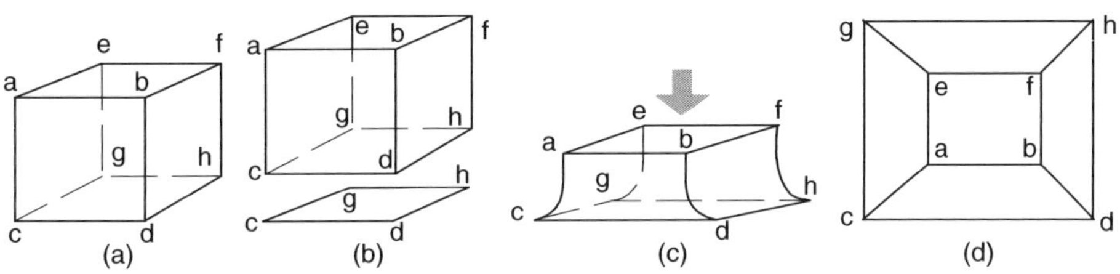

Fig. 6.11 Making of a planar graph of a cube: (a) initial cube; (b) removal of one face; (c) flattening the rest of the surface; (d) final planar graph.

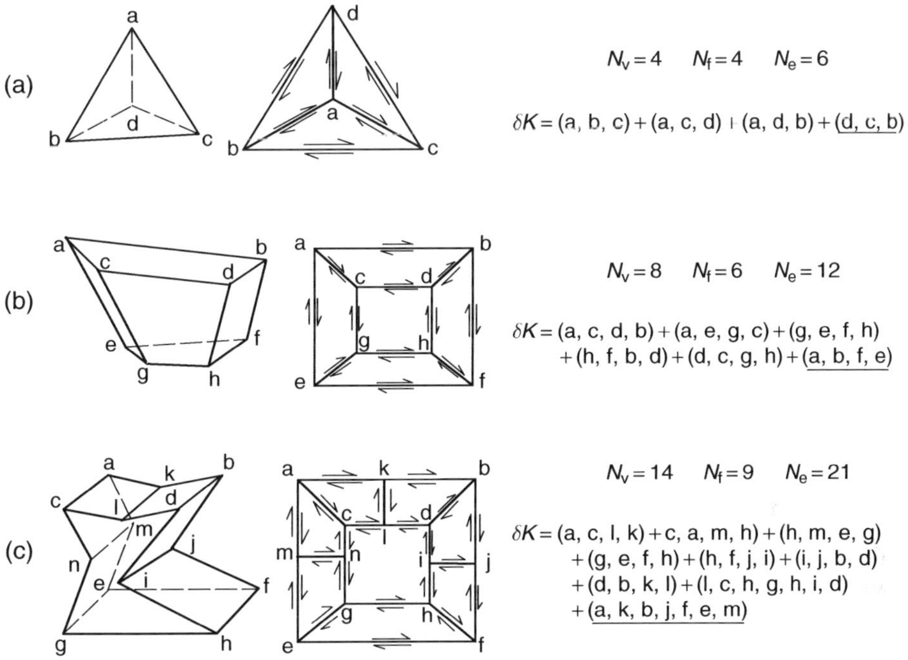

Fig. 6.12 Boundary representation of a polyhedron homeomorphic to a sphere. The arrows indicate positive orientation of the removed face (underlined).

face underlined in the chain of faces in each case is the imagined removed face for the boundary representation. In the case that a polyhedron has some of its faces not simple, i.e., the faces are defined by more than one chain of edges (corresponding to a multiply-connected planar figure), auxiliary edges should be added on these non-simple faces to make them simple, see Fig. 6.13. The addition of edges to a complex does not affect the characteristics of the polyhedron (Fréchet and Fan, 1967; Henle, 1974).

Figure 6.14a shows a cylindrical open surface – a cylinder without top and bottom faces. By topological transformation, it can be transformed into an annulus (Fig. 6.14b), i.e., an open cylindrical surface is homeomorphic to an annulus. An annulus is, however, not a simplex of one boundary, but two boundaries.

Another representation is to cut the cylindrical surface along one of its generation lines $\overline{pa}$ with the orientation from vertex p to vertex a (Fig. 6.14c), and then flatten it into a rectangle with two edges (p, a) but of opposite orientations (Fig. 6.14d). This rectangle (p, a, a, p) with the orientation of (p, a) is then the planar graph of the cylindrical surface. A special topology is then defined on this rectangle so that two opposite edges (p, a) are *topologically identified*, i.e., they are actually the same edge.

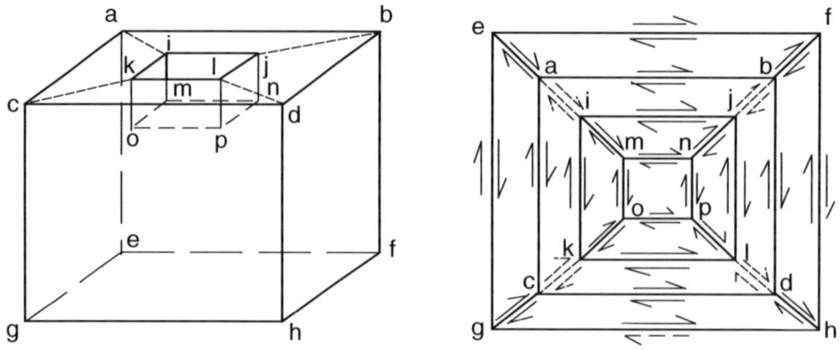

$N_v = 16$ $N_f = 14$ $N_e = 28$ ------ Auxiliary edge

δK = (a, e, g, c) + (c, g, h, d) + (d, h, f, b) + (b, f, e, a) + (i, a, c, k) + (k, c, d, l) + (l, d, b, j) + (j, b, a, i) + (m, i, k, o) + (o, k, l, p) + (p, l, j, n) + (n, j, i, m) + (n, m, o, p) + (e, f, h, g)

Fig. 6.13 Boundary representation of a polyhedron homeomorphic to a sphere, but where one of its faces (the top face) is not simplicial.

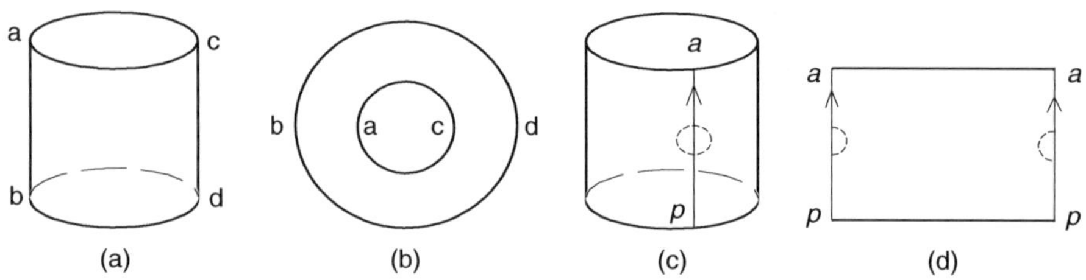

Fig. 6.14 Boundary representation of a cylindrical surface. (a) A cylinder without top and bottom; (b) an annulus homeomorphic to the cylinder; (c) making of a new extra edge (p, a); (d) topological identification of edge (p, a) and the resultant planar graph.

A topological identification of two edges has the same effect as gluing them together. Figure 6.15 illustrates topological identifications used for a sphere and a torus. The planar graph for the sphere is formed by two curvilinear edges that are topologically identified. The points on two topologically identified edges have a unique and reversible one-to-one correspondence and they are to be superimposed during the imaginary gluing operation and the sense of orientation of the two edges are opposite. If a small neighborhood is defined at a point on the introduced new edge (as the small disk around a point in Figs 6.14 and 6.15), this neighborhood is then shared by the identified edges. Below is a more precise definition of the topological identification.

Let P be a set of polygons and a_i $(i = 1, 2, \ldots, n)$ is a set of edges from these polygons. These edges are termed *topologically identified* when a new topology is defined on P such that:

(1) Each edge has an orientation from one endpoint to another and is placed in topological correspondence with a unit real interval [0, 1] such that the initial endpoints are corresponding to 0 and the end points are corresponding to 1.

(2) The points on the edges a_i $(i = 1, 2, \ldots, n)$ that are corresponding to the same value from the unit interval are treated as the same point.

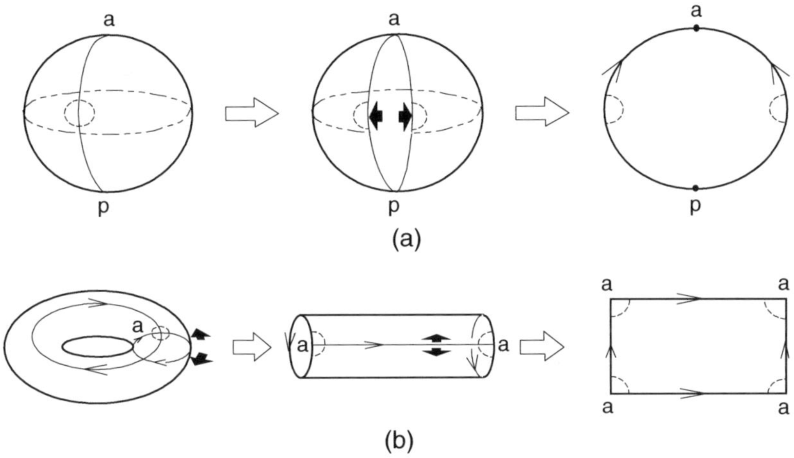

Fig. 6.15 Planar graph representation of (a) a sphere and (b) a torus by topological identification.

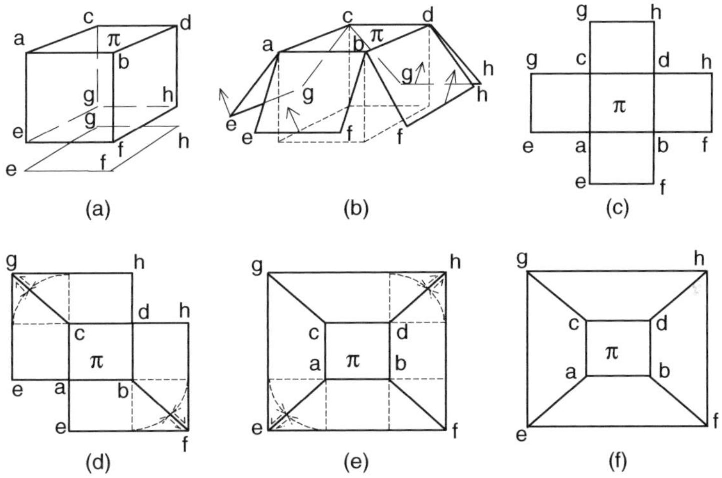

Fig. 6.16 Successive projection and edge identification for making a planar graph of a cube. (a) Removal of one face and select one face as the projection plane π; (b) projection of other faces on the π plane; (c) resultant projection; (d) and (e) identification of edges (cg), (bf), (ae) and (dh); (f) final planar graph.

(3) The neighborhood of the new topology on P are the disks entirely contained in a single polygon and the union of half disks (or quarterly disks or sectors) whose diameters match the interval values (0 to 1) around corresponding points on the edges a_i $(i = 1, 2, \ldots, n)$ (Henle, 1974).

The computer algorithm for making planar graphs for polyhedra homeomorphic to a sphere is to use the successive projection and topological identification process in a straightforward manner, see Fig. 6.16. The procedure used in Fig. 6.16 can be applied for more complicated polyhedra homeomorphic to a sphere. It is, however, much more difficult to make planar graphs for polyhedra homeomorphic to a torus, because two additional edges need to be introduced along which topological identifications are required.

For practical problems, the fractures are generally rough, not planar, and the blocks are of irregular shapes. The usual procedure is first to assume that fractures are smooth and planar (or piecewise planar,

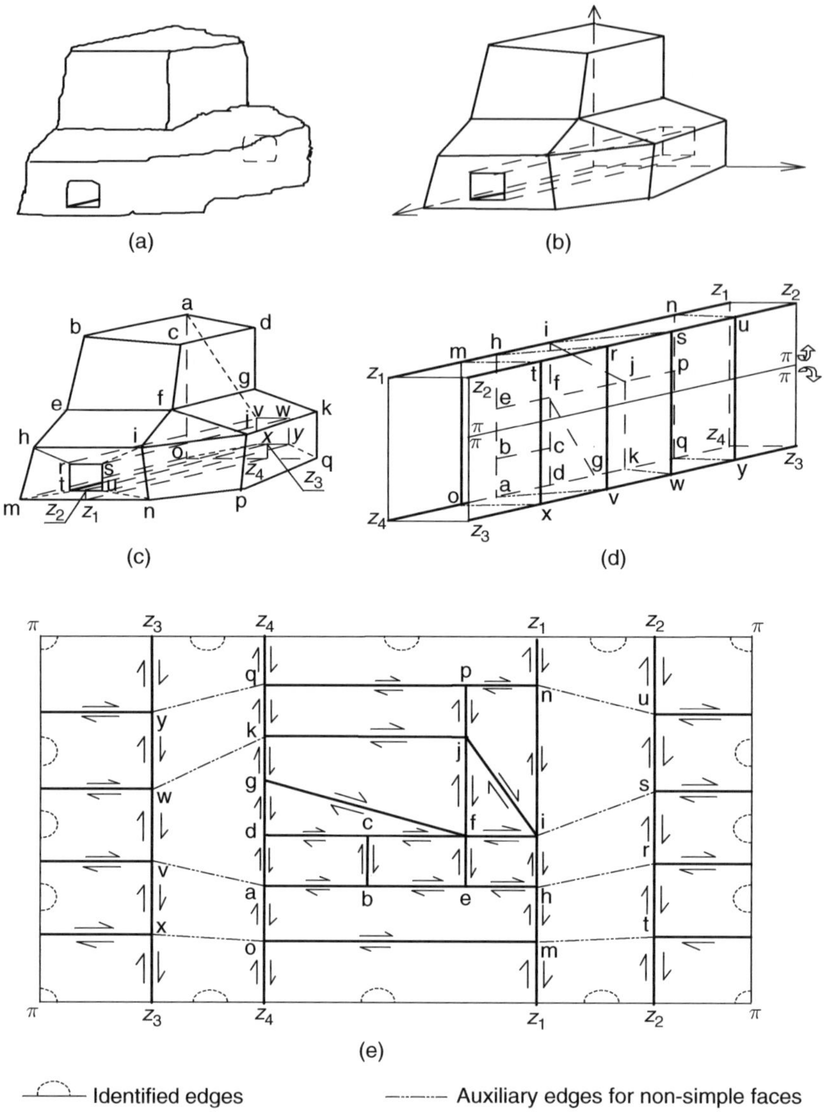

Fig. 6.17 Planar graph of a general block. (a) Initial irregular block; (b) polyhedron approximation to the initial block in a metric space; (c) labeling of vertices of the polyhedron in a topological space; (d) boundary representation of the polyhedron with orientation by a planar graph in a topological space. $N_v + N_f - N_e = 24 + 22 - 46 = 0$.

although this is not necessary for their topological analysis) and then represent the block surfaces by their corresponding planar faces (Fig. 6.17a and b). The resultant rock blocks are seen as polyhedral and their vertices are labeled (Fig. 6.17c). These operations are carried out in a three-dimensional metric space in which the locations of vertices and dimensions of edges and faces are important. The task of block generation is then to trace the individual blocks and their numbers of vertices, edges and faces using boundary operators. Planar graphs can be made (Fig. 6.17d and e) to help in analyzing these relations, but it is not necessary to calculate the metric information in the block-tracing process.

The combinatorial equations (6.1) and (6.18) do not require any metric properties of the polyhedra and are therefore carried out entirely in a topological space.

6.5 Data Sets for Boundary Representation of Polyhedra

The topological properties of polyhedra are useful for designing a proper data structure to represent the geometry of the polyhedra so that storage and retrieval of information are most convenient and economic in terms of computational efficiency. Different data structures for this purpose have been created, chief among them are the linked-list structure and arrays. The linked-list data structure has the advantage of occupying smaller computer memory but needs extensive use of pointers for information storage and retrieval and is relatively slow in computer implementation. The array structure needs more computer memory but is faster in retrieval and storage of information. The difference is relative and computational efficiency depends very much on the programming skills.

Let integer sets $M_v = (1, 2, \ldots, N_v)$, $M_e = (1, 2, \ldots, N_e)$, and $M_f = (1, 2, \ldots, N_f)$ be the sets of vertex numbers, edge numbers and face numbers, respectively. The set $V(X)$ records the coordinates of vertices, i.e.,

$$V(X) = \{v_i(x_i, y_i, z_i), i \in M_v, (x_i, y_i, z_i) \in R^3\}$$

The set $E(V)$ records the pairs of vertex numbers defining each edge and set $F(V)$ records the lists of closed vertex loops defining each face in a counter-clockwise manner, written as

$$E(V) = \{(v_j, v_k)_i, j \in M_v, k \in M_v, i \in M_e\}$$
$$F(V) = \{(v_j, v_k, \ldots, v_l, v_j)_i, j \in M_v, k \in M_v, l \in M_v, i \in M_f\}$$

To make the data structure more efficient for computer implementation, three more sets of data can also be defined to represent the topology of the block system. One set is called $V(E)$ recording the cyclic order of edges connected to each vertex; another is called $V(F)$ recording the cyclic order of faces around each vertex; and the third is called $F(E)$ recording the lists of closed edge loops defining each face (cf. Fig. 6.18 for a cube and its data sets):

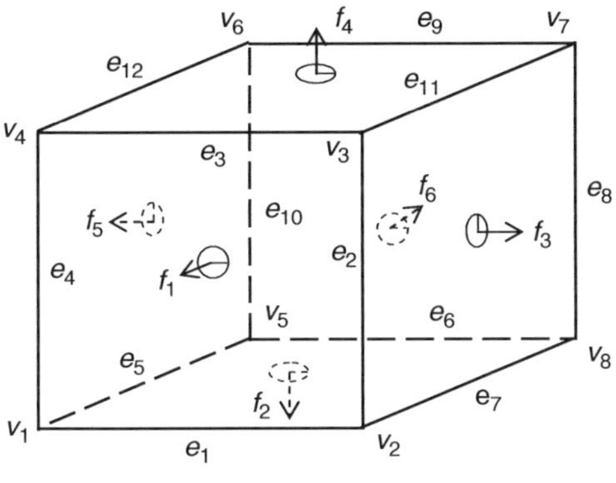

	$V(x)$
v_1	x_1, y_1, z_1
v_2	x_2, y_2, z_2
v_3	x_3, y_3, z_3
v_4	x_4, y_4, z_4
v_5	x_5, y_5, z_5
v_6	x_6, y_6, z_6
v_7	x_7, y_7, z_7
v_8	x_8, y_8, z_8

	$E(x)$
e_1	v_1, v_2
e_2	v_2, v_3
e_3	v_3, v_4
e_4	v_4, v_1
e_5	v_1, v_5
e_6	v_5, v_8
e_7	v_8, v_2
e_8	v_8, v_7
e_9	v_7, v_6
e_{10}	v_6, v_5
e_{11}	v_7, v_3
e_{12}	v_6, v_4

	$F(E)$
f_1	e_1, e_2, e_3, e_4
f_2	e_5, e_6, e_7, e_1
f_3	e_7, e_8, e_{11}, e_2
f_4	e_3, e_{11}, e_9, e_{12}
f_5	e_4, e_{12}, e_{10}, e_5
f_6	e_{10}, e_9, e_8, e_6

	$F(V)$
f_1	v_1, v_2, v_3, v_4, v_1
f_2	v_1, v_5, v_8, v_2, v_1
f_3	v_2, v_8, v_7, v_3, v_2
f_4	v_3, v_7, v_6, v_4, v_3
f_5	v_1, v_4, v_6, v_5, v_1
f_6	v_5, v_6, v_7, v_8, v_5

	$V(E)$
v_1	e_1, e_4, e_5
v_2	e_1, e_7, e_2
v_3	e_2, e_{11}, e_3
v_4	e_3, e_{12}, e_4
v_5	e_5, e_6, e_{10}
v_6	e_3, e_9, e_{12}
v_7	e_8, e_9, e_{11}
v_8	e_6, e_8, e_7

	$V(F)$
v_1	f_1, f_5, f_2
v_2	f_1, f_2, f_3
v_3	f_1, f_3, f_4
v_4	f_1, f_4, f_5
v_5	f_2, f_6, f_5
v_6	f_4, f_6, f_5
v_7	f_3, f_6, f_4
v_8	f_2, f_6, f_3

Fig. 6.18 Data sets for a cube.

$$V(E) = \{(e_j, e_k, \ldots, e_l)_i, j, k, l \in M_e, i \in M_v\},$$
$$V(F) = \{(f_j, f_k, \ldots, f_l)_i, j, k, l \in M_f, i \in M_v\},$$
$$F(E) = \{(e_j, e_k, \ldots, e_l)_i, j, k, l \in M_e, i \in M_f\},$$

The data sets $V(X)$, $E(V)$ and $F(V)$ are basic sets to represent the topology of a polyhedron. The sets $V(E)$, $V(F)$ and $F(E)$ are induced sets that can be derived from the basic sets. Theoretically, they provide redundant information; however, for computer execution, it may require much less computational time to have this redundant information readily accessible, than obtaining it every time by calculation. Considering the extensive use of faces, vertices and edges during contact detection in discrete element methods, these computational efforts may be very time-consuming. Therefore, it might be more efficient to have these data sets explicitly in both the local (block-wise) and global data structures, either by linked-lists or arrays.

6.6 Block Tracing Using Boundary Operators

The ordered data sets representing the topology of a polyhedron are generated during the formation of block assemblages involving space sub-division and block tracing.

The task of space sub-division is to introduce all fractures at once so that they will divide the domain of interest (a sub-space with a specific set of artificial boundary surfaces) into an assemblage of a finite number of blocks (polyhedra). The faces, edges and vertices of the blocks are defined by the intersections of the introduced fractures. After removing the 'dangling' edges and faces that do not form a closed surface (this is called regularization), the individual blocks are then traced out one by one through successive applications of the Euler–Poincáre formula and boundary operators, represented by Eqns (6.1)–(6.3) and (6.18).

To illustrate how to apply these governing relations to trace a polyhedron, a cube with six faces, eight vertices and 12 edges is generated gradually face by face, see Fig. 6.19, assuming all faces, vertices and

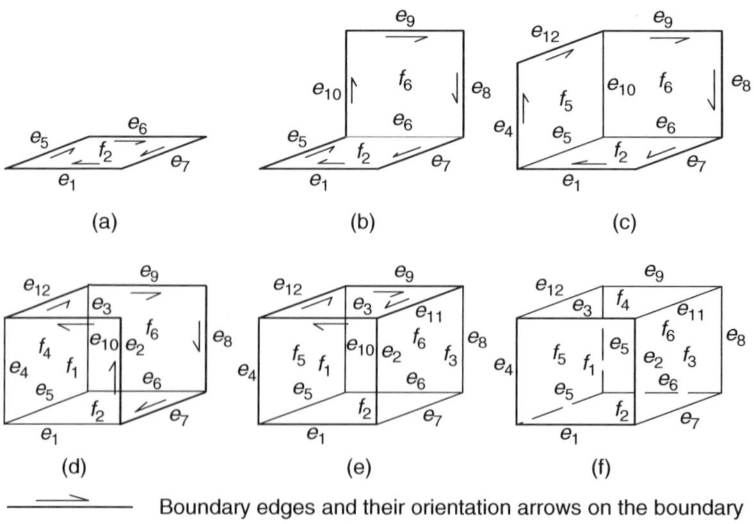

Fig. 6.19 Boundary operators during a cubic block generation.

edges are known. Denoting the cube as a complex K with its six 3-simplexes (faces) forming its algebraic boundary face loop,

$$\partial K = (f_1) + (f_2) + (f_3) + (f_4) + (f_5) + (f_6) \tag{6.21}$$

which is a 2-chain of K. The tracing starts with any vertex of a face, say face 2, (f_2) in this example in Fig. 6.19. The 2-chain $\partial K = (f_2)$ has just one face. Application of the boundary operator (Eqn (6.18)) to the 2-chain leads to a non-zero boundary of ∂K, with a boundary edge loop represented by $\partial\partial K = -(e_1) + (e_5) + (e_6) + (e_7)$, indicating this as an open surface, Fig. 6.19a. Finding (f_6) to be the next face connected to (f_2), the boundary of the 2-chain becomes $\partial\partial K = -(e_1) + (e_5) + (e_{10}) + (e_9) + (e_8) + (e_7)$, still a non-zero boundary (Fig. 6.19b). Continuing this operation at each stage of the block tracing when a new face has been recognized for the cube, the boundary operator $\partial\partial K$, is calculated. If $\partial\partial K$ is not zero, then the complex is still an open surface, not a polyhedron, and the tracing continues with the boundary edges until $\partial\partial K = 0$ is achieved. Then the complex represents a polyhedron and the block tracing is completed. The natural order of faces defining the block is the order of appearance of faces during the tracing. For the cube illustrated in Fig. 6.19, the order of faces in the data structure is $f_2 \to f_6 \to f_5 \to f_1 \to f_3 \to f_4$. Application of Eqn (6.1) leads to $N_v + N_f - N_e = 2$, and indicates that the block is homeomorphic to a sphere. The boundary operators at each step in Fig. 6.19 are given as follows:

(a) The base face 2, $\partial\partial K = -(e_1) + (e_5) + (e_6) + (e_7)$.

(b) Add face 6, $\partial\partial K = -(e_1) + (e_5) + (e_{10}) + (e_9) + (e_8) + (e_7)$.

(c) Add face 5, $\partial\partial K = -(e_1) - (e_4) + (e_{12}) + (e_9) + (e_8) + (e_7)$.

(d) Add face 1, $\partial\partial K = (e_2) + (e_3) + (e_{12}) + (e_9) + (e_8) + (e_7)$.

(e) Add face 3, $\partial\partial K = (e_3) + (e_{12}) + (e_9) + (e_{11})$.

(f) Add face 4, $\partial\partial K = 0$, block tracing completed.

The block-tracing algorithm depends on the availability of data sets of faces, edges and vertices and their connectivity. These data sets are generated during the sub-division of computational space by fractures.

References

Cundall, P. A. and Strack, O. D. L., A distinct numerical model for granular assemblies. Geomechanique, 1979;29:47–65.

Fréchet, M. and Fan, K. Y., Initiation to combinatorial topology. Prindle, Weber and Schmidt, Inc., Boston, 1967.

Henle, M., A combinatorial introduction to topology. Freeman, San Francisco, 1974.

Itasca Consulting Group Ltd., The 3DEC Manual, 2000.

Lin, D., Fairhurst, C. and Starfield, A. M., Geometrical identification of three-dimensional rock block systems using topological techniques. International Journal of Rock Mechanics and Mining Sciences and Geomechanics Abstracts, 1987;24(6):331–338.

Mäntylä, M., An introduction to solid modeling. Computer Science Press. Rockville, Maryland, USA, 1998.

7 NUMERICAL TECHNIQUES FOR BLOCK SYSTEM CONSTRUCTION

7.1 Introduction

Problems of fractured rocks are rarely, if ever, 2D, simply because of the very 3D nature of the fracture system geometry and the anisotropy and inhomogeneity of the rock matrix and fracture properties. Two-dimensional simplifications have been used extensively and they do have significant theoretical roles to play in research and development. However, it is the 3D models and their solutions that are the ultimate objectives for numerical solutions of rock engineering problems. The achievement of this goal depends significantly on the representation of 3D fracture–block systems, especially for DEM models.

Discrete representation of fracture networks have been used for simulations of groundwater and other Newtonian fluid flow since the early 1980s, with notably the well-known codes such as FRACMAN/MAFIC developed by Golder Associates (1995) and NAPSAC developed by AEA Technology (Herbert, 1996). The former can be used on various PCs and stations and the latter can be applied for large-scale problems, using mainframe computers. The construction of the fracture system geometry is straightforward, and the solutions of the flow equations are presented in Chapter 10. Figure 7.1 provides an illustrative example of a fracture system in 3D (Niemi et al., 2000).

As shown in Fig. 7.1, the realizations of the fracture systems are random and irregular. The tracing of the solid blocks from these irregular fractures by their intersections is not an easy numerical task, and it requires face and edge regularizations. The basic requirement of the fracture system regularization is that the 'dangling' and isolated faces in 3D networks and the 'dangling' and isolated edges in 2D networks should be removed, since they do not contribute to block building and the Euler–Poincaré formula for the topological relations between face, edge and vertices will not be upheld if they are included (Figs. 7.2 and 7.3).

A regularized fracture network has the following properties:

(1) Each edge has two and only two vertices that are intersections with other fractures.

(2) Each face is shared by two and only two polyhedra (3D blocks).

(3) Denoting the number of independent sub-networks as N_{sn}, the number of vertices (intersections) as N_V, the number of edges as N_e and number of polygons (face) as N_f, then these topological measures follow a relation defined by

$$N_v + N_f - N_e = 1 + N_{sn} \tag{7.1}$$

Equation (7.1) is an extended Euler–Poincaré formula for polygonal systems in the combinatorial topology. A completely connected network must satisfy Eqn (7.1). In other words, a network not satisfying

Fig. 7.1 An illustrative sub-division of space by fracture systems (Niemi et al., 2000). An example network realization used in the 10-m scale well test (at the center) simulations. The model dimensions are 30 m × 30 m × 30 m.

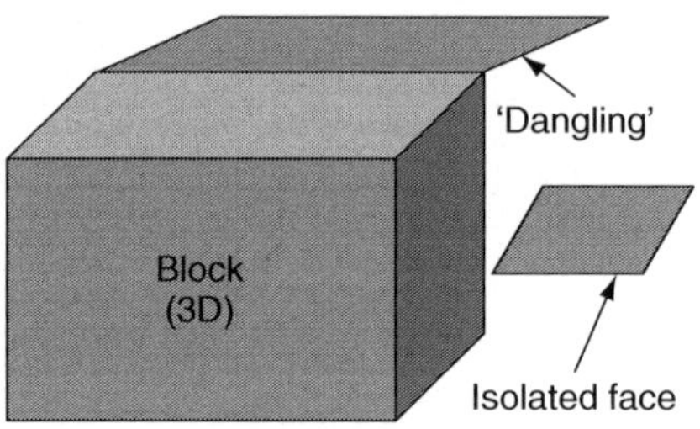

Fig. 7.2 'Dangling' and isolated fractures that need to be removed by regularization, the 3D case.

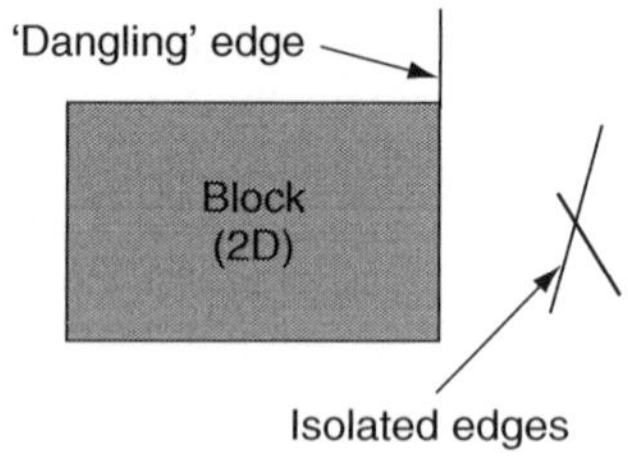

Fig. 7.3 'Dangling' and isolated fractures that need to be removed by regularization, the 2D case.

Eqn (7.1) must contain irregular elements (dead ends, isolated fractures or single connected fractures). Equation (7.1) can, therefore, serve as the criterion for network regularization processes. Tracing blocks (either convex, concave or multiply-connected) and flow paths, and detecting the disconnected sub-networks, can be performed by using a boundary operator (Jing and Stephansson, 1994a,b). Caution, however, needs to be taken regarding DFN networks in 3D for fluid flow analysis. A fracture system that

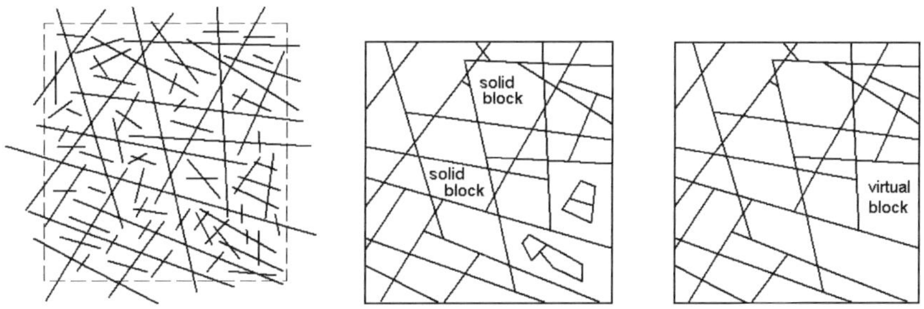

(a) (b) (c)

Fig. 7.4 Regularization of fracture networks: (a) original fracture network, (b) regularized fracture network for block tracing, (c) regularized fracture network after elimination of independent sub-networks for flow analysis (Jing and Stephansson, 1996).

may not form a complete solid block may, however, form part of the fluid-conducting pathways, and therefore should not be removed for fluid flow analysis. For stress and deformation or motion analysis of blocks systems using DEM, on the other hand, they must be removed since they do not contribute to block building. Different numerical techniques should therefore be used according to the objectives.

Figure 7.4 provides an illustrative example of a 2D fracture network and its regularization (Jing and Stephansson, 1996). Figure 7.4a shows the original fracture system with dead end segments and isolated fractures with 126 intersections (vertices). The regularization results in 34 blocks (faces), 146 edges (parts of fractures) and 92 vertices (intersections), which leads to three networks (one global and two extra independent sub-networks) as shown in Fig. 7.4b, and the Euler–Poincaré formula is verified by

$$N_v + N_f - N_e = 1 + N_{sn} = 92 + 57 - 146 = 1 + 2 = 3 \tag{7.2a}$$

Removing the two smaller independent sub-networks, only the one global network contains 130 fracture edges, 53 faces and 78 vertices and the Euler–Poincaré formula is verified again by (Fig. 7.4c)

$$N_v + N_f - N_e = 1 + N_{sn} = 78 + 53 - 130 = 1 + 0 = 1 \tag{7.2b}$$

The construction of the block system geometry, however, is not as straightforward as that of the fracture system. The difficulty is the tracing of the blocks from the intersected fractures, which are complex due to complexities in both shape and size distributions. Unless the fracture sizes are very large and shapes are uniformly regular, very complex block geometry may be generated. Block tracing in 3D remains a difficult and time consuming computational problem. This is partially the reason why, in many 3D DEM codes, either the fractures are assumed to be infinitely large or simple block geometry is used. The topological method of block construction using boundary operator and the Euler–Poincaré formula can help to overcome or reduce this difficulty. The 3D block tracing using boundary operator algorithms for natural fracture system models of general complexity in shape and size distributions are presented in detail in this chapter. The theoretical foundation and the basic algorithms are similar in many respects to the 2D counterparts.

7.2 Block System Construction in 2D Using a Boundary Operator Approach

Construction of a fracture–block system in two dimensions is usually a straightforward operation, and therefore has not appeared much in the literature. However, the computational efficiency is very different, depending on the algorithm used. The technique presented in this chapter uses the same combinatorial topology theory and 2D boundary operators, together with the extended Euler–Poincaré formula for polyhedra (Eqn (7.1)). The latter can be used since a polyhedron can be represented by a planar graph and the topological relation (6.2) is still valid, by removing one face (or including an exterior face compassing the whole model). The principle and algorithms are presented mostly in Shi (1988), Lin (1992), Jing and Stephansson (1994a,b, 1996). The main reference for graph theory is Wang (1987).

The following assumptions are adopted to simplify the problem:

(1) all fractures are straight- and smooth-line segments with two end vertices and finite lengths (Fig. 7.5a);

(2) blocks are the generally shaped polygons formed by edges that are formed by intersections (vertices) of fractures (Fig. 7.5b). The curved fractures can be represented as a series of a finite number of connected straight-line segments, and the developed algorithm in this chapter can be readily extended to consider them without undue difficulties.

The shapes of blocks can be convex, concave or multiply-connected. The blocks can be classified into either exterior blocks, which represent internal holes in an infinitely large region (Fig. 7.6a) or interior blocks, which have a finite volume, bounded by a set of ordered edges and vertices (Fig. 7.6b,c). If a block has one or more interior holes, it is called a multiply-connected block, see Fig. 7.6c.

As stated in Chapter 6, all edges and faces are oriented simplexes. An edge has a positive orientation from its starting to the ending vertex, and a face has a positive orientation if its edge and vertex loops follow a counter-clockwise arrangement (cf. Fig. 7.5b). The imaginary boundaries of the domain of interest are treated as artificial fractures.

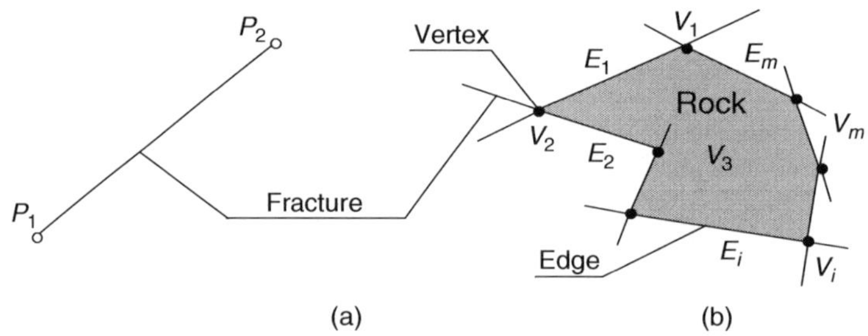

Fig. 7.5 (a) A straight fracture with two vertices (P_1, P_2); (b) A block with its vertices (V_i) and edges (E_i) arranged counter clockwise for a positive orientation.

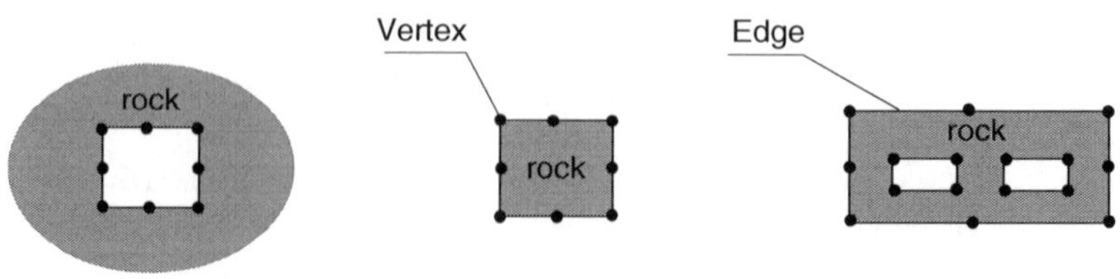

Fig. 7.6 Topological types of 2D blocks. (a) An exterior block; (b) an interior block and (c) a multiply-connected interior block.

7.2.1 Fracture Intersection and Edge Set Formation

Let d_i and d_j be two fractures of half-length r_i and r_j, and the coordinates of center points be $c_i(\mathrm{x}_i^{\mathrm{c}}, \mathrm{y}_i^{\mathrm{c}})$ and $c_j(\mathrm{x}_j^{\mathrm{c}}, \mathrm{y}_j^{\mathrm{c}})$ in a global O–XY coordinate system. The unit normal vectors of these two fractures are $\mathbf{n}_i(n_x^i,n_y^i) = (\cos\theta_i,\ \sin\theta_i)$, $\mathbf{n}_j(n_x^j,n_y^j) = (\cos\theta_j,\ \sin\theta_j)$, respectively. The distance between c_i and c_j is given by

$$d_{ij}^{\mathrm{c}} = \sqrt{(\mathrm{x}_i^{\mathrm{c}} - \mathrm{x}_j^{\mathrm{c}})^2 + (\mathrm{y}_i^{\mathrm{c}} - \mathrm{y}_j^{\mathrm{c}})^2} \tag{7.3}$$

If one of the following two conditions

$$d_{ij}^{\mathrm{c}} > r_i + r_j \tag{7.4a}$$

$$(\mathbf{n}_i) \cdot (\mathbf{n}_j) = n_x^i n_x^j + n_y^i n_y^j + n_z^i n_z^j = \pm 1 \tag{7.4b}$$

is satisfied, then the two fractures have no intersection. They are either too far away, or are parallel with each other, or are partially overlapped (in such a case, they should be merged). Otherwise, two fractures may (but are not certain to) have an intersection, depending on their relative positions.

Let $P_1^i(x_1^i, y_1^i)$, $P_2^i(x_2^i, y_2^i)$, $P_1^j(x_1^j, y_1^j)$, $P_2^j(x_2^j, y_2^j)$ be the starting and ending points of the ith and jth fractures, respectively, Fig. 7.7a. A vertex $V(x_s, y_s)$ is defined by the intersection of the two segments (Fig. 7.7b) if it exists. The parametric equations of the two intersection segments can be written

$$\begin{cases} x = x_1^i + (x_2^i - x_1^i)t_i \\ y = y_1^i + (y_2^i - y_1^i)t_i \end{cases} \quad \text{and} \quad \begin{cases} x = x_1^j + (x_2^j - x_1^j)t_j \\ y = y_1^j + (y_2^j - y_1^j)t_j \end{cases} \tag{7.5}$$

where $0 \le t_i \le 1$ and $0 \le t_j \le 1$ are the length parameters. The solution of Eqn (7.5) leads to

$$\begin{cases} x_s = \dfrac{1}{\Delta}\left[(x_1^j y_2^j - x_2^j y_1^j)(x_2^i - x_1^i) - (x_1^i y_2^i - x_2^i y_1^i)(x_2^j - x_1^j)\right] \\ y_s = \dfrac{1}{\Delta}\left[(x_1^i y_2^i - x_2^i y_1^i)(y_1^j - y_2^j) - (x_1^j y_2^j - x_2^j y_1^j)(y_1^i - y_2^i)\right] \end{cases} \tag{7.6a}$$

and

$$\Delta = \begin{vmatrix} y_1^i - y_2^i, & x_2^i - x_1^i \\ y_1^j - y_2^j, & x_2^j - x_1^j \end{vmatrix}, \qquad t_i = \frac{x_s - x_1^i}{x_2^i - x_1^i}, \qquad t_j = \frac{x_s - x_1^j}{x_2^j - x_1^j} \tag{7.6b}$$

If conditions $\Delta \ne 0$, $0 \le t_i \le 1$ and $0 \le t_j \le 1$ are satisfied simultaneously, the two segments have an intersection with coordinates (x_s, y_s), which defines a new vertex; otherwise, no intersection exits.

The segments between two adjacent vertices are called edges. The next step is to trace out all faces defined by single loops of edges on each fracture – after intersections between all fracture segments, including the artificial fractures defining the domain of interest, are determined. The pairs of adjacent vertices $(v_{i,j}, v_{i,j+1})$ ($j = 1,2,\ldots,M$) represent the adjacent edges on the ith fracture. This ordering can

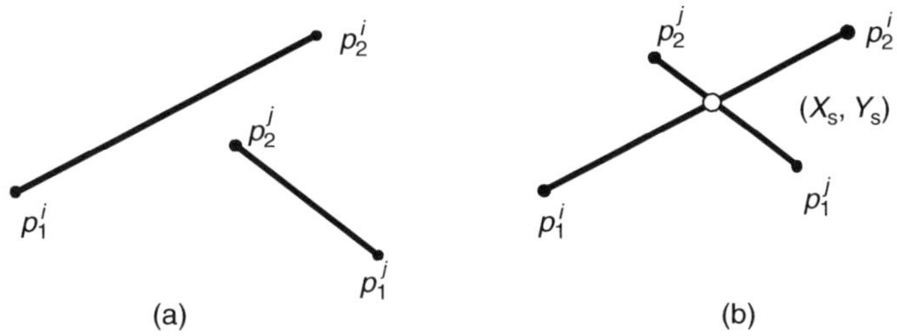

Fig. 7.7 Determination of a vertex as the intersecting point between two fracture segments. (a) Two segments have no intersecting point; (b) two segments have an intersecting point as a vertex.

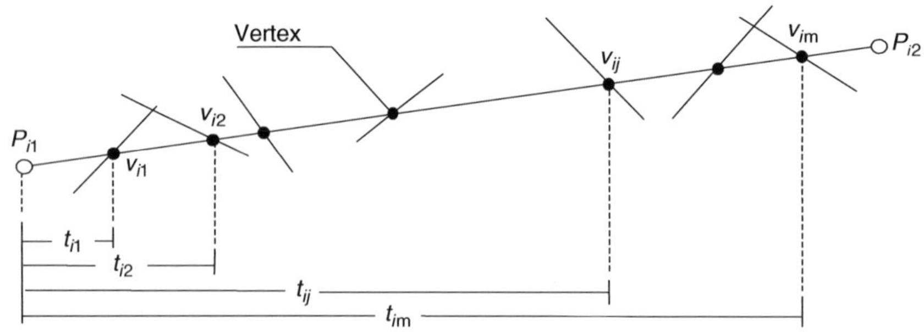

Fig. 7.8 Ordered vertices along the *i*th discontinuity with increasing values of length parameter *t* (Jing and Stephansson, 1994a).

be easily done by use of the length parameter values *t* in Eqn (7.6b) for each fracture, starting from the start vertex (Fig. 7.8).

Three sets of data should be generated at the end of the operation for the fracture intersection (vertex) determination: vertex coordinate *V*(*X*); edges defined by ordered pairs of vertices *E*(*V*); and fracture-edge connectivity *D*(*E*):

$V(X) = \{(x_i, y_i), i = 1, 2, \ldots, N_v\}$, N_v – Total number of vertices;

$E(V) = \{(v_i^s, v_i^e), i = 1, 2, \ldots, N_e\}$, N_e – Total edge number, v_i^s, v_i^e – the starting and ending vertex numbers;

$F(V) = \{(v_{i,1}, v_{i,2}, \ldots, v_{i,M}), i = 1, 2, \ldots, N_d, \}$, N_d – Total number of fractures, $v_i^j, j = 1, 2, \ldots, M$ – The total number of vertices arranged in a sequential order from the starting to the ending vertex of the *i*th fracture *i*.

Figure 7.9 shows a network of 19 fractures with 25 vertices and its corresponding fracture–vertex connectivity matrix *F*(*V*) of the fracture system. We will use this example to illustrate the block-tracing

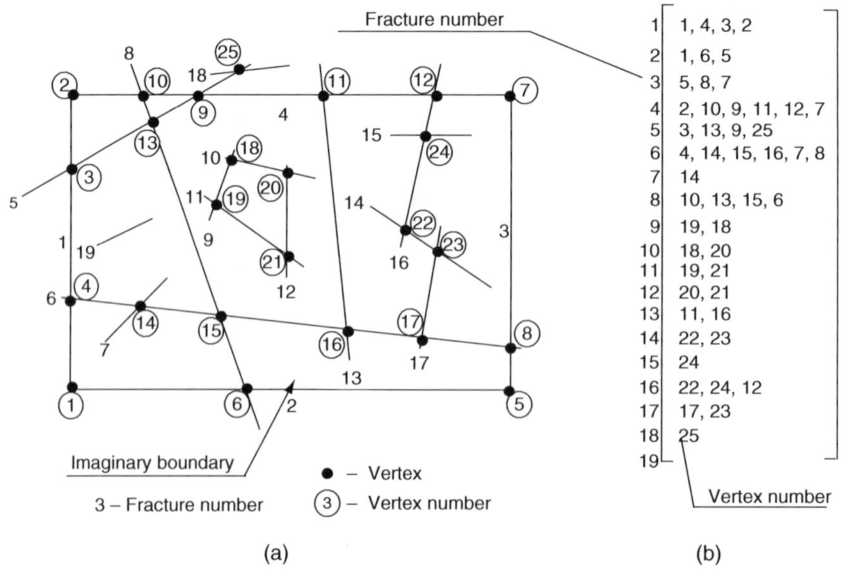

Fig. 7.9 A fracture network example: (a) fracture network and its vertices; (b) its initial fracture–vertex matrix *F*(*V*) (Jing and Stephansson, 1994a).

algorithm. The tracing algorithm is based on the intersections of fractures as calculated and recorded above.

7.2.2 Edge Regularization

The edge regularization is performed by removing the isolated fractures that have no vertex (no intersection with other fractures) or 'dangling' fracture segments that have only one vertex. The resultant set of edges is called the regularized set of edges. The regularized network of edges and vertices form a completely connected planar directed graph. Each edge is defined by exactly two vertices (the starting and ending vertices), shared by exactly two adjacent polygons (blocks) that have two opposite orientations, counter-clockwise (positive) and clockwise (negative). Therefore, edge $\overline{v_{ij}v_{i,j+1}}$ and edge $\overline{v_{i,j+1}v_{ij}}$ refer to the same edge, but with opposite orientations. A graph with the orientation of edges is called a digraph and an oriented complex in combinatorial topology, denoted by K, which can be taken as a planar schema of a polyhedron in a 3D space, after removing one face (cf. Chapter 6). Figure 7.10 shows the complex (digraph) of the fracture system in Fig. 7.9, after edge set regularization and the updated F(E) matrix.

An auxiliary matrix $\mathbf{O} = \lfloor o_{ij} \rfloor$, called an edge orientation index matrix, is useful to keep track of the states of orientations of edges during the identification process.

Corresponding to the ith row of $F(V) = [v_{ij}]$ with r elements (vertices), $v_{i1}, v_{i2}, \ldots, v_{ir}$, there are $(r-1)$ elements in the ith row of matrix $\mathbf{O}$, $o_{i1}, o_{i2}, \ldots, o_{i,r-1}$, corresponding to $(r-1)$ edges formed by adjacent pairs of vertices v_{ij} and $v_{i,j+1}$ (or $v_{i,j-1}$) with values:

$o_{ij} = 2$ (initial state of edge $\overline{v_{ij}, v_{i,j+1}}$, both orientations of the edge unused);

$o_{ij} = 1$ (intermediate state of edge $\overline{v_{ij}, v_{i,j+1}}$, the orientation from vertex v_{ij} to vertex $v_{i,j+1}$ unused);

$o_{ij} = -1$ (intermediate state of edge $\overline{v_{ij}, v_{i,j+1}}$, the orientation from vertex v_{ij} to vertex $v_{i,j+1}$ unused);

$o_{ij} = 0$ (final state of edge $\overline{v_{ij}, v_{i,j+1}}$, both orientations of the edge used).

Figure 7.10c shows the **O** matrix for the example fracture network shown in Fig. 7.9 after regularization. All its elements have the initial value 2, indicating an initial state of block identification. At the end of identification, all elements in matrix **O** should be zero.

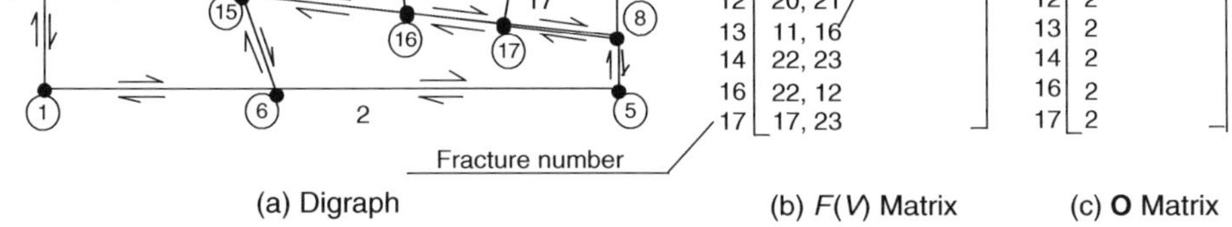

Fig. 7.10 (a) The oriented complex (digraph) of the fracture network; (b) the fracture–vertex connectivity matrix $F(V)$ and (c) the edge orientation index matrix **O** of the fracture system in Fig. 7.9 after edge set regularization (Jing and Stephansson, 1994a).

7.2.3 Boundary Operators of 2D Complexes

The boundary operator, ∂, of a 2D complex K produces a set of oriented edges of the complex, denoted as $\partial(K)$ and is called its 1-chain in the context of combinatorial topology. They represent the boundary edge loop of the complex. For closed edge loops (faces), the boundary operator $\partial(\partial K) = 0$. Block tracing is based on the boundary operator of the set of edges.

Regarding an oriented edge–vertex network as a planar complex, a boundary operator for the edge set is needed for face tracing algorithm. To define such a boundary operator, the concept of decomposed and signed vertices is used. Let $E_B = (v_1, v_2, \ldots, v_n, v_1)$ be a closed loop of vertices defining a polygon in which the adjacent pairs of the vertices define the edges in a counter-clockwise order. If v is a vertex of an edge, then it is denoted as v^+ if it is the starting vertex of the edge along the positive orientation of a face, or v^- if it is the ending vertex on the edge (Fig. 7.11). The pair (v^+, v^-) refers to the same vertex v but belongs to two adjacent edges connected at vertex v.

For a pair of decomposed and signed vertices (v^+, v^-), we define a topological operation (summation rule) such that

$$v^+ + v^- = 0 \tag{7.7}$$

This relation means that when the signed starting and ending vertices of the same vertex meet, they cancel each other and their topological sum is zero. This relation will enable us to define the starting and ending vertices of a set of connected edges. The boundary vertices defined by the pair of vertices (v_i, v_j) are the signed vertices (v_i^+) and (v_j^-). For a single edge, the boundary operator leads to

$$\partial(v_i, v_j) = (v_i^+) + (v_j^-) \tag{7.8}$$

For a set of connected edges, the boundary operator results in

$$\partial\left(\sum (v_i, v_j)\right) = \sum \partial(v_i, v_j) = \sum [(v_i^+) + (v_j^-)] \tag{7.9}$$

For a closed loop of edges defining a block, the boundary operator of the edge set leads to an empty set of vertices,

$$\partial\left(\sum (v_i, v_{i+1})\right) = 0 \tag{7.10}$$

Equation (7.10) is the 2D boundary operator that is used to check if a block tracing is completed. Together with the Euler-Poincaré formula, they are the basis of the 2D block identification algorithm. Figure 7.12 shows the process of successive application of the boundary operator for an edge set to trace a block of six edges, (a^+, b^-), (b^+, c^-), (c^+, d^-), (d^+, e^-), (e^+, f^-) and (f^+, a^-) with signed vertices for tracing a single block.

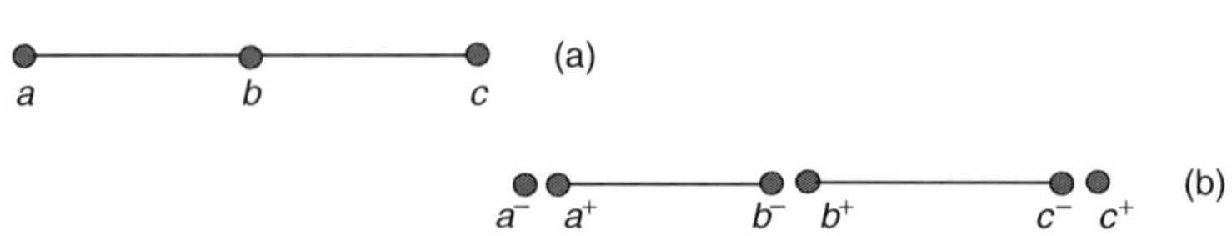

Fig. 7.11 Decomposed and signed vertex concept: (a) original chain of vertices; (b) decomposed and signed vertices.

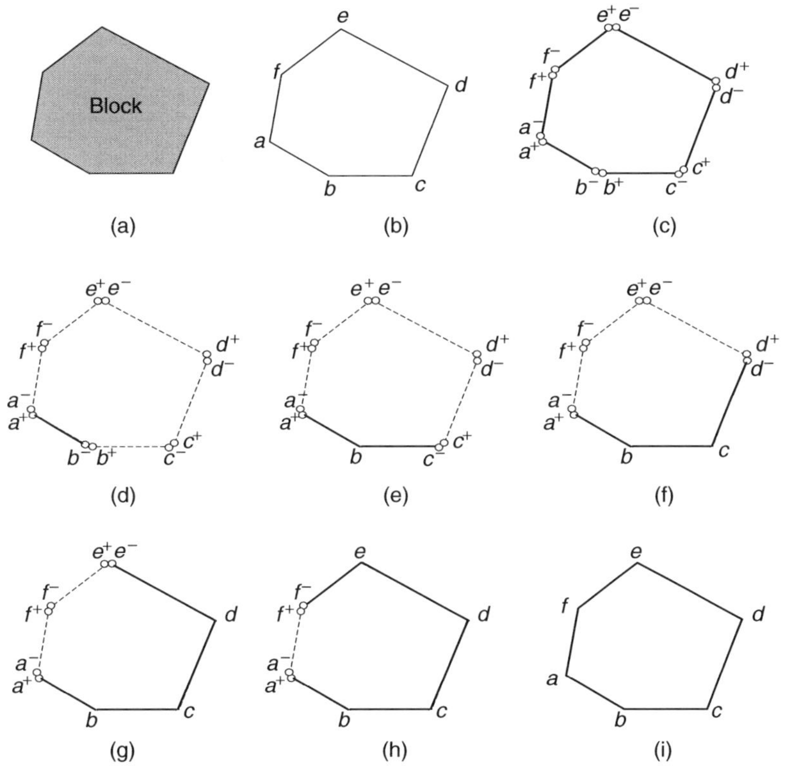

Fig. 7.12 Tracing an example block of six edges by successive application of the boundary operator for edge set (Jing and Stephansson, 1994b).

a) the block;
b) vertex labeling;
c) the oriented edges with their decomposed and signed vertex pairs;
d) $\partial(a, b) = a^+ + b^-$;
e) $\partial[(a, b) + (b, c)] = a^+ + c^-$;
f) $\partial[(a, b) + (b, c) + (c, d)] = a^+ + d^-$;
g) $\partial[(a, b) + (b, c) + (c, d) + (d, e)] = a^+ + e^-$;
h) $\partial[(a, b) + (b, c) + (c, d) + (d, e) + (e, f)] = a^+ + f^-$;
i) completion of the block tracing, with $\partial[(a, b) + (b, c) + (c, d) + (d, e) + (e, f) + (f, a)] = a^+ + a^- = 0$.

7.2.4 Block Tracing in 2D

7.2.4.1 The principle of 'minimum turning-left-angle'

The block tracing starts with any vertex and any one of the edges connected at this vertex. All edges converged on this vertex can be found from the fracture-edge connectivity matrix F(E). Taking the current edge as a base line and rotating it clockwise about the current vertex, the edge that forms the smallest angle with the base line should be chosen as the next edge defining the block. In other words the criterion for determination of the next edge is to keep turning left with the smallest angle while advancing from the current vertex (Fig. 7.13). An auxiliary array is needed to record the number of times each edge has been processed (since each edge has exactly two orientations defining two different faces,

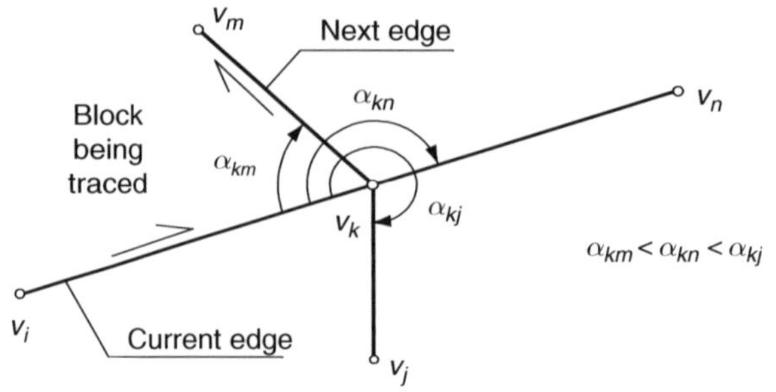

Fig. 7.13 Determination of the next edge connected with the current edge at its end vertex (Jing and Stephansson, 1994a).

each edge needs to be processed exactly twice). The tracing should be completed when all edges are processed twice.

7.2.4.2 Determination of vertex loops of interior holes

The block-tracing algorithm presented here will trace out all types of blocks (interior, exterior, singly-or multiply-connected). Determination of them requires use of the following corollaries.

Corollary 1: If a point P is located inside a convex polygon with n vertices connected by n straight edges, and the vertices of the polygon are arranged with counter-clockwise rotation orientation, then all n triangles formed by point P and n pairs of adjacent vertices have positive values of area (Fig. 7.14) calculated as

$$A^i = \frac{1}{2}\begin{vmatrix} 1, & x_p, & y_p \\ 1, & x_i, & y_i \\ 1, & x_j, & y_j \end{vmatrix} \tag{7.11}$$

where $i = 1, n$ and $j = i + 1$. $j = 1$ if $i = n$.

The point P (x_p, y_p) is called the reference point. All faces (blocks) of positive area are interior blocks. $A > 0$ indicates that the loop of vertices of the block identified is arranged counter-clockwise and encloses a solid interior block. $A < 0$ means that the edges are arranged clockwise, representing either an internal opening of a larger interior block or the exterior block whose edge loop is located entirely along the artificial boundary fractures (because of their clockwise ordering). The edge loops of internal openings need to be

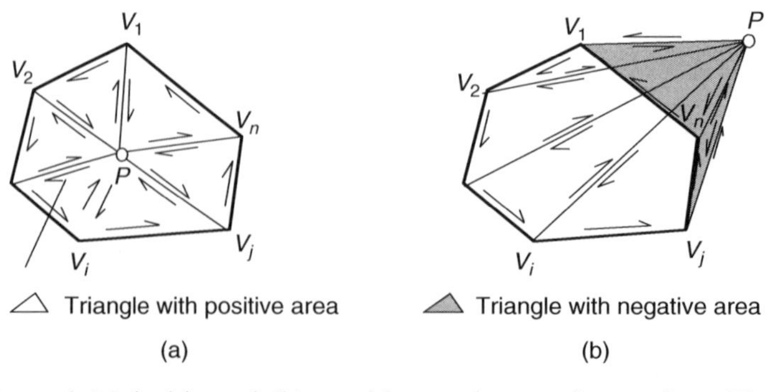

Fig. 7.14 A point located (a) inside and (b) outside a polygon of n vertices (Jing and Stephansson, 1994a).

associated with their parent interior blocks to define the multiply-connected interior blocks. Detection of these multiply-connected blocks requires knowledge of whether one vertex of a negatively oriented edge loop of an internal opening is located inside another positively oriented edge loop (block).

Corollary 2: If a point P is located outside a convex polygon with n vertices connected by n straight edges, and the vertices of the polygon are arranged in a positive orientation with counter-clockwise rotation, then among all n triangles formed by point P and n pairs of adjacent vertices, at least one of them has a negative value of area calculated by Eqn (7.11), see Fig. 7.14b.

The above corollary holds also for the simplest convex polygon – the triangles ($n=3$). The general polygons, however, could be either convex or concave and the above corollary is not valid for concave polygons. To examine if a point P is located inside or outside a general polygon of m vertices without requiring the knowledge that a polygon is convex or concave, the polygon is triangularized into $m-2$ triangles by connecting a specific vertex (for example, the first vertex in the loop) to the remaining $(m-2)$ vertices (see triangles $\Delta V_1 V_i V_{i+1}$ $(i=3, m-1)$ in Fig. 7.15).

Let point P be the reference point, apply the above corollary to examine if point P is located inside or outside the triangles and define an index I_c^k as

$$I_c^k = 1 \quad \text{(the point } P \text{ inside the } k\text{th triangle of a positive value of area)} \tag{7.12a}$$

$$I_c^k = 0 \quad \text{(the point } P \text{ outside the } k\text{th triangle)} \tag{7.12b}$$

$$I_c^k = -1 \quad \text{(the point } P \text{ inside the } k\text{th triangle of a negative value of area)} \tag{7.12c}$$

Then, the algebraic sum of index I_c^k over $(m-2)$ triangles will indicate if the point P is inside or outside the polygon by

$$\sum_{k=1}^{m-2} I_c^k > 0 \quad \text{(the point } P \text{ is inside the polygon)} \tag{7.13a}$$

$$\sum_{k=1}^{m-2} I_c^k \leq 0 \quad \text{(the point } P \text{ is outside the polygon)} \tag{7.13b}$$

The criteria (7.13a,b) are valid for generally shaped polygons: i.e., convex, concave or multiply-connected.

Because of the uniqueness and completeness of the regularized set of edges forming a planar graph, if one of the vertices of an edge loop is inside a polygon, then all vertices of this loop are also inside this polygon (Jing and Stephansson, 1994a).

For the network example shown in Figs. 7.9 and 7.10, the block tracing process is shown in Fig. 7.16. The resultant blocks identified for the example network are listed in Table 7.1.

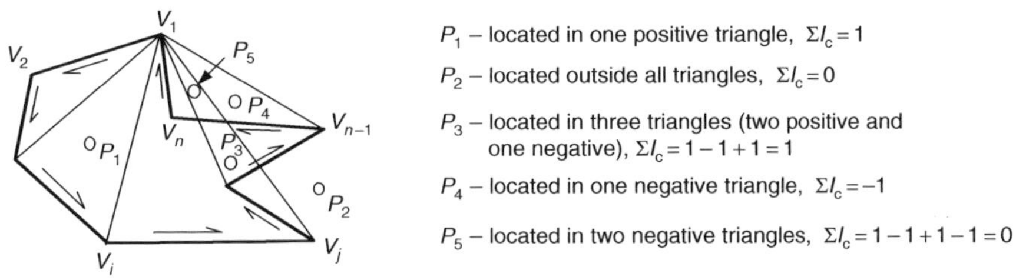

Fig. 7.15 Points located in positive or negative triangles of a general polygon (Jing and Stephansson, 1994a).

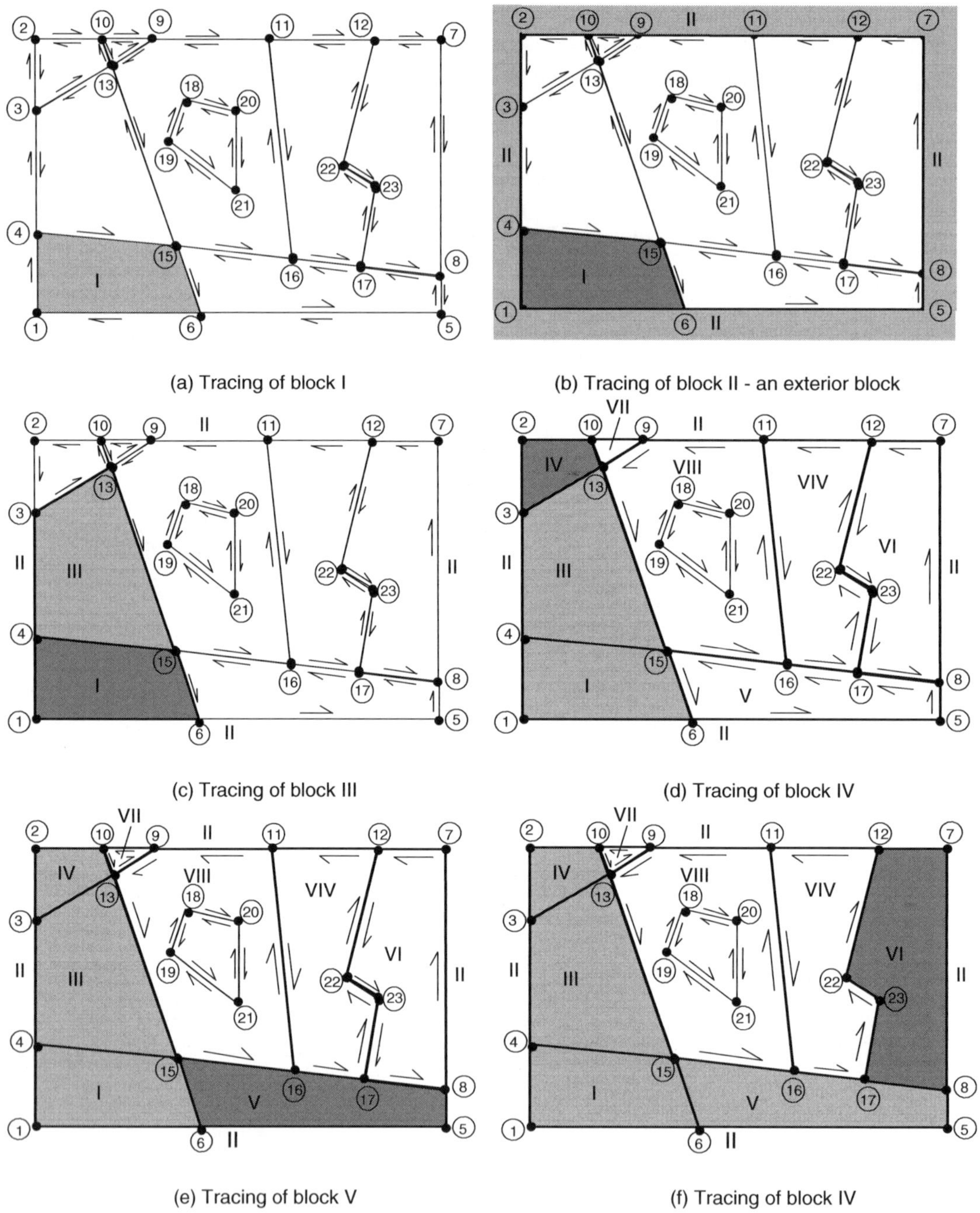

Fig. 7.16 Identification of blocks for the example network. Blocks I–VI (Jing and Stephansson, 1994a). (a) $\partial\partial(K) = (1,6) + (6,15) + (15,4) + (4,1)$; (b) $\partial\partial(K) = (6,15) + (15,4) + (4,3) + (3,2) + (2,10) + (10,9) + (9,11) + (11,12) + (12,7) + (7,8) + (8,5) + (5,6)$; (c) $\partial\partial(K) = (6,15) + (15,13) + (13,3) + (3,2) + (2,10) + (10,9) + (9,11) + (11,12) + (12,7) + (7,8) + (8,5) + (5,6)$; (d) $\partial\partial(K) = (6,15) + (15,13) + (13,10) + (10,9) + (9,11) + (11,12) + (12,7) + (7,8) + (8,5) + (5,6)$; (e) $\partial\partial(K) = (15,13) + (13,10) + (10,9) + (9,11) + (11,12) + (12,7) + (7,8) + (8,17) + (17,16) + (16,15)$; (f) $\partial\partial(K) = (15,13) + (13,10) + (10,9) + (9,11) + (11,12) + (12,22) + (22,23) + (23,17) + (17,16) + (16,15)$.

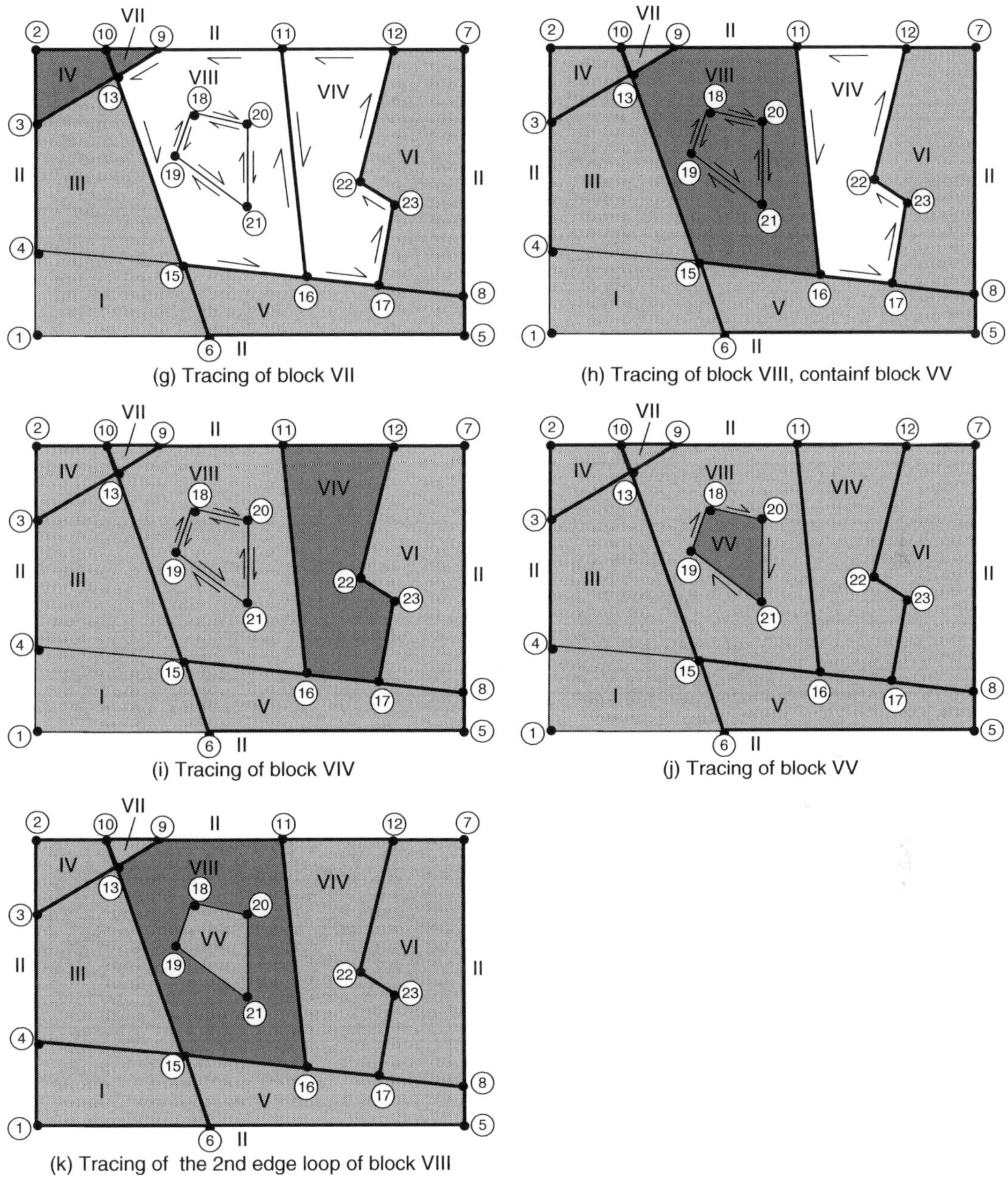

(g) Tracing of block VII

(h) Tracing of block VIII, containf block VV

(i) Tracing of block VIV

(j) Tracing of block VV

(k) Tracing of the 2nd edge loop of block VIII

Fig. 7.16 (Continued)
(g) $\partial\partial(K) = (15,13) + (13,9) + (9,11) + (11,12) + (12,22) + (22,23) + (23,17) + (17,16) +$ $(16,15)$; (h) $\partial\partial(K) = (16,11) + (11,12) + (12,22) + (22,23) + (23,17) + (17,16)$; (i) $\partial\partial(K) = 0$, but the remaining edges are not used; (j) $\partial\partial(K) = (18,19) + (19,21) + (21,20) + (20,18)$; (k) $\partial\partial(K) = 0$ and no edge remains are unused. Block tracing completed.

7.2.5 Representation of Flow Path and Mechanical Contacts of Blocks

At the completion of block tracing, two tasks remain for fracture–block model construction: the establishment of flow paths through the regularized fracture network and the establishment of contact

Table 7.1 Loops of vertices defining blocks (Jing and Stephansson, 1994a)

Block	Loops of vertices	Loops of edges	Block Type
I	1,6,15,4,1	1,4,21,15	Interior
II	1,4,3,2,10,9,11,12,7,8,5,6,1	1,2,3,8,9,10,11,12,7,6,5,4	Exterior
III	3,4,15,13,3	2,15,20,13	Interior
IV	2,3,13,10,2	3,13,19,8	Interior
V	6,5,8,17,16,15,6	5,6,18,17,16,21	Interior
VI	8,7,12,22,23,17,8	7,12,28,27,29,18	Interior
VII	9,10,13,9	9,19,14	Interior
VIII	Loop 1: 11,9,13,15,16,11	Loop 1: 10,14,26,16,26	Multiply
	Loop 2: 19,18,20,21,19	Loop 2: 22,23,25,24	connected
VIV	12,11,16,17,23,22,12	11,26,17,29,27,28	Interior
VV	18,19,21,20,18	22,24,25,23	Interior

relations between blocks. The former is required for fluid flow analyses and the latter for stress/deformation/motion analyses of blocks. Both are required for coupled hydro-mechanical analyses.

7.2.5.1 Flow-path connectivity and percolating fracture system identification

A graph showing the connectivity of fractures without dead-end segments can be readily constructed by using the F(E) matrix, see Fig. 7.17a. The graph can be represented by a square and symmetric matrix $C_V = \lfloor c_{ij}^V \rfloor$ with its elements defined by (Fig. 7.17b)

$$c_{ij}^V = \begin{cases} 0 \text{ (vertices } i \text{ and } j \text{ not connected)} \\ 1 \text{ (vertices } i \text{ and } j \text{ connected)} \end{cases}, \quad c_{ij}^V = c_{ji}^V \tag{7.14}$$

This graph also represents the flow path of the block system if only fluid flow through fractures is concerned (the flow field in the intact rock blocks is assumed to be negligible). The continuity equations of fluid flow can be written at each vertex, and the boundary conditions should be specified at the vertices provided by the loop of vertices of the exterior block II (cf. Table 1 and Fig. 7.15b).

The isolated interior fracture system formed by vertices 18, 19, 20 and 21 (corresponding to edges 22, 23, 24 and 25) should be discarded since they are not connected to the overall flow system, and therefore do not contain percolating fluids. In other words the overall fracture system may become 'compartmentalised' into connected and isolated compartments and only the compartments connected to the boundary of the domain of interest are needed for flow analyses, assuming no flow in the rock matrix.

7.2.5.2 Block extraction for mechanical analyses

For mechanical analysis, the vertex–edge loops defined at the completion of block tracing cannot be directly used for DEM analysis, since all blocks are 'locked-up' by sharing common vertices and edges. Therefore, an operation of block extraction is needed to extract individual blocks with distinct numbering of vertices and edges but maintaining the coordinates of the vertices. This block extraction and vertex–edge renumbering can be done in many different ways, and all of them are not difficult. The basic technique is to add some extra vertices at some of the vertex locations in the graph at the end of block tracing so that the total vertex number at that location equals the total block number connected at that location. Renumbering of the vertices and edges and redefining the loops of vertices and edges of blocks

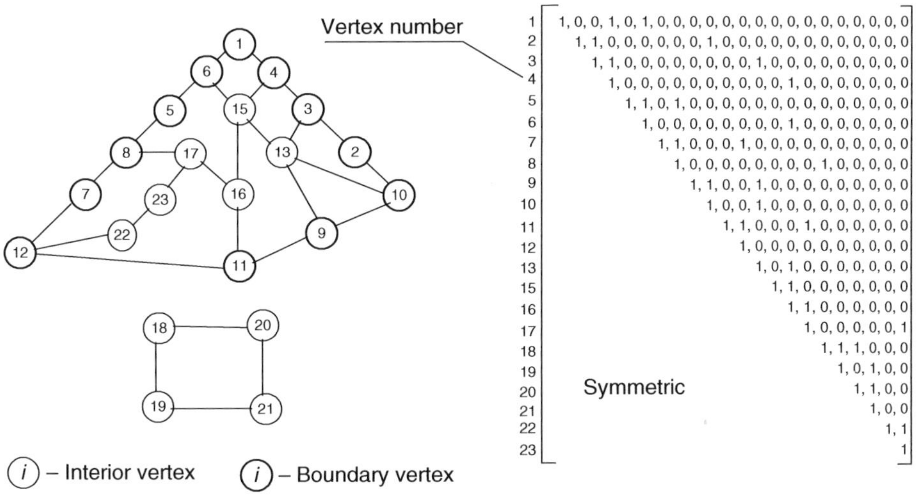

(a) Graph of vertex connectivity (b) Matrix C_V

Fig. 7.17 (a) Graph of vertex connectivity and (b) its matrix representation for the flow path (Jing and Stephansson, 1994a).

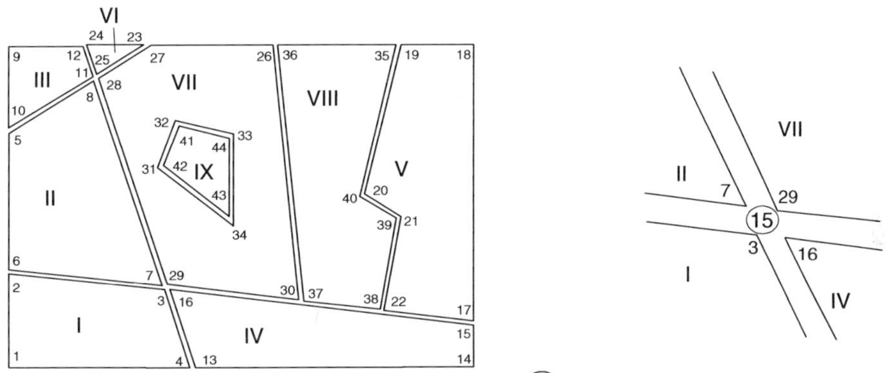

I, II, III – New block number; 1, 2, 3 – New vertex number; (15) – Old vertex number

Fig. 7.18 Renumbering of vertices and block extraction (left) and addition of new vertices to an old one (right) (Jing and Stephansson, 1994a).

with the new numbering system should follow. Figure 7.18 shows an example, where vertex clusters (2,6), (5,10), (4,13), (3,7,29,16) share the same coordinates with the old vertices 4, 3, 6, 15, … and the edges are also renumbered accordingly.

For mechanical analysis, the initial contact relations between blocks can also be represented as a graph, and its matrix representation, $C_B = [c_{ij}^B]$, can be written as (Fig. 7.19)

$$c_{ij}^B = \begin{cases} 0 \text{ (blocks } i \text{ and } j \text{ not in contact)} \\ 1 \text{ (blocks } i \text{ and } j \text{ in contact)} \end{cases}, \quad c_{ij}^B = c_{ji}^B \tag{7.15}$$

The block contact matrix is most conveniently constructed using the edge loops defining the blocks at the completion of block tracing, before block extraction for mechanical analysis. It is helpful when

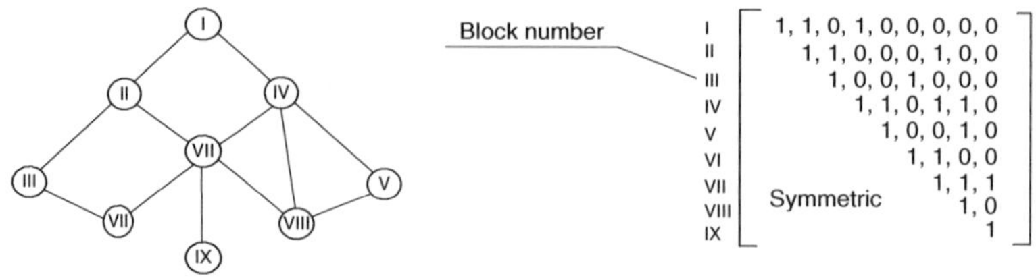

(a) Contact graph (b) Contact matrix

Fig. 7.19 (a) Block contact graph and (b) its matrix representation (Jing and Stephansson, 1994a).

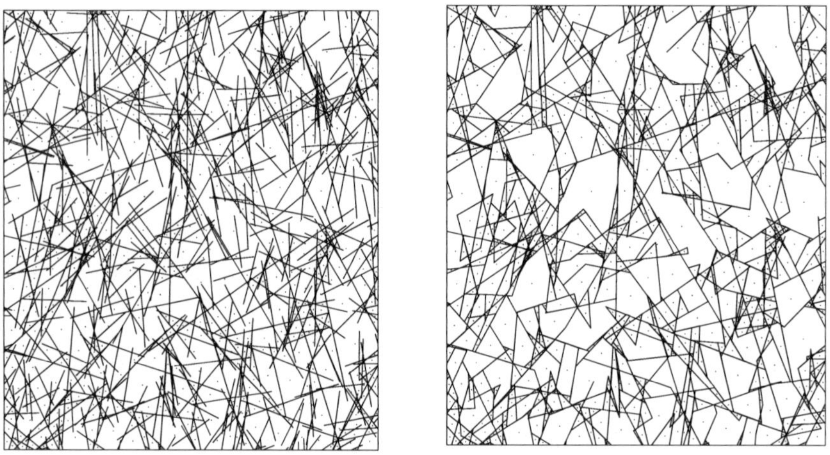

(a) (b)

Fig. 7.20 An example of block identification from a fracture network realization, using a 2D DEM code UDEC (Itasca, 2000). (a) Generated fracture system realization before edge regularization and (b) block systems after the edge regularization.

implicit solution methods such as DDA are used to solve the equations of motion of a discrete rigid block system. An element stiffness matrix $[k_{ij}] \neq [0]$ resulting from an energy minimization process can only be possible if $c_{ij}^{\mathrm{B}} \neq 0$.

The block-tracing algorithm using the boundary-operator approach is not the only algorithm suitable for 2D problems. More straightforward algorithms also exist so long as the ‘minimum left-turn-angle’ principle is applied. However, using the boundary operator algorithm can ensure the topological correctness of the block system construction without the need for extra checking of potential errors that can be identified using Eqn (7.1). Figure 7.20 shows an example of block identification using the UDEC code, without using the boundary-operator approach.

7.3 Block System Construction in 3D Using the Boundary Operator Approach

For block construction in 3D, the space sub-division is based on the determination of fracture intersection lines that form the edges of the polyhedral blocks. The fractures are assumed to be planar and smooth, but with finite size and polygonal shapes, and are generated realizations through deterministic or stochastic processes. The fractures are generated with their mass centers located within a regularly shaped domain, on whose boundaries specific boundary conditions can be readily specified.

For analysis of physical problems, a finite thickness can be assigned to the fractures as a property corresponding to the initial aperture of the fractures. The mathematical principle is described in Jing (2000) for topological block identification from connected random fractures and in Mäntylä (1988) for solid modeling using boundary representation techniques based on regular surface shapes most often encountered in mechanical engineering.

7.3.1 Fracture Representation and Coordinate Systems

For construction of block or DFN models, the orientation of a fracture is uniquely determined in a 3D space (noted as R^3) by its dip direction ω and dip angle ρ. To represent a fracture in R^3, besides the dip direction and dip angle, the coordinates of a reference point on the fracture plane, e.g., its mass or geometrical center, and parameters describing its shape and size (e.g., its radius for a circular fracture or side length for a rectangular fracture) are also required. Therefore, a seven-parameter set $d_i(x_i, y_i, z_i, \omega_i, \rho_i, a_i, b_i)$ can uniquely determine the orientation, location, shape and dimension of an elliptical, circular or rectangular fracture plane in R^3, but cannot describe geometry and size of a generally shaped irregular fracture. The latter has to be represented by a finite number of M vertices $d_i\{(v_{i,1}, v_{i,2}, \ldots, v_{i,M}), i = 1, 2, \ldots, N_\mathrm{d}\}$, with known coordinates. The symbol N_d stands for the total number of fractures. All these vertices must locate on the same fracture plane.

On the other hand, for elliptical and circular fractures (with $a_i = b_i = r_i$ in the data string, r_i- radius), a fracture is computationally represented also by a finite number of M vertices whose coordinates are calculated from the fracture equation. For rectangle fractures (including the square ones with $a_i = b_i =$ constant side length), $M = 4$. Therefore, fracture shapes, sizes and locations can always easily be represented by the data strings of $d_i\{(v_{i,1}, v_{i,2}, \ldots, v_{i,M}), i = 1, 2, \ldots, N_\mathrm{d}\}$ approximating its boundary. The only difference is how the number M is defined: by calculation for regular fractures and by direct input or Monte Carlo simulation for irregular fractures. For simplicity and clarity of demonstrating the algorithm, the fractures are assumed to be circular in shape in this chapter and characterized by a six-parameter string $d_i(x_i, y_i, z_i, \omega_i, \rho_i, r_i)$ for the derivation of equations. For simplicity in computations, fractures are always represented by their strings of vertices $d_i\{(v_{i,1}, v_{i,2}, \ldots, v_{i,M}), i = 1, 2, \ldots, N_\mathrm{d}\}$ where the vertex number M may vary from fracture to fracture.

Block tracing depends on the intersections of fractures, which is performed with a dual global–local coordinate system O–XYZ and o–n–s–t, as shown in Fig. 7.21. The local frame is defined on each fracture

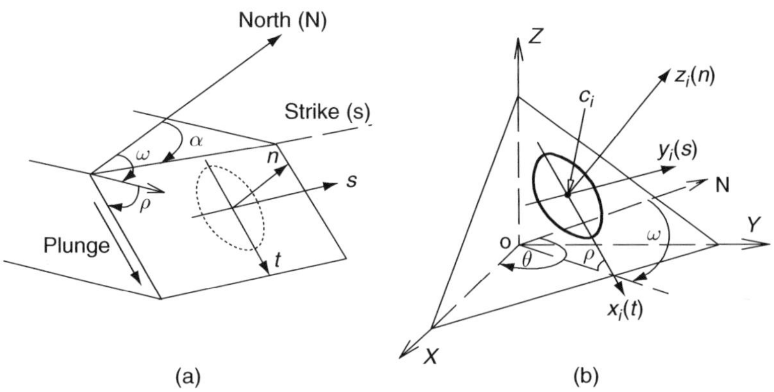

Fig. 7.21 Global and local coordinate systems: (a) definition of local coordinate system (n–s–t) on a fracture; (b) relation between the global coordinate system O–X–Y–Z and the local coordinate system of fracture $(x_1(t) - y_1(s) - z_1(n))$ (Jing, 2000).

plane as shown in Fig. 7.21a, which is uniquely specified as a right-handed system with n–s–t axes. The n-axis points along the outward normal direction of the fracture, the s-axis along the strike and the t-axis along the dip, respectively. Assuming that the global coordinates of the center of the ith fracture are $c_i(x_i^c, y_i^c, z_i^c)$, the relation between the global and local coordinate systems of the ith fracture, (see Fig. 7.21b) is written as

$$\begin{Bmatrix} t^i \\ s^i \\ n^i \end{Bmatrix} = \begin{Bmatrix} x_i \\ y_i \\ z_i \end{Bmatrix} = \begin{bmatrix} t_{11}^i & t_{12}^i & t_{13}^i \\ t_{21}^i & t_{22}^i & t_{23}^i \\ t_{31}^i & t_{32}^i & t_{33}^i \end{bmatrix} \begin{Bmatrix} x \\ y \\ z \end{Bmatrix} - \begin{Bmatrix} x_i^c \\ y_i^c \\ z_i^c \end{Bmatrix} \tag{7.16}$$

where $\mathbf{T}^i = \left[t_{kl}^i\right]$, $(k, l = 1, 2, 3)$, is the transformation matrix between the local coordinate system of the ith fracture and the global coordinate system. The transformation matrix is uniquely determined by the angle between the North and X-axis, θ (a quantity determined by the choice of global coordinate system, but independent of fracture fractures), angle of dip direction, ω^i, and dip angle, ρ^i, of the ith fracture, written as

$$\mathbf{T}^i = \left[t_{kl}^i\right] = \begin{bmatrix} \cos(\rho^i)\cos(\theta - \omega^i), & -\sin(\theta - \omega^i), & \sin(\rho^i)\cos(\theta - \omega^i) \\ \cos(\rho^i)\sin(\theta - \omega^i) & \cos(\theta - \omega^i), & \sin(\rho^i)\sin(\theta - \omega^i) \\ -\sin(\rho^i) & 0, & \cos(\rho^i) \end{bmatrix} \tag{7.17}$$

The matrix $\mathbf{T}^i$ is an orthogonal matrix with $(\mathbf{T}^i)^{-1} = (\mathbf{T}^i)^{\mathrm{T}}$ and $\left|(\mathbf{T}^i)\right| = \left|(\mathbf{T}^i)^{-1}\right| = 1$. It is a right-hand to right-hand system transformation.

7.3.2 Intersection Lines Between Fractures – the Initial Step

Let d_i and d_j be two fractures of radii r_i and r_j, respectively, with $c_i(x_i^c, y_i^c, z_i^c)$ and $c_j(x_j^c, y_j^c, z_j^c)$ as the coordinates of their respective geometric centers in the global O–XYZ system. The normal directions of these two fractures in the global system (O–XYZ) are $\mathbf{n}_i(n_x^i, n_y^i, n_z^i) = \mathbf{n}_i(t_{13}^i, t_{23}^i, t_{33}^i)$ and $\mathbf{n}_j(n_x^j, n_y^j, n_z^j) = (t_{13}^j, t_{23}^j, t_{33}^j)$, respectively. The distance between c_i and c_j is given by

$$d_{ij}^c = \sqrt{(x_i^c - x_j^c)^2 + (y_i^c - y_j^c)^2 + (z_i^c - z_j^c)^2} \tag{7.18}$$

If one of the following two conditions

$$d_{ij}^c > r_i + r_j \tag{7.19}$$

$$(\mathbf{n}_i) \cdot (\mathbf{n}_j) = n_x^i n_x^j + n_y^i n_y^j + n_z^i n_z^j = \pm 1 \tag{7.20}$$

is satisfied, then the two fractures have no intersection. They are either too far away, or parallel with each other, or partially overlapped. Otherwise two fractures may (but not certainly as in the 2D case) have an intersection trace line, called an edge here, defined by two vertices, as shown by an edge denoted as e_k in Fig. 7.22. The existence and coordinates of the two vertices defining the edge are determined by the local systems of the fractures.

The equations of planes containing the ith and jth fractures in the global system are written as

$$\{x - x_i^c, \quad y - y_i^c, \quad z - z_i^c\} \cdot \begin{Bmatrix} n_x^i \\ n_y^i \\ n_z^i \end{Bmatrix} = \left(\begin{Bmatrix} x \\ y \\ z \end{Bmatrix} - \begin{Bmatrix} x_i^c \\ y_i^c \\ z_i^c \end{Bmatrix} \right)^T \cdot \begin{Bmatrix} n_x^i \\ n_y^i \\ n_z^i \end{Bmatrix} = 0 \tag{7.21}$$

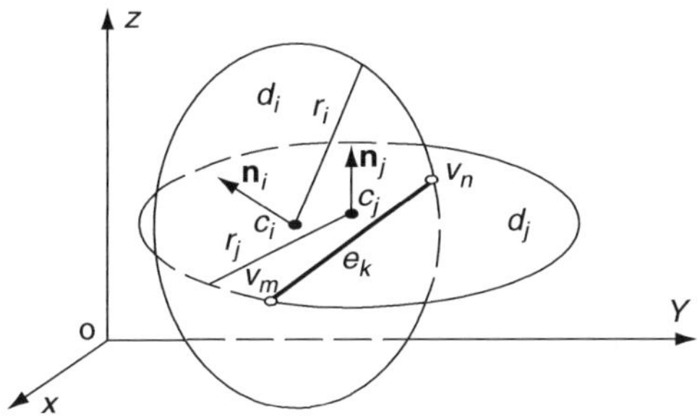

Fig. 7.22 The intersection of two circular fractures in R^3 (Jing, 2000).

for the plane containing the ith fracture and

$$\{\mathrm{x}-x_j^c,\ \mathrm{y}-y_j^c,\ \mathrm{z}-z_j^c\}\cdot\begin{Bmatrix} n_x^j \\ n_y^j \\ n_z^j \end{Bmatrix} = \left(\begin{Bmatrix} x \\ y \\ z \end{Bmatrix} - \begin{Bmatrix} x_j^c \\ y_j^c \\ z_j^c \end{Bmatrix}\right)^{\mathrm{T}} \cdot \begin{Bmatrix} n_x^j \\ n_y^j \\ n_z^j \end{Bmatrix} = 0 \tag{7.22}$$

for the plane containing the jth fracture, respectively. Rearrangement of Eqn (7.16) leads to

$$\begin{Bmatrix} x \\ y \\ z \end{Bmatrix} = \begin{bmatrix} t_{11}^i & t_{21}^i & t_{31}^i \\ t_{12}^i & t_{22}^i & t_{32}^i \\ t_{13}^i & t_{23}^i & t_{33}^i \end{bmatrix} \begin{Bmatrix} x_i + x_i^c \\ y_i + y_i^c \\ z_i + z_i^c \end{Bmatrix} \tag{7.23}$$

which transform the global co-ordinate system into a local system defined on the ith fracture plane with the origin located at $c_i(x_i^c, y_i^c, z_i^c)$. Substituting Eqn (7.23) into Eqn (7.22) leads to the equation of the plane containing the jth fracture in the local system defined on the ith fracture,

$$\left(\begin{Bmatrix} x \\ y \\ z \end{Bmatrix} - \begin{Bmatrix} x_j^c \\ y_j^c \\ z_j^c \end{Bmatrix}\right)^{\mathrm{T}} \cdot \begin{Bmatrix} n_x^j \\ n_y^j \\ n_z^j \end{Bmatrix} = \left(\begin{bmatrix} t_{11}^i & t_{21}^i & t_{31}^i \\ t_{12}^i & t_{22}^i & t_{32}^i \\ t_{13}^i & t_{23}^i & t_{33}^i \end{bmatrix} \begin{Bmatrix} x_i + x_i^c \\ y_i + y_i^c \\ z_i + z_i^c \end{Bmatrix} - \begin{Bmatrix} x_j^c \\ y_j^c \\ z_j^c \end{Bmatrix}\right)^{\mathrm{T}} \cdot \begin{Bmatrix} n_x^j \\ n_y^j \\ n_z^j \end{Bmatrix} = 0 \tag{7.24}$$

The equation of the ith fracture in its own coordinate system is written as

$$\begin{cases} (x_i)^2 + (y_i)^2 = (r_i)^2 \\ z_i = 0 \end{cases} \tag{7.25}$$

The solution of the combined Eqns (7.24) and (7.25) leads to

$$\begin{cases} (x_i)^2 + (y_i)^2 = (r_i)^2 \\ Ax_i + By_i + C = 0 \end{cases} \tag{7.26}$$

with

$$A = t_{11}^i n_x^j + t_{12}^i n_y^j, \quad B = t_{21}^i n_x^j + t_{22}^i n_y^j, \quad C = (A - n_x^j)x_i^c + (B - n_y^j)y_i^c \tag{7.27}$$

If condition $\Delta = 4A[B^2(r_i)^2 + A(r_i)^2 - C^2] > 0$ holds, the plane has two intersection points with the fractures, with coordinates,

$$(x_{i1}, y_{i1}, z_{i1}) = \left(\frac{B^2C - B\sqrt{\Delta}}{(B^2 + A)},\ \frac{-BC + \sqrt{\Delta}}{(B^2 + A)},\ 0\right) \tag{7.28}$$

$$(x_{i2}, y_{i2}, z_{i2}) = \left(\frac{B^2C + B\sqrt{\Delta}}{(B^2 + A)} + C, \ \frac{-BC - \sqrt{\Delta}}{(B^2 + A)}, \ 0 \right) \tag{7.29}$$

The global coordinates of these two points are

$$\begin{Bmatrix} x_1 \\ y_1 \\ z_1 \end{Bmatrix} = \begin{bmatrix} t^i_{11} & t^i_{21} & t^i_{31} \\ t^i_{12} & t^i_{22} & t^i_{32} \\ t^i_{13} & t^i_{23} & t^i_{33} \end{bmatrix} \begin{Bmatrix} x_{i1} + x^c_i \\ y_{i1} + y^c_i \\ z_{i1} + z^c_i \end{Bmatrix} \quad \text{and} \quad \begin{Bmatrix} x_2 \\ y_2 \\ z_2 \end{Bmatrix} = \begin{bmatrix} t^i_{11} & t^i_{21} & t^i_{31} \\ t^i_{12} & t^i_{22} & t^i_{32} \\ t^i_{13} & t^i_{23} & t^i_{33} \end{bmatrix} \begin{Bmatrix} x_{i2} + x^c_i \\ y_{i2} + y^c_i \\ z_{i2} + z^c_i \end{Bmatrix} \tag{7.30}$$

If $\Delta = 4A[B^2(r_i)^2 + A(r_i)^2 - C^2] \leq 0$, then the ith fracture has no intersection with the plane containing the jth fracture, i.e., no intersection between the two fractures.

The same process for the local system defined on the jth fracture will produce either another two intersection points, (x_3, y_3, z_3) and (x_4, y_4, z_4) on the jth fracture with the plane containing the ith fracture, or no intersection at all. In the latter case, the four intersection points $P_i(x_i, y_i, z_i)$, $i = 1, 2, 3, 4$, are located on the same line in the global space and the actual intersection trace line is determined by the common portion of finite length between segments $\overline{P_1P_2}$ and $\overline{P_3P_4}$ (Fig. 7.23). In Fig. 7.23a,b, the two trace line segments have either no common portion or the common portion is a single point. The two fractures have no intersection. In Fig. 7.23c,d, the common portion between the two trace line segments is defined by the non-zero segments $\overline{P_3P_2}$ and $\overline{P_3P_4}$, respectively, representing the intersection trace line segment between the two fractures. This trace line segment has the potential to be one of the edges defining a block and is called an intersection edge.

A matrix can be created to record the connectivity relations between fractures. Assuming the total number of fractures for a problem is N_d, then a square matrix of rank $(N_d \times N_d)$, $\mathbf{C}^d = [c^d_{ij}]$, $i, j = 1, 2, \ldots, N_d$, is defined as

$$c^d_{ij} = \begin{cases} 0, & \text{the fracture } i \text{ and } j \text{ not intersected} \\ 1, & \text{the fracture } i \text{ and } j \text{ intersected} \end{cases} \tag{7.31}$$

and

$$c^d_{ii} = 0 \tag{7.32}$$

for the diagonal elements in matrix $\mathbf{C}^d$.

The matrix $\mathbf{C}^d$ is a symmetric matrix (because, if the ith disk intersects with the jth disk, then the jth disk also intersects with the ith disk) and is called the disk connectivity matrix. The summation of its elements in the ith row of $\mathbf{C}^d$, M^e_i, is the number of intersection segments of the i-th disk with other disks.

The matrix $\mathbf{C}^d$ can become very large if conventional matrix storage techniques are used. Therefore, for large-scale problems with numerous fractures involved, special compression techniques or a linked-

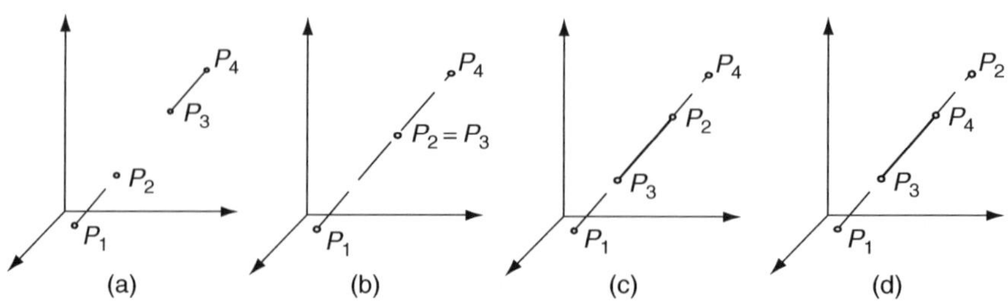

Fig. 7.23 An intersection edge defined as the common portion between two intersection trace line segments on the intersection line (Jing, 2000).

list structure should be used to record and keep track of the fracture connectivity, which plays an important role in flow analysis and block tracing.

The initial data sets, both inputted and generated, include:

(1) Fracture edge data:$D(E) = d_i\{(e_{i,1}, e_{i,2}, \ldots, e_{i,M+K}), i = 1, 2, \ldots, N_d\}$, where the loops of edges $e_{i,j}$ ($j = 1, 2, \ldots, M+K$) are the M boundary edges defining the geometry of the ith fracture and K interior edges as the intersection edges with other fractures.

(2) Vertex data: $V(X) = \{(x_i, y_i, z_i), i = 1, 2, \ldots, N_v\}$, where N_v is the current total number of vertices and (x_i, y_i, z_i) are their global coordinates, including all vertices defining all fracture boundary edges and all interior intersection edges.

(3) Edge data: $E(V) = \{(v_{i,1}, v_{i,2}), i = 1, 2, \ldots, N_e\}$, where N_e is the current total number of edges, including both the edges defining the fracture boundaries and intersections between fractures.

(4) Fracture connectivity data: the current fracture connectivity matrix $\mathbf{C}^d$ or its equivalent linked-list counterparts.

By the definition of 'complex', its sets of faces and edges must be complete and closed sets. 'Dangling' and isolated faces in the model should be discarded (Fig. 7.2) because they do not form blocks (polyhedra) as relevant simplexes, similar to the edge regularization in the 2D case. However, in the 3D case, both edge and face regularizations are needed.

7.3.3 Face and Edge Regularizations

The faces of blocks in a block system are defined by the 2D polygons (discs representing fractures) and the intersection edges between them. Face set regularization is performed by eliminating those polygons that have less than three intersection edges with other disks (because triangles are the simplest polygons that may form a face of a potential polyhedron). The initial data sets of edge ($E(V)$), Vertex ($V(X)$) and fracture ($D(V)$) need to be updated by including new additional vertices, on each fracture plane, as the intersecting points between two intersection edges between this fracture and all other fractures.

The edges of a block (polyhedron) are defined by intersection edges between disks, including the artificial boundary surfaces defining the computational model. These intersection edges define a network of vertices and edges on a particular fracture disk, forming a set of polygonal faces. The task of edge set regularization is to determine the coordinates of vertices as the intersecting points of the intersection edges on each disk, eliminate the isolated and 'dangling' segments and form a complete planar graph of edges and vertices on each disk.

Let $P_1^i(x_1^i, y_1^i, z_1^i)$, $P_2^i(x_2^i, y_2^i, z_2^i)$, $P_1^j(x_1^j, y_1^j, z_1^j)$ and $P_2^j(x_2^j, y_2^j, z_2^j)$ be the starting and ending points of the ith and jth intersection segments, respectively, on a particular fracture, Fig. 7.24a. A new vertex $V(x_s, y_s, z_s)$ is defined by the intersection of the two segments (Fig. 7.24b) if it exists. The parametric equations of the two intersection segments can be written as

$$\begin{cases} x = x_1^i + (x_2^i - x_1^i)t_i \\ y = y_1^i + (y_2^i - y_1^i)t_i \\ z = z_1^i + (z_2^i - z_1^i)t_i \end{cases} \quad \text{and} \quad \begin{cases} x = x_1^j + (x_2^j - x_1^j)t_j \\ y = y_1^j + (y_2^j - y_1^j)t_j \\ z = z_1^j + (z_2^j - z_1^j)t_j \end{cases} \tag{7.33}$$

where $0 \le t_i \le 1$ and $0 \le t_j \le 1$ are parameters. The solution of the Eqns in (7.33) leads to

$$x_s = x_1^i + (x_2^i - x_1^i)\frac{\Delta_x}{\Delta}, \quad y_s = y_1^i + (y_2^i - y_1^i)\frac{\Delta_x}{\Delta}, \quad z_s = z_1^i + (z_2^i - z_1^i)\frac{\Delta_x}{\Delta} \tag{7.34}$$

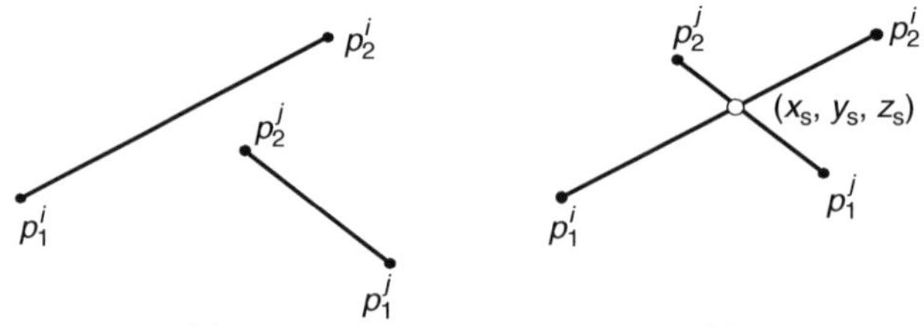

Fig. 7.24 Determination of vertices as the intersecting points between two intersection edges on a fracture: (a) two intersection edges have no intersecting point; (b) two intersection edges have an intersecting point as an additional vertex (Jing, 2000).

where

$$\Delta = \begin{vmatrix} x_2^i - x_1^i, & -(x_2^j - x_1^j) \\ y_2^i - y_1^i, & -(y_2^j - y_1^j) \end{vmatrix}, \quad \Delta_x = \begin{vmatrix} x_1^j - x_1^i, & -(x_2^j - x_1^j) \\ y_1^j - y_1^i, & -(y_2^j - y_1^j) \end{vmatrix}, \quad \Delta_y = \begin{vmatrix} x_2^i - x_1^i, & x_1^j - x_1^i \\ y_2^i - y_1^i, & y_1^j - y_1^i \end{vmatrix} \tag{7.35}$$

and

$$t_i = \frac{x_s - x_1^i}{x_2^i - x_1^i} = \frac{\Delta_x}{\Delta}, \quad t_j = \frac{x_s - x_1^j}{x_2^j - x_1^j} = \frac{\Delta_y}{\Delta} \tag{7.36}$$

If conditions $\Delta \neq 0$, $0 \leq t_i \leq 1$ and $0 \leq t_j \leq 1$ are satisfied simultaneously, the two intersection edges have a new intersection of coordinates (x_s, y_s, z_s), which defines a new vertex with a new label number. The two intersection edges then should be divided into four edges with proper starting and ending vertex numbers. The operation should be repeated between all intersection edges, including the edges defining the fracture boundaries, treating them like other interior intersection edges.

Edge set regularization is achieved by removing those intersection edge segments which have no intersection or only one intersection point (vertex) with all other segments on a particular disk. The resultant network of edges and vertices form a completely connected planar graph which itself can be regarded as a planar complex. This means that on each disk, its vertices and edges determined by the intersecting edges with other disks should satisfy the following conditions: (a) each vertex connects at least two edges; (b) each edge is defined by exactly two vertices and is shared by exactly two faces as their common edge; (c) the minimum numbers for the vertices and edges is three.

Edge set regularization should be performed in combination with face set regularization and done iteratively. The removal of segments with no, or only one, vertex means removal of a disk connected with this particular disk that should also be done in the face connectivity matrix $\mathbf{C}^{\mathrm{d}}$, with the consequent changes in its other elements. The combined face and edge set regularization should continue iteratively until all isolated faces and edges, faces with less than three intersecting segments and edges with less than two vertices, are removed.

The initial data sets then should be updated so that new vertex numbers and their coordinates are included in edge data $E(V)$ and vertex data $V(X)$, and the new edges, both interior and boundary, in $D(E)$. Two new data sets are created on each fracture plane. One is edge segment matrix $S(V)_j = \{(v_{i,1}, v_{i,2}, \ldots, v_{i,k})_j, \ i = 1, 2, \ldots, N_s; \ j = 1, 2, \ldots, N_d\}_j$ recording the sequential order of vertices along each intersection edge on the jth disk, similar to the $F(V)$ data in the 2D case (cf. Fig. 7.8). The other is the vertex-edge connectivity matrix $V(E)_j = \{(e_{i,1}, e_{i,2}, \ldots, e_{i,k})_j, i = 1, 2, \ldots, N_v;$ $j = 1, 2, \ldots, N_d\}$ recording the edges sharing a common vertex i.

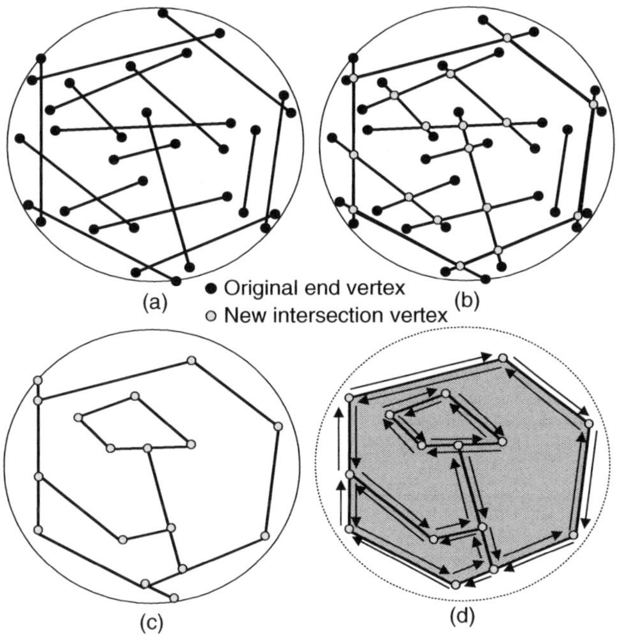

Fig. 7.25 Face/edge regularization for a circular fracture disk: (a) initial intersection trace map; (b) intersection trace map after adding new intersection points between traces; (c) intersection trace map after regularization; (d) final trace map after removing the two outer most 'dangling' faces.

An edge connectivity matrix, $\mathbf{C}^{e}_{kl}$, $(l = 1, M_v;\ \ k = 1, M^{e}_{i})$, is created to perform the edge set regularization on a particular disk, where M_v is the maximum number of vertices along an edge on this disk. The elements $v_{k1}, v_{k2}, \ldots, v_{kn}$ in each row of $\mathbf{C}^{e}_{kl}$ are the vertex numbers ordered according to their relative distance to the starting point of the corresponding segment (from small to large). This edge connectivity matrix reflects the connection relations between edges on the disk. The vertex number v_{kl} appears in both the k- and lth rows of the matrix because it connects both the k- and lth segments. Figure 7.25 shows the sequence of face/edge regularizations on an example fracture disk. Figure 7.25a shows the original map of fracture intersection traces on the plane of the fracture disk concerned. The coordinates of the new intersection points between the traces are calculated and added in the data structure (Fig. 7.25b).

After applying the edge regularization, the 'dangling' trace segments are removed and some of the end points become isolated vertices if they were not located on the fracture boundary. These isolated vertices are also then removed. The trace map then becomes a complete planar graph (Fig. 7.25c), composed of a number of polygonal faces by single loops of edges (alternated by the same number of vertices), after the identification of edges and vertices along intersection lines. The initial face regularization is performed by eliminating those faces which have less than three non-collinear edges along which connection with other faces are identified (because triangles are the simplest polygon on a fracture plane which may form a face of a potential polyhedron). Faces with one or more edges not shared by other faces are 'dangling' faces and are removed, such as the outermost faces containing the outer boundary of the circular disk. After assigning orientations of the edges (and therefore faces), the resultant fracture intersection trace map is an oriented planar complex (digraph), as shown in Fig. 7.25d, which can be used for block tracing.

The algorithm used for tracing these 3D faces is essentially the same as in the 2D block tracing, but the operations are performed on each fracture disk in their embedded local frames, then transformed back to the global frame. The directions of the edges are assigned using the decomposed and assigned vertex concept as introduced in the 2D case. Counter- clockwise edge loop directions are chosen since this will yield a positive area value of the face they outline for interior faces and a negative area value for exterior faces.

The data generated at the end of face and edge regularizations are then updated and represent the topological structure of faces and edges of the whole model, based on the vertices (and their coordinates) recorded in arrays $F(E)$, $E(V)$ and $V(X)$, with

$$F(E) = [\{f_i\}] = \left\{e_i^1, e_i^2, \ldots, e_i^{N_e}; i = 1, 2, \ldots, N_F\right\}$$

where N_F is the updated label number of faces and N_e the number of edges defining the face f_i and e_i^k the assigned edge number;

$$F(V) = [\{f_i\}] = \left\{v_i^1, v_i^2, \ldots, v_i^{N_V}, i = 1, 2, \ldots, N_F\right\}$$

where N_F is the updated label number of faces and N_V the number of vertexes defining the face f_i in a counter-clockwise orientation and e_i^k the assigned vertex number;

$$V(X) = \{v_i\} = \{(x_i, y_i, z_i),\ i = 1, 2, \ldots, N_v\}$$

where N_v is the updated number of vertices; and

$$E(V) = \{e_i\} = \left\{(v_{i,1}, v_{i,2}),\ i = 1, 2, \ldots, N_e\right\}$$

where N_e is the updated number of edges.

These three data sets are the primary sets, and other types of data can be induced from them. Although $F(E)$ and $F(V)$ are redundant, i.e., one can be readily derived from the other, they are both useful for block tracing operations.

Of particular usefulness is an edge–face connectivity matrix and a face–face connectivity matrix, defined as

$$\begin{aligned} \mathbf{C}_F^E &= [\{e_i\}] = \left[f_{i,1}, f_{i,2}, \ldots, f_{i,m}\right], \quad (i = 1, 2, \ldots, N_e) \\ \mathbf{C}_F^F &= [\{f_i\}] = \left[f_{i,1}, f_{i,2}, \ldots, f_{i,n}\right], \quad (i = 1, 2, \ldots, N_f) \end{aligned}$$

where m is the total number of faces connected at edge i, n the total number of faces connected to face I, N_e the total number of edges, N_f the total number of faces and $f_{i,j}$ the label number of faces for each case. Other sets of data, which may also be needed to speed up the data retrieval process, can all be readily calculated from the primary sets.

An index array $\{I_F\} = \{(I_i),\ i = 1, 2, \ldots, N_F\}$ is created to facilitate the block tracing. This index array is used to record the state of a face during the block tracing. The value assigned to the elements I_F during the block tracing is given as

$I_i = 2$ (initial state, face i has not been used during block tracing);

$I_i = 2$ (intermediate state, face i with its negative normal direction has been used);

$I_i = -1$ (intermediate state, face i with its positive normal direction has been used);

$I_i = 0$ (residual state, both the positive and the negative normal directions of face i have been used).

These data are all useful for block tracing. Due to their irregular dimensions, the linked-list data structure is usually most effective and memory-efficient.

The block (polyhedron) tracing is performed with the combined application of boundary operators, Euler–Poincaré formula, face connectivity matrix $\mathbf{C}_{ij}^{FF}$, edge–face connectivity matrix $\mathbf{C}_F^{EF}$ and the index

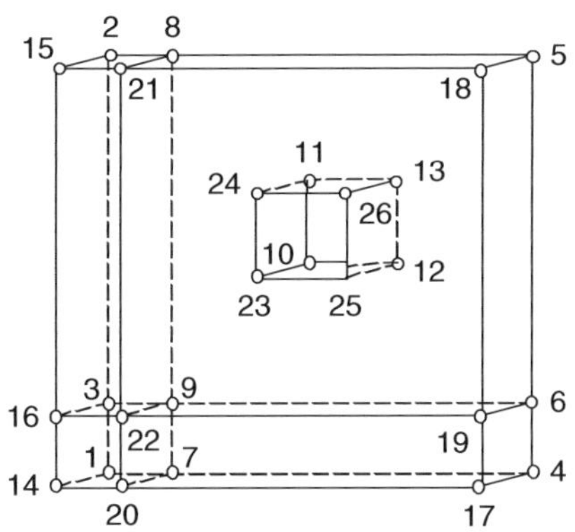

	$F(V)$		$F(E)$		$\{I_F\}$
1	3, 1, 7, 9, 3	1	−1, −6, −10, 12	1	2
2	2, 3, 9, 8, 2	2	−2, −12, −9, 8	2	2
3	4, 6, 9, 7, 4	3	3, 11, 10, −5	3	2
4	6, 5, 8, 9, 6	4	4, 7, 9, −11	4	2
	10, 11, 13, 12, 10		13, 16, 15, 14		2
5	11, 10, 12, 13, 11	5	−13, −14, −15, −16	5	2
6	15, 16, 22, 21, 15	6	17, 27, −25, 24	6	2
7	16, 14, 20, 22, 16	7	18, 21, −26, −27	7	2
8	17, 19, 22, 20, 17	8	19, −28, 26, 22	8	2
9	19, 18, 21, 22, 19	9	20, 23, 25, 28	9	2
	23, 24, 26, 25, 23		−29, −32, −31, −30		2
10	24, 23, 25, 26, 24	10	29, 30, 31, 32	10	2
11	3, 1, 14, 16, 3	11	−1, 33, −18, −35	11	2
12	2, 3, 16, 15, 2	12	−2, 35, −17, 34	12	2
13	4, 6, 19, 17, 4	13	3, −38, −19, 36	13	2
14	6, 5, 18, 19, 6	14	4, 37, −20, 38	14	2
15	4, 7, 20, 17, 4	15	5, −39, 22, 36	15	2
16	7, 1, 14, 20, 7	16	6, 33, 21, 39	16	2
17	5, 8, 21, 18, 5	17	7, 40, −23, −37	17	2
18	8, 2, 15, 21, 8	18	8, −34, −24, −40	18	2
19	8, 9, 22, 21, 8	19	9, −41, −25, −40	19	2
20	9, 7, 20, 22, 9	20	10, −39, −26, 41	20	2
21	6, 9, 22, 19, 6	21	11, −41, 28, 38	21	2
22	9, 3, 16, 22, 9	22	12, 35, 27, 41	22	2
23	10, 11, 24, 23, 10	23	13, 42, 29, −43	23	2
24	12, 10, 23, 25, 12	24	14, 43, 30, 44	24	2
25	13, 12, 25, 26, 13	25	15, −44, 31, −45	25	2
26	11, 13, 26, 24, 11	26	16, 45, 32, −42	26	2

Fig. 7.26 An example of a block system and its data sets generated for block tracing (Jing, 2000).

array $\{I_F\}$. The block tracing starts with face 1 in the index array $\{I_F\}$ and ends when all elements in the index array $\{I_F\}$ become zero.

Figure 7.26 is an example of a data set structure. The model region is defined by 12 rectangular fractures with data sets *F(E)*, *F(V)* and $\{I_F\}$ used for block tracing. Figure 7.27 shows the edge–vertex composition (Fig. 27a) and edge–face connectivity (Fig. 27b) matrices of the demonstrative example.

It should be noted that the above algorithms and data structure are definitely not the optimal choices for computer implementation but have the advantage of clearer illustration of the topological relations between the elementary variables of vertices, edges, faces and polyhedra/polygons. Removal of redundant data sets is needed during code development.

7.3.4 Block Tracing in 3D

7.3.4.1 The principle

The block tracing in 3D is performed by using the boundary chain operators with successive addition of relevant faces until the sum of its boundary 2-chain becomes zero. The Euler–Poincaré formula for polyhedra is then applied to ensure the correctness of the operation and detect the type of the polyhedron (homeomorphism to sphere or torus). Note that each face is shared by exactly two polyhedra with two opposite normal directions. Therefore one face must be processed twice, for both its negative and positive normal directions since they define two different blocks.

Similar to their 2D counterparts the 3D polyhedra have three types: (i) The exterior polyhedron of an infinitely large volume and an inner boundary defined by a finite number of faces (Fig. 7.28a); (ii) Simple (singly connected) interior polyhedron of an outer boundary defined by a finite number of faces (Fig. 7.28b) and (iii) Multiply-connected interior polyhedron (homeomorphic to a torus of one or a finite number of inner openings) of one outer boundary and a finite number of inner boundaries (Fig. 7.28c).

The determination of the volume of a polyhedron of n faces is given by

$$V = \sum_{i=1}^{n} V_i = \frac{1}{6} \sum_{i=1}^{n} \sum_{j=1}^{m_i-2} \begin{vmatrix} x_p, & y_p, & z_p, & 1 \\ x_j, & y_j, & z_j, & 1 \\ x_{j+1}, & y_{j+1}, & z_{j+1}, & 1 \\ x_{j+2}, & y_{j+2}, & z_{j+2}, & 1 \end{vmatrix} \tag{7.37}$$

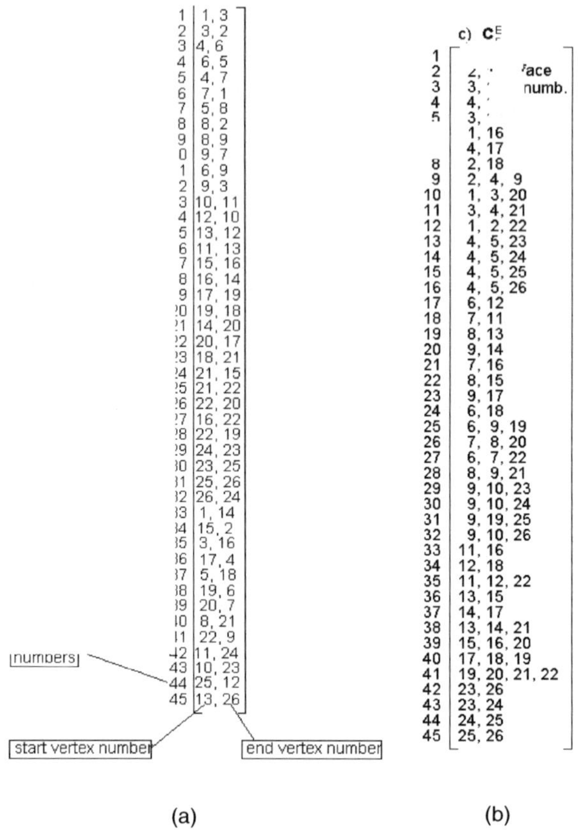

Fig. 7.27 Additional data set for the example in Fig. 7.26: (a) edge–vertex composition matrix and (b) edge–face connectivity matrix.

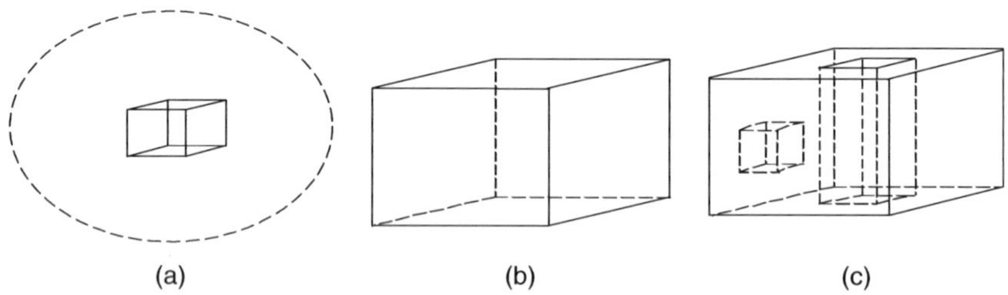

Fig. 7.28 Three types of polyhedral: (a) an exterior polyhedron; (b) an interior polyhedron; (c) a multiply-connected polyhedron (Jing, 2000).

where V_i is the volume of a cone formed by a reference point (x_p, y_p, z_p) and face i of the polyhedron. The polygonal face i of m_i vertices can be divided into $(m_i - 2)$ triangles forming $(m_i - 2)$ tetrahedra with the reference point (Fig. 7.29). If the vertices on each face of the polyhedron are arranged in a counter-clockwise orientation, the volume V calculated by formula (7.37) is positive for interior (singly or multiply-connected) polyhedron and negative for exterior polyhedron and interior openings of multiply-connected polyhedron, respectively. The exterior block is usually discarded from the block data structures, but it also provides the boundary faces for the model, useful for implementation of boundary conditions in numerical analyses.

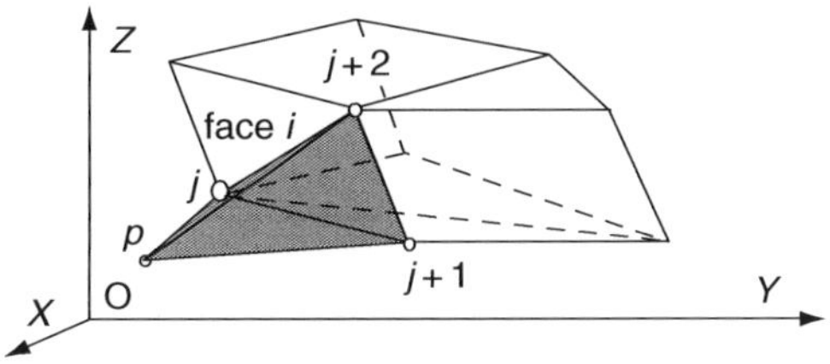

Fig. 7.29 The algorithm for calculating the volume of a tetrahedron formed by the reference point *P* and three adjacent vertices on a face of a polyhedron (Jing, 2000).

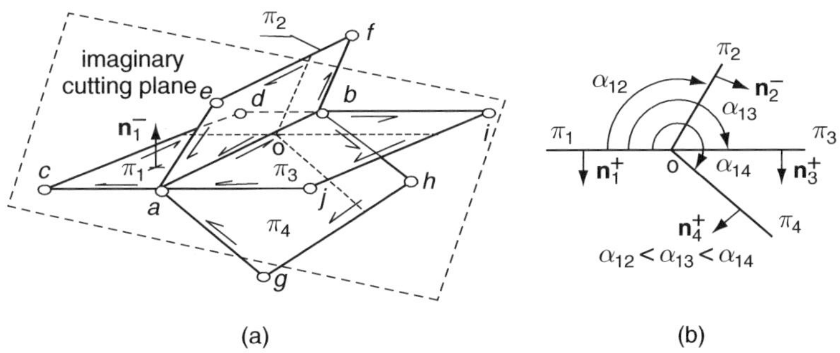

Fig. 7.30 Determination of the next face connected with base face: (a) a bundle of faces connected with the base face at a starting edge (*a,b*); (b) determination of the minimum angle between the base face and other connected faces (Jing, 2000).

Similar to the 2D face tracing the next face connected with the current face during block tracing is also determined by the minimum angle between the current face (called the base face) and other faces connected with it at a common edge (Fig. 7.30). If a bundle of faces is connected at a common edge (edge (*a,b*) in Fig. 7.30a) and one of them is the base face (face π_1 in Fig. 7.30a), then the next face which forms a solid block with the base face has the minimum angle, in a clockwise rotational sense, with the base face (Fig. 7.30b). It is important here to notice the orientations of the faces. Because each face has two orientations with its positive and negative unit normal vectors, corresponding to the counter-clockwise and clockwise rotational order of vertices (edges), respectively. Since the orientation of the base face is determined previously, so that the orientation of the common edge is also determined as the one following the same orientation of the base face. The orientations of the other faces connected at this edge should be chosen so that the orientation of the common edge in their loops of edges is opposite to the one in the edge loop of the base face. For example, in Fig. 7.30, the base face is the positive π_1 of the unit normal vector $\mathbf{n}_1^+$, vertex loop (*a,c,d,b,a*) and edge loop $\{(a,c)+(c,e)+(e,b)+(b,a)\}$. The chosen orientation of the common edge is therefore (*b,a*). The orientations of the other three faces should therefore be those corresponding to vertex loops π_2 (*a,b,f,e,a*) ($\mathbf{n}_2^-$), π_3 (*a,b,i,j,a*) ($\mathbf{n}_3^+$), and π_4 (*a,b,h,g,a*) ($\mathbf{n}_4^+$), respectively. The clockwise rotational angles between the base face and other three faces are calculated as α_{12}, α_{13} and α_{14}, respectively. Because $\alpha_{12} < \alpha_{13} < \alpha_{14}$, the face π_2 should be chosen as the next face.

Throughout the remaining parts of this chapter, the convention is used to denote the faces and edges with their orientations as f_i^+, f_i^- for faces and e_i^+, e_i^- for edges, where *i* is the number of faces or edges.

The general procedure of block tracing in 3D is:

(1) select a starting edge on a base face;

(2) determine the next candidate face by the 'minimum angle' principle from the bundle of faces connected along this edge;

(3) update the list of face orientations data set;

(4) add the candidate face to the complex and apply the boundary operator to see if the boundary edge chain is equal to zero;

(5) if the boundary edge chain is non-zero, the first edge of the boundary edge set is selected as the starting edge and the face containing this edge, already identified in the face list of the complex, is chosen as the next base face, and steps 2–4 are repeated;

(6) if the boundary edge chain equals zero, check with the Euler–Poincaré formula to ensure the correctness of the operation and calculate the block volume to determine the topological type of the block. The process should be repeated until every face is processed exactly twice, once for its positive orientation and once for its negative orientation, as shown in Fig. 6.19 for an example of tracing of a rectangular block with boundary chain operators, illustrating the steps and orientation of the boundary edge loops by arrows.

Block tracing is a complex operation and additional data sets, especially concerning connectivity information (face–face, edge–face, edge–edge, vertex–edge, vertex–face, etc.) are often needed to help the storage and retrieval of the dynamic data status.

7.3.4.2 An example

To demonstrate the 3D block tracing process using the boundary operator, the block system shown in Fig. 7.26 is used (Jing, 2000).

Tracing starts with f_1^+ (the base face) of the edge loop $(e_1^- + e_6^- + e_{10}^- + e_{12}^+) = \{(3,1) + (1,7) + (7,9) + (9,3)\}$ (from $\{F_{\mathrm{E}}\}$) and its first edge $e_1^- = (3,1)$. Changing $I_1 = -1$ in $\{I_{\mathrm{F}}\}$ (because f_1^+ is used) and checking the row 1 of the edge–face connectivity matrix $C_{\mathrm{F}}^{\mathrm{E}}$, the only face connected at edge 1 with f_1 is f_{11} of an edge loop $(e_1^- + e_{33}^+ + e_{18}^- + e_{35}^-) = \{(3,1) + (1,14) + (14,16) + (16,3)\}$ for f_{11}^+. Face f_{11}^- is then the next assigned face (because its edge (1,3) has the opposite orientation to that of edge (3,1) in f_1^+) with edge loop $(e_1^+ + e_{35}^+ + e_{18}^+ + e_{33}^-) = \{(1,3) + (3,16) + (16,14) + (14,1)\}$. The complex so far is composed of $K = (f_1^+ + f_{11}^-)$ and its boundary of 2-chains are calculated as a formal sum (Fig. 7.31a)

$$\partial K = \partial(f_1^+ + f_{11}^-) = \partial(f_1^+) + \partial(f_{11}^-) = (e_1^- + e_6^- + e_{10}^- + e_{12}^+) + (e_1^+ + e_{35}^+ + e_{18}^+ + e_{33}^-) =$$
$$(e_6^- + e_{10}^- + e_{12}^+ + e_{35}^+ + e_{18}^+ + e_{33}^-) = (1,7) + (7,9) + (9,3) + (3,16) + (16,14) + (14,1) \neq 0$$

where only $e_1^- + e_1^+ = 0$. I_{11} in the index array $\{I_F\}$ should be changed to 1 according to the rules mentioned above. Because that the set of boundary edges of the complex is not empty, i.e., a block (polyhedron) has not been formed, the tracing continues with the current complex of faces and boundary edges as the base, with additions of connected faces until finally we add face f_{10}^- (Fig.7.31a) as the last closing face, which leads to

$$\begin{aligned}\partial K &= \partial(f_1^+ + f_{11}^- + f_{16}^+ + f_3^+ + f_2^+ + f_{12}^- + f_7^- + f_{15}^+ + f_{13}^- + f_4^+ + f_{18}^- + f_6^- + f_8^- + f_{14}^- + f_{17}^- + f_9^- + f_5^+ + f_{10}^-) \\ &= \partial(f_1^+ + f_{11}^- + f_{16}^+ + f_3^+ + f_2^+ + f_{12}^- + f_7^- + f_{15}^+ + f_{13}^- + f_4^+ + f_{18}^- + f_6^- + f_8^- + f_{14}^- + f_{17}^- + f_9^- + f_5^+) + \partial(f_{10}^-) \\ &= (e_{30}^+ + e_{31}^+ + e_{32}^+ + e_{29}^+) + (e_{32}^- + e_{31}^- + e_{30}^- + e_{29}^-) = 0\end{aligned}$$

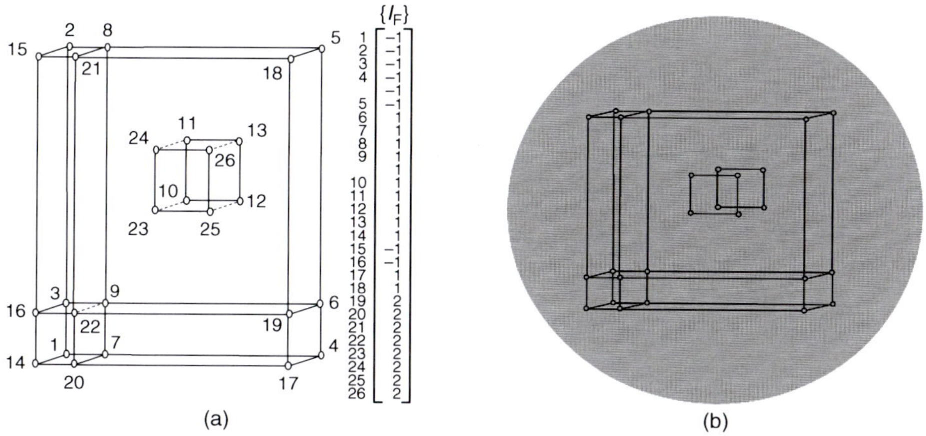

Fig. 7.31 (a) Tracing of block B_0 and (b) the exterior block enclosing a hollow space bounded by all identified and connected faces.

Because $\partial K = 0$ is achieved, so the tracing of one block is complete. By calculating the volume of the block by using formula (7.37), it can be found that the volume is negative, indicating the block is either an internal opening inside another interior block, or it is the exterior block enclosing the whole model (Fig. 7.31b). By checking that all faces of this block are located on the boundary disks of the model, it can be determined that this is the exterior block enclosing the whole model. This block is usually discarded from the block data structures, but it also provides the boundary faces useful for other applications (such as specifying boundary conditions). The block is denoted $B_0(K)$.

Continuation of the similar block tracing process using the boundary operator ∂K until condition $\partial K = 0$ is satisfied yields a progressive identification of all other interior blocks $B1$–$B5$, as shown in Fig. 7.32–7.36, respectively.

The oriented faces and their lists of oriented edges that form the above identified blocks for this example are listed in Table 7.2, in which the symbol '-' before the numbers of faces and edges stands for the reverse

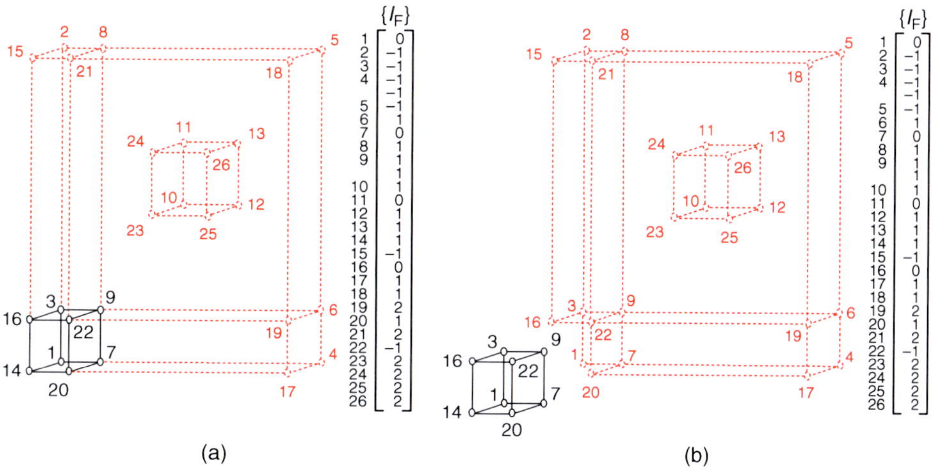

Fig. 7.32 (a) Completion of tracing $B_1(K)$; (b) separate presentation of $B_1(K)$ in solid lines, the remaining unidentified blocks are in dotted lines. The complex and boundary operator of block B_1 are $K_1 = (f_1^- + f_{22}^+ + f_{20}^- + f_{16}^- + f_{11}^+ + f_7^+)$ and $\partial K_1 = \partial(f_1^- + f_{22}^+ + f_{20}^- + f_{16}^- + f_{11}^+ + f_7^+) = 0$.

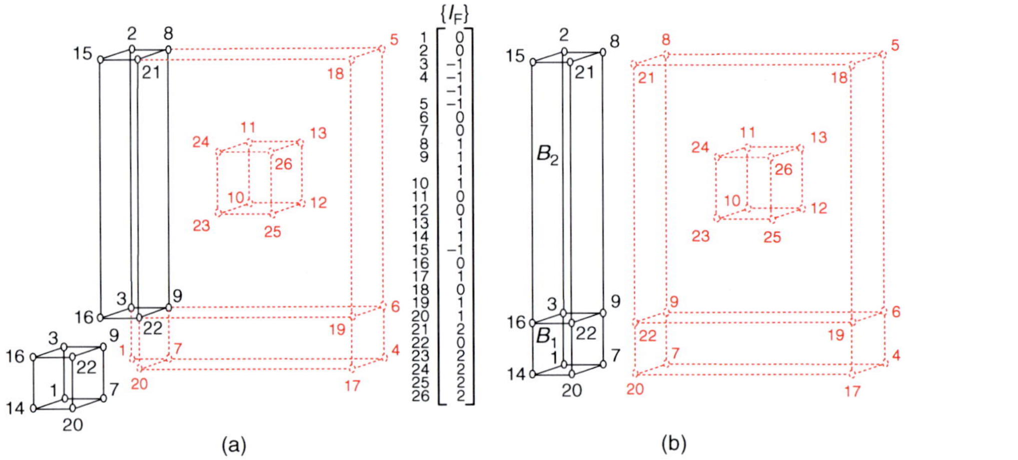

Fig. 7.33 (a) Completion of $B_2(K)$; (b) the separate representation of identified blocks $B_1(K)$ and $B_2(K)$. The complex and boundary operator of block B_2 are $K_2 = (f_2^- + f_{18}^+ + f_{19}^- + f_{22}^- + f_{12}^+ + f_6^+)$ and $\partial K_2 = \partial(f_2^- + f_{18}^+ + f_{19}^- + f_{22}^- + f_{12}^+ + f_6^+) = 0$.

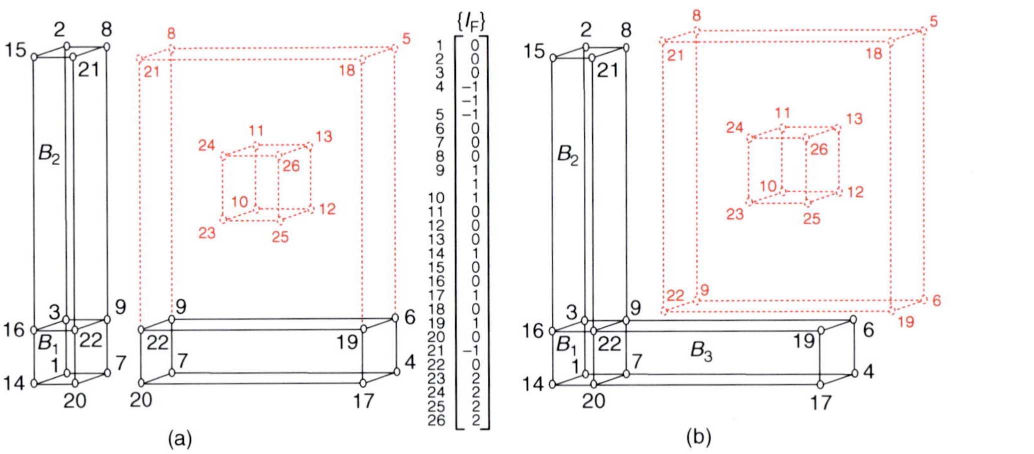

Fig. 7.34 (a) Completion of tracing $B_3(K)$, and separate representation of identified blocks $B_1(K)$, $B_2(K)$ and $B_3(K)$, (b) complex and boundary operator of block B_3 are $K_3 = (f_3^- + f_{15}^- + f_{20}^+ + f_{21}^+ + f_{13}^+ + f_8^+)$ and $\partial K_3 = \partial(f_3^- + f_{15}^- + f_{20}^+ + f_{21}^+ + f_{13}^+ + f_8^+) = 0$.

orientations of those faces and edges as originally defined during the system generation. Numbers of faces and edges without any symbol mean that they have the original orientations (as positive). It can be verified that, for each block, the total sum of the numbers of the directed (assigned) edges is zero.

7.4 Summary Remarks

Warburton (1983) was the first person to develop a rock block identification algorithm using infinitely large fractures. This technique, as commented on in the introduction of this chapter, is not suitable to represent complex block shapes in rock masses and the blocks thus generated are limited to convex shapes. A similar algorithm was also reported in Heliot (1988), and was implemented in some

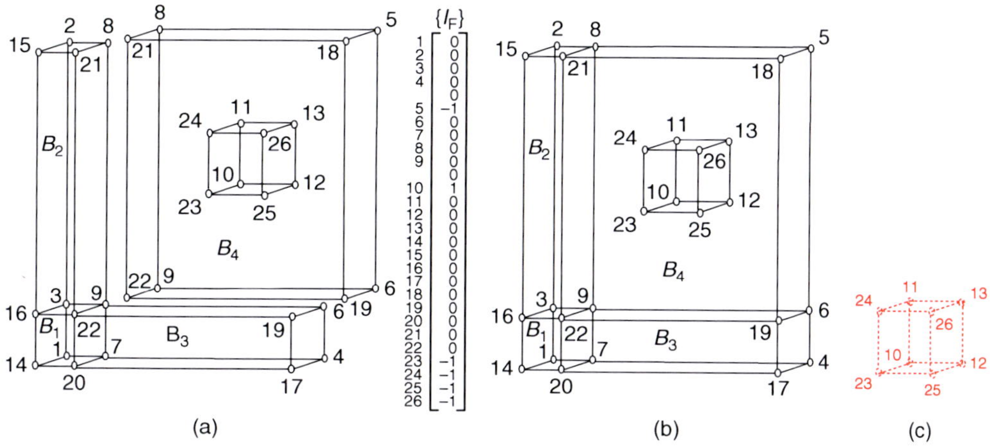

Fig. 7.35 (a) Completion of tracing block $B_4(K)$, a multiply-connected block of ten faces; (b) separate representation of identified blocks $B_1(K)$, $B_2(K)$, $B_3(K)$ and $B_4(K)$ and (c) the remaining unidentified block. Complex and boundary operator of block B_4 are $K_4 = (f_4^- + f_{21}^- + f_{19}^+ + f_{17}^+ + f_{14}^+ + f_{24}^+$ $f_{25}^+ + f_{26}^+ + f_{23}^+ + f_9^+)$ and $\partial K_4 = \partial(f_4^- + f_{21}^- + f_{19}^+ + f_{17}^+ + f_{14}^+ + f_{24}^+ + f_{25}^+ + f_{26}^+ + f_{23}^+ + f_9^+) = 0$.

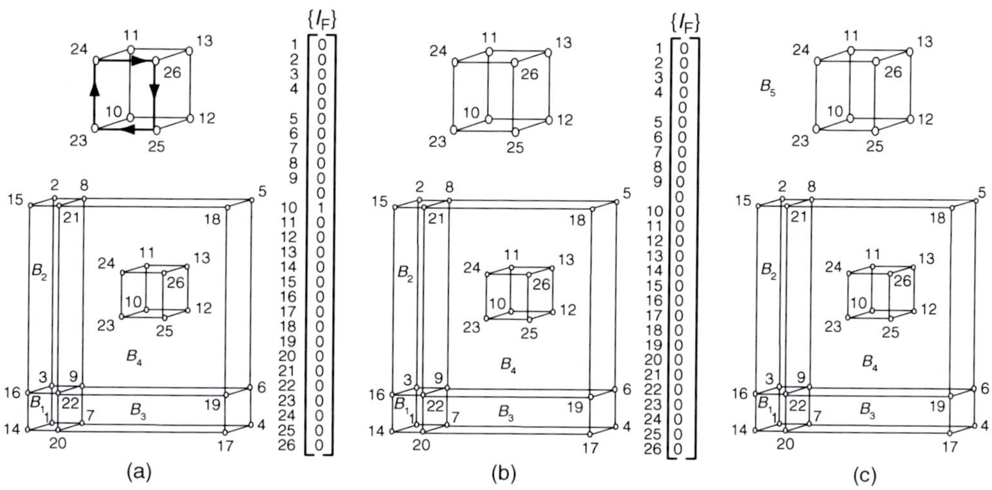

Fig. 7.36 (a) Adding the final face to B_5, (b) completion of tracing block B_5 and (c) the final presentation of five blocks. The complex and boundary operator of block B_5 are $K_5 = (f_5^- + f_{26}^- + f_{25}^- + f_{24}^- + f_{23}^- + f_{10}^+)$ and $\partial K_5 = (f_5^- + f_{26}^- + f_{25}^- + f_{24}^- + f_{23}^- + f_{10}^+) = 0$.

DEM codes. The extended part of fractures introduced by this assumption need to be treated as artificial parts of the infinite fractures. In DEM models, artificial fractures with fictitious mechanical properties are also needed to define excavation boundaries. For mechanical analysis, such artificial fractures (or the extended parts of the infinitely large fractures) may not affect the motion and deformation process very much, but may lead to non-negligible effects on stress concentrations, especially near the natural–artificial and artificial–artificial fracture intersections. The most important uncertainty in using such a simple technique is the fact that the fracture system's connectivity becomes much overestimated, which may lead to unknown effects on fluid flow processes if proper steps are not taken regarding the transmissivity of the artificial fractures (or the extended artificial parts of the real fractures). The boundary operator approach introduced in this chapter overcomes this difficulty with the price paid

Table 7.2 Face numbers and their edge loops of the $B_1(K)$–$B_5(K)$

$B_1(K)$	$B_2(K)$	$B_3(K)$	$B_4(K)$	$B_5(K)$
−1 (−12,10,6,1)	−2 (−8,9,12,2)	−3 (5,−10,−11,−3)	−4(11,−9,−7,−4; −14,−15,−16,−13)	−5 (16,15,14,13)
22 (12,35,27,41)	18 (8,−34,−24,−40)	−15 (−36,−22,39,−5)	−21 (−38,−28,41,−11)	−26 (42,−32,−45,−16)
−20 (−41,26,39,−10)	−19 (40,25,41,−9)	20 (10,−39,−26,41)	19 (9,−41,−25,−40)	−25 (45,−31,44,−15)
−16 (−39,−21,−33,−6)	−22 (−41,−27,−35,−12)	21 (11,−41,28,38)	17 (7,40,−23,−37)	−24 (−44,−30,−43,−14)
11 (−1,33,−18,−35)	12 (−2,35,−17,34)	13 (3,−38,19,36)	14 (4,37,−20,38)	−23 (43,−29,−42,−13)
7 (18,21,−26,−27)	6 (17,27,−25,24)	8 (19,−28,26,22)	24 (14,43,30,44)	10 (29,30,31,32)
			25 (15,−44,31,−45)	
			26 (16,45,32,−42)	
			23 (13,42,29,−43)	
			9 (20,23,25,28; −29,−32,−31,−30)	

for the more challenging block tracing algorithms, which, however, is paid once and for all at the pre-processing stage of the problem solution.

As pointed out in Lu (2002), the topological concepts of simplex and complex may not be necessary in practice for block system identifications. Using directional graph theory or oriented polygons can also achieve the same objective. On the other hand the combinatorial topology provides a more fundamental basis for such algorithms, and the Euler-Poincaré relation is useful for checking the correctness of the tracing operations. In programming, simple vector algebra is often enough to carry out the operations without resorting to complex set operations as in Lu (2002) and Lin (1992). The key issue is the establishment of properly chosen data structures for the tracing operations. The data structure and tracing algorithm introduced above cannot be said to be the most effective since it has a large degree of redundancy, but more efficient algorithms can be developed by more skillful code developers.

In developing fracture–block system algorithms for DEM methods, an important issue to be kept in mind is that the needs for modeling the coupling effects of fluid flow, fracture deformation and block motion/deformation must be considered at the same time. This is necessary because the fluid flow process becomes a more and more important issue in rock engineering, and one that cannot be discarded in many practical problems concerning environmental impacts. Such an objective (not object)-oriented development strategy must be adopted at all levels of the DEM methods since fracture systems are often the major fluid-conducting pathways whose more realistic representations in the models are decisive factors for obtaining reliability of results.

References

Golder Associates, FracMan Manual, 1995.

Heliot, D., Generating a blocky rock mass. International Journal of Rock Mechanics and Mining Sciences and Geomechanics Abstract, 1988;25(3):127–138.

Herbert, A. Modeling approaches for discrete fracture network flow analysis. In: Stephansson, O., Jing, L. and Tsang, C.-F. (ed.), Coupled thermo-hydro-mechanical processes of fractured media. Developments in geotechnical engineering. Vol. 79, pp. 213–229. Elsevier Sciences B. V. 1996.

Itasca Consulting Group Ltd., The UDEC Manual, 2000.

Jing, L., Block system construction for three-dimensional discrete element models of fractured rocks. International Journal of Rock Mechanics and Mining Sciences, 2000;37(4):645–654.

Jing, L. and Stephansson, O., Topological identification of block assemblage for jointed rock masses. International Journal of Rock Mechanics and Mining Sciences and Geomechanics Abstract, 1994a;31(2):163–172.

Jing, L. and Stephansson, O., Identification of block topology for jointed rock masses using boundary operators. Proc. Int. ISRM Symp. on Rock Mechanics, Santiago, Chile, Vol. I, 19–29, 1994b.

Jing, L. and Stephansson, O. Network topology and homogenization of fractured rocks. In: Jamtveit, B and Yardley, B. (eds), Fluid flow and transport in rocks: mechanisms and effects, pp. 191–202. Chapman and Hall, 1996.

Lin, D., Elements of rock block modeling. Ph.D. Thesis, University of Minnesota, Minneapolis, 1992.

Lu, J., Systematic identification of polyhedral rock blocks with arbitrary joints and faults. Computers and Geotechnics, 2002;29(1):49–72.

Mäntylä, M., An introduction to solid modeling. Computer Science Press, Rockville, Maryland, 1988.

Niemi, A., Kontio, K. and Kuusela-Lahtinen, A, Hydraulic characterization and upscaling of fracture networks based on multiple-scale well test data. Water Resources Research, 2000;36(12): 3481–3497.

Shi, G., Discontinuous deformation analysis – a new numerical model for the statics and dynamics of block systems. Ph.D. Thesis, University of California, Berkeley, CA, 1988.

Wang, C., Graph theory. Press of Beijing University of Technology, 399 pp. (in Chinese), 1987.

Warburton, P. M., Application of a new computer model for reconstructing blocky rock geometry – analysing single block stability and identifying keystones. In: Proc. 5th Int. Cong. on Rock Mechanics (preprints), Melbourne, 1983, Section F, pp. F225–F230. Brown Prior Anderson Pty Ltd., 1983.

Part Three
DEM Approaches

8 EXPLICIT DISCRETE ELEMENT METHOD FOR BLOCK SYSTEMS – THE DISTINCT ELEMENT METHOD

8.1 Introduction

The distinct element method is an explicit DEM based on finite difference principles, originated in the early 1970s by a landmark work on the progressive movements of rock masses as 2D rigid block assemblages (Cundall, 1971a,b). The work was extended later into a code RBM written in machine language (Cundall, 1974). The method and the RBM code later progressed by first approximating the deformation of blocks of complex 2D geometry by a constant strain tensor, with the code translated into FORTRAN language and called SDEM (Cundall et al., 1978). A separate version of the SDEM code, called CRACK, was created to consider fracturing, cracking and splitting of intact blocks under loading, based on a tensile failure criterion. However, the representation of 'simply deformable blocks' causes incompatibility between the complex block geometry and uniform strain tensor. This difficulty was overcome later by using full internal discretization of blocks by finite difference meshes with triangular elements, leading to the early versions of the UDEC and 3DEC codes (Cundall, 1980; Cundall and Hart, 1985). The principle of simulating the large-scale deformation of elasto-plastic materials using finite difference schemes by Wilkins (1963) was used. The method and codes were then developed further by coupling heat conduction and viscous fluid flow through fractures (treated as interfaces between block boundaries) for 2D problems (Itasca, 1993) and coupled thermo-mechanical simulations for 3D problems (Cundall, 1988; Hart et al., 1988; Itasca, 1994). The hybrid technique of combining distinct element and BEMS was also developed (Lemos, 1983) to treat the far-field effects, most efficiently for 2D problems.

Typical computer codes are the 2D code UDEC and the 3D code 3DEC, developed by Cundall and his colleagues (Itasca, 1993, 1994). The method has also been developed for simulating the mechanical behavior of granular materials (Cundall and Strack, 1979a,b,c, 1982), with a typical early code BALL (Cundall, 1978) which later evolved into the codes of the PFC group for 2D and 3D problems of particle systems (Itasca, 1995).

Through continuous developments and extensive applications over the last three decades, there has accumulated a great body of knowledge and a rich field of literature about the distinct element method. Besides the works represented in the developments and applications of the UDEC/3DEC/PFC code groups, different formulations and numerical codes based on the same principles have been developed and applied to various problems, such as code BLOCKS for rigid block motion (Taylor, 1982), code CICE (Williams et al., (1986); Williams and Mustoe, 1987; Mustoe, 1992) and the BSM techniques by Kawai (1977a,b, 1979; Kawai et al., 1978), Wang and Garga (1993), Li and Wang (1998), Li and Vance (1999) and Hu (1997). However the main trend in the development and application of the method in rock engineering is represented by the history and results of the code groups UDEC/3DEC. In this chapter, therefore, the main features of the distinct element method are presented in the context of the UDEC/3DEC codes, followed by a brief discussion on the basic characteristics of the other codes. Since the 'simply deformable block' has only historical values and is no longer applied in practice, it is not included.

As presented in Chapter 1, the distinct element method was created to simulate the interactions between blocks in an assemblage that simulates the blocky structure of a fractured rock mass. The generation of the geometry of the block assemblage created by the fracture network has been presented in Chapters 5–7 for both 2D and 3D problems. In the present chapter, the emphasis is on the presentation of the block geometry, deformation, contact, damping, solution techniques and the data structure. The main mathematical techniques of dynamic relaxation are presented in detail and other solution techniques or formulations are briefly presented at the end of the chapter.

Before the presentations, it is necessary to note that there are basically three most important tasks in a distinct element code for mechanical analyses:

- to create the block assemblage, record the block topology with an appropriate data structure plus updating of the record throughout the entire deformation process;
- to select a proper form of the equations of motion and constitutive models for the rock and fractures, and the solution techniques and
- to determine and update the geometry and mechanical behavior of contacts between the blocks during the deformation process.

If fluid flow (through fractures) and heat conduction are also involved, the management of fracture spaces through contact detection and fracture deformation, and heat diffusion through the blocks are also essential tasks.

Most of the available distinct element codes use the linked-list data structure to create and update the geometry of the block systems. The central FDM is applied to integrate the Newton–Euler equations of motion for rigid block systems or Cauchy equations of motion for deformable block systems based on either the dynamic or static relaxation techniques. The contact detection and updating is performed based on the 'contact overlap' concept, by which the amount of interpenetration depth between two blocks at a contact point is regarded as relative normal deformation of the two blocks at the point. These three main aspects of the distinct element method are presented in detail in this chapter, together with other important issues regarding damping, numerical stability and alternative formulations. Before going into the detailed presentation of the method, some theoretical concepts relating to finite difference approximation to functions and their derivatives and the relaxation technique are introduced.

The distinct element method can be classified into two categories: those using static relaxation principles and those using dynamic relaxation principles. Although the latter is more widely applied in practice at present, the former has distinct values in both conceptualization and theoretical development. We introduce both in this chapter.

The FDM for stress analysis of solids is the basis of the distinct element method, besides the contact analysis. The forerunner of this approach can be traced back to the finite difference scheme for calculating large-scale elasto-plastic deformation of material for both 1D and 2D problems, as developed in Wilkins (1963). Otter et al. (1966) also developed similar algorithms for dynamic relaxation solutions of elasticity problems, and similar algorithms were also used later in different DEM methods for plastic flow studies of geological materials (Marti and Cundall, 1982; Cundall and Board, 1988).

The formulation of the distinct element method for motion and deformation of block systems is based on three fundamental formulation aspects:

(1) the internal discretization of the blocks with finite difference zones or finite volume elements when the blocks are treated as deformable;

(2) the dynamic relaxation technique for stress analysis of the deformable blocks and

(3) an efficient detection and representation of contacts between the blocks.

Fluid flow through the connected fracture networks formed by the interfaces between the blocks forms an additional basis if coupled hydro-mechanical analysis is required.

In the following sections, the above fundamental aspects of the formulation of the distinct element method are presented in different degrees of detail. The presentation starts with the principles of static and dynamic relaxation, and is then followed by internal discretization schemes, the representation for deformation and stress analyses, contacts and treatment of fluid flow and heat transfer of block systems. Some of the fundamental formulation aspects are similar or identical to that in the codes UDEC and 3DEC, such as stress and strain analyses and the solution of equations of motion, but some aspects, such as internal discretization, may not be the same. However, the principles are the same and a good understanding of the approaches presented here will lead to easier understanding of other alternative approaches adopted in different DEM codes. The partitioning of the modeling space by dividing the domain of interest into block systems is not presented, because the topic is adequately discussed in previous chapters.

8.2 Finite Difference Approximations to Derivatives

8.2.1 Regular Meshes of Rectangular Elements

When a function *f(x)* and its derivatives are single valued, finite and continuous functions of variable x, its Taylor's expansion is given by

$$\begin{cases} f(x+\Delta x) = f(x) + \dfrac{\mathrm{d}f(x)}{\mathrm{d}x}(\Delta x) + \dfrac{1}{2}\dfrac{\mathrm{d}^2 f(x)}{\mathrm{d}x^2}(\Delta x)^2 + \cdots + \dfrac{1}{n!}\dfrac{\mathrm{d}^n f(x)}{\mathrm{d}x^n}(\Delta x)^n + \cdots \\ f(x-\Delta x) = f(x) - \dfrac{\mathrm{d}f(x)}{\mathrm{d}x}(\Delta x) + \dfrac{1}{2}\dfrac{\mathrm{d}^2 f(x)}{\mathrm{d}x^2}(\Delta x)^2 - \cdots - \dfrac{(-1)^n}{n!}\dfrac{\mathrm{d}^n f(x)}{\mathrm{d}x^n}(\Delta x)^n + \cdots \end{cases} \tag{8.1}$$

The addition of the two expressions in Eqn (8.1) then results in

$$f(x+\Delta x) + f(x-\Delta x) = 2f(x) + (\Delta x)^2 \frac{\mathrm{d}^2 f(x)}{\mathrm{d}x^2} + \mathrm{O}[(\Delta x)^4] \tag{8.2}$$

where $\mathrm{O}[(\Delta x)^4]$ denotes terms containing the fourth and higher powers of (Δx). Assuming that these are negligible compared with the lower powers of (Δx), it follows that

$$\frac{\mathrm{d}^2 f(x)}{\mathrm{d}x^2} \approx \frac{1}{(\Delta x)^2}[f(x+\Delta x) - 2f(x) + f(x-\Delta x)] \tag{8.3}$$

with a leading error on the right-hand-side of order $(\Delta x)^2$. Subtraction between the two expressions of Eqn (8.1) and neglecting terms of order $(\Delta x)^3$ leads to

$$\frac{\mathrm{d}f(x)}{\mathrm{d}x} \approx \frac{1}{2(\Delta x)}[f(x+\Delta x) - f(x-\Delta x)] \tag{8.4}$$

also with a leading error on the right-hand-side of order (Δx).

Equation (8.4) approximates the slope of the tangent of *f(x)* at *P* by the slope of the chord *AB* (Fig. 8.1a), and is called a central-difference approximation.

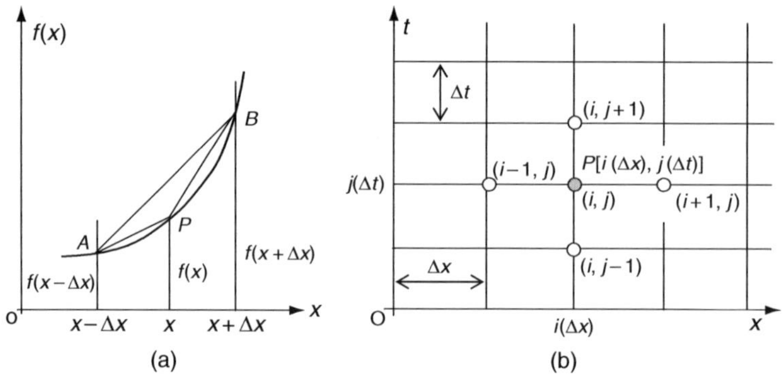

Fig. 8.1 Central finite difference schemes for approximations to derivatives of functions: (a) 1D problem; and (b) 2D problems.

The approximation by

$$\frac{\mathrm{d}f(x)}{\mathrm{d}x} \approx \frac{1}{\Delta x}\left[f(x+\Delta x)+f(x)\right] \tag{8.5}$$

is called the forward-difference approximation to d*f(x)*/d*x*, represented by the slope of chord *PB* and

$$\frac{\mathrm{d}f(x)}{\mathrm{d}x} \approx \frac{1}{\Delta x}\left[f(x)-f(x-\Delta x)\right] \tag{8.6}$$

is called the backward-difference approximation to d*f(x)*/d*x*, represented by the slope of chord *AP*.

Equations (8.5) and (8.6) can be derived from Eqns (8.1) and (8.2) directly by neglecting second and higher powers of (Δx), with a leading error, therefore, of O(Δx).

For functions of multiple variables $(x_1, x_2, ..., x_n)$, the technique is similar. The variable space is divided into sets of equal sub-spaces of sides $(\Delta x_1, \Delta x_2, ..., \Delta x_n)$, followed by the same approximation to the Taylor expansion of the function by neglecting higher powers of the sub-spacing increments. Figure 8.1b provides an example for 2D problems of the variable (*x*, *t*).

The *x*–*t* plane is divided into sets of equal rectangles of sides Δx and Δt. Let the coordinates (*x*, *t*) of a representative point *P* be $x = i(\Delta x)$, $t = j(\Delta t)$ where *i* and *j* are integers and the value of the function *f*(*x*, *t*) *at* P *be* $f_p(x,t) = f(i\Delta x, j\Delta t) = f_{i,j}$, then by Eqn (8.3),

$$\left[\frac{\partial^2 f(x,t)}{\partial x^2}\right]_{i,j} \approx \frac{f[(i+1)(\Delta x), j(\Delta t)] - 2f[i(\Delta x), j(\Delta t)] + f[(i-1)(\Delta x), j(\Delta t)]}{(\Delta x)^2} \tag{8.7}$$

$$\left[\frac{\partial^2 f(x,t)}{\partial t^2}\right]_{i,j} \approx \frac{f[i(\Delta x), (j+1)(\Delta t)] - 2f[i(\Delta x), j(\Delta t)] + f[i(\Delta x), (j-1)(\Delta t)]}{(\Delta t)^2} \tag{8.8}$$

or simply

$$\left[\frac{\partial^2 f(x,t)}{\partial x^2}\right]_{i,j} \approx \frac{f_{i+1,j} - 2f_{i,j} + f_{i-1,j}}{(\Delta x)^2}, \quad \left[\frac{\partial^2 f(x,t)}{\partial t^2}\right]_{i,j} \approx \frac{f_{i,j+1} - 2f_{i,j} + f_{i,j-1}}{(\Delta x)^2} \tag{8.9}$$

with a leading error of order $(\Delta x)^2$ and $(\Delta t)^2$, respectively. Similarly one can derive

$$\left[\frac{\partial f(x,t)}{\partial x}\right]_{i,j} \approx \frac{f[(i+1)(\Delta x), j(\Delta t)] - f[(i-1)(\Delta x), j(\Delta t)]}{2(\Delta x)} = \frac{f_{i+1,j} - f_{i-1,j}}{2(\Delta x)} \tag{8.10}$$

$$\left[\frac{\partial f(x,t)}{\partial t}\right]_{i,j} \approx \frac{f[i(\Delta x),(j+1)(\Delta t)] - f[i(\Delta x),(j-1)(\Delta t)]}{2(\Delta t)} = \frac{f_{i,j+1} - f_{i,j-1}}{2(\Delta t)} \tag{8.11}$$

for the central-difference scheme with leading error of order O(Δx) and O(Δt), respectively.

8.2.2 Meshes with Generally Shaped Elements – the Finite Volume Scheme

The scheme for the FEM with regular rectangular meshes presented above requires that the shape of the elements must be regular rectangles aligned to the directions of the coordinate axes (although one can have different interval lengths in different coordinate directions). This often causes numerical difficulties for discretization if the shape of the domain of interest is complex and irregular, especially near the boundaries.

An alternative scheme can be used to overcome this difficulty. This is the so-called Finite Volume Scheme (or sometimes control volume scheme) of FDMs, often used for fluid mechanics problems. The scheme depends on an integral definition of partial derivatives of a continuous function, using the Gauss divergence theorem, over domains (or elements) of generally shaped polygons or polyhedra (for example the hexagonal, quadrilateral and triangular elements in Fig. 8.2 for 2D problems).

Assume that a continuous (or piecewise continuous) function is defined on a domain (element) of volume A that is enclosed by boundary surface S of arbitrary shape of N plane faces (or straight edges in the 2D case), then the definitions of the partial derivatives in the directions of the coordinate axes are given by

$$\begin{cases} \dfrac{\partial f}{\partial x} = \dfrac{\oint_S f n_x \, \mathrm{d}S}{\operatorname{Lim}_{A\to 0} A} \approx \dfrac{1}{A}\oint_S f n_x \, \mathrm{d}S \approx \dfrac{1}{A}\sum_{m=1}^{N} f_m n_x \Delta S_m \\ \dfrac{\partial f}{\partial y} = \dfrac{\oint_S f n_y \, \mathrm{d}S}{\operatorname{Lim}_{A\to 0} A} \approx \dfrac{1}{A}\oint_S f n_y \, \mathrm{d}S \approx \dfrac{1}{A}\sum_{m=1}^{N} f_m n_y \Delta S_m \\ \dfrac{\partial f}{\partial z} = \dfrac{\oint_S f n_z \, \mathrm{d}S}{\operatorname{Lim}_{A\to 0} A} \approx \dfrac{1}{A}\oint_S f n_z \, \mathrm{d}S \approx \dfrac{1}{A}\sum_{m=1}^{N} f_m n_z \Delta S_m \end{cases} \tag{8.12}$$

where f_m is the mean value of function f on the mth face (or edge) of the boundary S, ΔS_m and $\mathbf{n} = (n_x, n_y, n_z)$ is the outward unit normal vector of surface S. The relation between $\mathbf{n}$ and the unit vectors of the coordinate axes is given by

$$\mathbf{n} = n_x\mathbf{i} + n_y\mathbf{j} + n_z\mathbf{k} = \frac{\partial x}{\partial n}\mathbf{i} + \frac{\partial x}{\partial n}\mathbf{j} + \frac{\partial x}{\partial n}\mathbf{k} \tag{8.13}$$

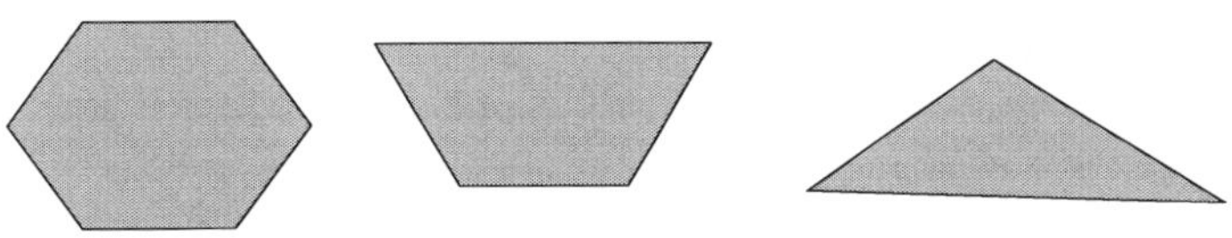

Fig. 8.2 Different element shapes for finite volume schemes: hexagonal elements, quadrilateral and triangular elements.

where **i**, **j**, **k** are the unit vectors of the coordinate axes, x, y and z, respectively. For 2D cases, Eqn (8.13) can be rewritten simply as

$$\mathbf{n} = \frac{\partial x}{\partial n}\mathbf{i} + \frac{\partial x}{\partial n}\mathbf{j} = \frac{\partial y}{\partial S}\mathbf{i} - \frac{\partial x}{\partial S}\mathbf{j} \tag{8.14}$$

The partial derivatives $\partial f/\partial x$ and $\partial f/\partial y$ in 2D cases can then be written as

$$\begin{cases} \dfrac{\partial f}{\partial x} \approx \dfrac{1}{A}\oint\limits_{\Gamma} f n_x \, \mathrm{d}S = \dfrac{1}{A}\oint\limits_{\Gamma} f\dfrac{\partial x}{\partial n}\, \mathrm{d}S = \dfrac{1}{A}\oint\limits_{\Gamma} f\dfrac{\partial y}{\partial S}\, \mathrm{d}S \approx \dfrac{1}{A}\sum_{m=1}^{N} f_m \dfrac{\Delta y_m}{\Delta S_m}\Delta S_m = \dfrac{1}{A}\sum_{m=1}^{N} f_m \Delta y_m \\ \dfrac{\partial f}{\partial y} \approx \dfrac{1}{A}\oint\limits_{\Gamma} f n_y \, \mathrm{d}S = \dfrac{1}{A}\oint\limits_{\Gamma} f\dfrac{\partial y}{\partial n}\, \mathrm{d}S = -\dfrac{1}{A}\oint\limits_{\Gamma} f\dfrac{\partial x}{\partial S}\, \mathrm{d}S \approx -\dfrac{1}{A}\sum_{m=1}^{N} f_m \dfrac{\Delta x_m}{\Delta S_m}\Delta S_m = -\dfrac{1}{A}\sum_{m=1}^{N} f_m \Delta x_m \end{cases} \tag{8.15}$$

where Δx_m and Δy_m are the differences in x and y coordinates between the starting and ending vertices of the mth edge. For the example of the quadrilateral element given in Fig.1b, the partial derivatives can be written as

$$\begin{cases} \dfrac{\partial f}{\partial x} \approx \dfrac{1}{A}\sum_{m=1}^{4} f_m \Delta y_m = \dfrac{1}{A}[f_{12}(y_2 - y_1) + f_{23}(y_3 - y_2) + f_{34}(y_4 - y_3) + f_{41}(y_1 - y_4)] \\ \dfrac{\partial f}{\partial y} \approx -\dfrac{1}{A}\sum_{m=1}^{4} f_m \Delta x_m = -\dfrac{1}{A}[f_{12}(x_2 - x_1) + f_{23}(x_3 - x_2) + f_{34}(x_4 - x_3) + f_{41}(x_1 - x_4)] \end{cases} \tag{8.16}$$

where f_{ij} $(i, j = 1, 2, 3, 4)$ are the mean values of function f along the mth edge defined by vertex i and j. Assuming that $f_{ij} = 1/2(f_i + f_j)$, where f_i and f_j are values of function f at vertices i and j, respectively, the derivatives can be calculated explicitly as

$$\begin{cases} \dfrac{\partial f}{\partial x} \approx \dfrac{1}{2A}[(f_1 - f_3)(y_2 - y_4) + (f_2 - f_4)(y_3 - y_1)] \\ \dfrac{\partial f}{\partial y} \approx -\dfrac{1}{2A}[(f_1 - f_3)(x_2 - x_4) + (f_2 - f_4)(x_3 - x_1)] \end{cases} \tag{8.17}$$

8.3 Dynamic and Static Relaxation Techniques

8.3.1 General Concepts

Relaxation is a classical solution method in structural and stress analysis problems (Southwell, 1935, 1940) and was later extended to general problems in physics and engineering sciences (Southwell, 1956). The concept was, however, first used by Cross (1932) to solve moment distributions of continuous beams without solving simultaneous equations. Its basic concept is to 'relax', i.e., move or load, part or the whole of an initially unloaded and constrained system (continuum or discrete) progressively in small steps, calculate the interaction stresses/forces and strains/displacements with neighboring elements, and remove the initial and artificial constraints to the members in proper locations according to the governing equations used, until the total stored internal (strain) energy of the system is minimized. The relaxation is therefore a progressive process through usually a time marching scheme. By a member-by-member relaxation process (called block relaxation by Southwell) the need for forming and solving a large number of simultaneous equations such as that used in FEM is eliminated. On the contrary, combinations of members can also be used as entities of relaxation called group relaxation. The problem of producing influence coefficients of a framework in structural analysis is a typical relaxation technique.

Applied to a rigid block system of rocks, this technique can be described by the following steps:

1. The block system is initially unloaded and constrained in its initial position, without causing any internal interactions.
2. When any loading effect is applied (by introducing boundary loads/displacements or gravity, for example), the effect is not considered simultaneously for all blocks but initially considered separately for one block after another.
3. For one block which is relaxed, i.e., which is affected by the loads and therefore it will have translational and rotational movements according to the governing equations (equations of motion in this case) while all other blocks remain with their initial positions and conditions, we say the constraints on the current block are removed, i.e., the block is relaxed.
4. Reaction forces between the relaxed block and its neighboring blocks are calculated according to the contact laws and contact positions. Thus, the effect of the loading is transferred from the relaxed block to the system;
5. The blocks neighboring the current 'relaxed' block are, in turn, relaxed, i.e., moved according to the interaction forces/moments they received through their contacts with the currently 'relaxed' block, and their initial constraints are removed. Their relaxation is not performed simultaneously but one by one. Therefore, no set of simultaneous equations is formed.
6. Steps 1–5 are repeated for all blocks affects by boundary loads (or all blocks by gravity).
7. The resultant out-of-balance forces and moments of all blocks are calculated and the relaxation continues for those blocks whose resultant out-of-balance forces and moments are compared with a preset criterion. Convergence is reached when the total out-of-balance of forces and moments of the whole system is minimized.

The above operation is performed using small time steps to achieve convergence in a reasonable period of time without causing numerical instabilities, by avoiding the production of too large overlaps at the contact points so that contact detection algorithm will not fail to work. The convergence of the calculation can be detected by thresholds of variables, such as the difference between the maximum velocity or the total magnitude of the out-of-balance forces of the whole system between two successive steps. The name 'relaxation' represents the nature of the operations of removing the constraints of the blocks with marching of the time steps. To illustrate the principle of the relaxation technique for block systems, an example by Stewart (1981) is presented below.

The system consists of three rigid blocks with one degree of freedom, connected by springs with different stiffness and loaded by self-weight. The motion is restricted to the vertical direction only. The blocks are numbered as i, $i+1$ and $i-1$, with self-weight W_i, W_{i+1}, W_{i-1} and connected by springs of stiffness K_i, K_{i+1}, K_{i+2} and K_{i-1}, respectively, to the fixed reference system (Fig. 8.3). The vertical displacement of block i at the kth iteration is denoted by u_i^k and the forces in spring i and $i+1$ are denoted by F_i and F_{i+1}, respectively.

The unit problem for the static equilibrium of the block i, after k iterations, is written as

$$F_i + F_{i+1} = W_i \tag{8.18a}$$

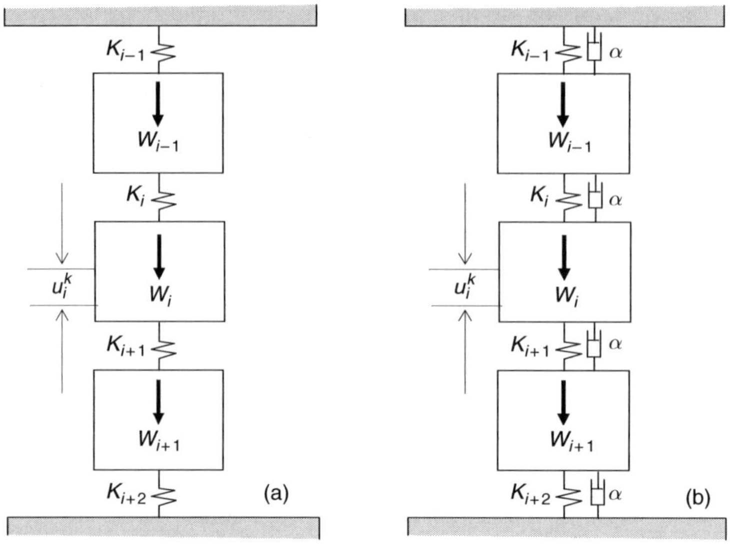

Fig. 8.3 An example of a three-rigid-block dynamic relaxation problem: (a) without mass damping and (b) with mass damping (Stewart, 1981).

with the spring forces given by

$$F_i = K_i[(u_i^{k+1} - u_{i-1}^{k+1}) + (u_i^k - u_{i-1}^k) + \cdots + (u_i^1 - u_{i-1}^1)] = K_i \sum_{j=1}^{k+1} (u_i^j - u_{i-1}^j) \tag{8.18b}$$

$$F_{i+1} = K_{i+1}[u_i^{k+1} + (u_i^k - u_{i+1}^k) + \cdots + (u_i^k - u_{i+1}^1)] = K_{i+1}\left[u_i^{k+1} + \sum_{j=1}^{k} (u_i^j - u_{i+1}^j)\right] \tag{8.18c}$$

Assuming that $W_{i-1} = W_{i+1} = 12$, $W_i = 6$, $K_{i-1} = K_i = K_{i+1} = K_{i+2} = 1$, the results are listed in a relaxation table – Table 8.1. The approximation of a 0.1% discrepancy is achieved by 10 iterations.

Table 8.1 Relaxation table for the example shown in Fig. 8.2 (Stewart, 1981).

Number of iterations	Block $i-1$		Block i		Block $i+1$	
	F_{i-1}	F_i	F_i	F_{i+1}	F_{i+1}	F_{i+2}
1	6	6	0	6	3	9
2	9	3	3	3	0	12
3	12	0	3	3	−1.5	13.5
4	13.5	−1.5	3	3	−2.25	14.25
5	14.25	−2.25	3	3	−2.625	14.625
6	14.625	−2.625	3	3	−2.8125	14.8125
7	14.8125	−2.8125	3	3	−2.9063	14.9063
8	14.9063	−2.9063	3	3	−2.9531	14.9531
9	14.9531	−2.9531	3	3	−2.9766	14.9766
10	14.9766	−2.9766	3	3	−2.9883	14.9883
Analytical solution	15	−3	3	3	−3	15

8.3.2 Dynamic Relaxation Method for Block Systems

Based on the principle proposed by Southwell (1940), Otter et al. (1966) proposed a 'dynamic relaxation' technique to solve stress analysis problems of elasticity, based on a finite difference formulation. This is a significant development because it represents a forerunner of the early formulations for the distinct element methods developed by Cundall (1971a), Parekh (1976), Hocking (1977), Ozgenoglu (1978) and Cundall and Strack (1979a,b,c). The technique has the following features:

- It is a step-by-step numerical integration of the dynamic equations of motion (vibration equation) of an elastic continuum as an initial-boundary value problem. The inertial term is therefore included in the calculations.
- It uses critical viscous damping to achieve a steady-state solution, with the purpose of absorbing the excessive kinetic energy caused by using spring models of contacts.
- Stresses and displacements are calculated using central FDMs, using an internal discretization with a triangular element mesh.
- The loads/stresses on each block/element by its neighboring blocks/elements are updated (i.e., the constraints of the block/element are removed, or the block/element relaxed) only after all other blocks/elements have also been relaxed in the same iteration cycle, i.e., at the end of each complete iteration cycle for each time step.
- The dynamic equations of motion are formed using finite-difference approximations to derivatives and solved element by element; therefore, no formation and solution of simultaneous equations in the conventional finite-difference method are needed.
- An appropriate method of detecting contacts between the blocks and proper constitutive models of different contacts (point contacts, edge contacts and face contacts) are needed to determine the reactive forces/stresses of blocks.

The technique is called dynamic relaxation because the inertial terms (products of mass and acceleration) are presented in the equations of motion even for static problems with the use of viscous damping to achieve steady state solutions. The method is, in essence, a numerical integration technique of the equations of motion of a discrete block system, using two sets of governing equations (Fig. 8.4): the equations of motion and equations of constitutive laws. The equations of motion are solved to determine the increments of kinematic quantities such as translational and rotational displacements,

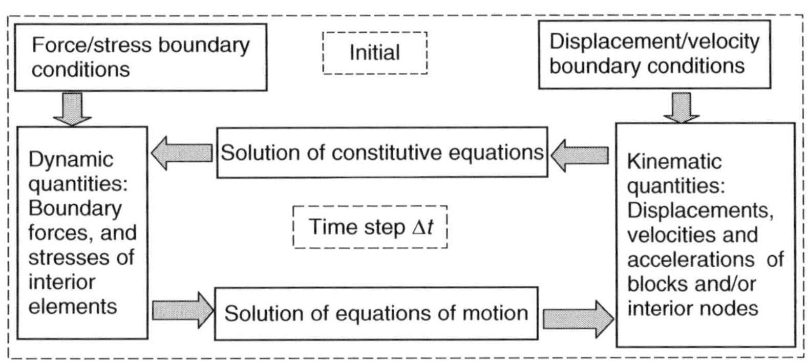

Fig. 8.4 Dynamic relaxation calculation cycles.

velocities and accelerations of each block (or each element for deformable blocks with FDM discretization), together with increments of relative displacements at contacts on the block boundaries and strain increments at interior elements. The constitutive equations for both contacts at the block boundaries and rock material (if the blocks are treated as deformable and discretized with FDM meshes) are solved in order to produce increments of dynamic quantities such as reactive forces/stresses at the boundary contacts according to the available increments of relative displacements, using constitutive models prescribed for contacts, and/or stress increments at interior FDM elements according to corresponding strain increments.

The solution is performed by iteration with a small enough time step until the sum of the total out-of-balance forces (or the maximum velocity) of the system is minimized without causing numerical instabilities. To demonstrate the basic features of the dynamic relaxation technique, the system of three rigid blocks shown in Fig. 8.3 is used as an example, with addition of a mass-proportional viscous damping term, characterized by a coefficient α (Fig. 8.3b).

For block i, the equation of motion for the vertical movements can be written for the kth iteration as

$$m_i \frac{\mathrm{d}^2 u_i^k}{\mathrm{d}t^2} + \alpha m_i \frac{\mathrm{d}u_i^k}{\mathrm{d}t} + (K_i - K_{i+1})u_i^k = W_i - K_i \sum_{j=1}^{k-1}(u_i^j - u_{i-1}^j) - K_{i+1} \sum_{j=1}^{k-1}(u_{i+1}^j - u_i^j) = f_i^{k-1} \quad (8.19)$$

where f_i^{k-1} represents the accumulated resultant force just before the current iteration.

Application of the central difference scheme as shown by Eqns (8.3) and (8.4) leads to an approximation of Eqn (8.19) as a difference equation with time step Δt instead of the general marching step of the variable Δx

$$\ddot{u}_i^k = \frac{1}{(\Delta t)^2}(u_i^{k+1} - 2u_i^k + u_i^{k-1}), \quad \dot{u}_i^k = \frac{1}{2(\Delta t)}(u_i^{k+1} - u_i^{k-1}) \quad (8.20)$$

Substitution of Eqn (8.20) into Eqn (8.19) and rearrangement of the terms result in

$$u_i^{k+1} = \left(1 + \frac{\Delta t}{2}\alpha\right)^{-1}\left\{f_i^{k-1}\frac{(\Delta t)^2}{m} + \left[2 - (K_i - K_{i+1})\frac{(\Delta t)^2}{m}\right]u_i^k - \left(1 - \frac{\Delta t}{2}\alpha\right)u_i^{k-1}\right\} \quad (8.21a)$$

or

$$u_i^k = \left(1 + \frac{\Delta t}{2}\alpha\right)^{-1}\left\{f_i^{k-2}\frac{(\Delta t)^2}{m} + \left[2 - (K_i - K_{i+1})\frac{(\Delta t)^2}{m}\right]u_i^{k-1} - \left(1 - \frac{\Delta t}{2}\alpha\right)u_i^{k-2}\right\} \quad (8.21b)$$

The displacement calculation is therefore always one time step ahead of the velocity and acceleration calculations.

Since all the displacement values before the $(k+1)$th iteration are known, including the initial value u_i^0 when $k=1$, the current value of u_i^{k+1} can be directly determined by the scheme without resort to solution of simultaneous sets of equations. The method is therefore called 'explicit'. The solution is straightforward with the only complexity to update the reactive forces at contacts (represented by the spring forces in Eqn (8.21)), according to the accumulated relative displacements at the contacts. The laws of the contact force/stress-relative displacements are different for different problems and codes and will be described in detail later for different formulations.

In general, for 2D rigid block systems, the standard equations of motion at time t for a block of mass m and a mass-proportional viscous damping coefficient α can be written as

$$m\ddot{u}_x^t + \alpha m\dot{u}_x^t = F_x, \quad m\ddot{u}_y^t + \alpha m\dot{u}_y^t = F_y, \quad I\ddot{\theta}^t + \alpha I\dot{\theta}t = T \quad (8.22)$$

where F_x and F_y are the resultant force components in the x- and y-axis directions, T and I are the resultant torque and moment of inertia of the block about the imaginary z-axis perpendicular to the x–y

plane, (u_x^t, u_y^t) are the displacement components at the mass center of the block at time t, and θ^t is the rotational displacement (rotation angle) of the block. The resultant forces and torque include contributions from all external forces such as boundary conditions, reaction forces at contacts and body forces such as self-weight due to gravity and electro-magnetic or centrifugal forces if present. The application of the standard dynamic relaxation scheme to Eqn (8.22) with time step Δt leads to

$$\begin{cases} u_x^{t+\Delta t} = \left(1 + \frac{\Delta t}{2}\alpha\right)^{-1} \left[\frac{(\Delta t)^2}{m} F_x^{t-\Delta t} + 2u_x^t - \left(1 - \frac{\Delta t}{2}\alpha\right) u_x^{t-\Delta t} \right] \\ u_y^{t+\Delta t} = \left(1 + \frac{\Delta t}{2}\alpha\right)^{-1} \left[\frac{(\Delta t)^2}{m} F_y^{t-\Delta t} + 2u_y^t - \left(1 - \frac{\Delta t}{2}\alpha\right) u_y^{t-\Delta t} \right] \\ \theta^{t+\Delta t} = \left(1 + \frac{\Delta t}{2}\alpha\right)^{-1} \left[\frac{(\Delta t)^2}{I} T^{t-\Delta t} + 2\theta^t - \left(1 - \frac{\Delta t}{2}\alpha\right) \theta^{t-\Delta t} \right] \end{cases} \tag{8.23}$$

Recall Eqn (8.20), the velocity and acceleration components are given by

$$\dot{u}_x^t = \frac{u_x^{t+\Delta t} - u_x^{t-\Delta t}}{2(\Delta t)}, \quad \dot{u}_y^t = \frac{u_y^{t+\Delta t} - u_y^{t-\Delta t}}{2(\Delta t)}, \quad \dot{\theta}^t = \frac{\theta^{t+\Delta t} - \theta^{t-\Delta t}}{2(\Delta t)} \tag{8.24a}$$

$$\ddot{u}xt = \frac{u_x^{t+\Delta t} - 2u_x^t + u_x^{t-\Delta t}}{(\Delta t)^2}, \quad \ddot{u}_y^t = \frac{u_y^{t+\Delta t} - 2u_y^t + u_y^{t-\Delta t}}{(\Delta t)^2}, \quad \ddot{\theta}^t = \frac{\theta^{t+\Delta t} - 2\theta^t + \theta^{t-\Delta t}}{(\Delta t)^2} \tag{8.24b}$$

Since all terms on the right-hand sides of Eqns (8.23) and (8.24) are known, the determination of the unknowns on the left-hand sides of the equations is explicit and straightforward. The integration of the equations of motion can therefore be performed block by block at each successive time step without the need to form and solve simultaneous equations numerically. The second advantage is that any arbitrarily complex constitutive laws at contacts can be implemented without any undue complexities, such as those encountered in FEM or BEM procedures where iterative processes must be carried out to ensure that the correct stress paths are followed according to prescribed constitutive laws.

The relaxation procedure is a staggered one with displacement calculations being one time step ahead of velocities and accelerations. The constitutive relations describing the force/stress-relative displacements at contacts, however, depend on the types of contacts (point-point, edge-edge, surface-surface, etc.) and the mathematical platforms used (simple spring models, spring-dashpot combinations, plasticity theories, damage mechanics, contact mechanics, etc.) and are therefore approach/code-specific. They will be presented at appropriate places in the later sections of this chapter.

8.3.3 Static Relaxation Method for Rigid Block Systems in DEM

An alternative method in the relaxation approach is called static relaxation, developed first by Stewart (1981) and followed by Wei (1992), Chen et al. (1994), Chen (1998) and others, specifically for 2D discrete models of fractured rocks. The method is based also on the principles of Southwell (1935, 1940), and is an explicit formulation. The main features of the method are as follows:

- The same iterative interaction process of the solution of equations of motion and application of the constitutive equations as shown in Fig. 8.4 is used, with all blocks/elements being relaxed at the end of each time step.
- Compared with the dynamic relaxation method that relaxes blocks/elements one-by-one successively, static relaxation can also relax all blocks/elements simultaneously in order to eliminate the non-physical path-dependency introduced during successive relaxation due to the use of complex

non-linear constitutive laws of both the rock blocks and the fractures/contacts. The latter is termed 'simultaneous relaxation' (Stewart, 1981). This will induce a reduced rate of perturbation propagation and convergence compared with the successive relaxation approach, but just by a small fraction, with the added advantage of being more robust for complex rock/fracture behavior.

- Only the static equilibrium of forces and torque of rigid blocks are considered without the use of inertial forces and viscous damping effects in the original formulation of the method by Stewart (1981); hence the term 'static relaxation'. The only mechanism to absorb the dissipated energy of the system is through the friction along the interfaces between blocks. This will inevitably result in small-scale continuing vibrations of adjacent blocks relative to each other, and thus exact equilibrium may not be completely achievable in theory. The technique to overcome this difficulty is to control the rate at which the loads are applied and to use smaller time steps.
- In the original formulation of the static relaxation method for rigid rock block systems (Stewart, 1981), the contacts between blocks are represented by two types: (1) vertex-edge (point) contact characterized by three parameters, the normal stiffness K_n, the shear stiffness K_s, and the friction angle ϕ, all constants; and (2) edge-edge contacts (interfaces) represented by equivalent continuum layer elements following Goodman's joint model (Goodman, 1976).
- The solution starts with the application of external loads (boundary loads and gravity, for example) to the related blocks in turn at the first time step, thus producing out-of-balance increments of forces and torques, and increments of translational and rotational displacements to these blocks. The affected blocks are then relaxed (imaginary constraints removed) and moved to the new positions according to the incremental displacement components. This is different compared to dynamic relaxation that uses the total rather than the out-of-balance force increments during the solution. The imaginary constraints are then applied (i.e., the blocks are temporarily fixed at their new positions until the next iteration). The interaction forces between the initially affected blocks and their neighbors are then determined according to contact types and constitutive laws, thus creating increments of displacement to the neighboring blocks in turn. The perturbation thus continues until all blocks having non-zero out-of-balance torques/forces move to their new positions. The process is repeated if the total out-of-balance force increments are larger than the prescribed threshold for the next iteration.

8.3.3.1 Successive static relaxation method

For a block in contact with a finite number of other blocks, at vertices or along edges (Fig. 8.5), under the action of resultant out-of-balance forces and torques existing at the end of the previous iteration $(k-1)$, the center of the block undergoes translational and rotational rigid body displacement increments $(\Delta u_x^c, \Delta u_y^c, \Delta\theta^c)$. Because of the rigid body assumption, the corresponding displacements at any point i of the block can be approximated by

$$\begin{cases} \Delta u_x^i = \Delta u_x^c + \Delta\theta^c(x^i - x^c) \\ \Delta u_y^i = \Delta u_y^c + \Delta\theta^c(y^i - y^c) \end{cases} \tag{8.25}$$

after a linear truncation of the rotational displacement field, where (x^i, y^i) and (x^c, y^c) are the global coordinates of the arbitrary field point in a block and the geometric center of the block, respectively. Assuming that the field point is allocated at M contact points of the typical block, the normal and shear stiffness of the contact are K_n and K_t, and the inclination angles of each contact with respect to the horizontal axes are $\alpha_i, i = 1, 2, ..., M$, respectively, the transformation between the normal/tangential and global axial components of displacement and force increments are written as

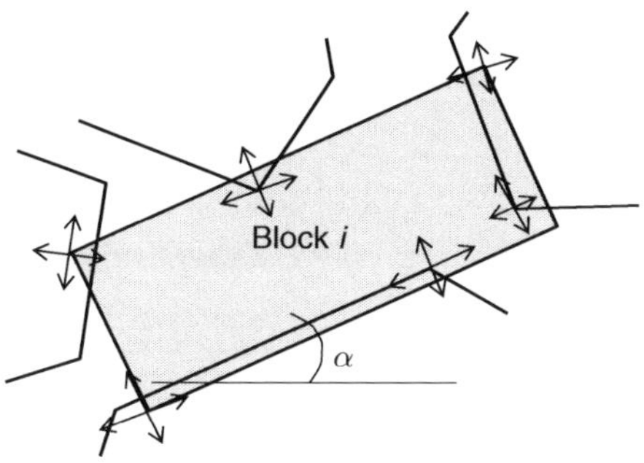

Fig. 8.5 Illustrative example of a block i in contact with several other blocks at vertices and along edges (Stewart, 1981). The arrows indicate the vector of contact force vectors at contact points for both action and reaction components.

$$\begin{cases} \Delta u_n^i = -\Delta u_x^i \sin\alpha_i + \Delta u_y^i \cos\alpha_i \\ \Delta u_t^i = \Delta u_x^i \cos\alpha_i + \Delta u_y^i \sin\alpha_i \end{cases} \tag{8.26}$$

$$\begin{cases} \Delta F_x^i = -\Delta F_n^i \sin\alpha_i + \Delta F_t^i \cos\alpha_i \\ \Delta F_y^i = \Delta F_n^i \cos\alpha_i + \Delta F_t^i \sin\alpha_i \end{cases} \tag{8.27}$$

$$\begin{cases} \Delta F_{\mathrm{n}}^i = K_{\mathrm{n}} \Delta u_{\mathrm{n}}^i \\ \Delta F_{\mathrm{t}}^i = K_{\mathrm{t}} \Delta u_{\mathrm{t}}^i \end{cases} \tag{8.28}$$

The incremental contact forces will then be given by

$$\begin{cases} \Delta F_x^i = -K_{\mathrm{n}}(\Delta u_x^i \sin\alpha_i - \Delta u_y^i \cos\alpha_i)\sin\alpha_i + K_{\mathrm{t}}(\Delta u_x^i \cos\alpha_i - \Delta u_y^i \sin\alpha_i)\cos\alpha_i \\ \Delta F_y^i = K_{\mathrm{n}}(\Delta u_x^i \sin\alpha_i - \Delta u_y^i \cos\alpha_i)\cos\alpha_i + K_{\mathrm{t}}(\Delta u_x^i \cos\alpha_i - \Delta u_y^i \sin\alpha_i)\sin\alpha_i \end{cases} \tag{8.29}$$

and the torques due to the contact forces at contact point i with respect to the block center are

$$\Delta T_{\mathrm{c}}^i = \Delta F_x^i (y^i - y^{\mathrm{c}}) + \Delta F_y^i (x^i - x^{\mathrm{c}}) \tag{8.30}$$

All the above relations are written for contact point i, ($i = 1, \ldots, M$).

The static relaxation solution requires that the sum of the contact forces acting at all M contact points of the block satisfies the static equilibrium equation without considering inertial and viscous terms, i.e.,

$$\sum_{i=1}^{M} (F_x^i + \Delta F_x^i) + b_x + F_x^{\mathrm{e}} = 0 \tag{8.31a}$$

$$\sum_{i=1}^{M} (F_y^i + \Delta F_y^i) + b_y + F_y^{\mathrm{e}} = 0 \tag{8.31b}$$

$$\sum_{i=1}^{M} (F_y^i + \Delta F_y^i)[x^i - \Delta\theta^{\mathrm{c}}(y^i - y^{\mathrm{c}}) - x^{\mathrm{c}}] - \sum_{i=1}^{M} (F_x^i + \Delta F_x^i)[y^i - \Delta\theta^{\mathrm{c}}(x^i - x^{\mathrm{c}}) - y^{\mathrm{c}}] + T^{\mathrm{e}} = 0 \tag{8.31c}$$

where (F_x^i, F_y^i) are the known residual resultant force components and $(F_x^{\mathrm{e}}, F_y^{\mathrm{e}}, T^{\mathrm{e}})$ the known resultant external loads (forces and torques), at the end of the (k−1)th iteration, respectively. Substitution of

Eqns (8.25)–(8.30) into Eqn (8.31) and solving for the unknown displacement components at the center of the block yields simultaneous equations for a typical block:

$$\begin{cases} k_{11}\Delta u_x^{\rm c} + k_{12}\Delta u_y^{\rm c} + k_{13}\Delta\theta^{\rm c} + F_x = 0 \\ k_{21}\Delta u_x^{\rm c} + k_{22}\Delta u_y^{\rm c} + k_{23}\Delta\theta^{\rm c} + F_y = 0 \\ k_{31}\Delta u_x^{\rm c} + k_{32}\Delta u_y^{\rm c} + k_{33}\Delta\theta^{\rm c} + T = 0 \end{cases} \tag{8.32}$$

where

$$F_x = \sum_{i=1}^{M} F_x^i + F_x^{\rm e} \tag{8.33a}$$

$$F_y = \sum_{i=1}^{M} F_y^i + F_y^{\rm e} \tag{8.33b}$$

$$T = \sum_{i=1}^{M} (F_y^i)[x^i - x^{\rm c}] - \sum_{i=1}^{M} (F_x^i)[y^i - y^{\rm c}] + M^{\rm e} \tag{8.33c}$$

$$k_{11} = \sum_{i=1}^{M} (-K_{\rm t}\cos^2\alpha_i - K_{\rm n}\sin^2\alpha_i) \tag{8.33d}$$

$$k_{12} = k_{21} = \sum_{i=1}^{M} \cos\alpha_i \sin\alpha_i (K_{\rm n} - K_{\rm t}) \tag{8.33e}$$

$$\begin{aligned} k_{13} = k_{31} = & \sum_{i=1}^{M} \{K_{\rm t}[\cos^2\alpha_i (y^i - y^{\rm c}) - \sin\alpha_i \cos\alpha_i (x^i - x^{\rm c})] \\ & + K_{\rm n}[\cos\alpha_i \sin\alpha_i (x^i - x^{\rm c}) + \sin^2\alpha_i (y^i - y^{\rm c})]\} \end{aligned} \tag{8.33f}$$

$$k_{22} = \sum_{i=1}^{M} (-K_{\rm t}\sin^2\alpha_i - K_{\rm n}\cos^2\alpha_i) \tag{8.33g}$$

$$\begin{aligned} k_{23} = k_{32} = & \sum_{i=1}^{M} \{K_{\rm t}[\sin\alpha_i \cos\alpha_i \ (y^i - y^{\rm c}) - \sin^2\alpha_i \ (x^i - x^{\rm c})] \\ & - K_{\rm n}[\cos^2\alpha_i \ (x^i - x^{\rm c}) + \sin\alpha_i \cos\alpha_i \ (y^i - y^{\rm c})]\} \end{aligned} \tag{8.33h}$$

$$k_{33} = \sum_{i=1}^{M} [k_{23}(x^i - x^{\rm c}) - k_{13}(y^i - y^{\rm c})] \tag{8.33i}$$

are evaluated at the last $(k-1)$th iteration. The matrix form of Eqn (8.32) is

$$\begin{bmatrix} k_{11} & k_{12} & k_{13} \\ k_{21} & k_{22} & k_{23} \\ k_{31} & k_{32} & k_{33} \end{bmatrix} \begin{Bmatrix} \Delta u_x^{\rm c} \\ \Delta u_y^{\rm c} \\ \Delta\theta^{\rm c} \end{Bmatrix} + \begin{Bmatrix} F_x \\ F_y \\ T \end{Bmatrix} = \begin{Bmatrix} 0 \\ 0 \\ 0 \end{Bmatrix} \tag{8.34}$$

and matrix $[k_{ij}], i,j = 1,2,3$ is called the contact stiffness matrix of the block.

Equation (8.34) can therefore be made into formulae of recurrence for successive iterations by evaluating the contact stiffness matrix and force vector at the $(k-1)$ iteration and solving the unknown block displacement vector for the kth iteration, i.e.,

$$\begin{Bmatrix} \Delta u_x^{c} \\ \Delta u_y^{c} \\ \Delta \theta^{c} \end{Bmatrix}_k = \begin{bmatrix} k_{11} & k_{12} & k_{13} \\ k_{21} & k_{22} & k_{23} \\ k_{31} & k_{32} & k_{33} \end{bmatrix}_{k-1}^{-1} \begin{Bmatrix} F_x \\ F_y \\ T \end{Bmatrix}_{k-1} \tag{8.35}$$

until all blocks in the system are relaxed and out-of-balance forces(torques) are minimized.

8.3.3.2 *Group static relaxation method*

Stewart's successive static relaxation can be seen as an explicit approach (since all unknowns can be directly obtained at the end of previous iteration without need for the solution of large matrix equations). An implicit approach with group relaxation was presented in Chen et al. (1994) and more comprehensively formulated in Chen (1998). The technique does not require the one-by-one successive relaxation of blocks as in Stewart (1981). Relaxation occurs for all blocks simultaneously, without considering inertial and viscous terms. The main difference between the successive and group relaxation techniques is that, for the former, the contact stiffness matrix is written for one block only, with consideration of the effects of all forces at all contact points of the block. However, for the latter, a sub-matrix of contact stiffness is written for each of the contact points between two blocks individually and then all such sub-matrices are assembled into a large global stiffness matrix, in a similar fashion to that of the FEM. The summation over all contacts ($i = 1, \ldots, M$) in Eqns (8.31) (therefore Eqn (8.32) by default) is replaced by distributing contributions from each contact point to the appropriate locations in the global contact stiffness matrix.

Assume that block m and block n have a vertex-edge contact at point i on one edge of block m with an inclination angle α, as shown in Fig. 8.6. Similar to that used in the successive relaxation by Stewart (1981), the incremental displacement at point i, under small displacement assumption, is given by the incremental displacements of the centers of block m and block n, respectively, as

$$\begin{Bmatrix} \Delta u_{x,m}^{i} \\ \Delta u_{y,m}^{i} \end{Bmatrix} = \begin{bmatrix} 1 & 0 & x^i - x_m^{c} \\ 0 & 1 & -(y^i - y_m^{c}) \end{bmatrix} \begin{Bmatrix} \Delta u_{x,m}^{c} \\ \Delta u_{y,m}^{c} \\ \Delta \theta_m^{c} \end{Bmatrix} \tag{8.36}$$

$$\begin{Bmatrix} \Delta u_{x,n}^{i} \\ \Delta u_{y,n}^{i} \end{Bmatrix} = \begin{bmatrix} 1 & 0 & x^i - x_n^{c} \\ 0 & 1 & -(y^i - y_n^{c}) \end{bmatrix} \begin{Bmatrix} \Delta u_{x,n}^{c} \\ \Delta u_{y,n}^{c} \\ \Delta \theta_n^{c} \end{Bmatrix} \tag{8.37}$$

where $(\Delta u_{x,m}^{c}, \Delta u_{y,m}^{c}, \Delta \theta_m^{c})^{T}$ and $(\Delta u_{x,n}^{c}, \Delta u_{y,n}^{c}, \Delta \theta_n^{c})^{T}$ are the increments of translational and rotational displacements at the centers of blocks m and n, respectively and $(\Delta u_{x,m}^{i}, \Delta u_{y,m}^{i})^{T}$ and $(\Delta u_{x,n}^{i}, \Delta u_{y,n}^{i})^{T}$ are

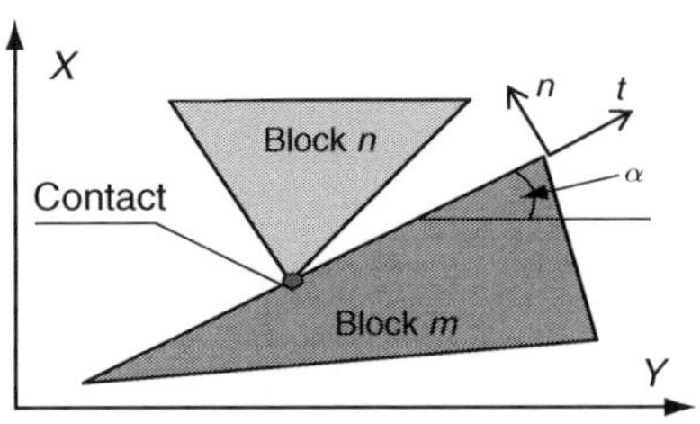

Fig. 8.6 Contact between two blocks (m, n) at point i for the group relaxation method.

the displacement increments at point i caused by displacements of block m and n, respectively. The relative displacements at point i, $(\Delta u_x^i, \Delta u_y^i)^{\mathrm{T}}$, due to movement of blocks m and n is therefore

$$\begin{Bmatrix} \Delta u_x^i \\ \Delta u_y^i \end{Bmatrix} = \begin{Bmatrix} \Delta u_{x,m}^i \\ \Delta u_{y,m}^i \end{Bmatrix} - \begin{Bmatrix} \Delta u_{x,n}^i \\ \Delta u_{y,n}^i \end{Bmatrix} = \begin{bmatrix} 1 & 0 & x^i - x_m^{\mathrm{c}} & -1 & 0 & -(x^i - x_m^{\mathrm{c}}) \\ 0 & 1 & y^i - y_m^{\mathrm{c}} & 0 & -1 & -(y^i - y_m^{\mathrm{c}}) \end{bmatrix} \begin{Bmatrix} \Delta u_{x,m}^{\mathrm{c}} \\ \Delta u_{y,m}^{\mathrm{c}} \\ \Delta \theta_m^{\mathrm{c}} \\ \Delta u_{x,n}^{\mathrm{c}} \\ \Delta u_{y,n}^{\mathrm{c}} \\ \Delta \theta_n^{\mathrm{c}} \end{Bmatrix} \tag{8.38}$$

Recalling the transformation relations (8.27) and (8.30), the combined matrix form can be written as

$$\begin{Bmatrix} \Delta F_{x,m}^i \\ \Delta F_{y,m}^i \\ \Delta T_m^i \end{Bmatrix} = \begin{bmatrix} -\sin \alpha_i & \cos \alpha_i \\ \cos \alpha_i & \sin \alpha_i \\ -(y_m^{\mathrm{c}} - y^i) \sin \alpha_i + (x_m^{\mathrm{c}} - x^i) \cos \alpha_i & (y_m^{\mathrm{c}} - y^i) \sin \alpha_i + (x_m^{\mathrm{c}} - x^i) \cos \alpha \end{bmatrix} \begin{Bmatrix} \Delta F_{\mathrm{n}}^i \\ \Delta F_{\mathrm{t}}^i \end{Bmatrix} \tag{8.39}$$

for block m. Substitution of relations (8.26), (8.28) and (8.38) into Eqn (8.39) leads to the contribution to the contact stiffness matrix at point i from the block m,

$$\begin{Bmatrix} \Delta F_{x,m}^i \\ \Delta F_{y,m}^i \\ \Delta T_m^i \end{Bmatrix} = \begin{bmatrix} k_{11} & k_{12} & k_{13} & k_{14} & k_{15} & k_{16} \\ k_{21} & k_{22} & k_{23} & k_{24} & k_{25} & k_{26} \\ k_{31} & k_{32} & k_{33} & k_{34} & k_{35} & k_{36} \end{bmatrix} \begin{Bmatrix} \Delta u_{x,m}^{\mathrm{c}} \\ \Delta u_{y,m}^{\mathrm{c}} \\ \Delta \theta_m^{\mathrm{c}} \\ \Delta u_{x,n}^{\mathrm{c}} \\ \Delta u_{y,n}^{\mathrm{c}} \\ \Delta \theta_n^{\mathrm{c}} \end{Bmatrix} = [[k_{mm}] \ [k_{mn}]] \begin{Bmatrix} \{\Delta u_m^{\mathrm{c}}\} \\ \{\Delta u_n^{\mathrm{c}}\} \end{Bmatrix} \tag{8.40}$$

where $\{\Delta u_m^{\mathrm{c}}\} = (\Delta u_{x,m}^{\mathrm{c}}, \Delta u_{y,m}^{\mathrm{c}}, \Delta \theta_m^i)^{\mathrm{T}}$ and $\{\Delta u_n^{\mathrm{c}}\} = (\Delta u_{x,n}^{\mathrm{c}}, \Delta u_{y,n}^{\mathrm{c}}, \Delta \theta_n^i)^{\mathrm{T}}$, and the elements in $[k_{mm}], (m = 1,2,3)$, are identical to those in Eqn (8.32), but without the summation symbols, i.e.,

$$k_{11} = -K_{\mathrm{t}} \cos^2 \alpha_i - K_{\mathrm{n}} \sin^2 \alpha_i \tag{8.41a}$$

$$k_{12} = k_{21} = \cos \alpha_i \ \sin \alpha_i (K_{\mathrm{n}} - K_{\mathrm{t}}) \tag{8.41b}$$

$$\begin{aligned} k_{13} = k_{31} = K_{\mathrm{t}}[\cos^2 \alpha_i (y^i - y_m^{\mathrm{c}}) - \sin \alpha_i \ \cos \alpha_i (x^i - x_m^{\mathrm{c}})] \\ + K_n[\cos \alpha_i \ \sin \alpha_i (x^i - x_m^{\mathrm{c}}) + \sin^2 \alpha_i (y^i - y_m^{\mathrm{c}})] \end{aligned} \tag{8.41c}$$

$$k_{22} = -K_{\mathrm{t}} \sin^2 \alpha_i - K_{\mathrm{n}} \cos^2 \alpha_i \tag{8.41d}$$

$$k_{33} = k_{23}(x^i - x_m^{\mathrm{c}}) - k_{13}(y^i - y_m^{\mathrm{c}}) \tag{8.41e}$$

$$\begin{aligned} k_{23} = k_{32} = K_{\mathrm{t}}[\sin \alpha_i \ \cos \alpha_i (y^i - y_m^{\mathrm{c}}) - \sin^2 \alpha_i (x^i - x_m^{\mathrm{c}})] \\ - K_{\mathrm{n}}[\cos^2 \alpha_i (x^i - x_m^{\mathrm{c}}) + \sin \alpha_i \ \cos \alpha_i (y^i - y_m^{\mathrm{c}})] \end{aligned} \tag{8.41f}$$

The elements of $[k_{mn}], (n = 4,5,6; m = 1,2,3)$, are given by

$$k_{14} = -k_{11} = K_{\mathrm{t}} \ \cos^2 \alpha_i + K_{\mathrm{n}} \ \sin^2 \alpha_i \tag{8.42a}$$

$$k_{15} = -k_{12} = \cos \alpha_i \ \sin \alpha_i (K_{\mathrm{t}} - K_{\mathrm{n}}) \tag{8.42b}$$

$$k_{16} = K_t[-\cos^2 \alpha_i(y^i - y^c_n) + \sin \alpha_i \cos \alpha_i(x^i - x^c_n)] \\ -K_n[\cos \alpha_i \sin \alpha_i(x^i - x^c_n) + \sin^2 \alpha_i(y^i - y^c_n)] \tag{8.42c}$$

$$k_{24} = k_{15} = \cos \alpha_i \sin \alpha_i(K_t - K_n) \tag{8.42d}$$

$$k_{25} = -k_{22} = K_t \sin^2 \alpha_i + K_n \cos^2 \alpha_i \tag{8.42e}$$

$$k_{26} = -k_{23} = K_t[-\sin \alpha_i \cos \alpha_i(y^i - y^c_n) + \sin^2 \alpha_i(x^i - x^c_n)] \\ + K_n[\cos^2 \alpha_i(x^i - x^c_n) + \sin \alpha_i \cos \alpha_i(y^i - y^c_n)] \tag{8.42f}$$

$$k_{34} = k_{24}(x^i - x^c_m) - k_{14}(y^i - y^c_m) \tag{8.42g}$$

$$k_{35} = k_{25}(x^i - x^c_m) - k_{15}(y^i - y^c_m) \tag{8.42h}$$

$$k_{36} = k_{26}(x^i - x^c_m) - k_{16}(y^i - y^c_m) \tag{8.42i}$$

For block n, the contact forces are just the negative pair of those for block m, and therefore the contribution of block m to the contact stiffness matrix is similarly

$$\begin{Bmatrix} \Delta F^i_{x,n} \\ \Delta F^i_{y,n} \\ \Delta T^i_n \end{Bmatrix} = \begin{bmatrix} k_{41} & k_{42} & k_{43} & k_{44} & k_{45} & k_{46} \\ k_{51} & k_{52} & k_{53} & k_{54} & k_{55} & k_{56} \\ k_{61} & k_{62} & k_{63} & k_{64} & k_{65} & k_{66} \end{bmatrix} \begin{Bmatrix} \Delta u^c_{x,m} \\ \Delta u^c_{y,m} \\ \Delta \theta^c_m \\ \Delta u^c_{x,n} \\ \Delta u^c_{y,n} \\ \Delta \theta^c_n \end{Bmatrix} = [\,[k_{nm}] \quad [k_{nn}]\,] \begin{Bmatrix} \{\Delta u^c_m\} \\ \{\Delta u^c_n\} \end{Bmatrix} \tag{8.43}$$

with the elements in the stiffness matrix given by

$$k_{41} = -k_{11}, \quad k_{42} = -k_{12}, \quad k_{43} = -k_{13}, \quad k_{44} = -k_{14}, \quad k_{45} = -k_{15}, \quad k_{46} = -k_{16} \tag{8.44a}$$

$$k_{51} = -k_{21}, \quad k_{52} = -k_{22}, \quad k_{53} = -k_{23}, \quad k_{54} = -k_{24}, \quad k_{55} = -k_{25}, \quad k_{56} = -k_{26} \tag{8.44b}$$

$$k_{61} = k_{51}(x^i - x^c_n) - k_{41}(y^i - y^c_n), \quad k_{62} = k_{52}(x^i - x^c_n) - k_{42}(y^i - y^c_n) \tag{8.44c}$$

$$k_{63} = k_{53}(x^i - x^c_n) - k_{43}(y^i - y^c_n), \quad k_{64} = k_{54}(x^i - x^c_n) - k_{44}(y^i - y^c_n) \tag{8.44d}$$

$$k_{65} = k_{55}(x^i - x^c_n) - k_{45}(y^i - y^c_n), \quad k_{66} = k_{56}(x^i - x^c_n) - k_{46}(y^i - y^c_n) \tag{8.44e}$$

Denoting $\{\Delta F^i_m\} = (\Delta F^i_{x,m}, \Delta F^i_{y,m}, \Delta T^i_m)^T$ and $\{\Delta F^i_n\} = (\Delta F^i_{x,n}, \Delta F^i_{y,n}, \Delta T^i_n)^T$, as the vectors of incremental contact forces, the combination of Eqns (8.40) and (8.41) leads to the complete contact stiffness matrix at point i between blocks m and n,

$$\begin{Bmatrix} \Delta F^i_{x,m} \\ \Delta F^i_{y,m} \\ \Delta T^i_m \\ \Delta F^i_{x,n} \\ \Delta F^i_{y,n} \\ \Delta T^i_n \end{Bmatrix} = \begin{bmatrix} k_{11} & k_{12} & k_{13} & k_{14} & k_{15} & k_{16} \\ k_{21} & k_{22} & k_{23} & k_{24} & k_{25} & k_{26} \\ k_{31} & k_{32} & k_{33} & k_{34} & k_{35} & k_{36} \\ k_{41} & k_{42} & k_{43} & k_{44} & k_{45} & k_{46} \\ k_{51} & k_{52} & k_{53} & k_{54} & k_{55} & k_{56} \\ k_{61} & k_{62} & k_{63} & k_{64} & k_{65} & k_{66} \end{bmatrix} \begin{Bmatrix} \Delta u^c_{x,m} \\ \Delta u^c_{y,m} \\ \Delta \theta^c_m \\ \Delta u^c_{x,n} \\ \Delta u^c_{y,n} \\ \Delta \theta^c_n \end{Bmatrix} \tag{8.45a}$$

or

$$\begin{Bmatrix} \{\Delta F^i_m\} \\ \{\Delta F^i_n\} \end{Bmatrix} = \begin{bmatrix} [k_{mm}] & [k_{mn}] \\ [k_{nm}] & [k_{nn}] \end{bmatrix} \begin{Bmatrix} \{\Delta u^c_m\} \\ \{\Delta u^c_n\} \end{Bmatrix} \tag{8.45b}$$

where the elements in the sub-matrices are given by

$$[k_{mm}] = \begin{bmatrix} k_{11} & k_{12} & k_{13} \\ k_{21} & k_{22} & k_{23} \\ k_{31} & k_{32} & k_{33} \end{bmatrix}, \quad [k_{mn}] = \begin{bmatrix} k_{14} & k_{15} & k_{16} \\ k_{24} & k_{25} & k_{26} \\ k_{34} & k_{35} & k_{36} \end{bmatrix} \tag{8.46a}$$

and

$$[k_{nm}] = \begin{bmatrix} k_{41} & k_{42} & k_{43} \\ k_{51} & k_{52} & k_{53} \\ k_{61} & k_{62} & k_{63} \end{bmatrix}, \quad [k_{nn}] = \begin{bmatrix} k_{44} & k_{45} & k_{46} \\ k_{54} & k_{55} & k_{56} \\ k_{64} & k_{65} & k_{66} \end{bmatrix} \tag{8.46b}$$

All these sub-matrices are symmetric.

Let $\{F^e_m\} = (F^e_{x,m}, F^e_{y,m}, T^e_m)^T$ and $\{F^e_n\} = (F^e_{x,n}, F^e_{y,n}, T^e_n)^T$ be the external force/torque vectors acting on blocks m and n, respectively, then the contribution to the static equilibrium equation for blocks m and n due to contact at point i is written as

$$\begin{Bmatrix} \{\Delta F^i_m\} \\ \{\Delta F^i_n\} \end{Bmatrix} + \begin{Bmatrix} \{F^e_m\} \\ \{F^e_n\} \end{Bmatrix} = \begin{bmatrix} [k_{mm}] & [k_{mn}] \\ [k_{nm}] & [k_{nn}] \end{bmatrix} \begin{Bmatrix} \{\Delta u^c_m\} \\ \{\Delta u^c_n\} \end{Bmatrix} + \begin{Bmatrix} \{F^e_m\} \\ \{F^e_n\} \end{Bmatrix} = \begin{Bmatrix} \{0\} \\ \{0\} \end{Bmatrix} \tag{8.47}$$

The global equation of static equilibrium of a system of N blocks is then given by

$$\begin{bmatrix} [k_{11}] & [k_{12}] & [k_{13}] & \dots & [k_{1N}] \\ [k_{21}] & [k_{22}] & [k_{23}] & \dots & [k_{2N}] \\ [k_{31}] & [k_{32}] & [k_{33}] & \dots & [k_{3N}] \\ \dots & \dots & \dots & \dots & \dots \\ [k_{N1}] & [k_{N2}] & [k_{N3}] & \dots & [k_{NN}] \end{bmatrix} \begin{Bmatrix} \{\Delta u^c_1\} \\ \{\Delta u^c_2\} \\ \{\Delta u^c_3\} \\ \dots \\ \{\Delta u^c_N\} \end{Bmatrix} + \begin{Bmatrix} \{F^e_1\} \\ \{F^e_2\} \\ \{F^e_3\} \\ \dots \\ \{F^e_N\} \end{Bmatrix} = \begin{Bmatrix} \{0\} \\ \{0\} \\ \{0\} \\ \dots \\ \{0\} \end{Bmatrix} \tag{8.48}$$

where all sub-matrices $[k_{ij}]$, $(i, j = 1, 2, \dots, N)$ have rank 3×3 and are given by the sub-matrices in Eqn (8.46), according to their labels and whether they are in contact. If block i and j are not in contact, then $[k_{ij}] = [0]$.

The technique of assembly of the global contact stiffness matrix is similar to that in the FEM, i.e., the sub-matrices $[k_{ij}]$ are allocated at the ith row and the jth column of the global stiffness matrix of rank $3N \times 3N$. The solution of the global equilibrium Eqn (8.48) yields simultaneously displacement increments of all blocks (at their centers) according to the current external loads. The equation can be made into a recurrence formulae form by evaluating the elements of stiffness matrix and external load vector at the $(k-1)$th iteration, and the solution determines the incremental block displacements for the kth iteration, i.e.,

$$\begin{Bmatrix} \{\Delta u^c_1\} \\ \{\Delta u^c_2\} \\ \{\Delta u^c_3\} \\ \dots \\ \{\Delta u^c_N\} \end{Bmatrix}_k = - \begin{bmatrix} [k_{11}] & [k_{12}] & [k_{13}] & \dots & [k_{1N}] \\ [k_{21}] & [k_{22}] & [k_{23}] & \dots & [k_{2N}] \\ [k_{31}] & [k_{32}] & [k_{33}] & \dots & [k_{3N}] \\ \dots & \dots & \dots & \dots & \dots \\ [k_{N1}] & [k_{N2}] & [k_{N3}] & \dots & [k_{NN}] \end{bmatrix}^{-1}_{k-1} \begin{Bmatrix} \{F^e_1\} \\ \{F^e_2\} \\ \{F^e_3\} \\ \dots \\ \{F^e_N\} \end{Bmatrix}_{k-1} \tag{8.49}$$

The convergence of the solution is determined by either the minimization of the out-of-balance forces of the whole system, or meeting a prescribed tolerance threshold between the results of two successive iterations.

The extension to the dynamic relaxation scheme is straightforward by adding the inertial and viscous terms as additional RHS vectors, see Chen (1998). The above technique applies only to rigid block systems.

8.3.4 Dynamic Relaxation Method for Fluid Flow in Porous Media

The dynamic relaxation method can also be used for seepage problems for porous media, as presented by Day (1965) and Parekh (1976). The objective is to seek a final steady state solution of a damped dynamic wave equation for fluid flow through a porous medium,

$$\frac{\partial}{\partial x}\left(k_x\frac{\partial h}{\partial x}\right)+\frac{\partial}{\partial y}\left(k_y\frac{\partial h}{\partial y}\right)+\frac{\partial}{\partial z}\left(k_z\frac{\partial h}{\partial z}\right)+s=c\frac{\partial h}{\partial t}+\rho\frac{\partial^2 h}{\partial t^2} \tag{8.50}$$

where h is the head, (k_x, k_y, k_z) the anisotropic permeability of the medium, s the source term, c the damping coefficient and ρ fluid density.

Denoting

$$v=\frac{\partial h}{\partial t}\quad\text{and}\quad q=\frac{\partial}{\partial x}\left(k_x\frac{\partial h}{\partial x}\right)+\frac{\partial}{\partial y}\left(k_y\frac{\partial h}{\partial y}\right)+\frac{\partial}{\partial z}\left(k_z\frac{\partial h}{\partial z}\right)+s$$

Eqn (8.50) can be simplified to

$$q=cv+\rho\frac{\partial v}{\partial t} \tag{8.51}$$

The explicit finite difference formulation of Eqn (8.51) for a typical node o at time t (Fig. 8.7) is

$$q_{o,t}=\frac{c}{2}\left[v_{o,t+\Delta t/2}+v_{o,t-\Delta t/2}\right]+\frac{\rho}{\Delta t}\left[v_{o,t+\Delta t/2}-v_{o,t-\Delta t/2}\right] \tag{8.52}$$

Solving for $v_{o,t+\Delta t/2}$ yields

$$v_{o,t+\Delta t/2}=\left(\frac{\rho}{\Delta t}+\frac{c}{2}\right)^{-1}\left[\left(\frac{\rho}{\Delta t}-\frac{c}{2}\right)v_{o,t-\Delta t/2}+q_{o,t}\right] \tag{8.53}$$

where q can be defined by the finite difference approximation as

$$q_{o,t}=\frac{k_x}{(\Delta x)^2}[h_1-2h_o+h_2]_t+\frac{k_y}{(\Delta y)^2}[h_3-2h_o+h_4]_t+\frac{k_z}{(\Delta z)^2}[h_5-2h_o+h_6]_t+s_{o,t} \tag{8.54}$$

and the value of h at time $t+\Delta t$ at point o is then calculated by

$$h_{o,t+\Delta t}=h_{o,t}+(\Delta t)v_{o,t+\Delta t/2} \tag{8.55}$$

Equations (8.53) and (8.55) provide the required recurrence scheme for iterative solution of the Eqn (8.51) (therefore Eqn (8.50)) when the initial values $h_{o,t=0}$ are known (which is the usual case). The stability criterion for the solution is given by a critical time step,

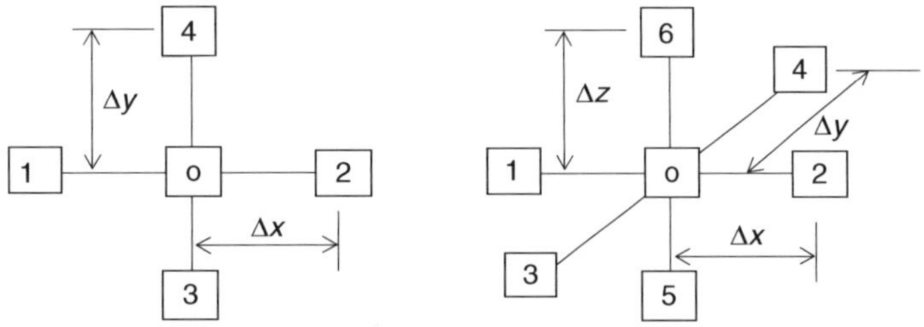

Fig. 8.7 Finite difference meshes for 2D and 3D cases (Parekh, 1976).

$$\Delta t \leq \Delta t^{c} = \frac{1}{2} \frac{c}{[\max(k_x, k_y, k_z)]} [\min(\Delta x, \Delta y, \Delta z)]^2 \tag{8.56}$$

For constant permeability k and a square mesh, the criterion is simplified to

$$\Delta t \leq \Delta t^{c} = \frac{1}{2} \frac{c}{k} [\Delta x]^2 \tag{8.57}$$

The choice of the damping coefficient c also affects the rate of convergence of the solution process. A critical damping coefficient should be chosen so that the damped oscillations of the head should be quickly reduced to zero in order to achieve a steady state solution. The coefficient is a function of the fundamental frequency of the undamped system. Accurate evaluation of the fundamental frequency in closed form is not practical for most of the practical problems, but it can be achieved by approximations. One method is to run the problem as undamped and observe the oscillation of values of the head h at a representative node. If the number of iterations required to complete one cycle of h variation is N, then the fundamental frequency of the system in consideration is given by

$$\omega = \frac{2\pi}{N(\Delta t)} \quad \text{(unit rad/s)} \tag{8.58}$$

and the critical damping coefficient is then determined by

$$c = 2\rho\omega = \frac{4\rho\pi}{N(\Delta t)} \tag{8.59}$$

The second method is to observe the oscillation of the kinetic energy of the whole system. Since the velocity at each node is known, the sum of $K = \sum (v_i)^2$ can be readily calculated as a measure of the system kinetic energy. The number of iterations required to reach the first true maximum of the K curve is equivalent to one quarter of a cycle of the fundamental mode of the system. If the iteration number is M for reaching a true maximum of K, then the fundamental frequency is given by

$$\omega = \frac{2\pi}{4M(\Delta t)} \tag{8.60}$$

and the critical damping coefficient is calculated by

$$c = 2\rho\omega = \frac{\rho\pi}{M(\Delta t)} \tag{8.61}$$

8.4 Dynamic Relaxation Method for Stress Analysis of Deformable Continua

With the finite difference scheme introduced above, the calculations of the stresses, strains and their rates (over time) and increments for deformable continuous materials are straightforward operations. The use of rectangular meshes was developed in the early 1940s and will not be repeated here. The presentation is confined to the Finite Volume Approach. The shape of the elements can be general polygons in 2D and polyhedra in 3D, and can be different at different regions within the domain of interest. An example is given below for general quadrilateral elements (Fig. 8.8) to demonstrate the

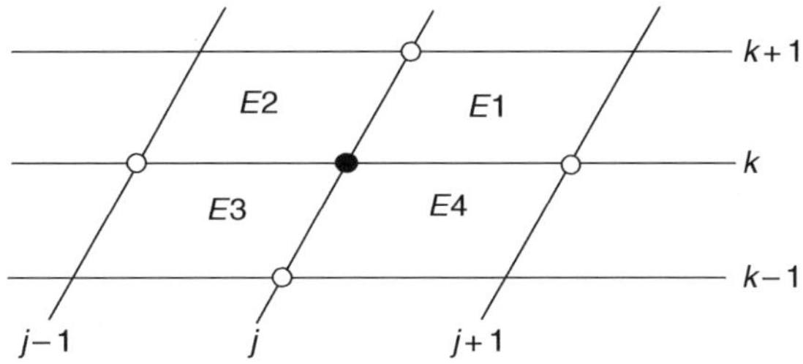

Fig. 8.8 A finite volume scheme with quadrilateral elements.

formulations. The assumption is that the density of the material and area of the element are constant during the deformation (although the shape of the element may change) and isotropic linear elasticity is the assumed behavior of the material with constant Young's modulus E and Poisson's ratio v.

(a) Strain gradients at the center of element i (of nodes 1, 2, 3 and 4 arranged counter-clockwise)

$$(\dot{\varepsilon}_{xx})_i = \left(\frac{\dot{x}}{x}\right)_i = \frac{1}{2A}[(\dot{x}_1 - \dot{x}_3)(y_2 - y_4) + (\dot{x}_2 - \dot{x}_4)(y_3 - y_1)] \tag{8.62a}$$

$$(\dot{\varepsilon}_{yy})_i = \left(\frac{\dot{y}}{y}\right)_i = \frac{-1}{2A}[(\dot{y}_1 - \dot{y}_3)(x_2 - x_4) + (\dot{y}_2 - \dot{y}_4)(x_3 - x_1)] \tag{8.62b}$$

$$(\dot{\varepsilon}_{xy})_i = \frac{1}{2}\left(\frac{\partial \dot{y}}{\partial x} + \frac{\partial \dot{x}}{\partial y}\right)_i = \frac{1}{4A}[(\dot{y}_1 - \dot{y}_3)(y_2 - y_4) + (\dot{y}_2 - \dot{y}_4)(y_3 - y_1) + (\dot{x}_1 - \dot{x}_3)(x_2 - x_4) + (\dot{x}_2 - x_4')(x_3 - x_1)] \tag{8.62c}$$

(b) Strain increments at the center of element i

$$(\Delta\varepsilon_{xx})_i = (\dot{\varepsilon}_{xx})_i\Delta t, \quad (\Delta\varepsilon_{yy})_i = (\dot{\varepsilon}_{yy})_i\Delta t, \quad (\Delta\varepsilon_{xy})_i = (\dot{\varepsilon}_{xy})_i\Delta t \tag{8.63}$$

(c) Stress increments at the center of element i

$$(\Delta\sigma_{xx})_i = \frac{E(1-\nu)}{(1+\nu)(1-2\nu)}\left(\Delta\varepsilon_{xx} + \frac{\nu}{1-\nu}\Delta\varepsilon_{yy}\right)_i \tag{8.64a}$$

$$(\Delta\sigma_{yy})_i = \frac{E(1-\nu)}{(1+\nu)(1-2\nu)}\left(\Delta\varepsilon_{yy} + \frac{\nu}{1-\nu}\Delta\varepsilon_{xx}\right)_i \tag{8.64b}$$

$$(\Delta\sigma_{xy})_i = \frac{E}{2(1+\nu)}(\Delta\varepsilon_{xy})_i \tag{8.64c}$$

(d) Stress rotation correction terms at the center of element i

If an element rotates by an angle ω during a time step Δt, its stresses must be adjusted to its new position relative to the global x–y coordinate system, with the transformation

$$\begin{Bmatrix} \sigma'_{xx} \\ \sigma'_{yy} \\ \sigma'_{xy} \end{Bmatrix} = \begin{bmatrix} \cos^2\omega & \sin^2\omega & -2\sin\omega\ \cos\omega \\ \sin^2\omega & \cos^2\omega & 2\ \sin\omega\ \cos\omega \\ \sin\omega\ \cos\omega & -\sin\omega\ \cos\omega & \cos^2\omega - \sin^2\omega \end{bmatrix} \begin{Bmatrix} \sigma^0_{xx} \\ \sigma^0_{yy} \\ \sigma^0_{xy} \end{Bmatrix} \quad (8.65)$$

where σ'_{ij} and σ^0_{ij} $(i, j = x, y)$ are the stresses after and before the rotation, with the angle ω given by

$$\begin{aligned} \sin\omega &= \frac{1}{2}[\nabla\times(x\mathbf{i}+y\mathbf{j})] = \frac{\Delta t}{2}\left(\frac{\partial\dot{y}}{\partial x} - \frac{\partial\dot{x}}{\partial y}\right) \\ &= \frac{\Delta t}{8A}[(\dot{y}_1-\dot{y}_3)(y_2-y_4)+(\dot{y}_2-\dot{y}_4)(y_3-y_1)-(\dot{x}_1-\dot{x}_3)(x_2-x_4) \\ &\quad +(\dot{x}_2-\dot{x}_4)(x_3-x_1)] \end{aligned} \quad (8.66)$$

The corresponding stress correctors can then be calculated as

$$(\delta_{xx}) = (\sigma'_{xx}) - (\sigma^0_{xx}) = [(\sigma^0_{yy}) - (\sigma^0_{xx})]\sin^2\omega - (\sigma^0_{xy})\sin\ 2\omega \quad (8.67a)$$

$$(\delta_{yy}) = (\sigma'_{xx}) - (\sigma^0_{xx}) = [(\sigma^0_{xx}) - (\sigma^0_{yy})]\sin^2\omega + (\sigma^0_{xy})\sin\ 2\omega \quad (8.67b)$$

$$(\delta_{xy}) = (\sigma'_{xy}) - (\sigma^0_{xy}) = -2(\sigma^0_{xy})\sin^2\omega + \frac{1}{2}[(\sigma^0_{xx}) - (\sigma^0_{yy})]\sin\ 2\omega \quad (8.67c)$$

The stress updating at time step $n+1$ for element i is then given by the sum of the initial stresses, stress increments and the rotation correction terms,

$$(\sigma_{xx})_i^{n+1} = (\sigma_{xx})_i^n + (\Delta\sigma_{xx})_i^n + (\delta_{xx})_i^n \quad (8.68a)$$

$$(\sigma_{yy})_i^{n+1} = (\sigma_{yy})_i^n + (\Delta\sigma_{yy})_i^n + (\delta_{yy})_i^n \quad (8.68b)$$

$$(\sigma_{xy})_i^{n+1} = (\sigma_{xy})_i^n + (\Delta\sigma_{xy})_i^n + (\delta_{xy})_i^n \quad (8.68c)$$

(e) Finite Volume scheme for solving the equations of motion

For the mesh of quadrilateral elements shown in Fig. 8.8, the equations of motion are written for nodes (grid points) (j, k) connecting four elements $E1$, $E2$, $E3$ and $E4$, and four grid points $(j, k-1)$, $(j+1, k)$, $(j, k+1)$ and $(j-1, k)$, respectively. The equations of motion for node (j, k) are then written for a typical time step $(n, n+1)$ as

$$\begin{aligned} \dot{x}_{j,k}^{n+1} = \dot{x}_{j,k}^n - \frac{\Delta t}{2M_m}&[(\sigma_{xx})_{E1}^n(y_{j+1,k}^n - y_{j,k+1}^n) + (\sigma_{xx})_{E2}^n(y_{j,k+1}^n - y_{j-1,k}^n) \\ &+ (\sigma_{xx})_{E3}^n(y_{j-1,k}^n - y_{j,k-1}^n) + (\sigma_{xx})_{E4}^n(y_{j,k-1}^n - y_{j+1,k}^n) - (\sigma_{xy})_{E1}^n(x_{j+1,k}^n - x_{j,k+1}^n) \\ &- (\sigma_{xy})_{E2}^n(x_{j,k+1}^n - x_{j-1,k}^n) - (\sigma_{xy})_{E3}^n(x_{j-1,k}^n - x_{j,k-1}^n) - (\sigma_{xy})_{E4}^n(x_{j,k-1}^n - x_{j+1,k}^n)] \end{aligned} \quad (8.69a)$$

$$\begin{aligned} \dot{y}_{j,k}^{n+1} = \dot{y}_{j,k}^n - \frac{\Delta t}{2M_m}&[(\sigma_{yy})_{E1}^n(x_{j+1,k}^n - x_{j,k+1}^n) + (\sigma_{yy})_{E2}^n(x_{j,k+1}^n - x_{j-1,k}^n) \\ &+ (\sigma_{yy})_{E3}^n(x_{j-1,k}^n - x_{j,k-1}^n) + (\sigma_{yy})_{E4}^n(x_{j,k-1}^n - x_{j+1,k}^n) - (\sigma_{xy})_{E1}^n(y_{j+1,k}^n - y_{j,k+1}^n) \\ &- (\sigma_{xy})_{E2}^n(y_{j,k+1}^n - y_{j-1,k}^n) - (\sigma_{xy})_{E3}^n(y_{j-1,k}^n - y_{j,k-1}^n) - (\sigma_{xy})_{E4}^n(y_{j,k-1}^n - y_{j+1,k}^n)] \end{aligned} \quad (8.69b)$$

where

$$M_m = (M_{E1} + M_{E2} + M_{E3} + M_{E4})/4 \tag{8.69c}$$

is the attributed mass associated to node (j, k). The displacements are then calculated as

$$x_{j,k}^{n+1} = x_{j,k}^n + \dot{x}_{j,k}^n \Delta t, \quad y_{j,k}^{n+1} = y_{j,k}^n + \dot{y}_{j,k}^n \Delta t \tag{8.70}$$

8.5 Representation of Block Geometry and Internal Discretization

Blocks are represented as convex polyhedra in 3D with each face being a planar convex polygon having a finite number of rectilinear edges (Fig. 8.9b), due to the use of the SSD method (see Chapter 7). Their 2D counterparts are general polygons with a finite number of straight edges (Fig. 8.9a). The 2D polygons can be either convex or concave, but the 3D polyhedra are convex. These blocks are formed by fractures that are represented in the problem domain either individually (for larger scale fractures) or by a fracture sets generator (for sets of smaller scale fractures) using random distributions of dip angles, dip directions, spacing and apertures of the fracture sets. The vertices (corners), edges and faces of individual blocks and their connectivity relations are identified during the block generation process, using the oriented faces concept. The topological algorithm and the Euler–Poincaré principle are not used. This is possible since all fractures in a 3D space are assumed to be infinitely large (although algorithms for hidden blocks or hidden surfaces are used to terminate fracture extensions by other fractures or boundaries).

The deformable blocks are further divided into a finite number of internal elements for stress, strain and displacement calculations. These elements are either constant strain triangles or linear strain quadrilateral elements in 2D and constant strain tetrahedra in 3D, see Fig. 8.10. They form a mesh of finite volume (or finite difference) elements (also called zones in the UDEC/3DEC code group).

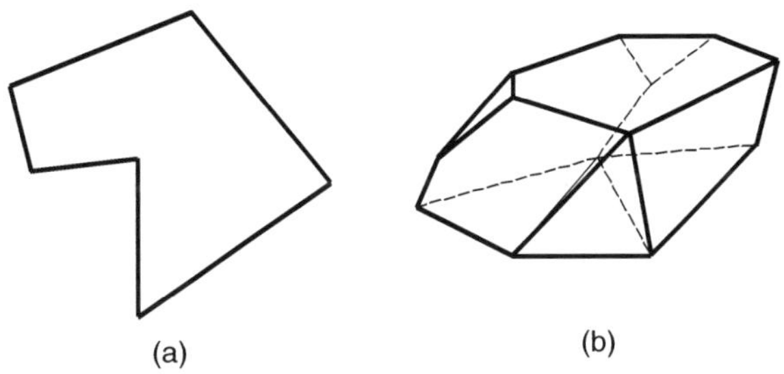

Fig. 8.9 Blocks in the distinct element method in (a) 2D and (b) 3D.

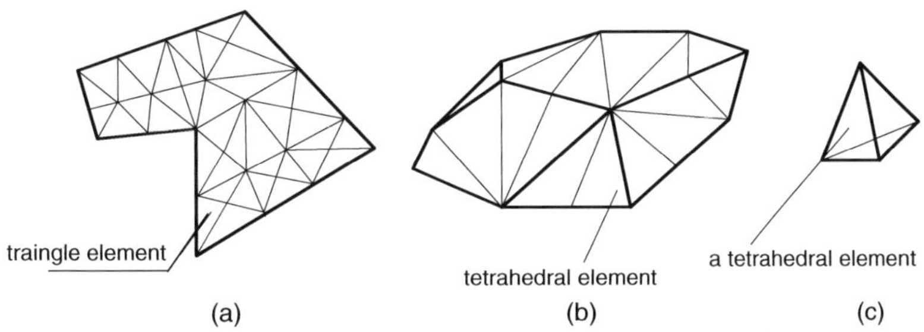

Fig. 8.10 Discretization of blocks by (a) constant strain triangles, (b) constant strain tetrahedra and (c) a typical tetrahedral element.

An explicit, large strain Lagrangian formulation for the constant strain elements is used for both cases. The displacement field of each element varies linearly, and the faces or edges of the zones will then remain as planar surfaces or straight line segments. Higher order elements may also be used, but curved boundary surfaces (or edges) may be obtained, which may in turn complicate the contact detection algorithm. The zone generation can use the same grid generation techniques in the FEM or FDM, by triangularization. The important issue in discretization is to ensure the topological compatibility of the mesh regarding the number of elements, nodes (grid points), edges, faces and boundary surfaces of the blocks. The topological relations relating these topological features described in Chapter 6 still applies and can be used to check the compactness of the meshes in the DEM.

8.5.1 Internal Triangulation and Voronoi Grids

The triangulation of 2D polygons or 3D polyhedra by automatic generation of triangular/tetrahedral elements is standard in mesh generation operations in numerical modeling, in many existing FEM codes and commercially available pre-processing program packages, and with many different techniques, such as the body-center-radiation method (Kamel and Eisenstein, 1971), drag-line or drag-surface method (Park and Washam, 1979), Delaunay triangulation (Watson, 1981; Cavendish et al., 1985; Schroeder and Shephard, 1988) and advance-front method based on the Delaunay triangulation principles (Lo, 1985, 1989, 1991a,b). Mathematical principles of the general grid generation can be seen in Knupp and Steinberg (1993). The Delaunay triangulation is the most commonly used mesh generation technique with the potential for automatic operation. The key requirement for an automatic mesh generation algorithm is the compactness of the generated mesh, i.e., the Euler–Poincáre relation between the vertices, edges and faces should be met exactly so that no loose vertex or gangling edges or faces remain at the end of the generation process.

The automatic mesh generation method adopted in the UDEC/3DEC code group is similar to the Delaunay triangulation. This is a technique to triangulate a 2D or 3D domain by generating a set of points inside and on the boundary of the domain according to the desired mesh density, then connecting these points to form planar triangular or quadrilateral elements or 3D tetrahedral elements.

Another technique is not to connect the points but to form irregular local polygons around each interior point and smooth them through iterations (if required) until a desired mesh of smooth polygons is achieved. This is the so-called Voronoi process and is implemented in the UDEC code (Itasca, 1993).

In this section, the principles of Delaunay triangulation and Voronoi mesh generations are briefly reviewed without going into too much detail.

8.5.2 Two-Dimensional Delaunay Triangulation Scheme

The Delaunay mesh generation process is based on two definitions:

Definition 1: Given a set $\{P\}$ of m unique, random points in a 2D or 3D volume, associated with each point $P_i \in \{P\}$, there exists a local sub-region V_i such that

$$V_i = \{X : ||X - P_i|| < ||X - P_j||, \quad \text{for all } j \neq i\} \tag{8.71}$$

The collection of sub-regions, $\{V\} = \{V_i, i = 1, ..., m\}$ is called a Dirichlet tessellation of the volume. This definition implies that the volume can be divided into a compact collection of sub-regions surrounding each of the randomly generated m interior points. Each sub-region V_i can be shown to be the convex intersection of the open half-spaces (lines or planes) separating the point P_i and P_j ($j \neq i$). This is called a random partition of space by a compact set of convex hulls associated with a set of randomly generated interior points. It can be proved that the Euler–Poincáre relation (5.1) is satisfied by this Dirichilet tessellation of convex hulls (also called tiles) and each interior edge or face of V_i is shared by two and only two tiles. The tiles may or may not be triangular (Fig. 8.11a and b).

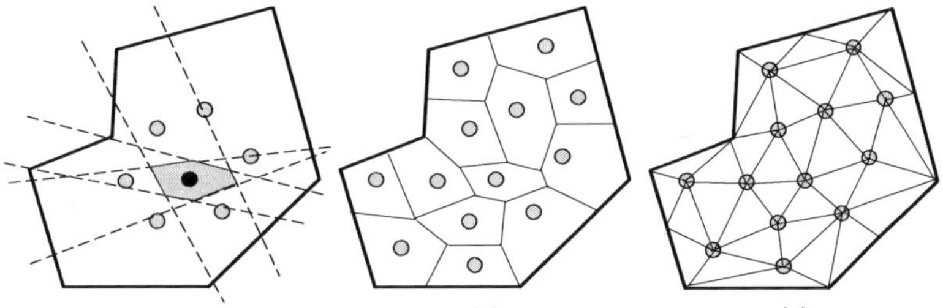

Fig. 8.11 Dirichlet tessellation and Delaunay triangulation. (a) convex hull formation around randomly generated interior points by half planes (or spaces); (b) Dirichlet tessellation by convex hulls (tiles); (c) Delaunay triangulation. Boundary points are not shown.

Definition 2: Given a Dirichlet tessellation $\{V\} = \{V_i, i = 1, ..., m\}$, with each tile V_i containing one and only one interior point P_i, the line segments connecting neighboring interior points P_i and P_j $(j \neq i)$, where V_i and V_j $(j \neq i)$ are neighboring tiles, form a Delaunay triangulation of the volume. The sub-regions formed by connecting line segments are triangular in 2D and tedrahedral in 3D spaces (Fig. 8.11c). This is actually nothing other than a simplex partition of the volume as described in Chapter 6, with triangles and tetrahedra as the simplexes for two and three dimensions.

An efficient automatic mesh generation algorithm can be designed according to either of the above two definitions, with the following five key components:

(1) determination of the grid density (or point spacing) according to user-supplied average linear dimension (usually the mean edge size) of the elements;

(2) placing of boundary nodes according to the desired node spacing on external and internal boundaries (representing excavations, for example);

(3) generation of interior points according to the desired point spacing;

(4) create elements as either Dirichlet tessellations or Delaunay triangulations;

(5) smoothing of the generated mesh for improved homogeneity of the grid.

Step 1 – generation and placing of nodes on external and internal boundaries

Assume that the desired mean edge length is $\bar{l}_e$, and the external and internal boundaries are represented by M straight-line segments with their lengths represented by L_i, $(i = 1, ... M)$. The number of nodes on each of the segments, $N_i (i = 1, ..., M)$ is then given by

$$N_i = \left[\frac{L_i}{\bar{l}_e}\right] + 1 \tag{8.72}$$

where [ ] is the integer operator removing the real fraction of $L_i/\bar{l}_e$. The new mean edge length on each of the boundary segments is

$$l_e^i = \frac{L_i}{N_i} \quad (i = 1, ..., M) \tag{8.73}$$

The generation of coordinates of the nodes on each of the boundary segments is simple and straightforward. However, the ordering of the nodes on the external boundary should be counter-clockwise and ordering on the internal boundaries should be clockwise. Care should also be taking to avoid double nodes with identical coordinates at corners, as a result of repeated generations of nodes at the ends of the line segments defining the boundaries of the domain. (However, double node representations of corners may also be useful since the unit normal vectors at corners are not uniquely determined. One solution is to use two nodes having one unique unit normal vector. This is sometimes used in BEM).

Step 2 – generation and placing of interior nodes inside the solid material

The generation of interior points (nodes) can be implemented randomly, as used by Fukuda and Suhara (1972) and Cavendish (1974), by superimposing on the given domain a square grid just large enough to cover the whole domain, then generating interior points randomly and iteratively so that each sub-square contains just one point. Shaw and Pitchen (1978) abandoned the random trials for point generation, and used a rectangular (rather than square) grid superimposed over the interested domain, with nodes placed at one of the diagonals of each rectangular grid, which can be connected to form triangular elements. The above methods all have shortcomings with undesired shapes of meshes near external or internal boundaries of the domain.

The method used by Lo (1985) for interior point generation is more efficient and straightforward for achieving homogeneous meshes. The algorithm can be described by the following operations:

(1) Determine the top and bottom of the domain of interest with coordinates Y_{max} and Y_{min} , and place imaginary horizontal lines between the top and bottom of the domain, according to the desired edge length of the mesh and relations (8.72) and (8.73) by replacing L_i with $Y_{max} - Y_{min}$. The spacing between the imaginary horizontal lines are the modified edge lengths after Eqn (8.73) in the vertical direction. Each line is represented by its two intersection nodes with the external boundary of the domain.

(2) If the imaginary line does not intersect any internal boundaries, the line is taken as just one continuous segment and the interior points are placed along the line between the two extreme nodes defining its boundary, according to the desired edge length and Eqns (8.72) and (8.73). The two extreme nodes are not included.

(3) If the imaginary line intersects one or more internal boundaries, see Fig. 8.12, the line is divided into several segments with nodes representing intersections between the line and domain boundaries. Interior points are then placed along each segment according to the desired edge length and Eqns (8.72) and (8.73), without including the intersection nodes between the imaginary line and domain boundaries (since all boundary nodes are already placed at Step 1).

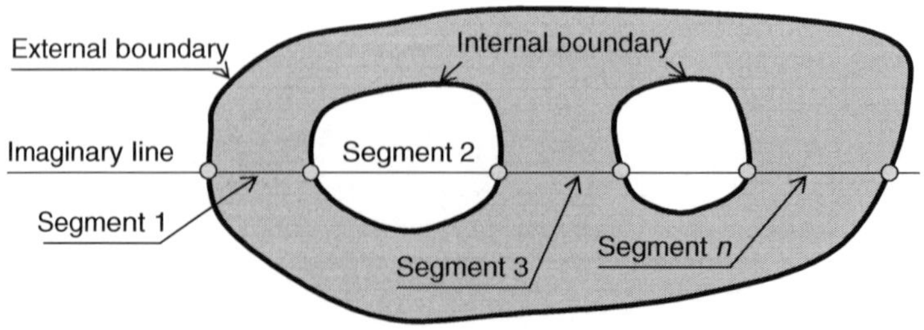

Fig. 8.12 Placing of imaginary horizontal lines across the domain for interior point placing (Lo, 1985), with slight modifications.

(4) Assuming that a typical straight line segment defining part of the external or internal boundaries of the domain has the coordinates of its two end points as (x_i, y_i) and (x_j, y_j), and the height of the kth imaginary horizontal line is y_k, then there will be no intersection if the following conditions apply:

$$\begin{cases} (y_i - y_k)(y_j - y_k) < 0 & \text{or} \\ (y_i - y_k)(y_j - y_k) = 0 & \text{and} \quad (y_k > y_i \text{ or } y_k > y_j) \end{cases} \tag{8.74}$$

otherwise the line intersects the boundary segment at the coordinates

$$\left(x_i + \frac{(y_k - y_i)(x_j - x_i)}{y_j - y_i}, y_k \right) \tag{8.75}$$

Step 3 – forming of internal elements – Delaunay triangulation

By virtue of the counter-clockwise ordering of the exterior boundary nodes and clockwise ordering of interior boundary nodes as described at Step 1, the solid area to be discretized is situated always on the left side of the advancing direction of the boundary nodes. It will be situated on the right hand side of the advancing front if the opposite ordering convention is adopted. Denoting Γ as the generation front representing the topological boundary of the set of generated triangular elements, and Λ as the set of all interior nodes generated at Step 2, the generation of elements is performed by connecting interior nodes with those on the generation front Γ, whose initial position is that of the boundary (external and internal). The generation front Γ is therefore updated continuously whenever a new triangular element is formed, and is formed by a set of segments representing edges of generated elements (or initial boundary segments between adjacent nodes at the beginning). The generation front is therefore the $(n-1)$ chain of the set of the generated mesh in n-dimensions ($n = 2$ or 3).

Delaunay triangulation starts by selecting the last segment $AB \in \Gamma$. The aim is to find an interior point $C \in (\Gamma \bigcup \Lambda)$ such that C lies to the left side of the oriented line segment AB and the triangle ΔABC thus formed is as close as to an equilateral triangle as possible if there are several candidate corners at the vicinity. The criterion is to choose the candidate interior node which provides the minimum value of the norm $||AC||^2 + ||BC||^2$. Assuming that there are two candidate nodes $C_1 \in (\Gamma \bigcup \Lambda)$ and $C_2 \in (\Gamma \bigcup \Lambda)$, Fig. 8.13, and defining the following parameters

$$\alpha_1 = \frac{\Delta ABC_1}{||AB||^2 + ||BC_1||^2 + ||C_1A||^2}, \quad \beta_1 = \frac{\Delta C_1BC_2}{||C_1B||^2 + ||BC_2||^2 + ||C_1C_2||^2} \tag{8.76a}$$

$$\eta_1 = \frac{\Delta AC_1C_2}{||AC_1||^2 + ||C_1C_2||^2 + ||C_2A||^2}, \quad \lambda_1 = \max\,(\beta_1, \eta_1) \tag{8.76b}$$

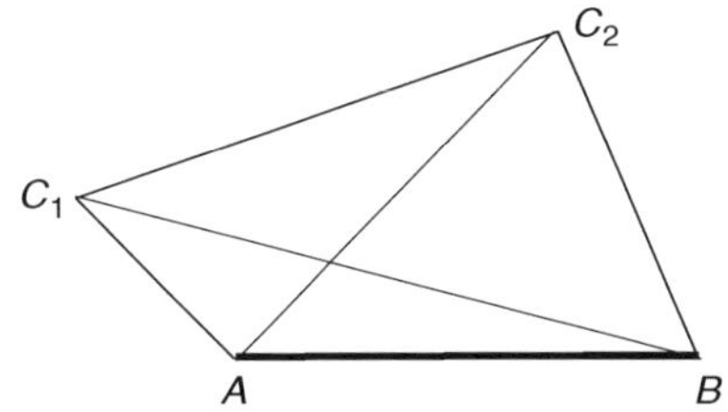

Fig. 8.13 Selection between two candidate corners C_1 and C_2 for Delaunay triangulation with edge AB (Lo, 1985).

for node C_1 and

$$\alpha_2 = \frac{\Delta ABC_2}{||AB||^2 + ||BC_2||^2 + ||C_2A||^2}, \quad \beta_2 = \frac{\Delta C_2BC_1}{||C_2B||^2 + ||BC_1||^2 + ||C_1C_2||^2} = -\beta_1 \tag{8.76c}$$

$$\eta_2 = \frac{\Delta AC_2C_1}{||AC_2||^2 + ||C_1C_2||^2 + ||C_1A||^2} = -\eta_2, \quad \lambda_2 = \max(\beta_2, \eta_2) \tag{8.76d}$$

for node C_2, respectively. Node C_1 will be selected if

$$\alpha_1\lambda_1 > \alpha_2\kappa_2 \tag{8.77}$$

and vice versa. The symbols ΔABC_1 etc. in Eqn (8.76) represent the area of the triangle formed by the associated nodes A, B, and C_1, etc. The algorithm can be extended for any number of candidate nodes. At the end of the triangulation, both the generation front Γ and interior node set Λ are reduced to zero sets. This also identifies the end of the triangulation process. The check with Euler–Poincaré formula is not necessary using Lo's Delaunay triangulation since it is automatically satisfied. The technique is called the 'advance front method' because of the fundamental nature of the generation front.

Step 4 – smoothing

Smoothing is sometimes necessary to produce a more homogeneous mesh. For each interior node, the procedure connects several triangles generated at Step 3. These triangles form a generally shaped polygon. The aim of the smoothing is to shift the coordinates of the interior nodes toward the shape-center of the polygon, through an iterative process.

8.5.3 Two-Dimensional Voronoi Process

The Delaunay triangulation is an automatic generation algorithm of a simplex mesh and is based on Definition 2 of the process. The end result is that all elements are triangular which forms the simplest constant strain elements. On the contrary, a mesh with general multiedged polygons can also be generated, based on Definition 1, with a Dirichlet tessellation. This is also called a Voronoi tessellation or process. The principle of the Voronoi process is simple and is as described in Definition 2 and by Fig. 8.11, by direct formation of polygonal elements followed by a smoothing process. The interior nodes can be generated either randomly or directly according to the desired edge dimension and density of the elements. A different technique is to combine the triangular elements generated by using the Delaunay triangulation to form polygonal elements around each interior point. The latter technique is the so-called Voronoi mesh generation implemented in the UDEC code. The algorithm includes following steps:

Step 1 – expand the tessellation area slightly over the boundary of the domain of interest to reduce the boundary effect (same as Step 1 of the Delaunay triangulation).

Step 2 – distribute nodes along the domain boundaries (external or internal) according to the desired edge length of elements (same as Step 2 of Delaunay triangulation).

Step 3 – generate interior nodes randomly until the desired density of interior nodes is achieved.

Step 4 – smoothing the position of interior nodes so that a more uniform grid of nodes can be obtained by an iteration process (the more iteration, the more uniform the grid).

Step 5 – connect interior nodes to form the intermediate triangular element mesh (can use the same algorithm as that of Step 3 of Delaunay triangulation).

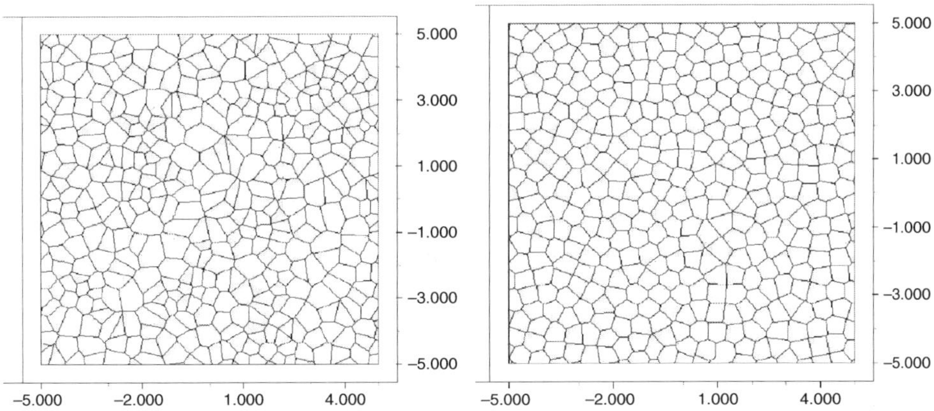

Fig. 8.14 An example of a Voronoi mesh created by the UDEC code: (a) initial mesh; (b) smoothed mesh after 50 iterations.

Step 6 – form the Voronoi polygonal elements by constructing perpendicular bisectors of all triangles which share a common edge. The polygonal elements are truncated at the boundaries of the initial tessellation region.

Figure 8.14 shows an example of a Voronoi mesh generation process with 50 iterations, using the UDEC code.

The Voronoi meshes have some advantages when applied for modeling intact rock blocks with more complex internal structural features and their interactions with large deformation, stress and fluid flow though element interfaces, in a similar way as that of the particle mechanics approach. The additional requirement is, however, that the properties of the elements/particles and their contacts may need special measurement techniques or numerical treatment algorithms.

8.5.4 Three-Dimensional Delaunay Triangulation – Tetrahedral Elements

The advance front technique, based on the principle of Delaunay triangulation, was also developed for 3D discretization of polyhedra, with slightly different algorithms for computer implementation. The basics of the method were presented in Peraire et al. (1988) and Lo (1991a,b), and can be seen as a direct extension of the 2D Delaunay triangulation process described above.

The process starts at Delaunay triangulation of the boundary surfaces (including both external and internal boundaries) of the polyhedral volume of interest. The collection of the triangular elements representing the boundary surfaces is the initial advance front of the 3D Delaunay process, to be updated continuously with the addition of each new tetrahedral element which is constructed by connecting a properly chosen interior node to the candidate triangular facet of the generation front surface. All interior nodes are generated before the triangulation. As in the 2D case, correct topological orientation of all tetrahedral element facets is essential, therefore the ordering of all tetrahedral elements generated must follow the principles of oriented surfaces presented in Chapters 6 and 7. A special difficulty associated with 3D Delaunay triangulation is the fact that interior nodes may be located on the candidate triangular facets on the generation front surface, thus causing topological inconsistencies, or generation of ill-shaped tetrahedra with very small volumes. An example is given below for a demonstration of the difficulty, see Fig. 8.15.

Let ΔABC be the candidate triangular facet on a generation front and D is an interior node. It is possible in a 3D space that the perpendicular distance, d, between D and the plane formed by A, B and

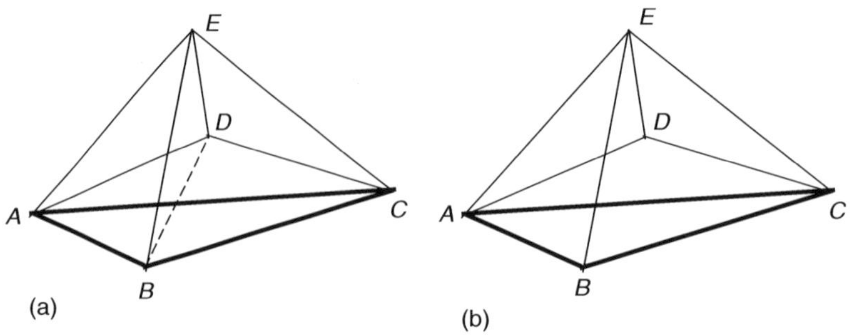

Fig. 8.15 Typical difficulties in 3D Delaunay triangulation (Lo, 1991a).

C is very small so that the new tetrahedron *ABCD* will have a very small volume, which may cause extra difficulties in hydro-mechanical calculations (e.g., forcing a very small time step). However, the triangular facets Δ*ABD*, Δ*DCB* and Δ*ACD* may all be well proportioned. In addition, if the vertical distance *d* is of the same order of magnitude as that of the accumulated computer truncation error, this error may be propagated to the next stage of triangulation with a new interior node *E*, which may form two new tetrahedra *ABDE* and *BCDE* with their base facets Δ*ABD* and Δ*BCD* not exactly on the same plane (Fig. 8.18a). In fact, a better solution is to use the interior *E* ahead of interior node *D* to form two new tetrahedral elements *ABCE* and *ACDE*, both being perfectly well proportioned for all facets. Although a series of user interferences and built-in checks (e.g., with user-supplied minimum element volume associated with desired edge length) may be implemented, the technique is not fully automatically optimal in any sense.

Lo (1991b) presented an algorithm of 3D Delaunay triangulation for arbitrary polyhedral solids, which can reduce the above mentioned intrinsic difficulty to some extent. The technique is presented in the following steps.

Step 1 – Delaunay triangulation of the boundary surfaces of the polyhedron

An arbitrary polyhedron in 3D space, which represents a domain of interest (e.g., a domain of a computational model) can be singly- or multiply-connected with only an external or with a finite number of additional internal boundary surfaces (representing, e.g., excavation boundaries). The triangulation is performed by using the same algorithm as presented before according to the desired edge length of the tetrahedral elements, which results in a set of triangular facets representing the complete boundary surface of the domain. This set of triangular facets serves as the initial generation front Γ. Similarly, care should be taken with the orientation (therefore numbering of nodes) of all these triangular facets so that their unit normal vectors always point outward, defined by the right-hand rule.

Step 2 – generation of interior nodes

Similar to the imaginary horizontal lines of a uniform spacing equal to or near the desired edge length for 2D Delaunay triangulation, imaginary horizontal planes of uniform spacing equal to or near the desired edge length of the tetrahedral elements are placed across the domain, therefore cutting the domain into a series of horizontally placed layers (Fig. 8.16). On each cross-sectional plane formed by the cutting planes and the interior volume of the domain, additional imaginary parallel straight lines are placed, and interior nodes are generated and placed along these imaginary lines in the same way as in its 2D counterparts.

Because the boundary of the solid domain is represented by the set of triangular facets, the contour of the cross section of the solid domain and the cutting plane is determined by the intersection line segments

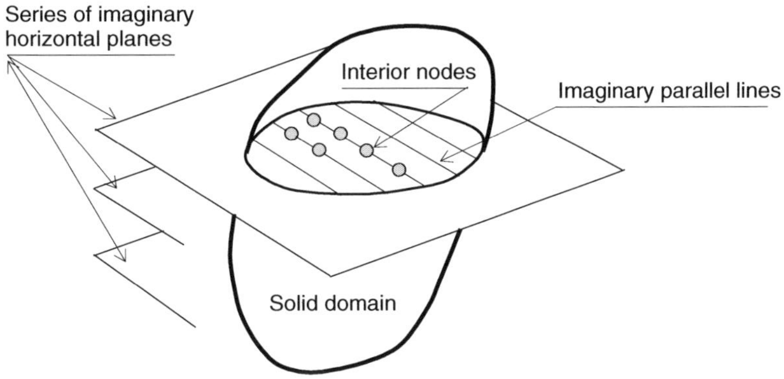

Fig. 8.16 Volume domain cut by a series of imaginary horizontal layers (Lo, 1991a).

of potential triangular facets of the boundary surface with the imaginary cutting plane. Let the vertical coordinate of the kth imaginary plane be z_k (k = 1, ..., M), and set $\{\Delta\} = \{\Delta A_iB_iC_i; i = 1, ..., N\}$ represents all triangular facets of the boundary surface, where N is the total number of boundary facets. The typical triangular facet $\Delta A_iB_iC_i$ has three nodes A_i, B_i and C_i (i = 1, ..., N) positively oriented, with vertical coordinates (x_A^i, y_A^i, z_A^i), (x_B^i, y_B^i, z_B^i), (x_C^i, y_C^i, z_C^i), respectively. The vertical coordinates are z_A^i, z_B^i and z_C^i. The intersection between the horizontal plane of height z_k with $\Delta A_iB_iC_i$ is determined by intersections of the plane with two of the three edges of the facet, A_iB_i, B_iC_i and C_iA_i. for edge A_iB_i, the intersection exists if the following conditions are met (Fig. 8.17)

$$\begin{cases} (z_A^i - z_k)(z_B^i - z_k) < 0 \quad \text{or} \\ (z_A^i - z_k)(z_B^i - z_k) = 0 \quad \text{and} \quad (z_k > z_A^i \ \text{ or } \ z_k > z_B^i) \end{cases} \tag{8.78}$$

The point of intersection $P_{AB}(u, v)$ is given by

$$\begin{cases} u = (1-t)x_A^i + tx_B^i \\ u = (1-t)y_A^i + ty_B^i \end{cases} \quad \text{with } t = (z_k - z_A^i)/(z_k - z_A^i) \tag{8.79}$$

Similar calculations can be performed for the other two edges, B_iC_i and C_iA_i, to determine the second intersection point $Q_{BC}(r, s)$ or $Q_{CA}(m, n)$. The line segment PQ will be the line segment between the imaginary horizontal plane at $z = z_k$ with triangular facet $\Delta A_iB_iC_i$. The set of all intersection segments of the imaginary horizontal plane at $z = z_k$ with the boundary triangular facet set $\{\Delta\} = \{\Delta A_iB_iC_i; i = 1, ..., N\}$ will lead to the complete contour segment $\{S\} = \{P_jQ_j; i = 1, ..., K\}$, whose ordering is related to the numbering of facets set $\{\Delta\}$ and defines a closed loop of line segments.

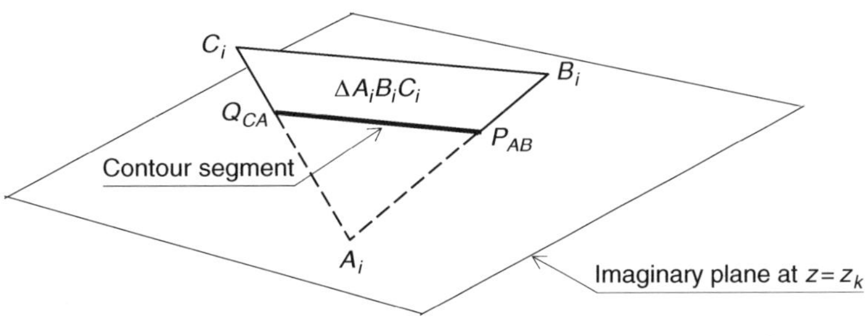

Fig. 8.17 Intersection segment between a typical triangular facet $\Delta A_iB_iC_i$ on the boundary and an imaginary horizontal cutting plane at $z = z_k$ (Lo, 1991a).

Parallel horizontal lines (e.g., along the x or y directions) with uniform spacing (equal to or near the desired edge length of elements) can then be placed on the cross section defined by the contour segment set $\{S\} = \{P_jQ_j\}$, and the interior nodes can be placed along these lines, using the algorithm described for the 2D Delaunay triangulation, with caution taken in regard to the distance of the nodes close to the boundary. To judge whether a potential node X close to the boundary contour segments is proper, a measure of the quality of the tetrahedron formed by node X and the closest triangular facet ΔABC (containing the line segment closest to the node X) can be defined by

$$\begin{aligned}\gamma &= \frac{(72\sqrt{3})V_{\Delta ABC}}{(\text{sum of the squares of edges})^{1.5}} \\ &= \frac{(12\sqrt{3})(AC \times AB \cdot AX)}{(||AC||^2 + ||BC||^2 + ||AB||^2 + ||AX||^2 + ||BX||^2 + ||CX||^2)^{1.5}}\end{aligned} \tag{8.80}$$

Note that X will be an acceptable node if $\gamma > \gamma_{\min}$, where $\gamma_{\min}$ is a user determined tolerance. For example, Lo (1991b) suggested that $\gamma_{\min} = 0.5\alpha$ where α is the shape factor of a triangle (cf. Eqns (8.76a) and (8.76c)).

Step 3 – forming of tetrahedral elements – advance front approach

After generation of the interior nodes, and taking the boundary triangle facets as the initial generation front $\Gamma_0 = \{\Delta\} = \{\Delta A_iB_iC_i; i = 1, ..., N\}$, the construction of interior tetrahedral elements is achieved by connecting proper interior nodes to the current candidate triangle facet (Fig. 8.18a), starting with the last one in the set of Γ_0. By virtue of the correct topological orientation of facets in $\{\Delta\}$, the solid domain to be triangulated has an interior volume which lies always opposite the unit normal vectors of the triangular facets. While boundary triangular facets set $\{\Delta\}$ remain unchanged, the generation front is updated every time when a new tetrahedral element is constructed, causing a change to the generation front, $\Delta\Gamma$, $\Gamma_{i+1} = \Gamma_i + \Delta\Gamma$ $(i = 1, 2, \ldots, M)$. The process continues until $\Gamma_M = 0$, resulting in M elements in total and all interior nodes are used.

Three conditions must be satisfied for a properly chosen interior node X for constructing a new element with the current triangular facet ΔABC on Γ:

(1) ensure that the tetrahedron to be formed lies inside the interior of the solid domain (i.e., the volume of the tetrahedron is positive), ensured by satisfying the relation

$$AC \times AB \cdot AX > 0 \tag{8.81}$$

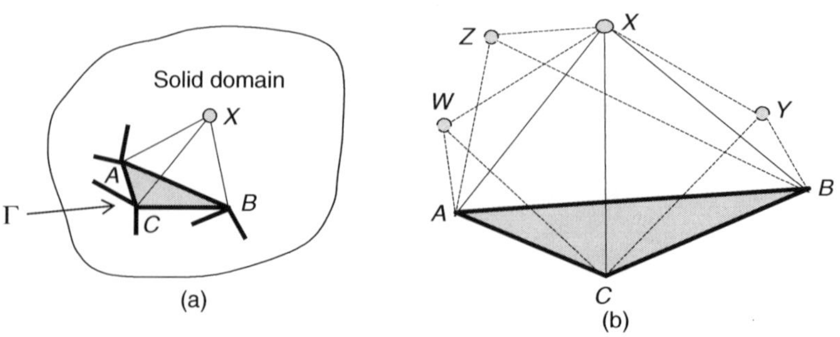

Fig. 8.18 (a) Construction of tetrahedral elements and (b) selection of the locally optimal interior node (Lo, 1991b).

(2) ensure that the proposed tetrahedron does not cut cross the generation front Γ. This is ensured by satisfying the following condition

$$AX \cap \Gamma \in \{A, \{A, X\}, AX\}; BX \cap \Gamma \in \{B, \{B, X\}, BX\}; CX \cap \Gamma \in \{C, \{C, X\}, CX\} \tag{8.82}$$

(3) ensure that the shape of the new tetrahedral element is optimized. This is not easy to achieve, but can be approximated by taking two steps:

(i) to ensure that the γ-value given by Eqn (8.80) of the new element is the maximum, i.e.,

$$\gamma_X = \gamma(ABC; X) = \max\{\gamma(ABC; X_i), i = 1, 2, ..., N_\Lambda\} \tag{8.83}$$

where N_Λ is the current total number of unused interior nodes.

(ii) condition (i) is necessary but not sufficient, especially for very irregular boundaries. Additional consideration is given to the shape quality of the triangular facets of the new tetrahedron ΔXBC, ΔXCA and ΔXAB (Fig. 8.21b). Imagine that three new tetrahedral elements are constructed with candidate facets ΔXBC, ΔXCA and ΔXAB as the best-shaped elements with nearby interior nodes Y, W and Z, and that their respective γ-values are γ_Y, γ_W, and γ_Z. Should the imaginary nodes Y, W or Z be located on the generation front, then we set $\gamma_Y = 1.0$, $\gamma_W = 1.0$, respectively. Then the parameter $\lambda_X = \gamma_X\gamma_Y\gamma_W\gamma_Z$ represents the quality of the tetrahedron $(ABC;X)$ and future tetrahedra $(XBC;Y)$, $(XCA;W)$ and $(XAB;Z)$. Similarly one can construct parameters λ_Y, λ_W, and λ_Z, using nodes Y, W and Z as the primary interior node, respectively Therefore the candidate node $P \in \{X, Y, W, Z\}$should be chosen so that the λ-value is the local maximum, i.e.,

$$\lambda_P = \max\ \{\lambda_X, \lambda_Y, \lambda_W, \lambda_Z\} \tag{8.84}$$

Updating the generation front Γ is achieved by removing the final candidate basic facet after the optimal selection of the interior node and addition of the remaining other three facets to Γ.

Step 4 – smoothing

The quality of the generated tetrahedral elements can be improved through a smoothing process after the Delaunay triangulation, in a similar way to the 2D case. The smoothing is done by shifting the location of an interior node toward the center of the polyhedron formed by the collection of tetrahedral elements connected to the considered interior node, through an iteration process.

Assume that there are N tetrahedral elements connected to an interior node P, and each has a γ-value, γ_i, $(i = 1, 2, \ldots, N)$. The total γ-value at P is then $\gamma_P = \sum_{i=1}^{N} \gamma_i$. The correct direction of the smoothing process is measured by the increase of the γ_P value and should stop when it reaches its maximum, or the difference in the γ_P value between two adjacent iterations becomes equal to or smaller than a prescribed tolerance.

8.5.5 Higher Order Elements

The triangular elements in 2D and tetrahedral elements in 3D are the simplest interior elements with constant strains. They are also the most flexible elements to fit complex and irregular boundaries. Higher order elements can, of course, be constructed using standard mesh generation techniques in FEM, FDM or FVM, and the latter is used in the 2D DEM code UDEC to construct quadrilateral elements to enhance non-linear stress analysis capacity of the code. In this book, the triangular and tetrahedral elements are considered.

8.6 Strain and Stress Calculations for the Internal Elements

The shapes of the internal finite difference elements or zones are not necessarily regular rectangles or cubes. A finite volume scheme for the FDM similar to that developed by Wilkins (1963) is employed to calculate the displacements and strains of the internal elements of general shapes, with the principle given below.

Gauss's theorem for a function, f, and its gradient defined on a body of volume V with boundary S, may be written in the form

$$\int_S f n_i \, \mathrm{d}s = \int_V \frac{\partial f}{\partial x_i} \mathrm{d}v \tag{8.85}$$

where n_i is the unit normal of S, pointing out of V. From Eqn (8.85), the average value of the derivative of the function f over volume V may be written

$$\left\langle \frac{\partial f}{\partial x_i} \right\rangle = \frac{1}{V} \int_V \frac{\partial f}{\partial x_i} \mathrm{d}v = \frac{1}{V} \int_S f n_i \, \mathrm{d}s \tag{8.86}$$

where $\langle \bullet \rangle$ stands for the average of function. For a polygon with N straight edges or a polyhedron with N planar boundary faces, Eqn (8.86) can be written in the form

$$\left\langle \frac{\partial f}{\partial x_i} \right\rangle \approx \frac{1}{V} \sum_{k=1}^{N} \bar{f} n_i^k \Delta S^k \tag{8.87}$$

where ΔS^k is the area (or length) of the kth boundary face (or edge) with unit normal n_i^k, and $\bar{f}$ is the mean value of f over ΔS^k . The summation extends over the N faces (or edges). Taking a 2D example, replacing function f by velocity v_i, and taking the value of v_i of an edge bounded by two consecutive nodes, a and b, as mean values of v_i^a and v_i^b (see Fig. 8.19), the velocity gradients can then be written as

$$\left\langle \frac{\partial v_i}{\partial x_j} \right\rangle = \frac{1}{2V} \sum_{k=1}^{N} \left[(v_i^a + v_i^b) n_j \Delta S^k \right] \approx \frac{\partial v_i}{\partial x_j} \tag{8.88}$$

The increments of strain of the element can be written as

$$\Delta\varepsilon_{ij} = \frac{1}{2} \left(\frac{\partial v_i}{\partial x_j} \pm \frac{\partial v_i}{\partial x_i} \right) \Delta t \tag{8.89}$$

where the sign '+' is used if $i = s\, j$, otherwise the sign '−' is used. Δt is the time step.

The stress increments are obtained by invoking the constitutive equations for the block materials.

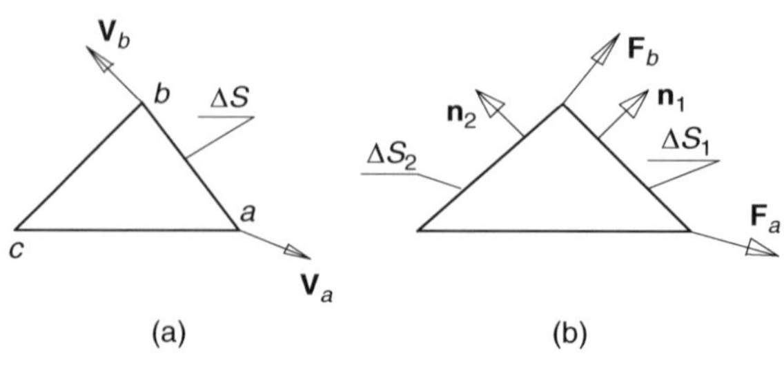

Fig. 8.19 Triangle elements with (a) nodal velocity vectors and (b) nodal force vectors.

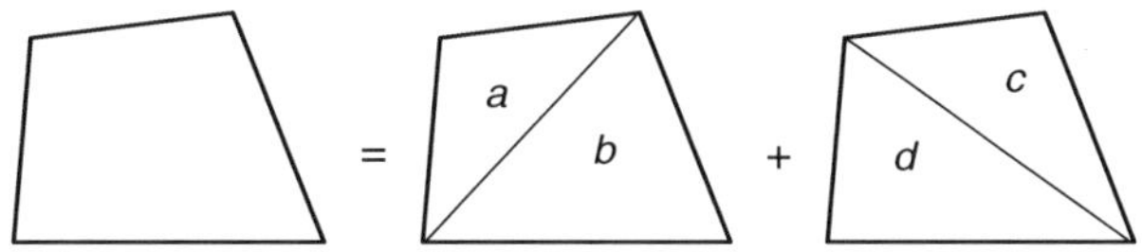

Fig. 8.20 A quadrilateral element in the UDEC code and its division into two pairs of triangular sub-elements.

The forces acting at nodes (grid points) of an element are obtained from the element stress, σ_{ij} (which is constant over the element due to constant strains), by

$$f_i^z = \int_S \sigma_{ij} n_j \, ds \approx \sigma_{ij} \sum_{k=1}^{N} (n_j^k \Delta S^k) \tag{8.90}$$

If one node (grid point) is shared by M such elements, the total force, F^z, acting at this node is the summation of f^z from each element connected to this node

$$F^z = \sum_{l=1}^{M} (f_i^z)_l = \sum_{l=1}^{M} \left[(\sigma_{ij}) \sum_{k=1}^{N} (n_j^k \Delta S^k) \right]_l \tag{8.91}$$

Quadrilateral elements were also developed for the 2D DEM code UDEC (Cundall and Board, 1988), with algorithms similar to that presented in Wilkins (1963). For a general quadrilateral element, it is divided into two overlaying pairs of constant strain triangular elements, *a*/*b*, and *c*/*d* (Fig. 8.20).

The stress components of each triangle element are calculated independently using the algorithms presented later in this chapter. There are, therefore, twelve stress components in total calculated and stored for the quadrilateral element. The stress components of each quadrilateral element are then determined as the average values of the same stress components of the four sub-elements. The advantage of this overlaying and averaging of triangle sub-elements is to eliminate any possible mesh-induced stress and strain anisotropy and to ensure the stress symmetry. For each triangle sub-element, the increments of strain components are determined by

$$(\Delta\epsilon_{ij})_{k=a,b,c,d} = \frac{1}{2} \left[\left(\frac{\partial \dot{u}_i}{\partial x_j} \right)_{k=a,b,c,d} + \left(\frac{\partial \dot{u}_j}{\partial x_i} \right)_{k=a,b,c,d} \right] \Delta t \quad (i,j = x, y) \tag{8.92}$$

The volumetric strain increment is then calculated by averaging over each pair of the triangle sub-elements, using the mixed discretization scheme (Marti and Cundall, 1982)

$$(\Delta\epsilon_m)_{j,k} = \frac{1}{2} [(\Delta\varepsilon_{11})_q + (\Delta\varepsilon_{22})_q + (\Delta\varepsilon_{11})_b + (\Delta\varepsilon_{11})_b] \tag{8.93}$$

and the stresses of each of the two pairs of sub-elements are also equalized to ensure isotropic stress of the whole quadrilateral element k

$$(\sigma_{ij})_k^{(a)} = (\sigma_{ij})_k^{(b)} = \left[\frac{(\sigma_{ij})_k^{(a)} A^{(a)} + (\sigma_{ij})_k^{(b)} A^{(b)}}{A^{(a)} + A^{(b)}} \right] \tag{8.94}$$

where superscripts (a) and (b) stand for triangle element a and b and A is the triangle element's area. Similar operations are performed for triangle elements c and d, as well for final averaging.

The quadrilateral elements are more suitable for analyses of non-linear material behavior, such as plastic flow, but are not as flexible as the triangular elements for matching complex block shapes. It is therefore more desirable to have mixed meshes with quadrilateral elements used only for cases where large, non-linear deformation may occur.

8.7 Representation of Block Contacts

Block contacts are determined by the smallest distance between adjacent blocks. When this distance is within a prescribed threshold, a potential contact between these two blocks is numerically established. The contact detection algorithm in the distinct element method programs determines the contact type (vertex-to-edge, edge-to-edge, face-to-face, etc., if two blocks touch), the maximum gap (if two blocks do not touch but are separated by a gap close to the preset tolerance) and the unit normal vector defining the tangential plane on which sliding can take place.

Mechanically the interaction between two contacting blocks is characterized by a finite stiffness (spring) in the normal direction and a finite stiffness and friction angle (spring-'slipping board' in series) in the tangential direction relative to the fracture surface (contact plane, see Fig. 8.21a). Interaction forces developed at contact points are determined by the deformations of the springs and slipping boards (i.e., the relative movements of blocks at contact points) and resolved into normal and tangential components, F_n and F_t. The interaction forces are assumed to be linearly proportional to the relative displacements, u_n and u_t, between the two blocks. This relation may be expressed in an incremental form as

$$\begin{cases} \Delta F_n = K_n \Delta u_n \\ \Delta F_t = K_t \Delta u_t \end{cases} \text{(without slipping) or } \begin{cases} \Delta F_n = K_n \Delta u_n \\ \Delta F_t = (\Delta F_n)\tan\phi \end{cases} \text{(with slipping)} \tag{8.95}$$

where K_n and K_t are the normal and shear stiffness of the point contacts with unit of force/length between the vertices or the vertex and edges.

An edge–edge (or face–face) contact represents a fracture, and can be decomposed into combinations of vertex-to-edge (or vertex-to-face) contacts. A relation between the increments of stresses (rather than forces) and displacements is then required to characterize the behavior of fractures. It is assumed that before or after slipping takes place between the two opposite surfaces of the fracture represented by an edge-edge contact, a linear or non-linear relation can be written between contact stresses and relative displacements in the form

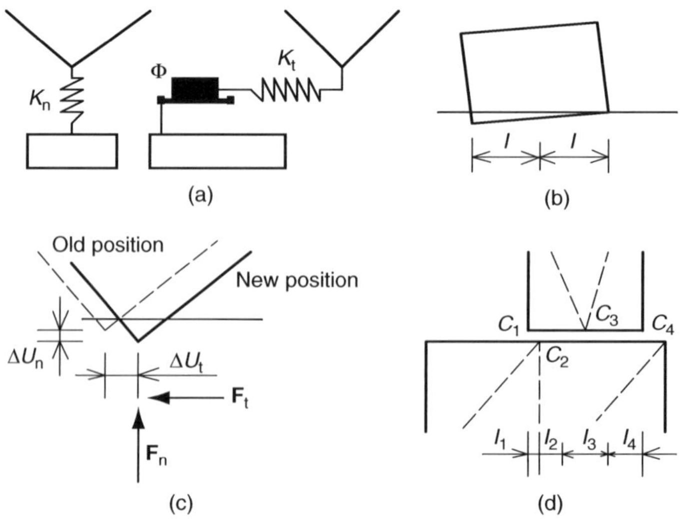

Fig. 8.21 Mechanical representation of contact in the UDEC code (Itasca, 1993).

$$\begin{cases} \Delta\sigma_n = k_n \Delta u_n \\ \Delta\sigma_t = k_t \Delta u_t \end{cases} \text{(without slipping) or } \begin{cases} \Delta\sigma_n = k_n \Delta u_n \\ \Delta\sigma_t = (\Delta\sigma_n)\tan\phi \end{cases} \text{(with slipping)} \tag{8.96}$$

The stiffness parameters k_n and k_t have the unit of stress/length rather than force/length as for the point contact in the previous case. The increment Δu_n in relations (8.95) and (8.96) stands for the amount of contact overlap (i.e., interpenetration between two blocks) at the contact point in the normal direction and is required to develop interaction forces, see Fig.8.21c. Cohesion is assumed to be zero when slipping occurs.

When the deformable blocks are discretized into a finite number of finite difference elements, a number of grid points will be placed along the boundaries of blocks. These grid points will be taken as the new vertices of the blocks and are called sub-contacts. The same relation (8.96) is used for these sub-contacts with the sub-contact length defined as shown in Fig. 8.21d for 2D problems. At each time step, the stress increments are calculated by Eqn (8.96), added to the current total stresses of the elements and checked against the constitutive law specified for the fracture. Similar concepts are used for 3D problems. The contact overlap is checked against a prescribed tolerance to determine if the calculation should continue.

Equation (8.95) and (8.96) are simple constitutive laws at contacts, with Coulomb friction law for slipping. More comprehensive constitutive models for fractures have been established in code UDEC and 3DEC, such as Mohr-Coulomb model, continuously-yielding model, BB-model, etc., see Itasca (1993, 1994).

8.8 Numerical Integration of the Equations of Motion

An explicit central difference scheme is applied in the distinct element method to integrate the equations of motion of the block system, as opposed to the implicit approach utilized in other continuum-based numerical methods. The unknown variables (contact forces or stresses) on the block boundary or in the internal elements are determined locally at each time step from the known variables on the boundaries, in the elements and their immediate neighbors. There is no need to set up and solve the matrix form of the equations of motion. The non-linearity in the material behavior (of the fractures or intact blocks) can be handled in a straightforward manner.

For rigid blocks, the central difference scheme in the time domain for the translational and angular velocities (v_i and ω_i) and accelerations at time t can be written as

$$v_i^{(t)} = \frac{1}{2}\left[v_i^{(t+\Delta t/2)} + v_i^{(t-\Delta t/2)}\right], \quad \omega_i^{(t)} = \frac{1}{2}\left[\omega_i^{(t+\Delta t/2)} + \omega_i^{(t-\Delta t/2)}\right] \tag{8.97}$$

and

$$\frac{dv_i^{(t)}}{dt} = \frac{v_i^{(t+\Delta t/2)} - v_i^{(t-\Delta t/2)}}{\Delta t}, \quad \frac{d\omega_i^{(t)}}{dt} = \frac{\omega_i^{(t+\Delta t/2)} - \omega_i^{(t-\Delta t/2)}}{\Delta t} \tag{8.98}$$

The equations of motion for a rigid block can then be written as

$$v_i^{(t+\Delta t/2)} = v_i^{(t-\Delta t/2)} + \left[\frac{\sum f_i}{m} + b_i\right]\Delta t, \quad \omega_i^{(t+\Delta t/2)} = \omega_i^{(t-\Delta t/2)} + \frac{\sum M_i}{I}\Delta t \tag{8.99}$$

where m is the block mass, I the moment of inertia, b_i the volume force components of the block and M_i the components of the resultant moment. The displacement at the next time step is then given by

$$u_i^{(t+\Delta t)} = u_i^{(t)} + v_i^{(t+\Delta t/2)}\Delta t, \quad \theta_i^{(t+\Delta t)} = \theta_i^{(t)} + \omega_i^{(t+\Delta t/2)}\Delta t \tag{8.100}$$

where θ_i is the angular displacement of the block.

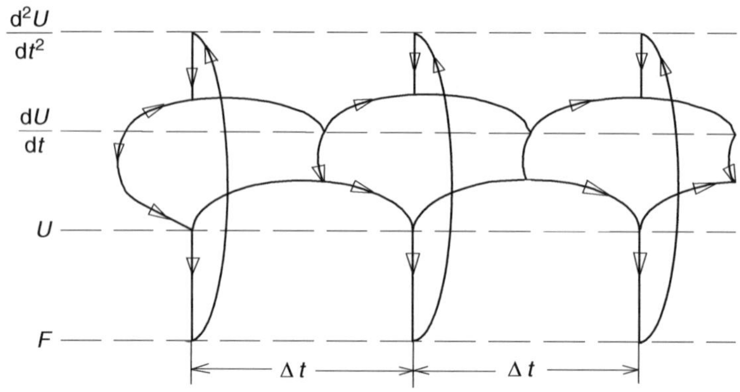

Fig. 8.22 Interlaced calculation order in the distinct element method (Itasca, 1993).

For deformable blocks, the equations of motion are written for grid points – the vertices of internal difference elements. The central difference scheme is similar to the first equation in Eqns (8.99) with only one modification to the resultant out-of-balance force,

$$f_i = f_i^{c} + \sum_{k=1}^{N} \sigma_{ij}(n_j^k \Delta S^k) \tag{8.101}$$

where f_i^{c} are the resultant contact forces if the grid point is on the boundary of the block. The symbol N denotes the number of difference elements connected by this grid point.

At each time step, the kinematic quantities (velocities, displacements and accelerations) are first calculated, and the dynamic quantities (contact forces or stresses, as well as the internal stresses of the elements) are then obtained through invoking the constitutive relations for contacts and internal materials. Figure 8.22 shows the central difference scheme with the order of calculation indicated by arrows, for analysis involving rigid blocks as an example.

In the general calculation procedure, two basic tasks are performed in turn. The kinematic quantities are updated first, followed by invoking the constitutive relations to provide the corresponding forces and stresses, see Fig. 8.23.

The time step Δt used in the integration of equations of motion by a central difference algorithm should not be larger than a critical value to avoid numerical instability. This critical time step is determined by two components: one required for the stability of the calculation of contact between blocks, Δt_{b} and another required for the material deformation of blocks, Δt_{n} (Itasca, 1993)

$$\Delta t_{\mathrm{b}} = 2\beta \left[\frac{m_{\min}^{(b)}}{k_{\max}^{(d)}} \right]^{1/2}, \quad \Delta t_{\mathrm{n}} = 2\min \left[\frac{m_i^{(z)}}{k_i^{(z)}} \right]^{1/2} \tag{8.102}$$

where $m_{\min}^{(\mathrm{b})}$ and $k_{\max}^{(\mathrm{d})}$ are the minimum mass of individual blocks and the largest normal (or shear) stiffness of fractures, and $m_i^{(z)}$ and $k_i^{(z)}$ are the mass lumped at the grid point i and the measure of stiffness of elements associated with grid point i.

Parameter $0 < \beta < 1.0$ is a user-defined fraction constant which accounts for the fact that a single block is in contact with several other blocks. A typical value of β is 0.2. The value of $k_i^{(z)}$ can be estimated by

$$k_i^{(z)} = \sum k_k^{(z)} + \max \quad (k_{\mathrm{n}}^i, k_{\mathrm{t}}^i) \tag{8.103}$$

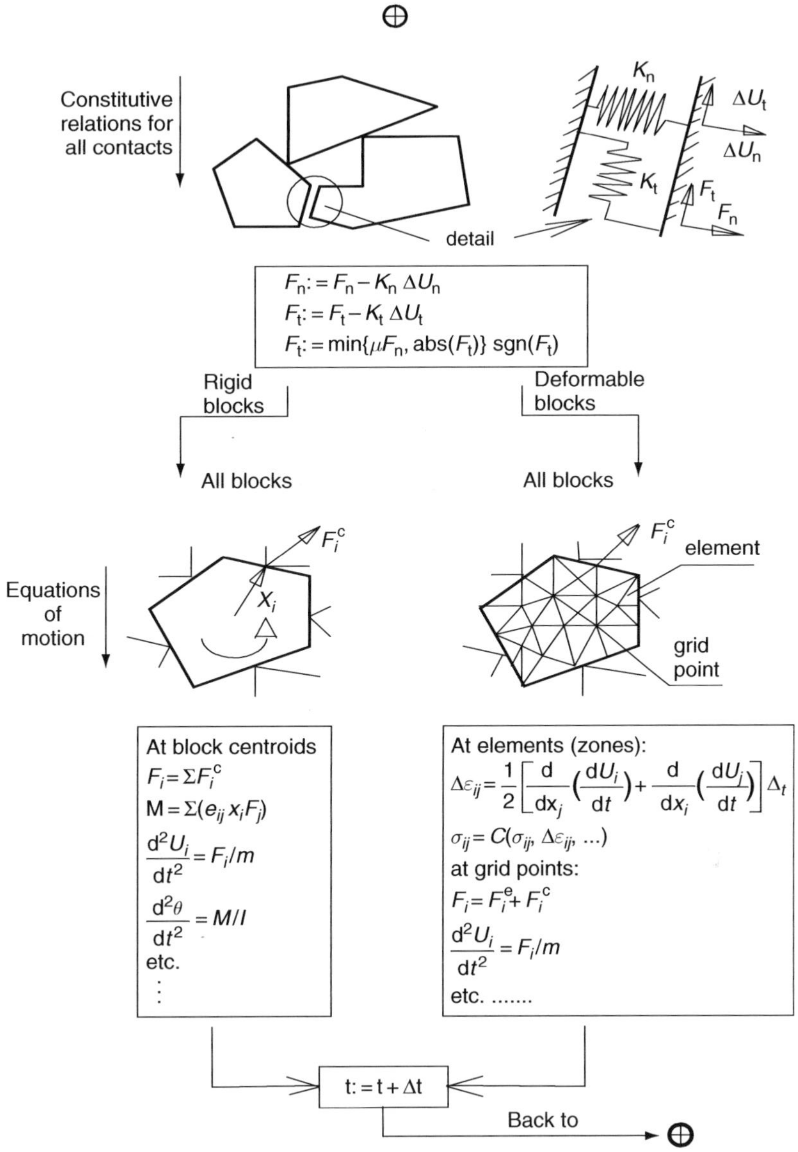

Fig. 8.23 Calculation cycles in the distinct element method (Hart, 1993).

where k_n^i and k_t^i are the normal and shear stiffness at grid point i if it is located on the block boundary, and $\sum k_k^{(z)}$ represents the sum of contributions of deformation stiffness of all elements associated with the grid point i, estimated as

$$k_k^{(z)} = \frac{8}{3}\left(K + \frac{4}{3}G\right)\frac{(b_{\max})^2}{h_{\min}} \tag{8.104}$$

where K and G are the bulk and shear elastic moduli of the block material, $b_{\max}$ is the maximum length of zone edge and $h_{\min}$ is the minimum height of the triangular zone. The critical time step for a computational model, Δt^c, is taken as the lesser one of Δt_b and Δt_n

$$\Delta t^{c} = \min(\Delta t_{b}, \Delta t_{n}) \tag{8.105}$$

When stiffness-proportional damping is used, this time step restriction may not guarantee numerical stability. Belytschko (1983) points out that the time step in this case can be estimated in two ways as given by

$$\Delta t^{c} = \frac{2}{\omega_{\max}} = \xi \frac{2}{\sqrt{\lambda_{\max}}} \tag{8.106}$$

$$\Delta t^{c} = \frac{A_{e}}{c_{p} \Delta l_{\max}} \tag{8.107}$$

where $\omega_{\max}$ and $\lambda_{\max}$ are the maximum frequency and eigenvalue of the DEM mesh, respectively, evaluated using the mass and stiffness matrices of the system (Zienkiewicz and Taylor, 2000), $\xi \leq 1.0$ is an empirical parameter, c_p the p-wave velocity of the rock material, A_e the area of a triangular element and $\Delta l_{\max}$ the maximum linear dimension of the all elements.

In the case of a non-uniform mesh with a large difference between the minimum and maximum element sizes, the final time step determined by the numerical stability requirement is usually very small according to Eqn (8.98). This may lead to unnecessarily long computation time. To reduce such limitations, two numerical techniques were developed for quasi-static problems: the adaptive density-scaling (Cundall, 1982) and dynamic time-stepping (Unterberger et al., 1997), developed for explicit finite difference solution of stress analysis problems.

The basic concept of dynamic time stepping is to determine the time step of each grid point and each element individually and store them as their own properties, instead of using a constant global time step, determined using criteria from Eqns (8.104)–(8.107), regardless of different element sizes. During the process of cycling (iteration), the time step of each grid point/element is used when the relaxation is carried out for that specific grid point/element. Denoting the time step determined by using relations (8.102)–(8.107) as the global time step, Δt_G, the grid-point time step, Δt_{gp}, is then taken as the product of an integer multiplier M_{gp} and Δt_G,

$$\Delta t_{gp} = M_{gp} \Delta t_{G} \tag{8.108}$$

where the multiplier $M_{gp} \leq 2^n$ ($n \leq 5$) is determined by the algorithm shown in Fig. 8.24. The reason for choosing powers of 2 for the multiplier is to ensure that the force contributions from all elements connected at a specific grid point. Arbitrary integral multipliers will not ensure such a characteristic. For special grid points such as those on the boundary, structure elements or null elements (representing excavated spaces), the multiplier is equal to 1. The multiplier for the element time step, M_e, is chosen as the minimum of the grid point multipliers of the element, i.e., where $\omega_{\max}$ is the highest eigenfrequency

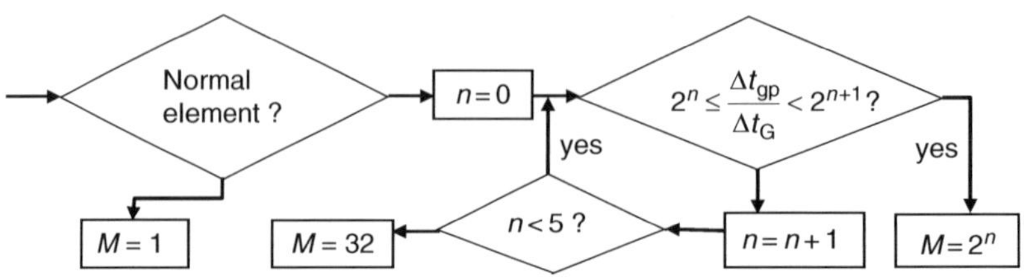

Fig. 8.24 Flowchart for determining the grid-point time-step multiplier (Unterberger et al. 1997).

of the block system and λ the fraction of damping at this frequency. This can be used as user-defined damping parameters.

An alternative expression for the critical time step is given as (Unterberger et al., 1997)

$$M_e = \min\left\{M^i_{gp}, i \in l, k, m\right\} \tag{8.109}$$

for triangular elements with three vertices l, k and m or

$$M_e = \min\left\{M^i_{gp}, i \in l, k, m, n\right\} \tag{8.110}$$

for quadrilateral elements with four vertices l, k, m and n.

With the determination of the gridpoint and element multipliers, the local time steps for grid-point operations (such as calculation of new velocities and displacements) and element operations (such as calculations of new strains, stresses and sum of grid-point forces) are $M_{gp}\Delta t_G$ and $M_e\Delta t_G$, respectively, i.e., the calculations performed only every M_{gp} cycles for the grid points and M_e cycles for the elements.

It was stated by Unterberger et al. (1997) that the above dynamic time steping is accurate for dynamic simulations that represent stress wave frequencies well below the natural frequencies of the individual elements, because the minimum wavelength is ensured to be at least ten times the maximum linear dimension of the element (since the element multiplier is the minimum of the grid-point multiplier). Some errors, however, may be induced for simulations involving high wave frequencies due to the fact that velocities used for updating strain rates are not defined in time at the center of the interval Δt if the element multiplier does not equal the grid-point multiplier, therefore a departure from the second order accuracy of the central finite difference scheme occurs. For such cases, the global time step should be used as an option to overcome such local difficulties. It was reported by Unterberger et al. (1997) that the speed of dynamic simulations could be increased by up to 3–5 times compared with the original global time step – a significant improvement.

Another numerical technique adopted in explicit DEM codes for speeding up the calculation process is density-scaling (or sometimes called mass-scaling) (Cundall, 1982). The basic concept of this approach is that for quasi-static processes, the inertial terms in the equations of motion (the product of mass and acceleration) serve only as an accelerator of the converging process to a final static (steady state) solution, and the psuedo-dynamic nature of the process plays no physical role at all. Therefore the density of the rock material (or mass of elements or blocks) can be artificially adjusted to achieve an optional value of time step that is larger than the conventional global time steps determined by Eqns (8.102)–(8.107), therefore speeding up the calculation process. The condition is that the gravitational forces should be kept correctly represented. The reason for choosing density for scaling up the time step is that the automatic time-step determination is also positively proportional to the material density ρ

$$\Delta t \propto \sqrt{\rho} \tag{8.111}$$

The question is then to find an optimal artificial density so that a maximum Δt can be used while numerical stability is retained. This is achieved by an automatic adjustment algorithm in the code to increase or decrease the density by fractions incrementally so that a near-optimal time-step value is maintained. The adopted density scaling also affects the damping schemes used, see Section 8.9 for detailed comments.

8.9 Contact Types and Detection in the Distinct Element Method

For arbitrary polygonal or polyhedral blocks, the total numbers of different contact types are three and six, respectively, for 2D and 3D problems, see Table 8.2.

Table 8.2 Types of contacts for 2D polygons and 3D polyhedral blocks

Block shape	Contact type
General 2D polygons (convex or concave, singly or multiply connected)	Vertex-to-vertex Vertex-to-edge Edge-to-edge
Convex 3D polyhedra	Vertex-to-vertex Vertex-to-edge Vertex-to-face Edge-to-edge Edge-to-face Face-to-face

For 2D polygons, the vertex-to-edge contact is the basic type and the other two types of contacts can be deduced from this basic type. For example, a vertex-to-vertex contact can be decomposed into two vertex-to-edge contacts occurring at the same contact point. An edge-to-edge contact can be decomposed into two vertex-to-edge contacts.

For 3D convex polyhedra, the basic contact types are the vertex-to-face and edge-to-edge. The other four types of contact can be similarly decomposed as combinations of these two basic types.

(1) Vertex-to-vertex: equivalent to three or more vertex-to-face contacts existing at the same location.

(2) Vertex-to-edge: equivalent to two vertex-to-face contacts coinciding for two adjacent faces.

(3) Edge-to-face: equivalent to two vertex-to-face contacts existing between two adjacent vertices and a face.

(4) Face-to-face: equivalent to three or more vertex-to-face or two or more edge-to-edge contacts.

The objective of the contact detection algorithm in a distinct element method is to identify the potential contact making blocks and their possible types so that the correct physical laws can be applied to determine the contact forces (or stresses). The contact detection algorithm should also provide a unit normal vector which uniquely defines, at the contact points, a normal direction along which the normal component of the contact force will apply and a tangential plane along which the shear force can be applied and sliding may take place. This may not always be possible for some cases, for example, a vertex-to-vertex contact has no such plane, theoretically, because it cannot be defined.

In the distinct element method a contact is kinematically established whenever an overlap occurs between vertices, edges or faces. The gap between two blocks also needs to be calculated so that potential contact making blocks may be established. The algorithm must be very efficient because the bulk of the computation time in the whole calculation process is spent in contact detection and updating.

The direct search method is to test all possibilities of contacts between two blocks. If block A and block B are in contact, the total number of possible contact combinations, N, is (Cundall, 1988)

$$N = (V_A + E_A + F_A)(V_B + E_B + F_B) \tag{8.112}$$

for 3D polyhedra and

$$N = (V_A + E_A)(V_B + E_B) \tag{8.113}$$

for 2D polygons. Variables V_i, E_i and F_i $(i = A, B)$ are the numbers of vertices, edges and faces of block A and block B, respectively. If only the basic types of contact are detected, the total test number for contact becomes

$$N = V_A F_B + F_A V_B + E_A F_B + F_A E_B \tag{8.114}$$

for 3D polyhedra and

$$N = V_A E_B + E_A V_B \tag{8.115}$$

for 2D polygons.

The direct search method is the simplest and most reliable contact detection algorithm. It is efficient for 2D polygons, but too time consuming for the 3D case. For example, for two cubic blocks, the total test number is $N = 676$ by Eqn (8.112) and 240 by Eqn (8.114). To enhance the contact detection algorithm for 3D problems, the concept of a common plane is introduced (Cundall, 1988) (Fig. 8.25).

A common plane is simply a nominal tangential plane which bisects the void space (or overlapped portions) between the two blocks in contact, like a rigid plate (without thickness) loosely held between two blocks (Fig. 8.25). An algorithm is created for this common plane so that it can be translated and rotated with the relative movements of the two blocks. With this common plane in position, the contact test between two blocks can be performed by testing the contact between the common plane (a nominal face) and each block separately. The common plane will always take up a position midway between two approaching blocks, at a maximum distance from both, and defines the sliding plane for two touching blocks of any shape and orientation. The contact detection logic is greatly simplified because only the simple vertex-to-the-plane (a face) tests need to be performed for each block. The face-to- face and edge-to-edge contacts between two convex blocks can be recognized by counting the number of vertex-to-the common plane overlaps for both blocks (three overlaps for face contacts and two overlaps for edge contacts).

The number of tests for contact is also greatly reduced because tests relating to contact between the common plane and vertices of blocks A and B can be carried out separately for the two approaching blocks. The number of tests is

$$N = V_A + V_B \tag{8.116}$$

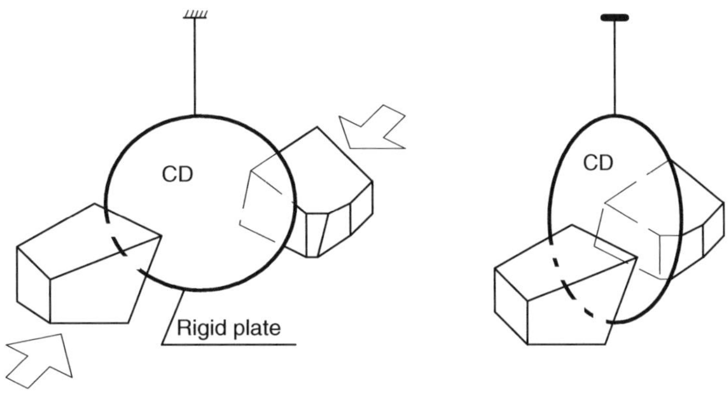

Fig. 8.25 Visualization of the common plane rotating with block movements (Cundall, 1988).

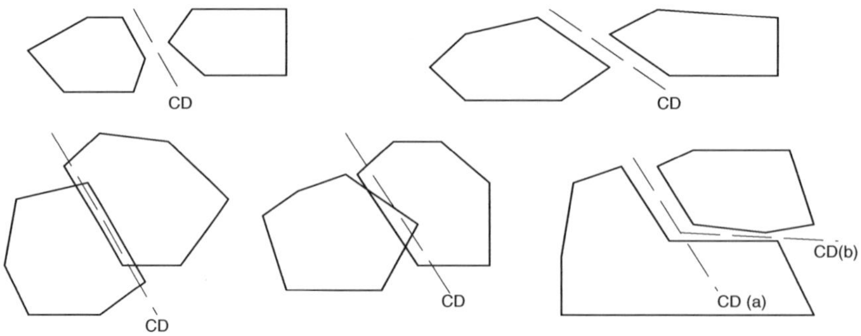

Fig. 8.26 The common plane bisecting the space between closest vertices of two blocks (Cundall, 1988)

i.e., it depends linearly on the number of vertices, compared to the quadratic dependence before in the direct search method. For two cubes, the number of tests becomes $N = 16$.

The location and movement of the common plane between blocks is updated, by rotation and translation, at each time step, in parallel with the mechanical calculation. The concept for locating the common plane between two approaching blocks is to move (by shift and rotation) the common plane so that the gaps between the two closest vertices on the two approaching blocks and the common plane are equal. The space between the two closest vertices is therefore bisected by the common plane, see Fig. 8.26. For overlapping blocks, 'gaps' becomes negative, but the same logic applies.

Associated with the common plane, the contact types between two blocks can be represented through counting the numbers of vertices touching the common plane, as listed in Table 8.3. Contact type is important because it determines the mechanical response of the contact.

There is no need to test if two blocks are potential contact makers. Because the common plane bisects the space between two blocks then, if two blocks touch the common plane, they must contact each other; otherwise, they will not touch the common plane. The unit normal of the common plane is also the unit normal of the contact point, and is uniquely defined, and can change smoothly with block movements. The minimum gap between two approaching blocks is just the algebraic sum of the distances of the two blocks from the common plane.

The number of operations necessary to establish and maintain a common plane for a pair of approaching blocks depends linearly on the number of vertices. The number of operations for translation correction can be given as

Table 8.3 Contact types associated with the common plane

Number of vertices in contact, Block A	Number of vertices in contact, Block B	Contact types
0	0	null
1	1	vertex-to-vertex
1	2	vertex-to-edge
1	>2	vertex-to-face
2	1	edge-to-vertex
2	2	edge-to-edge
2	>2	edge-to-face
>2	1	face-to-vertex
>2	2	face-to-edge
>2	>2	face-to-face

$$M_{\mathrm{T}} = V_A + V_B \tag{8.117}$$

and the number of operations for rotation correction is given by

$$M_{\mathrm{R}} = 4I(V_A + V_B) \tag{8.118}$$

where I is the number of iterations. The total operation number is therefore

$$M = M_{\mathrm{T}} + M_{\mathrm{R}} = (4I + 1)(V_A + V_B) \tag{8.119}$$

This number cannot be directly compared with the total number of contact tests in the direct search scheme (cf. relations (8.114)–(8.115)), because the number of iterations is not constant. However, combined with other advantages (smooth movements of the common plane which also supply the tangential plane for sliding), the common plane logic is a more suitable choice for rock engineering problems with relatively tightly packed block systems. For loosely packed block systems (e.g., flying blocks in blasting or explosions) or granular materials, the direct search scheme may still be the effective method.

The concept of contact 'overlap', though physically inadmissible in block kinematics because blocks should not interpenetrate each other, may be accepted as a mathematical means to represent the deformability of the contacts. However, it does present a numerical shortcoming that is difficult to overcome when the normal forces or stresses at contact points are very large. In this case even with very high normal stiffness, the 'overlap' may develop too excessively to be acceptable and the calculation has to stop in order to implement some remedial measure (for example, increase the normal stiffness) and start all over again. It also presents a problem for fluid flow calculation in which the apertures of fractures may become negative if 'overlap' occurs at contact points. The mathematical representation of the contacts is not fully compatible with the physical reality of the concept of overlap.

Theoretically to identify the potential contact-making blocks (i.e., the neighboring blocks) for a particular block, the search should encompass all other blocks in the problem domain. This direct searching scheme is, however, prohibitive in terms of computer time (it increases quadratically with the number of blocks) and not necessary for the relatively tightly packed block systems in rock mechanics in which most of the blocks initially in contact will probably remain in contact during the deformation process, except the blocks in the neighborhood of excavations or the ground surface, where large movements of blocks may cause frequent changes of the contact patterns. It is therefore desirable to distinguish different areas in the problem domain so that the contact detection work is concentrated on the 'active' areas. 'Cell mapping' is such a technique (Fig. 8.27).

By cell mapping, the problem domain is divided into a small number of regular shaped 'cells' whose boundaries are parallel with the problem boundaries in the directions of the coordinate axes. Each block is enveloped by its 'envelope space' which is defined by the minimum rectangle containing the block and whose boundaries are parallel with the coordinate directions. The block is then mapped into one or more of these cells that overlap with its envelope space, see Fig. 8.27a. Once all blocks are mapped into corresponding cells, the identification of the neighboring blocks of one particular block becomes much easier: they are contained in the same cell. The neighboring cells have an overlapped area between themselves via a tolerance (see Fig. 8.27b) so that all blocks within the given tolerance can be found. The cell mapping technique is also helpful to develop parallel processing algorithms for programs run on parallel computers. The computer time necessary to perform the mapping and searching functions for each block depends on the shape and size of the block. The total computer time is the time spent on the management of cells (mapping and re-mapping of blocks during calculation) and the time spent on the actual searching of neighboring blocks. The time for cell management will increase with an increase in the number of cells (or decrease of cell volumes), but the time spent on neighboring block searching will decrease with the increase of the number of cells. Therefore there might be an optimal cell number (or volume) at which the sum of the time components for cell management

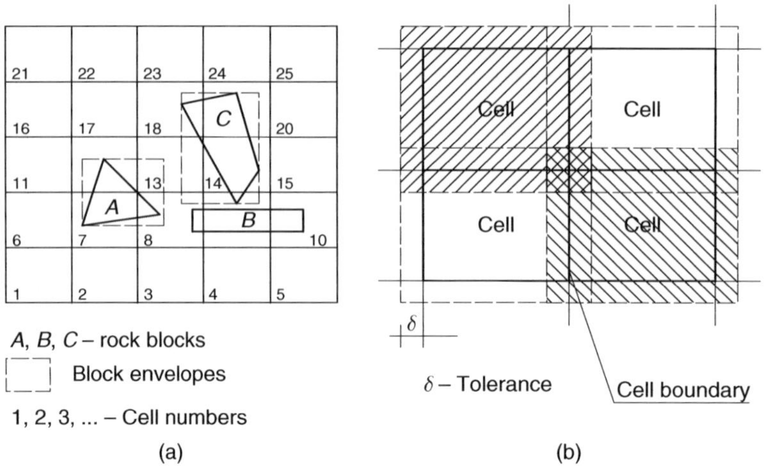

Fig. 8.27 Cell mapping in the distinct element method. (a) Cell mapping of blocks; (b) overlapped tolerance area between neighboring cells (modified, based on Cundall, 1988).

and neighboring blocks searching is the minimum for a particular problem. An analytical expression for this optimal cell volume versus block number seems to be difficult because of the great variety of block shapes involved in rocks. Cundall (1988) suggested that this optimal cell volume should be in the order of one cell per block, i.e., the cell number equal to the block number.

The re-mapping of cells is triggered whenever one block moves outside its cell space. The contact updating, however, may be different for different programs and different problems. For tightly packed block systems under static loading, for example, most of the blocks retain their contacts (which may move during simulation), except for those 'active' areas mentioned above. Therefore, the contact updating in the 'active' areas should be undertaken more frequently than for the rest of the problem domain where contacts remain more or less static. For loosely packed block systems under dynamic loading (for example, flying blocks during blasting or explosions), on the other hand, the whole problem domain may be 'active' and so the contact patterns will change rapidly. A much smaller cell volume is needed and the time spent on the cell management may become more than that on the direct search scheme. In this case, the cell mapping technique is no longer useful.

8.10 Damping

Damping is used in the explicit distinct element method to dissipate the excessive energy in the block system due to the use of linear springs between blocks to represent the contacts. Two types of damping are used: mass-proportional damping and stiffness-proportional damping.

The mass-proportional damping applies, at the contact points, a force proportional to the velocity of the centers of rigid blocks or grid points of deformable blocks, but in the opposite direction. It has an effect similar to that of immersing the block system in a viscous liquid, i.e., to damp the absolute motion relative to the inertial frame of reference.

The stiffness proportional damping applies, at the contact points, a force proportional to the force or stress increments at the contacts, in the same sense. It is physically equivalent to adding a viscous dashpot across block contacts (in both normal and shear directions) so that the relative motions between blocks are damped.

The mass proportional damping is effective in reducing low frequency motion, where the whole block system 'sloshes' from side to side. The stiffness proportional damping is more effective in removing the high frequency noise of individual block's 'rattling' against their neighbors. Either form of numerical damping can be used separately or in combination. The combined use of two damping forms is usually termed Rayleigh damping for elastic continuous systems (Bathe and Wilson, 1976).

The mass-proportional damping force term, d_i^m, for translational degrees of freedom in the momentum equation takes the form

$$d_i^{\mathrm{m}} = -\alpha \frac{\partial u_i}{\partial t} m \tag{8.120}$$

where m is the block mass or lumped mass at a grid point and α a constant. The stiffness-proportional damping force term, d_i^{s}, is expressed as

$$d_i^{\mathrm{s}} = \beta k_{ij} \frac{\partial u_j}{\partial t} \tag{8.121}$$

where β is a constant and k_{ij} the contact stiffness tensor. The velocity $\partial u_j / \partial t$ in this case is the relative velocity at the contact. This damping is used to eliminate the extra vibration energy generated at contacts between blocks or within internal zones. This damping should be 'switched off' if slip occurs at contacts or failure occurs because frictional dissipation should provide natural damping.

For multiple degree-of-freedom systems, the choice of α and β cannot be made analytically with certainty. However, the critical damping ratio, λ, at any natural (angular) frequency of the system, ω, can be given as (Bathe and Wilson, 1976)

$$\lambda = \frac{1}{2}\left(\frac{\alpha}{\omega} + \beta\omega\right) \tag{8.122}$$

The relation between α, β and λ is plotted in Fig. 8.28.

The level of damping is shown to be frequency dependent. The values of α and β should be chosen so as to provide a suitable fraction of critical damping. The minimum of λ with respect to the frequency ω is given by

$$\lambda_{\min} = \sqrt{\frac{\alpha}{\beta}} = \omega_{\min} \tag{8.123}$$

The fundamental frequency of the system, $f_{\min}$, is then defined as

$$f_{\min} = \frac{\omega_{\min}}{2\pi} \tag{8.124}$$

where the unit is cycles/second. The values of $\lambda_{\min}$ and $f_{\min}$ are required input parameters in explicit distinct element method computer programs. Their values can be experimentally determined by trial-and-error processes.

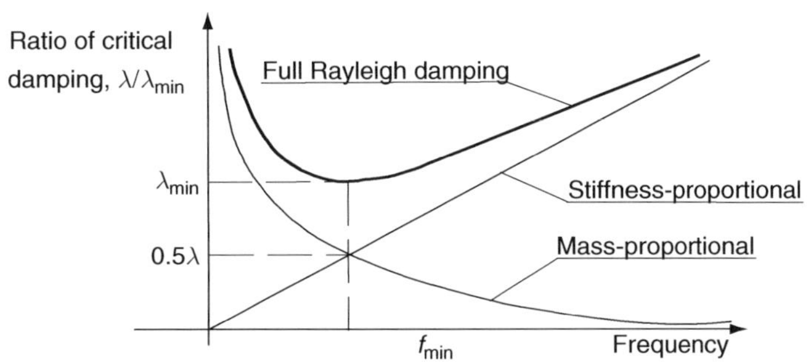

Fig. 8.28 Rayleigh damping (Itasca, 1993).

Theoretically damping can be used for quasi-dynamic problems because only the final steady state or equilibrium state of the system is of interest. The time marching scheme can be taken as just an iteration process to achieve the convergence of the solution. However, for dynamic problems, the damping parameters represent the physical behavior of the system for the combined effects of natural energy dissipation mechanisms, such as the friction and slipping along fractures of various scale (faults, joints, grain boundaries), the initiation and propagation of micro-fractures, the damage of rock materials and the interactions between rocks/fractures and other environmental agents (fluids, gases, thermal gradients, etc.). It is, however, very difficult to define explicitly and quantitatively the damping parameters in practice and care should be taken when using damping for dynamic problems.

Care should also be taken concerning the density scaling when dynamic problems are considered. Since density affects mass, therefore it also affects the mass-proportional damping property α in Eqns (8.120), (8.122), (8.123). Therefore the density scaling scheme as described before cannot be used for dynamic problems since the inertial terms are physically meaningful and significant for the starting, intermediate and final states of the system behavior. The only restriction on the time step is the numerical stability of the process, which is controlled by the smallest element size. Therefore block and element generations should be carefully performed to avoid too small blocks and elements so that the resultant time step is reasonable regarding the allocated computing time and resources.

For quasi-dynamic problems, on the other hand, damping is also an artificial device to achieve final steady state solutions by absorbing the excessive kinetic energy of the system. For such cases, it is desirable, if can be achieved, to use density scaling to modify the damping parameters so that an optimal damping scheme can be obtained. This is called adaptive damping. However, it is impossible in practice to derive optimal adaptive damping parameters in closed-form for block systems of complex shapes, sizes and heterogeneous material properties of the block/element systems. A numerical scheme of adaptive damping was developed (Cundall, 1982) based on the density scaling algorithm presented before.

In order to evaluate the effects of damping, the ratio R representing the rate of energy dissipation by dashpots, $\dot{D}$, and rate of change of kinetic energy, $\dot{E}$,

$$R = \frac{\dot{\mathrm{D}}}{\dot{\mathrm{E}}} \tag{8.125}$$

is evaluated to measure the proportion of energy the dashpots extract from the periodic energy flux of an oscillating block system, with energy alternating between storage of strain energy and kinetic energy. It was found by numerical experiments that $R = 1.0$ provides an effect similar to that of using critical damping. $R < 1.0$ represents 'under-damping' and $R > 1.0$ represents 'over-damping'. The numerical algorithm for the adaptive damping is therefore identical to that of density scaling with the substitution of the damping parameter α for the density ρ and substitution of the ratio R in Eqn (8.121) for the ratio of balance-of-force, r

$$r = \frac{\text{maximum out-of-balance force}}{\text{representative grid-point force}} \tag{8.126}$$

8.11 Linked-list Data Structure

A 'linked-list' data structure is used in distinct element method programs to store and retrieve all data elements. It is suitable for the hierarchical structure of the data required to represent the rock structure (problem domain, blocks, faces, edges, vertices, etc.). The variables in the program are grouped into different data blocks representing different physical items: blocks, contacts, fluid domains (voids

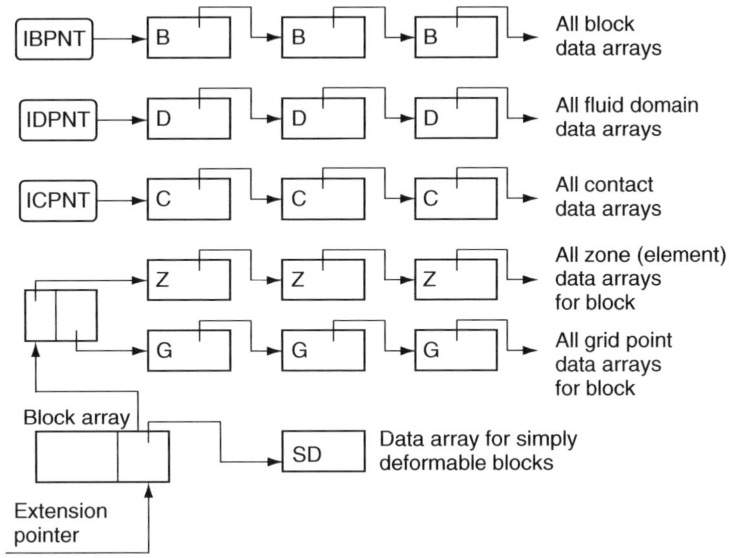

(a)

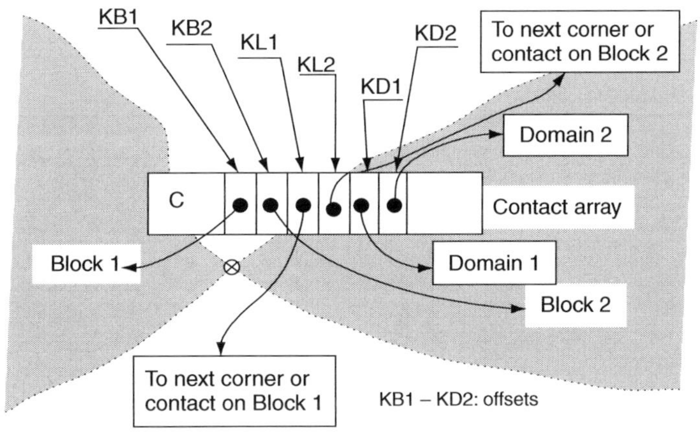

(b)

Fig. 8.29 (a) Linked-list for the main data arrays and (b) the pointers used for contact registration (Itasca, 1993).

between contacting blocks at the intersection of fractures), zones (finite difference elements), grid points, etc. All data are stored in a main array and arranged by their 'addresses' (the first memory location in the main array for a particular item, for example, a block or a contact) and 'offsets' describing the memory locations of particular variables associated with the particular item. Figure 8.29a shows the linked-list for the main data array used for the UDEC code. Each data block is accessed via a pointer, which can be accessed via the previous block data array. Global pointers are given for different data blocks, as IBPNT for all block data arrays and ICPNT for all contact data arrays. The convention used for pointers within a contact array is shown in Fig. 8.29b.

Figure 8.30 shows the hierarchical data structure for rigid polyhedral blocks in the 3D DEM code 3DEC. Each element in the data structure, although drawn separately in the figure, is embedded in the main data array and connected by the pointers. Blocks are accessed via a global pointer IBPNT which

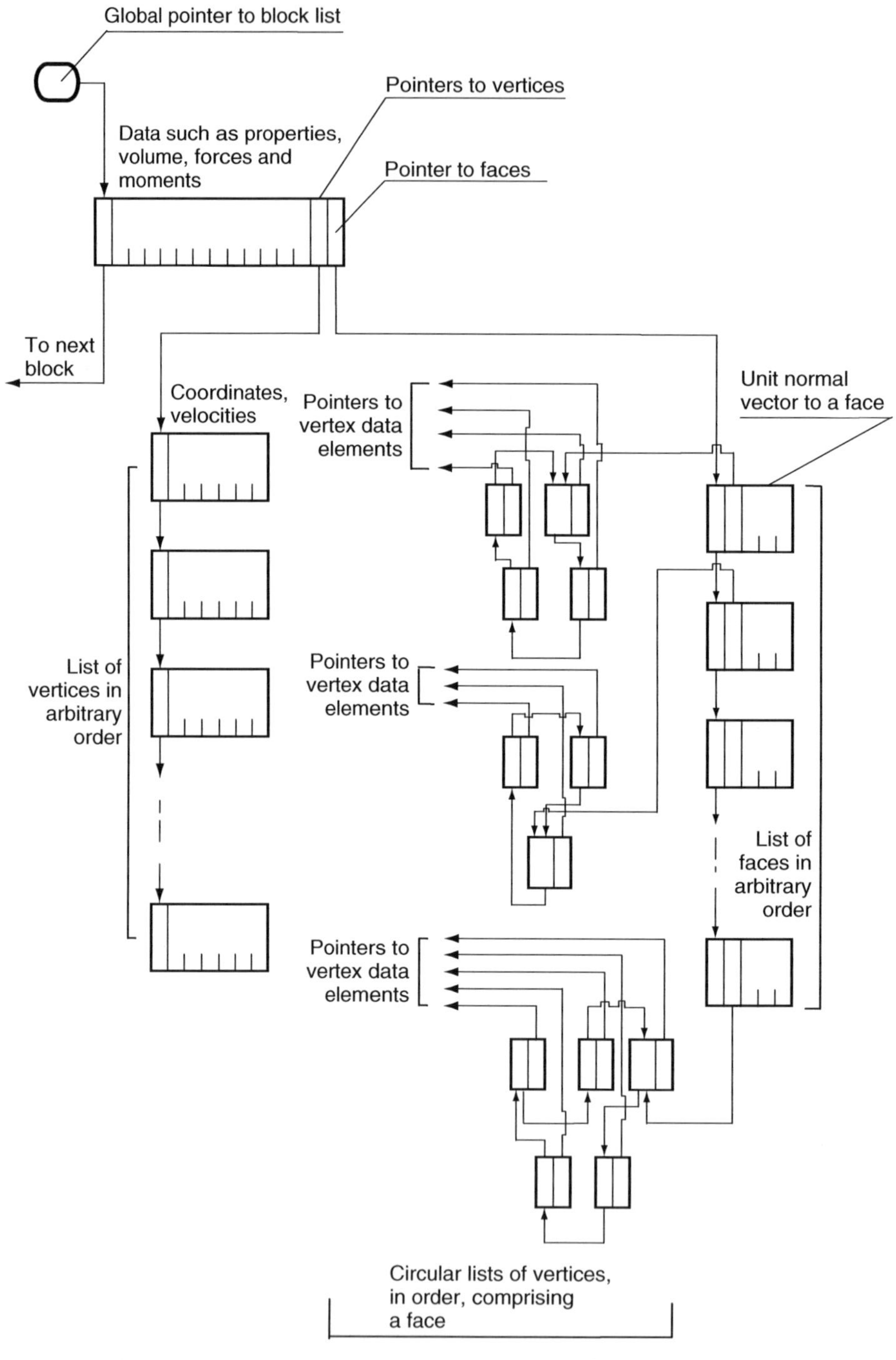

Fig. 8.30 Simplified data structure associated with a rigid polyhedral block (modified, based on Cundall, 1988).

provides entry to a list of all blocks in arbitrary order. Each block data array contains a pointer providing access to lists of vertices and faces. Each face data array, in turn, contains a pointer that gives access to a circular list that contains addresses of vertices that make up the face, arranged in a certain order. For fully deformable blocks, the data structure is similar, but each original polygonal face is further discretized

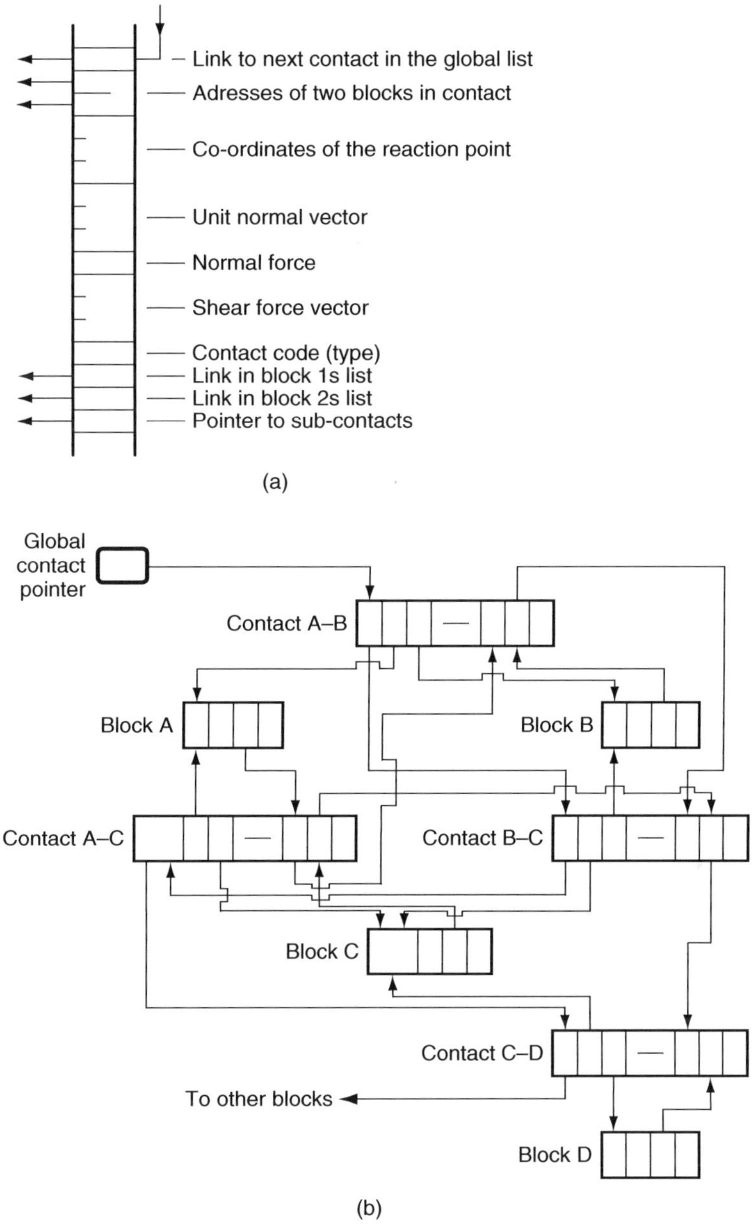

Fig. 8.31 (a) Main items in a contact data element and (b) Global and local links for a four-block system (modified, based on Cundall, 1988).

into triangular sub-faces, in accordance with the discretization of the block interior into tetrahedra. The data structure associated with a sub-face is exactly like that associated with a regular face. An extra pointer is added for deformable blocks to point to a list of all internal tetrahedral elements.

A similar hierarchical structure is used for other data arrays. The major items in a contact data array are shown in Fig. 8.31. A data element is assigned in the main data array for each pair of blocks in contact. This element contains relevant information, such as friction, shear and normal forces, etc., see Fig. 8.31a. Each contact element is linked globally to all other contacts as well as the two blocks making up the contact. The form of the linked data structure for such contact data elements is shown in Fig. 8.31b for an example of four blocks. Contacts can be accessed in several ways, depending on the requirement. When all contact forces are updated during the main calculation cycle, all contacts are scanned one at a time, via the global pointer for contact data array (ICPNT). During contact detection and updating, the existing contacts of constituent blocks are accessed and thread through all blocks.

8.12 Coupled Thermo-Hydro-Mechanical Analysis

Many efforts were made to develop capacities in DEM codes for handling coupled stress, flow and heat transfer analysis for fractured rocks for different kinds of engineering applications, especially concerning underground radioactive waste repositories. However, only a few DEM codes are able to simulate such complex physical processes – and with considerable assumptions regarding the physical behavior of the materials (rocks fractures, fluids of different phases), geometry (fracture system at the macroscopic level and fracture surface at the microscopic level) and problem size (2D versus 3D). There are also difficult physical processes to handle numerically (such as energy conversion from mechanical work (or dissipated energy) into heat energy, fluid phase change and buoyancy, interaction between fracture flow and matrix flow, scale effects, etc.) and also runtime computing costs. Regarding the above limitations, the most representative DEM code for handling coupled THM processes for problems in fractured rocks is UDEC for 2D problems, with the following assumptions:

(1) Flow is conducted only through connected fractures, based on the Cubic Law (with possible simple corrective terms for effects of roughness of the fracture surfaces);

(2) Transient heat conduction through the rock matrix and heat convection by flow through fracture spaces are the main heat transfer modes considered;

(3) No fluid phase change and buoyancy force due to heating are considered and

(4) No energy conversion between mechanical work and heat (e.g., that caused by friction) is considered.

The coupled processes considered in UDEC code are shown in Fig. 8.32, with one-directional coupling for thermo-mechanical and thermo-hydraulic aspects. However, it has been shown by many theoretical studies and experimental measurements (Jing et al., 1994a,b) that such simplifications are very often realistic for practical problems without causing significant errors. It is especially true for hard rocks (such as granites) where the volume of water in the rock mass is small compared to the volume of rock, and that the flow velocity is generally very small. Except in the very close vicinity of fractures or heat sources, heat convection plays a minor role in the overall temperature field.

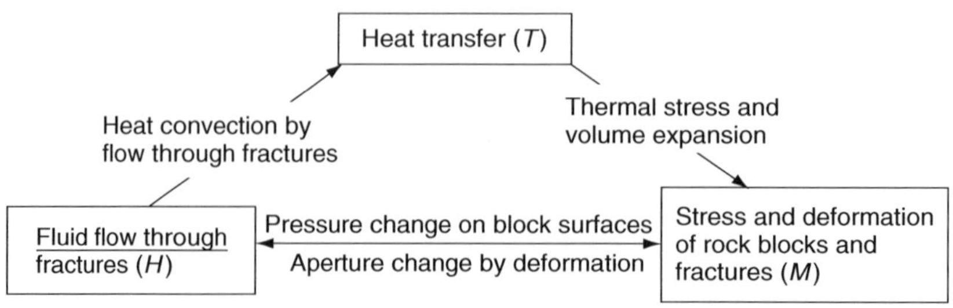

Fig. 8.32 Coupled THM processes simulated in the UDEC code.

The coupling mechanisms shown in Fig. 8.32 show only the major coupling mechanisms at the process level, i.e., between stress/deformation, fluid flow and heat transfer. There are also minor coupling mechanisms at a lower level – the property level, such as temperature dependency of hydro-mechanical properties, stress/deformation dependency of thermo-hydraulic properties and fluid dependency of thermo-mechanical properties. Typical examples of such interdependency of physical properties are the change of density and viscosity of fluids, shear strength of fractures and other mechanical properties of rock materials with temperature variations, and the changes of porosity, permeability and thermal conductivity of rock materials with stresses. These changes may become very significant under high temperature conditions, especially with the occurrence of phase changes of liquids and rocks (such as melting), which may involve other physico-chemical processes. Such complexities are not considered at present in the DEM codes due mainly to the assumption of impermeable rock blocks, except for the change of fluid viscosity μ and density ρ_f with temperature, given in a general form

$$\begin{cases} \rho_f = \rho_f^0[1 - \alpha(T - T^0)] \\ \dfrac{1}{\mu} = \dfrac{1}{\mu_0}[1 + \beta(T - T^0)] \end{cases} \tag{8.127}$$

where ρ_f^0 and μ_0 are the initial density and viscosity of the fluid at the reference temperature T^0, and α and β are material constants to be determined by laboratory experiments.

8.12.1 Flow and Hydro-Mechanical Analysis Technique Using a Domain Structure

The solution of the equation of fluid flow based on the classic Cubic Law, with possible corrections for non-parallel wedge-shaped fractures in UDEC is based on a 'domain' structure. A domain is a 2D area defined to represent the void spaces created at fracture intersections by contact points around that spot and the void spaces along the fractures (for edge–edge contact type). Hydraulic contacts are also established between flow domains to facilitate the connectivity of domains so that fluid flow can be conveniently and correctly calculated from domain to domain according to the flow equations, following the time steps. The domains, therefore, operate as if they are 'flow elements' along the fractures and at intersections where the shape may be very irregular, see Fig. 8.33.

A hydraulic contact is a separator between the domains, which is represented by mechanical contacts along a fracture between a vertex (a block corner or a vertex of a finite difference element) on its one surface with an edge or vertex on its opposite surface (Fig. 8.34). Mechanical contacts always therefore accompany the hydraulic separators. Thus a fracture formed between two block edges may be separated into a number of hydraulic domains, based on the internal discretization by the finite difference elements. Therefore, the speed and accuracy of flow analysis depends also on the discretization of blocks for mechanical analysis in UDEC.

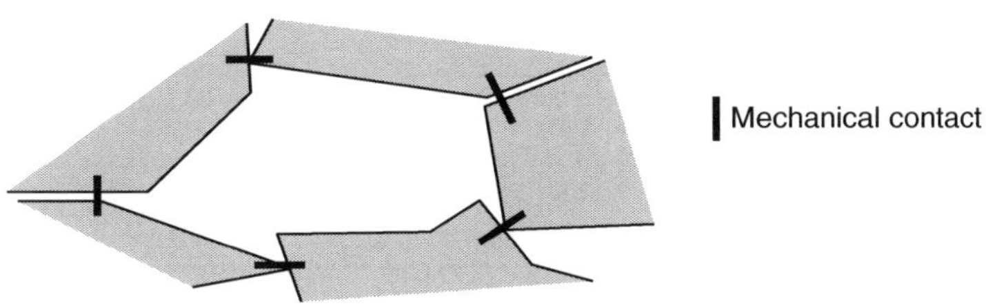

Fig. 8.33 An irregularly shaped domain at a fracture intersection for a UDEC model.

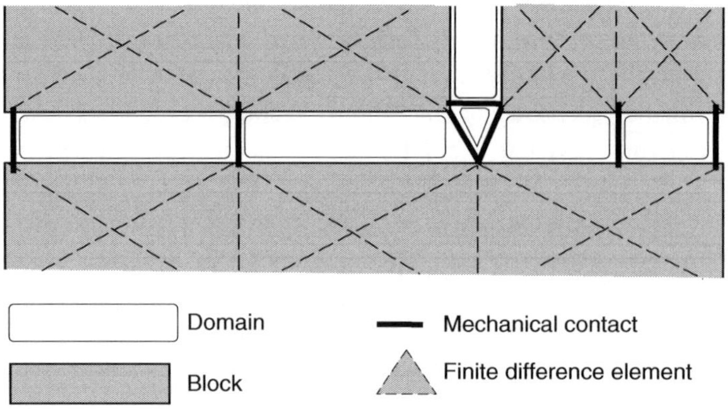

Fig. 8.34 Domains defined both at intersections and along fractures with edge–edge contact types in a UDEC model (after Ahola et al., 1996, with slight modifications to the legends).

The governing equations and solution techniques for the fluid flow in 2D fracture networks in UDEC are basically the same as those presented in Chapter 4 and Chapter 10, using the Cubic Law. However, due to different numerical implementation techniques, some particular fluid–solid coupling aspects in the flow analysis of the UDEC code need to be described:

(1) Fluid pressure is usually assumed to be uniform (constant) over a domain when gravity is absent, and or to be linearly variable according to the hydrostatic gradient when the gravity is considered. The mechanical effect of the fluid pressure is therefore calculated as (Fig. 8.35a)

$$F_i = pn_iL \tag{8.128}$$

which is applied on the boundaries of the solid block (elements) for movements and deformation analysis.

(2) Hydraulic aperture (e) variation with mechanical deformation/displacement of blocks (or elements) is considered through updating by

$$e = e_0 + \Delta e \tag{8.129}$$

where e_0 and Δe are the initial aperture and its variation due to mechanical deformation and movement of blocks, usually taken as the normal displacement of the fracture u_n (with opening taken as positive) (Fig. 8.35b).

(3) Generation of pressure gradient at domains representing intersections (Fig. 8.35c)

$$\Delta p = \frac{K_f}{V}\Delta V = \frac{K_f}{V}(\Delta V_f - \Delta V_m) = \frac{K_f}{V}\left(\sum_{k=1}^{M_i} q_k \Delta t - \Delta V_m\right) \tag{8.130}$$

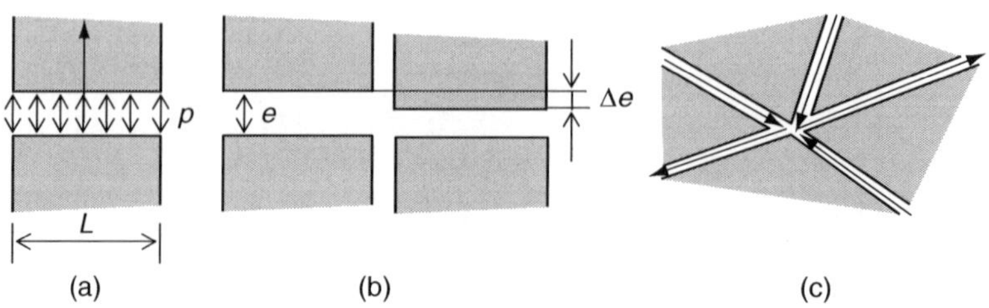

Fig. 8.35 Fluid–rock interactions in fractures considered in the UDEC code (after Itasca, 1994).

where K_f is the bulk modulus of the fluid, $M_i > 2$ the number of fractures connected at domain i, ΔV_f the change of fluid volume at the domain, ΔV_m the volume change of the domain due to mechanical deformation, calculated from updated contact position vectors, and V is the current domain volume. The first term in the bracket in (8.130) represents the fluid volume transferred from the upstream domain to the downstream domain, and the second term in the bracket represents the reduction (or enlargement) of domain volume due to the mechanical deformation of movements.

(4) Generation of pressure between two adjacent domains (i and $i + 1$) between two opposite fracture surfaces

$$\Delta p = p_{i+1} - p_i + \rho_f g(y_{i+1} - y_i) \tag{8.131}$$

where ρ_f is fluid density, g gravity acceleration, y_i and y_{i+1} are the vertical coordinates of the center of domain i and $i + 1$, respectively.

(5) The flow is governed by pressure difference between the adjacent domains and the calculation of flow rate is different for different contact types. For vertex-to-edge (point) contact, the flow rate is determined by

$$q = -k_e \Delta p \tag{8.132}$$

where k_e is a point contact permeability factor. For edge-to-edge contact representing a fracture, the flow rate is given by

$$q = -\frac{e^3}{12\mu}\frac{\Delta p}{L} = -k_j \Delta p \tag{8.133}$$

where μ fluid viscosity and L the assigned contact length between the two adjacent domains.

(6) At each mechanical time step, the apertures of the fractures are updated according to the current block (element) vertex (boundary grid points) coordinates. This will then facilitate the determination of the flow rates, using Eqns (8.132) and (8.133), with the domain pressures updated by the relation below

$$p^{i+1} = p^i + \Delta p^i \tag{8.134}$$

where i is the previous iteration number and Δp^i the pressure increment determined by using Eqns (8.130) or (8.131), depending on the domain types.

(7) Numerical stability of the explicit flow analysis method depends on the time step, whose critical value should be governed by the following criterion

$$\Delta t_f \leq \Delta t_f^c = \min\left\{ V_i \Big/ \left(K_f \sum_{k=1}^{M_c^i} k_k \right); i = 1, 2, .., N_d \right\} \tag{8.135}$$

where N_d is the total number of domains, V_i the volume of the ith domain, M_c^i the number of contacts connected to the ith domain and k_k the permeability of the kth contact of domain i, its value being determined as $k_k = k_e$ or $k_k = k_j$, depending on the contact types, using either Eqn (8.132) or Eqn (8.133). It is clear that hydraulic time step depends largely on the minimum domain volume.

Figure 8.36 summarizes and illustrates the transient hydro-mechanical interactive analysis algorithm presented above, using the dynamic relaxation technique based on the domain structure. The algorithm is implemented by performing a series of iterations (termed cycles in the UDEC code) for fluid flow with the time step defined by the user but controlled by Eqn (8.135). For a typical cycle i, the following tasks are carried out in the given order:

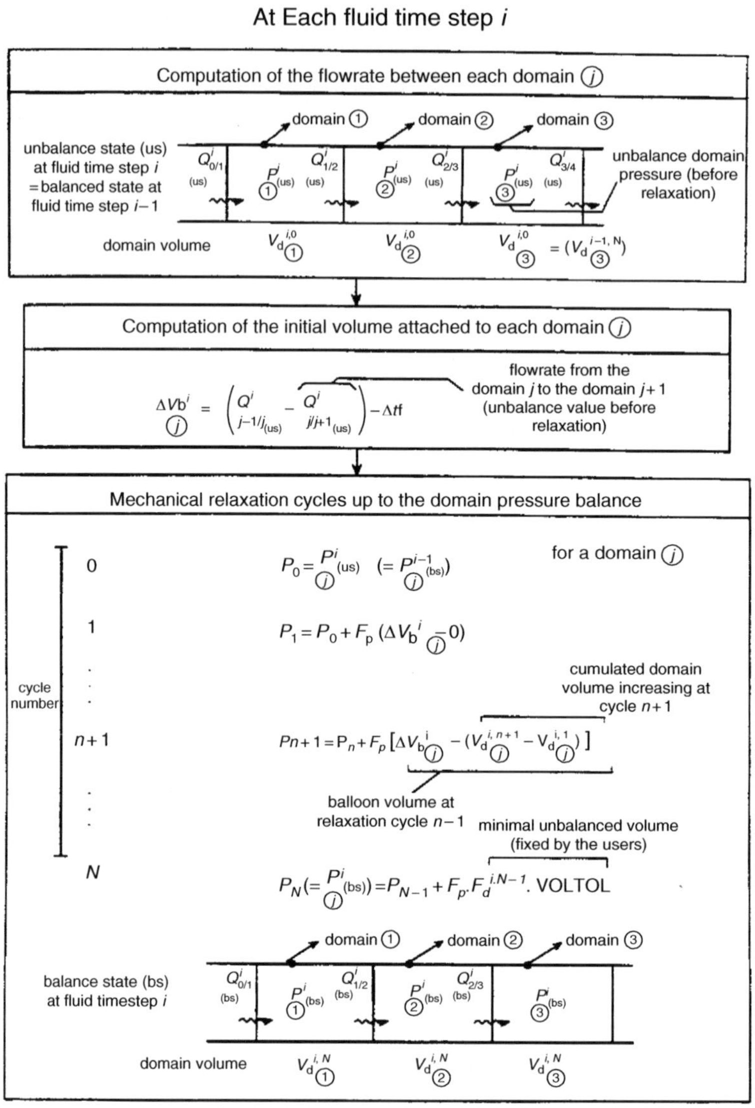

Fig. 8.36 Summary of the coupled hydro-mechanical coupling algorithm in UDEC for transient problems (Ahola et al., 1996).

(1) Computing the flow rate $Q_{i/i+1}$ from domain i to domain $i+1$ as a function of the unbalanced pressure p_i and p_{i+1} in each domain.

(2) Computing the initial volume of each domain, $Vd^{i,0}$.

(3) Performing a series of mechanical relaxation steps until a continuous flow in each domain is reached. Assuming incompressibility of the fluid, the net volume of fluid flowing into a domain during a hydraulic time step must equal the total volume change of the domain. The unbalanced fluid volume, being the difference between the two, is gradually reduced during the relaxation process.

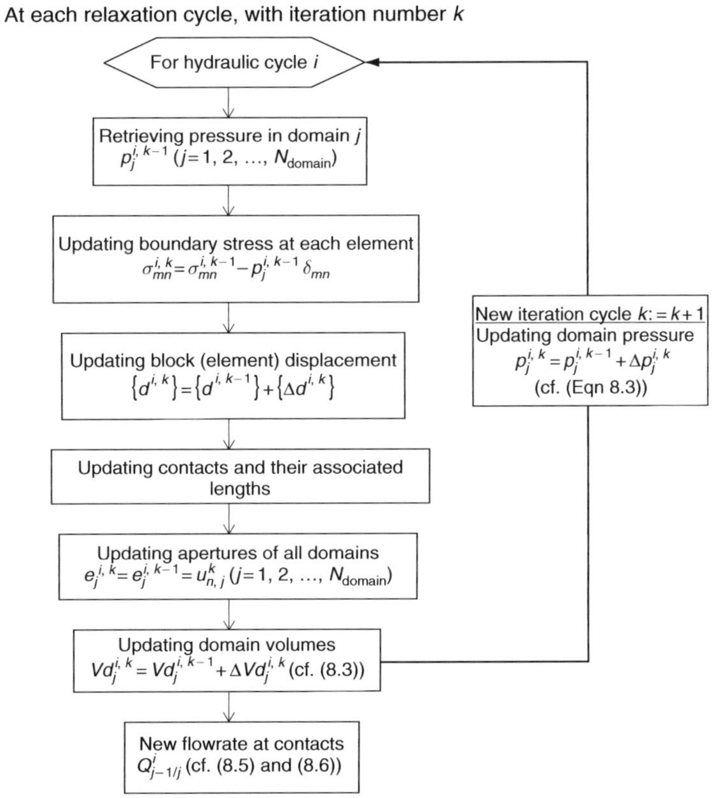

Fig. 8.37 The dynamic relaxation scheme for transient stress-flow analysis in UDEC (Ahola et al., 1996).

(4) The domain pressure is changed in accordance with the unbalanced volume of each domain, according to Eqn (8.130), following an adaptive change of domain volume, until a stable pressure value is obtained.

Figure 8.37 shows graphically the above dynamic relaxation scheme.

8.12.2 Heat Conduction and Thermo-Mechanical Analysis in the UDEC Code

For coupled thermo-mechanical processes with an elasticity assumption for the rock material, the governing equations are the combined equations of elastic deformation and heat conduction, without considering the conversion of mechanical work into thermal energy, namely,

$$\left(\begin{array}{c} Gu_{i,jj}+(\lambda+G)u_{j,ji}+\rho b_i+(3\lambda+2G)\alpha(T-T_0)_{,j}\delta_{ji}=\rho\ddot{u}_i \\ (k_k T_{,k})_{,k}+s^{\mathrm{h}}=\rho c_p\dfrac{\partial T}{\partial t} \end{array}\right. \tag{8.136}$$

Therefore the heat conduction is not coupled with mechanical deformation and can be simulated independently, but the thermal stress and strain increments must be included in the stress/deformation analysis as indicated by the extended Navier equations of motion. Solution of Eqn (8.136) follows the similar dynamic relaxation scheme, based on the FDM through an explicit time marching process.

For inelastic responses of rock material, the elastic small stress–strain relation no longer holds, and the total mechanical strain is assumed to be defined by the sum,

$$\varepsilon_{ij} = \varepsilon_{ij}^{\mathrm{e}} + \varepsilon_{ij}^{\mathrm{ne}} + \varepsilon_{ij}^{\mathrm{T}} \tag{8.137}$$

where ε_{ij} is the total mechanical strain, $\varepsilon_{ij}^{\mathrm{e}}$ the elastic strain representing the reversible elastic responses of the material under loading, $\varepsilon_{ij}^{\mathrm{ne}}$ the non-elastic part of the strain representing the irreversible strain of the material, such as plasticity, damage, etc. and $\varepsilon_{ij}^{\mathrm{T}}$ the thermal strain component caused by thermal expansion of the rock volume. The reversible part of the strain is usually taken as an elastic response given by

$$\varepsilon_{ij}^{\mathrm{e}} = C_{ijkl}\sigma_{kl} = [\lambda\delta_{ij}\delta_{kl} + \mu(\delta_{ik}\delta_{jl} + \delta_{il}\delta_{jk})]\sigma_{kl} \tag{8.138}$$

where λ and μ are the Lame's elasticity constants. The tensor C_{ijkl} is called the compliance tensor of the material and is the inverse of the commonly used stiffness tensor.

In rock mechanics, as well in the UDEC code, the most common source of the non-linearity of the rock material is taken to be the plastic deformation. The yield function F and the plastic potential Q are usually taken as the same or as different forms of the Mohr–Coulomb model or Hoek–Brown model.

The governing equations, in such a case, are better given in terms of stresses, plus separate definitions of strains and constitutive laws, together with the heat conduction equation, namely, the combination of the following equations:

(1) Extended equations of motion with damping

$$\frac{\partial \sigma_{ij}}{\partial xi} + b_i = \rho\frac{\partial^2 u_i}{\partial t^2} + c\frac{\partial u_i}{\partial t} \tag{8.139a}$$

(2) Geometry equation – strain definition

$$\varepsilon_{ij} = \varepsilon_{ij}^{\mathrm{e}} + \varepsilon_{ij}^{\mathrm{ne}} + \varepsilon_{ij}^{\mathrm{T}} \tag{8.139b}$$

(3) Constitutive equation, e.g., plasticity models, etc.,

$$f\left(\varepsilon_{ij}, \dot{\varepsilon}_{ij}, \sigma_{ij}, \dot{\sigma}_{ij}\right) = 0 \tag{8.139c}$$

(4) Heat conduction equation, the second equation in Eqn (8.136)

For analysis of underground radioactive waste repository problems, a method proposed by St. John (1985) was adopted to determine the radius of influence of a single heat source or a waste container on rock temperature as a function of time, in order to determine the size of the problem required in a model for heat transfer analysis. The equation of temperature delay for a single point source, expressed as the temperature change ΔT as a function of a moving distance from the heat source, R, and the initial heat intensity, Q_0, (Christiansson, 1979) is

$$\Delta T = \frac{Q_0}{\pi^{3/2}} \exp\left(-At\right)\frac{\sqrt{\pi}}{4k} \exp\left(-\frac{R^2}{4\kappa t}\right) Re\left[w\left(\sqrt{At} + \frac{\mathrm{i}R}{\sqrt{4\kappa t}}\right)\right] \tag{8.140}$$

where i $= \sqrt{-1}$, A is thermal constant, κ thermal diffusivity, t time, $w = w(z)$ complex error function of the complex variable z and Re the real part of a complex function. The temperature therefore has an exponential decay proportional factor $\exp(-R^2/4\kappa t)$. St. John (1985) proposed that the minimum problem size L for a model should be checked against the following criterion (where t should be taken in years)

$$L \geq 4\sqrt{\kappa t} \tag{8.141}$$

In the UDEC code, continuous heat sources, such as line or area sources, are divided into discrete point source arrays with lumped heat intensities, so that the overall temperature behavior of the whole model will be the same as caused by continuous heat sources. However, this lumped point source replacement will cause certain errors in temperature distribution in close vicinities of the lumped points sources. For all practical problems in rock masses, the effect of these errors is not significant and can be reduced by refined source arrays.

Because of the one-way coupling for the stress–temperature analysis, the solution of Eqn (8.136) or (8.139) is greatly simplified, since the heat conduction can be solved separately. Therefore only the induced thermal stresses at the end of each thermal time step need to be introduced into the mechanical relaxation time steps to affect the stress and deformation of the rocks.

The heat conduction simulation in code UDEC is straightforward, based on the element mesh created for the block deformation, Fourier's law and the numerical solution of the transient heat conduction equation. The algorithm is given below for the example of triangular elements.

Similar to the strain calculations, for each triangular element of area A_e bounded by a boundary S, the temperature gradients $\partial T/\partial x_i$ $(x_1 = x,\ x_2 = y)$ are given by

$$\frac{\partial T}{\partial x_i} = \frac{1}{A_e}\int_S T n_i \, ds \approx \frac{1}{A_e}\sum_{i=1}^{3} \bar{T}_i e_{ij} \Delta x_j^m \tag{8.142}$$

where n_i is the unit normal vector of the boundary S, e_{ij} the 2D permutation tensor with $e_{11} = e_{22} = 0, e_{12} = 1$ and $e_{21} = -1$. Symbol $\bar{T}_i$ stands for the mean temperature along the ith edge of the element and Δx_j^m the difference in x_j (i.e., in x or y) between the two grid points of the edge (Fig. 8.38).

For simplicity, one can have

$$\bar{T}_{lm} = \frac{1}{2}(T_l + T_m), \quad \bar{T}_{mk} = \frac{1}{2}(T_m + T_k), \quad \bar{T}_{kl} = \frac{1}{2}(T_k + T) \tag{8.143a}$$

$$\Delta x_{lm} = (x_m - x_l), \quad \Delta x_{mk} = (x_m - x_k), \quad \Delta x_{kl} = (x_k - x_l) \tag{8.143b}$$

$$\Delta y_{lm} = (y_m - y_l), \quad \Delta y_{mk} = (y_m - y_k), \quad \Delta y_{kl} = (y_k - y_l) \tag{8.143c}$$

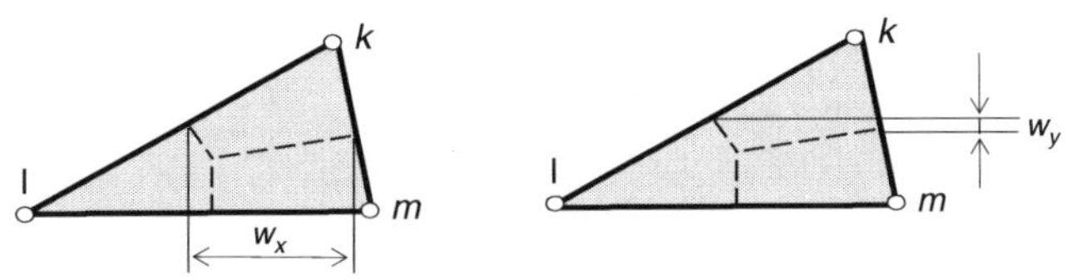

Fig. 8.38 Heat flux into grid point k of a triangular element (zone) (Itasca, 1993).

for the side defined by grid point (vertex) l, m and k, where side number $1 = lm$, $2 = mk$ and $3 = kl$. Defining w_x and w_y as the representative width of the line across which a heat flux flows in the element in the respective direction, and the heat flux vector of the element is given by Fourier's law

$$Q_i = -k_{ij}\frac{\partial T}{\partial x_i} \quad (i = x, y) \tag{8.144}$$

where the temperature gradient is given by Eqn (8.142); the heat flux at each vertex (grid point) of the element is calculated as

$$F = w_j Q_j = w_x Q_x + w_y Q_y \tag{8.145}$$

Equation (8.145) represents the contribution to the total heat flux at each grid point from the element they define. If a grid point k connects N elements, then the total heat flux consisting of contributions from all connected elements is

$$F_k = \sum_{i=1}^{N} F_i = \sum_{i=1}^{N}(w_x^i Q_x^i + w_y^i Q_y^i) \tag{8.146}$$

where w_x^i, w_y^i, Q_x^i, Q_y^i are the representative width and heat flux components of the element i connecting to grid point k. The temperature change at the grid point k is then given by

$$\Delta T = \frac{F_k}{C_p M}\Delta t \tag{8.147}$$

where M is the lumped mass at the grid point k (the sum of one-third of the areas of all elements connected at point k).

The above scheme is an explicit algorithm and the magnitude of Δt depends entirely on the numerical stability requirement. An implicit scheme can also be derived, using different finite difference solutions to the heat conduction equation, see Itasca (1993) for details.

A staggered iteration scheme for the coupled thermo-mechanical calculations is implemented in UDEC (Board, 1989), as shown in Fig. 8.39.

8.12.3 Heat Convection along Fractures and Coupled Thermo-Hydraulic Processes

For a rock fracture formed by two heated rock surfaces and filled with moving fluids, the major coupled THM variables are the fracture deformation, temperature and flow rate or velocity. These variables are mutually interdependent or two-way coupled. The individual processes of fracture deformation, fluid flow, heat conduction and principles of the thermo-mechanical and hydro-mechanical couplings implemented in the UDEC code have been presented above. One remaining major mechanism is the thermo-hydraulic coupling in and along a fracture – the thermal convection by fluid flow in fractures – and is presented below. The principle and the computer implementation rest on the work of Abdaliah et al. (1995) that was also summarized in Ahola et al. (1996). The heat convection by fluid flow inside rock blocks is not considered because of the assumption of impermeable rock blocks. Therefore the convective heat transfer is limited to that through fluid flow in fractures, in DEM and in the UDEC code. Such assumptions have often proved to be acceptable in practice, especially for hard rocks, as mentioned before.

The theoretical model for convective heat transfer due to fluid flow along a rock fracture developed by Abdaliah et al. (1995) considers the total heat balance of an elementary volume in a fracture formed

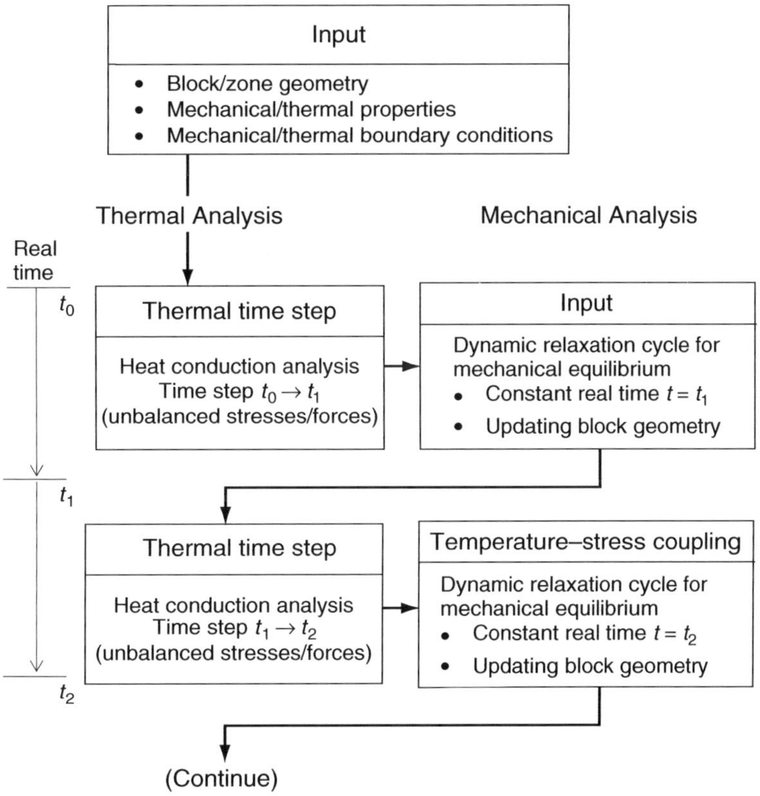

Fig. 8.39 Staggered time marching scheme for stress–temperature analysis in UDEC, based on Board (1989) with slight modifications.

by two parallel smooth surfaces of aperture, e, and flow displacement increment, dx, in the flow direction, as the area $ABCD$ shown in Fig. 8.40, for 2D problems. The mechanisms of heat transfer considered are fluid–fluid (domain–domain) heat convection, fluid–fluid (domain–domain) heat conduction and fluid–rock (domain–element) heat convection.

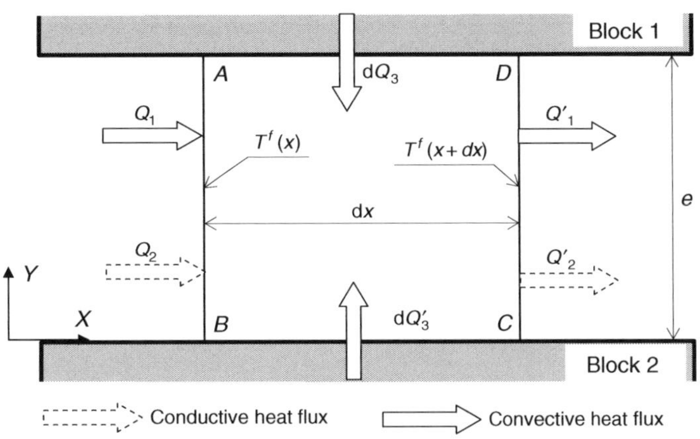

Fig. 8.40 Heat balance of a elementary fluid volume in a fracture (Abdaliah, 1995).

The elementary volume *ABCD* has a unit thickness in the out-of-plane direction and a surface area of $e(\mathrm{d}x)$ in the flow plane. There are conductive $Q_1, Q_1^{'}, Q_3, Q_3^{'}$ and convective $(Q_2, Q_2^{'})$ heat fluxes entering and leaving the elementary volume across its fluid–fluid (*AB*, *CD*) and/or fluid–rock (*AD*, *BC*) boundaries. The temperatures in the rock and in the fluid are denoted by T^{r} and T^{f}, and the density and specific heat of the fluid are represented by the symbols ρ_{f} and C_p^{f}, respectively. The interfaces *AB* and *CD* represent the hydraulic contacts connecting two adjacent domains defined in the UDEC code. They are established at the locations of mechanical contacts with either vertex-to-edge or vertex-to-vertex contacts and are used to facilitate the 'domain element' data structure for flow analysis.

(1) Fluid–fluid heat convection

Across the fluid–fluid interfaces (*AB* and *CD*), the convective heat flux increment is given by

$$\mathrm{d}Q_x^{\text{convection-fluid/fluid}} = \mathrm{d}Q_1 = Q_1 - Q_1^{'} = -\int_0^e \left[\rho_{\mathrm{f}} C_p^{\mathrm{f}} \left(v_x \frac{\partial T^{\mathrm{f}}}{\partial x} + T^{\mathrm{f}} \frac{\partial v_x}{\partial x} \right) \mathrm{d}x \right] \mathrm{d}y \tag{8.148}$$

where v_x is the fluid velocity in the x direction, along the fracture. This increment of heat flux is caused by the velocity of the fluid flow.

(2) Fluid–fluid heat conduction

Heat conduction occurs also between fluid particles, for the elementary volume in Fig. 8.42; this is given by

$$\mathrm{d}Q_x^{\text{conduction-fluid/fluid}} = \mathrm{d}Q_2 = Q_2 - Q_2^{'} = \int_0^e \left[\kappa^{\mathrm{f}} \frac{\partial}{\partial x} \left(\frac{\partial T^{\mathrm{f}}}{\partial x} \right) \mathrm{d}x \right] \mathrm{d}y \tag{8.149}$$

where κ^{f} is the thermal conductivity of the fluid (*W/m K*) and is assumed to be a constant.

(3) Rock–fluid heat convection

Heat convection occurs also between the rock–fluid interface, following Newton's cooling law. The general expression of the law is given by

$$\mathrm{d}Q_x^{\text{convection-rock/fluid}} = h(T^{\mathrm{r}} - T^{\mathrm{f}})\mathrm{d}A \tag{8.150}$$

where h is called the thermal exchange coefficient $(W/m^2 K)$ between the rock surface and fluid, T^{r} the temperature at the rock surface and $\mathrm{d}A = \mathrm{d}x \cdot 1$ the differential area in the out-of-plane direction. For the elementary volume in Fig. 8.40

$$\mathrm{d}Q_x^{\text{convection-rock/fluid}} = \mathrm{d}Q_3 + \mathrm{d}Q_3^{'} = h(T^{\mathrm{r}1} - T^{\mathrm{f}})\mathrm{d}x + h(T^{\mathrm{r}2} - T^{\mathrm{f}})\mathrm{d}x \tag{8.151}$$

where $T^{\mathrm{r}1}$ and $T^{\mathrm{r}2}$ are the surface temperature of the rock blocks 1 and 2, respectively. The heat energy balance equation can then be written as

$$\begin{aligned} mC_p^{\mathrm{f}} \frac{\partial T^{\mathrm{f}}}{\partial t} &= \mathrm{d}Q_x^{\text{convection-fluid/fluid}} + \mathrm{d}Q_x^{\text{conduction-fluid/fluid}} + \mathrm{d}Q_x^{\text{convection-rock/fluid}} \\ &= \mathrm{d}Q_1 + \mathrm{d}Q_2 + \mathrm{d}Q_3 + \mathrm{d}Q_3^{'} \end{aligned} \tag{8.152}$$

where m and C_p^{f} are the fluid mass of domain and specific heat of the fluid, respectively.

It is assumed by Abdaliah et al. (1995) that the effect of conductive heat flux between fluids is very small and therefore it may be safely taken that $dQ_2 \approx 0$. The simplified Eqn (8.152) then becomes

$$mC_p^{\rm f}\frac{\partial T^{\rm f}}{\partial t} = {\rm d}Q_1 + {\rm d}Q_3 + {\rm d}Q_3' = -\int_0^e \left\{ \left[\rho_{\rm f} C_p^{\rm f} \left(v_x \frac{\partial T^{\rm f}}{\partial x} + T^{\rm f} \frac{\partial v_x}{\partial x} \right) \right] {\rm d}x \right\} {\rm d}y + h(T^{\rm r1} + T^{\rm r2} - 2T^{\rm f}){\rm d}x \tag{8.153}$$

A further assumption is that the heat energy change at fracture intersections is also negligible. The heat transfer in the fracture is confined to the convective heat transfer between the fluid–rock surfaces and across imaginary domain boundaries along the fracture by flow of fluids (for 2D problems). Based on these assumptions, a numerical scheme was proposed to solve Eqn (8.153) with the following algorithms.

As shown in Fig. 8.41, a typical flow domain (i) is defined by four 'hydraulic corners' *A*, *B*, *C* and *D*, two 'hydraulic contacts' (imaginary fluid interfaces) *AB* and *CD* and two rock–fluid interfaces *AD* and *BC*. The domain i connects with adjacent domain $i-1$ and domain $i+1$ through the two hydraulic contacts *AB* and *CD*. The lengths of the domains are l_i, l_{i+1} and l_{i-1}, the mean temperature and the mean fluid velocity in the x direction and heat exchange coefficient of the domains are denoted by $T_i^{\rm f}, T_{i-1}^{\rm f}, T_{i+1}^{\rm f}, v_x^i, v_x^{i-1}, v_x^{i+1}$ and h_i, h_{i-1}, h_{i+1}, respectively.

Based on the above geometrical discretization, the heat fluxes in Eqn (8.153) are also discretized as given by

$$\Delta Q_1 \approx -(\rho_f C_p^{\rm f})(e_i l_i)\left[v_x^i \frac{T_i^{\rm f} - T_{i-1}^{\rm f}}{l_i} + T_i^f \frac{v_x^{CD} - v_x^{AB}}{l_i} \right] \tag{8.154}$$

$$\Delta Q_3 \approx h_i l_i \left(\frac{T_A^{\rm r} + T_D^{\rm r}}{2} - T_i^{\rm f} \right) \tag{8.155}$$

$$\Delta Q_3' \approx h_i l_i \left(\frac{T_B^{\rm r} + T_C^{\rm r}}{2} - T_i^{\rm f} \right) \tag{8.156}$$

$$mC_p^{\rm f}\frac{\partial T^{\rm f}}{\partial t} \approx (\rho_{\rm f} C_p^{\rm f})(e_i l_i)\frac{\Delta T_i^{\rm f}}{\Delta t} \tag{8.157}$$

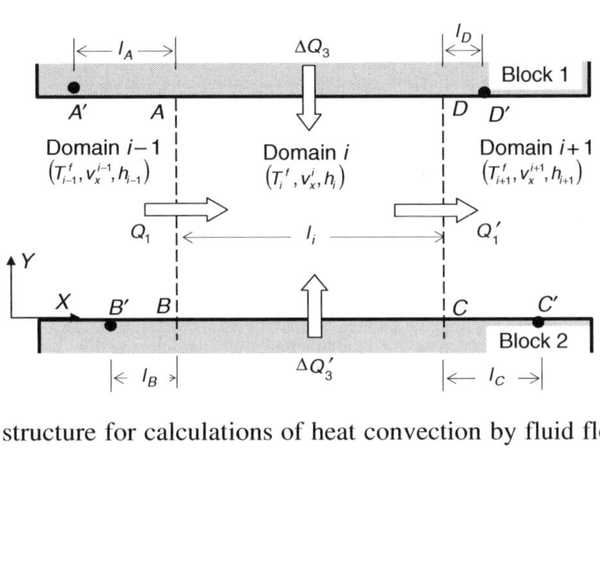

Fig. 8.41 Domain data structure for calculations of heat convection by fluid flow along fractures (after Abdaliah et al., 1995).

where no summation is intended over index i. In Eqn (8.154), the term $(T_i^{\mathrm{f}} - T_{i-1}^{\mathrm{f}})$ is used instead of the more accurate term $(T_D^{\mathrm{f}} - T_A^{\mathrm{f}})$ for the sake of numerical stability, as proposed by Patankar (1980).

Substitution of Eqns (8.154)–(8.157) into Eqn (8.147) and rearranging the terms lead to the equation for calculating temperature change during a time step:

$$\Delta T_i^{\mathrm{f}} = \left[\frac{2h_i}{\rho_f C_p^f e_i} \left(\frac{T_A^{\mathrm{r}} + T_B^{\mathrm{r}} + T_C^{\mathrm{r}} + T_D^{\mathrm{r}}}{4} - T_i^f \right) - \left(v_x^i \frac{T_i^{\mathrm{f}} - T_{i-1}^{\mathrm{f}}}{l_i} + T_i^f \frac{v_x^{CD} - v_x^{AB}}{l_i} \right) \right] \Delta t \tag{8.158}$$

where no summation over index i is intended. This equation represents the algorithm in UDEC for the combined free heat convection between linear fluid domains along the fractures and the forced heat convection between fracture surfaces and the moving fluid.

Due to the heat convection, the temperature on the rock fracture surfaces in the rock is changed. These changes are approximated by temperature changes at the domain corners. For example at corner A, the temperature change is calculated by

$$\Delta T_A^{\mathrm{r}} = h_i \left(\frac{l_i}{2} \right) (T_i^{\mathrm{f}} - T_A^{\mathrm{r}}) C_p^A \Delta t \quad \text{(no summation over } i\text{)} \tag{8.159}$$

where symbol C_p^A stands for the thermal capacity at the hydraulic corner A, which is calculated using the thermal capacities of its nearest mechanical grid point A' (Fig. 8.41)

$$C_p^{A'} = (\rho_{\mathrm{r}} C_p^{\mathrm{r}} S_{A'})^{-1} \tag{8.160}$$

and

$$C_p^A = \left(\frac{C_p^{D'} - C_p^{A'}}{l_A + l_i + l_D} \right) l_A + C_p^{A'} \tag{8.161}$$

where l_A and l_D are the distances between hydraulic corner A to mechanical grid point A', and between hydraulic corner D and mechanical grid point D' (Fig. 8.41). The symbol $S_{A'}$ stands for a third of the sum of the surface areas of all elements connected at grid A' (each element has three grid points for triangle elements). The temperature change at grid point A', B', C' and D' are obtained by interpolation between two adjacent hydraulic corners belonging to the same block. A nested iterative loop algorithm for the calculation of fluid and rock temperature, obtained from the respective convective and conductive heat transfer, shown in Fig. 8.42, was implemented in UDEC for the calculations. Figure 8.43 shows the complete calculation cycles of the UDEC code for coupled THM processes, where the symbol N stands for the total cycle number, Δt the computing time step,

$$T = N_{\mathrm{T}} \Delta t_{\mathrm{T}} = N_{\mathrm{HM}} \Delta t_{\mathrm{HM}} \tag{8.162}$$

where N_{T} is the cycle number for thermal calculations, N_{HM} the cycle number for hydro-mechanical calculations, Δt_{T} the magnitude of time step for thermal calculations (in seconds) and Δt_{HM} the magnitude of time step for hydro-mechanical calculations.

An alternative method for evaluating the net heat flux $\mathrm{d}Q_1$ in Eqn (8.148) can be derived using closed-form integration. The net heat flux can be written as

$$\mathrm{d}Q_1 = -\frac{\rho_{\mathrm{f}} C_p^{\mathrm{f}} g}{2\nu} \int_0^e \left(y^2 - \frac{e^2}{4} \right) \left\{ \frac{\partial}{\partial x} \left[(h - b_x x) T^{\mathrm{f}} \right] \right\} \mathrm{d}x \, \mathrm{d}y \tag{8.163}$$

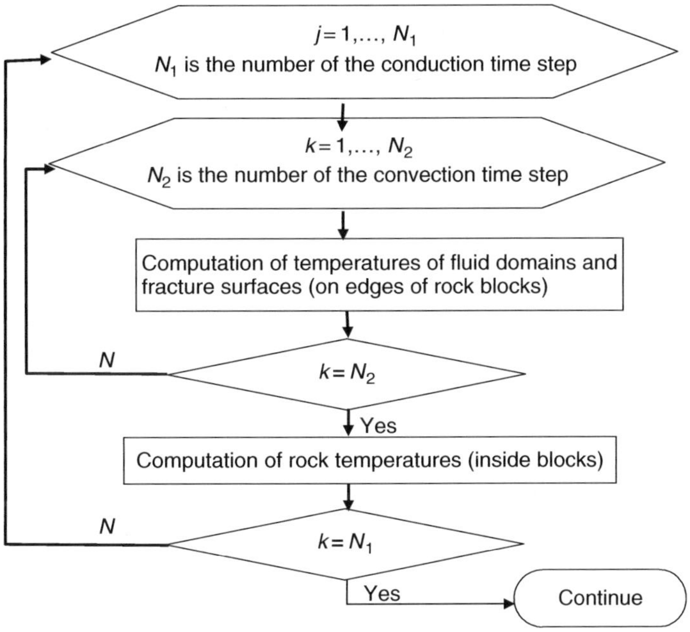

Fig. 8.42 The nested loops for calculations of temperatures in fluid domains and rock blocks (on surfaces and inside) by convective and conductive heat transfer.

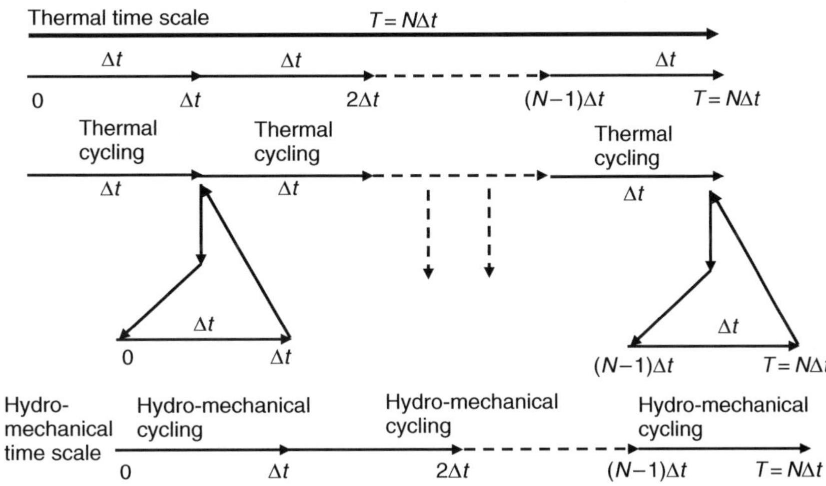

Fig. 8.43 The time-marching relaxation scheme of the coupled THM processes in UDEC (Abdaliah et al., 1995).

Assuming further that the head and fluid temperature gradients do not vary much in the y direction (across the fracture), which is also implied in the Abdaliah et al. (1995) when deriving Eqns (8.154)–(8.156), the integration about y in Eqn (8.163) can be evaluated as

$$dQ_1 = -\frac{\rho_f C_p^f g}{2\nu}\left(\frac{e^3}{12}\right)\left\{\frac{\partial}{\partial x}\left[(h - b_x x)T^f\right]\right\}dx = \left[-\frac{\rho_f C_p^f g}{2\nu}\left(\frac{e^3}{12}\right)\right]d\left[(h - b_x x)T^f\right] \tag{8.164}$$

The net heat flux of the domain i in Fig. 8.41 is then given by

$$\Delta Q_1 = -\frac{\rho_f C_p^f g}{2\nu}\left(\frac{e^3}{12}\right)\int_0^{l_i}\left\{\frac{\partial}{\partial x}\left[(h-b_x x)T^f\right]\right\}dx = -\frac{\rho_f C_p^f g}{2\nu}\left(\frac{e^3}{12}\right)(h_{CD}T^f_{CD} - h_{AB}T^f_{AB} + b_x l_i) \tag{8.165}$$

where $h_{AB}, h_{CD}, T^f_{AB}, T^f_{CD}$ are the heads and fluid temperatures at fluid domain boundary faces AB and CD, respectively. The fluid head is used instead of velocity.

8.12.4 Treatment of Coupled Processes in the Distinct Element Method

In modeling the coupled T-H-M-C processes of fractured rocks, especially the hard crystalline rocks, the modeling of coupled THM processes has been progressed steadily in the past decades since 1970's, with focus mainly on coupled HM processes, i.e., interactions between fluid flow and deformation/stress. With the distinct element method, also the DDA as presented later in Chapter 9, the fluid flow is considered in fractures and the rock matrix is generally assumed to be impermeable. The heat transfer is usually treated as a conduction process without considering heat convection due to fluid flow. This may be a reasonable assumption for some engineering applications, due to the fact that the very low porosity of the hard crystalline rocks makes the fluid flow velocity very slow and the fluid volume, compared with rock volume, is very small. Therefore simple heat conduction algorithms included in the DEM codes can usually satisfy temperature analysis requirements, at least for far-field problems, without causing unacceptable error margins, as established by both *in situ* measurements and numerical analysis using both continuum and discrete approaches (Jing et al., 1996). For near-field problems, such as a nuclear waste repository with both a natural barrier (rock) and an engineered barrier (bentonite), convective heat transfer and fluid flow driven by temperature gradients may become important (Rutqvist et al., 2001; Tsang et al., 2005). Heat convection and the fluid phase change (such as evaporation and condensation) will be much more important if a high temperature environment is involved. For other engineering applications requiring heat transport by fluid flow in connected fracture systems, such as geothermal energy reservoirs in crystalline rocks, convective heat transfer in fracture systems become an important mechanism for the heat production and cannot be ignored.

One of the special issues that has not been investigated in depth in both research and applications is the stress/deformation-induced anisotropy and channeling in individual fractures and clustered fluid flow paths in fractured rocks, especially by fracture shear (Yeo et al., 1998; Koyama, 2005; Koyama et al., 2004, 2006). This issue may have a fundamental influence on both DEM and DFN methods since fluid flow behavior in individual fractures is the basis of understanding the coupled HM processes. The effect of stress can be considered in DEM only for transmissivity (or aperture) changes, not shear-induced flow anisotropy. On the contrary, the effect of stress on fluid flow (and transport) processes cannot be considered in DFN models at all, see Chapter 10.

For the coupled HM processes in fractured rocks, it was demonstrated in Min et al. (2004a,b,c) that, when a critical stress state is encountered in fractured rock masses, many localized fluid flow paths may be formed which may change the original permeability of the rock mass considerably, using the DEM approach (cf. Chapter 12). The work represents an important application of the DEM approach for coupled HM processes of fractured rocks and derivation of equivalent hydraulic and mechanical properties, and quantitative estimation of stress on rock permeability, using the UDEC code, through a numerical homogenization and upscaling process.

Chemical processes in rock fractures, such as mineral dissolution and precipitation, solute sorption and diffusion and general water–rock interaction, have long been central issues in the fields of reactive mass transport, geochemistry and geochemical engineering, and are central scenarios in the safety assessment analysis of nuclear waste repositories and geothermal energy extraction problems (Neretnieks, 1981; Tsang, 1991; Yasuhara and Elsworth, 2004). The associated results, however, have not been incorporated in any DEM approaches or codes.

Considering coupled processes certainly makes the DEM models/codes much more complex, thus adding additional computational costs, especially regarding the high demands for computer memory and time for contact detection and updating. Therefore for solving large-scale practical problems, hybrid approaches are often adopted, such as a DEM representation of a small, near-field region with extensive fractures embedded in a much larger, but basically continuous, far-field region represented via the FEM or BEM, as briefly introduced in Section 8.13.

8.13 Hybrid DEM–FEM/BEM Formulations

As mentioned in Chapter 1, the advantages of both discrete-and continuum-based methods can be best realized if the two methods are used in combination – with discrete and explicit representations of fractures and rock blocks in the smaller near-field region and an equivalent continuum representation of the much larger far-field region. Figure 8.44 shows a hybrid DEM–FEM model. The DEM region is highly fractured and contains the excavation (but this is not restrictive since the excavation could also be present in the FEM region), and the FEM region is continuous (although this may also be relaxed to contain a few specific large-scale fractures simulated by special fracture elements). The key issue is to ensure displacement compatibility at the interface, which connects both the DEM and FEM regions. Different coupling techniques can be applied to formulate the hybrid model. The combined representations are often termed hybrid models and are useful for large-scale practical problems.

The main trend of such hybrid discrete–continuum representations is using DEM–BEM for mechanical or hydro-mechanical analyses, but DEM–FEM solutions have also been developed, mainly for deformation analysis. In this section the techniques of the hybrid discrete–continuum representations are presented briefly to complete the presentation on discrete models and to highlight the advantageous use of them for rock engineering problems.

When dynamic relaxation or individual static relaxation is used for the analysis of the DEM region, no global matrix equation is required. Instead, an iterative procedure is used, following the time-marching process of the relaxation scheme. The determination of the induced contact forces in the DEM region permits direct determination of the induced nodal forces of the interfaces with the FEM region, and this, in turn, is used to update the residual force vectors used for the FEM analysis, resulting in the induced nodal displacement vector on the interface. The interface displacement is then fed into the DEM analysis to produce incremental force vectors, and the nested iterative loops continues until the process converges. An example of such a hybrid model was given by Pan and Reed (1991), using the static relaxation technique proposed by Stewart (1981) and further developed in Pan (1988), with

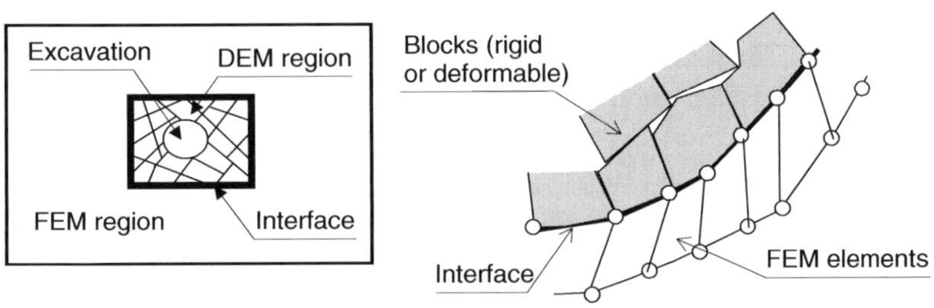

Fig. 8.44 Hybrid DEM–FEM model and interface representation.

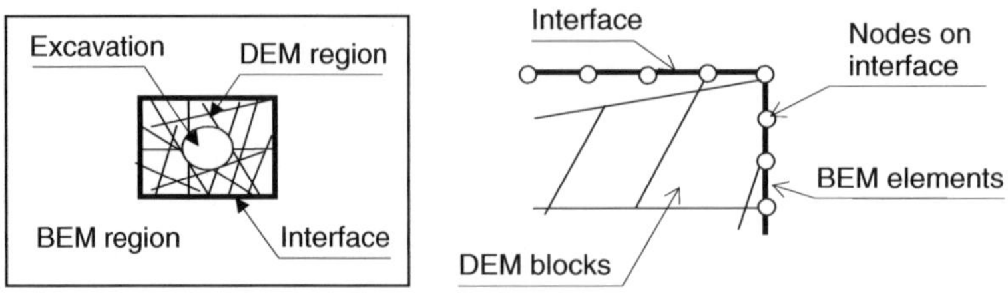

Fig. 8.45 Hybrid DEM–BEM model and interface representation.

implementation of Goodman's model for fracture behavior (Goodman, 1976) and with the assumption of rigid blocks. The FEM technique is a 2D plane-strain FEM formulation using isoparametric 8-noded elements with large-scale deformation analysis capabilities based on the updated Lagrangian formulation of elasto-viscoplastic material behavior developed in Owen and Hinton (1980). It should be noted that eight-noded isoparametric FEM elements are not compatible with the rigid block assumption of the DEM region in Pan and Reed's work, since the curved boundaries of the FEM elements along the interface conflict with the straight edges of the rigid blocks on the DEM side. In addition, there are also difficulties in contact detection when curved boundaries are used. Using bi-linear quadrilateral elements will overcome this difficulty, but more elements may be required in the FEM region. Similar work was also performed by Dowding et al. (1983).

The main trend in the hybrid models is, however, the DEM-BEM approach with a DEM region embedded in a larger (or infinite or half space) elastic domain, because the far-field rock can be assumed to be linearly elastic if the DEM region surrounding the excavation is large enough (Fig. 8.45). Much work has been performed in this direction, notably that by Lorig and Brady (1982) and Lorig et al. (1986) for mechanical analysis, and Wei (1992) and Wei and Hudson (1998) for hydraulic and hydro-mechanical analysis. The coupling formulation is implemented in the DEM code group UDEC and 3DEC (Itasca, 1993, 1994).

The advantage of DEM–BEM coupling is the fact that only the interface between the DEM and BEM regions needs to be discretized with BEM elements. The conditions to be satisfied on the interface are the displacement continuity and stress equilibrium, without separation and slipping between the two regions.

Details of the hybrid DEM–FEM or DEM–BEM will not be given in this book since no new insights into the DEM models can be found through such hybrid models.

8.14 An Example of Comparative Modeling Using the FEM and DEM

A full-scale model of part of a cloister façade at the Sao Vicente de Fora Monastery in Lisbon (Fig. 8.46a) was constructed (Fig. 8.46b) and tested at the ELSA Laboratory of the Joint Research Center (Fig. 8.46c), featuring three stone block columns, two complete arches and two half arches, with stone blocks connected by mortar. The upper part of the model is made of masonry and held by pre-tensioned bars. Vertical loads were applied to the pillars and panels to simulate the action of the missing upper floors (Fig. 8.46c).

Two types of numerical models were used to simulate the full model test: the FEM with the codes ABAQUS and CASTEM 2000; and the DEM with the code UDEC. The approach using smeared cracks was adopted by the ABAQUS model. The isoparametric elements for mortar joint, stone blocks and bricks were used by the CASTEM 2000 model, and deformable blocks/bricks were adopted in the UDEC model. The model geometries are shown in Fig. 8.47. It was found that the DEM approach overcame the numerical difficulties encountered in the FEM approach in terms of equivalent material properties required by the ABAQUS model and the large number of joint elements in a FEM mesh in the CASTEM 2000 model, especially as regards compatibility between the brick/block matrix and mortar joints which

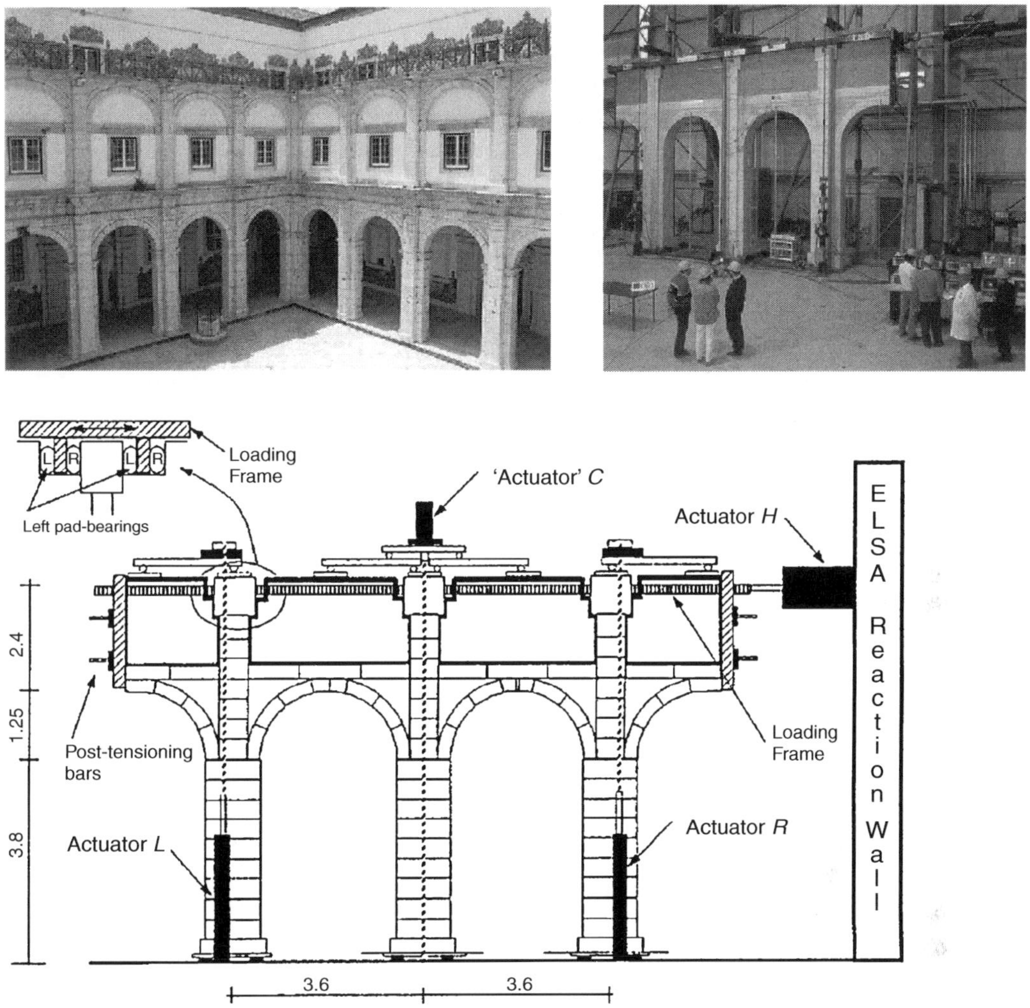

Fig. 8.46 Physical model test of the historical masonry structure of Sao Vicente de Fora Monastery (Giordano et al., 2002). The Cloister façade (top-left), the site for the model test (top-right) and the test setup (below).

had to be enforced, due to the fact that the DEM model can more readily handle the large number of deformable brick/blocks and mortar joints with a simple Mohr–Coulomb elasto-plastic constitutive model with three parameters: normal and shear stiffness and a friction angle. The measured and numerically simulated vertical loads plotted against displacements are shown in Fig. 8.48, with the similar results illustrating the equal applicability of both the equivalent continuum and discrete numerical modeling approaches for such structures.

8.15 Summary Remarks

Due mainly to its conceptual attraction for the explicit representation of fractures, the distinct element method has been enjoying a wide application ranges in rock engineering. A large quantity of associated publications has been published, in both journal papers and conference proceedings, these becoming an

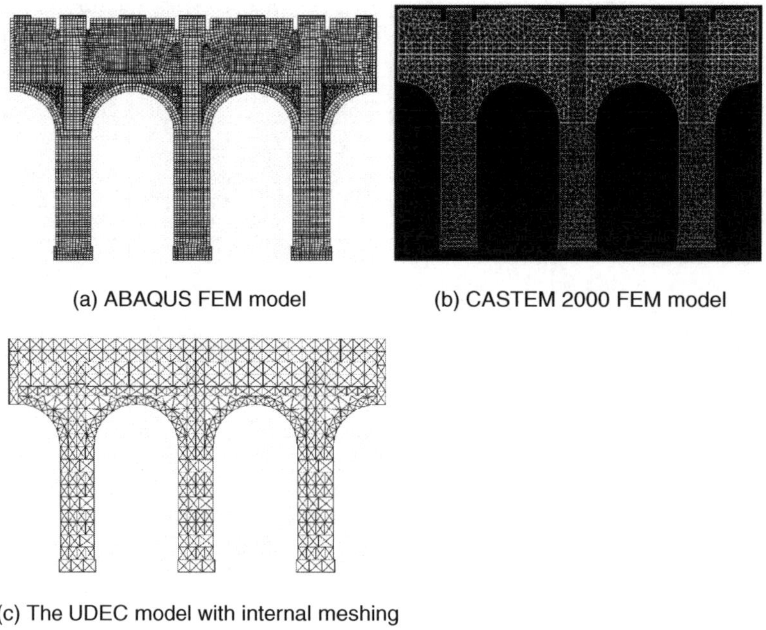

Fig. 8.47 An example of numerical modeling of historical masonry structure using FEM and DEM. (a) The FEM model using ABAQUS code; (b) The FEM model using CASTEM 2000 code; (c) UDEC model with internal meshing (Giordano et al., 2002).

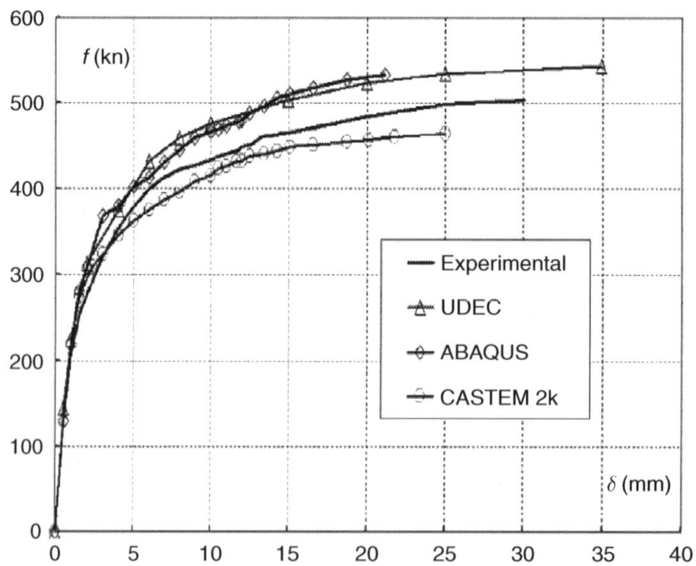

Fig. 8.48 Comparison of the measured and modeled load-displacements curves at one of the monitoring points (Giodano et al., 2002).

important subject in engineering education. It is not practical to list all relevant literature, even at a moderate level, in this book. Therefore a few representative references, mainly in international journals, are given here to demonstrate the wide range of the applicability of the method. A few of the references are presented in more detail in Chapter 12 as case studies of particular interest:

- Tunneling, underground excavations and mining: Barton (1991), Jing and Stephansson (1991), Hanssen et al. (1993), McNearny and Abel (1993), Lorig et al. (1995), Nordlund et al. (1995), Bhasin et al. (1996), Kochen and Andrade (1997), Chryssanthakis et al. (1997), Souley et al. (1997a,b), Shen and Barton (1997), Sofianos and Kapenis (1998), Bhasin and Høeg (1998), Diederichs and Kaiser (1999), Senseny and Pučik (1999), Dowding et al. (2000), Konietzky et al. (2001), Monsen and Barton (2001), Nomikos et al. (2002a,b), Kamata and Mashimo (2003), Sapigni et al. (2003);
- Rock dynamics and blasting: Kim et al. (1997a,b), Ma and Brady (1999), Zhao et al. (1999), Chen et al. (2000) and Cai and Zhao (2000);
- Nuclear waste repository design and performance assessment: Chan et al. (1995), Hansson et al. (1995), Jing et al. (1995, 1997), Hökmark (1998), Rejeb and Bruel (2001), Hutri and Antikainen (2002), Min et al. (2005a);
- Reservoir simulations: Brignoli et al. (1997), Gutierrez and Makurat (1997);
- Geophysical investigations: Harper and Last (1989, 1990a,b), Jing (1990), Hu et al. (1997), Homberg et al. (1997), Su and Stephansson (1999), Hu et al. (2001), Su (2004);
- Rock and soil slopes and landslides: Kim et al. (1997a,b), Allison and Kimber (1998), Zhu et al. (1999), Zhang et al. (2001), Eberhardt et al. (2004);
- Laboratory test simulations and constitutive model development for hard rocks: Jing et al. (1993, 1994a,b), Makurat et al. (1995), Lanaro et al. (1997);
- Stress-flow coupling: Liao and Hencher (1997);
- Hard rock reinforcement: Lorig (1985);
- Well and borehole stability: Santarelli et al. (1992), Rawlings et al. (1993), Zhang et al. (1999);
- Historical structures: Psycharis et al. (2000), Giordano et al. (2002), Jiang and Esaki (2002), Papantonopoulos et al. (2002);
- Hydropower and road structures: Zhang et al. (1997), Hashash et al. (2002), Huang et al. (2003);
- Structural geology: Zhang and Sanderson (1996), Pascal (2002), Finch et al. (2003);
- Derivation of equivalent hydro-mechanical properties of fractured rocks: Zhang and Sanderson (1995), Zhang et al. (1996), Min and Jing (2003, 2004), Min et al. (2001; 2004a,b,c, 2005b,c).

A recent book by Sharma et al. (2001) includes reference to a collection of DEM application papers for various aspects of rock engineering. A book was also written by Mohammadi (2003) which presents the principles of DEM and DDA at a fundamental level.

The applications of the distinct element method concentrate mainly on hard rock problems, but the coupled hydro-mechanical behavior of fractured rocks is also an important field because of the dominating effects of the rock fractures in this regard, when explicit representation of fractures is necessary.

For the softer and weaker rocks, equivalent continuum models are more applicable because there is less difference between the deformability of the fractures and the rock matrix.

Compared with the attractiveness of conceptual simplicity, the largely unknown geometry of the hidden rock fractures limits the more wide and in-depth applications of DEM models. The geometry of fracture systems in rock masses cannot be fully known and can only be roughly estimated. The adequacy of the DEM results depends on capturing the rock reality that in turn is therefore dependent on the interpretation of the *in situ* fracture system geometry, which cannot be even moderately validated in practice. Of course, the same problem applies to the continuum models, such as the FEM or FDM, but the requirement for explicit fracture geometry representation in the DEM highlights its potential limitation. The Monte Carlo approach to fracture simulation may assist in reducing the level of uncertainty, paying the price of much increased computational efforts. A prime subject for research, therefore, is establish how to increase the quality of the rock fracture system characterization with more advanced and affordable means, possibly using more reliable and high resolution geophysical exploration techniques. This issue is, however, not specific to the DEM but applies to all numerical approaches for fractured rocks where there is a requirement to include the detailed fracture pattern.

References

Abdaliah, G., Thoraval, A. Sfeir A. and Piguet, J. P. Thermal convection of fluid in fractured media. International Journal of Rock Mechanics and Mining Sciences and Geomechanics Abstracts 1995;32(5):481–490.

Ahola, M. P., Thoraval, A. and Chowdhury, A. H., Distinct element models for the coupled T-H-M processes: theory and implementation. In: Stephansson, O., Jing, L. and Tsang, C.-F. (eds), Coupled thermo-hydro-mechanical processes of fractured media, development in geotechnical engineering, Volume 79, pp. 181–211. Elsevier Science B.V., Rotterdam, 1996.

Allison, R. J. and Kimber, O. G., Modeling failure mechanisms to explain rock slope change along the Isle of Purbek coast, UK. Earth Surface Processing and Landforms, 1998;23(8):731–750.

Barton, N. Modeling jointed rock behaviour and tunnel performance. World Tunnelling, 1991;4(7): 414–416.

Bathe, K.-J. and Wilson, E. L., Numerical methods in finite element analysis. Prentice-Hall, Englewood Cliffs, NJ, 1976.

Bhasin, R. and Høeg, K., Parametric study for a large cavern in jointed rock using a distinct element model (UDEC-BB). International Journal of Rock Mechanics and Mining Sciences, 1998:35(1): 17–29.

Bhasin, R. K., Barton, N., Grimstad, E., Chryssanthankis, P. and Shende, F. P., Comparison of predicted and measured performance of a large cavern in the Himalayas. International Journal of Rock Mechanics and Mining Sciences and Geomechanics Abstracts, 1996;33(6):607–626.

Board, M., UDEC (Universal Distinct Element Code) Version ICGI.5 Vols. 1–3, NUREG/CR-6021, Washington, D.C., Nuclear Regulatory Commission, 1989.

Belytschko, T., An overview of semi-discretization and time integration procedures. In: Belytschko, T. and Hughes, T. J. R., (eds), Computational methods for transient analysis, North Holland, New York, NY, Computational Methods in Mechanics Series, Vol. 1, pp.1–65, 1983.

Brignoli, M., Pellegrino, A., Santarelli, F. J., Musso, G. and Barla, B., Continuous and discontinuous deformation above a compacting reservoir: consequences upon the lateral extension of the subsidence bowl. International Journal of Rock Mechanics and Mining Sciences. 1997;34(3/4):449, Paper no. 39.

Cai, J.G. and Zhao, J., Effects of multiple parallel fractures on apparent attenuation of stress waves in rock mass. International Journal of Rock Mechanics and Mining Sciences. 2000;37(4):661–682.

Cavendish, J. C., Automatic triangulation of arbitrary planar domains for the finite element method. International Journal for Numerical Methods in Engineering, 1974;8(4):679–696.

Cavendish, J. C., Field, D. A. and Frey, W. H., An approach to automatic three-dimensional finite element mesh generation. International Journal for Numerical Methods in Engineering. 1985;21(2):329–347.

Chan, T., Khair, K., Jing, L., Ahola, M., Noorishad, J. and Vuillold, E., International comparison of coupled thermo-hydro-mechanical models of a multiple-fracture bench mark problem: DECOVALEX Phase I, Bench Mark Test 2. International Journal of Rock Mechanics and Mining Sciences and Geomechanics Abstracts, 1995;32(5):435–452.

Chen, W., Research on relaxation numerical methods in rock and solid mechanics and its appplication. Ph.D. Thesis, Institute of Rock and Soil Mechanics, Chinese Academy of Sciences, Wuhan, China, pp. 142, 1998. (in Chinese).

Chen, S. G., Cai, J. G., Zhao, J. and Zhou, Y. X., Discrete element modeling of an underground explosion in a jointed rock mass. Geotechnical and Geological Engineering, 2000;18(2):59–78.

Chen, W., Gu, X. and Ge. X., A new discrete element method and its application in engineering. In: Siniwardane, H. J. and Zaman, M. M. (eds), Proc. of 8th conference on computer methods and advances in geomechanics, pp. 883–887. Balkema, Rotterdam, 1994.

Christiansson, M., TEMP3D: A computer program for determining temperatures around single or arrays of constant or decaying heat sources. Users' guide and manual. Minneapolis, MN. University of Minnesota, Department of Civil and Mineral Engineering, 1979.

Chryssanthakis, P, Barton, N., Lorig, L. and Christiansson, M., Numerical simulation of fibre reinforced shotcrete in a tunnel using the distinct element method. International Journal of Rock Mechanics and Mining Sciences, 1997;34(3–4):590, Paper no. 054.

Cross, H., Analysis of continuous frames by distributing fixed-end moments. Transactions of the American Society of Civil Engineers. 1932;96(1793):1–10.

Cundall, P. A., A computer model for simulating progressive, large scale movements in blocky rock systems. Proc. Int. Symp. Rock Fracture, ISRM, Nancy, Vol. 1, Paper II–8, 1971a.

Cundall, P. A., The measurement and analysis of acceleration in rock slopes. Ph.D. Thesis, Imperial College of Science and Technology, University of London, 1971b.

Cundall, P. A., Rational design of tunnel supports: a computer model for rock mass behaviour using interactive graphics for the input and output of geomaterial data. Technical Report MRD-2-74, Missuri River Division, U.S. Army Corps of Engineers, NTIS Report No. AD/A-001 602, 1974.

Cundall, P. A., BALL – A program to model granular media using the distinct element method. Dames and Moore Advanced Technology Group, London, 1978.

Cundall, P. A., UDEC – A generalized distinct element program for modelling jointed rock. Report PCAR-1-80, Peter Cundall Associates, European Research Office, U.S. Army Corps of Engineers, 1980.

Cundall, P. A., Adaptive density-scaling for time-explicit calculations. Proc. 4th Int. Conf. Num. Meth. in Geomech, Edmonton, May–June, pp. 21–26, 1982.

Cundall, P. A., Formulation of a three-dimensional distinct element model - Part I: A scheme to detect and represent contacts in a system composed of many polyhedral blocks. International Journal of Rock Mechanics and Mining Sciences and Geomechanics Abstracts, 1988;25(3):107–116.

Cundall, P. A. and Board, M., A microcomputer program for modeling large-strain plasticity problems. In: Swoboda, G. (ed.), Numerical methods in geomechanics (Innsbruk, 1988), pp. 2101–2108. A. A. Balkema, Rotterdam, 1988.

Cundall, P. A. and Hart, R. D., Development of generalized 2-D and 3-D distinct element programs for modelling jointed rock. Itasca Consulting Group, Misc. Paper SL-85-1. U.S. Army Corps of Engineers, 1985.

Cundall, P. A. and Strack, O. D. L., A discrete numerical model for granular assemblies. Géotechnique, 1979a;29(1):47–65.

Cundall, P. A. and Strack, O. D. L., The development of constitutive laws for soil using the distinct element method. In: Wittke, W. (ed.), Numerical methods in Geomechanics, Aachen, pp. 289–298. A. A. Balkema, Rotterdam, 1979b.

Cundall, P. A. and Strack, O. D. L., The distinct element method as a tool for research in granular media. Report to NSF concerning grant ENG 76-20711, Part II, Dept. Civ. Engng, University of Minnesota, 1979c.

Cundall, P. A. and Strack, O. D. L., Modelling of microscopic mechanisms in granular material. Proc. US-Japan Sem. New Models Const. Rel. Mech. Gran. Mat., Itasca, 1982.

Cundall, P. A., Marti, J., Beresford, P. J., Last, N. C. and Assgian, M. M., Computer modelling of jointed rock masses. Technical Report N-78-4. U.S. Army Corps of Engineers, Waterways Experiment Station, Vicksburg, Mississipi, 1978.

Day, A. S., An introduction to dynamic relaxation. The Engineer, London, 1965;219(5688):218–221.

Diederichs, M. S. and Kaiser, P. K., Stability of large excavations in laminated rock masses: the voussoir analogue revisited. International Journal of Rock Mechanics and Mining Sciences, 1999;36(1):97–117.

Dowding, C. H., Belytschko, T. B. and Yen, H. J., A coupled finite element – rigid block method for transient analysis of rock caverns. International Journal for Numerical and Analytical Methods in Geomechanics, 1983;7(1):117–127.

Dowding, C. H., Belytschko, T. B. and Dmytryshyn, O. Dynamic response of million block cavern models with parallel processing. Rock Mechanics and Rock Engineering, 2000;33(3):207–214.

Eberhardt, E., Stead, D. and Coggan, J. S., Numerical analysis of initiation and progressive failure in natural rock slopes – the 1991 Randa rock slide. International Journal of Rock Mechanics and Mining Sciences, 2004;41(1):69–87.

Finch, E., Hardy, S. and Gawthorpe, R., Discrete element modeling of contractional fault-propagation folding above rigid basement fault blocks. Journal of Structural Geology, 2003;25(4):515–528.

Fukuda, J. and Suhara, J., Automatic mesh generation for finite element analysis. In: Oden, J. T., Clough, R. W. and Yamamoto, Y. (eds), Adbances in computational methods in structural mechanics and designs, UAH Press, Huntsvile, Alabama, 1972.

Giordano, A., Mele, E., de Luca, A., Modeling historical masonry structures: comparison of different approaches through a case study. Engineering Structures, 2002;24(8):1057–1069.

Goodman, R. E., Methods of geological engineering in discontinuous rocks. West Publishing Company, San Francisco, 1976.

Gutierrez, M. and Makurat, A., Coupled THM modelling of cold water injection in fractured hydrocarbon reservoirs. International Journal of Rock Mechanics and Mining Sciences, 1997;34(3–4):Paper 113.

Hanssen, F.H., Spinnler, L. and Fine, J., A new approach for rock mass cavability. International Journal of Rock Mechanics and Mining Sciences and Geomechanics Abstracts, 1993;30(7):1379–1385.

Hansson, H., Jing, L. and Stephansson, O., 3-D DEM modelling of coupled thermo-mechanical response for a hypothetical nuclear waste repository. Proc. NUMOG V – Int. Symp. on Numerical Models in Geomechanics, Davos, Switzerland, pp. 257–262. Balkema, Rotterdam, 1995.

Harper, T.R. and Last, N.C., Interpretation by numerical modeling of changes of fracture system hydraulic conductivity induced by fluid injection. Géotechnique, 1989;39(1):1–11.

Harper, T.R. and Last, N.C., Response of fractured rocks subjected to fluid injection. Part II: characteristic behaviour. Techtonophysics, 1990a;172(1–2):33–51.

Harper, T.R. and Last, N.C., Response of fractured rocks subjected to fluid injection. Part III: practical applications. Techtonophysics, 1990b;172(1–2):53–65.

Hart, R.D., An introduction to distinct element modelling for rock engineering. In: Hudson, J. A. (ed. in-chief), Comprehensive rock engineering, Vol. 2, pp. 245–261. Pergamon Press, Oxford, 1993.

Hart, R. D., Cundall, P. A. and Lemos, J. V. Formulation of a three-dimensional distinct element method – Part II: Mechanical calculations for motion and interaction of a system composed of many polyhedral blocks. International Journal of Rock Mechanics and Mining Sciences and Geomechanics Abstracts, 1988;25(3):117–125.

Hashash, Y. M. A., Cording, E. J. and Oh, J., Analysis of shearing of a rock ridge, International Journal of Rock Mechanics and Mining Sciences, 2002;39(8):945–957.

Hocking, G., Development and application of the boundary integral and rigid block method for geotechnics. Ph.D. Thesis, University of London, 1977.

Homberg, C., Hu, J. C., Angelier, J., Bergerat, F. and Lacombe, O., Characterization of stress perturbations near major fault zones: insight from 2-D distinct-element numerical modelling and field studies (Jura Mountains). Journal of Structural Geology, 1997;19(5):703–718.

Hu, Y., Block-spring-model considering large displacements and non-linear stress-strain relationships of rock joints. In: Yuan, J.-X. (ed.), Computer methods and advances in geomechanics, Vol. 1. pp. 507–512. Balkema, Rotterdam, 1997.

Hu, J. C., Angelier, J. and Yu, S.-B., An interpretation of the active deformation of southern Taiwan based on numerical simulation and GPS studies. Tectonophysics, 1997;274(1–3):145–169.

Hu, J. C., Angelier, J., Homberg, C., Lee, J.-C., and Chu, H.-T., Three-dimensional modelling of the behavior of the oblique convergent boundary of southeast Taiwan: friction and strain partitioning. Tectonophysics, 2001;333(1–2):261–276.

Huang, Z. Q., Jiang, T., Yue, Z. Q., Lee, C. F. and Wang, S. J., Deformation of the central pier of the permanent ship-lock, Three-Gorge project, China: an analysis case study. International Journal of Rock Mechanics and Mining Sciences, 2003;40(6):877–892.

Hutri, K.-L. and Antikainen, J., Modeling of the bedrock response to glacial loading at the Olkilouto site, Finland. Engineering Geology, 2002;67(1–2):39–49.

Hökmark, H., Numerical study of the performance of tunnel plugs. Engineering Geology, 1998;49(3):327–335.

Itasca Consulting group, Inc., UDEC Manual, 1993.

Itasca Consulting Group, Inc., 3DEC Manual, 1994.

Itasca Consulting Group, Inc., PFC-2D and PFC-3D Manuals, 1995.

Jiang, K. and Esaki, T., Quantitative evaluation of stability changes in historical stone bridges in Kagoshina, Japan, by weathering. Engineering Geology, 2002;63(1–2):83–91.

Jing, L., Numerical modelling of jointed rock masses by distinct element method for two and three-dimensional problems. Ph.D. Thesis, 1990:90 D, Luleå University of Technology, Luleå, Sweden, 1990.

Jing, L and Stephansson,O., Distinct element modelling of sublevel stoping. Proc. 7th Int. Congress of ISRM, Aachen, Germany, Vol.1, pp. 741–746, 1991.

Jing, L., Stephansson, O. and Nordlund, E., Study of rock joints under cyclic loading conditions. Rock Mechanics and Rock Engineering, 1993;26(3):215–232.

Jing, L., Nordlund, E. and Stephansson, O., A 3-D constitutive model for rock joints with anisotropic friction and stress dependency in shear stiffness. International Journal of Rock Mechanics and Mining Sciences and Geomechanics Abstracts, 1994a;31(2):173–178.

Jing, L., Rutqvist, J., Stephansson, O., Tsang, C.-F. and Kautsky, F., DECOVALEX – Mathematical models of coupled T-H-M processes for nuclear waste repositories. Report of Phase II. SKI Report 94:16, Swedish Nuclear Power Inspectorate, Stockholm, Sweden, 1994b.

Jing, L., Tsang, C.-F. and Stephansson, O., DECOVALEX – an international co-operative research project on mathematical models of coupled THM processes for safety analysis of radioactive waste repositories. International Journal of Rock Mechanics and Mining Sciences and Geomechanics Abstracts, 1995;32(5):389–398.

Jing, L., Tsang, C.-F., Stephansson, O. and Kautsky, F., Validation of mathematical models against experiments for radioactive waste repositories-DECOVALEX experience. In: Stephansson, O. Jing, L. and Tsang, C.-F. (eds), Coupled thermo-hydro-mechanical processes of fractured media. Developments in geotechnical engineering, Vol.79, pp. 25–56. Elsevier, Rotterdam, 1996.

Jing, L., Hansson, H., Stephansosn, O. and Shen, B., 3D DEM study of thermo-mechanical responses of a nuclear waste repository in fractured rocks – far and near-field problems. In: Yuan, J.-X. (ed.), Computer Methods and Advances in Geomechanics, Vol. 2, pp.1207–1214. Balkema, Rotterdam, 1997.

Kamata, H. and Mashimo, H., Centrifuge model test of tunnel face reinforcement by bolting. Tunneling and Underground Space Technology, 2003;18(2–3):205–212.

Kamel, H. A. and Eisenstein, K. Automatic mesh generation in two and three-dimensional inter-connected domains. Les Congré et Colloques de l'Université de Liège, 1971.

Kawai, T., New discrete structural models and generalization of the method of limit analysis. Proc. Int. Conf. on Finite Elements in Nonlinear Solid and Structural Mechanics, Norway, Vol. 2, G04.1–G04.20, 1977a.

Kawai, T., New element models in discrete structural analysis. Japan Society of Naval Architects, Japan, 1977b;141:174–180.

Kawai, T., Collapse load analysis of engineering structures by using new discrete element models. IABSE Colloquium, Copenhagen, 1979.

Kawai, T., Kawabata, K. Y., Kondou, I. and Kumagai, K., A new discrete model for analysis of solid mechanics problems. Proc. 1st Conf. numerical methods in fracture mechanics, pp. 26–27. Swansea, UK, 1978.

Kim, M. K., Kim, S. E., Oh, K. H. and Kim, W. J., A study on the behavior of rock mass subjected to blasting using modified discrete element method. International Journal of Rock Mechanics and Mining Sciences, 1997a;34(3–4), p. 629, Paper no. 156.

Kim, J. S., Kim, J. Y. and Lee, S. R., Analysis of soil nailed slope by discrete element method. Computers and Geotechnics, 1997b;20(1):1–14.

Knupp, P. and Steinberg, S., Fundamentals of grid generation. CRC Press, Boca Raton, 1993.

Kochen, R. and Andrade, C., Predicted behaviour of a subway station in weathered rock. International Journal of Rock Mechanics and Mining Sciences, 1997;34(3–4):587.

Konietzky, H., te Kamp, L., Hammer, H., Niedermeyer, S. Numerical modeling of in-situ stress conditions as an aid in route selection for rail tunnels in complex geological formations in South Germany. Computers and Geotechnics, 2001;28(67):495–516.

Koyama, T., Numerical modelling of fluid flow and particle transport in rock fractures during shear. Licentiate Thesis, Royal Institute of Technology, Stockholm, Sweden, 2005

Koyama, T., Fardin, N. and Jing, L., Shear induced anisotropy and heterogeneity of fluid flow in a single rock fracture with translational and rotary shear displacements – a numerical study, International Journal of Rock Mechanics and Mining Sciences, 2004;41(3):426, Paper No. 2A08.

Koyama, T., Fardin, N., Jing, L. and Stephansson, O., Numerical simulation of shear induced flow anisotropy and scale dependent aperture and transmissivity evolutions of fracture replicas. International Journal of Rock Mechanics and Mining Sciences, 2006;43(1):89–106.

Lanaro, F., Jing, L., Stephansson, O. and Barla, G., DEM modelling of laboratory tests of Block toppling. International Journal of Rock Mechanics and Mining Sciences, 1997;34 (3–4):506–507.

Lemos, J. V., A hybrid distinct element computational model for the half-plane. M.Sc Thesis, Department of Civil Engineering, University of Minnesota, 1983.

Li, G. and Vance, J., A 3-D block-spring model for simulating the behaviour of jointed rocks. In: Amadei, Kranz, Scott and Smeallie (eds), Rock mechanics for industry, Vol. 1, pp. 141–146. Balkema, Rotterdam, 1999.

Li, G. and Wang, B., Development of a 3-D block-spring model for jointed rocks. In: Rossmanith, H. P. (ed.), Mechanics of jointed and faulted rock, pp. 305–309. Balkema, Rotterdam, 1998.

Liao, Q. H. and Hencher, S. R., Numerical modeling of the hydro-mechanical behaviour of fractured rock masses. International Journal of Rock Mechanics and Mining Sciences, 1997;34(3–4):Paper 177.

Lo, S. H. A new mesh generation scheme for arbitrary planar domains. International Journal for Numerical Methods in Engineering, 1985;21:1403–1426.

Lo, S. H., Delaunay triangulation of non-convex planar domains. International Journal for Numerical Methods in Engineering, 1989;28(11):2695–2707.

Lo, S. H., Volume discretization into tetrahedra – I. Verification and orientation of boundary surfaces. Computers and Structures, 1991a;39(5):493–500.

Lo, S. H., Volume discretization into tetrahedra – II. 3D triangulation by advancing front approach. Computers and Structures, 1991b;39(5):501–511.

Lorig, L., A simple numerical representation of fully bounded passive rock reinforcement for hard rock. Computers and Geotechnics, 1985;1(2):79–97.

Lorig, L. and Brady, B. H. G., A hybrid discrete element-boundary element method of stress analysis. In: Issues in rock mechanics, Proc. 23rd US Symp. on Rock Mech., Berkeley, pp. 628–636, 1982.

Lorig, L., Brady, B.H.G. and Cundall, P. A., Hybrid distinct element-boundary element analysis of jointed rock. International Journal of Rock Mechanics and Mining Sciences and Geomechanics Abstracts, 1986;23(4):303–312.

Lorig, L., Gibson, W., Alvial, J. and Cuevas, J., Gravity flow simulations with the particle flow code (PFC). ISRM News Journal, 1995;3(1):18–24.

Ma, M. and Brady, B. H., Analysis of the dynamic performance of an underground excavation in jointed rock under repeated seismic loading. Geotechnical and Geological Engineering, 1999;17(1):1–20.

Makurat, A., Ahola, M., Khair, K., Noorishad, J., Rosengren, L. and Rutqvist, J., The DECOVALEX test case one. International Journal of Rock Mechanics and Mining Sciences and Geomechanics Abstracts, 1995;32(5):399–408.

Marti, J. and Cundall, P. A., Mixed discretization procedure for accurate modeling of plastic collapse. International Journal for Numerical and Analytical Methods in Geomechanics, 1982;6(1):129–139.

McNearny, R.L. and Abel, J.F., Large-scale two-dimensional block caving model tests. International Journal of Rock Mechanics and Mining Sciences and Geomechanics Abstracts, 1993;30(2):93–109.

Min, K.-B. and Jing, L., Numerical determination of the equivalent elastic compliance tensor for fractured rock masses using the distinct element method. International Journal of Rock Mechanics and Mining Sciences, 2003;40(6):795–816.

Min, K.-B. and Jing, L., Stress-dependent mechanical properties and bounds of Poisson's ratio for fractured rock masses investigated by a DFN-DEM technique. International Journal of Rock Mechanics and Mining Sciences, 2004;41(Suppl 1):Paper 2A 13.

Min, K. B., Mas-Ivars, D. and Jing, L., Numerical derivation of the equivalent hydro-mechanical properties of fractured rock masses using distinct element method. In: Elsworth, D., Tinucci, J. P. and Heasley, P. E. (eds), Rock mechanics in the national interest, pp.1469–1476. Swets and Zeitlinger Lisse, 2001.

Min, K.-B., Jing, L. and Stephansson, O., Fracture system characterization and evaluation of the equivalent permeability tensor of fractured rock masses using a stochastic REV approach. International Journal of Hydrogeology, 2004a;12(5):497–510.

Min, K.-B., Rutqvist, J., Tsang, C.-F. and Jing, L., Stress-dependent permeability of fractured rock masses: a numerical study, International Journal of Rock Mechanics and Mining Sciences, 2004b;41(7):1191–1210.

Min, K.-B., Rutqvist, J., Jing, L. and Tsang, C.-F., A block-scale stress-permeability relationship of fractured rock by numerical experiments. In: Stephansson, O., Hudson, J. A. and Jing, L. (eds), Coupled thermo-hydro-mechanical-chemical processes in geo-systems-fundamentals, modeling, experiments and applications. pp. 269–274. Elsevier, Amsterdam, 2004c.

Min, K.-B., Rutqvist, J., Tsang, C.-F. and Jing L., Thermo-mechanical impacts on performance of a nuclear waste repository in fractured rock masses – a far-field study using an equivalent continuum approach International Journal of Rock Mechanics and Mining Sciences, 2005a;42(5–6):765–780.

Min, K.-B., Jing, L., Rutqvist, J., Tsang, C.-F. and Stephansson, O., Representation of fractured rock masses as equivalent continua using a DFN-DEM approach. Proc. Int. Conf. Int. Ass. Comp. Modell. Anal. Geomech. (IACMAG), Vol. 2, pp. 531–536. Turin, Italy, 2005b.

Min, K.-B., Stephansson, O. and Jing, L., Effect of stress on mechanical and hydraulic rock mass properties – application of DFN-DEM approach for the site investigation at Forsmark, Sweden. Int. Symp. Rock Mech. in Europe (EUROCK05), pp. 389–395. Brno, Czech Republic, 2005c

Mohammadi, S., Discontinuum mechanics using finite and discrete elements. WIT Press, Southampton, 2003.

Monsen, K. and Barton, N., A numerical study of cryogenic storage in underground excavations with emphasis on the rock joint response. International Journal of Rock Mechanics and Mining Sciences, 2001;38(7):1035–1045.

Mustoe, G. G. M., A generalized formulation of the discrete element method, Engineering Computations, 1992;9(2):181–190.

Neretnieks, I., Migration of radionuclides in fissured rock: The influence of micropore diffusion and longitudinal dispersion. Journal of Geophysical Research, 1981;86(B5):3749–3758.

Nomikos, P. P., Sofianos, A. I. and Tsoutrelis, C. E., Symmetric wedge in the roof of a tunnel excavated in an inclined stress field . International Journal of Rock Mechanics and Mining Sciences, 2002a;39(1):59–67.

Nomikos, P. P., Sofianos, A. I. and Tsoutrelis, C. E., Structural response of vertically multi-jointed roof rock beams. International Journal of Rock Mechanics and Mining Sciences, 2002b;39(1):79–94.

Nordlund, E., Rådberg, G. and Jing, L., Determination of failure modes in jointed pillars by numerical modelling. In: Myer, L. R., Cook, N. G. W., Goodman, R. E. and Tsang, C.-F. (eds), Fractured and jointed rock masses, pp. 345–350. Balkema, Rotterdam, 1995.

Otter, J. R. H., Cassell, A. C. and Hobbs, R. E., Dynamic relaxation. Proceedings of Institute of Civil Engineers, 1966;35:633–635.

Owen, D. R. J. and Hinton, Finite element in plasticity. Pineridge Press, Swansea, 1980.

Ozgenoglu, A., The analysis of toppling failure using models and numerical methods. M.Phil. Thesis, University of London, 1978.

Pan, X. D., Numerical modelling of rock movements around mine openings. Ph.D. Thesis, Imperial College, University of London, 1988.

Pan, X. D. and Reed, M. B. A coupled distinct element – finite element method for large deformation analysis of rock masses. International Journal of Rock Mechanics and Mining Sciences and Geomechanics Abstracts, 1991;28(1):93–99.

Patankar, S. V., Numerical heat transfer and fluid flow. Hemisphere, Washington, D.C., 1980.

Papantonopoulos, C., Psycharis, I. N., Papastamatiou, D. Y., Lemos, J. V. and Mouzakis, H. P., Numerical prediction of the earthquake response of classical columns using distinct element method. Earthquake Engineering and Structural Dynamics, 2002;31(9):1699–1717.

Parekh, C. J., Dynamic relaxation solution of seepage problems. Proc. of ASCE Engineering Fundation Conference on Numerical Methods in Geomechanics, pp.1133–1144. Virginia, 1976.

Park, S. and Washam, C. J., Drag method as a finite element mesh generation scheme. Computers and Structures. 1979;10(1–2):343–346.

Pascal, C., Interaction of faults and perturbation of slip: influence of anisotropic stress state in the presence of fault friction and comparison between Wallece-Bott and 3D distinct element models. Tectonophysics, 2002;356(4):307–322.

Peraier, J., Ppeiro, J., Formaggia, L., Morgan, K. and Zienkiewicz, O., Finite element Euler computations in three dimensions, International Journal of Numerical Methods in Engineering, 1988;26(10): 2135–2159.

Psycharis, I. N., Papastamatiou, D. Y., and Alexandris, A. P., Parametric investigation of the stability of classical columns under harmonic and earthquake excitations. Earthquake Engineering and Structural Dynamics, 2000;29(8):1093–1109.

Rawlings, C. G., Barton, N. R., Bandis, S. C., Addis, M. A. and Gutierrez, M. S., Laboratory and numerical discontinuum modeling of wellbore stability. Journal of Petroleum Technology, 1993; 45(11):1086–1092.

Rejeb, A. and Bruel, D., Hydromechanical effects of shaft sinking at the Sellafield site. International Journal of Rock Mechanics and Mining Sciences, 2001;38(1):17–29.

Rutqvist, J., Börgesson, L., Chijimatsu, M., Nguyen, T. S., Jing, L., Noorishad, J. and Tsang, C.-F., Coupled thermo-hydro-mechanical analysis of a heater test in fractured rock and bentonite at Kamaishi Mine-comparison of field results to predictions of four finite element codes. International Journal of Rock Mechanics and Mining Sciences, 2001;38(1):129–142.

Santarelli, F.J., Dahen, D., Baroudi, H. and Sliman, K.B., Mechanisms of borehole instability in heavily fractured rock. International Journal of Rock Mechanics and Mining Sciences and Geomechanics Abstracts, 1992;29(5):457–467.

Sapigni, M., La Barbera, G. and Ghirotti, M., Engineering geological characterization and comparison of predicted and measured deformations of a cavern in Italian Alps. Engineering Geology, 2003; 69(1–2):47–62.

Schroeder, W. J. and Shephard, M. S. Geometry-based fully automatic mesh generation and the Delaunay triangulation. International Journal for Numerical Methods in Engineering. 1988; 26(11):2503–2515.

Senseny, P. E. and Pučik, T. A., Development and validation of computer models for structures in jointed rock. International Journal of Numerical and Analytical Methods in Geomechanics, 1999;23(8): 751–778.

Sharma, V.M., Saxena, K.R. and Woods, R.D. (eds), Distinct element modelling in geomechanics. A. A. Balkema, Rotterdam, 2001.

Shaw, R. D. and Pitchen, R. G., Modification to the Suhara-Fukuda method of network generation. International Journal for Numerical Methods in Engineering, 1978;12(1):93–99.

Shen, B. and Barton, N., The disturbed zone around tunnels in jointed rock masses. International Journal of Rock Mechanics and Mining Sciences, 1997;34(1):117–125.

Sofianos, A.I. and Kapenis, A.P., Numerical evaluation of the response in bending of an underground hard rock voissoir beam roof. International Journal of Rock Mechanics and Mining Sciences, 1998;35(8):1071–1086.

Souley, M., Hommand, F. and Thoraval, A., The effect of joint constitutive laws on the modelling of an underground excavation and comparison with in-situ measurements. International Journal of Rock Mechanics and Mining Sciences, 1997a;34(1):97–115.

Souley, M., Hoxha, D. and Hommand, F., The distinct element modelling of an underground excavation using a continuum damage model. International Journal of Rock Mechanics and Mining Sciences, 1997b;35(4–5):Paper 026.

Southwell, R. V., Stress calculation in frameworks by the method of systematic relaxation of constraints. Parts I and II. Proceedings of the Royal Society. A, 1935;151:p.56–95.

Southwell, R. V., Relaxation methods in engineering sciences – a treatise on approximate computation, Oxford University Press, London, 1940.

Southwell, R. V., Relaxation methods in theoretical physics – a continuation of the treatise: relaxation methods in engineering science. Vol. I and II, Oxford University Press, London 1956.

St. John, C. M., Thermal analysis of spent nuclear fuel disposal in vertical displacement boreholes in a welded tuff repository. SAND84-7207, Albuquerque, NM. Sandia National Laboratories, 1985.

Stewart, I. J., Numerical and physical modeling of underground excavations in discontinuous rock. Ph.D. Thesis, University of London (Imperial College), 1981.

Su, S., Effect of fractures on in situ rock stresses studied by the distinct element method. International Journal of Rock Mechanics and Mining Sciences, 2004;41:159–164.

Su, S. and Stephansson, O., Effect of a fault on in situ stresses studied by the distinct element method. International Journal of Rock Mechanics and Mining Sciences, 1999;36(8):1051–1056.

Taylor, L. M., BLOCKS – a block motion code for geomechanics studies. SANDIA Report, SAND82-2373, DE83 009221. Sandia National Laboratories, Albuquerque, New Mexico, 1982.

Tsang, C.-F., Coupled thermomechanical and hydrochemical processes in rock fractures. Review of Geophysics, 1991;29(4):537–551.

Tsang, C.-F., Jing L., Stephansson, O. and Kautsky, F., Numerical modeling for the coupled THMC processes in underground nuclear waste disposal systems. International Journal of Rock Mechanics and Mining Sciences, 2005;42(5–6):563–610.

Unterberger, W., Cundall, P. A. and Zettler, A. H., Dynamic substepping – increasing the power of explicit finite difference modeling. In: Yuan, J.-X. (ed.), Computer Methods and Advances in Geomechanics, pp. 409–412. Balkema, Rotterdam, 1997.

Wang, B. and Garga, V.K., A numerical method for modeling large displacements of jointed rocks – Part I: Fundamentals. Canadian Geotechnical Journal, 1993;30(1):96–108.

Watson, D. F., Computing the n-dimensional Delaunay tessellation with application to Voronoi polytopes. Computer Journal, 1981;24(2):167–172.

Wei, L. Numerical studies of the hydro-mechanical behaviour of jointed rocks. Ph.D. Thesis, Imperial College of Science, Technology and Medicine, University of London, 1992.

Wei, L. and Hudson, J. A., A hybrid discrete-continuum approach to model hydro-mechanical behaviour of jointed rocks. Engineering Geology, 1998;49(3–4):317–325.

Wilkins, M. L., Calculation of elastic-plastic flow. UCRL-7332, Rev. I, TID-4500, UC-34, Physics. Lawrence Radiation Laboratory, University of California, Livermore, 1963.

Williams, J. R. and Mustoe, G. G. M., Modal methods for the analysis of discrete systems. Computers and Geotechnics, 1987;4(1):1–19.

Williams, J. R., Hockings, G. and Mustoe, G. G. M., The theoretical basis of the discrete element method. Proc. of the NUMETA'86 Conf., Swansea, 7–11 January, pp. 897–906, 1986.

Yasuhara, H. and Elsworth, D., Evolution of permeability in a natural fracture: Significant role of pressure solution. Journal of Geophysical Research, 2004;109(B3):B03204.

Yeo, I. W., De Freitas, M. H. and Zimmerman, R. W., Effect of shear displacement on the aperture and permeability of a rock fracture, International Journal of Rock Mechanics and Mining Sciences, 1998;35(8):1051–1070

Zhang, X. and Sanderson, D. J., Anisotropic features of geometry and permeability in fractured rock masses. Engineering Geology, 1995;40(1–2):65–75.

Zhang, X. and Sanderson, D. J., Numerical modeling of the effects of fault slip on fluid flow around extensional faults. Journal of Structural Geology, 1996;18(1):109–119.

Zhang, X., Sanderson, D.J., Harkness, R.M. and Last, N.C., Evaluation of the 2-D permeability tensor for fractured rock masses. International Journal of Rock Mechanics and Mining Sciences and Geomechanics Abstracts, 1996;33(1):17–37.

Zhang, C., Pekau, O. A., Feng, J. And Wang, G., Application of distinct element method in dynamic analysis of high rock slopes and blocky structures. Soil Dynamics and Earthquake Engineering, 1997;16(6):385–394.

Zhang, X., Last, N., Powrie, W., Harkness, R., Numerical modeling of wellbore behavior in fractured rock masses. Journal of Petroleum Science and Engineering, 1999;23(2):95–115.

Zhang, J.-H., He, J.-D. and Fan, J.-W., Static and dynamic stability assessment of slopes or dam foundations using a rigid body-spring element method. International Journal of Rock Mechanics and Mining Sciences, 2001;38(8):1081–1090.

Zhao, J., Zhou, Y.X., Hefny, A.M., Cai, J.G., Chen, S.G., Li, H.B., Liu, J.F., Jain, M., Foo, S.T. and Seah, C.C., Rock dynamics research related to cavern development for ammunition storage. Tunnelling and Underground Space Technology, 1999;14(4):513–526.

Zhu, W., Zhang, Q. and Jing, L., Stability analysis of the ship-lock slopes of the Three-Gorge project by three-dimensional FEM and DEM techniques. In: Amadei, B. (ed.), Proc. of the 3rd Int. Conf. of discontinuous deformation Analysis (ICADD-3). Vail, USA, pp. 263–272. American Rock Mechanics Association (ARMA), Alexandria, USA, 1999.

Zienkiewicz, O. and Taylor, R., The finite element method, Volume 1, The Basis, 5th edition, Butterworth-Heinemann, Oxford, 2000.

9 IMPLICIT DISCRETE ELEMENT METHOD FOR BLOCK SYSTEMS – DISCONTINUOUS DEFORMATION ANALYSIS (DDA)

The formulation of an implicit DEM is similar to, or parallel with, that of the FEM. Both methods use displacements as the basic variables of unknowns, and the energy minimization principle is used to derive the equation of motion of the system in a matrix form. The fundamental difference between the implicit DEM and FEM is the different treatment of material discontinuities. The FEM assumes that the overall computational domain is a single continuous body, and the interelement relation is the displacement compatibility condition along the boundaries of adjacent elements. The discontinuities (e.g., fractures in rocks) are treated as element boundaries, represented as special fracture elements. These fracture elements undergo deformations both along and across the fractures with the condition that these deformations should not exceed the overall magnitude of deformation of their continum neibourhoods so that continuous deformation assumption is not violated. Complete detachment of individual elements or opening-up of fracture elements is not allowed. Therefore the connectivity of the elements in FEM is fixed throughout the computation. The adjacent elements should have the same order of magnitude of displacements along their common boundaries so the continuum assumption is satisfied. The DEM, on the other hand, assumes that the overall problem domain is discontinuous and comprised of a finite number of discrete blocks as the basic components, and the interblock relations are via the mechanical contacts between neighboring, but independent, blocks. The coefficient matrix in the DEM is composed of contact stiffness sub-matrices between blocks as mechanical contacts, with the addition of deformational stiffness sub-matrices for intact blocks if deformable blocks are considered. An internal discretization scheme can be used to discretize the discrete blocks, using standard FEM techniques. The implicit DEM is therefore parallel with the FEM in formulation, but can be made more general.

Historically the development of DDA originated from a back analysis to determine the best fit to the deformed configuration of a block system from measured displacements and deformations (Shi and Goodman, 1985), with a coupling between the FEM and a rigid block system. The work was further developed, leading to the complete deformation analysis of a block system called Discontinuous Deformation Analysis (DDA) (Shi, 1988). The early formulation of the DDA technique concentrated on simply deformable blocks, which coupled a constant strain tensor with the rigid body movement mode of an arbitrarily shaped block. Fully deformable blocks with internal discretization using a finite element mesh of triangular or quadrilateral elements were developed by Shyu (1993), so that the deformability of arbitrarily shaped blocks could be more rigorously represented. More strict contact formulation regarding frictional effects was reported in Jing (1993, 1998) and non-linear material behavior was considered in Chang (1994). The method was also extended to enable a more user-friendly coding structure and environment (Chen et al., 1996; Doolin and Sitar, 2001), to cover problems of rigid block systems (Koo and Chern, 1998; Cheng and Zhang, 2000), fractured rock masses (Lin et al., 1996) and coupled flow-stress processes (Ma, 1999; Kim et al., 1999; Jing et al., 2001) with fluid conducted only in fractures, as evaluated in Cheng (1998) and Jing (2003). The extension to 3D problems (Shi, 2001; Yeung et al., 2003; Jiang and Yeung, 2004), the use of higher order elements (Hsiung, 2001) and the use of more comprehensive fracture models (Zhang and Lu, 1998) have also been reported. The applications focus mainly on slopes, tunneling, caverns, mining, fracturing and fragmentation processes of geological and structural materials, masonry structures, earthquake effects and powder transportation (Yeung, 1993; Chang et al., 1996; Yeung & Leong, 1997; Hatzor & Benary, 1998; Chiou et al., 1998, 1999; Pearce et al., 2000; Cai et al., 2000; MacLauphlin et al., 2001; Mortazavi and Katsabanis, 1998, 2000, 2001; Hatzor and Feintuch, 2001; Thavalingam et al., 2001; Hsiung and Shi, 2001; Soto-Yarritu and De Andres Martinez, 2001; Kong and Liu, 2002). A comprehensive review is given in Ohnishi et al. (2006) which summarized key numerical issues addressed in DDA, parameters used and how they are determined, and treatment of damping and time-step issues.

The fundamentals of the DDA approach are presented in this chapter in some detail, covering basically the rigid block, triangular element and quadrilateral element representations. The formulation of simply deformable blocks, originally developed by Shi (1988, 1992a) and Shi and Goodman (1985, 1989), has already been fully presented in a book (Shi, 1993) and will not be presented again here. The block generation algorithm is similar to that introduced in Chapter 7 for the 2D problem and will also not be presented.

9.1 Energy Minimization and Global Equilibrium Equations

By the second law of thermodynamics, a mechanical system under loading (external and/or internal) must move or deform in a direction which results in the minimum total energy of the whole system. The total energy consists of the potential energy due to external loads, system constraints and internal deformations (strain energy) of the bodies, the kinetic energy due to block mass and the energy absorbed by the system (dissipated irreversible energy in the system, energy dissipated through friction and heat generation, for example). The minimization of the system energy will produce an equation of motion of the system. This is the so-called energy minimization principle, as used in FEM.

Let $\mathbf{U}_i$ be the potential energy due to different deformation mechanisms (external loads, strain energy, etc.), $\mathbf{K}$ be the kinetic energy and $\mathbf{W}$ be the dissipated energy in the system, then the total energy, Π, is given as the sum

$$\prod = \sum(\mathbf{U}_i) + \mathbf{K} + \mathbf{W} \tag{9.1}$$

The minimization of the total energy is performed by first order differentiation with respect to the displacement vector, $\mathbf{d} = \{d\}$, written as

$$\frac{\partial \Pi}{\partial \mathbf{d}} = \left[\sum \partial(\mathbf{U}_i) + \partial \mathbf{K} + \partial \mathbf{W}\right] / \partial\{\mathbf{d}\} = 0 \tag{9.2}$$

Equation (9.2) will yield a weak form of the equilibrium equation describing the motion and/or deformation of the block system. The differentiation can be carried out separately for individual energy mechanisms, thus producing local equations of motion due to these individual mechanisms.

The first step in energy minimization is to define the energy functional, Π, as a function of the nodal displacement vector $\mathbf{d}_i$ of a block or element i, $\Pi = F(\mathbf{d}_i)$ for a specific energy mechanism. The minimization operator $\partial \Pi / \partial \mathbf{d}_i = 0$ then leads to

$$\mathbf{k}_{ii}\mathbf{d}_i + \mathbf{f}_i = 0 \text{ (no summation over } i) \tag{9.3}$$

when only one block (element) i is involved. On the other hand, if two blocks i and j (or two elements i and j belong to two different blocks) are in contact, then the resultant equation becomes

$$\begin{cases} (\mathbf{k}_{ii} + \mathbf{k}_{ij})\mathbf{d}_i + \mathbf{f}_i = 0 \\ (\mathbf{k}_{ji} + \mathbf{k}_{jj})\mathbf{d}_j + \mathbf{f}_j = 0 \end{cases} \text{(no summation over } i \text{ and } j) \tag{9.4}$$

These local equations of motion are then assembled together to yield the final, global, equation of motion in the same manner as in the FEM. In the case of N blocks, each with m_i primary variables (such as displacement) per block, the minimization will yield $(N \times N)$ simultaneous equation blocks in total, written symbolically as

$$\begin{bmatrix} \mathbf{k}_{11} & \mathbf{k}_{12} & \mathbf{k}_{13} & \dots & \mathbf{k}_{1N} \\ \mathbf{k}_{21} & \mathbf{k}_{22} & \mathbf{k}_{23} & \dots & \mathbf{k}_{2N} \\ \mathbf{k}_{31} & \mathbf{k}_{32} & \mathbf{k}_{33} & \dots & \mathbf{k}_{3N} \\ \dots & \dots & \dots & \dots & \dots \\ \mathbf{k}_{N1} & \mathbf{k}_{N2} & \mathbf{k}_{N3} & \dots & \mathbf{k}_{NN} \end{bmatrix} \begin{Bmatrix} \mathbf{d}_1 \\ \mathbf{d}_2 \\ \mathbf{d}_3 \\ \dots \\ \mathbf{d}_N \end{Bmatrix} = \begin{Bmatrix} \mathbf{f}_1 \\ \mathbf{f}_2 \\ \mathbf{f}_3 \\ \dots \\ \mathbf{f}_N \end{Bmatrix} \tag{9.5}$$

where $\mathbf{k}_{ij}$ are $(m_i \times m_i)$ sub-matrices, $\mathbf{d}_i$ $(m_i \times 1)$ sub-vectors of the displacement variables of block i and $\mathbf{f}_i$ the $(m_i \times 1)$ sub-vectors of the resultant general forces acting on block (element) i, with $i, j = 1$ to N. The diagonal terms, $\mathbf{k}_{ii}$, usually contain the inertial and deformation sub-matrices of block i ($i = 1, N$) and the off-diagonal sub-matrices $\mathbf{k}_{ij}$ $(i \neq j)$ contain the sub-matrices of contact stiffnesses between blocks i and j. The final $\mathbf{k}_{ij}$ and $\mathbf{f}_i$ are the matrix or vectorial sums of a number of $(m_i \times m_i)$ sub-matrices $[k_{ij}]$ and $(m_i \times 1)$ vectors $\{f_i\}$, corresponding to the different energy minimization mechanisms. The value of m_i depends on how the deformability of the blocks is represented, such as rigid block assumption or fully deformable blocks with FEM meshes with triangular or quadrilateral elements, respectively.

Equation (9.5) can be simply written as

$$[\mathbf{K}]\{\mathbf{D}\} = \{\mathbf{F}\} \tag{9.6}$$

in the same fashion as the global stiffness matrix equation in the FEM, and $[\mathbf{K}]$ can also be called the global 'stiffness matrix', which bears a resemblance with the global connectivity matrix $[\mathbf{C}]$. $\mathbf{k}_{ij} = 0$ if and only if $\mathbf{c}_{ij} = 0$ because $\mathbf{k}_{ij}$ describes the contact stiffness between block i and block j when $i \neq j$.

The stiffness matrices $\mathbf{k}_{ij}$ are obtained by energy minimization of the different mechanisms according to the basic assumptions concerning the material behavior, loading cases, types of initial boundary conditions, etc. Some of these considerations are listed below as examples:

(1) Potential energy due to deformation (strain energy);

(2) Potential energy due to external loads (work done by point forces, line forces and volume forces);

(3) Potential energy due to bolting between blocks (work done by bolting forces);

(4) Kinetic energy due to block mass (kinetic energy due to inertia);

(5) Potential energy due to displacement constraints at boundaries (boundary conditions);

(6) Potential energy due to block contacts (interactions between blocks).

The above items by no means exhaust the complete spectrum of the energy contribution mechanisms. Other mechanisms can also be separately formulated and minimized, and the local equilibrium equations obtained can be assembled into the global equation of motion (9.5) so that their effects are included in the analysis.

9.2 Contact Types and Detection

During a loading process, blocks move according to the forces acting on them and some of the blocks may contact each other at vertices or along edges. Interactive forces (contact forces) are developed at these contact points or edges. Like in any other DEM, contact detection is the essential part of DDA.

9.2.1 Least Distance between Two Approaching Blocks

Contact detection is achieved on the basis of the least distance between a pair of neighboring blocks. Two blocks may come into contact for the next time step only if they are very close at the current time step. The measure of the closeness of two blocks is defined by their least separation distance compared with a prescribed threshold.

The least distance between block i and block j is defined as the minimum distance, η_{ij}, between any two points $p_1(x_1, y_1)$ in block i and point $p_2(x_2, y_2)$ in block j (see Fig. 9.1) with

$$\eta_{ij} = \min \left\{ \sqrt{(x_2 - x_1)^2 + (y_2 - y_1)^2}, \quad \forall (x_1, y_1) \in B^i, \forall (x_2, y_2) \in B^j \right\} \tag{9.7}$$

where B^i and B^j are the sets of all points belonging to block i and block j, respectively.

By this definition, block i and block j are overlapped if $\eta_{ij} < 0$, at contact if $\eta_{ij} = 0$ and separated (not in contact) if $\eta_{ij} > 0$. Let the maximum resultant displacement increment, ρ, at the current time step, be given by

$$\rho = \max \left\{ \sqrt{\Delta u_x^2(x, y) + \Delta u_y^2(x, y)}, \quad \forall (x, y) \in B^k, \ k = 1, N \right\} \tag{9.8}$$

where $\Delta u_x(x, y)$ and $\Delta u_y(x, y)$ are two components of displacement increment at point (x,y) at current time step, and N is the number of blocks. It is reasoned that if

$$\eta_{ij} < 2\rho \tag{9.9}$$

then block i and block j are in potential contact for the next time step (Fig. 9.2). However, it is impossible to compute the least distance between two approaching blocks by equation (9.7) because the number of points in any one block is infinite. A more practical approach is to express the least distance between the vertices and line segments of block i and block j, as given by

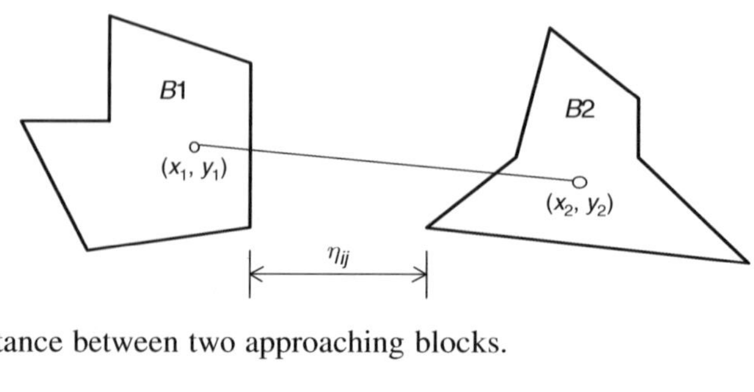

Fig. 9.1 The least distance between two approaching blocks.

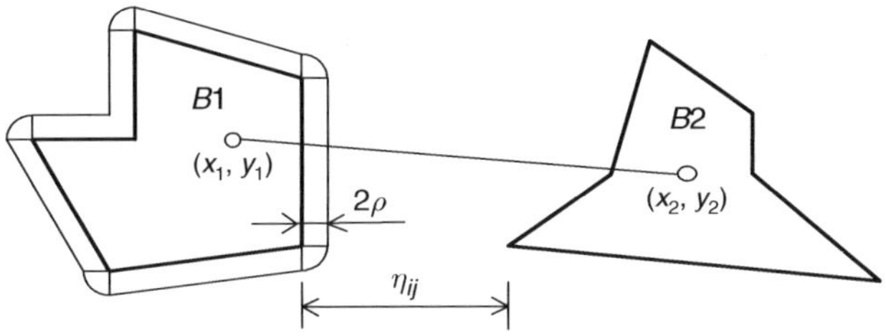

Fig. 9.2 Two blocks not in contact at the next time step.

$$\eta_{ij} = \min\left[\sqrt{(x_2 - x_1)^2 + (y_2 - y_1)^2}, \quad \text{with} \begin{cases} \text{either} & \forall(x_1, y_1) \in V^i, \forall(x_2, y_2) \in B^i \\ \text{or} & \forall(x_1, y_1) \in B^i, \forall(x_2, y_2) \in V^i \end{cases} \right] \tag{9.10}$$

where B^i and B^j are the sets of boundary segments of block i and block j, and V^i and V^j are the sets of vertices of block i and block j, respectively.

The boundary of a block consists of line segments of straight or curved edges represented as straight line segments. The minimum distance from a point $p_1(x_1, y_1)$ to a line segment $\overline{p_2p_3}$ defined by points $p_2(x_2, y_2)$ and $p_3(x_3, y_3)$ can be calculated by the following algorithm. The parametric equation of line segment $\overline{p_2p_3}$ is given by

$$\begin{cases} x = x_2 + (x_3 - x_2)t \\ y = y_2 + (y_3 - y_2)t \end{cases} \quad (0 \leq t \leq 1) \tag{9.11}$$

The distance from point p_1 to an arbitrary point (x,y) on a line segment $\overline{p_2p_3}$ is given by

$$\eta = \sqrt{(x - x_1)^2 + (y - y_1)^2} = \sqrt{[(x_2 - x_1) + (x_3 - x_2)t]^2 + [(y_2 - y_1) + (y_3 - y_2)t]^2} \tag{9.12}$$

with condition $(0 \leq t \leq 1)$. Using $\partial\eta/\partial t = 0$ to minimize the distance leads to

$$[(x_2 - x_1) + (x_3 - x_2)t](x_3 - x_2) + [(y_2 - y_1) + (y_3 - y_2)t](y_3 - y_2) = 0 \tag{9.13}$$

The solution of this equation yields

$$t = \frac{(x_3 - x_2)(x_1 - x_2) + (y_3 - y_2)(y_1 - y_2)}{(x_3 - x_2)^2 + (y_3 - y_2)^2} \tag{9.14}$$

If $0 < t < 1$, the point $(\bar{x}, \bar{y})$ with co-ordinates

$$\begin{cases} \bar{x} = x_2 + (x_3 - x_2)t \\ \bar{y} = y_2 + (y_3 - y_2)t \end{cases} \tag{9.15}$$

on the line segment $\overline{p_2p_3}$ has the minimum distance $\bar{\eta}$ to point p_1 with

$$\bar{\eta} = \sqrt{(\bar{x} - x_1)^2 + (\bar{y} - y_1)^2} \tag{9.16}$$

If $t \leq 0$ or $t \geq 1$, then one of the vertices on the line segment $\overline{p_2p_3}$ has the minimum distance with point p_1

$$\bar{\eta} = \min\{|p_1p_2|, |p_1p_3|\} \tag{9.17}$$

Equations (9.10), (9.16) and (9.17) can be used to calculate the distances between two blocks, expressed as distances between the vertices of one block to the boundary line segments of another block.

Table 9.1 Possible contact types between vertices and edges in 2D

	Convex vertex	Concave vertex	Edge
Convex vertex	Possible	Possible	Possible
Concave vertex	Possible	Not possible	Not possible
Edge	Possible	Not possible	Possible

9.2.2 Contact Types and Detection Algorithm

For 2D polygons, because a vertex can be either convex (of an internal solid vertex angle $\leq 180^{\circ}$) or concave (of an internal solid vertex angle $> 180^{\circ}$), the complete types of contacts can be summarized as: convex vertex-to-convex vertex, convex vertex-to-concave vertex, convex vertex-to-edge, concave vertex-to-concave vertex, concave vertex-to-edge and edge-to-edge. The fact that concave vertex-to-concave vertex and concave vertex-to-edge contacts are not physically admissible (see Table 9.1) reduces the possibilities of contacts into four types: convex vertex-to-convex vertex, convex vertex-to-concave vertex, convex vertex-to-edge and edge-to-edge. Among these four possible contact types, the convex vertex-to-edge contact (see Fig. 9.3a) is the basic type and the other three types can be decomposed as combinations of this basic type. A convex vertex-to-convex vertex or a convex vertex-to-concave vertex contact can be decomposed as two simultaneous convex vertex-to-edge contacts at the same position (Fig. 9.3b,c, the vertex p_1 contacts with two adjacent edges $\overline{p_2p_3}$ and $\overline{p_3p_4}$ simultaneously at vertex p_3), an edge-to-edge contact can be decomposed as two simultaneous convex vertex-to-edge contacts at the same edge (see Fig. 9.3d, the contact between edge $\overline{p_1p_2}$ and edge $\overline{p_3p_4}$ can be decomposed as two vertex-to-edge contacts: (vertex) p_1 to (edge) $\overline{p_3p_4}$ and (vertex) p_4 to (edge) $\overline{p_1p_2}$.

For a possible contact, the geometrical relations between vertices and edges should be kinematically admissible. Two kinematic conditions must be satisfied for a contact:

(1) The distance between the approaching vertices and edges must satisfy relation (9.9).

(2) No large overlapping of solid material, except for a small neighborhood at the approaching vertices due to small numerical truncation errors, should occur when the vertices or edges translate (no rotation) to the contact position.

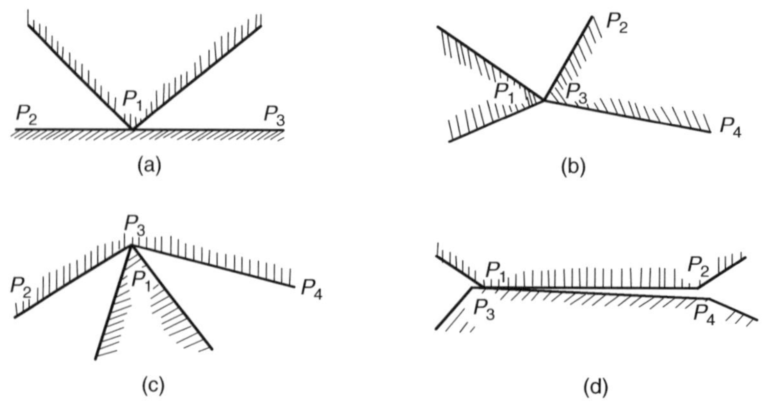

Fig. 9.3 Four types of contacts. (a) convex vertex-to-edge; (b) convex vertex-to-convex vertex; (c) convex vertex-to-concave vertex; (d) edge-to-edge.

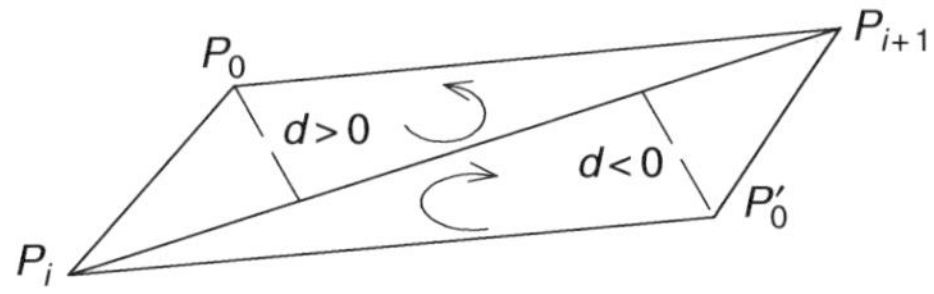

Fig. 9.4 Determination of the interpenetration depth, *d*.

While the first condition is determined largely by the time-stepping schemes used in DDA for calculating block/element displacement and deformations, the second condition is automatically satisfied, no matter what types of vertex–edge contacts are being considered. The impossible concave vertex–edge contact is prevented by earlier convex vertex–edge contacts so that solid material overlapping will never occur, as long as the time step is well within the critical time-step criterion. Therefore only the convex vertex needs to be considered for contact detection.

Contacts are detected by a vertex–edge inference detection algorithm in DDA. Taking a vertex as a point and an edge as a reference line, and recalling that the vertex-to-edge contact is the basic type of contact from which other contact types can be composed of, a contact is made whenever the vertex of one block (or element) moves across an edge of another block (or element belonging to another block) with a non-zero penetration depth, *d*, usually called an interpenetration of distance *d*.

The criterion for occurrence of interpenetration is established between a moving convex vertex and an edge (see Fig. 9.4 with p_0 as the moving vertex and segment $\overline{p_i p_{i+1}}$ as the edge), belonging to two different blocks (or elements from two different blocks).

Let the initial positions of p_0, p_i and p_{i+1} be (x_0, y_0), (x_i, y_i) and (x_{i+1}, y_{i+1}), respectively. During a time step the incremental displacements of vertices p_0, p_i and p_{i+1} are $(\Delta u_x^0, \Delta u_y^0)$, $(\Delta u_x^i, \Delta u_y^i)$ and $(\Delta u_x^{i+1}, \Delta u_y^{i+1})$, respectively. Arranging the three vertices in the order of $p_1 \to p_i \to p_{i+1}$, and denoting Δ as the value of the determinant

$$\Delta = \begin{vmatrix} 1, & x_0 + \Delta u_x^0, & y_0 + \Delta u_y^0 \\ 1, & x_i + \Delta u_x^i, & y_i + \Delta u_y^i \\ 1, & x_{i+1} + \Delta u_x^{i+1}, & y_{i+1} + \Delta u_y^{i+1} \end{vmatrix} \tag{9.18}$$

an interpenetration between two blocks occurs whenever the moving point p_0 (a vertex on one block) crosses the reference line $\overline{p_i p_{i+1}}$ (an edge on another block) with $\Delta < 0$. The interpenetration depth is given by

$$d = \frac{\Delta}{\sqrt{(x_{i+1} - x_i)^2 + (y_{i+1} - y_i)^2}} \tag{9.19}$$

Whenever $d < 0$ occurs, an interpenetration is detected and a contact between the participating blocks (elements) is established at the points of contacting vertices and edges. The corresponding physical laws are then applied to the different types of contact to calculate the contact forces which, in turn, determine the movement (deformation) of the blocks (elements) at the next time step. The interpenetration is minimized in DDA through an iteration process because it is not admissible physically. The most common physical models of contacts in DDA are very similar to that of the distinct element method, using constant (Shi, 1988; Shyu, 1993; Chang, 1994) or variable (Chen and Ohnishi, 1996) normal stiffness and constant shear stiffness to represent point (vertex–edge) contacts. An edge–edge contact is composed of two vertex–edge contacts.

9.3 The Rigid Block Formulation

The simplest assumption regarding the deformability of the rock blocks in DEM is the rigid body assumption, which implies that the distance between any two interior points within a block remains unchanged under loading conditions and block movements. The main driving mechanisms for the movements of a rigid block system are therefore the external and internal loading (gravity, inertial forces, water pressure, bolting forces, etc.), boundary conditions (boundary forces and displacement constraints) and the laws determining the reaction forces between the block contacts. Because of the rigidity assumption, the distances between the mass center and the boundary vertices of a rigid block, and the relative positions between its vertices, remain the same during any movements. The position vectors (i.e., co-ordinates) of the vertices are therefore used to characterize the block geometry and the co-ordinates of the mass center of the block are used to represent the block movement. The characterizing information for a rigid block system is therefore the labels (numbering) of the blocks, their mass centers and vertices which are generated automatically during block system formation using the algorithms presented in Chapters 6 and 7. For the example of the three block system shown in Fig. 9.5, the available charactering data are

Block i: mass center of co-ordinates x_i^c, y_i^c,
boundary vertices of co-ordinates $P_1^i(x_1^i, y_1^i)$, $P_2^i(x_2^i, y_2^i)$, $P_3^i(x_3^i, y_3^i)$, $P_4^i(x_4^i, y_4^i)$.

Block j: mass center of co-ordinates (x_j^c, y_j^c),
boundary vertices of co-ordinates $P_1^j(x_1^j, y_1^j)$, $P_2^j(x_2^j, y_2^j)$, $P_3^j(x_3^j, y_3^j)$, $P_4^j(x_4^j, y_4^j)$.

Block k: mass center of co-ordinates (x_k^c, y_k^c),
boundary vertices of co-ordinates $P_1^k(x_1^k, y_1^k)$, $P_2^k(x_2^k, y_2^k)$, $P_3^k(x_3^k, y_3^k)$, $P_4^k(x_4^k, y_4^k)$ and $P_5^k(x_5^k, y_5^k)$

The key technique is to represent the movements of a rigid block by the movements of its mass center, including both translational and rotational components. A rigid body has, for 2D problems, three degrees of freedom (therefore three displacement components as primary variables): two translational displacements in the x and y directions, u_x^c and u_y^c, and a rotation, θ^c, about the mass center of the block, Fig. 9.6. The vector

$$\{d\} = \begin{Bmatrix} u_x^c \\ u_y^c \\ \theta^c \end{Bmatrix} \tag{9.20}$$

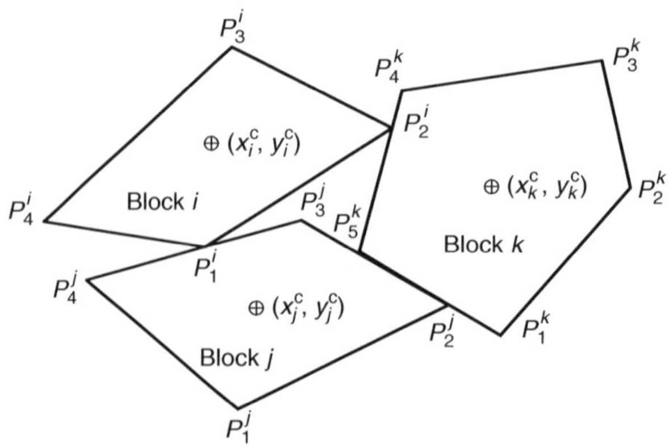

Fig. 9.5 A rigid block system with labeled vertices and mass centres.

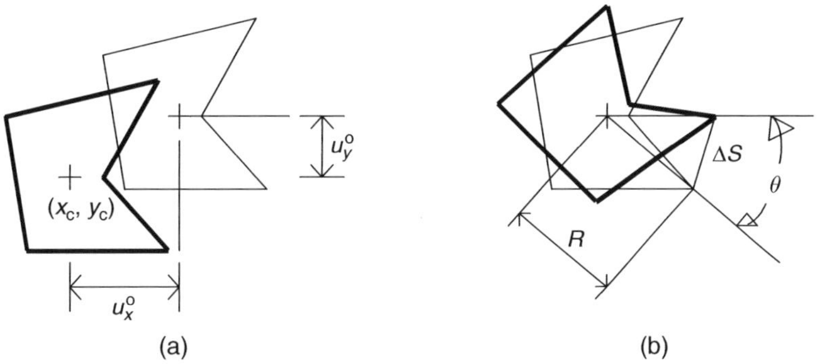

Fig. 9.6 Rigid body movements of a block: (a) translational displacements; (b) rotational displacement.

is used as the displacement vector of the rigid block, representing translational and rotational displacements at every point in the block.

The contributions made by the rigid body translation of a block to the displacement at an arbitrary point (x, y) in the block can be written as (see Fig. 9.6a)

$$\begin{Bmatrix} u_x \\ u_y \end{Bmatrix}_1 = \begin{bmatrix} 1 & 0 \\ 0 & 1 \end{bmatrix} \begin{Bmatrix} u_x^c \\ u_y^c \end{Bmatrix} \tag{9.21}$$

The contributions from the rigid rotation of the block about its mass center to the displacement at an arbitrary field point (x,y) within the block, when the displacement is small (if the time step is small enough), can be given by the equations given below (see Fig. 9.6b). The distance R from an arbitrary point (x,y) in the block to the rotation center x_c, y_c is given by

$$R = \sqrt{(x - x_c)^2 + (y - y_c)^2} \tag{9.22}$$

so that

$$\begin{cases} \cos \quad (\theta^c) = (x - x_c)/R \\ \sin \quad (\theta^c) = -(y - y_c)/R \end{cases} \tag{9.23}$$

When θ^c is small (which is the usual case when the time step is small enough), the arc length, Δs, in Fig. 9.6b can be taken as approximately equal to $R\theta^c$. Thereby, using relation (9.23), one can obtain

$$\begin{Bmatrix} u_x \\ u_y \end{Bmatrix}_2 = \begin{Bmatrix} [R\theta^c](\sin(\theta^c)) \\ [R\theta^c](\cos(\theta^c)) \end{Bmatrix} = \begin{Bmatrix} -(y - y_c) \\ (x - x_c) \end{Bmatrix} \theta^c \tag{9.24}$$

The total displacement vector at point (x,y) is then the sum of the displacement contributions from the two modes of block displacement

$$\begin{Bmatrix} u_x \\ u_y \end{Bmatrix} = \begin{Bmatrix} u_x \\ u_y \end{Bmatrix}_1 + \begin{Bmatrix} u_x \\ u_y \end{Bmatrix}_2 = \begin{bmatrix} 1 & 0 & -(y - y_c) \\ 0 & 1 & (x - x_c) \end{bmatrix} \begin{Bmatrix} u_x^c \\ u_y^c \\ \theta^c \end{Bmatrix} \tag{9.25}$$

or simply

$$\{u\} = [T]\{d\} \tag{9.26}$$

where matrix $[T]$ is called the displacement matrix of the ith block and is given by

$$[T] = \begin{bmatrix} 1 & 0 & -(y - y_c) \\ 0 & 1 & (x - x_c) \end{bmatrix} \tag{9.27}$$

To be consistent with the formulations using triangular and quadrilateral elements in the following sections, Eqn (9.26) is modified to

$$\{u\} = [T][X]\{d\} \tag{9.28}$$

where the matrix $[X]$ is an identity matrix

$$[X] = \begin{bmatrix} 1 & 0 & 0 \\ 0 & 1 & 0 \\ 0 & 0 & 1 \end{bmatrix} \tag{9.29}$$

The matrix $[H]$ denotes the integration of the product $[T]^{\mathrm{T}}[T]$ over the block domain

$$[H] = \iint_{\Omega} [T^i]^{\mathrm{T}}[T]\mathrm{d}x \ \mathrm{d}y = \begin{bmatrix} S_{00} & 0 & 0 \\ 0 & S_{00} & 0 \\ 0 & 0 & S_{20} + S_{02} - y_c S_{01} - x_c S_{10} \end{bmatrix} \tag{9.30}$$

and is useful later for stiffness matrix evaluations. For a generally shaped polygon of N vertices, the integral expressions for the inertial properties $S_{00}, S_{10}\ S_{01}, S_{02}$ and S_{20} are respectively given by

$$S_{00} = \iint_{\Omega} \mathrm{d}x \ \mathrm{d}y = \frac{1}{2}\sum_{i=1}^{N} \begin{vmatrix} 1 & x_c & y_c \\ 1 & x_i & y_i \\ 1 & x_{i+1} & y_{i+1} \end{vmatrix} = A \quad \text{(the block area)} \tag{9.31}$$

$$S_{10} = \iint_{\Omega} x \ \mathrm{d}x \ \mathrm{d}y = \frac{A}{3}\sum_{i=1}^{N}(x_c + x_i + x_{i+1}) \tag{9.32}$$

$$S_{01} = \iint_{\Omega} y \ \mathrm{d}x \ \mathrm{d}y = \frac{A}{3}\sum_{i=1}^{N}(y_c + y_i + y_{i+1}) \tag{9.33}$$

$$S_{11} = \iint_{\Omega} xy \ \mathrm{d}x \ \mathrm{d}y = \frac{A}{12}\sum_{i=1}^{N}(2x_c y_c + 2x_i y_i + 2x_{i+1}y_{i+1} + x_i y_c + x_{i+1}y_c + x_c y_i + x_c y_{i+1} + x_i y_{i+1} + x_{i+1}y_i) \tag{9.34}$$

$$S_{20} = \iint_{\Omega} x^2 \ \mathrm{d}x \ \mathrm{d}y = \frac{A}{6}\sum_{i=1}^{N}(x_c^2 + x_i^2 + x_{i+1}^2 + x_c x_i + x_c x_{i+1} + x_i x_{i+1}) \tag{9.35a}$$

$$S_{02} = \iint_{\Omega} y^2 \ \mathrm{d}x \ \mathrm{d}y = \frac{A}{6}\sum_{i=1}^{N}(y_c^2 + y_i^2 + y_{i+1}^2 + y_c y_i + y_c y_{i+1} + y_i y_{i+1}) \tag{9.35b}$$

Note that the vertices of the block should be labeled counter-clockwise and $(x_{N+1}, y_{N+1}) = (x_1, y_1)$. The term S_{11} is not used here but later for deformable blocks.

For a much simplified formulation for rigid block motions, we consider here an example of energy minimization for the damping effect, assuming that the damping coefficient is given by the product of a damping coefficient α and the rock density ρ, which then produces a force term related to the block velocity

$$\begin{Bmatrix} f_x \\ f_y \end{Bmatrix}_d = -\alpha\rho \begin{Bmatrix} \dfrac{\partial u_x}{\partial t} \\ \dfrac{\partial u_y}{\partial t} \end{Bmatrix} \tag{9.36}$$

The potential energy due to the damping for block k is then written as

$$\Pi_D = -\iint_\Omega \{u_x, u_y\} \begin{Bmatrix} f_x \\ f_y \end{Bmatrix} dx\ dy = \iint_\Omega \partial\rho_k \{u_x, u_y\} \begin{Bmatrix} \dfrac{\partial u_x}{\partial t} \\ \dfrac{\partial u_y}{\partial t} \end{Bmatrix} dx\ dy = \iint_\Omega \partial\rho_k \{d_k\}^T [T_k]^T [T_k] dx\ dy \tag{9.37}$$

Recalling Eqn (9.30), minimizing the potential energy Π_D then leads to

$$[k_{kk}] = \frac{\alpha\rho_i}{\Delta t} [H_k] \tag{9.38}$$

More detailed information on rigid block motions using the DDA approach can be seen in Koo and Chern (1998) and Cheng and Zhang (2000).

We omit the DDA formulations about cable support, damping effects and coupling with fluid flow through fractures in the rigid block systems, since most of these issues are either more fully covered in DDA formulations with deformable blocks as described in the following sections or simple algorithms that were fully presented in literature.

9.4 Deformable Blocks with a FEM Mesh of Triangular Elements

The rigid block assumption may capture the gross, also often the large scale, movements of rock block systems but lacks the capacity to consider stress, strength, failure and non-linear deformation of the rock materials. The latter items are often important issues in rock engineering. The formulations using simply deformable blocks in the early stage of the DDA method, which characterizes the blocks' deformability using only one constant strain tensor regardless of the block geometry complexity, also has limitations in this regard. Both are not suitable for considering coupling mechanisms of stress, deformation, fluid flow and heat transport in fractured rocks.

The important contribution by Shyu (1993) of discretizing the blocks with finite element meshes provided the DDA method with not only a greater stress and deformation analysis capacity for block systems but also the possibility to include coupled hydro-mechanical analysis with considerations of both fracture flow and fracture-and-matrix interaction for flow problems. Figure 9.7 provides an illustrative example of a system of three blocks with a triangular FEM element mesh, with the shaded elements being in contact with other elements.

Using FEM meshes to simulate the deformation of individual blocks introduces several additional tasks for a DDA formulation:

(1) selection of the element types used for the problem;

(2) detection of contacts between not only blocks but also elements;

(3) evaluation of the element stiffness matrices for deformation, contacts and other loading and constraint energy mechanisms, considering proper constitutive laws for contacts and rock material behavior;

(4) assemblage of element stiffness matrices into the global stiffness matrix; and

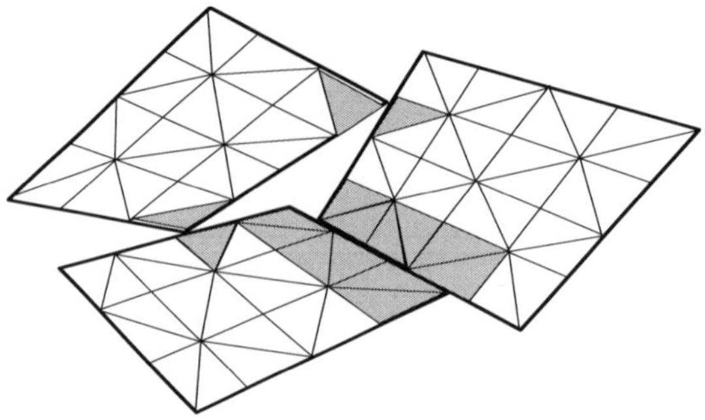

Fig. 9.7 Internal discretization of blocks by triangular FEM elements. The shaded elements are in contact.

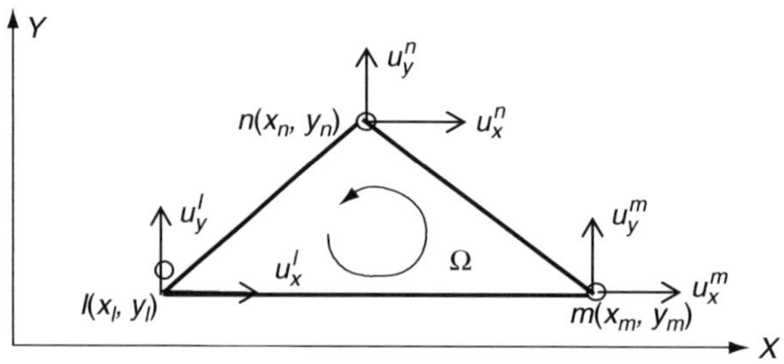

Fig. 9.8 A typical constant strain triangular element.

(5) solution of the global matrix equation to obtain the values of nodal displacements and element stresses in the same manner as in FEM.

All the above tasks can, however, be performed with standard FEM techniques without undue difficulty. In this book, homogeneous, isotropic and linear elastic properties are assumed for the behavior of all rock materials, characterized by the Young's modulus, E, and Poisson's ratio, v.

The discretization of an arbitrary 2D block into a number of triangular elements is performed in the same way as in the FEM. Each element has three nodes (l, m, n) and six displacement variables in two orthogonal directions, written in a (6×1) vector $\{d\} = (u_x^l, u_y^l, u_x^m, u_y^m, u_x^n, u_y^n)^{\mathrm{T}}$, where u_x^i, u_y^i $(i{=}l,m,n)$ represent the displacements in the x and y directions, respectively (Fig. 9.8). The element is positively oriented when the three nodes l–m–n are arranged counter-clockwise (as indicated by the arrow in Fig. 9.8). The displacement field of a triangular element with constant strain is given by

$$\begin{cases} u_x = a_0 + a_1 x + a_2 y \\ u_y = b_0 + b_1 x + b_2 y \end{cases} \tag{9.39}$$

where (a_0, a_1, a_2) and (b_0, b_1, b_2) are the coefficient to be determined by using the displacement components at the three nodes l, m and n. Substitution of nodal displacements into (9.39) then results in

$$\begin{Bmatrix} u_x^l \\ u_x^m \\ u_x^n \end{Bmatrix} = \begin{bmatrix} 1 & x_l & y_l \\ 1 & x_m & y_m \\ 1 & x_n & y_n \end{bmatrix} \begin{Bmatrix} a_0 \\ a_1 \\ a_2 \end{Bmatrix} \quad \text{and} \quad \begin{Bmatrix} u_y^l \\ u_y^m \\ u_y^n \end{Bmatrix} = \begin{bmatrix} 1 & x_l & y_l \\ 1 & x_m & y_m \\ 1 & x_n & y_n \end{bmatrix} \begin{Bmatrix} b_0 \\ b_1 \\ b_2 \end{Bmatrix} \tag{9.40}$$

Solutions for coefficients (a_0, a_1, a_2) and (b_0, b_1, b_2) lead to

$$\begin{Bmatrix} a_0 \\ a_1 \\ a_2 \end{Bmatrix} = \begin{bmatrix} 1 & x_l & y_l \\ 1 & x_m & y_m \\ 1 & x_n & y_n \end{bmatrix}^{-1} \begin{Bmatrix} u_x^l \\ u_x^m \\ u_x^n \end{Bmatrix} = \begin{bmatrix} n_{11} & n_{12} & n_{13} \\ n_{21} & n_{22} & n_{23} \\ n_{31} & n_{32} & n_{33} \end{bmatrix} \begin{Bmatrix} u_x^l \\ u_x^m \\ u_x^n \end{Bmatrix} \tag{9.41a}$$

and

$$\begin{Bmatrix} b_0 \\ b_1 \\ b_2 \end{Bmatrix} = \begin{bmatrix} 1 & x_l & y_l \\ 1 & x_m & y_m \\ 1 & x_n & y_n \end{bmatrix}^{-1} \begin{Bmatrix} u_y^l \\ u_y^m \\ u_y^n \end{Bmatrix} = \begin{bmatrix} n_{11} & n_{12} & n_{13} \\ n_{21} & n_{22} & n_{23} \\ n_{31} & n_{32} & n_{33} \end{bmatrix} \begin{Bmatrix} u_y^l \\ u_y^m \\ u_y^n \end{Bmatrix} \tag{9.41b}$$

Substitution of (9.41a) and (9.41b) into (9.39), and rearrangement, produces

$$\begin{Bmatrix} u_x \\ u_y \end{Bmatrix} = \begin{bmatrix} 1 & 0 & x & 0 & y & 0 \\ 0 & 1 & 0 & x & 0 & y \end{bmatrix} \begin{bmatrix} 1 & 0 & x_l & 0 & y_l & 0 \\ 0 & 1 & 0 & x_l & 0 & y_l \\ 1 & 0 & x_m & 0 & y_m & 0 \\ 0 & 1 & 0 & x_m & 0 & y_m \\ 1 & 0 & x_n & 0 & y_n & 0 \\ 0 & 1 & 0 & x_n & 0 & y_n \end{bmatrix} \begin{Bmatrix} u_x^l \\ u_y^l \\ u_x^m \\ u_y^m \\ u_x^n \\ u_y^n \end{Bmatrix} \tag{9.42}$$

or simply

$$\{u\} = [T][X]\{d\} \tag{9.43}$$

where

$$[T] = \begin{bmatrix} 1 & 0 & x & 0 & y & 0 \\ 0 & 1 & 0 & x & 0 & y \end{bmatrix} \tag{9.44}$$

$$[X] = \begin{bmatrix} 1 & 0 & x_l & 0 & y_l & 0 \\ 0 & 1 & 0 & x_l & 0 & y_l \\ 1 & 0 & x_m & 0 & y_m & 0 \\ 0 & 1 & 0 & x_m & 0 & y_m \\ 1 & 0 & x_n & 0 & y_n & 0 \\ 0 & 1 & 0 & x_n & 0 & y_n \end{bmatrix} \tag{9.45}$$

Let the matrix $[\delta]$ stand for a differential operator

$$[\delta] = \begin{bmatrix} \partial/\partial x & 0 & \partial/\partial y \\ 0 & \partial/\partial y & \partial/\partial x \end{bmatrix}^{\mathrm{T}} \tag{9.46}$$

then the differentiation of the matrix $[T]$ leads to a matrix $[B]$ given by

$$[B] = [\delta][T] = \equiv \begin{bmatrix} 0 & 1 & 0 & 0 & 0 & 0 \\ 0 & 0 & 0 & 0 & 0 & 1 \\ 0 & 0 & 1 & 0 & 1 & 0 \end{bmatrix} \tag{9.47}$$

The stress vector $\{\sigma\} = (\sigma_x, \sigma_y, \tau_{xy})_i^{\mathrm{T}}$ and strain vector $\{\varepsilon\} = (\varepsilon_x, \varepsilon_y, \gamma_{xy})_{\mathrm{i}}^{\mathrm{T}}$ can then be expressed as functions of the nodal displacement vector for elastic deformation

$$\{\varepsilon\} = [\delta]\{u\} = [\delta][T][X]\{d\} = [B][X]\{d\} \tag{9.48}$$

and

$$\{\sigma\} = [E]\{\varepsilon\} = [E][B][X]\{d\} \tag{9.49}$$

where

$$[E] = \frac{E}{1-\nu^2}\begin{bmatrix} 1 & \nu & 0 \\ \nu & 1 & 0 \\ 0 & 0 & (1-2\nu)/2 \end{bmatrix} \tag{9.50}$$

is the elasticity tensor for plane strain problems, with E being the Young's modulus and ν the Poisson's ratio. Two more matrices given below are useful later in evaluation of the stiffness matrices of triangular elements, given by

$$[H] = \iint_{\Omega} [T]^{\mathrm{T}}[T]\mathrm{d}x\ \mathrm{d}y = \begin{bmatrix} S_{00} & 0 & S_{10} & 0 & S_{01} & 0 \\ 0 & S_{00} & 0 & S_{10} & 0 & S_{01} \\ S_{10} & 0 & S_{20} & 0 & S_{11} & 0 \\ 0 & S_{10} & 0 & S_{20} & 0 & S_{11} \\ S_{01} & 0 & S_{02} & 0 & S_{02} & 0 \\ 0 & S_{01} & 0 & S_{11} & 0 & S_{02} \end{bmatrix} \tag{9.51}$$

$$[D] = \iint_{\Omega} [B]^{\mathrm{T}}[E][B]\mathrm{d}x\ \mathrm{d}y = \frac{S_0 E}{1-\nu^2}\begin{bmatrix} 0 & 0 & 0 & 0 & 0 & 0 \\ 0 & 0 & 0 & 0 & 0 & 0 \\ 0 & 0 & 1 & 0 & 0 & \nu \\ 0 & 0 & 0 & (1-\nu)/2 & (1-\nu)/2 & 0 \\ 0 & 0 & 0 & (1-\nu)/2 & (1-\nu_i)/2 & 0 \\ 0 & 0 & \nu & 0 & 0 & 1 \end{bmatrix} \tag{9.52}$$

9.5 Deformable Blocks with a FEM Mesh of Quadrilateral Elements

The quadrilateral elements have four nodes and bilinear strain fields over the element domain. Due to the higher order of the strain interpolation function, the accuracy of calculation using quadrilateral elements can be improved to a great extent, compared with that for a mesh of triangular elements. The meshing is performed in the same manner as in the FEM (Fig. 9.9).

Similar to triangular elements, the displacement at an arbitrary point (x, y) within a quadrilateral element, (u_x, u_y), is given by

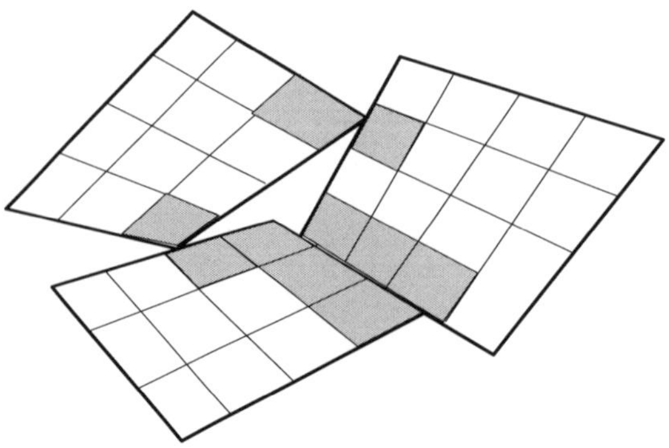

Fig. 9.9 Internal discretization of blocks by quadrilateral FEM elements. The shaded elements are in contact.

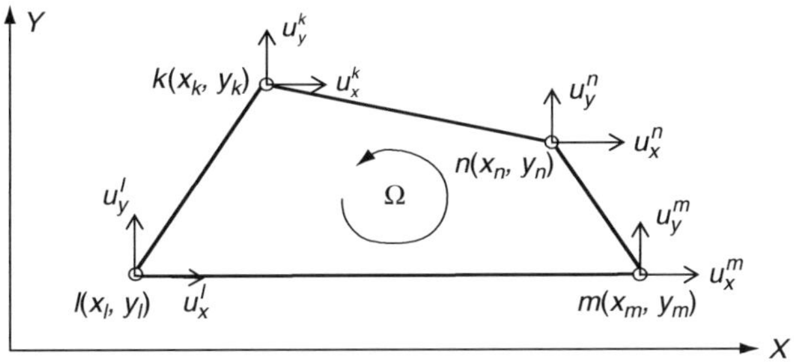

Fig. 9.10 A typical bilinear strain quadrilateral element.

$$\begin{cases} u_x = c_1 + c_2x + c_3y + c_4xy \\ u_y = d_1 + d_2x + d_3y + d_4xy \end{cases} \tag{9.53}$$

The coefficients c_i, d_i $(i = 1,2,3,4)$ in Eqn (9.53) are determined by using the displacements at the element nodes. Let the displacement components of the four nodes k, l, m, n be (u_x^k, u_y^k), (u_x^l, u_y^l), (u_x^m, u_y^m) and (u_x^n, u_y^n), and their co-ordinates be (x_k, y_k), (x_l, y_l), (x_m, y_m) and (x_n, y_n), respectively, as shown in Fig. 9.10. Substitution of their co-ordinates and displacement components into (9.53) leads to

$$\begin{Bmatrix} u_x^k \\ u_x^l \\ u_x^m \\ u_x^n \end{Bmatrix} = \begin{bmatrix} 1 & x_k & y_k & x_ky_k \\ 1 & x_l & y_l & x_ly_l \\ 1 & x_m & y_m & x_my_m \\ 1 & x_n & y_n & x_ny_n \end{bmatrix} \begin{Bmatrix} c_1 \\ c_2 \\ c_3 \\ c_4 \end{Bmatrix} \quad \text{and} \quad \begin{Bmatrix} u_y^k \\ u_y^l \\ u_y^m \\ u_y^n \end{Bmatrix} = \begin{bmatrix} 1 & x_k & y_k & x_ky_k \\ 1 & x_l & y_l & x_ly_l \\ 1 & x_m & y_m & x_my_m \\ 1 & x_n & y_n & x_ny_n \end{bmatrix} \begin{Bmatrix} d_1 \\ d_2 \\ d_3 \\ d_4 \end{Bmatrix} \tag{9.54}$$

Equation (9.54) can be rewritten as

$$\begin{Bmatrix} c_1 \\ c_2 \\ c_3 \\ c_4 \end{Bmatrix} = \begin{bmatrix} 1 & x_k & y_k & x_ky_k \\ 1 & x_l & y_l & x_ly_l \\ 1 & x_m & y_m & x_my_m \\ 1 & x_n & y_n & x_ny_n \end{bmatrix}^{-1} \begin{Bmatrix} u_x^k \\ u_x^l \\ u_x^m \\ u_x^n \end{Bmatrix} \quad \text{and} \quad \begin{Bmatrix} d_1 \\ d_2 \\ d_3 \\ d_4 \end{Bmatrix} = \begin{bmatrix} 1 & x_k & y_k & x_ky_k \\ 1 & x_l & y_l & x_ly_l \\ 1 & x_m & y_m & x_my_m \\ 1 & x_n & y_n & x_ny_n \end{bmatrix}^{-1} \begin{Bmatrix} u_y^k \\ u_y^l \\ u_y^m \\ u_y^n \end{Bmatrix} \tag{9.55}$$

Substitution of Eqn (9.55) into Eqn (9.53) yields

$$u_x = \{1, x, y, xy\} \begin{bmatrix} n_{11} & n_{12} & n_{13} & n_{14} \\ n_{21} & n_{22} & n_{23} & n_{24} \\ n_{31} & n_{32} & n_{33} & n_{34} \\ n_{41} & n_{42} & n_{43} & n_{44} \end{bmatrix} \begin{Bmatrix} u_x^k \\ u_x^l \\ u_x^m \\ u_x^n \end{Bmatrix} \tag{9.56}$$

$$u_y = \{1, x, y, xy\} \begin{bmatrix} n_{11} & n_{12} & n_{13} & n_{14} \\ n_{21} & n_{22} & n_{23} & n_{24} \\ n_{31} & n_{32} & n_{33} & n_{34} \\ n_{41} & n_{42} & n_{43} & n_{44} \end{bmatrix} \begin{Bmatrix} u_y^k \\ u_y^l \\ u_y^m \\ u_y^n \end{Bmatrix} \tag{9.57}$$

with

$$[N] = [n_{ij}] = \begin{bmatrix} n_{11} & n_{12} & n_{13} & n_{14} \\ n_{21} & n_{22} & n_{23} & n_{24} \\ n_{31} & n_{32} & n_{33} & n_{34} \\ n_{41} & n_{42} & n_{43} & n_{44} \end{bmatrix} = \begin{bmatrix} 1 & x_k & y_k & x_ky_k \\ 1 & x_l & y_l & x_ly_l \\ 1 & x_m & y_m & x_my_m \\ 1 & x_n & y_n & x_ny_n \end{bmatrix}^{-1} \tag{9.58}$$

Denoting

$$[T] = \begin{bmatrix} 1 & 0 & x & 0 & y & 0 & xy & 0 \\ 0 & 1 & 0 & x & 0 & y & 0 & xy \end{bmatrix} \tag{9.59}$$

and

$$[X] = \begin{bmatrix} n_{11} & 0 & n_{12} & 0 & n_{13} & 0 & n_{14} & 0 \\ 0 & n_{11} & 0 & n_{12} & 0 & n_{13} & 0 & n_{14} \\ n_{21} & 0 & n_{22} & 0 & n_{23} & 0 & n_{24} & 0 \\ 0 & n_{21} & 0 & n_{22} & 0 & n_{23} & 0 & n_{24} \\ n_{31} & 0 & n_{32} & 0 & n_{33} & 0 & n_{34} & 0 \\ 0 & n_{31} & 0 & n_{32} & 0 & n_{33} & 0 & n_{34} \\ n_{41} & 0 & n_{42} & 0 & n_{43} & 0 & n_{44} & 0 \\ 0 & n_{41} & 0 & n_{42} & 0 & n_{43} & 0 & n_{44} \end{bmatrix} \tag{9.60}$$

the displacement vector at an arbitrary point (x,y) in a quadrilateral element is expressed as a function of the nodal displacements, given by

$$\begin{Bmatrix} u_x \\ u_y \end{Bmatrix} = \begin{bmatrix} 1 & 0 & x & 0 & y & 0 & xy & 0 \\ 0 & 1 & 0 & x & 0 & y & 0 & xy \end{bmatrix} \begin{bmatrix} n_{11} & 0 & n_{12} & 0 & n_{13} & 0 & n_{14} & 0 \\ 0 & n_{11} & 0 & n_{12} & 0 & n_{13} & 0 & n_{14} \\ n_{21} & 0 & n_{22} & 0 & n_{23} & 0 & n_{24} & 0 \\ 0 & n_{21} & 0 & n_{22} & 0 & n_{23} & 0 & n_{24} \\ n_{31} & 0 & n_{32} & 0 & n_{33} & 0 & n_{34} & 0 \\ 0 & n_{31} & 0 & n_{32} & 0 & n_{33} & 0 & n_{34} \\ n_{41} & 0 & n_{42} & 0 & n_{43} & 0 & n_{44} & 0 \\ 0 & n_{41} & 0 & n_{42} & 0 & n_{43} & 0 & n_{44} \end{bmatrix} \begin{Bmatrix} u_x^k \\ u_y^k \\ u_x^l \\ u_y^l \\ u_x^m \\ u_y^m \\ u_x^n \\ u_y^n \end{Bmatrix} \tag{9.61}$$

or simply

$$\{u\} = [T][X]\{d\} \tag{9.62}$$

where $\{d\} = (u_x^k, u_y^k, u_x^l, u_y^l, u_x^m, u_y^m, u_x^n, u_y^n)^{\mathrm{T}}$ is the nodal displacement vector.

Recall the differentiation operator matrix $[\partial]$ defined by Eqn (9.46), the matrix

$$[B] = [\partial][T] = \begin{bmatrix} \frac{\partial}{\partial x} & 0 \\ 0 & \frac{\partial}{\partial y} \\ \frac{\partial}{\partial y} & \frac{\partial}{\partial x} \end{bmatrix} \begin{bmatrix} 1 & 0 & x & 0 & y & 0 & xy & 0 \\ 0 & 1 & 0 & x & 0 & y & 0 & xy \end{bmatrix}$$

$$= \begin{bmatrix} 0 & 0 & 1 & 0 & 0 & 0 & y & 0 \\ 0 & 0 & 0 & 0 & 0 & 1 & 0 & x \\ 0 & 0 & 0 & 1 & 1 & 0 & x & y \end{bmatrix} \tag{9.63}$$

is used to define the strain of the element, given by

$$\{\varepsilon\} = [\partial]\{u\} = [\partial][T][X]\{d\} = [B][X]\{d\} \tag{9.64}$$

The product

$$[B][X] = \begin{bmatrix} n_{21}+n_{41}y & 0 & n_{22}+n_{42}y & 0 & n_{23}+n_{43}y & 0 & n_{24}+n_{44}y & 0 \\ 0 & n_{31}+n_{41}x & 0 & n_{32}+n_{42}y & 0 & n_{33}+n_{43}y & 0 & n_{34}+n_{44}y \\ n_{31}+n_{41}x & n_{21}+n_{41}y & n_{32}+n_{42}x & n_{22}+n_{42}y & n_{33}+n_{43}x & n_{23}+n_{43}y & n_{34}+n_{44}x & n_{24}+n_{44}y \end{bmatrix} \tag{9.65}$$

shows that all components of the normal strains have a linear distribution along the direction perpendicular to the direction of the strain component, and the shear strain varies linearly in both co-ordinate directions. Assuming an elasticity model for the material and denoting the vector $\{\sigma\} = \{\sigma_x, \sigma_y, \sigma_{xy}\}^{\mathrm{T}}$ as the stress vector of the element, the stress vector is given by

$$\{\sigma\} = [E]\{\varepsilon\} = [E][B][X]\{d\} \tag{9.66}$$

where $[E]$ is the elasticity tensor given by (9.50). Similar also to the case for triangular elements, two more useful properties of the four-noded elements can also be given in the same symbols as those in (9.28) and (9.30). The first is a matrix given by

$$[H] = \iint_{\Omega} [T]^{\mathrm{T}}[T]\mathrm{d}\Omega = = \begin{bmatrix} S_{00} & 0 & S_{10} & 0 & S_{01} & 0 & S_{11} & 0 \\ 0 & S_{00} & 0 & S_{10} & 0 & S_{01} & 0 & S_{11} \\ S_{10} & 0 & S_{20} & 0 & S_{11} & 0 & S_{21} & 0 \\ 0 & S_{10} & 0 & S_{20} & 0 & S_{11} & 0 & S_{21} \\ S_{01} & 0 & S_{11} & 0 & S_{02} & 0 & S_{12} & 0 \\ 0 & S_{01} & 0 & S_{11} & 0 & S_{02} & 0 & S_{12} \\ S_{11} & 0 & S_{21} & 0 & S_{12} & 0 & S_{22} & 0 \\ 0 & S_{11} & 0 & S_{21} & 0 & S_{12} & 0 & S_{22} \end{bmatrix} \tag{9.67a}$$

where

$$S_{mn} = \iint_{\Omega} x^m y^n \, \mathrm{d}\Omega \tag{9.67b}$$

and Ω is the integration domain defined by the co-ordinates of the four nodes of the element.

The second matrix is written as

$$[D] = \iint_{\Omega} [B]^{\mathrm{T}}[E][B]\mathrm{d}\Omega$$

$$= \frac{E}{1-\nu^2} \begin{bmatrix} 0 & 0 & 0 & 0 & 0 & 0 & 0 & 0 \\ 0 & 0 & 0 & 0 & 0 & 0 & 0 & 0 \\ 0 & 0 & S_{00} & 0 & 0 & S_{00}\nu & S_{01} & 0 \\ 0 & 0 & 0 & \frac{1-2\nu}{\nu}S_{00} & \frac{1-2\nu}{\nu}S_{00} & 0 & \frac{1-2\nu}{\nu}S_{10} & \frac{1-2\nu}{\nu}S_{01} \\ 0 & 0 & 0 & \frac{1-2\nu}{\nu}S_{00} & \frac{1-2\nu}{\nu}S_{00} & 0 & \frac{1-2\nu}{\nu}S_{10} & \frac{1-2\nu}{\nu}S_{01} \\ 0 & 0 & S_{00}\nu & 0 & 0 & S_{00} & S_{01}\nu & S_{10} \\ 0 & 0 & S_{01} & \frac{1-2\nu}{\nu}S_{10} & \frac{1-2\nu}{\nu}S_{10} & S_{01}\nu & S_{02}+\frac{1-2\nu}{\nu}S_{20} & \frac{(1-\nu)^2}{\nu}S_{11} \\ 0 & 0 & S_{10}\nu & \frac{1-2\nu}{\nu}S_{01} & \frac{1-2\nu}{\nu}S_{01} & S_{10} & \frac{(1-\nu)^2}{\nu}S_{11} & S_{20}+\frac{1-2\nu}{\nu}S_{02} \end{bmatrix} \tag{9.68}$$

The vectors of displacement of the rigid blocks, triangular elements and quadrilateral elements (cf. Eqns (9.28), (9.43) and (9.62)), as well as vectors of strain and stress for the triangular and quadrilateral elements (cf. Eqns (9.48), (9.49), (9.64) and (9.66)), are expressed in identical forms for the later uniform treatment of energy minimization. However, the expressions for the matrices $[T]$, $[B]$ and $[X]$, and nodal displacement vector $\{d\}$, are totally different, and their rank orders are 3, 6 and 8, respectively.

9.6 Evaluation of Element Stiffness Matrices and Load Vectors

The evaluation of stiffness matrices of blocks and elements in DDA is performed through the minimization of the energy functional related to certain physical mechanisms concerning deformation, contacts, external and internal loading and boundary constraints. The main concept is to represent the effect of a particular physical mechanism by energy functionals as functions of the element nodal displacement vector, then minimizing the functional with respect to the nodal displacement vector to derive the mathematical expressions for the required stiffness matrices and load vectors, as described in Section 9.1. The mathematical forms of the energy functionals are different for different mechanisms. In this book, only the following mechanisms are considered as demonstrative examples:

- Elastic deformation of the rock material – minimization of the strain energy – only for deformable blocks with triangular and quadrilateral meshes;
- Effect of mass inertia of the rock material – minimization of kinetic energy of the blocks and elements – for both rigid and deformable blocks;
- Effects of contacts between blocks (elements) for three different cases: pure normal contact, inclined contact with normal and tangential displacements without slipping, inclined contact with normal and tangential displacements with slipping – minimization of work done by contact forces;
- Effects of external loads: volume (body) forces, point forces, distributed forces and fluid pressure – minimization of work done by external loads;
- Effect of bolting with point anchor assumptions – minimization of work done by anchoring forces; and
- Effect of boundary constraints – minimization of potential energy due to specified displacement constraints.

The above mechanisms are the basic physical aspects involved in the deformation process of a block system, but they are not the complete set. Many important mechanisms, such as thermal effects, fluid flow, different reinforcement measures, dynamic effects, different combinations of stress (force) and displacement boundary conditions and fracturing, are crucial to the success of DDA applications for more general and complex practical problems. Evaluation of the stiffness matrices due to these effects can be performed using a similar technique as presented here, or they can also be evaluated directly using the standard FEM technique without using the formal energy minimization processes. Presentation of these formulations are, however, beyond the scope of this book, and they are, in any case, straightforward applications of existing techniques. Only the fluid flow in fractures is considered in Section 9.8.

9.6.1 Elastic Deformation of the Rock Material – Minimization of Strain Energy

The effect of elastic deformation of the rock material for a representative element i is considered by minimization of its elastic strain energy, and is considered for deformable blocks only. Recall expressions (9.48), (9.49) and (9.52) for stress, strain and matrix $[D]$ for triangular and quadrilateral elements as defined before, the functional of strain energy is written as

$$\Pi_e = \frac{1}{2}\iint_\Omega \{\varepsilon_i\}^{\mathrm{T}}\{\sigma_i\}\,\mathrm{d}\Omega = \frac{1}{2}\iint_\Omega (\{d_i\}^{\mathrm{T}}[X_i]^{\mathrm{T}}[B_i]^{\mathrm{T}})([E_i][B_i][X_i]\{d_i\})\mathrm{d}\Omega$$
$$= \frac{1}{2}\{d_i\}^{\mathrm{T}}[X_i]^{\mathrm{T}}\left(\iint_\Omega [B_i]^{\mathrm{T}}[E_i][B_i]\mathrm{d}\Omega\right)[X_i]\{d_i\} = \frac{1}{2}\{d_i\}^{\mathrm{T}}[X_i]^{\mathrm{T}}[D_i][X_i]\{d_i\} \tag{9.69}$$

The minimization by $\partial\Pi_e/\{d_i\} = 0$ leads to

$$[k_{ii}] = [X_i]^{\mathrm{T}}[D_i][X_i] \tag{9.70}$$

This is either a (6×6) or an (8×8) square matrix, depending on whether triangular or quadrilateral elements are selected. For the former, expressions in Eqns (9.45) and (9.52) should be used for $[X]$ and $[D]$ matrices and, for the latter, those from Eqns (9.60) and (9.68) should be used. The matrix represents the elastic deformability of the rock material contained in the element.

The deformation stiffness matrices by Eqn (9.70) are evaluated for every element and assembled in diagonal positions of the global stiffness matrix for the ith element in Eqn (9.5).

9.6.2 Mass Inertia – Minimization of Kinetic Energy

The differential form of the kinetic energy of a particle of mass m can be written as

$$\mathrm{d}K_{\mathrm{p}} = m\mathbf{v}\cdot\frac{\mathrm{d}\mathbf{v}}{\mathrm{d}t} = m\mathbf{v}\cdot\frac{\mathrm{d}^2\mathbf{u}}{\mathrm{d}t^2} \tag{9.71}$$

where $\mathbf{u} = (u_x, u_y)^{\mathrm{T}}$ is the displacement vector of the particle.

For a block or element of mass m^i, a more generalized form for the kinetic energy functional can be written as

$$\Pi_K = -m^i\iint_\Omega \left(\{u_x, u_y\}\begin{Bmatrix} \mathrm{D}^2u_x \\ \mathrm{D}^2u_y \end{Bmatrix}\right)\mathrm{d}\Omega = -m^i\iint_\Omega \left(\{u_x, u_y\}\mathrm{D}^2\begin{Bmatrix} u_x \\ u_y \end{Bmatrix}\right)\mathrm{d}\Omega$$
$$= -m^i\{d^i\}^{\mathrm{T}}[X^i]^{\mathrm{T}}\left(\iint_\Omega [T^i]^{\mathrm{T}}[T^i]\mathrm{d}\Omega\right)[X^i]\mathrm{D}^2(\{d^i\}) = -m^i\{di\}^{\mathrm{T}}[X^i]^{\mathrm{T}}[H^i][X^i]\mathrm{D}^2(\{d^i\}) \tag{9.72}$$

in which $\mathrm{D}^2(\ ) = \partial^2(\ )/\partial t^2$ and u_x and u_y are time-dependent displacement functions of the block or element. The matrix $[H^i]$ is a property matrix given by (9.30), (9.51) and (9.67) for rigid blocks, and triangular and quadrilateral elements, respectively. Equation (9.72) represents the kinetic energy due to the mass of the block or element i, which can also be regarded as the effects of the inertial forces given by

$$f_x^i = -m^i\frac{\partial^2 u_x}{\partial t^2},\quad f_y^i = -m^i\frac{\partial^2 u_y}{\partial t^2} \tag{9.73}$$

Expanding $\{d^i\}$ in a Taylor series for a neighborhood Δt at time $t = 0$ leads to

$$\{d^i\} = \{d^i\}_0 + \frac{\partial\{d^i\}}{\partial t}(\Delta t) + \frac{1}{2}\frac{\partial^2\{d^i\}}{\partial t^2}(\Delta t)^2 + \cdots \tag{9.74}$$

Since the Eulerian description of motion is being used, the initial displacement at $t = 0$ is zero, i.e., $\{d^i\}_0 = \{0\}$. Then one can assume that approximately

$$\{d^i\} \approx \frac{\partial\{d^i\}}{\partial t}(\Delta t) + \frac{1}{2}\frac{\partial^2\{d^i\}}{\partial t^2}(\Delta t)^2 \tag{9.75}$$

i.e.,

$$\frac{\partial^2(\{d^i\})}{\partial t^2} \approx \frac{2\{d^i\}}{(\Delta t)^2} - \frac{2}{\Delta t}\frac{\partial\{d^i\}}{\partial t} \tag{9.76}$$

Substitution of (9.76) into (9.72) then produces the kinetic energy functional of the element (or block)

$$\begin{aligned}\Pi_K &= -m^i\{d^i\}^{\mathrm{T}}[X^i]^{\mathrm{T}}[H^i][X^i]D^2(\{d^i\}) \\ &= -m^i\{d^i\}^{\mathrm{T}}[X^i]^{\mathrm{T}}[H^i][X^i]D^2(\{d^i\})\left(\frac{2\{d^i\}}{(\Delta t)^2} - \frac{2}{\Delta t}\frac{\partial\{d^i\}}{\partial t}\right) \\ &= -2m^i\left(\frac{\{d^i\}^{\mathrm{T}}[X^i]^{\mathrm{T}}[H^i][X^i]\{d^i\}}{(\Delta t)^2} - \frac{\{d^i\}^{\mathrm{T}}[X^i]^{\mathrm{T}}[H^i][X^i]\{v^0\}}{\Delta t}\right)\end{aligned} \tag{9.77}$$

where vector $\{v^0\}$ is the initial velocity of the element (or block) at the beginning of the time step, given by

$$\{v^0\} = \left.\frac{\partial\{d^i\}}{\partial t}\right|_{\Delta t=0} = \left.\left(\frac{\partial u_x^c}{\partial t}, \frac{\partial u_y^c}{\partial t}, \frac{\partial r^c}{\partial t}\right)^{\mathrm{T}}\right|_{\Delta t=0} \tag{9.78}$$

for rigid blocks,

$$\{v^0\} = \left.\frac{\partial\{d^i\}}{\partial t}\right|_{\Delta t=0} = \left.\left(\frac{\partial u_x^l}{\partial t}\ \frac{\partial u_y^l}{\partial t}\ \frac{\partial u_x^m}{\partial t}\ \frac{\partial u_y^m}{\partial t}\ \frac{\partial u_x^n}{\partial t}\ \frac{\partial u_y^n}{\partial t}\right)^{\mathrm{T}}\right|_{\Delta t=0} \tag{9.79}$$

for triangular elements and

$$\{v^0\} = \left.\frac{\partial\{d^i\}}{\partial t}\right|_{\Delta t=0} = \left.\left(\frac{\partial u_x^k}{\partial t}\ \frac{\partial u_y^k}{\partial t}\ \frac{\partial u_x^l}{\partial t}\ \frac{\partial u_y^l}{\partial t}\ \frac{\partial u_x^m}{\partial t}\ \frac{\partial u_y^m}{\partial t}\ \frac{\partial u_x^n}{\partial t}\ \frac{\partial u_y^n}{\partial t}\right)^{\mathrm{T}}\right|_{\Delta t=0} \tag{9.80}$$

for quadrilateral elements. Minimization of Π_K with respect to $\{d^i\}$ leads to

$$\frac{\partial\Pi_K}{\partial\{d^i\}} = -\left(\frac{2m^i}{(\Delta t)^2}\right)[X^i]^{\mathrm{T}}[H^i][X^i]\{d^i\} + \left(\frac{2m^i}{\Delta t}\right)[X^i]^{\mathrm{T}}[H^i][X^i]\{v^0\} = 0 \tag{9.81}$$

which represents the contribution of kinetic energy of the mass of element (block) i to the global equilibrium of the system. From Eqn (9.81), one can derive a stiffness matrix and a load vector from this mechanism as

$$[k^{ii}] = -\left(\frac{2m^i}{(\Delta t)^2}\right)[X^i]^{\mathrm{T}}[H^i][X^i] \tag{9.82}$$

$$\{f^i\} = -\left(\frac{2m^i}{\Delta t}\right)[X^i]^{\mathrm{T}}[H^i][X^i]\{v^0\} \tag{9.83}$$

and they should be added to the ith diagonal sub-matrix, $\mathbf{k_{ii}}$, and the ith element of the RHS, f_i, of the global equation of motion (9.5), respectively.

For dynamic analysis, the initial velocity for the next time step, $\{v^0\}$, is the velocity at the end of the current time step, i.e.,

$$\left\{v^0_{(t+\Delta t)}\right\} = \left\{v^0_{(t)}\right\} + (\Delta t)D^2\{d^i\} \tag{9.84}$$

For static analysis, the time step can still be used artificially for iteration purposes, but the initial velocity should reach zero at the end of the iteration at each time step.

9.6.3 Element (Block) Contacts

Because that the vertex-to-edge contact is the basic contact type and other types of contacts can be induced as combinations of this basic type, the formulation of the element (block) contact can therefore be based entirely on that of this basic type. During deformation at a vertex-to-edge contact, frictional forces may be mobilized at the vertex and on the edge so that slipping may take place. Supposing that the normal stiffness of the contact is K_n (equal to the stiffness of an imaginary spring acting at the contact point in the direction normal to the edge), and the artificial interpenetration depth is d, there are four possibilities for the combinations of normal and shear forces with different considerations for the physical possibilities.

(1) The normal component, F_{n}, of the contact force is tensile, i.e., $F_{\mathrm{n}} = K_{\mathrm{n}}d > 0$. The contact should be removed because no tensile force is admissible at contacts.

(2) The normal force is compressive, but no shear force, F_{s}, is developed, i.e., $F_{\mathrm{n}} = K_{\mathrm{n}}d \leq 0$ and $F_{\mathrm{s}} = 0$. The contact is formulated by considering work done over the normal displacement of the spring, i.e., the penetration depth d. The normal component of the contact force is compressive, but the shear component is not large enough to cause slipping.

(3) Assuming similarly that an elastic response also applies in the tangential direction at the contacts represented by a shear spring of shear stiffness K_{s}(unit: Force/Length), i.e.,

$$F_{\mathrm{n}} = K_{\mathrm{n}}d \leq 0, \quad F_{\mathrm{s}} = K_{\mathrm{s}}S < |F_{\mathrm{n}}|\tan(\phi) + C \tag{9.85}$$

where ϕ and C are the friction angle and cohesion at the contact point, and S is the shear displacement. Because no energy dissipation is involved, only the potential energy due to incremental displacement of the vertex, which is reversible and path-independent, needs to be minimized.

(4) The normal component of the contact force is compressive and the shear component along the edge satisfies the Mohr–Coulomb friction law for slipping, i.e.,

$$F_{\mathrm{n}} = K_{\mathrm{n}}d \leq 0, \quad F_{\mathrm{s}} = |F_{\mathrm{n}}|\tan(\phi) + C \tag{9.86}$$

Sliding of the vertex along the edge will occur. Both the potential energy for interpenetration, d, in the normal direction and the dissipated energy for sliding distance, S, on the edge need to be minimized.

All the above considerations are based on the detection of the interpenetration depth, d, and the normal and shear stiffness at the contact point.

Contacts are represented by springs at the contact points in DDA. Assume a contact is established between vertex p_1 of the block (element) i and an edge $\overline{p_2p_3}$ of the block (element) j with an interpenetration depth, d, which represents the deformation of springs at the end of the time step at the contact spot. Denote (x_i, y_i) and $\{u_{xi}, u_{yi}\}^{\mathrm{T}}$ $(i = 1, 2, 3)$ as the co-ordinates and incremental displacements of the vertices p_i $(i = 1, 2, 3)$, respectively, and recall the expression equations (9.24) and (9.25), the depth d is rewritten as (Fig. 9.11 and Fig. 9.12)

$$d = \frac{\Delta}{L} = \frac{1}{L}\begin{vmatrix} 1, & x_1 + u_{x1}, & y_1 + u_{y1} \\ 1, & x_2 + u_{x2}, & y_2 + u_{y2} \\ 1, & x_3 + u_{x3}, & y_3 + u_{y3} \end{vmatrix} \tag{9.87}$$

where $L = \sqrt{(x_3 - x_2)^2 + (y_3 - y_2)^2}$, the determinant in (9.87) can be decomposed as

$$\begin{aligned} \Delta &= \begin{vmatrix} 1, & x_1, & y_1 \\ 1, & x_2, & y_2 \\ 1, & x_3, & y_3 \end{vmatrix} + \begin{vmatrix} 1, & u_{x1}, & y_1 \\ 1, & u_{x2}, & y_2 \\ 1, & u_{x3}, & y_3 \end{vmatrix} + \begin{vmatrix} 1, & x_1, & u_{y1} \\ 1, & x_2, & u_{y2} \\ 1, & x_3, & u_{y3} \end{vmatrix} + \begin{vmatrix} 1, & u_{x1}, & u_{y1} \\ 1, & u_{x2}, & u_{y2} \\ 1, & u_{x3}, & u_{y3} \end{vmatrix} \\ &\approx \begin{vmatrix} 1, & x_1, & y_1 \\ 1, & x_2, & y_2 \\ 1, & x_3, & y_3 \end{vmatrix} + \begin{vmatrix} 1, & u_{x1}, & y_1 \\ 1, & u_{x2}, & y_2 \\ 1, & u_{x3}, & y_3 \end{vmatrix} + \begin{vmatrix} 1, & x_1, & u_{y1} \\ 1, & x_2, & u_{y2} \\ 1, & x_3, & u_{y3} \end{vmatrix} \end{aligned} \tag{9.88}$$

because the last term

$$\begin{vmatrix} 1, & u_{x1}, & u_{y1} \\ 1, & u_{x2}, & u_{y2} \\ 1, & u_{x3} & u_{y3} \end{vmatrix} \tag{9.89}$$

is second order of infinitesimal quantity and can be neglected.

The first term of Eqn (9.88) is the area of the triangle enclosed by the three vertices and is denoted as

$$a_0 = \begin{vmatrix} 1 & x_1 & y_1 \\ 1 & x_2 & y_2 \\ 1 & x_3 & y_3 \end{vmatrix} = (x_2y_3 - x_3y_2) - (x_1y_3 - x_3y_1) + (x_1y_2 - x_2y_1) \tag{9.90}$$

The determinant in (9.88) becomes

$$\begin{aligned} \Delta &= a_0 + \begin{vmatrix} 1, & u_{x1}, & y_1 \\ 1, & u_{x2}, & y_2 \\ 1, & u_{x3}, & y_3 \end{vmatrix} + \begin{vmatrix} 1, & x_1, & u_{y1} \\ 1, & x_2, & u_{y2} \\ 1, & x_3, & u_{y3} \end{vmatrix} \\ &= a_0 + u_{x1}(y_2 - y_3) + u_{y1}(x_3 - x_2) + u_{x2}(y_3 - y_1) + u_{y2}(x_1 - x_3) + u_{x3}(y_1 - y_2) + u_{y3}(x_2 - x_1) \\ &= a_0 + \{(y_2 - y_3), (x_3 - x_2)\}\begin{Bmatrix} u_{x1} \\ u_{y1} \end{Bmatrix} \\ &\quad + \{(y_3 - y_1), (x_1 - x_3)\}\begin{Bmatrix} u_{x2} \\ u_{y2} \end{Bmatrix} + \{(y_1 - y_2), (x_2 - x_1)\}\begin{Bmatrix} u_{x3} \\ u_{y3} \end{Bmatrix} \end{aligned} \tag{9.91}$$

Since vertex p_1 belongs to block i and vertices p_2 and p_3 belong to block j, the displacements at these three vertices can be expressed as

$$\begin{Bmatrix} u_{x1} \\ u_{y1} \end{Bmatrix} = [T^i_{(x_1,y_1)}][X^i]\{d^i\}, \quad \begin{Bmatrix} u_{x2} \\ u_{y2} \end{Bmatrix} = [T^j_{(x_2,y_2)}][X^j]\{d^j\}, \quad \begin{Bmatrix} u_{x3} \\ u_{y3} \end{Bmatrix} = [T^j_{(x_3,y_3)}][X^j]\{d^j\} \tag{9.92}$$

Denoting vectors $\{e^i\}$ and $\{g^j\}$ as

$$\{e^i\} = \frac{1}{L}\{(y_2 - y_3), (x_3 - x_2)\}[T^i_{(x_1,y_1)}] \tag{9.93}$$

$$\{g^j\} = \frac{1}{L}\left(\{(y_3 - y_1), (x_1 - x_3)\}[T^j_{(x_2,y_2)}] + \{(y_1 - y_2), (x_2 - x_1)\}[T^j_{(x_3,y_3)}]\right) \tag{9.94}$$

The penetration depth d (or length change of the spring in the normal direction) can then be written as

$$d = \frac{a_0}{L} + \{e^i\}^{\mathrm{T}}[X^i]\{d^i\} + \{g^j\}^{\mathrm{T}}[X^j]\{d^j\} \tag{9.95}$$

9.6.3.1 *Formulation of contacts under purely normal compression*

Under purely normal compressive loading, the contact is represented by a normal spring of stiffness K_{n}, the moving vertex for a vertex-to-edge contact from time t to $t + \Delta t$ may move toward the edge and cause a change of displacement of the spring, u_{n}, which is equal to the artificial interpenetration displacement, $u_n = d$, given by Eqn (9.95) (Fig. 9.11).

The normal contact force is then calculated as

$$F_{\mathrm{n}} = K_{\mathrm{n}} u_{\mathrm{n}} = K_{\mathrm{n}}\left(\{e^i\}^{\mathrm{T}}[X^i]\{d^i\} + \{g^j\}^{\mathrm{T}}[X^j]\{d^j\} + \frac{a_0}{L}\right) \tag{9.96}$$

The potential energy due to the normal deformation of the spring is given by,

$$\begin{aligned}
\Pi_{\mathrm{p}} &= \frac{F_{\mathrm{n}} u_{\mathrm{n}}}{2} = \frac{K_{\mathrm{n}}(u_{\mathrm{n}})^2}{2} = \frac{K_{\mathrm{n}}}{2}\left(\frac{a_0}{L} + \{e^i\}^{\mathrm{T}}[X^i]\{d^i\} + \{g^j\}^{\mathrm{T}}[X^j]\{d^j\}\right)^2 \\
&= \frac{K_{\mathrm{n}}}{2}\left(\left(\frac{a_0}{L}\right)^2 + \{d^i\}^{\mathrm{T}}[X^i]^{\mathrm{T}}\{e^i\}\{e^i\}^{\mathrm{T}}[X^i]\{d^i\} + \{d^j\}^{\mathrm{T}}[X^j]^{\mathrm{T}}\{g^j\}\{g^j\}^{\mathrm{T}}[X^j]\{d^j\}\right. \\
&\quad + 2\{d^i\}^{\mathrm{T}}[X^i]^{\mathrm{T}}\{e^i\}\{g^j\}^{\mathrm{T}}[X^j]\{d^j\} + 2\left(\tfrac{a_0}{L}\right)\{d^i\}^{\mathrm{T}}[X^i]^{\mathrm{T}}\{e^i\} \\
&\quad \left. + 2\left(\frac{a_0}{L}\right)\{d^j\}^{\mathrm{T}}[X^j]^{\mathrm{T}}\{g^j\}\right)
\end{aligned} \tag{9.97}$$

The minimization of Π_{p} with respect to $\{d^i\}^{\mathrm{T}}$ and $\{d^j\}^{\mathrm{T}}$ produces two respective equations of motion due to contact between block (element) i and block (element) j,

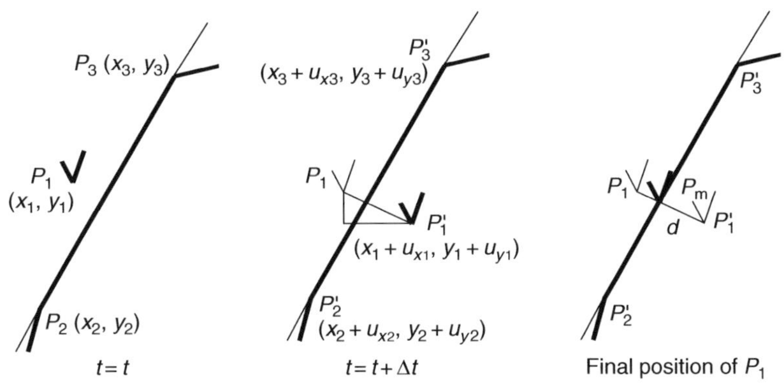

Fig. 9.11 Contact under pure normal compression (Jing, 1993).

$$(K_n[X^i]^{\mathrm{T}}\{e^i\}\{e^i\}^{\mathrm{T}}[X^i])\{d^i\}+(K_n[X^i]^{\mathrm{T}}\{e^i\}\{g^j\}^{\mathrm{T}}[X^j])\{d^j\}+\left(\frac{K_n a_0}{L}\{e^i\}\right)=0 \quad (9.98)$$

$$(K_n[X^j]^{\mathrm{T}}\{g^j\}\{e^i\}^{\mathrm{T}}[X^i])\{d^i\}+(K_n[X^j]^{\mathrm{T}}\{g^j\}\{g^j\}^{\mathrm{T}}[X^j])\{d^j\}+\left(\frac{K_n a_0}{L}\{g^j\}\right)=0 \quad (9.99)$$

which represent the contributions of the contact between block (element) i and block (element) j to the global equation of motion of the system, in the direction normal to the edge on the block (element) j. The derived stiffness matrices and load vectors can then be written as, respectively,

$$[k^{ii}] = K_n[X^i]^{\mathrm{T}}\{e^i\}\{e^i\}^{\mathrm{T}}[X^i] \quad (9.100a)$$

$$[k^{ij}] = K_n[X^i]^{\mathrm{T}}\{e^i\}\{g^j\}^{\mathrm{T}}[X^j] \quad (9.100b)$$

$$[k^{ji}] = K_n[X^j]^{\mathrm{T}}\{g^j\}\{e^i\}^{\mathrm{T}}[X^i], \quad (9.100c)$$

$$[k^{jj}] = K_n[X^j]^{\mathrm{T}}\{g^j\}\{g^j\}^{\mathrm{T}}[X^j] \quad (9.100d)$$

$$\{f^i\} = -\frac{K_n a_0}{L}[X^i]^{\mathrm{T}}\{e^i\} \quad (9.101a)$$

$$\{f^j\} = -\frac{K_n a_0}{L}[X^j]^{\mathrm{T}}\{g^j\} \quad (9.101b)$$

As before, they will be (3×3) sub-matrices and (3×1) vectors for rigid blocks, (6×6) sub-matrices and (6×1) vectors for triangular elements and (8×8) sub-matrices and (8×1) vectors for quadrilateral elements, using the corresponding $[X^i]$, $[X^j]$, $\{e^i\}$ and $\{g^j\}$ for their evaluations according to (9.93), (9.94), (9.100) and (9.101), respectively. These matrices and vectors should be assembled into sub-matrices $\mathbf{k}_{ii}$, $\mathbf{k}_{ij}$, $\mathbf{k}_{ji}$ and $\mathbf{k}_{jj}$ and the ith and jth vectorial elements of the RHS, $\mathbf{f}_i$ and $\mathbf{f}_j$ of the global equation (9.5), respectively. The above relations are valid only for $d < 0$. If $d > 0$, then the force is tensile and the contact is removed because no tensile force is allowed at a contact.

The contact stiffness matrices and generalized load vectors derived in this section represent the deformability of the contacts in the normal direction. The no-penetration condition is satisfied at the end of each time step by a numerical iteration process using sub-divided steps of the time-step interval to avoid the over-predicted magnitude of the predicted normal displacement so that the negative value of distance d by Eqn (9.95) will be within the pre-determined tolerance range. In theory, Lagrangian multiplier techniques can be applied to satisfy the non-penetration criterion. However, its numerical efficiency may or may not be better than this trial-and-error approach.

9.6.3.2 Contacts with mobilized shear force without slipping

Let vertex p_1 on block i move toward edge $\overline{p_2p_3}$ on block j with both normal and tangential displacements, thus mobilizing both normal and shear contact forces. However, the mobilized shear force is not large enough to cause slipping by the Mohr–Coulomb friction law (cf. Eqn (9. 86)). The contact is therefore represented by two springs: one in the normal and one in the tangential direction. Cohesion C is ignored here. The effect in the normal direction is treated in the same way as that presented in the previous section. The effect of the shear force is represented by an elastic shear spring of stiffness

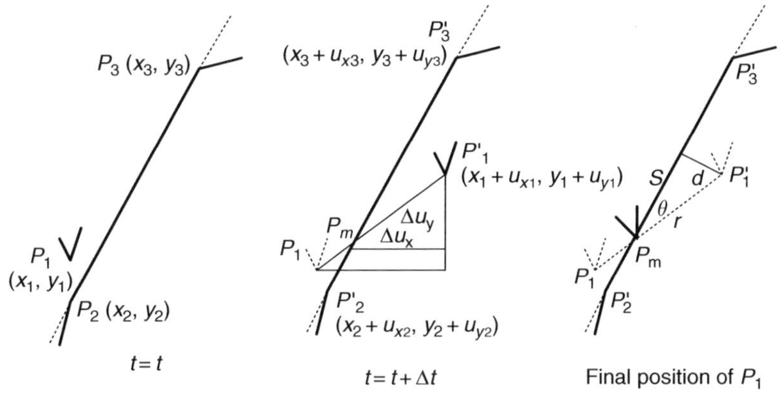

Fig. 9.12 Frictional contact between a vertex and an edge of shear distance S (Jing, 1993).

K_s, because slipping does not occur. The geometrical relation between moving vertex p_1 and edge $\overline{p_2p_3}$ is shown in Fig. 9.12. Let (x_m, y_m) be the target point of p_1 on the edge $\overline{p_2p_3}$ at the end of the time step, then the differential displacements can be written as

$$\begin{Bmatrix} \Delta u_x \\ \Delta u_y \end{Bmatrix} = \begin{Bmatrix} x_1 + u_{x1} - x_m - u_{xm} \\ y_1 + u_{y1} - y_m - u_{ym} \end{Bmatrix} = \begin{Bmatrix} u_{x1} \\ u_{y1} \end{Bmatrix} - \begin{Bmatrix} u_{xm} \\ u_{ym} \end{Bmatrix} + \begin{Bmatrix} x_1 - x_m \\ y_1 - y_m \end{Bmatrix}$$

$$= [T^i_{(x_1,y_1)}][X^i]\{d^i\} - [T^j_{(x_{m1},y_m)}][X^j]\{d^j\} + \begin{Bmatrix} x_1 - x_m \\ y_1 - y_m \end{Bmatrix} \tag{9.102}$$

From the geometrical relations, the angle θ between vector $\overline{p_2p_3}$ and $\bar{r} = \{\Delta u_x, \Delta u_y\}^{\mathrm{T}}$ and the shear distance S are given by

$$\cos(\theta) = \frac{\overline{p_2p_3} \cdot \bar{r}}{|\overline{p_2p_3}||\bar{r}|} \tag{9.103}$$

and

$$S = |\bar{r}| \cos(\theta) \tag{9.104}$$

From Eqns (9.102), (9.103) and (9.104), one obtains the shear distance (which is also the deformation of the shear spring) as

$$S = |\bar{r}| \cos(\theta) = \frac{\overline{p_2p_3} \cdot |\bar{r}|}{|\overline{p_2p_3}|} = \frac{\{(x_3 - x_2), (y_3 - y_2)\}\begin{Bmatrix} \Delta u_x \\ \Delta u_y \end{Bmatrix}}{\sqrt{(x_3 - x_2)^2 + (y_3 - y_2)^2}} = \left\{\frac{(x_3 - x_2)}{L}, \frac{(y_3 - y_2)}{L}\right\}\begin{Bmatrix} \Delta u_x \\ \Delta u_y \end{Bmatrix}$$

$$= \left\{\frac{x_3 - x_2}{L}, \frac{y_3 - y_2}{L}\right\}\left([T^i_{(x_1,y_1)}][X^i]\{d^i\} - [T^j_{(x_m,y_m)}][X^j]\{d^j\} + \begin{Bmatrix} x_1 - x_m \\ y_1 - y_m \end{Bmatrix}\right)$$

$$= \{m^i\}^{\mathrm{T}}[X^i]\{d^i\} - \{n^j\}^{\mathrm{T}}[X^j]\{d^j\} + X_m \tag{9.105}$$

with

$$\{m^i\}^{\mathrm{T}} = \left\{\frac{x_3 - x_2}{L}, \frac{x_3 - x_2}{L}\right\}[T^i_{(x_1,y_1)}] \tag{9.106a}$$

$$\{n^j\}^{\mathrm{T}} = \left\{\frac{x_3 - x_2}{L}, \frac{x_3 - x_2}{L}\right\}[T^j_{(x_m, y_m)}] \tag{9.106b}$$

$$X_m = \left\{\frac{x_3 - x_2}{L}, \quad \frac{y_3 - y_2}{L}\right\}\left\{\begin{matrix} x_1 - x_m \\ y_1 - y_m \end{matrix}\right\} \tag{9.106c}$$

The angle θ should always be in the range $0 \le \theta < 90^{\circ}$. It cannot be equal to 90° because in this case $S = 0$ so that no shear displacement occurs, which is contradictory to the assumption. The work done by the elastic shear force is then

$$\begin{aligned}\Pi_s &= -\frac{F_s S}{2} = -\frac{K_s}{2}S^2 = -\frac{K_s}{2}\left[\{m^i\}^T[X^i]\{d^i\} - \{n^j\}^{\mathrm{T}}[X^j]\{d^j\} + X_m\right]^2 \\ &= -\frac{K_s}{2}\left(\left(\frac{Xm}{L}\right)^2 + \{d^i\}^{\mathrm{T}}[X^i]^{\mathrm{T}}\{m^i\}\{m^i\}^{\mathrm{T}}[X^i]\{d^i\} + \{d^j\}^{\mathrm{T}}[X^j]^{\mathrm{T}}\{n^j\}\{n^j\}^{\mathrm{T}}[X^j]\{d^j\}\right. \\ &\quad -2\{d^i\}^{\mathrm{T}}[X^i]^{\mathrm{T}}\{m^i\}\{n^j\}^{\mathrm{T}}[X^j]\{d^j\} + 2\left(\frac{X_m}{L}\right)\{d^i\}^{\mathrm{T}}[X^i]^{\mathrm{T}}\{m^i\} \\ &\quad \left. -2\left(\frac{X_m}{L}\right)\{d^j\}^{\mathrm{T}}[X^j]^{\mathrm{T}}\{n^j\}\right)\end{aligned} \tag{9.107}$$

The minimization of the functional Π_s with respect to $\{d^i\}$ and $\{d^j\}$ then leads to

$$-(K_s[X^i]^{\mathrm{T}}\{m^i\}\{m^i\}^{\mathrm{T}}[X^i])\{d^i\} + (K_s[X^i]^{\mathrm{T}}\{m^i\}\{n^j\}^{\mathrm{T}}[X^j])\{d^j\} - K_s X_m[X^i]^{\mathrm{T}}\{m^i\}^{\mathrm{T}} = 0 \tag{9.108}$$

$$K_s[X^j]^{\mathrm{T}}\{n^j\}\{m^i\}^{\mathrm{T}}[X^i]\{d^i\} - (K_s[X^j]^{\mathrm{T}}\{n^i\}\{n^i\}^{\mathrm{T}}[X^j])\{d^j\} + K_s X_m[X^j]^{\mathrm{T}}\{n^i\}^{\mathrm{T}} = 0 \tag{9.109}$$

This leads to the following element stiffness matrices and load vectors

$$[k^{ii}] = -K_s[X^i]^{\mathrm{T}}\{m^i\}\{m^i\}^{\mathrm{T}}[X^i] \tag{9.110a}$$

$$[k^{ij}] = K_s[X^i]^{\mathrm{T}}\{m^i\}\{n^j\}^{\mathrm{T}}[X^j] \tag{9.110b}$$

$$[k^{ji}] = K_s[X^j]^{\mathrm{T}}\{n^j\}\{m^i\}^{\mathrm{T}}[X^i], \tag{9.110c}$$

$$[k^{jj}] = -K_s[X^j]^{\mathrm{T}}\{n^j\}\{n^j\}^{\mathrm{T}}[X^j] \tag{9.110d}$$

$$\{f^i\} = K_s X_m[X^i]^{\mathrm{T}}\{m^i\} \tag{9.111a}$$

$$\{f^j\} = -K_s X_m[X^j]^{\mathrm{T}}\{n^j\} \tag{9.111b}$$

These matrices and vectors should be assembled into sub-matrices $\mathbf{k}_{ii}$, $\mathbf{k}_{ij}$, $\mathbf{k}_{ji}$ and $\mathbf{k}_{jj}$ and the ith and jth vectorial elements of the RHS, $\mathbf{f}_i$ and $\mathbf{f}_j$ of the global equation (9.5), respectively.

Notice should be given to the different ranks and expressions of matrices $[X^i]$, $[X^j]$, $\{m^i\}$ and $\{n^j\}$ for rigid blocks, triangular elements or quadrilateral elements, respectively.

9.6.3.3 *Formulation of frictional contacts with slipping*

Using the same assumption as adopted in Section 9.6.3.2, but the shear force is now large enough to satisfy the friction law, the process is irreversible and the work done by the shear force over the sliding distance S is the dissipated energy.

Adopting a Mohr–Coulomb friction criterion and assuming that the shear force after slipping is given by

$$F_s = F_n \tan(\phi) + C = K_n \, d\tan(\phi) + C \tag{9.112}$$

and recall Eqns (9.95), (9.105) and (9.106), the work done by the shear force over a sliding distance S can then be written as

$$\begin{aligned}
\Pi_f &= -F_s S = -(F_n \tan\phi + C)S = -(K_n \tan\phi)(Sd) - CS \\
&= -(K_n \tan\phi)(\{m^i\}^T[X^i]\{d^i\} - \{n^j\}^T[X^j]\{d^j\} + X_m)\left(\frac{a_0}{L} + \{e^i\}^T[X^i]\{d^i\} + \{g^j\}^T[X^j]\{d^j\}\right) \\
&-C\left(\{m^i\}^T[X^i]\{d^i\} - \{n^j\}^T[X^j]\{d^j\} + X_m\right) \\
&= -(K_n \tan\phi)\Big(\{d^i\}^T[X^i]^T\{e^i\}\{m^i\}^T[X^i]\{d^i\} + \{d^i\}^T[X^i]^T\{m^i\}\{g^j\}^T[X^j]\{d^j\} \\
&-\{d^i\}^T[X^i]^T\{e^i\}\{n^j\}^T[X^j]\{d^j\} - \{d^j\}^T[X^j]^T\{g^j\}\{n^i\}^T[X^j]\{d^j\} + \frac{a_0}{L}\{d^i\}^T[X^i]^T\{m^i\} \\
&-\frac{a_0}{L}\{d^j\}^T[X^j]^T\{n^j\} + X_m\{d^i\}^T[X^i]^T\{e^i\} + X_m\{d^j\}^T[X^j]^T\{g^j\} + \frac{X_m a_0}{L}\Big) \\
&-C\{d^i\}^T[X^i]^T\{m^i\} + C\{d^j\}^T[X^j]^T\{n^j\} - X_m C
\end{aligned} \tag{9.113}$$

The minimization of Π_f with respect to $\{d^i\}$ and $\{d^j\}$ leads to two local equilibrium equations

$$\begin{aligned}
&-(K_n \tan\phi)\Big([X^i]^T\{e^i\}\{m^i\}^T[X^i]\{d^i\} + ([X^i]^T\{m^i\}\{g^j\}^T[X^j] - [X^i]^T\{e^i\}\{n^j\}^T[X^j])\{d^j\} \\
&+\frac{a_0}{L}[X^i]^T\{m^i\} + X_m[X^i]^T\{e^i\}\Big) - C[X^i]^T\{m^i\} = 0
\end{aligned} \tag{9.114}$$

$$\begin{aligned}
&-(K_n \tan\phi)\Big(([X^j]^T\{g^j\}\{m^i\}^T[X^i] - [X^j]^T\{n^j\}\{e^i\}^T[X^i])\{d^i\} - [X^j]^T\{g^j\}\{n^j\}^T[X^j]\{d^j\} \\
&-\frac{a_0}{L}[X^j]^T\{n^j\} + X_m[X^j]^T\{g^j\}\Big) + C[X^i]^T\{n^j\} = 0
\end{aligned} \tag{9.115}$$

These two equations represent the contribution of a sliding contact between block (element) i and block (element) j to the global equilibrium of the block system. The resultant stiffness matrices and load vectors therefore can be derived as

$$[k^{ii}] = -(K_n \tan\phi)[X^i]^T\{e^i\}\{m^i\}^T[X^i] \tag{9.116a}$$

$$[k^{ij}] = -(K_n \tan\phi)[X^i]^T\left(\{m^i\}\{g^j\}^T - \{e^i\}\{n^j\}^T\right)[X^j] \tag{9.116b}$$

$$[k^{ji}] = -(K_n \tan\phi)[X^j]^T\left(\{g^j\}\{m^i\}^T - \{n^j\}\{e^i\}^T\right)[X^i] \tag{9.116c}$$

$$[k^{jj}] = (K_n \tan\phi)[X^j]^T\{g^j\}\{n^j\}^T[X^j] \tag{9.116d}$$

$$\{f^i\} = (k_n \tan\phi)\left(-\frac{a_0}{L}[X^i]^T\{m^i\} - X_m[X^i]^T\{e^i\}\right) + C[X^i]^T\{m^i\} \tag{9.117a}$$

$$\{f^j\} = (k_n \tan\phi)\left(\frac{a_0}{L}[X^j]^T\{n^j\} - X_m[X^j]^T\{g^j\}\right) - C[X^j]^T\{n^j\} \tag{9.117b}$$

These matrices and vectors then are added into the respective sub-matrices, $\mathbf{k}_{ii}$, $\mathbf{k}_{ij}$, $\mathbf{k}_{ji}$ and $\mathbf{k}_{jj}$, and load vectors $\mathbf{f}_i$ and $\mathbf{f}_j$, of the global equilibrium equation (9.5), respectively. Similar care should be taken concerning the different ranks and expressions of the matrices $[X^i]$, $[X^j]$, $\{m^i\}$, $\{n^j\}$, $\{e^i\}$ and $\{g^j\}$ for rigid blocks, triangular elements or quadrilateral elements, respectively.

However, the sub-matrices $[k^{ii}]$, $[k^{ij}]$, $[k^{ji}]$ and $[k^{jj}]$ may become non-symmetric when frictional shear failure occurs. A special equation solver capable of handling asymmetrical matrix equations should be used.

9.6.4 External Forces

9.6.4.1 Point forces

Concentrated forces may lie either inside a block or on its boundary, and they can be either external loads or contact forces developed at contact points. Let $\{F_x, F_y\}^T$ be the point force vector acting at an arbitrary point (x, y) inside or on the boundary of a rigid block or element (triangular or quadrilateral) i (see Fig. 9.13), then the potential energy functional contributed by the external point force is given by

$$\Pi_p = (F_x u_x + F_y u_y) = \{F_x, F_y\}\begin{Bmatrix} u_x \\ u_y \end{Bmatrix}_{x=x_0, y=y_0} = \{F\}^T\{d^i\} = \{F\}^T[T^i(x_0, y_0)][X^i]\{d^i\} \tag{9.118}$$

The minimization of the functional is carried out to produce the following local equation of motion

$$\frac{\partial \Pi_p}{\partial\{d^i\}} = \frac{\partial}{\partial\{d^i\}}\left(\{F\}^T[T^i(x_0, y_0)]^T[X^i]\{d^i\}\right) = \{F\}^T[T^i(x_0, y_0)]^T[X^i] = 0 \tag{9.119}$$

which results in a load vector

$$\{f^i\} = -\{F\}^T[T^i(x_0, y_0)]^T[X^i] \tag{9.120}$$

representing the contribution of the point force vector to the equilibrium of the block system. It should be added to the ith element of the RHS vector of the global equation (9.5), $\mathbf{f}_i$, and different matrices $[T]$ and $[X]$ should be used for selecting rigid blocks, triangular elements or quadrilateral elements for the system.

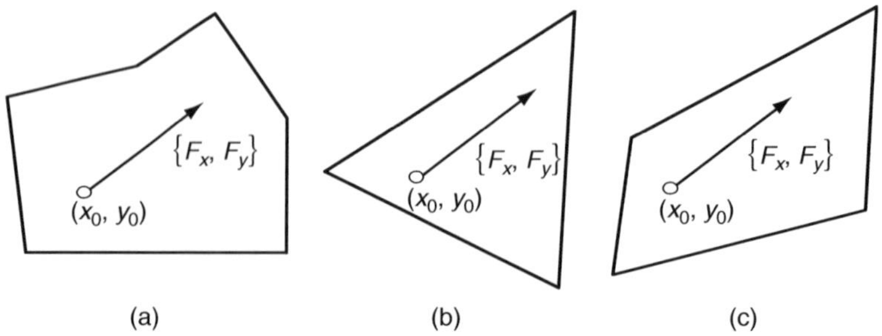

Fig. 9.13 Point forces on (a) a rigid block, (b) triangular element and (c) a quadrilateral element.

9.6.4.2 Distributed forces

Distributed forces are assumed to act along a straight line of length L from point (x_1, y_1) to point (x_2, y_2) in block or element i (Fig 9.14). A parametric equation of the line segment can be written as

$$\begin{cases} x = (x_2 - x_1)\omega + x_1 \\ y = (y_2 - y_1)\omega + y_1 \end{cases} \quad \text{and} \quad L = \sqrt{(x_2 - x_1)^2 + (y_2 - y_1)^2} \tag{9.121}$$

where $0 \leq \omega \leq 1$ is the parameter. Assuming that the intensity of the external force along the line is given by $\{F(\omega)\} = \{f_x(\omega), f_y(\omega)\}^{\mathrm{T}}$, the potential energy functional is written

$$\Pi_l = \int_0^L (f_x(\omega)u_x + f_y(\omega)u_y)\mathrm{d}l = \int_0^1 \{f_x(\omega), f_y(\omega)\}\begin{Bmatrix} u_x \\ u_y \end{Bmatrix} L \, \mathrm{d}\omega$$

$$= \int_0^1 \{F(\omega)\}^{\mathrm{T}}[T^i(\omega)][X^i]\{d^i\}L \, \mathrm{d}\omega = \left(L \int_0^1 \{F(\omega)\}^{\mathrm{T}}[T^i(\omega)][X^i]\mathrm{d}\omega \right)\{d^i\} \tag{9.122}$$

where matrix $[T^i]$ also becomes a function of parameter ω.

The minimization of energy functional Π_l with respect to the nodal displacement vector is given by

$$\frac{\partial \Pi_l}{\partial\{d^i\}} = \frac{\partial}{\partial\{d^i\}}\left(\left(L \int_0^1 \{F(\omega)\}^{\mathrm{T}}[T^i(\omega)][X^i]\mathrm{d}\omega \right)\{d^i\} \right) = L \int_0^1 \{F(\omega)\}^{\mathrm{T}}[T^i(\omega)][X^i]\mathrm{d}\omega = 0 \tag{9.123}$$

The result of this minimization is a load vector representing the contribution of distributed external forces to the global equilibrium of the block system, given by

$$\{f^i\} = -L\{F_T^i\}[X^i] \tag{9.124}$$

where

$$\{F_T^i\} = \int_0^1 \{F(\omega)\}^{\mathrm{T}}[T^i(\omega)]\mathrm{d}\omega \tag{9.125}$$

The resultant load vector should be added into the ith element of the RHS of the global equation (9.5), $\mathbf{f}_i$. The integration in Eqn (9.125) depends on the functional forms of $f_x(\omega)$ and $f_y(\omega)$. For simple forms of $f_x(\omega)$ and $f_y(\omega)$, the closed-form solutions may be obtained. Otherwise, numerical integration must be used.

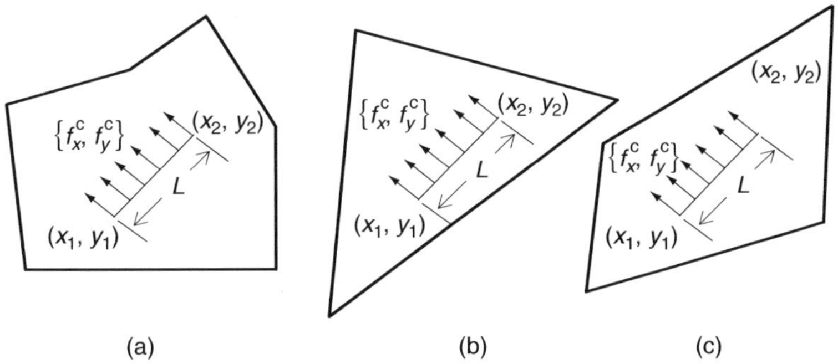

Fig. 9.14 Uniformly distributed forces on (a) a rigid block, (b) triangular element and (c) a quadrilateral element.

Recall Eqns (9.27), (9.44) and (9.59), the transformed matrix $[T^i(\omega)]$ is given by

$$[T^i(\omega)] = \begin{bmatrix} 1 & 0 & -(y_2 - y_1)\omega + (y_1 - y_c) \\ 0 & 1 & (x_2 - x_1)\omega + (x_1 - x_c) \end{bmatrix} \tag{9.126}$$

for rigid blocks,

$$[T^i(\omega)] = \begin{bmatrix} 1 & 0 & (x_2 - x_1)\omega + x_1 & 0 & (y_2 - y_1)\omega + y_1 & 0 \\ 0 & 1 & 0 & (x_2 - x_1)\omega + x_1 & 0 & (y_2 - y_1)\omega + y_1 \end{bmatrix} \tag{9.127}$$

for triangular elements and

$$[T^i(\omega)] = \begin{bmatrix} 1 & 0 \\ 0 & 1 \\ (x_2 - x_1)\omega + x_1 & 0 \\ 0 & (x_2 - x_1)\omega + x_1 \\ (y_2 - y_1)\omega + y_1 & 0 \\ 0 & (y_2 - y_1)\omega + y_1 \\ [(x_2 - x_1)\omega + x_1][(y_2 - y_1)\omega + y_1] & 0 \\ 0 & [(x_2 - x_1)\omega + x_1][(y_2 - y_1)\omega + y_1] \end{bmatrix}^{\mathrm{T}} \tag{9.128}$$

for quadrilateral elements.

Two simple cases are considered here: (a) $\{F(\omega)\}$ is uniform and (b) $\{F(\omega)\}$ is linearly distributed along the line.

(1) Uniformly distributed forces

In the case of $f_x(\omega) = f_x^0, f_y(\omega) = f_y^0$ and f_x^0 and f_y^0 are both constants (Fig. 9.14), the integration in (9.125) produces different vectors $\{F_{\mathrm{T}}^i\}$ for rigid body, triangular elements and quadrilateral elements. Using Eqns (9.126), (9.127) and (9.128), these vectors are given as

$$\{F_T^i\} = \left\{ f_x^c, \; f_y^c, \; \frac{f_y^c(x_2 - x_1) - f_x^c(y_2 - y_1)}{2} + f_y^c(x_1 - x_c) + f_x^c(y_1 - y_c) \right\}^{\mathrm{T}} \tag{9.129}$$

$$\{F_T^i\} = \left\{ f_x^c, \; f_y^c, \; f_x^c \frac{x_2 + x_1}{2}, \; f_y^c \frac{x_2 + x_1}{2}, \; f_x^c \frac{y_2 + y_1}{2}, \; f_y^x \frac{y_2 + y_1}{2} \right\}^{\mathrm{T}} \tag{9.130}$$

$$\{F_T^i\} = \begin{Bmatrix} f_x^c \\ f_y^c \\ f_x^c \dfrac{x_2 + x_1}{2} \\ f_y^c \dfrac{x_2 + x_1}{2} \\ f_x^c \dfrac{y_2 + y_1}{2} \\ f_y^c \dfrac{y_2 + y_1}{2} \\ f_x^c \left(\dfrac{(x_2 - x_1)(y_2 - y_1)}{3} + \dfrac{x_2 y_1 + x_1 y_2}{2} + x_1 y_1 \right) \\ f_y^c \left(\dfrac{(x_2 - x_1)(y_2 - y_1)}{3} + \dfrac{x_2 y_1 + x_1 y_2}{2} + x_1 y_1 \right) \end{Bmatrix} \tag{9.131}$$

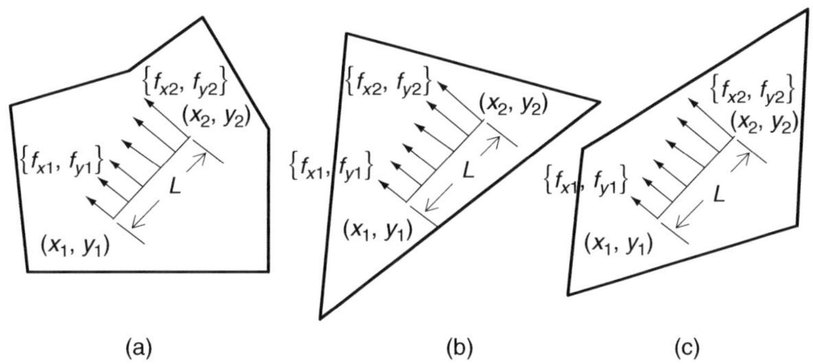

Fig. 9.15 Uniformly distributed forces on (a) a rigid block, (b) triangular element and (c) a quadrilateral element.

for rigid blocks, triangular elements and quadrilateral elements, respectively. The resultant load vectors for the three different cases are obtained by substitution of (9.129), (9.130) and (9.131) into (9.124), respectively, using the corresponding matrices $[X^i]$.

(2) Linearly distributed forces

If $\{F(\omega)\} = \{f_x(\omega), f_y(\omega)\}^{\mathrm{T}}$ varies linearly along the line of loading (Fig. 9.15), it can be expressed as a function of parameter ω as

$$\begin{cases} f_x(\omega) = (f_{x2} - f_{x1})\omega + f_{x1} \\ f_y(\omega) = (f_{y2} - f_{y1})\omega + f_{y1} \end{cases} (0 \le \omega \le 1) \tag{9.132}$$

Using Eqns (9.132), (9.126), (9.127), (9.128) and denoting $\Delta x_{21} = x_2 - x_1$, $\Delta y_{21} = y_2 - y_1$, $\Delta x_{1c} = x_1 - x_c$, $\Delta y_{1c} = y_1 - y_c$, $\Delta f_x = f_{x2} - f_{x1}$ and $\Delta f_y = f_{y2} - f_{y1}$, respectively, the integration in (9.125) becomes

$$\{F_T^i\} = \begin{Bmatrix} \dfrac{(f_{x2} + f_{x1})}{2} \\ \dfrac{(f_{y2} + f_{y1})}{2} \\ \dfrac{\Delta f_y \Delta x_{21}}{3} + \dfrac{\Delta f_y \Delta x_{1c}}{2} + \dfrac{f_{y1}\Delta x_{21}}{2} + f_{y1}\Delta x_{1c} - \dfrac{\Delta f_x \Delta y_{21}}{3} + \dfrac{\Delta f_x \Delta y_{1c}}{2} + f_{x1}\Delta y_{1c} - f_{x1}\dfrac{\Delta y_{21}}{2} \end{Bmatrix} \tag{9.133}$$

$$\{F_T^i\} = \begin{Bmatrix} \dfrac{\Delta f_x}{2} + f_{x1} \\ \dfrac{\Delta f_y}{2} + f_{y1} \\ \dfrac{\Delta f_x \Delta x_{21}}{3} + \dfrac{x_1 \Delta f_x + f_{x1}\Delta x_{21}}{2} + x_1 f_{x1} \\ \dfrac{\Delta f_y \Delta x_{21}}{3} + \dfrac{x_1 \Delta f_y + f_{y1}\Delta x_{21}}{2} + x_1 f_{y1} \\ \dfrac{\Delta f_x \Delta y_{21}}{3} + \dfrac{y_1 \Delta f_x + f_{x1}\Delta y_{21}}{2} + y_1 f_{x1} \\ \dfrac{\Delta f_y \Delta y_{21}}{3} + \dfrac{y_1 \Delta f_y + f_{y1}\Delta y_{21}}{2} + y_1 f_{y1} \end{Bmatrix} \tag{9.134}$$

$$\{F_T^i\} = \left\{ \begin{array}{c} \frac{\Delta f_x}{2} + f_{x1} \\ \frac{\Delta f_y}{2} + f_{y1} \\ \frac{\Delta f_x \Delta x_{21}}{3} + \frac{x_1 \Delta f_x + f_{x1} \Delta x_{21}}{2} + x_1 f_{x1} \\ \frac{\Delta f_y \Delta x_{21}}{3} + \frac{x_1 \Delta f_y + f_{y1} \Delta x_{21}}{2} + x_1 f_{y1} \\ \frac{\Delta f_x \Delta y_{21}}{3} + \frac{y_1 \Delta f_x + f_{x1} \Delta y_{21}}{2} + y_1 f_{x1} \\ \frac{\Delta f_y \Delta y_{21}}{3} + \frac{y_1 \Delta f_y + f_{y1} \Delta y_{21}}{2} + y_1 f_{y1} \\ \frac{\Delta f_x \Delta x_{21} \Delta y_{21}}{4} + \frac{\Delta f_x (y_1 \Delta x_{21} + x_1 \Delta y_{21}) + f_{x1} \Delta x_{21} \Delta y_{21}}{3} + \frac{\Delta f_x x_1 y_1 + f_{x1} (y_1 \Delta x_{21} + x_1 \Delta y_{21})}{2} + f_{x1} x_1 y_1 \\ \frac{\Delta f_y \Delta x_{21} \Delta y_{21}}{4} + \frac{\Delta f_y (y_1 \Delta x_{21} + x_1 \Delta y_{21}) + f_{y1} \Delta x_{21} \Delta y_{21}}{3} + \frac{\Delta f_y x_1 y_1 + f_{y1} (y_1 \Delta x_{21} + x_1 \Delta y_{21})}{2} + f_{y1} x_1 y_1 \end{array} \right\} \tag{9.135}$$

for rigid blocks, triangular elements and quadrilateral elements, respectively. Let $f_{x2} = f_{x1} = f_x^c$ and $f_{y2} = f_{y1} = f_y^c$, then $\Delta f_x = \Delta f_y = 0$, Eqn (9.133), (9.134) and (9.135) are reduced to the forms of (9.129), (9.130) and (9.131), respectively.

9.6.5 Body (Volume) Forces

Constant body (volume) forces are considered as another form of external loading, such as gravity. Assuming that the intensity of the body force is (b_x, b_y), then the potential energy due to this body force of the block is given by the functional

$$\Pi_b = \iint_\Omega \{b_x, b_y\} \begin{Bmatrix} u_x \\ u_y \end{Bmatrix} d\Omega = \{b^i\}^T \left(\iint_\Omega [T^i] d\Omega \right) [X^i]\{d^i\} = \{b^i\}^T [I^i][X^i]\{d^i\} \tag{9.136}$$

where $\{b^i\} = \{b_x, b_y\}^T$ and $[I^i] = \iint_\Omega [T^i] d\Omega$. Recall (9.27), (9.44) and (9.59), the matrix $[I^i]$ in (9.136) are evaluated as

$$[I^i] = \iint_\Omega \begin{bmatrix} 1 & 0 & -(y - y_c) \\ 0 & 1 & x - x_c \end{bmatrix} d\Omega = \begin{bmatrix} S_{00} & 0 & -S_{01} + y_c S_{00} \\ 0 & S_{00} & S_{10} - x_c S_{00} \end{bmatrix} = \begin{bmatrix} S_{00} & 0 & 0 \\ 0 & S_{00} & 0 \end{bmatrix} \tag{9.137}$$

for rigid blocks,

$$[I^i] = \iint_\Omega \begin{bmatrix} 1 & 0 & x & 0 & y & 0 \\ 0 & 1 & 0 & x & 0 & y \end{bmatrix} d\Omega = \begin{bmatrix} S_{00} & 0 & S_{10} & 0 & S_{01} & 0 \\ 0 & S_{00} & 0 & S_{10} & 0 & S_{01} \end{bmatrix} \tag{9.138}$$

for triangular elements and

$$[I^i] = \iint_\Omega \begin{bmatrix} 1 & 0 & x & 0 & y & 0 & xy & 0 \\ 0 & 1 & 0 & x & 0 & y & 0 & xy \end{bmatrix} d\Omega = \begin{bmatrix} S_{00} & 0 & S_{10} & 0 & S_{01} & 0 & S_{11} & 0 \\ 0 & S_{00} & 0 & S_{10} & 0 & S_{01} & 0 & S_{11} \end{bmatrix} \tag{9.139}$$

for quadrilateral elements.

The minimization of potential energy Π_b with respect to the nodal displacement vector is carried out as

$$\frac{\partial \Pi_b}{\partial \{d^i\}} = \{b^i\}^T [I^i][X^i] = 0 \tag{9.140}$$

and it leads to a (m × 1) vector

$$\{f^i\} = -\{b^i\}^T [I^i][X^i] \tag{9.141}$$

and should be added to the ith element of the RHS of the global equation (9.5), $\mathbf{f}_i$. The integer $m = 3$, 6 and 8 for rigid blocks, triangular elements and quadrilateral elements, respectively.

9.6.6 Displacement Constraints

The displacement constraints in specific directions are one of the two most commonly encountered boundary conditions (the another being force/stress boundary conditions which can be considered through point or distributed loads at the boundaries). More generally, an existing displacement δ in a certain direction at a point inside (or on the boundary of) a block or element can be treated as an input to the system (Fig. 9.16), using a very stiff artificial spring with the specified displacement as its pre-tension displacement.

Let (l_x, l_y) be the direction cosines of a specified displacement vector of magnitude δ, at a specified point (x_d, y_d) in the block (element) i. The specified displacement vector can be written as

$$\begin{Bmatrix} u_x^0 \\ u_y^0 \end{Bmatrix} = \delta \begin{Bmatrix} l_x \\ l_y \end{Bmatrix} \tag{9.142}$$

The differential displacement vector during the time step is therefore

$$\begin{Bmatrix} \Delta u_x \\ \Delta u_y \end{Bmatrix} = \begin{Bmatrix} u_x^d \\ u_y^d \end{Bmatrix} - \delta \begin{Bmatrix} l_x \\ l_y \end{Bmatrix} = \begin{Bmatrix} u_x^d \\ u_y^d \end{Bmatrix} - \begin{Bmatrix} u_x^0 \\ u_y^0 \end{Bmatrix} \tag{9.143}$$

Assuming that the deformation stiffness of the block (element) i at point (x_d, y_d) is k, which can be imagined as a strong spring acting in the opposite direction to the specified displacement, δ, then the imaginary spring force is

$$\begin{Bmatrix} f_x \\ f_y \end{Bmatrix} = k \begin{Bmatrix} \Delta u_x \\ \Delta u_y \end{Bmatrix} = k \left(\begin{Bmatrix} u_x^d \\ u_y^d \end{Bmatrix} - \begin{Bmatrix} u_x^0 \\ u_y^0 \end{Bmatrix} \right) \tag{9.144}$$

A large value of k is needed for displacement constraint.

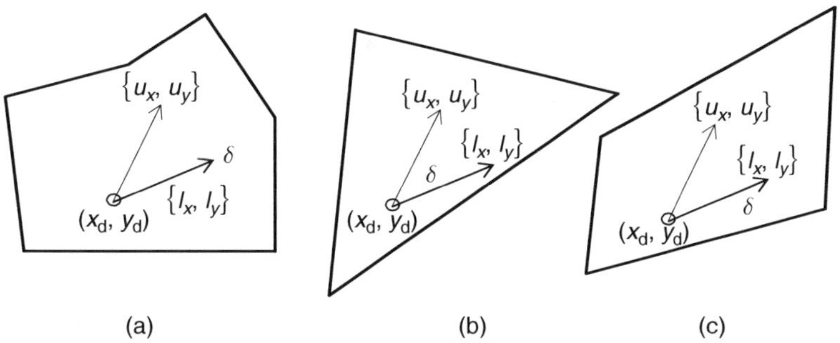

Fig. 9.16 Displacement constraint in a specific direction. (a) a rigid block; (b) a triangular element and (c) a quadrilateral element.

The potential energy caused by the prescribed displacement, δ can then be written as

$$\Pi_\delta = \frac{1}{2}\{f_x, f_y\}\begin{Bmatrix} \Delta u_x \\ \Delta u_y \end{Bmatrix} = \frac{k}{2}\left(\begin{Bmatrix} u_x^{\mathrm{d}} \\ u_y^{\mathrm{d}} \end{Bmatrix} - \begin{Bmatrix} u_x^0 \\ u_y^0 \end{Bmatrix}\right)^2$$

$$= \frac{k}{2}\left(\left\{u_x^{\mathrm{d}}, u_y^{\mathrm{d}}\right\}\begin{Bmatrix} u_x^{\mathrm{d}} \\ u_y^{\mathrm{d}} \end{Bmatrix} - 2\left\{u_x^0, u_y^0\right\}\begin{Bmatrix} u_x^{\mathrm{d}} \\ u_y^{\mathrm{d}} \end{Bmatrix} + \delta^2\right)$$

$$= \frac{k}{2}\left(\{d^i\}^{\mathrm{T}}[X^i]^{\mathrm{T}}[T^i(x_{\mathrm{d}}, y_{\mathrm{d}})]^{\mathrm{T}}[T^i(x_{\mathrm{d}}, y_{\mathrm{d}})][X^i](d^i)\right) - (k\delta)\{l_x, l_y\}[T^i(x_{\mathrm{d}}, y_{\mathrm{d}})][X^i]\{d^i\} + \frac{k\delta^2}{2} \tag{9.145}$$

because $l_x^2 + l_y^2 = 1$. The minimization of Π_δ with respect to the nodal displacement vector produces a local equation of motion

$$\frac{\partial \Pi_\delta}{\partial\{d^i\}} = k[X^i]^{\mathrm{T}}[T^i(x_{\mathrm{d}}, y_{\mathrm{d}})]^{\mathrm{T}}[T^i][X^i(x_{\mathrm{d}}, y_{\mathrm{d}})]\{d^i\} - k\left\{u_x^0, u_y^0\right\}[T^i(x_{\mathrm{d}}, y_{\mathrm{d}})][X^i] = 0 \tag{9.146}$$

which represents the contribution of a specified displacement of magnitude δ, at point $(x_{\mathrm{d}}, y_{\mathrm{d}})$, in the direction (l_x, l_y) to the global equation of the system. A $(m \times m)$ sub-matrix $[k^{ii}]$ and a $(m \times 1)$ vector $\{f^i\}$ can be derived as

$$[k^{ii}] = k[X^i]^{\mathrm{T}}[T^i(x_{\mathrm{d}}, y_{\mathrm{d}})]^{\mathrm{T}}[T^i(x_{\mathrm{d}}, y_{\mathrm{d}})][X^i] \tag{9.147a}$$

$$\{f^i\} = k\left\{u_x^0, u_y^0\right\}[T^i(x_{\mathrm{d}}, y_{\mathrm{d}})][X^i] \tag{9.147b}$$

with m = 3, 6 and 8 for rigid blocks, triangular elements or quadrilateral elements, respectively. They should be added to the ith diagonal sub-matrix, $\mathbf{k}_{ii}$, and the ith element of the RHS, $\mathbf{f}_i$, of the global equilibrium equation (9.5), respectively.

For the case in which $\delta = 0$, then $u_x^0 = u_y^0 = 0$, $\{f^i\} = \{0\}$ implying that the block (element) i is fixed in the specified direction, which can then be used as a fixed displacement boundary condition with respect to that specified direction, when applied to boundary blocks (elements). When $\delta \neq 0$, the formulation can be used as non-zero displacement boundary conditions to specify certain displacement paths which is desired for some particular blocks (elements), representing certain specific loading histories.

9.6.7 Rock Bolts

A simple scheme of bolting is implemented in the current version of the DDA code which connects two blocks (elements) i and j by a spring of constant stiffness. One anchoring point (x_1, y_1) is located inside block (element) i and another, (x_2, y_2), in block (element) j, see Fig. 9.17.

The bolt operates only in tension and no yielding of the bolt is considered at present. The respective displacements at the anchoring points are (u_{x1}, u_{y1}) and (u_{x2}, u_{y2}). The initial length of the bolt, L, and its direction cosines, (l_x, l_y), are written as

$$L = \sqrt{(x_2 - x_1)^2 + (y_2 - y_1)^2}, \quad l_x = \frac{x_2 - x_1}{L}, \quad l_y = \frac{y_2 - y_1}{L} \tag{9.148}$$

Let the stiffness of the bolt be $K_{\mathrm{b}} = EA_{\mathrm{b}}/L$, where E is the Young's modulus of the bolt material and A_{b} its cross-sectional area. The differential tensile displacement of the bolt during a time step is then

$$\Delta L = \{l_x, l_y\}\begin{Bmatrix} u_{x1} \\ u_{y1} \end{Bmatrix} - \{l_x, l_y\}\begin{Bmatrix} u_{x2} \\ u_{y2} \end{Bmatrix} \tag{9.149}$$

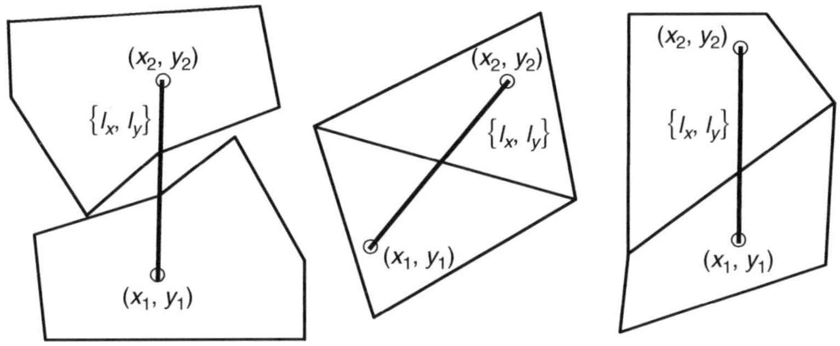

Fig. 9.17 Rock bolting between blocks and elements. (a) rigid blocks; (b) triangular elements and (c) quadrilateral elements.

Assume that the bolt has an initial tensional force F_T^0 and its incremental force due to ΔL is given by $K_b(\Delta L)$. The potential energy built up in the bolt is then

$$\begin{aligned}\Pi_m &= \frac{K_b}{2}(\Delta L)^2 + F_T^0(\Delta L)\\ &= \frac{K_b}{2}\left(\{l_x, l_y\}\begin{Bmatrix} u_{x1} \\ u_{y1}\end{Bmatrix} - \{l_x, l_y\}\begin{Bmatrix} u_{x2} \\ u_{y2}\end{Bmatrix}\right)^2 + F_T^0\left(\{l_x, l_y\}\begin{Bmatrix} u_{x1} \\ u_{y1}\end{Bmatrix} - \{l_x, l_y\}\begin{Bmatrix} u_{x2} \\ u_{y2}\end{Bmatrix}\right)\\ &= \frac{K_b}{2}\left(\begin{Bmatrix} u_{x1} \\ u_{y1}\end{Bmatrix}^T\begin{Bmatrix} l_x \\ l_y\end{Bmatrix}\begin{Bmatrix} l_x \\ l_y\end{Bmatrix}^T\begin{Bmatrix} u_{x1} \\ u_{y1}\end{Bmatrix} - 2\begin{Bmatrix} u_{x1} \\ u_{y1}\end{Bmatrix}^T\begin{Bmatrix} l_x \\ l_y\end{Bmatrix}\begin{Bmatrix} l_x \\ l_y\end{Bmatrix}^T\begin{Bmatrix} u_{x2} \\ u_{y2}\end{Bmatrix} + \begin{Bmatrix} u_{x2} \\ u_{y2}\end{Bmatrix}^T\begin{Bmatrix} l_x \\ l_y\end{Bmatrix}\begin{Bmatrix} l_x \\ l_y\end{Bmatrix}^T\begin{Bmatrix} u_{x2} \\ u_{y2}\end{Bmatrix}\right)\\ &\quad + F_T^0\left(\{l_x, l_y\}\begin{Bmatrix} u_{x1} \\ u_{y1}\end{Bmatrix} - \{l_x, l_y\}\begin{Bmatrix} u_{x2} \\ u_{y2}\end{Bmatrix}\right)\\ &= \frac{K_b}{2}\left(\{d^i\}^T[X^i]^T\{q^i\}\{q^i\}^T[X^i]\{d^i\} - 2\{d^i\}^T[X^i]^T\{q^i\}\{p^j\}^T[X^j]\{d^j\}\right.\\ &\quad \left. + \{d^j\}^T[X^j]^T\{p^j\}\{p^j\}^T[X^j]\{d^j\}\right) + F_T^0\left(\{q^i\}^T[X^i]\{d^i\} - \{p^j\}^T[X^j]\{d^j\}\right)\end{aligned} \tag{9.150a}$$

where

$$\{q^i\} = [T^i(x_1, y_1)]^T\begin{Bmatrix} l_x \\ l_y\end{Bmatrix}, \quad \{p^j\} = [T^j(x_2, y_2)]^T\begin{Bmatrix} l_x \\ l_y\end{Bmatrix} \tag{9.150b}$$

The minimization of Π_m with respect to $\{d^i\}$ and $\{d^j\}$ results in two local equations of motion

$$K_b([X^i]^T\{q^i\}\{q^i\}^T[X^i]\{d^i\} - [X^i]^T\{q^i\}\{p^j\}^T[X^j]\{d^j\} + F_T^0\{q^i\}^T[X^i] = 0 \tag{9.151}$$

$$K_b(-[X^j]^T\{p^j\}\{q^i\}^T[X^i]\{d^i\} + [X^j]^T\{p^j\}\{p^j\}^T[X^j]\{d^j\}) - F_T^0\{p^j\}^T[X^j] = 0 \tag{9.152}$$

which represent the contributions of the sum of the bolting and mobilized forces connecting blocks (elements) i and j to the global equation of motion of the system. Using the same technique as before, the resultant $(m \times m)$ stiffness matrices and $(m \times 1)$ load vectors are

$$[k^{ii}] = K_b[X^i]^T\{q^i\}\{q^i\}^T[X^i] \tag{9.153a}$$

$$[k^{ij}] = -K_b[X^i]^T\{q^i\}\{p^j\}^T[X^j] \tag{9.153b}$$

$$[k^{ji}] = -K_b [X^j]^T \{p^j\} \{q^i\}^T [X^i] \tag{9.153c}$$

$$[k^{jj}] = K_b [X^j]^T \{p^j\} \{p^j\}^T [X^j] \tag{9.153d}$$

$$\{f^i\} = -F_T^0 \{q^i\}^T [X^i] \tag{9.154a}$$

$$\{f^j\} = F_T^0 \{p^j\}^T [X^j] \tag{9.154b}$$

The values of rank m for these matrices and vectors, as before, are 3, 6 and 8, depending on whether rigid blocks, triangular elements or quadrilateral elements are used, respectively. They should be added to sub-matrices $\mathbf{k}_{ii}$, $\mathbf{k}_{ij}$, $\mathbf{k}_{ji}$ and $\mathbf{k}_{jj}$, and load vectors $\mathbf{f}_i$ and $\mathbf{f}_j$, of the global equilibrium equation (9.5), respectively.

9.7 Assembly of the Global Equations of Motion

The technique to assembly the global equations of motion (or equilibrium equation) is the same as the standard technique used in FEM, following a properly designed numbering system of blocks and elements. To simplify the assembly process and reduce the computer storage requirement, it is desirable to first number the blocks consecutively, as 1, 2, 3, 4,. . . , N. Then number the elements and nodes in each block consecutively, leaving no gaps between the blocks, i.e.,

Block 1: elements: 1, 2, . . . , M_E^1; nodes: 1, 2, . . . , M_N^1;

Block 2: elements: M_E^1+1, M_E^1+2, . . . , M_E^2; nodes: M_N^1+1, M_N^1+2, . . . , M_N^2;. . .

Block i: elements: M_E^{i-1}+1, M_E^{i-1}+2, . . . , M_E^i; nodes: M_N^{i-1}+1, M_N^{i-1}+2, . . . , M_N^i;. . .

Block N: elements: M_E^{N-1}+1, M_E^{N-1}+2, . . . , M_E^N; nodes: M_N^{N-1}+1, M_N^{N-1}+2, . . . , M_N^N.

To illustrate the assembly process, a simple example of a three block system is shown in Fig. 9.18. The solid deformation and block contact are the only two mechanisms considered in this example, using triangular elements. The formation of the deformation and contact stiffness sub-matrices of elements and blocks, and the resultant global stiffness matrix are shown in Fig. 9.19, in which each small non-blank square represents a (2 x 2) matrix corresponding to two orthogonal displacement components at a node.

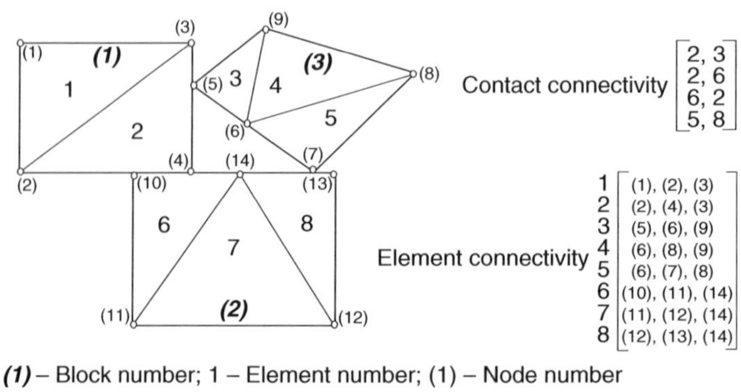

Fig. 9.18 A three block system and its element connectivity and block contact matrices (Jing, 1998).

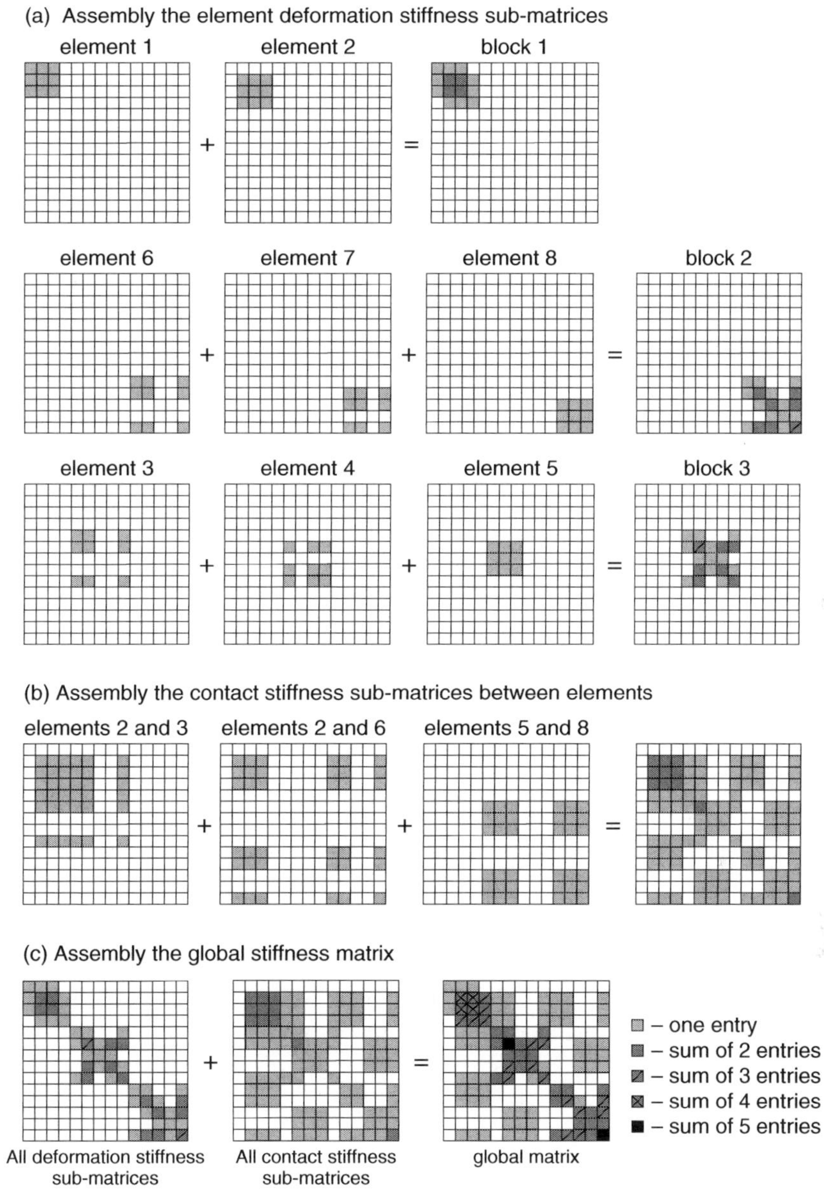

Fig. 9.19 Formation of the global stiffness matrix of a three block system. (a) Deformation stiffness sub-matrices for each block; (b) Contact stiffness sub-matrices of the three blocks; (c) The global stiffness matrix (Jing, 1998).

9.8 Fluid Flow and Coupled Hydro-Mechanical Analysis in DDA

Modeling fluid flow in the current DDA method follows the standard DFN technique to determine the pressure and flow-rate distributions along the fractures and at their intersections. The mathematical principle is well described in Wittke (1990) for both 2D and 3D problems, and is also summarized in Chapter 4. The capability of the DDA method for analysis of coupled fluid flow and rock deformation has been realized only recently for 2D problems, see Ma (1999), Kim et al. (2000) and Jing et al. (2001). A short presentation is given below.

The basic assumption is that the rock blocks are not permeable and the fluid flow is conducted entirely by the connected fracture network, i.e., the connected spaces formed between rock blocks. The flow in these fracture spaces, which are long and narrow with generally very small apertures, follows the Cubic Law as described in Chapter 4. The effects of roughness on the walls of fractures are ignored at present. The coupling of flow and deformation is presented in two ways:

(1) The effect of fluid pressure on the boundary surfaces of rock blocks as external loads, therefore affecting the stresses, strains and gross movements of rock blocks and

(2) The effect of strain and deformation of rock blocks on the flow-rate due to the change of fracture aperture during the process of deformation.

Note that the stress-deformation analysis technique is fully covered in early chapters of this book, and the governing equations for the viscous fluid flow through connected fracture networks are presented in Chapter 4 in detail. Therefore only the algorithms for coupled flow-deformation of discrete block systems are presented in this section. Only the steady-state fluid flow is considered.

9.8.1 Formulation of the Fluid Pressure-Block Deformation Coupling in DDA

The governing equations for the coupled flow-deformation analysis are the combined equations of (4.57) and (9.5)

$$\begin{cases} [T_{ij}]\{p_j\} = \{\hat{q}_j\} \\ [k_{ij}]\{d_j\} = \{f_j\} \end{cases} \tag{9.155}$$

where $\{p_j\}$ is the pressure vector (instead of the head vector in (4.57) but the two can be interchanged easily), $\{\hat{q}_j\}$ the flow rate vector, $[k_{ij}]$ the stiffness matrix of rock blocks (elements), $\{d_j\}$ the nodal displacement vector, $\{f_j\}$ the load vector and matrix $[T_{ij}]$ the conductivity matrix, given by an equivalent aperture e_{ij} and length L_{ij} of the fracture connecting intersection i and j (cf. Eqn (4.58)). Now because the aperture varies with the displacements of the nodes defining the fracture, and the load vector of blocks (elements) changes due to fluid pressure (or head), therefore the conductivity matrix becomes a function of the nodal displacement vector and the load vector becomes a function of fluid pressure or head, i.e.,

$$\begin{cases} [T_{ij}(d_j)]\{p_j\} = \{\hat{q}_j\} \\ [k_{ij}]\{d_j\} = \{f_j(\{p_j\})\} \end{cases} \tag{9.156}$$

The solution of this equation is through a time marching process using an appropriately selected time step Δt. The coupled analysis requires two operations at the end of each time step:

(1) Updating the conductivity matrix $[T_{ij}(\{d_j\})]$ according to current values of the nodal displacements, by re-calculating the equivalent aperture e_{ij} and length L_{ij} of the fracture connecting intersections i and j.

(2) Updating the load vector $\{f_j(p_j)\}$ according to the pressure distribution along the boundaries of blocks or boundary edges of elements forming fractures.

The first is a straightforward operation without any need for elaboration. The second requires extra formulations of the load vector due to pressure. The principle is given briefly in Chapter 4. Below is a

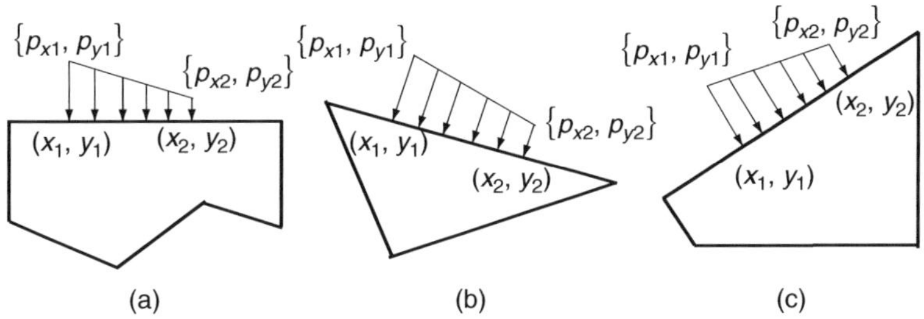

Fig. 9.20 Fluid pressure on (a) a rigid block, (b) triangular element and (c) a quadrilateral element.

more formal formulation of the load vector due to fluid pressure in terms of energy minimization used in DDA, for rigid blocks, triangular elements and quadrilateral elements (Fig. 9.20).

The formulation is similar to the linearly distributed loads. The convention here is that the pressure is always negative (compressive) pointing toward the boundary of blocks or elements, and is linearly distributed over a length L defined by two points (x_1, y_1) and (x_2, y_2). The co-ordinates of a moving point along the segment is defined by the two points and its length given by (cf. (9.121)), namely,

$$\begin{cases} x = (x_2 - x_1)\omega + x_1 \\ y = (y_2 - y_1)\omega + y_1 \end{cases} \text{ and } L = \sqrt{(x_2 - x_1)^2 + (y_2 - y_1)^2} \tag{9.157}$$

where $0 \leq \omega \leq 1$. The conversion between pressure p and head is given by

$$p = (h - z)\rho g \tag{9.158}$$

where ρ is the fluid density, g the gravitational acceleration and z the relative elevation. Noting that the vector $\{p(\omega)\} = \{p_x(\omega), p_y(\omega)\}^{\mathrm{T}}$ varies linearly along the line segment of pressure action (Fig. 9.20), it can be expressed as functions of parameter ω as

$$\begin{cases} p_x(\omega) = (p_{x2} - p_{x1})\omega + p_{x1} \\ p_y(\omega) = (p_{y2} - p_{y1})\omega + p_{y1} \end{cases} \tag{9.159}$$

The potential energy functional due to the action of pressure and material deformation is given by

$$\Pi_w = -\int_0^1 \{p(\omega)\}^{\mathrm{T}}\{u\}L \, \mathrm{d}\omega = -\int_0^1 \{p_x(\omega), p_y(\omega)\}\begin{Bmatrix} u_x \\ u_y \end{Bmatrix} L \, \mathrm{d}\omega \tag{9.160}$$

$$= -L\left(\int_0^1 \{p_x(\omega), p_y(\omega)\}[T^i(\omega)]\mathrm{d}\omega\right)[X^i]\{d^i\} = -L\{F_P^i\}^{\mathrm{T}}[X^i]\{d^i\} \tag{9.161}$$

Denoting $\Delta x_{21} = x_2 - x_1$, $\Delta y_{21} = y_2 - y_1$, $\Delta x_{1c} = x_1 - x_c$, $\Delta y_{1c} = y_1 - y_c$, $\Delta p_x = p_{x2} - p_{x1}$ and $\Delta p_y = p_{y2} - p_{y1}$, and using similar techniques as in Section 9.6.4, the vector $\{F_P^i\}$ in (9.161) becomes

$$\{F_P^i\} = \begin{Bmatrix} \dfrac{(p_{x2} + p_{x1})}{2} \\ \dfrac{(p_{y2} + p_{y1})}{2} \\ \dfrac{\Delta p_y \Delta x_{21}}{3} + \dfrac{\Delta p_y \Delta x_{1c}}{2} + \dfrac{p_{y1}\Delta x_{21}}{2} + p_{y1}\Delta x_{1c} - \dfrac{\Delta p_x \Delta y_{21}}{3} + \dfrac{\Delta p_x \Delta y_{1c}}{2} + p_{x1}\Delta y_{1c} - p_{x1}\dfrac{\Delta y_{21}}{2} \end{Bmatrix} \tag{9.162}$$

$$
\{F_P^i\} = \begin{Bmatrix} \dfrac{\Delta p_x}{2} + p_{x1} \\ \dfrac{\Delta p_y}{2} + p_{y1} \\ \dfrac{\Delta p_x \Delta x_{21}}{3} + \dfrac{x_1 \Delta p_x + p_{x1} \Delta x_{21}}{2} + x_1 p_{x1} \\ \dfrac{\Delta p_y \Delta x_{21}}{3} + \dfrac{x_1 \Delta p_y + p_{y1} \Delta x_{21}}{2} + x_1 p_{y1} \\ \dfrac{\Delta p_x \Delta y_{21}}{3} + \dfrac{y_1 \Delta p_x + p_{x1} \Delta y_{21}}{2} + y_1 p_{x1} \\ \dfrac{\Delta p_y \Delta y_{21}}{3} + \dfrac{y_1 \Delta p_y + p_{y1} \Delta y_{21}}{2} + y_1 p_{y1} \end{Bmatrix} \tag{9.163}
$$

$$
\{F_P^i\} = \begin{Bmatrix} \dfrac{\Delta p_x}{2} + p_{x1} \\ \dfrac{\Delta p_y}{2} + p_{y1} \\ \dfrac{\Delta p_x \Delta x_{21}}{3} + \dfrac{x_1 \Delta p_x + p_{x1} \Delta x_{21}}{2} + x_1 p_{x1} \\ \dfrac{\Delta p_y \Delta x_{21}}{3} + \dfrac{x_1 \Delta p_y + p_{y1} \Delta x_{21}}{2} + x_1 p_{y1} \\ \dfrac{\Delta p_x \Delta y_{21}}{3} + \dfrac{y_1 \Delta p_x + p_{x1} \Delta y_{21}}{2} + y_1 p_{x1} \\ \dfrac{\Delta p_y \Delta y_{21}}{3} + \dfrac{y_1 \Delta p_y + p_{y1} \Delta y_{21}}{2} + y_1 p_{y1} \\ \dfrac{\Delta p_x \Delta x_{21} \Delta y_{21}}{4} + \dfrac{\Delta p_x (y_1 \Delta x_{21} + x_1 \Delta y_{21}) + p_{x1} \Delta x_{21} \Delta y_{21}}{3} + \dfrac{\Delta p_x x_1 y_1 + p_{x1}(y_1 \Delta x_{21} + x_1 \Delta y_{21})}{2} + p_{x1} x_1 y_1 \\ \dfrac{\Delta p_y \Delta x_{21} \Delta y_{21}}{4} + \dfrac{\Delta p_y (y_1 \Delta x_{21} + x_1 \Delta y_{21}) + p_{y1} \Delta x_{21} \Delta y_{21}}{3} + \dfrac{\Delta p_y x_1 y_1 + p_{y1}(y_1 \Delta x_{21} + x_1 \Delta y_{21})}{2} + p_{y1} x_1 y_1 \end{Bmatrix} \tag{9.164}
$$

for rigid blocks, triangle elements and quadrilateral elements, respectively.

The minimization of the functional Π_w leads to a load vector of rank $(m \times 1)$

$$
\{f^i\} = L\{F_P^i\}^{\mathrm{T}}[X^i] \tag{9.165}
$$

which should be added to the ith element, $\mathbf{f}_i$ of the global load vector, representing the contribution of the fluid pressure to the global load system. The rank of the local vector, $\{f^i\}$, $m = 3$, 6 and 8 for rigid blocks, triangular elements and quadrilateral elements, respectively.

The above formulation is more general and flexible than that presented in Ma (1999) in which the fluid pressure was assumed to be distributed along the full length of an edge of an element. This algorithm has limitations when the free surface of the fluid cuts through part of the element boundaries during its final position determination through iteration. In such a case, only part of an element (or block) edge may be submerged inside the free surface and part of it is dry. For all submerged elements, one needs only to set the co-ordinates of (x_1, y_1) and (x_2, y_2) equal to the co-ordinates of intersections.

9.8.2 Solution Approach

Solution of the simultaneous equation (9.156) usually requires an iterative approach through a time marching process. The aim is to reach a steady state solution for both the mechanical and hydraulic variables, especially a stable hydraulic free surface and quasi-static distributions of fields of stress and strain. The technique applied for the flow analysis is to adjust, at each time step, the location of the hydraulic free surface (by changing the conductivity matrix), contact configurations and flow-rate vector (due to varied apertures, and therefore the changed conductivity matrix), using an initial flow-rate solution approach. The iteration scheme for flow analysis is then interwoven with a similar iteration process for the mechanical analysis so that aperture variations (due to changes in nodal displacements) can be included to adjust the conductivity matrix. In turn the changes in pressure along fractures due to aperture changes are included in re-evaluation of the load vectors as mechanical boundary conditions. The iteration continues until both a steady stress field and a near-zero velocity field for the block system is achieved, and a stable free surface for hydraulic analysis is obtained simultaneously. The approach can also be changed by inserting the iteration for flow analysis into the iteration process for mechanical calculations, which may cause a different convergence rate in the computer implementation.

In the remaining part of this section, the initial flow-rate approach for flow analysis is first presented, followed by the interwoven iteration scheme for coupled flow-stress calculations.

9.8.2.1 Residual flow method for unconfined flow problems

For confined flow problems, i.e., the complete boundary of the domain of interest is specified with either pressures or heads, or flow-rates, Eqn (9.155) is linear and the solution is straightforward. If, however, part of the boundary of the interested domain is a free surface with unknown location, then the flow is unconfined and the solution must be coupled with the determination of the free surface through an iteration procedure. An example of such a free surface is the location of the uppermost water table for seepage problems, which cannot be specified beforehand with certainty. There are two types of iteration approaches to achieve this aim: the residual flow method proposed first by Desai (1976, 1977) and the variable conductivity method as proposed by Bathe (1979). They are similar counterparts to the initial stress method and initial stiffness matrix method for the iterative solution of elasto-plastic deformation problems in solid mechanics. The techniques are well known and almost standard. Here, only the residual flow technique is presented since it was used in coupling with stress analysis using DDA in Jing et al. (2001), with discrete fracture network flows in two dimensions.

A free surface is characterized by the following features: (1) $h = z$, i.e., the head value at the free surface equals to the elevation of the point; (2) $p = 0$, i.e., the pressure is zero. This is the natural implication of (1); (3) $\partial h/\partial n = 0$, i.e., the flow rate across the free surface (in the direction normal to the surface) is zero.

The steps of the residual flow method are:

(1) Formation of initial conductivity matrix $\left[T_{ij}^0\right]$ using initial values of apertures of the fractures, which can be either known or assumed;

(2) Using an assumed initial location of the free surface (for example the top surface of the model), and calculating the initial head vector $\left\{h_j^0\right\}$, as the differences of the initial head value and elevations of all intersections of the connected fractures;

(3) Calculating the initial estimation of the flow-rate vector $\left\{\hat{q}_j^0\right\}$ by

$$\left\{\hat{q}_i^0\right\} = \left[T_{ij}^o\right]\left\{h_j^0\right\} \tag{9.166}$$

and using the appropriately defined boundary conditions as required by the problem.

(4) For iteration numbers i (i=1, 2, . . . , N), compare the elevations of the intersection points with their head values and connect those points at which $h = z$ conditions are satisfied. These points form the

current location of the free surface. Interpolation may be needed to locate the free surface more accurately across boundaries of boundary block (or element) edges.

(5) Check whether the current location of the free surface satisfies the condition $\partial h/\partial n = 0$. If the answer is 'Yes', the current location of the free surface is the final one and the iteration ends. If answer is 'No', its location should be corrected by adding a corrective term of the residual flow rate vector. The corrective term is formed by using the excessive flow across the free surface at those intersection points forming the current free surface, given by

$$\{\Delta q^i\} = -[T_{ij}^0]\left\{\Delta h_j^i\right\} \tag{9.167}$$

where

$$\left\{\Delta h_{ij}^i\right\} = \left\{h_j^i\right\} - \left\{z_j^{i-1}\right\} \tag{9.168}$$

(6) Correct the current flow vector by

$$\left\{\hat{q}_j^{i+1}\right\} = \left\{\hat{q}_j^i\right\} + \{\Delta q^i\} \tag{9.169}$$

and solve the equation

$$\left\{h_j^{i+1}\right\} = [T_{ij}^0]^{-1}\left\{\hat{q}_j^{i+1}\right\} \tag{9.170}$$

(7) Back to step (3) and start the next iteration.

For flow analysis, the conductivity matrix is unchanged throughout the iteration process, with the assumption that the initial values of apertures are representative values of the problem concerned and their variation due to mechanical deformation is ignored.

9.8.2.2 *Iterative solution of the coupled flow-stress calculations*

The solution of the coupled flow-stress equations (9.156) is by inserting the flow calculation cycles into the stress calculation cycles. This means that, at the end of each time step for stress analysis, the updated aperture from the new nodal displacement vector will be used to update the conductivity matrix, and the iteration for the new location of the free surface is performed until a stable free surface is found. The new pressures along the connected fractures are then updated according to the new head vector, and are used to re-evaluate the new load vector for stress analysis by considering new pressure distributions along edges of the relevant block (element) boundaries using Eqns (9.162), (9.163) or (9.164). The stress calculation is then continued to the next iteration step with the new RHS load vector to produce the new nodal displacement vector, and the process continues until final solutions for the mechanical variables are obtained. The free surface location is therefore updated at each mechanical time step due to changes in aperture, and the mechanical load vector for stress calculations is changed at the beginning of each time step by re-evaluating contributions due to fluid pressure along fracture surfaces. The calculation steps can therefore be summarized simply as follows:

(1) Formation of initial conductivity matrix $\left[T_{ij}^0\right]$, initial head vector $\left\{h_j^0\right\}$ and initial flow-rate vector $\left\{\hat{q}_j^0\right\}$ as described in steps (1), (2) and (3) in Section 9.8.2.1; formation of stiffness matrix $\left[k_{ij}\right]$,

which is kept constant throughout the process, and the RHS load vector $\left\{f_j^0\right\}$ according to initial head vector $\left\{h_j^0\right\}$ (after transforming it into pressure according to (9.158));

(2) Solution of Eqn (9.5) to obtain the initial nodal displacement vector $\left\{d_j^0\right\}$

$$\left\{d_j^0\right\} = [k_{ij}]^{-1}\left\{f_j^0\right\} \tag{9.171}$$

(3) For steps $l = 1, 2, \ldots, M$ for stress analysis and k (k=1, 2, . . ., N) for iterative flow analysis, determine the current location of the free surface by using the residual flow method equations

$$\{\Delta q^k\} = -[T_{ij}^{k-1}]\left\{\Delta h_j^k\right\} \tag{9.172}$$

$$\left\{\hat{q}_j^{k+1}\right\} = \left\{\hat{q}_j^k\right\} + \{\Delta q^k\} \tag{9.173}$$

and

$$\left\{h_j^{k+1}\right\} = [T_{ij}^k]^{-1}\left\{\hat{q}_j^{k+1}\right\} \tag{9.174}$$

where

$$\left\{\Delta h_j^k\right\} = \left\{h_j^k\right\} - \left\{z_j^{k-1}\right\} \tag{9.175}$$

until a new free surface is located, with updated $\left[T_{ij}^N\right], \left\{q_j^N\right\}$ and h_j^N

(4) Re-evaluate the RHS load vector by using the updated head vector at all intersections

$$\left\{f_j^{l+1}\right\} = \left\{f_j^l\right\} + \{\Delta f^j(h)\} \tag{9.176}$$

where the corrective term

$$\{\Delta f^j(h)\} = L\{\Delta F_P^j\}^{\mathrm{T}}[X^j] \tag{9.177}$$

and the term ΔF_P^j are obtained by replacing pressure components p_x and p_y with

$$\Delta p_x^l = p_x^l - p_x^{l-1}, \quad \Delta p_x^l = p_x^l - p_x^{l-1} \tag{9.178}$$

in Eqns (9.162), (9.163) or (9.164).

(5) Solution of mechanical equation of motion to produce new nodal displacement vector

$$\left\{d_j^{l+1}\right\} = [k_{ij}]^{-1}\left\{f_j^{l+1}\right\} \tag{9.179}$$

(6) Updating the conductivity matrix by correcting aperture values due to new nodal displacement vector

$$[T_{ij}^0] = [T_{ij}^N(\left\{d_j^{l+1}\right\})] \tag{9.180}$$

where the new initial conductivity matrix is formed using the new aperture data according to the new nodal displacement vector $\left\{d_j^{l+1}\right\}$, and using also

$$\{h_j^0\} = \{h_j^N\} \tag{9.181}$$

$$\{\hat{q}_j^0\} = \{\hat{q}_j^N\} \tag{9.182}$$

to renew all initial hydraulic variables.

(7) Go back to step (3) for the next step of stress analysis until a final solution is obtained.

9.9 Summary Remarks

The DDA method has emerged as an attractive model for geomechanical problems because its advantages cannot be replaced by continuum-based methods or explicit DEM formulations. Compared with the explicit approach of the DEM, the DDA method has four basic advantages over the explicit DEM:

(1) The equilibrium condition is automatically satisfied for quasi-static problems without the need for using excessive iteration cycles.

(2) The length of the time step can be larger, and without the risk of inducing numerical instability.

(3) Closed-form integrations for the element and block stiffness matrices can be performed without the need for Gaussian quadrature techniques.

(4) It is easy to convert an existing FEM code into a DDA code and include many mature FEM techniques without inheriting the limitations of the ordinary FEM, such as small deformation, continuous material geometry and reduced efficiency for dynamic analysis. However, matrix equations are produced and need to be solved, using the same FEM technique.

On the other hand, DDA also inherits some of the drawbacks of FEM as well. Chief among them is the much larger computer memory requirements for its matrix equation systems, which become even more sparsely populated due to contacts between blocks (elements). Unlike FEM, the extension of DDA to 3D and for coupled THM processes in porous and fractured porous rocks has not reached the same level of maturity and sophistication for more reliable and flexible simulations for rock engineering applications.

Formulations similar to DDA have also been developed and are reported in the literature, such as the discrete finite element approach for rock mechanics and geotechnical engineering problems in Ghaboussi (1988) and Barbosa and Ghaboussi (1990), and the combined finite-discrete method for fracturing and fragmentation processes of solids by Munjiza et al. (1995), Munjiza and Andrews (2000) and Munjiza and John (2002). The former used almost the same FEM approach for representing block deformation as that in DDA, and the later is fully described in a book by Munjiza (2004). These methods are, therefore, not repeated in this book.

In many aspects, DDA is also similar to the Manifold Method (MM) created by Shi (1991, 1992b, 1996) and further developed by Lin (1995, 2003) and Chen et al. (1998). The latter is in fact a more general and uniform treatment of FEM and DDA regarding rock matrix deformation, block movements, contacts and fractures, and DDA may be regarded as a special case of the MM. In practice, the fundamentals of DDA and MM are similar and linked. For reasons of space limitations, and the reason that at present at least DEM and DDA are the main stream of research and application of the discrete approaches in rock engineering, the details of the MM will not be presented in this book.

References

Barbosa, R. and Ghaboussi, J., Discrete finite element method for multiple deformable bodies. Journal of Finite Elements in Analysis and Design, 1990;7(2):145–158.

Bathe, K. J., Finite element free surface seepage analysis without mesh. International Journal for Numerical and Analytical Methods in Geomechanics, 1979;3(1):13–22.

Cai, Y., He, T. and Wang, R., Numerical simulation of dynamic process of the Tangshan earthquake by a new method-LDDA. Pure and Applied Geophysics, 2000;157(11–12):2083–2104.

Chang, Q., Non-linear dynamic discontinuous deformation analysis with finite element meshed block systems. Ph.D. Thesis, University of California, Berkeley, USA, 1994.

Chang, C.-T., Monteiro, P., Nemati, K. and Shyu, K., Behavior of marble under compression. Journal of Materials in Civil Engineering, 1996;8(3):157–171.

Chen, G., Miki, S. and Ohnishi, Y., Practical improvement on DDA. In: Salami, M. R. and Banks, D. (eds.), Discontinuous deformation analysis (DDA) and simulations of discontinuous media, pp. 302–309. TSI Press. Albuquerque, NM, 1996.

Chen, G., Ohnishi, Y. and Ito, T., Development of high-order manifold method. International Journal for Numerical Methods in Engineering, 1998;43(4):685–712.

Cheng, Y. M., Advancements and improvement in discontinuous deformation analysis. Computers and Geotechnics, 1998;22(2):153–163.

Cheng, Y. M. and Zhang, Y. H., Rigid body rotation and block internal discretization in DDA analysis. International Journal for Numerical and Analytical Methods in Geomechanics, 2000;24(6):567–578.

Chiou, Y.-J., Tzeng, J.-C. and Hwang, S.-C., Discontinuous deformation analysis for reinforced concrete frames infilled with masonry walls. Structural Engineering and Mechanics, 1998;6(2):201–215.

Chiou, Y.-J., Tzeng, J.-C. and Liou, Y.-W., Experimental and analytical study of masonry infilled frames. Journal of Structural Engineering, 1999;125(10):1109–1117.

Desai, C, S., Finite element residual schemes for unconfined flow. International Journal for Numerical Methods in Engineering, 1976;10(6):1415–1418

Desai, C. S., Flow through porous media. In: Numerical methods in geotechnical engineering. McGraw-Hill, New York, 1977.

Doolin, D. M. and Sitar, N., DDAML-discontinuous deformation analysis markup language. International Journal of Rock Mechanics and Mining Sciences 2001;38(3):467–474.

Ghaboussi, J., Fully deformable discrete element analysis using a finite element approach. Computers and Geotechnics, 1988;5(3):175–195.

Hatzor, Y. H. and Benary, R., The stability of a laminated Vousoir beam: back analysis of a historic roof collapse using DDA. International Journal of Rock Mechanics and Mining Sciences, 1998;35(2):165–181.

Hatzor, Y. H. and Feintuch, A., The validity of dynamic block displacement prediction using DDA. International Journal of Rock Mechanics and Mining Sciences, 2001;38(4):599–606.

Hsiung, S. M., Discontinuous deformation analysis (DDA) with *n*-th order polynomial displacement functions. In: Elsworth, D., Tinucci, J. P. and Heasley, P. E. (eds.), Rock mechanics in the national interest, Swets & Zeitlinger Lisse. pp. 1437–1444. 2001.

Hsiung, S. M. and Shi, G., Simulation of earthquake effects on underground excavations using discontinuous deformation analysis (DDA). In: Elsworth, D., Tinucci, J. P. and Heasley, P. E. (eds.), Rock mechanics in the national interest, Swets & Zeitlinger Lisse, pp. 1413–1420. 2001.

Jiang, Q. H. and Yeung, M. R., A model of point-to-face contact for three-dimensional discontinuous deformation analysis. Rock Mechanics and Rock Engineering, (2004);37(2):95–116.

Jing, L., Contact formulations via energy minimization. Proc. 2nd Int. Conf. DEM. IESL Publications, pp. 15–26. MIT, Boston, 1993.

Jing, L., Formulations of discontinuous deformation analysis (DDA) – an implicit discrete element model for block systems. Engineering Geology, 1998;49(3–4):371–381.

Jing, L., A review of techniques, advances and outstanding issues in numerical modelling for rock mechanics and rock engineering. International Journal of Rock Meehanics and Mining sciences, 2003;40(3):283–353.

Jing, L., Ma, Y. and Fang, Z., Modeling of fluid flow and solid deformation for fractured rocks with discontinuous deformation analysis (DDA) method. International Journal of Rock Mechanics and Mining Sciences, 2001;38(3):343–356.

Kim, Y., Amadei, B. and Pan, E., Modeling the effect of water, excavation sequence and rock reinforcement with discontinuous deformation analysis. International Journal of Rock Mechanics and Mining Sciences, 1999;36(7):949–970.

Kong, X. and Liu, J., Dynamic failure numeric simulations of model concrete-faced rock-fill dam. Soil Dynamics and Earthquake Engineering, 2002;22(9–12):1131–1134.

Koo, C. Y. and Chern, J. C., Modification of the DDA method for rigid block problems. International Journal of Rock Mechanics and Mining Sciences, 1998;35(6):683–693.

Lin, J. S., Continuous and discontinuous analysis using the manifold method, Proc. 1st Int. Int. Conf. on Analysis of Discontinuous Deformation (ICADD-I), Chungli, Taiwan, pp. 223–241. 1995.

Lin, J. S., A mesh based partition of unity method for discontinuity modeling. Computer Methods in Applied Mechanics and Engineering, 2003;192(11–12):1515–1532.

Lin, C. T., Amadei, B., Jung, J. and Dwyer, J., Extensions of discontinuous deformation analysis for jointed rock masses. Int. J. Rock Mech. Min. Sci. & Geomech. Abstr., 1996;33(7):671–694.

Ma, Y., Study on the DDA method for coupled fluid flow and rock deformation. Ph.D. Thesis, Beijing University of Science and Technology, Beijing, China, 1999.

MacLauphlin, M., Sitar, N., Doolin, D. and Aboot, T., Investigation of slope stability kinematics using discontinuous deformation analysis. International Journal of Rock Mechanics and Mining Sciences, 2001;38(5):753–762.

Mortazavi, A. and Katsabanis, P. D., Discontinuum modelling of blasthole expansion and explosive gas pressurization in jointed media. FRAGBLAST-International Journal for Blasting and Fragmentation, 1998;2:249–268.

Mortazavi, A.; Katsabanis, P. D., Modelling the effects of discontinuity orientation, continuity, and dip on the process of burden breakage in bench blasting. FRAGBLAST-International Journal for Blasting and Fragmentation, 2000;4(3–4):175–197.

Mortazavi, A. and Katsabanis, P. D., Modeling burden size and strata dip effects on the surface blasting process. International Journal of Rock Mechanics and Mining Sciences, 2001;38(4):481–498.

Munjiza, A., The combined finite-discrete element method, Wiley, New York. 2004.

Munjiza, A. and Andrews, K. R. F., Penalty function method for combined finite-discrete element systems comprising large number of separate bodies. International Journal of Numerical Methods in Engineering, 2000;49(11):1377–1396.

Munjiza, A. and John. N. W. M., Mesh size sensitivity of the combined FEM/DEM fracture and fragmentation algorithms. Engineering Fracture Mechanics, 2002;69(2):281–295.

Munjiza, A., Owen, D. R. J. and Bicanic, N., A combined finite-discrete element method in transient dynamics of fracturing solids. International Journal of Engineering Computations, 1995;12(2):145–174.

Ohnishi, Y., Nishiyama, S. and Sasaki, T., Development and application of discontinuous deformation analysis. In: Leung, C. F. and Zhou, Y. X. (eds.), Rock mechanics in underground constructions (Proc. of 4th Axia Rock Mechanics Symp. Nov. 8–10, 2006, Singapore), pp. 59–70. Keynote lecture. 2006.

Pearce, C. J., Thavalingam, A., Liao, Z. and Bicanic, N., Computational aspects of the discontinuous deformation analysis framework for modelling concrete fracture. Engineering Fracture Mechanics, 2000;65(2):283–298.

Shi, G., Discontinuous deformation analysis – a new numerical model for the statics and dynamics of block systems. Ph.D. Thesis, University of California, Berkeley, USA. 1988. Shi, G., Manifold method of material analysis. Trans, 9th Army Conf. on Applied Mathematics and Computing, Minneapolis, MN, pp. 57–76. 1991.

Shi, G., Discontinuous deformation analysis. A new numerical model for the statics and dynamics of deformable block structures. Engineering Computations, 1992a;9(2):157–168.

Shi, G., Modeling rock joints and blocks by manifold method, Proc. Of 33rd US Symp. on Rock Mechanics, Santa Fe, NM, pp. 639–648. 1992b.

Shi, G., Block system modeling by discontinuous deformation analysis. Computational Mechanics Publications. Boston. 1993.

Shi, G., Manifold method, In: Salami, M. R. and Banks, D. (eds.), Discontinuous deformation analysis and simulations of discontinuous media, pp. 52–262. TSI Press, Albuquerque, NM, 1996.

Shi, G., Three dimensional discontinuous deformation analysis. In: Elsworth, D., Tinucci, J. P. and Heasley, P. E. (eds.), Rock mechanics in the national interest, Swets & Zeitlinger Lisse. pp. 1421–1428. 2001.

Shi, G. and Goodman, R. E., Two dimensional discontinuous deformation analysis. International Journal for Numerical and Analytical Methods in Geomechanics, 1985;9:541–556.

Shi, G. and Goodman, R. E., Generalization of two-dimensional discontinuous deformation analysis for forward modelling. International Journal for Numerical and Analytical Methods in Geomechanics, 1989;13(4):359–380.

Shyu, K., Nodal-based discontinuous deformation analysis. Ph.D. Thesis, University of California, Berkeley, USA. 1993.

Soto-Yarritu, G. R. and Martinez, A., Computer simulation of granular material transport: Vibrating feeders. Powder Handling and Processing, 2001;13(2):181–184.

Thavalingam, A., Bicanic, N., Robinson, J. I. and Ponniah, D. A., Computational framework for discontinuous modeling of masonry arch bridges. Computers and Structures, 2001;79(19):1821–1830.

Wittike, W., Rock mechanics – theory and applications with case histories. Springer-Verlag, Berlin. 1990.

Yeung, M. R., Analysis of a mine roof using the DDA method. International Journal for Numerical and Analytical Methods in Geomechanics, 1993;30(7):1411–1417.

Yeung, M. R. and Loeng, L. L., Effects of joint attributes on tunnel stability. International Journal of Rock Mechanics and Mining Sciences, 1997;34(3–4):505.

Yeung, M. R., Jiang, Q. H. and Sun, N., Validation of block theory and three-dimensional discontinuous deformation analysis as wedge stability analysis method. International Journal of Rock Mechanics and Mining Sciences, 2003;40(2):265–275.

Zhang, X. and Lu, M. W., Block-interfaces model for non-linear numerical simulations of rock structures. International Journal of Rock Mechanics and Mining Sciences, 1998;35(7):983–990.

10 DISCRETE FRACTURE NETWORK (DFN) METHOD

10.1 Introduction

The connectivity of a fracture system determines the flow patterns in a fractured rock mass. The bulk volume of fluid is conducted through pathways formed by the connected fractures when the permeability of the rock matrix is negligible compared with that of fractures, especially for low-porosity rocks such as granites. In a population of distributed fractures, the isolated fractures (which have no intersection with any other fracture) and singly connected fractures (which have only one intersection with other fractures) do not contribute to the flow field, unless their propagation by external loading is included in the consideration. In addition, the flow field is very sensitive to the patterns of the fracture system connectivity when the system is near the critical percolation threshold. Under such a state, a small change of fracture connectivity (e.g., the addition of one small fracture) might lead to different flow patterns. On the other hand, the deformability, or strain fields of a fractured rock mass, depends more on the density and orientations of the fracture sets and less on the connectivity.

Both DEM and continuum-based numerical methods such as the FEM have been applied to simulate coupled hydro-mechanical processes in fractured rocks over the past decades (Lemos, 1988; Jing et al., 1995). When the DEM is applied, the first step is to set up a geometrical model describing the geometry of the connected fractures and the rock blocks formed by them. The solution of flow through the fractures can be achieved using special numerical or analytical methods to obtain the flow field in each fracture, thus leading to the DFN method (Long et al., 1982; Robinson, 1984). Simple flow laws can be used to describe the behavior of fractures when the rock matrix is assumed to be a rigid and impermeable material. To apply the continuum-based numerical methods, such as FEM or BEM, to fractured rocks, however, the rock mass properties must be homogenized in order to formulate the tensors of elastic stiffness (or elastic compliance) and permeability for the equivalent continua. This assumes that the fracture network is a percolating system. Whether or not the fracture network percolates can only be determined by a detailed analysis of the network topology.

The DFN method is a special discrete approach that considers fluid flow and transport processes in fractured rock masses through a system of connected fractures. The technique was created in the 1980s for both 2D and 3D problems (Long et al., 1982, 1985; Andersson, 1984; Endo, 1984; Robinson, 1984; Smith and Schwartz, 1984; Endo et al., 1984; Elsworth, 1986a,b; Andersson and Dverstop, 1987; Dershowtz and Einstein, 1987), and was continuously developed afterwards with many applications in civil, environmental and reservoir engineering and other geoscience and geoengineering fields.

The effects of the mechanical deformation and heat transfer in a rock mass on fluid flow and transport are difficult to model in DFN and are usually ignored or crudely approximated. Thus this method is most useful for the study of fluid flow and mass transport in fractured rocks for which an equivalent continuum model is difficult to establish or not necessary, or for the derivation of equivalent continuum flow and transport properties in the fractured rocks (Zimmerman and Bodvasson, 1996; Yu, et al., 1999). A large number of associated publications have been reported in journals and international symposia and conferences. Systematic presentation and evaluation of the DFN method have also appeared in books, such as Bear et al. (1993), Sahimi (1995), the US National Research Council (1996) and especially Adler and Thovert (1999). A most recent and comprehensive review on the issues relating to the DFN approach was given in Berkowitz (2002).

The DFN model is established on the understanding and representation of the two factors: fracture system geometry and aperture/transmissivity of individual fractures. The former is based on stochastic simulations of fracture systems, using probabilistic density functions of the geometric parameters of the fractures (density, orientation, size, aperture or transmissivity) formulated according to field mapping results, in addition to the assumption about fracture shape (circular, elliptical or generally polygonal). Direct mapping can only be conducted at surface exposures of limited size, boreholes of limited diameter/length/depth and on the walls of underground excavations (tunnels, caverns, shafts, etc.) of more limited measurement space and with cut-off limits for mapping. The reliability of fracture network information depends on the quality of mapping and sampling, and hence its adequacy and reliability is difficult to be evaluated. Equally difficult also is the determination of the aperture/transmissivity of the fracture population, due to the fact that *in situ* and laboratory tests can only be performed with a limited number of fracture samples from restricted locations, and the effect of sample size is difficult to determine.

Despite the above limitations the DFN model enjoys wide applications for fluid flow problems of fractured rocks, perhaps mainly due to the fact that it is to date an irreplaceable tool for modeling fluid flow and transport phenomena at both the 'near-field' and 'far-field' scales. The 'near-field' applicability is where the dominance of the fracture geometry at small and moderate scales makes the volume averaging principle used for continuum approximations unacceptable at such scales, and the 'far-field' applicability is where equivalent continuum properties of large rock volumes need to be approximated through upscaling and homogenization processes using DFN models with increasing model sizes. The latter is necessary when explicit representations of large numbers of fractures make the direct DFN models less efficient, and the continuum model with equivalent properties become more attractive, similar to the DEM.

There are many different DFN formulations and computer codes, but most notable are the approaches with the codes FRACMAN/MAFIC (Dershowitz et al., 1993) and NAPSAC (Stratford et al., 1990; Herbert, 1994, 1996; Wilcock, 1996) with many applications for rock engineering projects over the years. Some of the applications in the areas of hot-dry-rock (HDR), oil reservoir engineering, underground excavation and rock mass characterization are listed below as examples.

- HDR reservoir simulations: Layton et al. (1992), Ezzedine and de Marsily (1993), Watanabe and Takahashi (1995), Bruel (1995a,b), Kolditz (1995), Willis-Richards and Wallroth (1995),Willis-Richards (1995), Babadagli (2001);

- Characterization of the permeability of fractured rocks: Priest and Samaniego (1983), Dershowitz et al. (1992), Dershowitz (1993), Herbert and Layton (1995), Geier et al. (1995), Doe and Wallmann (1995), Barthélémy et al. (1996), Jing and Stephansson (1997), Margolin et al. (1998), Mazurek et al. (1998), Zhang and Sanderson (1999), Chen et al. (1999), Min et al. (2004);

- Hydrocarbon reservoir applications: Dershowitz and La Pointe (1994), Bruhn et al. (1997);

- Water effects on underground excavations and rock slopes: Rouleau and Gale (1987), Dverstorp and Andersson (1989), Xu and Cojean (1990), He (1997), Birkholzer et al. (1999).

In this chapter, we focus on saturated fracture systems in fractured rocks without considering matrix–fracture interaction, in terms of single-phase fluid flow, but the basic concepts of one hybrid DEM-BEM approach to fluid flow in fractured porous rocks with consideration of the matrix–fracture interaction are presented briefly. We will not cover the transport process and multiphase flow problems or the coupled T-H-M-C processes. Due to the large amount of literature and treatises published in this field, this chapter only attempts a brief summary and analysis of the basic concepts, the solution approaches for different methods and outstanding issues – so that the reader may achieve a basic understanding of the fundamental features and applicability of the DFN approach in rock engineering as a foundation for more advanced studies or applications.

10.2 Representation of Fracture Networks

10.2.1 Individual Fractures

The rock fractures are most commonly assumed to be pairs of smooth and parallel planar surfaces so that the Cubic Law can be readily applied. Such a simplification is particularly convenient for large-scale DFN models involving large numbers of fractures. In reality, however, the fracture surfaces are rough and the applicability of the Cubic Law may not be appropriate generally everywhere. Research efforts have been made to characterize the surface roughness and its effects on the flow and deformation of natural fractures, using probabilistic field theory, geostatistics and fractal geometry (Brown and Scholz, 1985; Keller and Bonner, 1985; Poon et al., 1992; Johns et al., 1993; Brown, 1995; Schmittbuhl et al., 1995; Lanaro et al., 1998, 1999; Fardin et al., 2001a,b, 2003). These approaches are based mostly on 2D or 3D profilometric measurements or X-ray computed tomography (CT) of the fracture surfaces (Duliu, 1999).

A repeatedly appearing finding is that fracture surfaces seem to exhibit fractal features, mostly characterized by the Hurst exponent of a power law, although the origin of such a consistent phenomenon has not been properly explained. This power law indicates the existence of a scaling effect which may have profound implications on the mathematical modeling of fractures. If this scaling effect exists at all scales with equal or similar degrees of importance, the physical properties of fractures must be functions of the fracture sizes in all ranges, which could then present an especially difficult challenge for the characterization of the physical properties of large-sized fractures beyond the laboratory scale, since a Representative Elementary Size (RES) of fractures would not then exist. However, it was found in Lanaro et al. (1998, 1999) and in Fardin et al. (2001a,b, 2003) that a stationarity threshold of surface roughness was reached with gradually increasing sampling area sizes for the tested rock fracture samples beyond which the scaling effect ceased to exist. This finding indicates that some 'dominant' scale of roughness may exist and in fact representative physical properties can be characterized at this scale.

Measurements of fracture surface topography using 2D or 3D profilometers, as for almost all other laboratory tests on fracture samples, are usually performed using small-scale samples of several to tens of centimeters in length, and in some cases a stationarity threshold may not be reached due to the fact that either the sample size is too small or 'structured non-stationarity' exists (i.e., the fracture surface is not nominally planar but has dominant undulations or inclinations). Detailed measurements of roughness on larger (field-scale) fractures are rarely reported, with one exception in Feng et al. (2003) using the total station technique, which indicated the presence of multiple-ordered fracture roughness and changed length correlations. This might be the reason for the fact that at field scales fracture surfaces might not appear as self-affine but nominally planar surfaces and a stationarity threshold may exist at certain representative fracture sizes. The difficulty is that, when this representative size is too large for laboratory testing, direct measurement of physical properties becomes difficult because field testing involves more uncertainties in controlling the initial and boundary conditions for testing. The current publications on fracture roughness threshold indicate that, for rock fractures generally encountered in rock engineering, the roughness stationarity threshold is found to be less than 1 m (usually 0.4–0.6 m). Regarding the fact that, in many (if not most of the) DFN models for practical applications, the lower cut-off limits for fracture sizes are much larger than 1 m, the assumption of planar and equivalent smooth fractures in DFN models is a relatively valid assumption.

Besides the scale effects due to roughness of fracture surfaces, for individual fractures in a DFN model another challenging issue is the definition and measurement of aperture for evaluating transmissivity. Different definitions of fracture aperture exist in the literature: geometric aperture, mechanical aperture and hydraulic aperture. Transmissivity is a function of the hydraulic aperture. It is important to note that all these apertures depend on the stresses and consequential deformation process of the fractures and are therefore variables, rather than constants, due to the fact that the stiffness (or deformability) of the fractures are functions of the stress and damage accumulation on the fracture surfaces because of the existence of roughness. For the same reason, apertures and transmissivity are stress-dependent, besides their dependence on size.

In common practice, hydraulic apertures of fractures are inferred (not directly measured) in laboratory tests or back-calculated in field tests by assuming the validity of the Cubic Law (Tsang, 1992) with known fracture geometry (size) and pressure gradient. The former technique has the drawback that the initial *in situ* conditions of the fracture samples are essentially irreversibly destroyed during the sampling process, therefore the calculated aperture value may not have a properly defined initial state, although it might be estimated from flow tests under estimated normal stresses at the sampling depth. The latter technique suffers the usual limitations in field testing with uncertain initial and boundary conditions for fluid flow, the uncertain geometry of the tested fracture and possible effects of unknown and hidden fractures connected to the tested fracture.

Fluid flow in individual fractures appears often in a 'channelized' mode (Tsang and Tsang, 1987; Tsang et al., 1988; Moreno and Neretnieks, 1993) due to the effects of roughness, like a corrugated iron roof. This channeling effect is enhanced by the stress and deformation processes (Yeo et al., 1998; Koyama et al., 2004, 2006).

In DFN practice, the aperture or transmissivity of individual fractures is taken either as a constant or as a stochastic distribution over individual fractures (Moreno et al., 1988; Nordqvist et al., 1996), with flow calculations using FEM, BEM or pipe network approaches (see Section 10.3). The FEM approach assumes that the aperture (or transmissivity) field of a fracture follows a certain probabilistic distribution according to measured data so that the aperture or transmissivity values can vary element-by-element. The pipe network model (or channel lattice model) represents the aperture field of a fracture by one or a set of connected pipes of effective hydraulic diameters according to the aperture (or transmissivity) values or distributions governed by results of measurements. Computational demand is much reduced when the pipe network models are used.

The shapes of individual fractures in most DFN codes are circular, rectangular or polygonal, mostly for purpose of convenience since the real shape of sub-surface fractures cannot be fully known. However, one argument is that, for large-scale DFN models with a very high density of fractures, the effect of fracture shape on the final results may be much reduced or diminished. On the other hand, shapes of individual fractures may become important in affecting fracture system connectivity if the population of fractures is not so large. The issue of fracture shape is an unresolved one and will remain so for the foreseeable future.

Geochemical processes operating in the minerals of fractured rocks may also be important in affecting fracture apertures, such as mineral dissolution and precipitation on fracture surfaces. This is an important issue for coupled T-H-M-C processes of rock fractures, especially when the time scale is large. In this chapter, we consider only single-phase fluid flow processes – without considering stress, heat and chemical effects.

10.2.2 Fracture Networks

Stochastic simulation of fracture systems is the geometric basis of the DFN approach and plays a crucial role in the performance and reliability of DFN models, in the same way as for the DEM. The key process is to create probabilistic density functions (PDFs) of geometric parameters of fracture sets relating to the densities, locations, orientations and sizes, based on field mapping results using borehole logging, surface mapping, window mapping or geophysical techniques (usually using seismic wave, electric resistance or MRI, i.e., magnetic resonance imaging, methods) (Balzarini et al., 2001). The generation of the realizations of the fractures systems according to these PDFs and assumptions about fracture shape (circular, elliptical or polygonal) (Dershowitz, 1984, 1993; Billaux et al., 1989) is then a straightforward inverse numerical process. A critical issue in this technique is the treatment of bias in estimation of the fracture densities, orientations and trace lengths measured via conventional 1D scanline or 2D window mappings. A notable recent development using circular mapping windows (Mauldon, 1998; Mauldon et al., 2001) provides an important step forward in this regard. Figure 10.1 shows two examples of generated fracture network realizations, one at Äspö Hard Rock Laboratory, a research facility for nuclear waste disposal in Sweden and another in South Korea for underground storage.

The connectivity of fractures is a critical feature controlling deformation, but especially fluid flow and transport processes in fractured rocks (Long and Billaux, 1987; Tsang and Tsang, 1987; Koudina et al., 1998; Margolin et al., 1998).

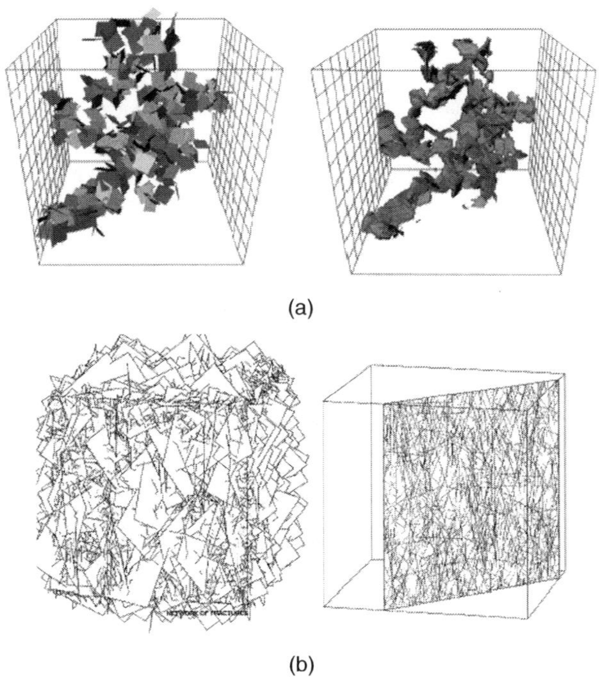

Fig. 10.1 Two examples of a DFN model: (a) one of the stochastic realizations of fracture networks generated for the Äspö HRL, Sweden (Svensson, 2001a): the total fracture system in the generation model (left) and the flow channels formed by the connected fractures from the total system (right); (b) A 3D DFN model generated using *in situ* borehole logging data (Park et al., 2002): the 3D fracture system model generated for modelling an underground storage facility in South Korea (left) and fracture traces on one of the vertical cross sections (right).

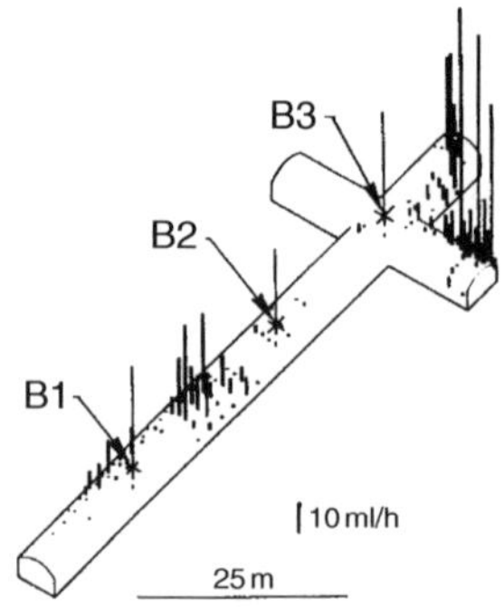

Fig. 10.2 Measured water flow into the experimental tunnels in the Stripa Mine, Sweden (Abelin et al., 1987).

One typical example is shown in Fig. 10.2, where the fluid flow in the test area of the Stripa Mine is highly compartmentalized and concentrated in a few locations, and is difficult for conventional continuum approaches to simulate. The DFN approach was applied to evaluate the experimental results using both the NAPSAC and FRACMAN/MAFIC codes which have served as one of the major drivers for further development of the DFN modeling techniques. Another example was reported in Long and Billaux (1987) concerning fracture-controlled fluid flow at the Fanay-Augères site in France. It was found that only about 0.1% of the fractures contributed to flow on a large scale. Similar phenomena have been discovered in many field test cases and underground excavation experiences where small portions of the fracture population have been found to dominate the flow. Even domains that appear to be heavily fractured may not, in fact, be well connected. Thus fracture connectivity is of prime importance, not only for DFN modeling but also for site characterization. However, numerical simulation of this phenomenon is not a trivial task because of the fact that realistically selecting a small number of conducting fractures from a well-connected fracture system is a subjective step, even when conditioning with test data is available. In a DFN model, after removing non-connected fractures, the connectivity of the remaining fractures is determined and is a function of the quality of the mapping techniques and extrapolation of the fracture system parameters from 1D and 2D data to 3D data.

The details of stochastic generation of fracture systems and the fluid flow equations of fractures in 2D are presented in Chapters 4 and 5 will not be repeated here.

10.3 Solution for the Flow Fields Within Fractures

Numerical techniques have been developed for the solution of flow fields in individual fractures using closed-form solutions, finite elements, boundary elements, simplified pipe networks and the channel lattice models. Closed-form solutions exist, at present, only for planar, smooth fractures with parallel surfaces of regular shape (i.e., circular or rectangular discs) for steady-state flow (Long, 1983) or for both steady-state and transient flow (Amadei and Illangaseekare, 1992). For fractures with more general shapes, numerical solutions must be used. The FEM is perhaps the most well-known techniques used in the DFN flow models and has been used in the DFN codes FRACMAN/MAFIC and NAPSAC. The basic concept is to impose a FEM mesh over the individual discs representing fractures in space (Fig. 10.3a) and solve the flow equations. The aperture or transmissivity field within the fracture can be either constant or randomly distributed. Similarly the BEM discretization can also be applied with the boundary elements defined only on the disc boundaries (Fig. 10.3b), with the fracture intersections treated as internal boundaries in the BEM solution. The compatibility condition

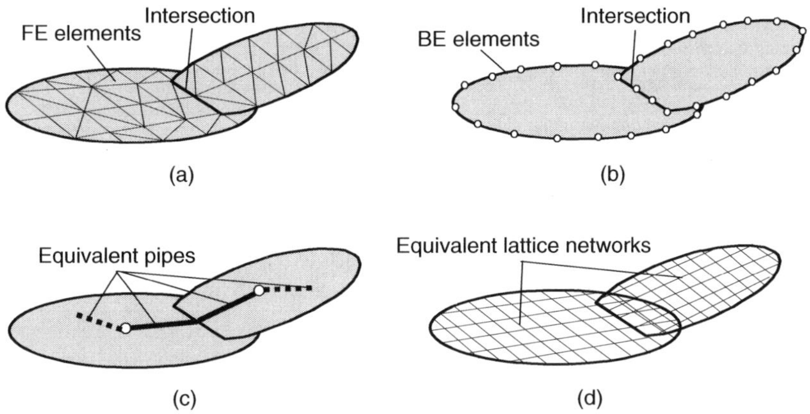

Fig. 10.3 Representation of rock fractures for solutions of fluid flow in DFN: (a) FEM; (b) BEM; (c) equivalent pipes and (d) channel lattice model (Jing, 2003).

is imposed at the intersections of the discs. See Elsworth (1986a,b) and Robinson (1986) for detailed formulations.

The pipe model represents a fracture as a pipe of equivalent hydraulic conductivity starting at the disc center and ending at the intersections with other fractures (Fig. 10.3c), based on the fracture transmissivity, size and shape distributions (Cacas et al., 1990). The channel lattice model represents the whole fracture by a network of regular channel networks (Fig. 10.3d). The pipe model leads to much simplified representation of the fracture system geometry and flow behavior, but may not be able to properly represent the behavior of large-sized fractures such as faults and fracture zones.

The channel lattice model is more suitable for simulating the complex flow behavior inside the fractures, such as the 'channel flow' phenomenon (Tsang and Tsang, 1987), and is computationally less demanding than the FEM and BEM models since the solutions of the flow fields through the channel elements are analytical.

In this chapter, we will focus on the FEM and BEM approaches. The pipe network and the channel lattice models are straightforward models using pipes/channels of equivalent hydraulic diameters replacing fracture volumes and will not be presented here

10.3.1 FEM Solution Techniques

10.3.1.1 General aspects

The general aspects of the FEM solution of fluid flow in a fracture discretized with a FEM mesh is briefly described first in this section, followed by a more detailed presentation concerning the numerical details.

The fluid flow within a fracture is assumed to be 2D Darcy flow parallel to the boundary surfaces of fractures that are assumed to be smooth. For a DFN model with a large number of fractures, the fluid flow equation of the ith fracture may be written as

$$S\frac{\partial H^i}{\partial t} = \nabla(T^i \nabla H^i) + W^i \tag{10.1a}$$

where the scalar S is the storativity, H^i the head, T^i the transmissivity and W^i the source term. No flow along the intersection or no head loss across the intersection is commonly adopted, but this assumption can be removed when the intersections are represented by 1D elements. The fluid flow in the fractures is assumed to be represented by the averaged velocity and pressure, with the parabolic distribution of velocity profile across the fracture ignored. The Galerkin FEM formulation of Eqn (10.1a) then leads to the FEM equation for the fluid flow in the ith fracture

$$\left\{\underset{m\times n}{[T^i]} + \frac{\underset{m\times n}{[S^i]}}{\Delta t}\right\}\underset{n\times 1}{[H^i]^k} = \left\{\underset{m\times 1}{[W^i]} + \frac{\underset{m\times n}{[S^i]}}{\Delta t}\right\}\underset{n\times 1}{[H^1]}^{k-1} \tag{10.1b}$$

where $[T^i]$ and $[S^i]$ are the transmissivity and storativity matrices, $[H^i]^k$ and $[H^i]^{k-1}$ the head vectors at time step k and $k-1$, respectively, and $[W^i]$ is the source term vector. The number n is the total number of FEM nodes of the ith fracture and m the number of nodes with unknown heads. The head continuity condition between connected fractures imposed at the intersection nodes, i.e., $H^i = H^j$ at all intersection nodes between fracture i and fracture j, serves to provide the links for assembling the global coefficient matrix.

Boreholes are commonly needed for DFN models and are usually assumed to be pipes of any shape with constant effective hydraulic diameters and are closed off except around intersecting fractures when the rock matrix is assumed to be impermeable. Boreholes can be at fixed pressures or have specified axial flow (pumping) rates.

The above fundamental assumptions lead to much simplified and straightforward models for flow calculations with the whole system of arbitrarily oriented fractures contained in a rectangular regional model for easier specification of boundary conditions, usually no-flow or fixed head conditions.

In order to solve the flow Eqn (10.1), discretizations at two levels – within single fractures and at the scale of the full fracture system – are sometimes used. The discretization within the fractures is totally separate from the discretization 'in the large'. Closed-form solutions within the single fractures can be directly used if exist. The approach described below is that from Robinson (1984, 1986).

10.3.1.2 The FEM approach in Robinson (1984, 1986)

The basic concept is a finite-element discretization at the full system scale with each fracture represented by a single element at the regional scale. The unknown in the system is usually the head P for easier specification of head boundary conditions, which would be impossible if the flux were to be chosen. Given the heads along the intersections, the heads within the fractures are determined by the flow Eqn (10.1). Therefore for the full system, we need only to discretize the intersections. The number of nodes along each intersection and the 1D Lagrangian shape functions associated with them can be found in any text book on the FEM. Let N_i be the number of nodes along intersection i, $\phi_j^{(i)}(\alpha)$ be the jth shape function at intersection i and α be a value varying from 0 to 1.0 along the intersection, then the shape functions have the following properties

$$\phi_j^{(i)}(\alpha_k) = \delta_{jk} \tag{10.2}$$

i.e., the jth shape function takes the values 1.0 at node j and 0 at the other nodes and

$$\sum_{j=1}^{N_i} \phi_j^{(i)}(\alpha) = 1 \tag{10.3}$$

i.e., the shape functions must sum to unity. The shape functions can be piecewise constant, piecewise linear or piecewise quadratic.

The governing equation represents conservation of mass at the intersections. A weak form of this equation in term of flux is written as

$$\int_{\alpha=0}^{1} Q^{(i)}(\alpha)\phi_j^{(i)}(\alpha)\mathrm{d}\alpha = 0, \quad j = 1, 2, \ldots, N_i, \text{ all } i \tag{10.4}$$

where $Q^{(i)}(\alpha)$ is the total flux into intersection i at position α along the 1D element of the intersection. Based on the linearity assumptions within the fractures, this flux can be written as a linear combination of the nodal heads, and equation (10.4) can be written as

$$\sum_s F_{rs}P_s = 0 \tag{10.5}$$

where matrix F_{rs} is the integral of the shape function associated with node r times the flux into the intersection containing r when $P_s = 1$, and all other heads are zero. F_{rs} is equal to zero unless r and s are on the same fracture (element). Denoting the contribution to F_{rs} from fracture (element) k as $F_{rs}^{(k)}$, an important property of this matrix is

$$\sum_s F_{rs}^{(k)} = 0 \tag{10.6}$$

indicating a no-flow state when the head is constant at all nodes. From (10.3), this property implies that the head is constant both along the boundary and in the interior of the fracture. The mass conservation within each fracture is given by the property

$$\sum_r F_{rs}^{(k)} = 0 \tag{10.7}$$

and the total flux-out is zero.

Another property of the matrix F_{rs} is that it is symmetric (Robinson, 1984)

$$F_{rs}^{(k)} = F_{sr}^{(k)} \tag{10.8}$$

and is true for any shaped fracture and any shape functions.

Equation (10.5) is to be solved for the full system with specified boundary conditions. From the resulting nodal heads, the head and flux everywhere can be calculated.

The matrix $F_{rs}^{(k)}$ needs to be calculated systematically for each fracture. When the aperture of a fracture is constant, one can, in principle, solve analytically for the head, given the fluxes. However, in reality this cannot be readily applied except in special cases (e.g., circular disc-shaped fractures) since the analytical solutions would consist of infinite sums whose convergence is very slow or not guaranteed, so that the analytical solutions would involve integrations which may still need numerical evaluations of matrix $F_{rs}^{(e)}$ and analytical solutions may not exist when the aperture within a fracture is not constant. Therefore the FEM method is needed.

Applying the conventional Galerkin weak form for Eqn (10.5), the equation for node i (here we refer to nodes of the fracture discretization not of the full system) is

$$A_{ij}P_j = Q_j \tag{10.9}$$

with

$$A_{ij} = \frac{e^3}{12\mu}\int\left(\frac{\partial\phi_i}{\partial x}\frac{\partial\phi_j}{\partial x} + \frac{\partial\phi_i}{\partial y}\frac{\partial\phi_j}{\partial y}\right)\mathrm{d}x\ \mathrm{d}y \tag{10.10}$$

where the term e is the fracture aperture, μ the fluid viscosity and Q_j the flux imposed at node j. At the intersections, these equations are replaced with the specified pressures. This leads to one RHS for each node of the full system on the fracture in question.

In order to calculate the fluxes, one can use the equations that were discarded for the nodes on the intersections. Using the coefficients to the solution for heads, one obtains the nodal fluxes consistent with head results. A weighted average of these nodal fluxes is then used to obtain the entry for $F_{rs}^{(k)}$ that satisfies properties Eqns (10.6) and (10.7) since the FEM will give exactly no-flow results when all the specified heads are equal and is globally conservative. Eqn (10.8) is satisfied simply because $A_{ij} = A_{ji}$ according to Eqn (10.10).

Solution of Eqn (10.5) with the two-level discretization is then a straightforward numerical procedure of the FEM.

A solution technique similar to the above technique is reported in de Druezy and Erhel (2003) without discretization of the fractures. Long (1983) proposed a semi-analytical FEM solution technique using matched images of sinks and sources to create no-flow boundary conditions at the exterior edges of fractures, but the resultant global matrix is not symmetric and conforms to a finite difference solution.

The FEM solution technique has the advantages of being able to consider varying aperture (or transmissivity) distributions within individual fractures to represent the effects of surface roughness. It requires, however, dense elements around the intersections to consider the sharp gradients of fluid velocity and head nearby, therefore leading to much increased total degrees of freedom of the matrix system.

10.3.2 BEM Solution Techniques

The BEM is another solution alternative to Eqn (10.1), under the same assumptions as for the FEM (Elsworth, 1986a,b). The system is envisaged as shown in Fig. 10.4a (Elsworth, 1986a), with a number of

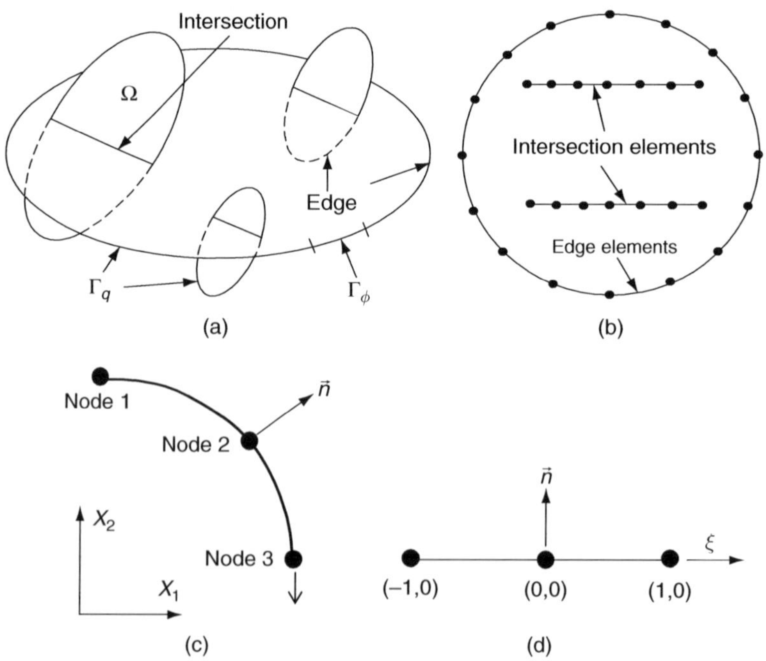

Fig. 10.4 BEM solution for DFN: (a) Intersecting fractures in a 3D space; (b) Discretization of an individual fracture with 1D boundary and intersection elements; (c) Isoparametric curved elements in global co-ordinates; (d) Isoparametric straight elements in local co-ordinates (Elsworth, 1986a).

connected fractures of any geometric shape. Each fracture is discretized on the edge and intersections with 1D isoparametric elements in the same way as that in the FEM (Fig. 10.4b).

For a linear potential flow within a fracture domain Ω and boundary Γ, the boundary integral equation is

$$c(p)\phi(p) + \int_{\Gamma} V(p,q)\phi(q)\mathrm{d}\Gamma = \int_{\Gamma} \Phi(p,q)v(q)\mathrm{d}\Gamma \tag{10.11}$$

where $V(p,q)$ and $\Phi(p,q)$ are the kernel functions of velocity and potential, respectively, at field point q on the boundary Γ of the fracture domain Ω due to a unit source at point p internally. The terms $v(q) = \frac{\partial \phi(q)}{\partial n}$ and $\phi(q)$ are the flow velocities and potentials (heads) at point q, which are either prescribed as boundary conditions or are unknowns requiring evaluation. Vector $\vec{n}$ is the outward normal of the boundary Γ, and $c(p)$ is a free term representing effects of the geometry at field point p, with $c(p) = 1.0$ when p is located entirely inside Ω and $c(p) = 1/2$ when p is located on the boundary Γ which is smooth.

For 2D linear flow problems with a line source of magnitude M within an infinite medium, the relevant kernel functions (fundamental solutions) are given as

$$\Phi(p,q) = \frac{M}{2\pi} \ln\ r \tag{10.12}$$

$$V(p,q) = -\frac{KM}{2\pi r} \tag{10.13}$$

where K is the fluid conductivity of the fracture and is determined as $K = (g/12\nu)e^2$, for laminar flow with a constant aperture e, and r is the radial distance from the field point p to the source point q. Terms g and ν are the gravitational acceleration and the kinematic viscosity of the fluid, respectively. Only the edges and intersections of fractures need to be discretized with standard curved or straight-line isoparametric elements (Fig. 10.4c, d), and the same quadrature techniques as used in FEM (Stroud and Secrest, 1966) are applied in the BEM for numerical integration of the kernel functions. The shape functions, in vector form $\mathbf{h}$, are given by

$$\mathbf{h} = \begin{Bmatrix} h_1 \\ h_2 \\ h_3 \end{Bmatrix} = \frac{1}{2}\begin{Bmatrix} (1-\xi) - (1-\xi^2) \\ (1+\xi) - (1-\xi^2) \\ 2(1-\xi^2) \end{Bmatrix} \tag{10.14}$$

where $\xi \in [-1,1]$ is the intrinsic co-ordinate along the isoparametric element. Using these shape functions, the geometric and physical variables defined in the global co-ordinate system can be mapped into the local intrinsic co-ordinate system of elements, such as

$$x_1 = \mathbf{h}^{\mathrm{T}}\mathbf{x}_1, \quad x_2 = \mathbf{h}^{\mathrm{T}}\mathbf{x}_2, \quad \phi = \mathbf{h}^{\mathrm{T}}\boldsymbol{\varphi}, \quad v \cdot \vec{n} = \mathbf{h}^{\mathrm{T}}\mathbf{v} \cdot \vec{\mathbf{n}} \tag{10.15}$$

where the vector terms on the right-hand sides are the values of the respective variables at the element nodes.

Dividing the boundary edge and the intersection traces into N elements and using the mapping of Eqn (10.15), Eqn (10.11) can be more readily integrated in the local intrinsic co-ordinate space over all elements as

$$c(p)\phi_i(p) + \sum_{j=1}^{N} \int_{\Gamma_j} V_i(p,q)\phi_j(q)\frac{\mathrm{d}\Gamma_j}{\mathrm{d}\xi}\mathrm{d}\xi = \sum_{j=1}^{N} \int_{\Gamma_j} \Phi_i(p,q)v_j(q)\frac{\mathrm{d}\Gamma_j}{\mathrm{d}\xi}\mathrm{d}\xi \tag{10.16}$$

or

$$c(p)\phi_i(p) + \sum_{j=1}^{N}\left[\int_{\Gamma_j} V_i(p,q)\frac{\mathrm{d}\Gamma_j}{\mathrm{d}\xi}\mathrm{d}\xi\right]\phi_j(q) = \sum_{j=1}^{N}\left[\int_{\Gamma_j} \Phi_i(p,q)\frac{\mathrm{d}\Gamma_j}{\mathrm{d}\xi}\mathrm{d}\xi\right]v_j(q) \tag{10.17}$$

after rearranging of the integrals. Denoting

$$G_{ij} = \int_{\Gamma_j} V_i(p,q)\frac{\mathrm{d}\Gamma_j}{\mathrm{d}\xi}\,\mathrm{d}\xi, \quad H_{ij} = \int_{\Gamma_j} \Phi_i(p,q)\frac{\mathrm{d}\Gamma_j}{\mathrm{d}\xi}\,\mathrm{d}\xi \tag{10.18}$$

Eqn (10.16) finally becomes

$$c(p)\phi_i(p) + \sum_{j=1}^{N} G_{ij}\phi_j(q) = \sum_{j=1}^{N} H_{ij}v_j(q) \tag{10.19}$$

The integrals in Eqn (10.18) can be evaluated using Gaussian quadrature techniques with the Jacobian of the mapping given by

$$\frac{\mathrm{d}\Gamma}{\mathrm{d}\xi} = \left[\left(\frac{\mathrm{d}x_1}{\mathrm{d}\xi}\right)^2 + \left(\frac{\mathrm{d}x_2}{\mathrm{d}\xi}\right)^2\right]^{1/2} \tag{10.20a}$$

with

$$\frac{\mathrm{d}x_1}{\mathrm{d}\xi} = \frac{\mathrm{d}(\mathbf{h}^{\mathrm{T}}\mathbf{x}_1)}{\mathrm{d}\xi} = \frac{\mathrm{d}(\mathbf{h}^{\mathrm{T}})}{\mathrm{d}\xi}\mathbf{x}_1 \tag{10.20b}$$

$$\frac{\mathrm{d}x_2}{\mathrm{d}\xi} = \frac{\mathrm{d}(\mathbf{h}^{\mathrm{T}}\mathbf{x}_2)}{\mathrm{d}\xi} = \frac{\mathrm{d}(\mathbf{h}^{\mathrm{T}})}{\mathrm{d}\xi}\mathbf{x}_2 \tag{10.20c}$$

$$\frac{\mathrm{d}\phi}{\mathrm{d}\xi} = \frac{\mathrm{d}(\mathbf{h}^{\mathrm{T}}\boldsymbol{\varphi})}{\mathrm{d}\xi} = \frac{\mathrm{d}(\mathbf{h}^{\mathrm{T}})}{\mathrm{d}\xi}\boldsymbol{\varphi} \tag{10.20d}$$

$$\frac{\mathrm{d}(v\cdot\vec{n})}{\mathrm{d}\xi} = \frac{\mathrm{d}(\mathbf{h}^{\mathrm{T}}\mathbf{v}\cdot\vec{n})}{\mathrm{d}\xi} = \frac{\mathrm{d}(\mathbf{h}^{\mathrm{T}})}{\mathrm{d}\xi}(\mathbf{v}\cdot\vec{n}) \tag{10.20e}$$

$$\frac{\mathrm{d}(\mathbf{h}^{\mathrm{T}})}{\mathrm{d}\xi} = [(\xi - 1/2), (\xi + 1/2), -2\xi] \tag{10.20f}$$

Equation (10.19) can be transformed into an algebraic matrix equation of order N

$$\underset{(N\times N)}{\left[G_{ij}(p,q)\right]}\ \underset{(N\times 1)}{\left\{\phi_j(q)\right\}} = \underset{(N\times N)}{\left[H_{ij}(p,q)\right]}\ \underset{(N\times 1)}{\left\{v_j(q)\right\}} \tag{10.21}$$

or simply

$$\mathbf{G}\boldsymbol{\varphi} = \mathbf{H}\mathbf{v} \tag{10.22}$$

where **G** and **H** are fully populated square matrices of order N, and $\mathbf{v}$ and $\boldsymbol{\varphi}$ the vectors of nodal normal (to the boundary) velocity and head, respectively. A total of N boundary conditions must be specified to yield meaningful solutions of Eqn (10.21).

Without losing generality, we may assume that vectors $\mathbf{v}$ and $\boldsymbol{\varphi}$ may be split into two parts, $\mathbf{v} = (\mathbf{v}_n,\mathbf{v}_m)$ and $\boldsymbol{\varphi} = (\boldsymbol{\varphi}_n,\boldsymbol{\varphi}_m)$, corresponding to n nodes on the outside edge and m nodes along the intersections, with $n + m = N$. Equation (10.22) can then be rewritten as

$$\begin{bmatrix} G_{nn} & G_{nm} \\ G_{mn} & G_{mm} \end{bmatrix} \begin{Bmatrix} \boldsymbol{\varphi}_n \\ \boldsymbol{\varphi}_m \end{Bmatrix} = \begin{bmatrix} H_{nn} & H_{nm} \\ H_{mn} & H_{mm} \end{bmatrix} \begin{Bmatrix} \mathbf{v}_n \\ \mathbf{v}_m \end{Bmatrix} \tag{10.23}$$

For fluid flow in fractures, we may choose the zero normal flow on the edges of fractures (i.e., $\mathbf{v}_n \cdot \vec{n} = 0$) and the unknown but constant heads along the intersections (i.e., $\boldsymbol{\varphi}_m$ = constant). Equation (10.23) can then be symbolically rewritten as

$$\begin{bmatrix} G_{nn} & -H_{nm} \\ G_{mn} & -H_{mm} \end{bmatrix} \begin{Bmatrix} \boldsymbol{\varphi}_n \\ \mathbf{v}_m \end{Bmatrix} = \begin{Bmatrix} -V_{mn}\boldsymbol{\varphi}_m \\ -V_{mm}\boldsymbol{\varphi}_m \end{Bmatrix} \tag{10.24}$$

Equation (10.24) may be solved directly, when the heads on the intersections ($\boldsymbol{\varphi}_m$) are all known, by a symbolic inversion

$$\begin{Bmatrix} \boldsymbol{\varphi}_n \\ \mathbf{v}_m \end{Bmatrix} = \begin{bmatrix} G_{nn} & -H_{nm} \\ G_{mn} & -H_{mm} \end{bmatrix}^{-1} \begin{Bmatrix} -V_{nm}\boldsymbol{\varphi}_m \\ -V_{mm}\boldsymbol{\varphi}_m \end{Bmatrix} \tag{10.25}$$

Otherwise a two-step procedure is required.

Equation (10.25) yields, in general, a subset equation purely for the intersection elements

$$\{\mathbf{v}\}_{m\times 1} = [A]_{m\times m}\{\boldsymbol{\varphi}\}_{m\times 1} \tag{10.26}$$

where $[A]_{m\times m}$ represents a tensor of geometric conductivity of the fracture without the effects of its outside edges. The raw $\mathbf{k}_i$ of $[A]_{m\times m}$ must be multiplied by the fracture aperture (e) to yield hydraulic conductivity to enforce flux continuity, through evaluation of the integral

$$\mathbf{k}_i = e\int_{-1}^{+1} \mathbf{h}^{\mathrm{T}} \frac{\mathrm{d}\Gamma_i}{\mathrm{d}\xi}\mathrm{d}\xi \tag{10.27}$$

and a hydraulic conductivity tensor $[K]_{m\times m}$ is thus obtained to relate flux and potential (head)

$$\{\mathbf{q}\}_{m\times 1} = [K]_{m\times m}\{\boldsymbol{\varphi}\}_{m\times 1} \tag{10.28}$$

where $\mathbf{q}$ is the flow rate.

Equation (10.28) is identical to a FEM statement and can then be assembled into a standard FEM formulation of a global DFN model for obtaining the heads along the intersections. The physical interpretation of the above reasoning is that of a homogeneous system of flux boundary conditions for an infinite domain truncated by the contour of the fracture edge.

The BEM approach has the advantage of a much reduced number of elements compared with FEM; therefore the size of the global coefficient matrix, which is, however, asymmetric and dense, requires efficient equation solvers to yield the final solution.

10.3.3 A BEM Approach Concerning Permeable Rock Matrix and Conducting Fractures

The influence of the rock matrix on flow in rock fractures is usually not considered in the DFN models. However, the related effects also need to be estimated when the permeability of the rock matrix cannot be ignored compared with that of the fractures, or the time scale of the problem is long enough so that matrix diffusion cannot be ignored. In such cases a fracture system embedded in a highly porous matrix needs to be properly represented, such as the FEM technique used by Sudicky and McLaren (1992). Dershowitz and Miller (1995) reported a simplified DFN technique for considering the matrix influence on flow in connected fracture systems, using a probabilistic technique. A more direct approach

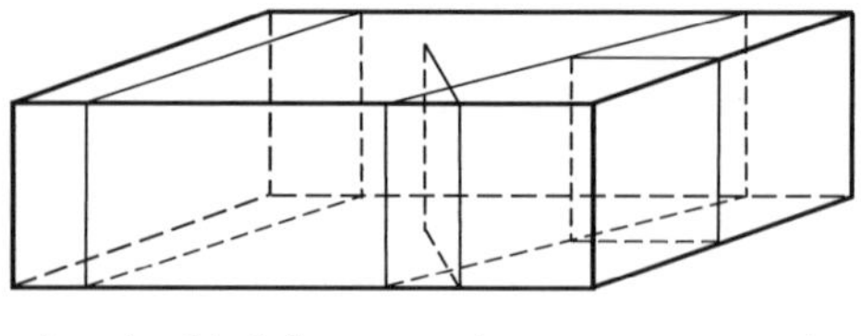

Fig. 10.5 A grid block model of embedded fractures in a porous matrix (Lough et al., 1998).

considering the coupling effects between fractures and rock matrix on fluid flow, using a BEM approach, was developed in Lough et al. (1998), aimed at the application of reservoir simulation at grid block levels. The original development was reported in Rasmussen et al. (1987) and was improved in Lough et al. (1998) for more efficient treatment of fractures in the BEM representation, which is summarized below mainly according to Lough et al. (1998).

The problem is envisaged as shown in Fig. 10.5 with a connected fracture system embedded in a porous matrix with fractures oriented randomly, and terminated either at the block boundary or inside the block.

The matrix and the fractures are treated as separate but interacting systems interfaced at the opposite pairs of surfaces of the fractures, (F_i^+, F_i^-), with $i = 1, 2, \ldots, N$, where N is the total number of fractures. The outside edge of the ith fracture is treated as a boundary F_i^{b} and its mean (central) plane is denoted as F_i which is parallel with F_i^+ and F_i^-. The aperture of the ith fracture, e_i, is the distance between F_i^+ and F_i^- in the direction of the unit normal of fracture pointing from F_i^- to F_i^+. The fluid flow in the fractures is treated as a 2D problem with the local co-ordinate system defined on F_i. The common interface contained in the grid block model is therefore the union of all fracture surfaces and edges, $\bigcup_{i=1}^{N}\left(F_i^+ \bigcup F_i^- \bigcup F_i^{\mathrm{b}}\right)$. In the total volume V of the block model, the volumes occupied by the matrix and fractures are labeled as V_m and V_i ($i = 1, 2, \ldots, N$), respectively.

In what follows, the boldface letters **x, y** and **z** refer to the position vectors of points in V and as such will have three components. Boldface Greek letters, such as ξ and ζ refer to position vectors of points on the mean plane of one of the fractures, F_i, referenced with respect to the 2D co-ordinate system on that fracture, with only two components.

Based on the assumption of averaged flow velocity in fractures and using $\mathbf{v}_i$ and p_i as representing flow velocity and pressure in fracture i with m_i intersection traces with other connected fractures, the flow equation can be written as

$$\mathbf{v}_i(\xi) = -k_i \nabla p_i(\xi) \tag{10.29a}$$

$$\nabla \cdot \mathbf{v}_i(\xi) = -\frac{1}{e_i} Q_i(\xi) + \sum_{j=1}^{m_i} \int_{L_i^j} q_i^j(\varsigma)\delta(\xi - \varsigma)\mathrm{d}l(\varsigma) \tag{10.29b}$$

where $Q_i(\xi)$ represents the source magnitude of the fluid flow from fracture to matrix and $\delta(\,)$ the 2D Dirac delta function. The intersections with other fractures are treated as sources or sinks for fluid flow, and the effects are accounted for by the line integrals in Eqn (10.29b) with trace length L_i^j ($j = 1, 2, \ldots, m_i$), where $q_i(\varsigma)$ is the magnitude of sources/sinks along the m_i intersection traces. The intrinsic permeability of fracture i in Eqn (10.29a) is then $k_i = (e_i)^2/12$. Note that the differentiation is restricted to the 2D space of the fracture. The boundary conditions for the flow in a fracture is given by zero normal flow flux if the fracture is located inside the boundary of the grid model (∂V) or pressure equal to the boundary pressure if the fracture intersects the grid model boundary,

$$\mathbf{v}_i(\xi)\cdot\mathbf{n}_i = 0 \quad (\text{if } \xi \text{ is inside } \partial\mathrm{V}) \tag{10.30a}$$

$$p_i(\xi)\cdot\mathbf{n}_i = p|_{\xi\in\partial V} \quad (\text{if } \xi \text{ is on } \partial\mathrm{V}) \tag{10.30b}$$

The fluid flow inside the porous matrix is assumed to be incompressible and to follow Darcy's law with permeability k_m, with its velocity $\mathbf{v}_m$ and pressure p_m, and is governed by

$$\mathbf{v}(\mathbf{x}) = -k_m\nabla p_m(\mathbf{x}) \tag{10.31a}$$

$$\nabla\cdot\mathbf{v}(\mathbf{x}) = \sum_{i=1}^{N}\int_{F_i} Q_i(\xi(\mathbf{x}))\delta(\mathbf{x}-\mathbf{z})\mathrm{d}A(\mathbf{z}) \tag{10.31b}$$

where dA( ) is the differential area element for fractures with respect to the global 3D co-ordinate space and Eqn (10.31b) represents the coupling between the fractures and matrix.

The usual potential or flux boundary conditions, or the combination of the two, are prescribed on ∂V. At the common interfaces between the matrix and fractures, the fluid pressure is taken as the fracture pressure, and the flow velocity depends on the source/sink magnitudes of the fracture,

$$p_m(\mathbf{x})|_{x\in F_i} = p_i\ \ (\xi(\mathbf{x}))|_{\xi\in F_i} \tag{10.32a}$$

$$(\mathbf{v}_m(\mathbf{x})|_{\mathbf{x}\in F_i^+} - \mathbf{v}_m(\mathbf{x})|_{\mathbf{x}\in F_i^-})\cdot\mathbf{n}_i = Q_i(\xi(\mathbf{x}))|_{\xi(\mathbf{x})\in F_i} \tag{10.32b}$$

Application of the fundamental solutions (cf. Eqns (10.12) and (10.13)) and Green's identity then leads to the boundary integral equation of the ith fracture

$$c_ip_i(\xi) + \int_{\partial F_i}\ln(|\varsigma-\xi|)\frac{\partial p_i(\varsigma)}{\partial n_i}\mathrm{d}l(\varsigma) = \int_{\partial \mathbf{F}_i}\frac{(\varsigma-\xi)}{|\varsigma-\xi|^2}\partial p_i(\varsigma)\mathrm{d}l(\varsigma) + \frac{1}{e_ik_i}\int_{F_i}\ln(|\varsigma-\xi|)Q_i(\varsigma)\mathrm{d}A(\varsigma)$$
$$-\frac{1}{k_i}\sum_{j=1}^{m_i}\int_{L_i^j}\ln(|\varsigma-\xi|)q_i^j(\varsigma)\mathrm{d}A(\varsigma) \tag{10.33}$$

where c_i is the free term and ∂F_i the outside edge of the mean plane F_i of the ith fracture. Similarly the boundary integral equation for flow in the matrix is written as

$$\begin{aligned} c_mp_m(\mathbf{x}) = &\int_{\partial V}\frac{1}{|\mathbf{y}-\mathbf{x}|}\frac{\partial p_m(\mathbf{y})}{\partial n}\mathrm{d}A(\mathbf{y}) + \int_{\partial V}\frac{(\mathbf{y}-\mathbf{x})\cdot\mathbf{n}(\mathbf{y})}{|\mathbf{y}-\mathbf{x}|^3}p_m(\mathbf{y})\mathrm{d}A(\mathbf{y}) \\ &+\frac{1}{k_m}\sum_{i=1}^{N}\int_{F_i}\frac{1}{|\mathbf{y}-\mathbf{x}|}Q_i(\xi(\mathbf{y}))\mathrm{d}A(\mathbf{y}) \end{aligned} \tag{10.34}$$

where c_m is the free term.

The boundary integral Eqns (10.33) and (10.34), the interface conditions (10.32a) and (10.32b) and the global boundary conditions (10.30a) and (10.30b) form the complete equation system. Application of standard BEM techniques using isoparametric elements and Gaussian quadrature can then transform the equations into a set of algebraic equations in the BEM formulation, with the primary unknown variables being pressure and normal velocity on the global grid model boundary, pressure and source magnitude on fracture planes, pressure and normal flux at the outside edges of fracture planes and the line source magnitude at the fracture intersection traces. This is a one-step solution technique compared with the two-step approach described in Section 10.3.2.

Let

- vector $\mathbf{U}$ hold values of either the pressure or the normal velocity at each node of the grid block boundary,
- vector $\mathbf{Q}$ hold the nodal values of the source magnitudes at the nodes of the fracture planes,
- vector $\mathbf{P}_{\mathrm{f}}$ hold the nodal values of the pressure of fracture planes,
- vector $\mathbf{P}_{\mathrm{fb}}$ hold the values of the pressure at the nodes on the fracture plane edges,
- vector $\mathbf{W}_{\mathrm{fb}}$ hold the values of the normal component of the velocity at the nodes on the fracture plane edges,
- vector $\mathbf{q}$ hold the values of the line source magnitudes at the nodes on the fracture intersections, and finally,
- vector $\mathbf{P}_{\mathrm{fi}}$ hold the corresponding values of the pressure at the fracture plane intersections.

Then the matrix–vector system of Eqns (10.33) and (10.34), from applying standard boundary element techniques, has the following more compact block-matrix form, with grouped equations into separate parts (blocks) for convenience,

$$\begin{bmatrix} A_1 & B_1 & 0 & 0 & 0 & 0 \\ A_2 & B_2 & C_2 & 0 & 0 & 0 \\ 0 & B_3 & C_3 & D_3 & E_3 & F_3 \\ 0 & 0 & C_4 & D_4 & 0 & 0 \\ A_5 & 0 & 0 & D_5 & E_5 & 0 \\ 0 & B_6 & 0 & D_6 & E_6 & F_6 \end{bmatrix} \begin{Bmatrix} \mathbf{U} \\ \mathbf{Q} \\ \mathbf{P}_{\mathrm{f}} \\ \mathbf{P}_{\mathrm{fb}} \\ \mathbf{W}_{\mathrm{fb}} \\ \mathbf{q} \end{Bmatrix} = \begin{Bmatrix} \mathbf{R}_1 \\ \mathbf{R}_2 \\ 0 \\ 0 \\ \mathbf{R}_5 \\ 0 \end{Bmatrix} \tag{10.35}$$

The first block results from collocating the matrix equations on the global model (grid block) boundaries and the second block results from collocating the equations on the fracture surfaces. The third block results from collocating the fracture equations on the fracture edges. The fourth block implies that the nodal values of the pressure in matrix and fracture must coincide at the fracture edges. The fifth block results from explicitly setting boundary conditions at the fracture edges. The sixth block results from equating the fracture pressures between each pair of intersecting fractures.

Some special features of the matrix blocks need to be noticed to simplify Eqn (10.35). If unknown variables in vector $\mathbf{U}$ are chosen as normalized pressure $\widehat{p}(\mathbf{x}) = p(\mathbf{x}) - \mathbf{x} \cdot \mathbf{J}$ (where $\mathbf{J}$ is the pressure gradient) and velocity $\widehat{\mathbf{v}}_m(\mathbf{x}) \cdot \mathbf{n} = \mathbf{v}_m(\mathbf{x}) \cdot \mathbf{n} + k_m \mathbf{J} \cdot \mathbf{n}$ on the global grid block model boundary, then $\mathbf{R}_1 = \mathbf{0}$ holds. Block D_4 is an identity matrix. Block C_2 represents the interpolation of nodal values on the fracture planes and its structure is quite predictable. The inverse of C_2 is equally straightforward. In addition, when periodic boundary conditions are assumed (this kind of boundary condition is especially applicable when the grid block serves as a standard cell of much larger models for reservoir characterization or derivation of representative behavior (properties) of fractured rocks), the fracture edges may be treated as being inside of the model boundary and therefore the normal velocity at these fracture edges can be set to zero, i.e. $\mathbf{W}_{\mathrm{fb}} = \mathbf{0}$. Application of all the above simplification then leads to a much reduced size of the equation

$$\begin{bmatrix} A_1 & B_1 & 0 \\ \widehat{A}_3 & \widehat{B}_3 & F_3 \\ \widehat{A}_6 & \widehat{B}_6 & F_6 \end{bmatrix} \begin{Bmatrix} \mathbf{U} \\ \mathbf{Q} \\ \mathbf{q} \end{Bmatrix} = \begin{Bmatrix} 0 \\ \mathbf{R}_3 \\ \mathbf{R}_6 \end{Bmatrix} \tag{10.36}$$

where

$$\widehat{A}_3 = -(C_3 - D_3C_4)C_2^{-1}A_2 \tag{10.37a}$$

$$\widehat{A}_6 = -D_6C_4C_2^{-1}A_2 \tag{10.37b}$$

$$\widehat{B}_3 = B_3 - (C_3 - D_3C_4)C_2^{-1}B_2 \tag{10.37c}$$

$$\widehat{B}_6 = B_6 + D_6C_4C_2^{-1}B_2 \tag{10.37d}$$

$$\mathbf{R}_3 = -(C_3 - D_3C_4)C_2^{-1}\mathbf{R}_2 \tag{10.37e}$$

$$\mathbf{R}_6 = D_6C_4C_2^{-1}\mathbf{R}_2 \tag{10.37f}$$

The above BEM approach also points a way to consider coupled stress-flow in the matrix so that the stress effect of the matrix on the fracture displacement and flow, and the fracture–matrix interaction in terms of fluid flow, can be included, with, of course, much increased computational efficiency challenges. Such a combination is an advantage for the BEM.

10.3.4 Pipe and Channel Lattice Models

Compared with FEM and BEM representations of the flow in fracture systems, the equivalent pipe network and channel lattice models are the simplest computationally, due to the fact that the fractures are replaced with single or a lattice of pipes of effective diameters determined by the volumes (i.e., areas and apertures) of fractures; also, modeling flow in pipes is straightforward without the need for numerical integration of partial differential equations, as in the cases of FEM and BEM approaches. However, this approach is applicable only for fluid flow and transport processes without considering the effects of heat and stress, therefore, cannot be applied for rock mechanics problems with such effects.

Pipes or channel lattice networks can be constructed as regular grids (Long, 1991) or as random networks following a Poisson process (Dershowitz, 1984). They have been conditioned either from large-scale field experiments (Long, 1991), or based on streamline distributions from continuum modeling (Herbert and Layton, 1992), in order to improve their applicability. A recent application of this approach is the channel lattice modeling of fluid flow for the fractured rocks of the RCF3 pump test at Sellafield, UK (Billaux and Detournay, 1997) for the international DECOVALEX II project for modeling and experiments of coupled THM processes of nuclear waste repositories (Jing et al., 1998). The random lattice model was also compared with Effective Medium Theory (EMT) in Adler and Berkowitz (2000) for equivalent continuum characterization of fractured media.

10.4 Alternative Techniques – Percolation Theory

Percolation theory is based on a random lattice model of conductors (fractures) for deriving fluid transport conditions (critical density) and properties (permeability), based on the system connectivity through a geometrical probability concept (Robinson, 1984; Hestir and Long, 1990; Berkowitz and Balberg, 1993; Sahimi, 1995). The theory provides a theoretical platform for understanding the geometric conditions for fluid conduction in fractured media in terms of a percolation threshold, expressed as a critical probability. Application of this theory in rock mechanics and engineering problems seems to concentrate on characterization of flow properties of fractured rocks. Some applications in this field are given below:

- Critical behavior of deformation and permeability of fractured rocks: Zhang and Sanderson (1998);
- Flow and transport in fracture networks: Mo et al. (1998);
- Microstructure and physical properties of rock: Guéquen et al. (1997); and
- Fluid, heat and mass transport in percolation clusters: Kimmich et al. (2001).

Percolation theory can be used for quantitative scaling of permeabilities of fractured media, usually in a form of a power law. It originated in the field of solid state physics, mainly semi-conductors, reported in the initial works by Broadbent and Hammersley (1957) and Essam (1980). Classical percolation theory starts with an infinite pairs of sites connected by bonds. The percolation models can be formulated as site percolation and bond-percolation models. The former concerns the probability $p^{(S)}$ of a site that is open, independently of other sites. The connected open sites then form paths. In bond percolation, all the sites are assumed to be open but there is a probability $p^{(B)}$ that a bond is open, again independently of all other bonds. A path is then formed by open sites connected by open bonds. Clusters of sites are formed in which any pair is connected by a path. The objective of the theory is to define whether a critical probability, $p_c^{(S)}$, exists, at which path connecting clusters of sites will reach from one part of a model to another. For the site percolation case if $p^{(S)} < p_c^{(S)}$ only finite size clusters (locally connected) exist, but for $p^{(S)} > p_c^{(S)}$ infinitely connected clusters exist and the whole system of clusters (sites) is conductive to electricity (or fluid flow). A similar position prevails in the bond percolation. Generally the systems considered are lattices with bonds connecting neighboring sites. Both site (Robinson, 1984) and bond (Adler and Thovert, 1999) percolation models have been used in characterizing fluid flow in fractured rocks and have essentially the same applicability. In this section we assume the site percolation models as described in Robinson (1984). The sites are taken to represent the fractures randomly generated with regard to position, size and orientation. Two fractures have a bond between them if they intersect. A cluster is a set of connected fractures.

There is no unique way of defining the critical density for a percolation model of a finite size. In most practice, a square (or cubic) region of fixed size is chosen and the random fractures are generated in increasing numbers until a cluster of connected fractures is formed and makes contact with all the sides of the region. The fracture system is then said to be percolating. The ratio of the number of fractures centered in the region divided by the area (or volume) of the region is taken as the percolating density of the system. This process is repeated for many realizations of the fracture system with the same statistics. The average percolating density of the multiple realizations is taken to be the critical density, also called the percolation threshold. The critical density thus defined is equivalent to the density at which half the square (or cubic) regions have a cluster contacting all four (or six) sides in 2D (or 3D).

An alternative definition is to require the clusters to connect either pair of opposite sides of the square, or only any pair of opposite faces in 3D. This is the same definition as the one mentioned above, but with relaxed demand on connectivity of the clusters at the boundary of the model. Another possible definition uses a periodic network with the square region as the basic cell. The intersections for such a system could easily be found but finding the infinite clusters would be computationally quite difficult and so the definition has not been often applied in practice.

Besides the critical fracture density, the number of intersections for each fracture, sometime also called the co-ordination number in literature, is another important variable in percolation theory. It can be calculated directly with the DFN algorithms as mentioned before and can also be predicted using fracture statistics from simple geometrical arguments. The number of fracture intersections at percolation is the continuum equivalent of $zp_c^{(S)}$ where z is the number of intersections to each fracture in a DFN model. It is found that, as the co-ordination number, z, increases $zp_c^{(S)} \rightarrow 4.5$ in two dimensions and $zp_c^{(S)} \rightarrow 2.7$ in three dimensions (Shante and Kirkpatrick, 1971).

Robinson (1984) presented an approach to define the relation between the density and mean number of intersections in general. Without losing generality, it is assumed that the geometry (location, size, orientation) of fractures in the system is described by a set of parameters s with known probability distributions, $f(s)$. Two fractures are connected by an intersection if their parameters satisfy some conditions, represented as $b(s_i, s_j) = 1$, and condition $b(s_i, s_j) = 0$ indicates no intersection. The mean number of intersections per fracture is then just the number of fractures times the integral over the parameter space of the connection function $b(\,)$ times the probability functions, written as

$$I = N \int \mathrm{d}s_i \int \mathrm{d}s_j b(s_i, s_j) f(s_i) f(s_j) \tag{10.38}$$

If we assume that the actual number of intersections on a particular fracture is distributed with a Poisson distribution with mean I^z, then the intersection number of fractures is given by the distribution

$$P_z = \frac{e^{-I} \cdot I^z}{z!} \tag{10.39}$$

This, however, does not apply to cases when there is a variation in fracture length.

An example is given here to demonstrate the validity of the above model, for a system of fractures with density ρ, a constant length $2l$ and oriented either horizontally or vertically with equal probability in a region of area A containing $N = \rho A$ fractures. A pair of fractures will intersect if they are orthogonal (for this example) and if the center of one lies within a square of side $2l$ around the other's center. Ignoring the effect of the boundaries, the probability that the two intersect is $p = (2l)^2/(2A)$. Therefore the probability that a given fracture intersects z other fractures is

$$p_z = \frac{(N-1)(N-2)\ldots(N-z)}{z!}(1-p)^{N-1} p^z (1-p)^{-z} \tag{10.40}$$

As the area A, and hence the total number of fractures N, tends to infinity, this gives

$$p_z = \frac{(2\rho l^2)^2}{z!} e^{-2\rho l^2} \tag{10.41}$$

which demonstrates that the distribution of the number of intersections is a Poisson distribution as claimed above and has a mean $I = 2\rho l^2$. The critical number of intersections per fracture for this case is therefore $I_c = 2\rho_c l^2$. Similar results can be derived for other fracture statistics. They are collected in the Table 10.1. On the other hand, the critical probability for bond percolation on a square lattice is $p_c = 1/2$.

The power law relations derived in percolation theory usually take the form (Berkowitz, 2002)

$$A \propto (N - N_c)^{-X} \tag{10.42}$$

where A is a geometrical or physically observable quantity (such as hydraulic conductivity), X an exponent specific to quantity A, N_c the critical number of fractures at the percolating threshold and N the total number of fractures in the system . Power laws of this form have been found to characterize geometrical characteristics such as the density and size of fractures necessary to ensure network connectivity, the size and extent of fracture 'clusters', and the likelihood of a fractured formation

Table 10.1 Relation between density and number of intersections (Robinson, 1984)

Fracture statistics	Number of intersections per fracture
Two orthogonal sets (all with length $2l$, half oriented each way and total density ρ)	$2\rho l^2$
Two sets with angle α between them (all with length $2l$, half oriented each way and total density ρ)	$2\rho l^2 \sin\alpha$
Uniformly distributed orientation (all with length $2l$ and total density ρ)	$8(\rho l^2)/\pi$
Orientation uniformly distributed between -0 and α (all with length $2l$ and total density ρ)	$\rho l^2(2/\alpha^2)(2\alpha - \sin 2\alpha)$
Any case with length uniformly distributed	As for case with fixed length equal to mean length, not a Poisson distribution

being hydraulically connected (Balberg et al., 1991; Berkowitz and Balberg, 1993; Sahimi, 1995; Bour and Davy, 1997, 1998, 1999). In practice, on the other hand, the typical power law for the trace length of the fractures is often given in a general form directly from measurements (fracture mapping at ground surface, lineament measurements by use of airplane or satellite survey maps) as

$$f \propto aL^{-X} \tag{10.43}$$

where f is the frequency density (a function of fracture trace length and model size), L the trace length, a a coefficient of proportionality (related to fracture density and model size) and the power law exponent X is generally in the range $1 < X < 3$ (Segall and Pollard, 1983; Scholz and Cowie, 1990; Davy, 1993). Such length distributions are widely reported for many crystalline and massive sedimentary rocks. The greatest uncertainty in characterizing such length distributions lies in definitions of the resolution limits of measurements, and lower cut-off lengths for data processing, but with large enough fracture sample population for a broad-band determination of coefficient a, and determination of correlations between surface and sub-surface features (e.g., layering or general change of lithology with depth considering fracturing patterns). Blind numerical fitting of power laws for estimates of a and X values regardless of site-specific conditions is likely to cause considerable uncertainty.

As pointed out above, percolating depends on both densities and connectivity (represented by the co-ordination number). Quite often seemingly 'dense' networks of fractures are not necessarily percolating due to the lack of connectivity and, conversely, sparse fracture systems may be percolating if they are well connected with respect to the model boundaries. However, dense systems may become close to a critical state so that small changes of local fracture geometry or density may cause drastic changes.

The DFN modeling does not require estimation of critical fracture density and co-ordination number of the system when the connectivity of the system is already numerically determined explicitly for each realization for systems of any complexity in terms of spatial correlation or anisotropy. Classical percolation theory is founded on the key assumption of purely random (uncorrelated, isotropic) systems for more or less regular fracture systems, and the power law relations are strictly valid only for fracture densities close to the percolation threshold. Its applicability for natural fracture systems with anisotropies, variable densities and/or spatial correlations may be limited and DFN modeling is more flexible in this regard. Even so, both DFN and percolation theory, either deterministic or stochastic, still have a long way to progress in order to capture the spatial organization, correlation properties and hence the connectivity of natural fracture patterns (Bour and Davy, 1999; Odling et al., 1999). The main hurdle is uncertainty in the relevant parameters that quantify connectivity and scaling properties and our inability to verify the reliability of the stochastic realizations for representing the real systems hidden in the sub-surface rock mass. Recent publications in Berkowitz et al. (2000) and Bour et al. (2002) with more general synthetic expressions for the fracture connectivity and length distribution considering the fractal nature of fracture density (Davy et al., 1990), as well as scales of measurement and resolution, are a step forward in overcoming such difficulties based on three basic parameters: the scaling exponent, the fractal dimension of the fracture center population and the density term.

10.5 Alternative Techniques – Combinatorial Topology Theory

In terms of the mathematical characteristics, there are two categories of geometrical properties of a fracture system. One category is the metric properties, such as orientation, spacing, aperture and size. These properties can be measured with certain physical units (degrees, meters, square or cubic meters, for example) and they may vary under certain conditions (e.g., during a deformation process). The other category is the topological properties, typically the fracture connectivity, which cannot be measured by using any physical units and does not change during continuous (and small) deformation processes. For example, two connected fractures will remain connected unless the block containing these two fractures is completely broken, which is, however, a discontinuous (and large) deformation process. The connectivity defines the geometrical interrelations between the fracture elements (shapes, boundaries and intersections) and the blocks so formed. Topological properties cannot be directly measured: they can only be represented in a combinatorial manner, i.e., counted.

This section presents a new technique for fracture system characterization, based on theories of combinatorial topology and percolation (Jing and Stephansson, 1996). The basic assumptions are that the permeability of the rock matrix is negligible (zero) and no fracture growth is considered. For simplicity and clarity of the demonstrations, the analysis is carried out only for 2D problems.

Dershowitz (1984), Low (1986) and Einstein (1993) used the following measures to characterize a fracture system:

(1) The density of fracture intersections (number of fracture intersections per unit area) in the regions of the rock mass concerned, denoted as $C1$.

(2) The percolation probability, i.e., the probability of a randomly selected fracture extending itself from one end of the region concerned to another, through its connections with other fractures, denoted as $C5$.

(3) The total lengths of fractures in the region concerned, denoted as $C8t$.

(4) The total length of fractures projected in the x-axis direction, denoted as $C8x$.

(5) The number of independent (disconnected) sub-networks in the region. An independent sub-network is formed by a portion of the total fracture population, but is disconnected with the remaining portion of the fractures and the global boundaries of the region.

Except for the percolating probability ($C5$), which is related to the density of fracture intersections ($C1$), the above measures are independent of each other and are defined by counting all fractures in the region concerned. Relations between these measures are not defined. For hydraulic characterization of fractured rocks, the isolated and singly connected fractures, which do not contribute to the permeability (or conductivity) of the whole region, should be treated by regularization first.

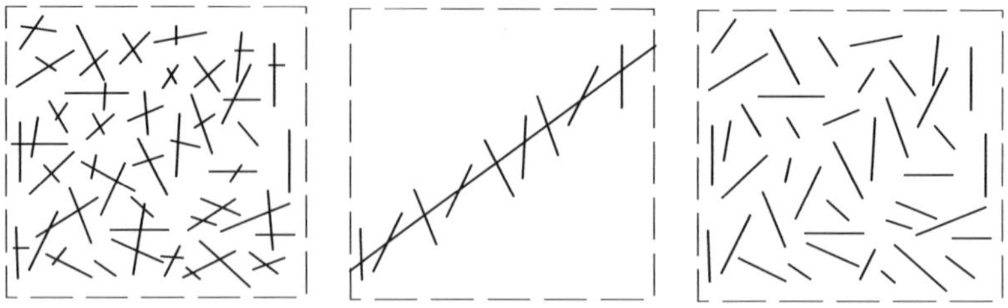

Fig. 10.6 Extreme situations of non-percolation: (a) System of singly connected fractures, (b) One multiple and nine singly connected fractures with equal values of C8*x* and (c) A completely disconnected fracture system (Jing and Stephansson, 1996).

The *C*1 measure is defined by counting all but the isolated fractures. If the singly connected fractures are included, a high value of *C*1 implies a high degree of fracture intersection, but this does not necessarily mean a high degree of fracture connectivity. Intersection is a geometrical relation between two (or several) fractures, but connectivity is a topological property for the whole fracture system (or network). In an extreme illustrative case of a fracture system as shown in Fig. 10.6a, the fractures are highly intersected with a high value of *C*1. However, the equivalent permeability of the rock mass (assuming impermeable rock matrix) is zero and the fracture system is not percolating since no connected fracture pathways are formed and there is no connection with the outer boundaries, i.e., zero connectivity.

Measure *C*5 is defined without the influence of isolated fractures, but may contain the effects of the singly connected fractures. In the second extreme example shown in Fig. 10.6b, the fracture system contains one persistent and multiply-connected fracture and many other singly connected fractures that do not have any contribution to the conductivity of the region. Again this shows that numbers of intersections alone do not automatically characterize connectivity properly.

The above two extreme examples demonstrate the necessity of the combined use of critical fracture density and intersection number in percolation theory, as described in Section 10.4

The measures *C*8*t* and C8*x* are metric measures and not topological properties of a fracture system. High values of *C*8*t* or *C*8*x* alone do not automatically indicate a high probability of conductivity since a completely disconnected discrete fracture system may also have very high *C*8*t* or *C*8*x* values without connectivity (Fig. 10.6c). Therefore *C*8*t* and *C*8*x* are not particularly useful for fracture network characterization, but might be useful for DFN software coding: the distributions of the fracture size and orientation must be considered.

Since connectivity is a topological property of a fracture system, it is natural to use the theory of combinatorial (or algebraic) topology for characterizing geometrical properties of fracture systems in DFN models (Aleksandrov, 1956; Henle, 1974). A special attraction of this theory is that it provides a simple tool for representing the algebraic relations between the vertices, edges and faces of polygonal systems or spatial polyhedral systems. A fracture network in rocks can be represented as 2D or 3D graphs defined by fracture planes (faces) and their intersections (edges and vertices). The topological relations defined in combinatorial topology concerning the vertices, edges and faces can then be adopted or modified to qualitatively characterize the connectivity of a fracture system. The extended Euler–Poincaré formula (7.1) (cf. Section 7.1 in Chapter 7) can then be used to characterize the network connectivity.

In Eqn (7.1), let $N_{sn} = 0$ (which holds for regulated global fracture networks in which no independent sub-networks remain) and dividing both sides of Eqn (7.1) by the total area of the region concerned, *A*, Eqn (7.1) can be rewritten as

$$d_v + \frac{1}{a_b} - d_e = \frac{1}{A} \tag{10.44}$$

or

$$d_v + d_p - d_e = \frac{1}{A} \tag{10.45}$$

where $d_v = N_v/A$ is the density of the intersections of regularized fractures and is equal to C1 minus the number of intersections formed by singly connected fractures in the original fracture system. The term $d_p = N_p/A$ is the density of blocks (polygons) and the term $a_b = A/N_p$ is the mean block (polygon) size. The term $d_e = N_e/A$ is the density of edges and is a function of the distribution properties of the trace length and spacing of the fracture sets in the region. Measures d_v, d_p and d_e are topological variables, and a_b and A are metric measures. Together, they characterize an average polygonal partition of the solution domain of area A.

Equation (10.45) represents an interrelation between key topological and metric parameters characterizing a completely connected fracture network. Therefore it is a useful model for fracture network characterization, especially when the hydraulic properties of fractured rock masses are concerned. Any fracture network with its topological measures satisfying Eqn (10.45) or (10.44) is a completely connected network that contains all possible pathways for fluid flow.

As introduced in Section 10.4, the critical fracture density, or the percolation threshold, is defined as the density of the fracture intersections in the domain when a global fracture cluster connects to the outer boundaries. Since the connectivity of the fracture network represents the topological relations between the intersections, using the density of fracture intersections is more appropriate for percolation characterization. The use of both combinatorial topology of a regularized fracture network and percolation theory thus provides a unique and precise definition of the percolation threshold of fracture intersection.

From only the statistical distributions of the fracture sets, there is no closed-form solution for determining the value of the percolation threshold for a general fracture system. The topological algorithms for fracture regularization and block-flow path tracing (Jing and Stephansson, 1994a,b) can be applied to determine whether the fracture system satisfies Eqn (7.1). Two conditions must be satisfied for a percolating fracture system: (i) Eqn (7.1) must be satisfied after the fracture network regularization; (ii) the regularized fracture system is connected to the outer boundaries of the solution domain. The critical density of the fracture intersections, denoted as d_{vc}, is the minimum of d_v when Eqn (7.1) is satisfied during the process of fracture generation and regularization while the distribution of the orientation, trace length and spacing of the fracture sets remains unchanged. The addition of more fractures, after the system becomes percolating, just makes the system more conductive, but will not change the fact that the system is already percolating – since the critical threshold d_{vc} has already been passed. Noting that $1/A \to 0$ for many practical problems, a percolation criterion, using the critical density of the fracture intersections as the basic measure, can be written as

$$\begin{cases} d_v = d_e - \dfrac{1}{a_b} \geq d_{vc} \quad \text{or} \quad d_v = d_e - d_p \geq d_{vc} \\ \max \quad (l_e^i, i = 1,\ 2,\ \ldots,\ N_e) < \min \quad (l_e^b) \end{cases} \tag{10.46}$$

where l_e^i (i = 1, 2, …,N_e) is the length of the edges and l_e^b the length of the sides defining the outer boundaries of the solution domain. This relation connects the topological properties of a fracture network with its percolating potential for fluid flow. It also represents a quantitative relation among the percolation threshold, area of the solution domain and block density. Equation (10.46) is a theoretical relation based on network topology and can be used to determine whether a fracture system is percolating, no matter whether the fracture system is generated directly or by using distribution functions of orientation, trace length and spacing, and can be directly combined with standard DFN modelling.

10.6 Summary Remarks

A few issues of importance relating to the DFN representation of fracture systems are discussed below regarding characterization methods, validity of the fractal or power law characterization and fracture system connectivity. The following summary remarks are based mainly on overviews in Odling (1997) and Berkowitz (2002).

10.6.1 Quality of Fracture Mapping and Data Estimation

In practice it is not usually possible to collect (spatial) data over more than two orders of magnitude. Truncation due to measurement resolution limits, censoring and other finite size effects often limits the scale range over which a power law can be meaningfully measured to less than one order of magnitude. Moreover the very definition of a fracture can be a function of resolution: a fracture that appears as a single, continuous trace on an aerial map often consists of a series of smaller, unconnected ('en echelon') fracture traces when viewed at ground level.

As demonstrated in Bonnet et al. (2001), blind application of analysis techniques and lack of objective evaluation of the suitability of a power law to characterize site-specific data, is likely to lead to unreliable fracture characteristics. To avoid such blind application, proper grouping or classification of a fracture population according to geological origins, formation history and sequence, and proper evaluation of the limitations of the mapping techniques are important issues in fracture system characterization and the subsequent generations of stochastic realizations. The difficulty is verification of the generated fracture system geometries. Published power law exponents and fractal dimensions for trace length and aperture vary widely. The vast majority of the data are based on 1D (scanline mapping and borehole logging) and 2D (surface trace mapping) measurements and their extrapolation to 3D systems is still problematic and depends largely on assumed geometric probabilities relating 2D and 3D features.

10.6.2 Representation of Scale Effects by a Fractal or Power Law

Using field measurements to estimate power law exponents and fractal dimensions (as well as other characteristic distributions) of fracture systems remains problematic. The fractal concept has been applied to fracture system characterization in order to consider the scale dependence of the fracture system geometry and for upscaling the permeability properties, using usually the box counting method or the Cantor dust model (Barton and Larsen, 1985; Chilés, 1988; Barton, 1992; Castaing et al., 1997; Wilson, 2001; Babadagli, 2002; Doughty and Karasaki, 2002). Power law relations have also been found to exist for trace lengths of fractures and have been applied for representing fracture system connectivity (Renshaw, 1999). A detailed analysis of these issues can be seen in Bonnet et al. (2001) for determining power law exponents and fractal dimensions, together with guidelines for their accurate and objective estimations.

As pointed out in Berkowitz (2002), an important consequence of systems that exhibit scaling behavior is that they do not possess any homogenization scale. If this scaling behavior is valid over all fracture size ranges, it means that no REV exists for fractured rocks and the equivalent continuum approach has no physical justification at all – since the elastic compliance and permeability tensors can only be defined for a continuum at a certain scale. On the other hand, continuum approaches often produce meaningful results at different scales with or without an explicit representation of the fractures, indicating that the equivalent continuum assumption may still be valid for problems at large enough scales. Therefore the scaling effect may have a certain range of validity, and homogenization may still be feasible to derive equivalent properties characterizing the overall properties of fractured rocks or fractured porous rocks (Cravero and Fidelibus, 1999; Barla et al., 2000; Svensson, 2001b; Park et al., 2002; Min et al., 2004). There may not exist universally valid criteria or approaches for the existence of REVs in fractured rocks, and the problems have to be treated as site-specific with careful evaluation of the ranges of applicability of different mapping and characterization techniques, and the conditionings enforced by local geology (e.g., the trace length restriction imposed by lithological layering) and the mechanical processes of fracture formation (e.g., orientation and growth rate of fractures in the rock matrix as restricted by tectonic stress processes), in order to obtain objective results (Genter et al., 1997; Gringarten, 1998; Meyer and Einstein, 2002).

10.6.3 Network Connectivity

Connectivity and other hydro-geological parameters of fracture networks depend on the geometrical structure of the generated fracture networks. The 2D stochastic realizations, although conceptually clear, and mathematically and computationally convenient, may not be reliable for interpreting the 3D situation since the connectivity in the third direction could be much different from that in the 2D plane of the DFN models. However, extrapolating 1D and 2D data to 3D may be the only available approach (Warburton, 1980a,b; Piggott, 1997). Berkowitz and Adler (1998) considered the extrapolation of fracture data from 1D to 2D, and from 2D to 3D, and systematically developed a series of analytical relations. In particular they studied the statistical characteristics of the intersections between a plane and a fracture network for several types of disk diameter distributions – uniform, power law, lognormal and exponential – as well as the influence of other parameters, such as the number of traces in the domain. However, since the direct measurement of the 3D fracture geometry in the sub-surface is not possible for large fracture populations of varying sizes in the foreseeable future, the validation for such extrapolation cannot be readily conducted.

In DFN practice, one approach to reduce the uncertainty is to use Monte Carlo realizations of synthetic 3D fracture networks to repeatedly produce hypothetical trace maps at mapping sites, contained in the model, until a suitable agreement of the mapped and generated trace maps is reached: in other words, a 'conditioning' with the available measurements. The reliability of such 'conditioning' is often limited due to the fact that the areas of mapping sites are often too small and too sparsely located on the ground surface or walls of excavations, compared with the much larger volumes of the DFN model regions, to have a dominating effect.

The above limitations and difficulties are not limited to DFN models alone but applicable to the whole set of DEM models since the geometrical characterization of the fracture system is the most fundamental basis of all DEM models.

It should also be noted that the fracture system at a specific site exists deterministically so that the concept of stochastic fracture systems is not physically correct. However, due to the limitation of measurement techniques for exploring the geometry and locations of sub-surface structures, we have to use stochastic approaches to expand the ranges of our predictions by including in our models as much information as possible according to the sparse available data measured at the limited locations. We have to live with such limitations and the associated uncertainties for the foreseeable future and try to provide our best estimations as objectively as possible for engineering purposes.

References

Abelin, H., Bergesson, L., Gidlund, J., Moreno, L., Neretnieks, I., Widen, I. and Ågren, T., 3-D migration experiment, Report 3, Performed experiments, Results and Evaluation, Stripa Project Technical Report TR 87-21, Swedish Nuclear Fuel and Waste Management Co. (SKB), Stockholm, 1987.

Adler, P. M. and Berkowitz, B., Effective medium analysis of random lattices. Transport in Porous Media, 2000;40(2):145–151.

Adler, P. M. and Thovert, J. F., Fractures and fracture networks. Kluwer Academic Publishers, Dordrecht, 1999.

Aleksandrov, P. S., Combinatorial topology, Graylock Press, Baltimore, MD, 1956.

Amadei, B. and Illangaseekare, T., Analytical solutions for steady and transient flow in non-homogeneous and anisotropic joints. International Journal of Rock Mechanics, Mining Sciences and Geomechanical Abstracts, 1992;29(6):561–572.

Andersson, J. A., Stochastic model of a fractured rock conditioned by measured information. Water Resources Research, 1984;20(1):79–88.

Andersson, J. and Dverstorp, B., Conditional simulations of fluid flow in three-dimensional networks of discrete fractures. Water Resources Research, 1987;23(10):1876–1886.

Babadagli, T., Fractal analysis of 2-D fracture networks of geothermal reservoirs in south-western Turkey. Journal of Volcanology and Geothermal Research, 2001;112(1–4):83–103.

Babadagli, T., Scanline method to determine the fractal nature of 2-D fracture networks. Mathematical Geology, 2002;34(6):647–670

Balberg, I., Berkowitz, B. and Drachsler, G., Application of a percolation model to flow in fractured hard rocks. Journal of Geophysical Research, 1991;96(B6):10015–10021.

Balzarini, M., Nicula, S., Mattiello, D. and Aliverti, E., Quantification and description of fracture network by MRI image analysis. Magnetic Resonance Imaging, 2001;19(3–4):539–541.

Barla, G., Cravero, M. and Fidelibus, C., Comparison methods for the determination of the hydrological parameters of a 2D equivalent porous medium. International Journal of Rock Mechanics and Mining Sciences, 2000;37(7):1133–1141.

Barthélémy, P., Jacquin, C., Yao, J., Thovert, J. F. and Adler, P. M., Hierarchical structures and hydraulic properties of a fracture network in the Causse of Larzac. Journal of Hydrology, 1996;187:237–258.

Barton, C. C., Fractal analysis of the scaling and spatial clustering of fractures in rock. In: Fractals and their application to geology. Proc. of 1988 GSA Annual Meeting, 1992.

Barton, C. C. and Larsen, E., Fractal geometry of two-dimensional fracture networks at Yucca Mountain, southwestern Nevada. In: Stephansson, O. (ed.), Proc. of Int. Symp. on Rock Joints, Bjorkliden, Sweden, 1985, CENTIK, pp. 77–84, 1985.

Bear, J., Tsang, C.-F. and de Marsily, G., Flow and contaminant transport in fractured rock. Academic Press, Inc., San Diego, 1993.

Berkowitz, B., Characterizing flow and transport in fractured geological media: A review. Advances in Water Resources, 2002;25(8–12):861–884.

Berkowitz, B. and Adler, P. M., Stereological analysis of fracture network structure in geological formations. Journal of Geophysical Research, 1998;103(B7):15339–15360.

Berkowitz, B. and Balberg, I., Percolation theory and its applications to groundwater hydrology. Water Resources Research, 1993;29(4):775–794.

Berkowitz, B., Bour, O., Davy, P. and Odling, N., Scaling of fracture connectivity in geological formations. Geophysical Research Letters, 2000;27(14):2061–2064.

Billaux, D. and Detournay, Ch., Groundwater flow modelling of the RCF3 pump test. Prepared for ANDRA by ITASCA Consultants, Report B RP ITA 96.014/A, 1997

Billaux, D., Chiles, J. P., Hestir, K. and Long, J. C. S., Three-dimensional statistical modeling of a fractured rock mass – an example from the Fanay-Augéres Mine. International Journal of Rock Mechanics and Mining Science and Geomechanics Abstracts, 1989; 26(3/4):281–299.

Birkholzer, J., Li, G., Tsang, C.-F. and Tsang, Y.-M., Modeling studies and analysis of seepage into drifts at Yucca Mountain. Journal of Contaminant Hydrology, 1999;38(1–3):349–384.

Bonnet, E., Bour, O., Odling, N. E., Davy, P., Main, I. and Cowie, P., Scaling of fracture systems in geological media. Reviews of Geophysics, 2001;39(3):347–83.

Bour, O. and Davy, P., Connectivity of random fault networks following a power law fault length distribution. Water Resources Research, 1997;33:1567–1583.

Bour, O. and Davy, P., On the connectivity of three-dimensional fault networks. Water Resources Research, 1998;34:2611–2622.

Bour, O. and Davy, P., Clustering and size distributions of fault patterns: Theory and measurements. Geophysical Research Letters, 1999;26:2001–4.

Bour, O., Davy, P., Darcel, C. and Odling, N. E., A statistical scaling model for fracture network geometry, with validation on a multi-scale mapping of a joint network (Hornelen Basin, Norway). Journal of Geophysical Research, 2002;17(6):ETG4-1–ETG4-12.

Broadbent, S. R. and Hammersley, J. M., Percolation process. I. Crystals and mazes. Proceedings of the Cambridge Philosophycal Society, 1957;53:629.

Brown, S. R., Simple mathematical model of a rough fracture. Journal of Geophysical Research, 1995;100(B4):5941–5952.

Brown, S. R. and Scholz, C. H., Broad bandwidth study of the topography of natural rock surfaces. Journal of Geophysical Research, 1985;90(B14):12575–12582.

Bruel, D., Heat extraction modeling from forced fluid flow from simulated fractured rock masses: Application to the Rosemanowes Hot Dry Rock Reservoir. Geothermics, 1995a;24(3):361–374.

Bruel, D., Modeling heat extraction from forced fluid flow through simulated fractured rock masses: Evaluation of the Soultz-Sous-Forets site potential. Geothermics, 1995b;24(3):439–450.

Bruhn, R. L., Bering, D., Bereskin, S. R., Magnus, C. and Fritsen, A., Field observations and analytical modeling of fracture network permeability in hydrocarbon reservoirs. International Journal of Rock Mechanics and Mining Sciences, 1997;34(3/4): 431. Paper no.41.

Cacas, M. C., Ledoux, E., de Marsily, G., Tille, B., Barbreau, A., Durand, E., Feuga, B. and Peaudecerf, P. Modeling fracture flow with a stochastic discrete fracture network: Calibration and validation. I. The flow model. Water Resources Research, 1990;26(3):479–489.

Castaing, C., Genter, A., Chilés, J. P., Bourgine, B. and Ouillon, G., Scale effects in natural fracture networks. International Journal of Rock Mechanics and Mining Sciences 1997;(34(3/4):389. Paper no. 45.

Chen, M., Bai, M. and Roegiers, J.-C., Permeability tensors of anisotropic fracture networks. Mathematical Geology, 1999;31(4):355–373.

Chilés, J. P., Fractal and geostatistical methods for modelling a fracture network. Mathematical Geology, 1988;20(6):631–654.

Cravero, M. and Fidelibus, C., A code for scaled flow simulation on generated fracture networks. Computers and Geosciences, 1999;25(2):191–195.

Davy, P., On the frequency–length distribution of the San Andreas fault system. Journal of Geophysical Research, 1993;98(B7):12141–12152.

Davy, P., Sornette, A. and Sornette, D., Some consequences of a proposed fractal nature of continental faulting. Nature, 1990;348:56–58.

de Druezy, J. R. and Erhel, J., Efficient algorithms for the determination of the connected fracture network and the solution to the steady-state flow equation in fracture network. Computers and Geosciences, 2003;29(1):107–111.

Dershowitz, W. S., Rock joint systems. Ph.D. Thesis, Massachusetts Institute of Technology, Boston, USA. 1984.

Dershowitz, W. S., Geometric conceptual models for fractured rock masses: Implications for groundwater flow and rock deformation. In: Sousa L. R. E., and Grossmann, N. E. (eds), Eurock' 93, Vol. 1, pp. 71–76. Balkema, Rotterdam, 1993.

Dershowitz, W. S. and Einstein, H. H., Three-dimensional flow modelling in jointed rock masses. In: Montreal, Canada. Herget, G. and Vongpaisal, S. (eds) Proc. of 6th Cong. ISRM, Vol. 1, pp. 87–92, 1987.

Dershowitz, W. S. and La Pointe P., Discrete fracture approaches for oil and gas applications. In: Nelson, P. P. and Laubach, S. E. (eds), Rock mechanics, pp. 19–30, Balkema, Rotterdam, 1994.

Dershowitz, W. and Miller, I., Dual porosity fracture flow and transport. Geophysical Research Letters, 1995;22(11):1441–1444.

Dershowitz, W. S., Wallmann, P. and Doe, T. W., Discrete feature dual porosity analysis of fractured rock masses: Applications to fractured reservoirs and hazardous waste. In: Tillerson and Wawersik (eds), Rock mechanics, pp. 543–550, Balkema, Rotterdam, 1992.

Dershowitz, W. S., Lee, G., Geier, J., Hitchcock, S. and La Pointe, P., User documentation: FracMan discrete feature data analysis, geometric modelling and exploration simulations. Golder Associates, Seattle, 1993.

Doe, T. W. and Wallmann, P. C., Hydraulic characterization of fracture geometry for discrete fracture modelling. Proc. of the 8th Cong. IRAM, Tokyo, pp. 767–772, 1995.

Doughty, C. and Karasaki, K., Flow and transport in hierarchically fractured rock. Journal of Hydrology, 2002;263(1–4):1–22.

Duliu, O. G., Computer axial tomography in geosciences: A review. Earth Science Reviews, 1999;48(4):265–281.

Dverstorp, B. and Andersson, J., Application of the discrete fracture network concept with field data: Possibilities of model calibration and validation. Water Resources Research, 1989;25(3),540–550.

Einstein, H. H., Modern developments in discontinuity analysis – the persistence-connectivity problem. In: Hudson, J. A. (ed.), Comprehensive rock engineering, Vol. 3, pp. 193–213, Pergamon Press, New York, 1993.

Elsworth, D., A hybrid boundary-finite element analysis procedure for fluid flow simulation in fractured rock masses. International Journal for Numerical and Analytical Methods in Geomechanics, 1986a;10(6):569–584.

Elsworth, D., A model to evaluate the transient hydraulic response of three-dimensional sparsely fractured rock masses. Water Resources Research, 1986b;22(13),1809–1819.

Endo, H. K., Mechanical transport in two-dimensional networks of fractures. Ph. D. Thesis, University of California, Berkeley, 1984.

Endo, H. K., Long, J. C. S., Wilson, C. K. and Witherspoon, P. A., A model for investigating mechanical transport in fractured media. Water Resources Research, 1984;20(10):1390–1400.

Essam, J.W., Percolation theory. Reports on Progress in Physics, 1980;43(7):833–912.

Ezzedine, S. and de Marsily, G., Study of transient flow in hard fractured rocks with a discrete fracture network model. International Journal of Rock Mechanics and Mining Science and Geomechanics Abstracts, 1993;30(7):1605–1609.

Fardin, N., Jing, L and Stephansson, O., Heterogeneity and anisotropy of roughness of rock joints. Proc. of EUROCK 2001, Finland., June, 2001, Balkema, Rotterdam, pp. 223–227, 2001a.

Fardin, N., Stephansson, O. and Jing, L., The scale dependence of rock joint surface roughness. International Journal of Rock Mechanics and Mining Sciences, 2001b;38(5):659–669.

Fardin, N., Stephansson, O. and Jing, L., Scale effect on the geometrical and mechanical properties of rock joints. International Society of Rock Mechanics, 10th Congress-Technology roadmap for rock mechanics. South African Institute of Mining and Metallurgy, 2003;1:319–324.

Feng, Q., Fardin, N., Jing, L. and Stephansson, O., A new method for in-situ non-contact roughness measurement of large rock fracture surfaces. Rock Mechanics and Rock Engineering, 2003;36(1): 3–25.

Geier, J., Dershowitz, W. S. and Doe, T. W., Discrete fracture modeling of in-situ hydrologic and tracer experiments. In: Myer, L. R., Cook, N. G. W., Goodman, R. E. and Tsang, C.-F. (eds), Fractured and jointed rock masses, pp. 511–518, Balkema, Rotterdam, 1995.

Genter, A., Castaing, C., Bourgine, B. and Chilès, J. P., An attempt to simulate fracture systems from well data in reservoirs. International Journal of Rock Mechanics and Mining Sciences 1997;(34(3/4):488. Paper no. 44.

Gringarten, E., FRACNET: Stochastic simulation of fractures in layered systems. Computers and Geosciences, 1998;24(8):729–736.

Guéquen, Y., Chelidze, T. and Le Ravalec, M., Microstructures, percolation thresholds, and rock physical properties. Tectonophysics, 1997;279(1–4):23–35.

He, S., Research on a model of seepage flow of fracture networks and modelling for coupled hydro-mechanical processes in fractured rock masses. In: Yuan, J.-X. (ed.), Computer methods and advances in geomechanics, Vol. 2, pp. 1137–1142, Balkema, Rotterdam, 1997.

Henle, M., Introduction to combinatorial topology, W. H. Freeman and Company, San Francisco, 1974.

Herbert, A. W., NAPSAC (Release 3.0) summary document. AEA D&R 0271 Release 3.0, AEA Technology, Harwell, UK, 1994.

Herbert, A. W., Modelling approaches for discrete fracture network flow analysis. In: Stephansson, O., Jing, L. and Tsang, C.-F. (eds), Coupled thermo-hydro-mechanical processes of fractured media-mathematical and experimental studies, pp. 213–229, Elsevier, Amsterdam, 1996.

Herbert, A. W. and Layton, G. W., Modeling tracer transport in fractured rock at Stripa. Stripa TR 92-01, SKB, Stockholm, Sweden, 1992

Herbert, A. W. and Layton, G. W., Discrete fracture network modeling of flow and transport within a fracture zone at Stripa. In: Myer, L. R., Cook, N. G. W., Goodman, R. E. and Tsang, C.-F. (eds), Fractured and jointed rock masses, pp. 603–609, Balkema, Rotterdam, 1995.

Hestir, K. and Long, J. C. S., Analytical expressions for the permeability of random two-dimensional networks based on regular lattice percolation and equivalent media theories. Journal of Geophysical Research, 1990;95(B13):21565–21581.

Jing, L. and Stephansson, O., Topological identification of block assemblages for jointed rock masses. International Journal of Rock Mechanics and Mining Science and Geomechanics Abstracts, 1994a;31(2):163–172.

Jing, L. and Stephansson, O, Identification of block topology for jointed rock masses using boundary operators. Proc. IV CSMR: Integrated approach to applied rock mechanics, Santiago, Chile, 10–14 May 1994, Published by SOCIEDAD CHILENA DE GEOTECNIA, San Martin No. 352, Sandiago, Chile, pp. 19–29, 1994b.

Jing, L,. and Stephansson, O., Network Topology and homogenization of fractured rocks. In: Jamtveit, B. and Yardley, B. (eds), Fluid flow and transport in rocks: Mechanisms and effects, pp. 191–202, Chapman and Hall, London, 1997.

Jing, L., Tsang, C.-F. and Stephansson, O., DECOVALEX – an International co-operative research project on mathematical models of coupled T-H-M processes for safety analysis of radioactive waste repositories. International Journal of Rock Mechanics and Mining Science and Geomechanics Abstracts, 1995;32(5):389–398.

Jing, L., Stephansson, O., Tsang, C.-F., Knight, L. J. and Kautsky, F., DECOVALEX II project – Technical Report-Task 1A and 1B. SKI report 98:39, Swedish Nuclear Power Inspectorate (SKI), Stockholm, Sweden, 1998.

Johns, R. A., Steude, J. S., Castenier, L. M. and Roberts, P. V., Non-destructive measurements of fracture aperture in crystalline rock cores using X-ray computed tomography. Journal of Geophysical Research, 1993;98(B2):1889–1990.

Keller K., and Bonner, B. P., Automatic, digital system for profiling rough surfaces. Review of Scientific Instruments, 1985;56:330–331.

Kimmich, R., Klemm, A. and Weber, M., Flow diffusion, and thermal convection in percolation clusters: NMR experiments and numerical FEM/FVM simulations. Magnetic Resonance Imaging, 2001;19(3–4):353–361.

Kolditz, O., Modelling flow and heat transfer in fracture rocks: Conceptual model of a 3-D deterministic fracture network. Geothermics, 1995;24(3):451–470.

Koudina, N., Gonzales-Garcia, R., Thovert, J. F. and Adler PM. Permeability of three-dimensional fracture networks. Physical Review E 1998;57(4):4466–4479.

Koyama, T., Fardin, N. and Jing, L., Shear-induced anisotropy and heterogeneity of fluid flow in a single rock fracture with translational and rotary shear displacements – a numerical study. International Journal of Rock Mechanics and Mining Sciences, 2004;41(Suppl.). Paper 2A 08.

Koyama, T., Fardin, N., Jing, L. and Stephansson, O., Numerical simulation of shear-induced flow anisotropy and scale-dependent aperture and transmissivity evolution of rock fracture replicas. International Journal of Rock Mechanics and Mining Sciences, 2006; 43(1), 89–106.

Lanaro, F., Jing, L. and Stephansson, O., 3-D laser measurements and representation of roughness of rock fractures. In: Rossmanith, H.-P. (ed.), Proc. of the 3rd Int. Conf. on Mechanics of Jointed and Faulted Rock (MJFR-3), April 6–9, Vienna, Austria, pp. 185–189, Balkema, Rotterdam, 1998.

Lanaro, F., Jing, L. and Stephansson, O., Scale dependency of roughness and stationarity of rock joints. Proc. 9th Cong. of International Society for Rock Mechanics, Paris, 1999. Vol. 2, pp. 1291–1295.

Layton, G. W., Kingdon, R. D. and Herbert, A. W., The application of a three-dimensional fracture network model to a hot-dry-rock reservoir. In: Tillerson, J. R. and Wawersik, W. R. (eds), Rock mechanics, pp. 561–570, Balkema, Rotterdam, 1992.

Lemos, J., A distinct element model for dynamic analysis of jointed rock with application to dam foundation and fault motion. Ph. D thesis, University of Minnesota, MN, USA, 1988.

Long, J. C. S., Investigation of equivalent porous media permeability in networks of discontinuous fractures. Ph.D. Thesis, Lawrence Berkeley Laboratory, University of California, Berkeley, CA, 1983.

Long, J. C. S., An inverse approach to the construction of fracture hydrology models conditioned by geophysical data: An example from the validation exercises at the Stripa Mine. International Journal of Rock Mechanics and Mining Science and Geomechanics Abstracts, 1991;28(2–3):121–142.

Long, J. C. S. and Billaux, D. M., From field data to fracture network modeling: An example incorporating spatial structure. Water Resources Research, 1987;23(7):1201–1216.

Long, J. C. S., Remer, J. S., Wilson, C. R. and Witherspoon, P.A., Porous media equivalents For networks of discontinuous fractures. Water Resources research, 1982;18(3):645–658.

Long. J. C. S., Gilmour, P. and Witherspoon, P. A., A model for steady fluid flow in random three dimensional networks of disc-shaped fractures. Water Resources Research, 1985;21(8):1105–1115.

Lough, M. F., Lee, S. H. and Kamath, J., An efficient boundary integral formulation for flow through fractured porous media. Journal of Computational Physics, 1998;143(2):462–483.

Low, L. S., Parametric studies in fracture geometry, M. A. Thesis, MIT, Cambridge, MA, 1986.

Margolin, G., Berkowitz, B. and Scher, H., Structure, flow and generalized conductivity scaling in fractured networks. Water Resources Research, 1998;34(9):2103–2121.

Mauldon, M., Estimating mean fracture trace length and density from observations in convex windows. Rock Mechanics and Rock Engineering, 1998;31(4):201–216.

Mauldon, M., Dunne, W. M. and Rohrbaugh, M. B., Jr, Circular scanlines and circular windows: new tools for characterizing the geometry of fracture traces. Journal of Structural Geology, 2001;23(2–3): 247–258.

Mazurek, M., Lanyon, G. W., Vomvoris, S. and Gautschi, A., Derivation and application of a geologic dataset for flow modeling by discrete fracture networks in low-permeability argillaceous rocks. Journal of Contaminant Hydrology, 1998;35(1–3):1–17.

Meyer, T. and Einstein, H. H., Geologic stochastic modelling and connectivity assessment of fracture systems in the Boston area. Rock Mechanics and Rock Engineering, 2002;35(1):23–41.

Min, K.-B., Jing, L. and Stephansson, O., Fracture system characterization and evaluation of the equivalent permeability tensor of fractured rock masses using a stochastic REV approach. International Journal of Hydrogeology, 2004;12(5):497–510.

Mo, H., Bai, M., Lin, D. and Roegiers, J. C., Study of flow and transport in fracture networks using percolation theory. Applied Mathematical Modeling, 1998;22(4–5):277–291.

Moreno, L. and Neretnieks, I., Fluid flow and solute transport in a network of channels. Journal of Contaminant Hydrology,1993;14(3–4):163–192.

Moreno L., Tsang, Y. W., Tsang, C.-F., Hale, F. V. and Neretnieks, I., Flow and tracer transport in a single fracture: A stochastic model and its relation to some field observations. Water Resources Research 1988;24(12):2033–2048.

Nordqvist, A. W., Tsang, Y. W., Tsang, C. F., Dverstorp, B. and Andersson, J., Effects of high variance of fracture transmissivity on transport and sorption at different scales in a discrete model for fractured rocks. Journal of Contaminant Hydrology, 1996;22(1–2):39–66.

Odling, N. E., Scaling and connectivity of joint systems in sandstone from western Norway. Journal of Structural Geology, 1997;19(10):1257–1271.

Odling, N. E., Gillespie, P. A., Bourgine, B., Castaing, C., Chilés, J. P. and Christiansen, N. P., Variations in fracture system geometry and their implications for fluid flow in fractured hydrocarbon reservoirs. Petroleum Geoscience, 1999;5:373–384.

Park, B. Y., Kim, K. S., Kwon, S., Kim, C., Bae, D. S., Hartley, L. J. And Lee, H. K., Determination of the hydraulic conductivity components using a three dimensional fracture network model in volcanic rock. Engineering Geology, 2002;66(1–2):127–141.

Piggott, A. R., Fractal relations for the diameter and trace length of disc-shaped fractures. Journal of Geophysical Research, 1997;102(B8):18121–18125.

Poon, C. Y., Sayles, R. S. and Jones, T. A., Surface measurement and fractal characterization of naturally fractured rocks. Journal of Physics D, 1992;25(8):1269–75.

Priest S. D. and Samaniego, A., A model for the analysis of discontinuity characteristics in two dimensions. Proc. of Int .Cong. on Rock Mechanics of ISRM, Melbourne, Australia, 1983, Vol. 2, pp. F199–F207.

Rasmussen, T. C., Yeh, T. C. J. and Evans, D. D., Effect of variable fracture permeability/matrix permeability ratios on three-dimensional fractured rock hydraulic conductivity. In: Bunton, B. E. (ed.), Proc. of the conference on geostatistical, sensitivity, and uncertainty methods for groundwater flow and radionuclide transport modelling. San Francisco, California, September 1987, p. 337, Batelle Press, Columbus, OH., 1987.

Renshaw, C. E., Connectivity of joint networks with power law length distribution. Water Resources Research, 1999;35(9):2661–2670.

Robinson, P. C., Connectivity, flow and transport in network models of fractured media. Ph.D. Thesis, St. Catherine's College, Oxford University, UK, 1984.

Robinson P. C. Flow modelling in three dimensional fracture networks. Research Report, AERE-R-11965, UK AEA, Harwell, 1986.

Rouleau, A. and Gale, J. E., Stochastic discrete fracture simulation of groundwater flow into underground excavation in granite. International Journal of Rock Mechanics and Mining Sciences and Geomechanics Abstracts, 1987;24(2):99–112.

Sahimi, M., Flow and transport in porous media and fractured rock. VCH Verlagsgesellschaft mbH, Weinheim, 1995.

Schmittbuhl, J., Schmitt, F. and Scholz, C. H., Scaling invariance of crack surfaces. Journal of Geophysical Research, 1995;100(B4):5953–73.

Scholz, C. H. and Cowie, P. A., Determination of total strain from faulting using slip measurements. Nature, 1990;346:837–839.

Segall, P. and Pollard, D. D., Joint formation in granitic rock of the Sierra Nevada. Geological Society of America Bulletin, 1983;94:563–571.

Shante, V. K. S. and Kirkpatrick, S., An introduction to percolation theory. Advances in Physics, 1971;20(85):325–357.

Smith, L. and Schwartz, F. W., An analysis of the influence of fracture geometry on mass transport in fractured media. Water Resources Research, 1984;20(9):1241–1252.

Stratford, R. G., Herbert, A. W. and Jackson, C. P., A parameter study of the influence of aperture variation on fracture flow and the consequences in a fracture network. In: Barton, N. and Stephansson, O. (eds), Rock Joints, pp. 413–422, Balkema, Rotterdam, 1990.

Stroud, A. H. and Secrest, D., Gaussian quadrature formulas. Prentice-Hall, Englewood Cliffs, NJ, 1966.

Sudicky, E. A. and McLaren, R. G., The Laplace transform Galerkin technique for large scale simulation of mass transport in discretely fractured porous formations. Water Resources Research, 1992;28(2):499–514.

Svensson, U., A continuum representation of fracture networks. Part I: Method and basic test cases. Journal of Hydrology, 2001a;250(1–4):170–186.

Svensson, U., A continuum representation of fracture networks. Part II: Application to the Äspö Hard Rock Laboratory. Journal of Hydrology, 2001b;250(1–4):187–205.

Tsang, Y. W., Usage of equivalent apertures for rock fractures as derived from hydraulic and tracer tests. Water Resources Research, 1992;28(5):1451–5.

Tsang, Y. W. and Tsang, C. F., Channel model of flow through fractured media. Water Resources Research, 1987;22(3):467–479.

Tsang, Y. W., Tsang, C.-F., Neretnieks, I. and Moreno, L., Flow and tracer transport in fractured media: A variable aperture channel model and its properties. Water Resources Research, 1988;24(12): 2049–2060.

US National Research Council, Rock fractures and fluid flow – contemporary understanding and applications. National Academy Press. Washington, DC, 1996.

Warburton, P. M., A stereological interpretation of joint trace data. International Journal of Rock Mechanics and Mining Sciences and Geomechanics Abstracts, 1980a;17(4):181–190.

Warburton, P. M., Stereological interpretation of joint trace data: Influence of joint shape and implication for geological surveys. International Journal of Rock Mechanics and Mining Sciences and Geomechanics Abstracts, 1980b;17(6):305–316.

Watanabe, K. and Takahashi, H. Parametric study of the energy extraction from hot dry rock based on fractal fracture network model. Geothermics, 1995;24(2):223–236.

Wilcock, P., The NAPSAC fracture network code. In: Stephansson, O., Jing, L. and Tsang, C.-F. (eds), Coupled thermo-hydro-mechanical processes of fractured media, pp. 529–538, Elsevier, Rotterdam, 1996.

Willis-Richards, J., Assessment of HDR reservoir simulation and performance using simple stochastic models. Geothermics, 1995;24(3):385–402.

Willis-Richards, J. and Wallroth, T., Approaches to the modeling of HDR reservoirs: A review. Geothermics, 1995;24(3):307–332.

Wilson, T. H., Scale transitions in fracture and active fault networks. Mathematical Geology, 2001;33(3):591-613.

Xu, J. and Cojean, R., A numerical model for fluid flow in the block interface network of three dimensional rock block system. In: Rossmanith (ed.), Mechanics of jointed and faulted rock, pp. 627–633, Balkema, Rotterdam, 1990.

Yeo, I. W., De Freitas, M. H. and Zimmerman, R. W., Effect of shear displacement on the aperture and permeability of a rock fracture. International Journal of Rock Mechanics and Mining Sciences 1998;35(8):1051–1070.

Yu, Q., Tanaka, M. and Ohnishi, Y., An inverse method for the model of water flow in discrete fracture network. Proc. of 34th Japan National Conf. on Geotechnical Engineering, Tokyo. pp. 1303–1304, 1999.

Zhang, X. and Sanderson, D. J., Numerical study of critical behaviour of deformation and permeability of fractured rocks. Marine and Petroleum Geology, 1998;15(6):535–548.

Zhang, X. and Sanderson, D. J., Scale up of two-dimensional conductivity tensor for heterogenous fracture networks. Engineering Geology, 1999;53(1):83–99.

Zimmerman, R. W. and Bodvarsson, G. S., Effective transmissivity of two-dimensional fracture networks. International Journal of Rock Mechanics and Mining Sciences and Geomechanics Abstracts, 1996;33(4):433–436.

11 DISCRETE ELEMENT METHODS FOR GRANULAR MATERIALS

11.1 Introduction

Although fractured rocks do not look like granular materials, the granular material models, however, are often applied to investigate the micro- or more correctly the quasi-microscopic mechanical behaviour of rocks and rock-like materials (such as concretes, ceramics and different composite material constituents), by assuming that at the micro- or quasi-microscopic scale the materials can be approximated as assemblies of particles of very small sizes bonded together by different models of contacts or cementation effects; thus, the overall mechanical behaviour can be evaluated through the collective contributions of particles under loading or unloading processes exhibiting motion, displacement, de-bonding or re-bonding, sliding and interparticle rotation. Heating effects and fluid pressure are also included in some of these models and the particles can be rigid or deformable, with smooth or rough surfaces of different shapes (mostly circular or elliptical in 2D and spherical or ellipsoidal in 3D). These models are important family members of the discrete element methods and are most often based on the distinct element method approach. In fact, the distinct element models for granular materials are forerunners of the discrete element methods and have been in the mainstream of the DEM literature since its more broad applicability in different scientific and engineering fields besides rock mechanics and rock engineering, such as in mineral engineering, chemical engineering, material science and engineering and geotechnical engineering.

The most striking characteristic of a granular material is its dual nature lying between a disjoint, discrete material and a continuum. Although the individual particles are solid, these particles are only partially connected at contact points. Under low normal stresses, the strength of the tangential bonds of most granular materials will be weak and the material may flow like a fluid under very small shear stresses. Therefore, the behaviour of granular material in motion can be studied as a fluid-mechanical phenomenon of particle flow where individual particles may be treated as 'molecules' of the flowing granular material. In many particle models in practice for geological materials, the number of particles contained in a typical domain of interest needs to be very large, similar to the large numbers of molecules that makes up a fluid or a gas. Figure 11.1 shows an example of applying the DEM code PFC2D for particle flow processes for simulating the microscopic deformation and damage (as tensile cracks) around a rock excavation.

A granular medium is distinguished from a continuous one by the existence of the particles that comprise the system and hence the contacts or interfaces between the particles. In simulating the deformation of a granular material with mathematical or numerical techniques, an important component is the formulation for representing contacts between individual particles. Cundall and Hart (1992) made the distinction between hard contacts and soft contacts. Interpenetration between two particles in a hard contact is prevented, although shear movement and opening between them can occur.

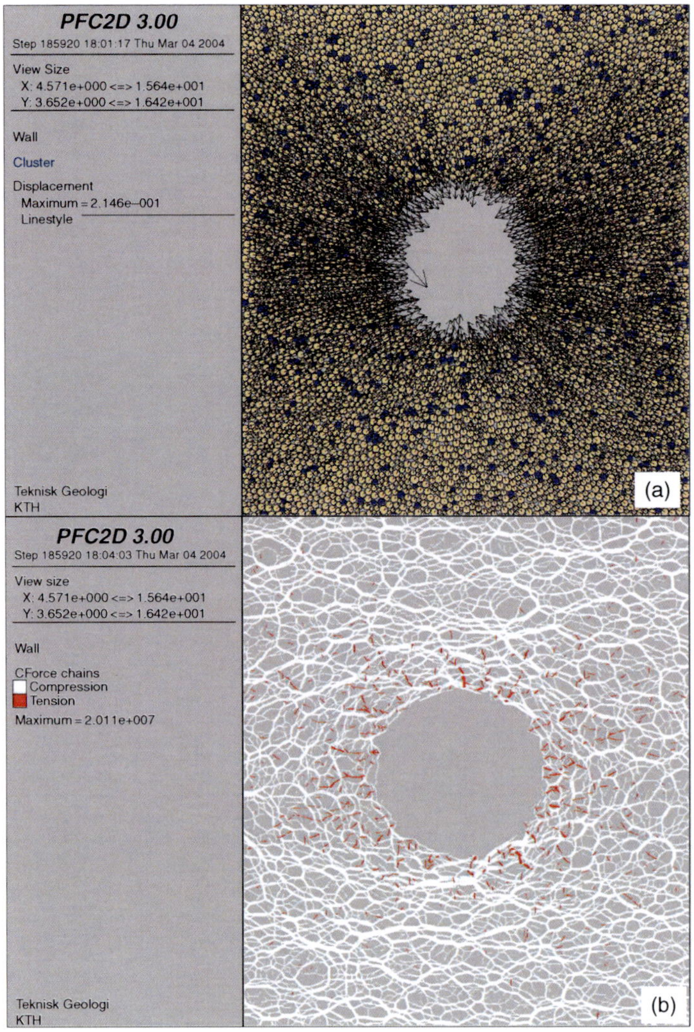

Fig. 11.1 Simulation of deformation and damage in the near field of a rock excavation: (a) displacement distribution (shown by arrows); (b) tensile crack distribution (shown in red).

A soft contact, on the other hand, uses interpenetration as a measure of the relative normal deformation that produces a contact force. As pointed out in Cundall and Hart (1992), although it seems inappropriate that two particles can penetrate each other in a mathematical sense, what it really represents is the relative deformation of the surface layers of the particles (especially when the surfaces are rough and have asperities) rather than real interpenetration as such. Figure 11.2 illustrates the logic for such soft contact of 3D particles (Maeda et al., 2003). An example of a 3D particle model with different mixing of hard and soft particles is illustrated in Fig. 11.3.

Shearing of two particles causes sliding and deformation and forces are generated at the point of contact. In the group of hard contacts, sliding occurs when the shear strength or friction criterion is exceeded. The simple law of friction, $T = \mu N$, where T and N are the tangential and normal forces, respectively, and μ is the interparticle friction coefficient, is the most common criterion used to simulate friction or bond strength.

The types of contact law selected to determine the contact stiffness and the decision for allowing or not allowing the particles to rotate are important subjects in the development of distinct element

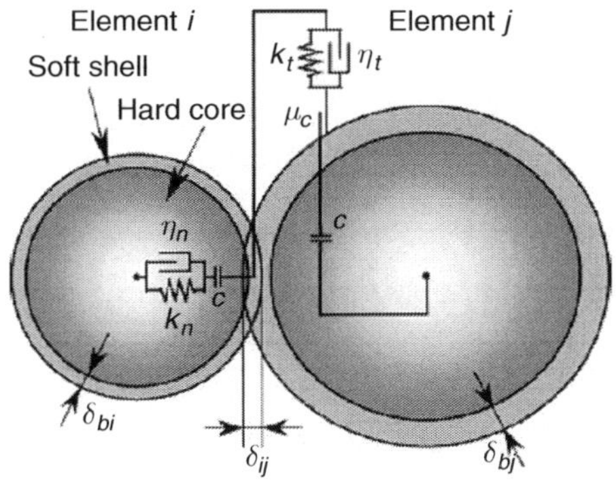

Fig. 11.2 Soft contact scheme for interpenetrations between particles with soft surface layers and hard cores for DEM granular models (Maeda et al., 2003).

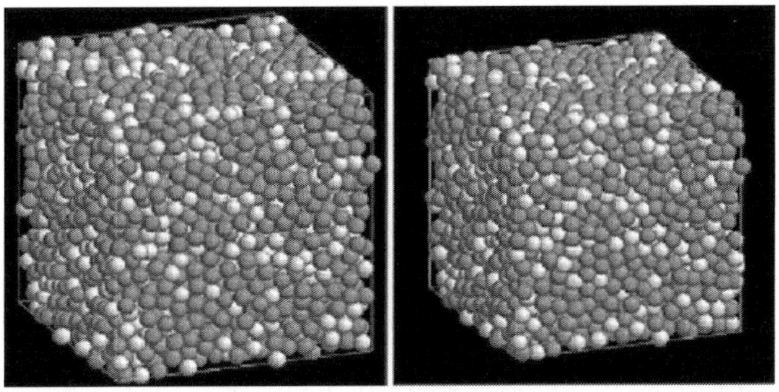

Fig. 11.3 Left: initial packing configuration with 30% hard particles (grey) and 70% soft particles (white) with relative density of 0.64; right: configuration with the same particle composition but with a different relative density of 0.90 (Martin and Bouvard, 2003).

formulations for granular materials. In principle, four different types of contact laws exist. The first and simplest one is to assume linear contact laws for normal compression, and constant shear stiffness and friction angle for the sliding. The other approaches assume the normal load–displacement response to be non-linearly elastic or to have force-dependent normal and shear stiffness. Note that in particle models, forces and displacements are used as primary variables – rather than stresses and strains, as for DEM models for block systems. Therefore, homogenization is often needed to evaluate the overall behavior of particle systems if equivalent continua are required. In a similar way as in the general DEM for block systems, damping is often used in DEM models for granular materials (such as the dashpot elements shown in Fig. 11.2) for absorbing excessive kinetic energy of the system in order to reach static or steady-state solutions or as representing physically meaningful energy dissipation mechanisms.

The original work of the DEM for granular materials for geomechanics and civil engineering application was reported in the series of papers by Cundall and Strack (1979a–d, 1982, 1983), which was based on an earlier work by Cundall (1978) and Strack and Cundall (1978). The term 'distinct element method' was coined in Cundall and Strack (1979a) to define the particular discrete scheme that uses deformable contacts and an explicit, time domain solution of the equations of motion for circular and rigid particles. The first presented distinct element computer program, BALL, is capable of describing the mechanical behavior of assemblies of particles in two dimensions. The method is based on the use of an explicit numerical scheme in which the interaction of the particles is monitored contact by contact and the motion of the particles is modeled particle by particle. For validation of the performance of the code BALL, results were compared with the results of photoelastic experiments conducted on an assembly of discs. The force vector diagrams obtained in the simulations with BALL closely resemble the measured ones and the method and program were demonstrated to be valid tools for research for the behaviour of granular materials. The three-dimensional extension to the code BALL (Cundall and Strack, 1979b) led to the program TRUBAL and these two codes became the most widely adopted platform for DEM code developments for granular material modeling in the 1980s and onwards, noticeably with Walton (1982, 1983), Oner (1984), Campbell and Brennen (1983), Hawkins (1983), Kawai and Takeuchi (1983), Thornton and Barnes (1986), Ting et al. (1986) and Ng (1989) as early runners in the 1980s, with focus on the fundamentals of the approach, coding techniques and application to the fundamental macroscopic mechanical behavior of granular materials, among them soils and sands, based on the microscopic behavior.

Cundall and Strack (1983) summarized the results of microscopic and macroscopic observations from experiments with the program BALL and they also presented the development of a partitioned stress tensor and a constraint ratio depending on the number of degrees of freedom and constraints, which are related to the stability of the assemblies of discs. The paper by Campbell and Brennen (1983) is among the first published for computer simulations of simple shear flows or Couette flow of granular materials using DEM. The simulation results are compared with experimental measurements of steady flows in order to establish appropriate models of the mechanics of particle–particle and particle–wall interactions. The intention of their modeling was to minimize the complexity of the geometry and the interactions so that steady flows of sufficiently long duration could be produced for rheological analysis.

The development of 'physically correct' models of granular material under shear deformations must recognize the discrete nature of the medium, as well as the mechanical properties of the constituent grains at the particle level. Typical developments in this subject were made by Rothenburg (1980), Thornton and Barnes (1986), Rothenburg and Bathurst (1989, 1991, 1992) and Bathurst and Rothenburg (1988a,b, 1990, 1992, 1994). In the paper by Rothenburg and Barthurst (1989), they developed the theoretical background for the so-called stress–force–fabric relation of granular materials and derived a general expression for the stress tensor for a group of disc contacts where the mean normal and tangential contact forces were included. The derived expressions of the principal components of the stress tensor imply that the principal direction of the stress tensor is coincident with the principal direction of anisotropy and contact forces, while principal stress rotations during loading cause small deviations between the direction of fabric anisotropy (orientation of contact vector) and principal stress directions. The algorithms for fluid–solid interactions in particle systems in the DEM approach were developed in Thornton et al. (1993) and Oda et al. (1993).

The 1980s and 1990s saw rapid development and a broad spread of the DEM for granular materials, and a number of codes were developed during this period for simulating granular material behavior with particle systems formed by 2D discs of circular, elliptical or polygonal shapes or 3D solid particles of spherical, ellipsoidal or polyhedral shapes, such as the programs SKRUBAL by Trent and Margolin (1992), MASOM by Issa and Nelson (1992), DISC by Bathurst and Rothenburg (1992), ELLIPSE2 by Ng and Lin (1993), ELLIPSE3 by Lin and Ng (1994), DEFORM by Saltzer (1993), JUMP by Zhang et al. (1993), FIBER by Fox et al. (1994) and POLY by Mirghasemi et al. (1994). The development and applications are mostly reported in journals and a series of proceedings of symposia and conferences, such as in Jenkins and Satake (1983), Satake and Jenkins (1988), Biarez and Gourvés (1989), Thornton (1993) and Siriwardane and Zaman (1994) in the field of micro-mechanics of granular media in general and in Mustoe et al. (1989), Williams and Mustoe (1993) and Shimizu et al. (2004) in the field of geomechanics in particular. The simulations of particle systems using the DDA approach appear mostly in ICADD conference series (e.g., Li et al., 1995; Salami and Banks, 1996; Ohnishi, 1997; Amadei, 1999; Bićanić, 2001; Hatzor, 2002).

In DEM for granular materials, the time step limit required for numerical stability is proportional to the square root of the mass divided by the spring constant. The simulation of small particles with high stiffness takes a significant number of time steps. In order to overcome this limitation, Meegoda (1997) and Kuraoka and Bosscher (1997) suggest the application of parallel computing to solve large problems. Meegoda (1997) developed a program for parallel computing termed 'Trubal for massively Parallel Machines' (TPM) based on the code TRUBAL. The TPM algorithm assigns each processor a multiple number of contact pairs existing within a consolidated assembly of spheres. To avoid excessive communication overheads, each sphere was joined with every other sphere in the granular material and the pair was placed in the same processor. This method allowed all computations to run simultaneously and speeded up the processing. An 800% increase in performance speed was reported for parallel computing on a supercomputer of 512 processors. Kuraoka and Bosscher (1997) developed a fully parallelized DEM program with a domain decomposition scheme and based on TRUBAL. A speedup of 900% was measured in a simulation with 400 balls using 16 processors of an parallel computer.

In addition to simulating granular flow and deformation of granular materials, the particle flow model can be used to simulate the behavior of solid materials. Groups of particles can be bonded together at the contact points in such a way that the groups behave as autonomous solid objects. This assembly of objects can also be regarded as a solid material that has elastic properties and is capable of fracturing once the bond strength is reached. The broken blocks can have arbitrary shape and can interact with one another.

The DEM method has been widely applied to many different fields, such as soil mechanics, the processing industry, non-metal material sciences and defense research. The most well-known codes in the field of rock engineering are the PFC (particle flow code in two- and three-dimension) codes for both 2D and 3D problems (Itasca, 1995a,b) and the DMC codes by Taylor and Preece (1989, 1992) and Preece et al. (2001). The program PFC^{2D} implements a particle flow model in terms of a collection of circular rigid particles or discs, and PFC^{3D} code simulates a collection of rigid spheres. These programs, PFC and DMC, are based on the idea that a rock mass can be represented as a large number of constituent particles whose contact stiffness and bounding behavior are simple in nature. A general particle flow model simulates the mechanical behavior of a system consisting of a collection of distinct, arbitrarily shaped particles that displace independently and interact only at contacts.

The PFC programs have several special features compared to the DEM programs UDEC and 3DEC for general block systems, as pointed out by Lorig et al. (1995a,b). Firstly, the PFC programs are potentially more efficient in contact detection, and calculation of contact forces between discs or spheres is much simpler than that in UDEC and 3DEC, chiefly due to the simplicity of the particle geometry. Secondly, breaking-up of blocks (formed by grouping of a number of small particles with special bonds) is allowed in PFC analysis, the so-called bonded-particle model for rock (Potyondy and Cundall, 2004).

In the PFC programs, the packing of granular material can be defined from statistical distributions of grain size and porosity, and the particles are assigned normal and shear stiffness and friction coefficients. Two types of bonds can be represented either individually or simultaneously; these bonds are referred to as contact and parallel bonds, respectively (Itasca, 1995a,b). Functions for fluid and heat flows were also added (Shimizu, 2006).

A general discrete element model simulates the mechanical behavior of a system comprised of an assembly of arbitrary shaped particles. The particles can displace independently from one another and interact only at contacts or interfaces between particles and between particles and boundaries. In terms of computational convenience, two-dimensional circular disk elements and three-dimensional spherical elements are the simplest shapes for representing granular materials and these comprised the geometry used in the first development of the method for the application to granular materials (Strack and Cundall, 1978; Cundall and Strack, 1979a,b). A single value, radius, defines the geometry of the particles and there is only one possible type of contact among particles, which can be detected easily. As a result, computer memory requirements and computer processing time are minimized with these particle shapes and in addition a large number of particles can be analyzed. The introduction of planar assemblies of elliptical particles, 3D ellipsoids and superquadricks provide more flexibility in particle characterization in the DEM (Ting and Corkum, 1988; Rothenburg and Bathurst 1991, 1992; Mustoe and DePoorter, 1993; Williams and Pentland, 1992; Ng and Lin, 1993; Ting et al., 1995; Ng, 1994; Sawada and Pradhan, 1994; Williams and O'Connor, 1995; Williams et al., 1995; Lin and Ng, 1997; Miyata et al., 2000), with extensions to polygonal or general shaped particles (Ghaboussi and Barbosa, 1990; Mirghasemi et al., 1994; Mustoe et al., 2000; Mustoe and Miyata, 2001). Bonded particle groups are also used to represent general shaped bodies in DEM models, such as in the PFC codes and as reported in Yamane et al. (1998).

Besides particle shapes, deformability of the particle themselves have also been considered for DEM modeling for granular materials. Oelfke et al. (1995) considered elastic particles in developing a Lagrangian DEM approach for ground control problems in underground mines. Thornton and Zhang (2003) applied a similar approach to investigate particle stress and deformation effects on shear-band behaviors of granular assemblies. However, the main thrust of development in this direction is the combined discrete and finite element methods, such as reported in Munjiza et al. (1995, 1996a,b, 2004), Mohammadi et al. (1998), Munjiza and Andrews (2000), Owen and Feng (2001), Owen et al. (2002), Komodromos and Williams (2002a,b) and Bangash and Munjiza (2003).

The following examples of publications are only a small part of the published works and are cited here to highlight the wide application range of the DEM for particle systems in the fields of rock and soil engineering:

- *Fracturing and fragmentation processes of rock blasting*: Preece (1990, 1994), Preece and Knudsen (1992), Preece et al. (1993, 2001), Preece and Scovira (1994), Donzé et al. (1997), Lee et al. (1997), Lin and Ng (1994);
- *Ground collapse and movements*: Iwashita et al. (1988), Zhai et al. (1997);
- *Hydraulic fracturing in rocks*: Thallak et al. (1991), Huang and Kim (1993), Kim and Yao (1994);
- *Sand production in oil reservoir engineering*: O'Connor et al. (1997);
- *Underground excavations*: Kiyama et al. (1991), Potyondy et al. (1996), Potyondy and Fairhurst (1998) and Cundall et al. (1996a,b), Potyondy and Cundall (1998);
- *Dam break*: Zhang et al. (1993);
- *Rock fracture and faulting*: Blair and Cook (1992), Saltzer (1993);
- *Soil/sand and granular geo-material behavior*: Anandarajah (1994), Ratnaweera and Meegoda (1993), Zhai et al. (1997); Jensen et al. (2001a,b), Zhang and Li (2006) and
- *Mining*: Lorig et al. (1995a,b).

In the following presentations, we present basics of DEM for granular media considering only rigid circular or spherical particles, in a sense of introduction to the approach. Some of the application studies using granular models of DEM are presented in Chapter 12.

11.2 Basic DEM Calculation Features for Granular Materials

The calculations conducted in the DEM for granular materials are almost identical to the DEM for rigid block systems but with much simplified contact detection and contact models (cf. Chapter 8) and alternate between invoking Newton's second law for the particles and force–displacement laws at the contacts through a time-stepping finite difference scheme. Newton's second law is used to determine the translation and rotation of each particle due to contact and body forces acting upon it. The force–displacement laws are used to update the contact forces arising from the relative motion at each contact. The presence of walls in the DEM models serves as user-specified boundary conditions and requires only that the force–displacement laws be specified for particle–wall contacts.

The solution scheme is similar to that used by the explicit finite difference method for analysis of continuum processes and is the same for rigid block systems in DEM. The time steps are chosen so small that, during a single time step, disturbance cannot propagate from any particle further than its closest neighbors. Then, the forces acting on a particle are determined by its interaction with the particles with which it is in contact. The use of an explicit numerical scheme for the time stepping makes it possible to study large number of particles without large computer memory requirements or the need for an iterative procedure. In the calculation cycle, the computation of motion according to Newton's law and the force–displacement law at each contact can be performed effectively in parallel (Sadd et al., 1993).

The algorithms for the DEM with rigid particles of circular shape are perhaps the most well known and have been described by many. The descriptions given in Mohammadi (2003) and Oda and Iwashita (1999) are both systematic presentations. Elliptical discs provide more flexibility in characterization of granular materials in the DEM modeling but, unlike circular ones, require the solution of fourth-order algebraic equations that can be achieved analytically. The key feature is the fact that the normal direction of the contact surface is now directed eccentrically and that will generate moments for rotation (or resistance to particle rotation) and this feature is therefore more suitable for loose granular materials such as soils and sands, with circular discs and spheres more suitable for quasi-solids. In this chapter, we introduce only the algorithms for 2D circular discs: the extensions to 3D spheres, 2D ellipses and 3D ellipsoids can be seen in the appropriate references (Lin and Ng, 1994; Ting et al., 1995; Mohammadi, 2003) and will not be repeated here.

Without losing generality, the algorithm associated with a two-disc system is shown in Fig. 11.4. Tables 11.1 and 11.2 list the algorithms for the calculation of relative velocities and displacements and

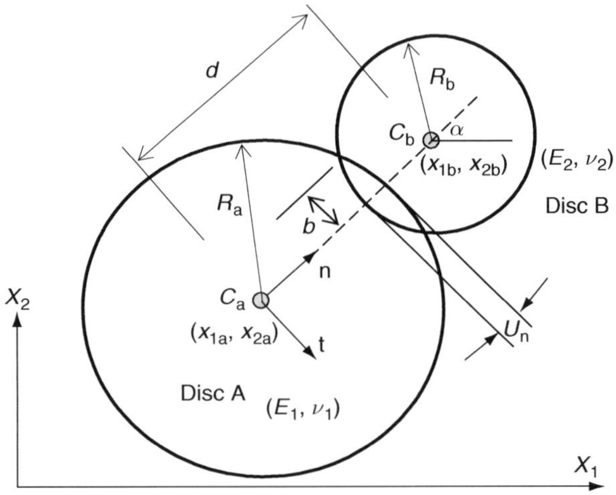

Fig. 11.4 Two circular particles in contact (Mohammadi, 2003).

Table 11.1 Sequence for contact force–displacement calculations of 2D disc pairs ($i = 1,2$)

Relative velocity	$v_i = (\dot{x}_{ia} - \dot{x}_{ib}) - (\dot{\theta}_a R_a + \dot{\theta}_b R_b) t_i$
Relative displacement increments in the normal and tangential directions of contacts	$\Delta U_n = (v_i n_i)\Delta t, \ \ \Delta U_t = (v_i t_i)\Delta t$
Contact force increments	$\Delta F_n = K_n \Delta U_n, \ \ \Delta F_t = K_t \Delta U_t$
Total force at time step j	$F_n^j = F_n^{j-1} + \Delta F_n, \ \ F_t^j = F_t^{j-1} + \Delta F_t$
Slip condition checking	$F_t \leq F_n \tan\phi + C$

Table 11.2 Sequence for particle motion calculations for 2D discs

Moment calculations (sum over clusters of particles in contacts)	$M_a = \sum F_a R_a, \ \ M_b = \sum F_b R_b$
Acceleration calculation (assuming constant force and moments from $t^{(j-1/2)}$ to $t^{(j+1/2)}$) at time step j	$\ddot{x}_i^j = \sum F_i/m, \ \ \ddot{\theta}^j = \sum M/I$
Velocity calculation	$\dot{x}_i^{(j+1/2)} = \dot{x}_i^{(j-1/2)} + \ddot{x}_i^j \Delta t, \ \ \dot{\theta}^{(j+1/2)} = \dot{\theta}^{(j-1/2)} + \ddot{\theta}^j \Delta t$
Displacement calculation (assuming constant velocities from time step t^j to t^{j+1})	$x_i^{j+1} = x_i^j + \dot{x}_i^{(j+1/2)} \Delta t, \ \ \theta^{j+1} = \theta^j + \dot{\theta}^{(j+1/2)} \Delta t$

contact forces. In the tables, t is time, θ the rotation angle, x_{ia} and x_{ib} the position vectors of the centers of the discs A and B of radius R_a and R_b, respectively; n_i and t_i are the unit directional vectors in the normal and tangential directions, respectively; U_n and U_t, are the normal and tangential displacements, respectively; F_n and F_t are the normal and tangential contact forces, respectively; K_n and K_t are the normal and tangential contact stiffness, respectively; ϕ and C are the friction angle and cohesion at the contacts, respectively; m is the particle mass; M and I are the resultant moment and moment of inertia of a particle, respectively. At the end of the above sequence, the time step is incremented and the whole process continues following the same sequence.

When the empirical contact stiffness, K_n and K_t in Table 11.1 are not used, the normal contact force with the viscous damping coefficient C_n for Hertzian contact is used as a sum of contact pressure P_n and the viscous term, given by

$$F_n = P_n + C_n v_{nr} \tag{11.1}$$

where v_{nr} is the relative velocity in the normal direction,

$$P_n = \frac{\pi}{4} \frac{1 + \left(\dfrac{K_1}{K_4}\right)^2}{(K_1 + K_2)} \frac{R_a + R_b}{R_a R_b} b^2 \tag{11.2a}$$

$$K_1 = \frac{1 - \nu_1}{G_1} \tag{11.2b}$$

$$K_2 = \frac{1 - \nu_2}{G_2} \tag{11.2c}$$

$$K_3 = \frac{2}{\pi}(K_1 + K_2) \tag{11.2d}$$

$$K_4 = \frac{1 - 2\nu_1}{G_1} - \frac{1 - 2\nu_2}{G_2} \tag{11.2e}$$

and the static constant normal stiffness from Hertz theory is given by

$$K_n = \frac{\pi h E_1 E_2}{2(E_1 + E_2)} \tag{11.2f}$$

with h being the thickness of a circular disk-shaped particle and E_1, E_2, G_1 and G_2 being the elastic and shear moduli of the two particles in contact (cf. Fig. 11.4). When the mass-proportional damping is adopted the coefficient C_n is given by

$$C_n = \alpha_m \left(2\sqrt{\frac{m_1 m_2}{m_1 + m_2} K_n} \right) \tag{11.3}$$

where α_m is an empirical constant. Similarly the tangential contact force is given by

$$F_t = P_t + C_t v_{tr} \tag{11.4}$$

where v_{tr} is the relative velocity in the tangential direction and the tangential contact, the pressure by Hertzian theory is given by

$$P_t = \tan \phi F_n \left(1 - \left(1 - \frac{v_{tr}}{\frac{8}{\pi} F_n (K_1 + K_2) \frac{R_a R_b}{R_a + R_b}} \right)^2 \right) \tag{11.5}$$

and the viscous term is similarly given by

$$C_t = \beta_m \left(2\sqrt{\frac{m_1 m_2}{m_1 + m_2} K_t} \right) \tag{11.6}$$

where β_m is an empirical coefficient.

The relative normal and tangential velocities are given by

$$v_{nr} = (\mathbf{v}_a - \mathbf{v}_b) \cdot \mathbf{n}, \quad v_{tr} = (\mathbf{v}_a - \mathbf{v}_b) \cdot \mathbf{t} \tag{11.7}$$

with the velocities obtained from the solution of the equations of motion of the two discs.

In the PFC codes (Itasca, 1995a,b), a modified Hertzian–Mindlin contact model (Mindlin and Deresiewicz, 1953) is adopted with the mean shear modulus $\langle G \rangle$, mean Poisson's ratio $\langle \nu \rangle$ and the normal and tangential contact stiffness are written as

$$K_n = \left(\frac{2\langle G \rangle \sqrt{2\langle R \rangle}}{3(1 - \langle \nu \rangle)} \right) \sqrt{U_n}, \quad K_t = \left(\frac{2(\langle G \rangle^2 3(1 - \langle \nu \rangle) \langle R \rangle)^{1/3}}{2 - \langle \nu \rangle} \right) |F_n| \tag{11.8}$$

$$\langle R \rangle = \frac{2R_a R_b}{R_a + R_b} \tag{11.9}$$

11.3 Demonstration Examples of the PFC Code Applications

Two examples are presented here to illustrate the applicability and the working procedure when the DEM models for granular materials are used for rock mechanics problems. The first example concerns simulating the mechanical behavior of rock samples of standard size in the laboratory under uniaxial compression until a complete stress–strain curve is obtained representing the structural breakdown of the sample, with and without a weak confining pressure, using a series of rock samples of diorite rock from the Äspö HRL (Hard Rock Laboratory) in Southern Sweden (Koyama and Jing, 2005). Figure 11.5 shows the test configurations for the two testing cases. The tests were assumed to be performed with a servo-controlled test machine with the axial strain rate as the controlling variable, with a loading velocity of 0.2 m. The model is a 2D simulation containing 3600 rigid particles of circular shape and uniform size. The stepwise changes of sample configuration, crack evolution and contact force development at different stages of the complete stress–strain curves are shown in Figs. 11.6–11.11. The calibrated rock properties and PFC model parameters are listed in Tables 11.3 and 11.4, respectively. It can be seen that, although small to moderate discrepancies between the measured and simulated particle system parameters exist, the PFC approach is capable of providing a quantitative estimation of the mechanical properties of the rocks and simulates the complete stress–strain curves, a difficult task for continuum models.

The above example demonstrates the advantages of the particle mechanics approach using the DEM. As pointed in the literature, micro-cracking in rock specimens occurs well before the peak uniaxial compressive strength is reached and often occurs at 70% of the peak strength. With continued loading, complete stress–strain curves were obtained for the case of a monotonic increase in strain as determined by the use of axial strain rate as the control variable, and as illustrated in Figs 11.6–11.11, with clear demonstration of progressive internal cracking pattern development and final failure at the different strain stages, so that after the peak strength the specimens lose their mechanical integrity and become failed structures instead of continuous solids.

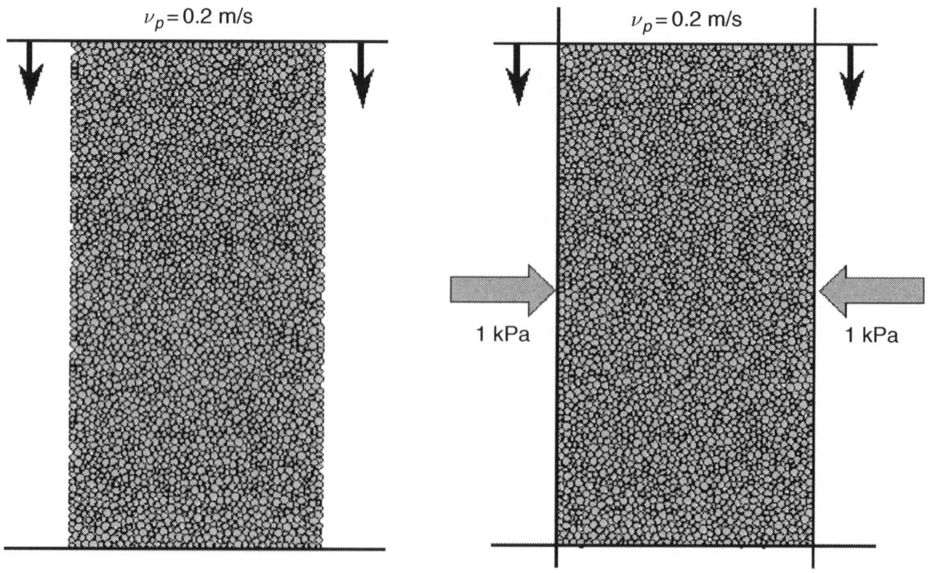

Fig. 11.5 Two PFC models simulating loading tests (Koyama and Jing, 2005). Case 1 (left): uniaxial compress test; case 2 (right): axial compression with a weak lateral confining stress of 1.0 kPa.

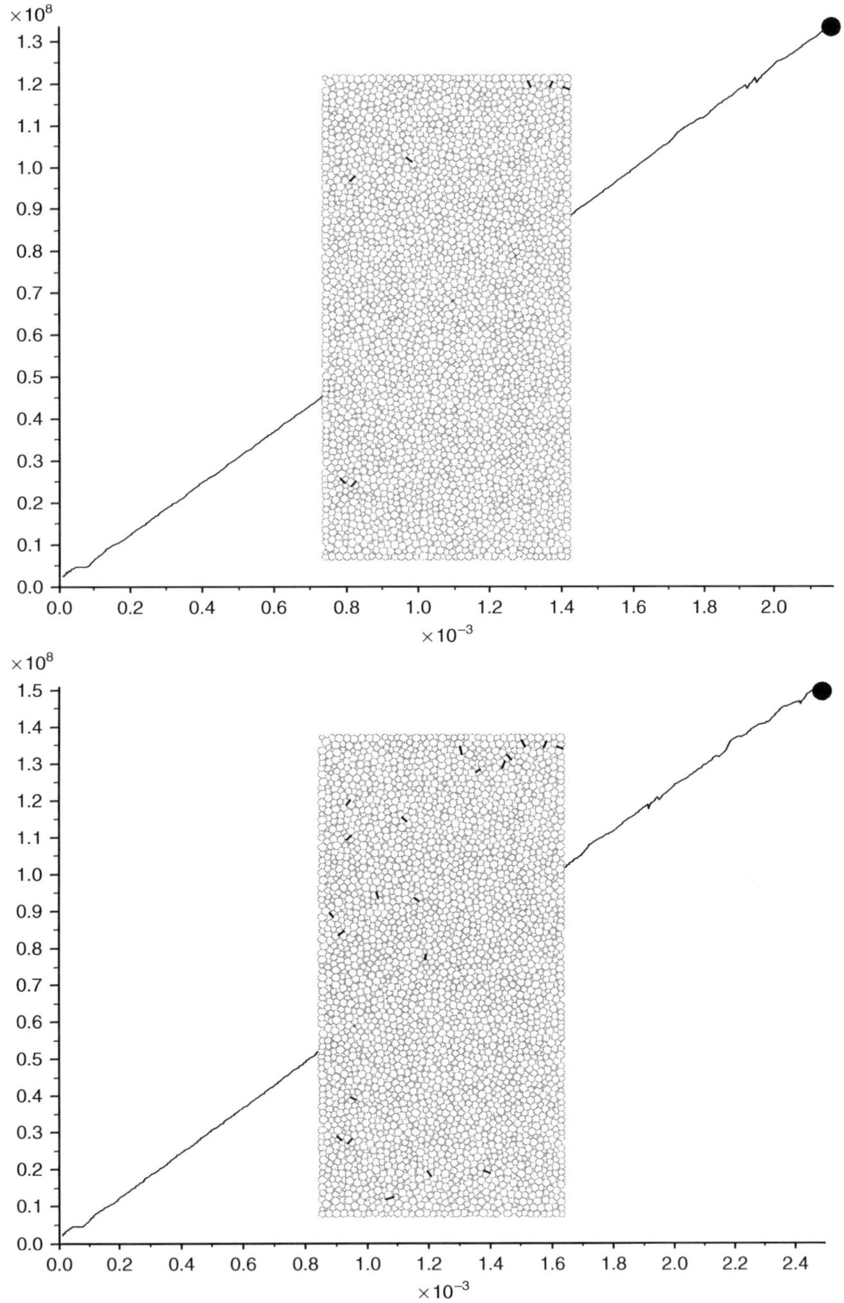

Fig. 11.6 Initiation of micro-cracks in the PFC model at 70% of the peak uniaxial compressive strength of the rock sample for boundary conditions case 1 (above) and case 2 (below). The black segments in the PFC model show the locations of crack initiation (either tensile or shear). The black dots at the end of stress–strain curves indicate the states of the samples on the stress–strain curves. The vertical axis is the averaged axial compressive stress and the horizontal axis is the averaged axial strain. Case 1 (above) has seven initiated micro-cracks and case 2 (below) has 21 initiated micro-cracks, distributed randomly and sparsely (Koyama and Jing, 2005).

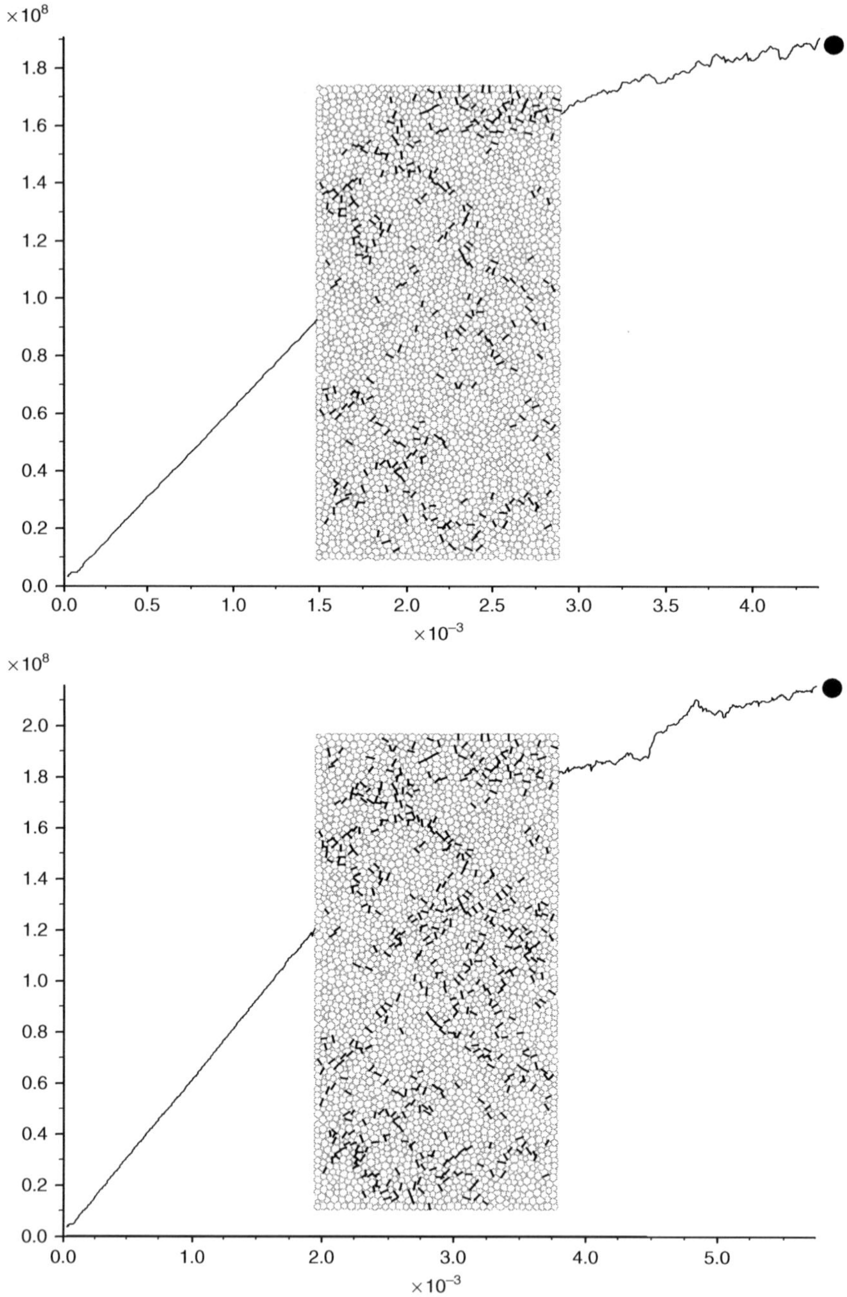

Fig. 11.7 Initiation of micro-cracks in the PFC model at the uniaxial compressive strength of the rock sample for boundary conditions case 1 (above) and case 2 (below). The black segments indicated in the PFC model show the locations of crack initiation (either tensile or shear). The black dots at the end of stress–strain curves indicate the states of the samples on the stress–strain curves. The vertical axis is the averaged axial compressive stress and the horizontal axis is the averaged axial strain. Case 1 (above) has 290 initiated micro-cracks and case 2 (below) has 496 initiated micro-cracks, no crack localization/concentration (Koyama and Jing, 2005).

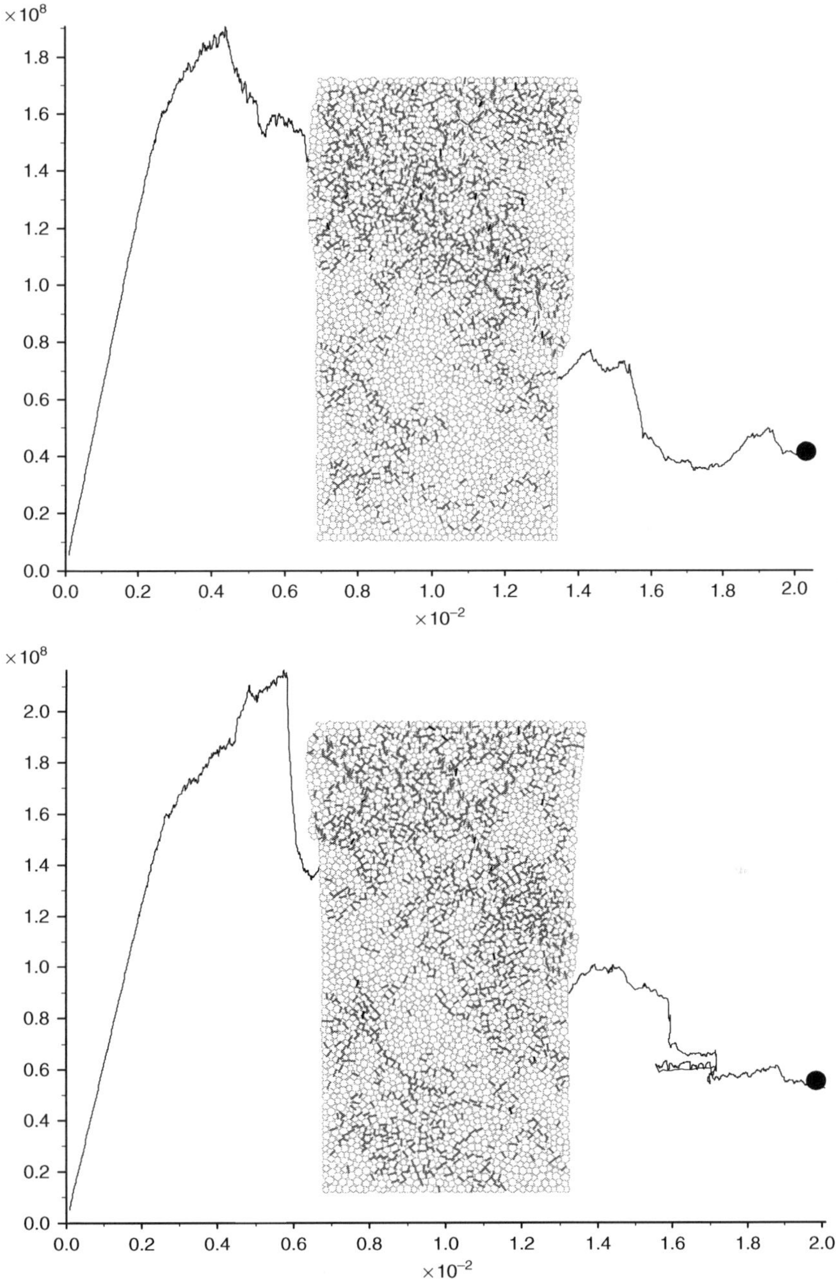

Fig. 11.8 Initiation of micro-cracks in the PFC model at the post-peak stage of the rock sample for boundary conditions case 1 (above) and case 2 (below). The black segments indicated in the PFC model show the locations of crack initiation (either tensile or shear). The black dots at the end of the stress–strain curves indicate the states of the samples on the stress–strain curves. The vertical axis is the averaged axial compressive stress and the horizontal axis is the averaged axial strain. Case 1 (above) has 1047 initiated micro-cracks and case 2 (below) has 1196 initiated micro-cracks distributed with clear localization (Koyama and Jing, 2005).

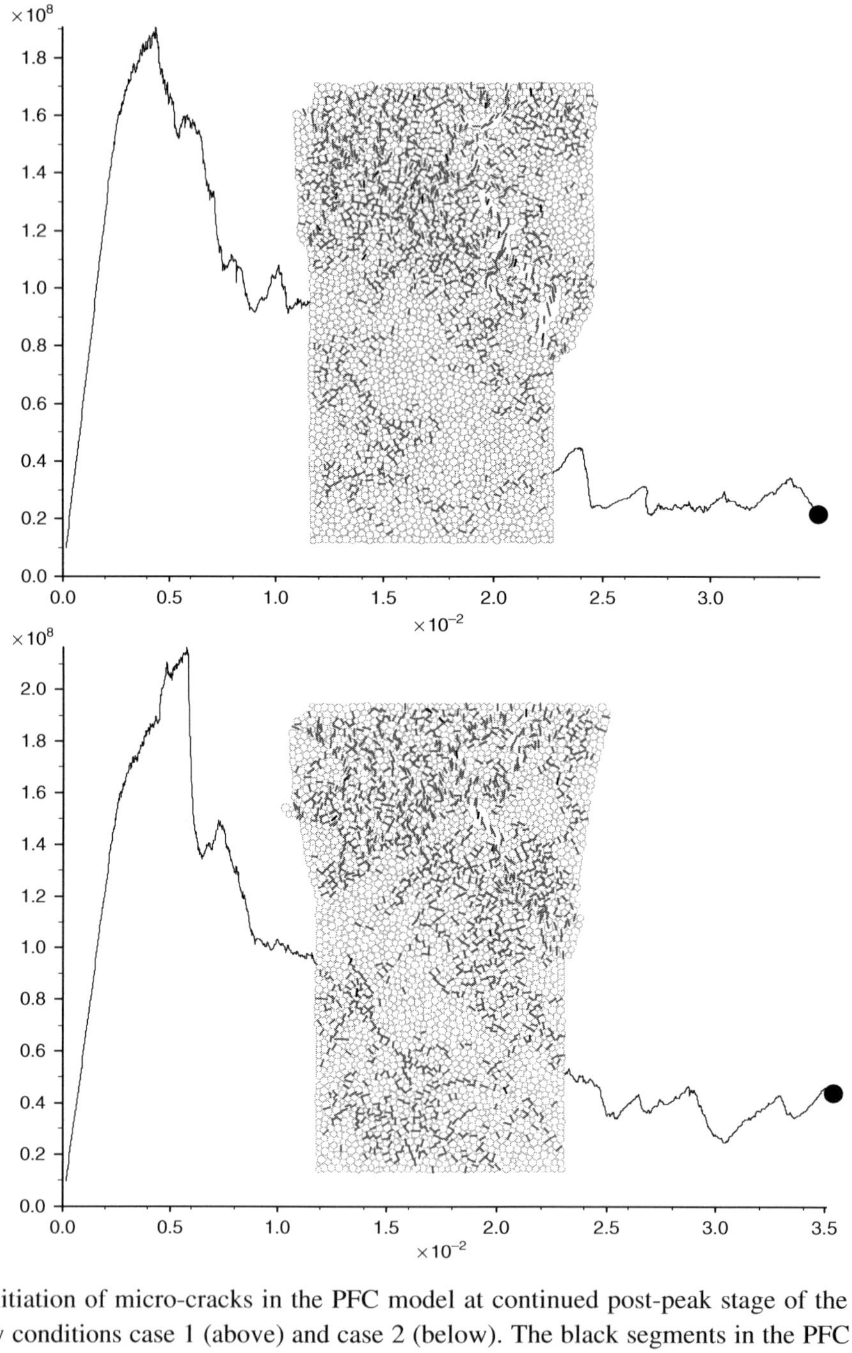

Fig. 11.9 Initiation of micro-cracks in the PFC model at continued post-peak stage of the rock sample for boundary conditions case 1 (above) and case 2 (below). The black segments in the PFC model show the locations of crack initiation (either tensile or shear). The black dots at the end of the stress–strain curves indicate the states of the samples on the stress–strain curves. The vertical axis is the averaged axial compressive stress and the horizontal axis is the averaged axial strain. Case 1 (above) has 1072 initiated micro-cracks and case 2 (below) has 1267 initiated micro-cracks with clear crack localization and ultimate failure (Koyama and Jing, 2005).

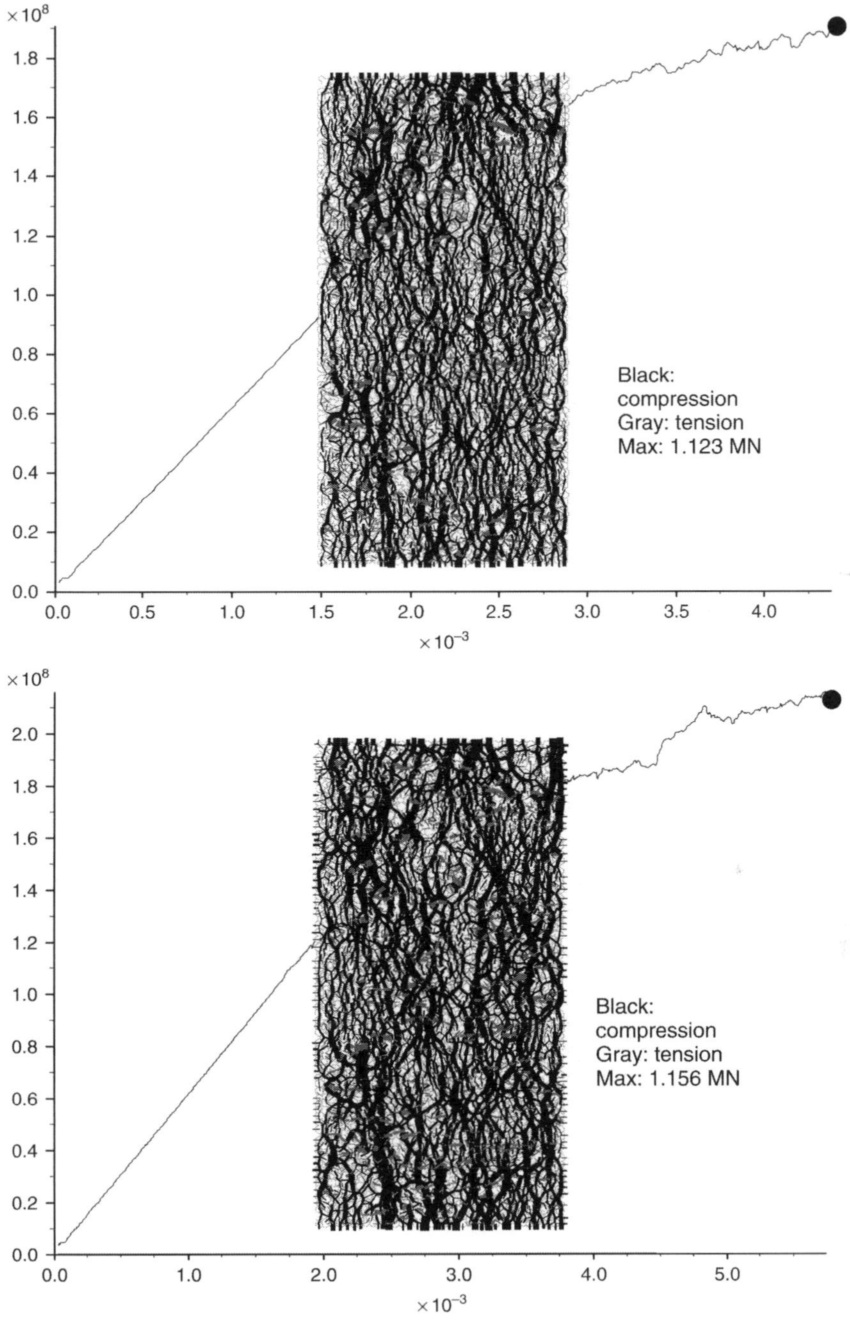

Fig. 11.10 Distributions of contact forces in the PFC model at the uniaxial compressive strength of the rock sample for boundary conditions case 1 (above) and case 2 (below). The black dots at the end of the stress–strain curves indicate the states of the samples on the stress–strain curves. The vertical axis is the averaged axial compressive stress and the horizontal axis is the averaged axial strain. Dominant compressive contact force with homogeneous distribution in the vertical direction without crack concentration (strain localization) (Koyama and Jing, 2005).

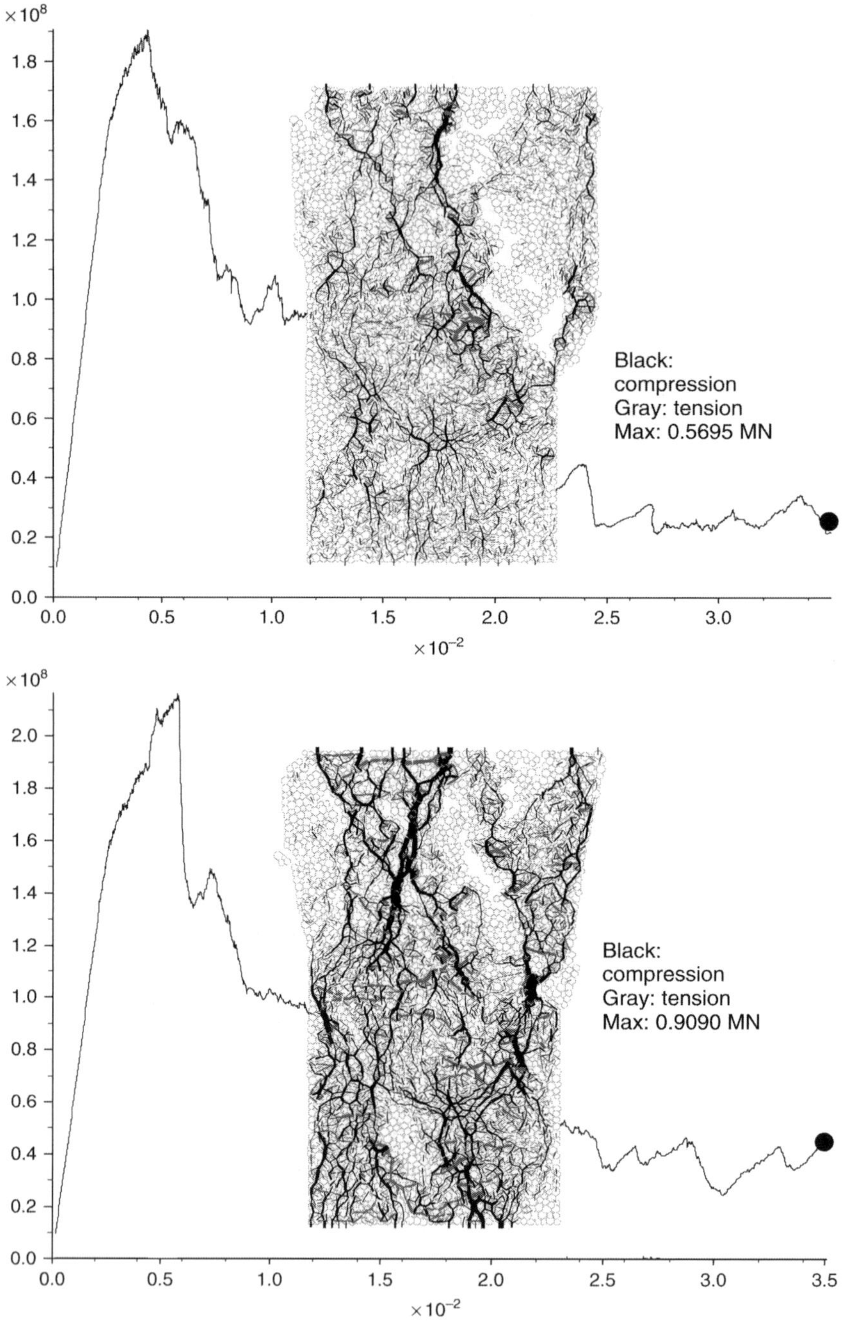

Fig. 11.11 Distributions of contact forces in the PFC model at the post-peak stage for boundary conditions case 1 (above) and case 2 (below). The black dots at the end of stress–strain curves indicate the states of the samples on the stress–strain curves. The vertical axis is the averaged axial compressive stress and the horizontal axis is the averaged axial strain. Dominant compressive contact force with homogeneous distribution in the vertical direction, clear crack concentration (strain localization) (Koyama and Jing, 2005).

Table 11.3 Calibrated rock properties through comparison with measured data for Äspö diorite rock (Koyama and Jing, 2005)

Macroscopic rock properties	Laboratory test data	PFC calibrated data	
		Case 1	Case 2
Young's modulus (GPa)	68	57.51	57.91
Poisson's ratio	0.24	0.27	0.27
Unconfined compressive strength (MPa)	214	190.67	216.47
Unconfined crack initiation stress (MPa)	121	131.14	136.82

Table 11.4 PFC model parameters corresponding to the calibrated rock properties in Table 11.3

Particle radius	Normal bonding strength	Contact modulus	Normal/shear stiffness ratio	Friction coefficient
0.5 – 0.83 mm	600 MPa	110 GPa	5	0.2

Number of particles used was 3600 with 15 880 time steps.

The stress–strain curve after the peak strength represents, therefore, not the so-called material 'softening' phenomenon but a structural behavior of the sample dominated by frictional properties between broken pieces, a clear bifurcation process.

The second example is a direct shear test of a rough rock fracture using the PFC^{2D} code, as shown in Fig. 11.12, with 9646 particles under normal stresses of 0.5, 1.0, 2.0, 5.0 and 10.0 MPa, respectively, with

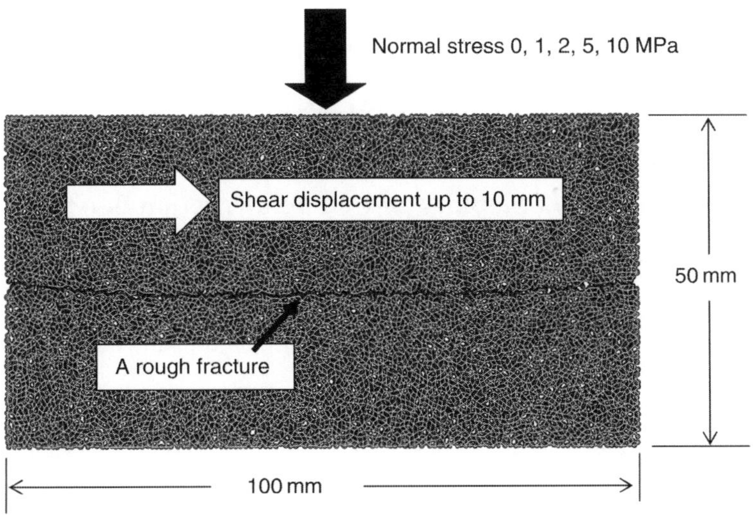

Fig. 11.12 PFC simulation of a direct shear test of a rough rock fracture under varying normal stresses (Koyama and Jing, 2005).

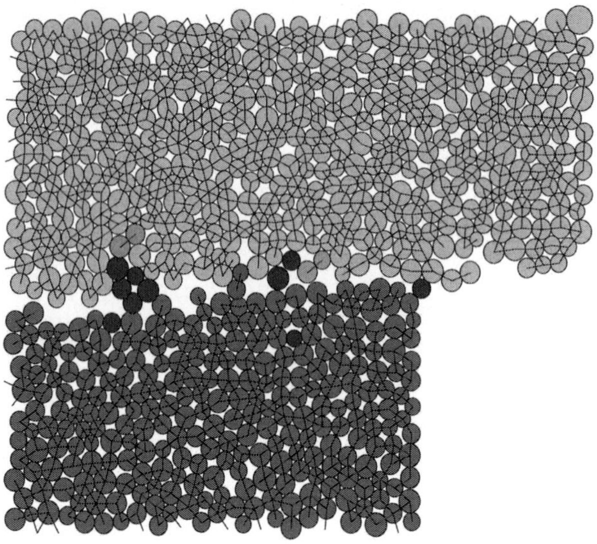

Fig. 11.13 Local detail of particle system deformation near the fracture sample edge at the end of a shear test under a normal stress of 5.0 MPa. Line segments in each particle indicate its contact number and directions (Koyama and Jing, 2005).

shear displacement up to 10 mm. The sizes of the particles vary from 0.25 to 0.53 mm. The other mechanical parameters are the same as those in Table 11.4.

Figure 11.13 shows the deformed particle pattern in detail at the end of a test under 5.0 MPa normal stress with darkened loose particles representing shear-induced gouge material between the fracture surfaces. The obtained complete shear stress–shear displacement curves of the fracture (averaged values) and the normal dilation versus shear displacement curves, under different normal loads, are shown in Fig. 11.14. Figure 11.15 illustrates the concentrated distribution of the contact forces at the contacted asperity tips (Fig. 11.15a) and the locations where micro-cracking is initiated as a consequence of the shear, representing the extent of the surface damage zone of the fracture. These figures are of course qualitative for this example but are similar to what one obtains from laboratory tests. Such details cannot be observed in continuum models for simulating conventional direct shear tests on rock fractures. Thus, PFC simulation approaches provide an alternative tool to deepen our understanding of the localized stresses, contact areas and damage development in fractures during shear.

11.4 Numerical Stability and Time Integration Issues

Numerical algorithms for integration of particle systems share similarities with that of FEM, and have been extensively reported in literature of particle mechanics problems. The recent review in O'Sullivan and Bray (2004) provides a very comprehensive treatment of the problem and is adopted as the basis of presentations in this section.

11.4.1 Analogue Between FEM Meshes and DEM Particle Systems

Both explicit and implicit DEM approaches have been applied to study the mechanical behavior of particle systems (e.g., granular materials), as reviewed in Kishino and Thornton (1999). In matrix form, the equations at discrete time intervals for the system of particles can be written as

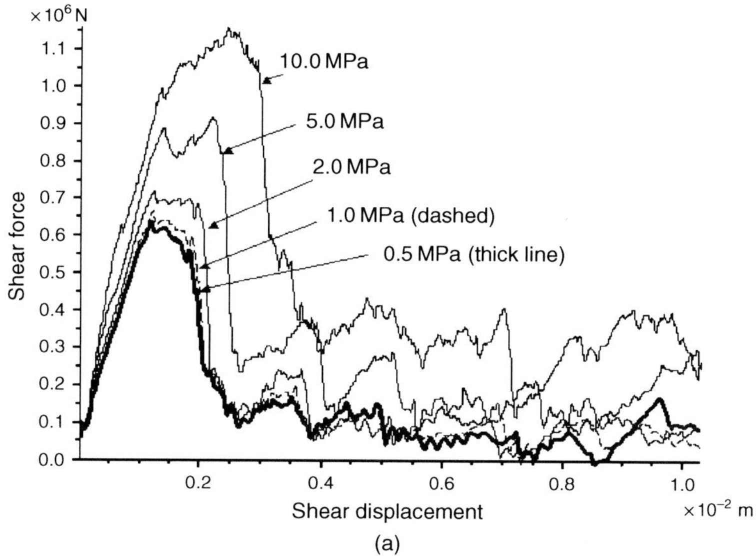

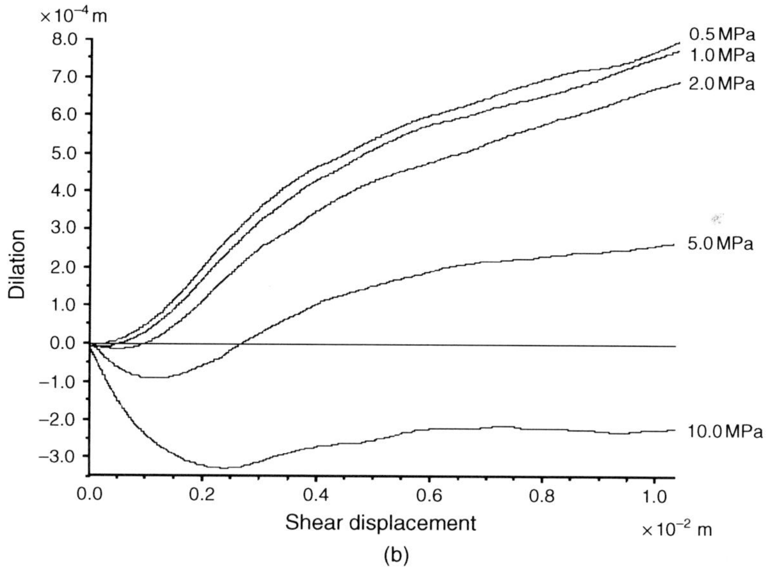

Fig. 11.14 (a) Shear force versus shear displacement curves under different normal stresses from the PFC model of direct shear test of a rough fracture (Koyama and Jing, 2005). (b) Normal dilation versus shear displacement curves under different normal stresses from the PFC model of direct shear test of a rough fracture (Koyama and Jing, 2005).

$$\mathbf{Ma} + \mathbf{Cv} + \mathbf{K}\Delta\mathbf{x} = \Delta\mathbf{f} \tag{11.10}$$

where $\mathbf{M}$ is the mass matrix, $\mathbf{a}$ the acceleration vector, $\mathbf{K}$ the stiffness matrix, $\Delta\mathbf{f}$ the incremental force vector, $\Delta\mathbf{x}$ the incremental displacement vector and $\mathbf{v}$ the velocity vector, and the damping matrix $\mathbf{C}$ is given by

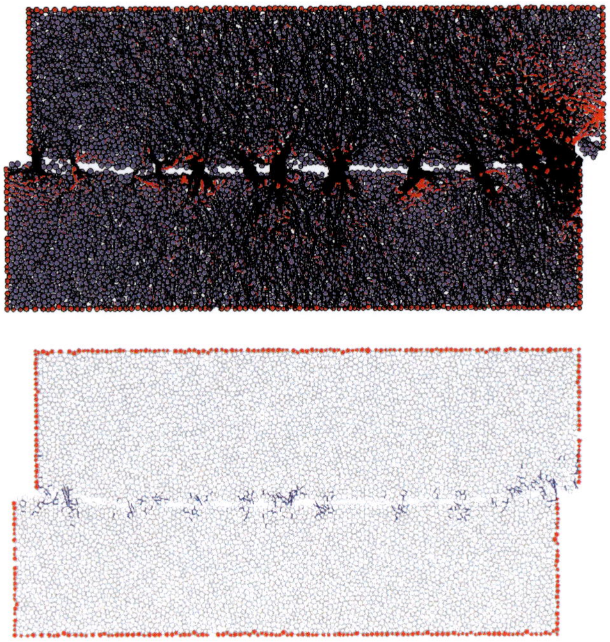

Fig. 11.15 Distribution of contact forces (top, with tensile forces in red and compressive forces in black), micro-crack development (bottom, the blue line segments indicating the locations of the shear-induced micro-cracking locations at a shear displacement of about 5.5 mm under a normal stress of 5 MPa (Koyama and Jing, 2005).

$$\mathbf{C} = \alpha\mathbf{M} + \beta\mathbf{K} \tag{11.11}$$

when a mixed mass-proportional and stiffness-proportional mode scheme is adopted, where α and β are constants determined by the problem-specific material and system configurations. Both the stiffness matrix **K** and the velocity vector **v** change with changing geometric distribution and the mechanical status of the particle contacts.

The above equation is similar to the FEM equations for dynamic analysis of continua. As stated in O'Sullivan and Bray (2001, 2004), there is an analogy between a DEM framework and a FEM framework, with the DEM particles corresponding to the FEM nodes and interparticle contacts corresponding to FEM elements as shown in Fig. 11.16. This analogy can help the DEM in estimating time integration issues, especially the critical time step determination, compared with that in Itasca (1998) for PFC models.

As stated in Belytschko et al. (2000), the performance of the time integration method depends mainly on the type of partial differential equation and the smoothness of the data. For hyperbolic equations, such as those for stress wave propagation and dynamic motion of particle systems with contacts, explicit time integration as adopted in PFC codes is well suited when small time steps are used to meet stability requirements. Numerical treatment for contact-caused changes in displacements, velocities and accelerations are straightforward without requiring a global stiffness matrix. The implicit discrete element formulations, typically DDA, require an iterative change in the stiffness matrix, such as that used in Shi (1988) and in Ke and Bray (1995). The DDA approach uses the state of contact of the particles (i.e., the contact stiffness matrix) at the end of the previous time step to predict the displacement increments at the current time step, followed by updating of the stiffness matrix. If convergence is not achieved, the interval of the time step is further divided and the

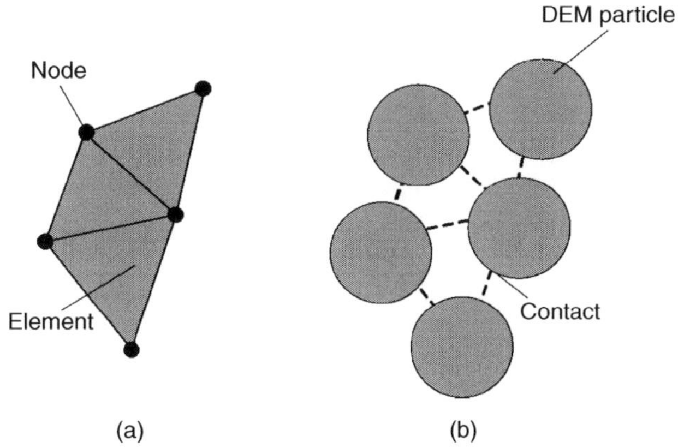

Fig. 11.16 Schematic diagram illustrating DEM–FEM analogy. (a) Elements and nodes (FEM); and (b) particles and contacts (DEM) (O'Sullivan and Bray, 2004).

calculation starts over again. This will involve increased computational costs for repeated matrix inversion operations. Such iteration is not needed in an explicit time integration scheme such as PFC codes for a particle system. This is an important advantage if large-scale DEM models with millions of discrete particles need to be simulated.

The central difference time integration scheme in DEM is conditionally stable and the critical time step for stable analyses can be estimated approximately. However, a universal criterion for estimating the critical time step for DEM models under general conditions does not exist, even for linearized dynamic systems. The maximum stable time increment (Δt_{crit}) is a function of the eigenvalues of the current stiffness matrix (Belytschko, 1983). For a linear system without damping the relation is given by Eqn (8.106) using the maximum frequency, $\omega_{\max}$, or the maximum eigenvalue, $\lambda_{\max}$, of the $\mathbf{M}^{-1}\mathbf{K}$ matrix defined in Eqn (11.10).

For a single degree of freedom system, i.e., considering a single DEM particle, for the central difference time integration scheme, the critical time step (Δt_{crit}) is simply given by (as similar to (8.102)),

$$\Delta t^{\mathrm{c}} = \frac{T}{\pi} = 2\sqrt{\frac{m}{K_{\mathrm{eff}}}} \tag{11.12}$$

where T is the period of free vibration of the degree of freedom, m the particle mass and K_{eff} the maximum spring stiffness of the particle contacts. In the PFC codes of Itasca (1998), a method for estimating the critical time step to avoid unstable behavior for an assembly of particles, the critical time step evaluated by considering a system of an infinite series of uniformly sized circular particles connected through springs (Fig. 11.17), is given by

$$\Delta t^{\mathrm{c}}_{\mathrm{particles}} = \sqrt{\frac{m}{K}} \tag{11.13}$$

where K is the contact spring stiffness and m the particle mass. A safety measure using a multiplier of 0.8 was also proposed.

For the quasi-static type of analysis that is frequently considered in geotechnical discrete element simulations, the time step constraint can be relaxed by using density or mass scaling (cf. Section 8.8), when the response of the system to inertia effects is assumed to be of no interest. If high frequency response is important, mass scaling should not be adopted.

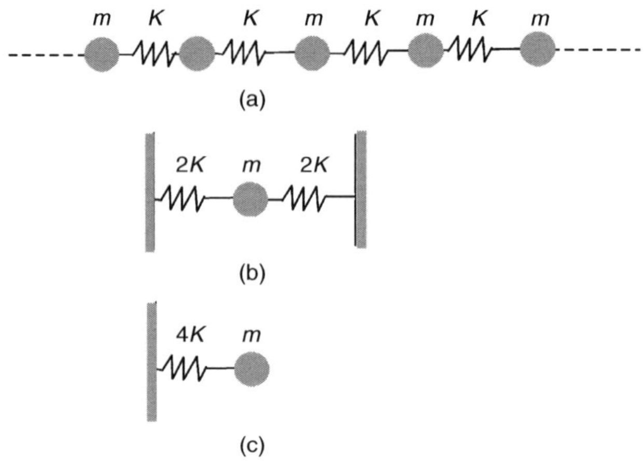

Fig. 11.17 Particle mass–spring systems in calculating the critical time step in PFC codes (O'Sullivan and Bray, 2004).

In O'Sullivan and Bray (2004) an approach for calculating the critical time steps for a circular disc or spherical particle systems is proposed, considering a direct analogy between the nodes and elements of a FEM mesh and the distribution of particles and contacts of a DEM assembly.

Considering the FEM mesh shown in Fig. 11.16a, the matrices ($\mathbf{M}^e$) and element stiffness matrices ($\mathbf{K}^e$) can be calculated for each element using standard FEM technique. For the particles in Fig. 11.16b, the contact stiffness matrix can be determined from the system configuration. Therefore, the geometrical distribution of particles and the contact modes (translational and rotational) at each contact determines the contact stiffness matrix. Therefore, direct calculations may be possible for regularly symmetric configurations of uniformly sized particles, not for general systems, since only in such simplified cases can the eigenvalues λ^e for the elements be determined from the element stiffness and mass matrices and the critical time step calculated.

11.4.2 Stiffness Matrices of Contact Elements

As shown in Fig. 11.18 and in two dimensions, the particle contacts in a DEM model are similar to strut elements used in structural analysis. Consequently, the element stiffness matrix for each contact is similar to the element stiffness matrix for a strut (Sack, 1989).

In a local frame that has an inclination angle θ with the global (inertial) frame, and for a pair of particles of equal size with a radius r, an increment of rotation, $\Delta\varphi_i$, will induce a moment increment $r^2 K_{s\Delta}\varphi_i$. When the rotational degrees of freedom are included in the analysis, the contact element's stiffness matrix has six degrees of freedom (Δu_n, Δu_s and $\Delta\phi$ for each particle in turn) and is given by (O'Sullivan and Bray, 2004)

$$\mathbf{K}^e_{\text{local}} = \begin{bmatrix} K_n & 0 & 0 & -K_n & 0 & 0 \\ 0 & K_s & 0 & 0 & -K_s & 0 \\ 0 & 0 & K_s r^2 & 0 & 0 & 0 \\ -K_n & 0 & 0 & K_n & 0 & 0 \\ 0 & -K_s & 0 & 0 & K_s & 0 \\ 0 & 0 & 0 & 0 & 0 & K_s r^2 \end{bmatrix} \tag{11.14}$$

and the coordinate transformation matrix (between local and global frames) is simply

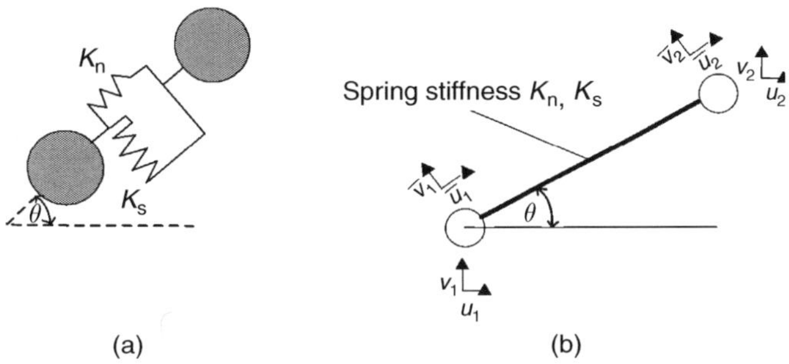

Fig. 11.18 (a) Normal and shear contact springs for DEM particles and (b) strut element used to determine the contact element stiffness matrix (O'Sullivan and Bray, 2004).

$$\mathbf{T} = \begin{bmatrix} \cos\theta & \sin\theta & 0 & 0 & 0 & 0 \\ -\sin\theta & \cos\theta & 0 & 0 & 0 & 0 \\ 0 & 0 & 1 & 0 & 0 & 0 \\ 0 & 0 & 0 & \cos\theta & \sin\theta & 0 \\ 0 & 0 & 0 & -\sin\theta & \cos\theta & 0 \\ 0 & 0 & 0 & 0 & 0 & 1 \end{bmatrix} \tag{11.15}$$

The local–global transformation of the element stiffness matrix is then

$$\mathbf{K}^{\mathrm{e}}_{\mathrm{global}} = \mathbf{T}^{\mathrm{T}}\mathbf{K}^{\mathrm{e}}_{\mathrm{local}}\mathbf{T} \tag{11.16}$$

In three-dimensional systems (cf. Chapter 2) of non-spherical particles, the moments of inertia around the three principal axes of inertia are not equal and the three rotational degrees of freedom (about each of the principal axes of inertia) are coupled. The central difference method cannot be used to integrate the rotational kinematic equations. Alternative time integration approaches have been proposed by Lin and Ng (1997) and by Munjiza et al. (2003), with the necessary additional computational costs.

For spherical particles, $I_{xx} = I_{yy} = I_{zz}$, no rotational coupling exists and the time integration using the central difference approach is applicable. For these non-coupled cases, the element stiffness matrix is given by

$$\mathbf{K}^{\mathrm{e}}_{\mathrm{local}} = \begin{bmatrix} K_n & 0 & 0 & 0 & 0 & 0 & -K_n & 0 & 0 & 0 & 0 & 0 \\ 0 & K_s & 0 & 0 & 0 & 0 & 0 & -K_s & 0 & 0 & 0 & 0 \\ 0 & 0 & K_s & 0 & 0 & 0 & 0 & 0 & -K_s & 0 & 0 & 0 \\ 0 & 0 & 0 & K_s r^2_{\mathrm{eff},1} & 0 & 0 & 0 & 0 & 0 & -K_s r^2_{\mathrm{eff},1} & 0 & 0 \\ 0 & 0 & 0 & 0 & K_s r^2_{\mathrm{eff},2} & 0 & 0 & 0 & 0 & 0 & -K_s r^2_{\mathrm{eff},2} & 0 \\ 0 & 0 & 0 & 0 & 0 & K_s r^2_{\mathrm{eff},3} & 0 & 0 & 0 & 0 & 0 & -K_s r^2_{\mathrm{eff},3} \\ -K_n & 0 & 0 & 0 & 0 & 0 & K_n & 0 & 0 & 0 & 0 & 0 \\ 0 & -K_s & 0 & 0 & 0 & 0 & 0 & K_s & 0 & 0 & 0 & 0 \\ 0 & 0 & -K_s & 0 & 0 & 0 & 0 & 0 & K_s & 0 & 0 & 0 \\ 0 & 0 & 0 & -K_s r^2_{\mathrm{eff},1} & 0 & 0 & 0 & 0 & 0 & K_s r^2_{\mathrm{eff},1} & 0 & 0 \\ 0 & 0 & 0 & 0 & -K_s r^2_{\mathrm{eff},2} & 0 & 0 & 0 & 0 & 0 & K_s r^2_{\mathrm{eff},2} & 0 \\ 0 & 0 & 0 & 0 & 0 & -K_s r^2_{\mathrm{eff},3} & 0 & 0 & 0 & 0 & 0 & K_s r^2_{\mathrm{eff},3} \end{bmatrix} \tag{11.17}$$

and the local–global transformation matrix is

$$\mathbf{T} = \begin{bmatrix} \cos\alpha_x & \cos\alpha_y & \cos\alpha_z & 0 & 0 & 0 & 0 & 0 & 0 & 0 & 0 & 0 \\ \cos\beta_x & \cos\beta_y & \cos\beta_z & 0 & 0 & 0 & 0 & 0 & 0 & 0 & 0 & 0 \\ \cos\gamma_x & \cos\gamma_y & \cos\gamma_z & 0 & 0 & 0 & 0 & 0 & 0 & 0 & 0 & 0 \\ 0 & 0 & 0 & \cos\alpha_x & \cos\alpha_y & \cos\alpha_z & 0 & 0 & 0 & 0 & 0 & 0 \\ 0 & 0 & 0 & \cos\beta_x & \cos\beta_y & \cos\beta_z & 0 & 0 & 0 & 0 & 0 & 0 \\ 0 & 0 & 0 & \cos\gamma_x & \cos\gamma_y & \cos\gamma_z & 0 & 0 & 0 & 0 & 0 & 0 \\ 0 & 0 & 0 & 0 & 0 & 0 & \cos\alpha_x & \cos\alpha_y & \cos\alpha_z & 0 & 0 & 0 \\ 0 & 0 & 0 & 0 & 0 & 0 & \cos\beta_x & \cos\beta_y & \cos\beta_z & 0 & 0 & 0 \\ 0 & 0 & 0 & 0 & 0 & 0 & \cos\gamma_x & \cos\gamma_y & \cos\gamma_z & 0 & 0 & 0 \\ 0 & 0 & 0 & 0 & 0 & 0 & 0 & 0 & 0 & \cos\alpha_x & \cos\alpha_y & \cos\alpha_z \\ 0 & 0 & 0 & 0 & 0 & 0 & 0 & 0 & 0 & \cos\beta_x & \cos\beta_y & \cos\beta_z \\ 0 & 0 & 0 & 0 & 0 & 0 & 0 & 0 & 0 & \cos\gamma_x & \cos\gamma_y & \cos ;\gamma_z \end{bmatrix} \tag{11.18}$$

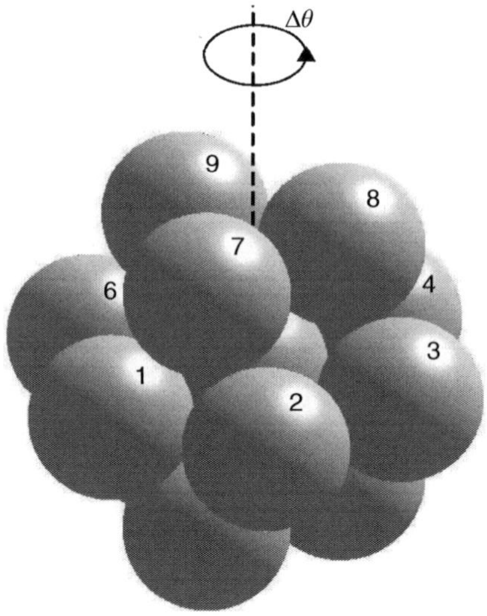

Fig. 11.19 Rotation of uniform spheres with rhombic packing configuration (O'Sullivan and Bray, 2004).

where $r_{\mathrm{eff},i}$ is the orthogonal distance from the axis of rotation i to the contact point on the surface of the sphere (O'Sullivan and Bray, 2004). This is different compared with the two-dimensional case. In three dimensions, $r_{\mathrm{eff}},i \neq r$ in all cases, where r is the sphere radius. Consider, for example, a rhombic (or hexagonal close) packing configuration and rotation about a vertical axis through the central sphere as shown in Fig. 11.19. For the contacts between the central sphere and spheres numbered 1–6, $r_{\mathrm{eff},i} = r$. However, for the contacts between the central sphere and spheres numbered 7–9, $r_{\mathrm{eff},i} = r/\sqrt{3}$ (O'Sullivan and Bray, 2004).

11.4.3 Mass Matrices of Contact Elements

In the FEM, 'lumped' mass matrices are typically used for constructing element mass matrices, since the approach yields a diagonal global mass matrix and is computationally efficient (Zienkiewicz and Taylor, 2000). The nodal mass values are the sum of the mass contributions from the elements contacting at that node and the calculation details depend on the element shape. For example, in two dimensions for a simple three node triangular element with two degrees of freedom at each node, the element mass matrix is given by

$$\mathbf{M}^{\mathrm{e}} = \frac{\rho A^{\mathrm{e}}}{3}\begin{bmatrix} 1 & 0 & 0 & 0 & 0 & 0 \\ 0 & 1 & 0 & 0 & 0 & 0 \\ 0 & 0 & 1 & 0 & 0 & 0 \\ 0 & 0 & 0 & 1 & 0 & 0 \\ 0 & 0 & 0 & 0 & 1 & 0 \\ 0 & 0 & 0 & 0 & 0 & 1 \end{bmatrix} \tag{11.19}$$

where ρ is the material density and A^{e} the element area. Then, the lumped mass at each node is given by

$$m_k = \frac{1}{3}\sum_{i=1}^{N_k} \rho_i A_j^{\mathrm{e}} \tag{11.20}$$

where m_k and n_k are the lumped mass and the number of elements meeting at node k, respectively. The FEM mass system is like an artificial particle system.

In a DEM particle system, the mass is naturally 'lumped' at the particles (nodes) already as an assigned value when the particle is generated. In a similar fashion as that for the stiffness matrix, the mass of a particle is proportionally distributed to the contacts it shares with other particles, so that a mass matrix can be developed for each contact element. Only for particle systems of uniform circular disks or spheres with a regular, symmetrical packing configuration, can the mass of the particles be distributed equally among the contacts to develop the element mass matrices and the closed-form expressions for mass matrix can be derived. If the particle mass is given by m_i, the particle rotational inertia is given by I_i and the number of contacts for particle i is n_i^{c}, the element mass matrices for the contact element linking particle i and j are given by a 6×6 matrix $\mathbf{M}_{\mathbf{ij}}^{\mathbf{e}}$ (O'Sullivan and Bray, 2004) with its all non-diagonal elements equal to zero and its diagonal elements given as

$$M_{11}^{\mathrm{e}} = M_{22}^{\mathrm{e}} = \frac{m_i}{n_i^{\mathrm{c}}}, \qquad M_{44}^{\mathrm{e}} = M_{55}^{\mathrm{e}} = \frac{m_j}{n_j^{\mathrm{c}}}, \qquad M_{33}^{\mathrm{e}} = \frac{I_i}{n_i^{\mathrm{c}}}, \qquad M_{66}^{\mathrm{e}} = \frac{I_j}{n_j^{\mathrm{c}}} \tag{11.21}$$

for two circular discs in contact for two-dimensional systems, for three degrees of freedom each (two translational and one rotational). Similarly, for two spheres in contact in three-dimensional systems, for six degrees of freedom each (three translational and three rotational), a 12×12 matrix $\mathbf{M}_{\mathbf{ij}}^{\mathbf{e}}$ is given in which all non-diagonal elements are zero and the diagonal elements are given by

$$M_{11}^{\mathrm{e}} = M_{22}^{\mathrm{e}} = M_{33}^{\mathrm{e}} = \frac{m_i}{n_i^{\mathrm{c}}}, \qquad M_{77}^{\mathrm{e}} = M_{88}^{\mathrm{e}} = M_{99}^{\mathrm{e}} = \frac{m_j}{n_j^{\mathrm{c}}} \tag{11.22a}$$

$$M_{44}^{\mathrm{e}} = \frac{I_x^i}{n_i^{\mathrm{c}}}, \quad M_{55}^{\mathrm{e}} = \frac{I_y^i}{n_i^{\mathrm{c}}}, \quad M_{66}^{\mathrm{e}} = \frac{I_z^i}{n_i^{\mathrm{c}}}, \quad M_{10,10}^{\mathrm{e}} = \frac{I_x^j}{n_j^{\mathrm{c}}}, \quad M_{11,11}^{\mathrm{e}} = \frac{I_y^j}{n_j^{\mathrm{c}}}, \quad M_{12,12}^{\mathrm{e}} = \frac{I_z^j}{n_j^{\mathrm{c}}} \tag{11.22b}$$

The mass matrix is a minimum (where n^{c} is a maximum) when the particle packing is the densest, and such cases can serve as limiting cases for mass matrix evaluations. For uniform disks (two dimensions), hexagonal packing is the densest configuration that can be attained (Fig. 11.20). This packing configuration is symmetrical so that the particle mass can be uniformly distributed among all six contacts.

In three dimensions, two packing configurations can achieve the maximum packing density for uniform spheres: the rhombic (or hexagonal close) and face-centred-cubic (FCC) packing configurations (Fig. 11.21). Although the void ratio and coordination number values are equivalent for both cases, however, the material fabric of the particles is different, resulting in different mechanical responses for these two spherical packing configurations.

Fig. 11.20 Hexagonally packed uniform circular disks, the lines between them showing the directions of contacts (O'Sullivan and Bray, 2004).

(a) Face-centered-cubic packing

(b) Rhombic packing

Fig. 11.21 Orthogonal views of uniform spheres with FCC (top) and rhombic (bottom) packing patterns (O'Sullivan and Bray, 2004).

The rhombic and FCC packing configurations are not perfectly symmetrical. Considering rotational symmetry, a lattice possesses an n-fold axis of rotational symmetry if it coincides with itself on rotation about the axis of $360°/n$. Taking the axis of symmetry to pass vertically through the central sphere (Fig. 11.21), the FCC packing has fourfold rotational symmetry, while the rhombic packing has threefold rotational symmetry.

Two approaches were then adopted to calculate the mass matrix for contact elements in the case of uniform spheres with rhombic and FCC packing configurations (O'Sullivan and Bray, 2004). In the first approach, the mass was simply distributed equally among all 12 contacts. In the second approach, the sphere was divided into three zones as shown in Fig. 11.22. The mass of the upper zone was then distributed equally among the upper row of contacts (four for the FCC case and three for the rhombic case). The mass of the central zone was distributed equally among the upper row of contacts and lower row of contacts (with a total of eight contacts for the FCC case and nine contacts for the rhombic case). Both mass matrices were used in calculating the critical time increments given below.

11.4.4 Eigenvalue Calculations

The calculation of eigenvalues of the matrix product $(\mathbf{M}^e)^{-1}(\mathbf{K}^e_{\text{global}})$ in the global frame is needed for calculating the critical time step for central difference integration of the particle system's equations of motion. As $\mathbf{M}^e$ is a diagonal matrix, then $\mathbf{M}^{e-1}$ is also diagonal and we have (O'Sullivan and Bray, 2004)

$$(\mathbf{M}^e)^{-1}\mathbf{T}^T(\mathbf{K}^e_{\text{local}})\mathbf{T} = \mathbf{T}^T(\mathbf{M}^e)^{-1}(\mathbf{K}^e_{\text{local}})\mathbf{T} \tag{11.23}$$

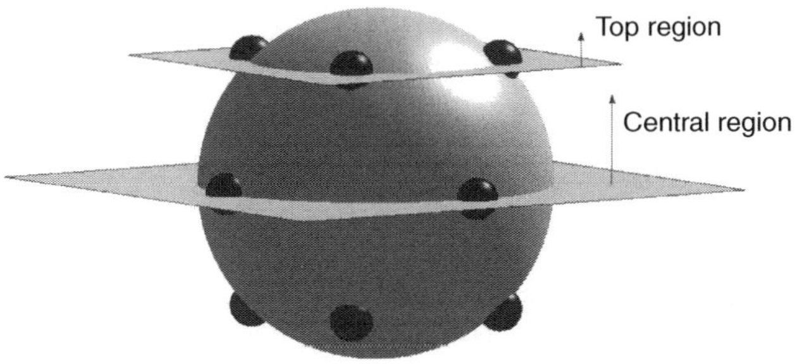

(a) Face-centered-cubic packing

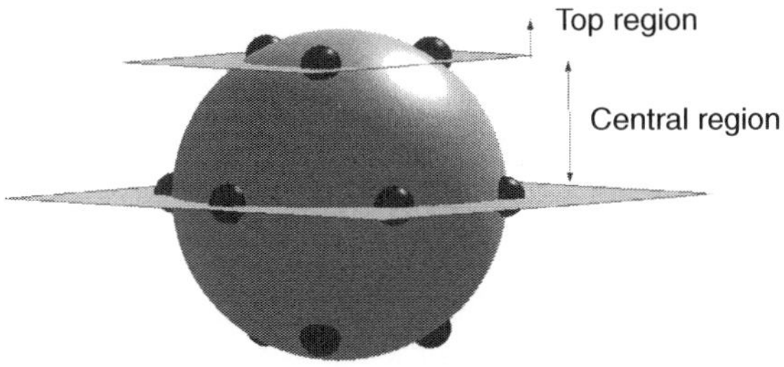

(b) Rhombic packing

Fig. 11.22 Illustration of division of spheres for non-uniform distribution of inertia values to contact elements (O'Sullivan and Bray, 2004).

Since again the matrix **T** is orthogonal so $\mathbf{T}^{\mathrm{T}} = \mathbf{T}^{-1}$, and the matrices $(\mathbf{M}^{\mathrm{e}})^{-1}(\mathbf{K}^{\mathrm{e}}_{\mathrm{local}})$ and $\mathbf{T}^{\mathrm{T}}(\mathbf{M}^{\mathrm{e}})^{-1}(\mathbf{K}^{\mathrm{e}}_{\mathrm{local}})\mathbf{T}$ are similar and have the same eigenvalues (Golub and Van Loan, 1983). Then, for each contact element, the calculation of the eigenvalues of $(\mathbf{M}^{\mathrm{e}})^{-1}(\mathbf{K}^{\mathrm{e}}_{\mathrm{global}})$ in the global frame is equivalent to calculating the eigenvalues of $(\mathbf{M}^{\mathrm{e}})^{-1}(\mathbf{K}^{\mathrm{e}}_{\mathrm{local}})$ in the local frame. Recognizing this equivalence and assuming that the normal and shear spring stiffness are equal (i.e., $K_{\mathrm{n}} = K_{\mathrm{s}} = K$), the eigenvalues were calculated for a number of symmetrical configurations of uniform disks and spheres. The critical time steps were then determined using eigenvalues of $(\mathbf{M}^{e})^{-1}(\mathbf{K}^{e}_{local})$ (cf. (Eqn (8.106)).

O'Sullivan and Bray (2004) show that for some symmetrical configurations of two-dimensional disks for both translational and rotational contacts, the minimum critical time increment in two dimensions for the case of translational motion only is $0.577\sqrt{m/K}$, while the minimum critical time increment is $0.408\sqrt{m/K}$ when particle rotation is also incorporated.

However, the critical time increments for three-dimensional analysis are more restrictive. If the particle inertia is distributed equally among all contacts, the minimum critical time step is $0.408\sqrt{m/K}$ for the case of translational motion only and the minimum critical time step is $0.258\sqrt{m/K}$ when rotation is also allowed.

If the particle inertia is distributed in a non-uniform manner according to Fig. 11.22, then the minimum critical time step is $0.348\sqrt{m/K}$ for translation only and $0.221\sqrt{m/K}$ if rotation is also included. It seems that the theoretical value of $\sqrt{m/K}$ is not adequately conservative. The examples above have shown that the critical time increment for rigid particles depends on packing density represented by the coordination number (the number of contacts per particle) of contacts. When this number increases, the critical time step decreases. Even when the packing density is the same, the material fabric also changes the critical time steps, such as for the examples of the rhombic and FCC packing with identical packing densities.

O'Sulllivan and Bray (2004) recommended a critical time step of $0.3\sqrt{m/K}$ in two dimensions and $0.17\sqrt{m/K}$ in three dimensions, where m is the minimum particle mass and K the maximum contact stiffness.

Estimation of critical time step value is important for maintaining numerical stability and result convergence. It also controls the computing time. Above theoretical estimations are applicable for only simple packing situations of uniform-sized particles. For irregular particle systems, multiple numerical experiments are often needed.

11.4.5 Energy Balance for Checking Numerical Instabilities

Numerical instabilities in explicit simulations can be detected by an energy balance check, indicated by the spurious generation of energy, which leads to a violation of the conservation of energy. Using the approach proposed by Belytschko et al. (2000), the energy balance requirement is given by

$$|W_{\text{kinetic}} + W_{\text{internal}} - W_{\text{external}}| \leq \varepsilon \max \quad (W_{\text{kinetic}}, W_{\text{internal}}, W_{\text{external}}) \tag{11.24}$$

where W_{kinetic} is the kinetic energy, W_{internal} the internal energy, W_{external} the external energy and ε is a constant on the order of 10^{-2}. An incremental approach was recommended to calculate the energy terms in (11.24) and is given in Itasca (1998) for DEM models.

The kinetic energy term, W_{kinetic}, is given by

$$W_{\text{kinetic}} = \frac{1}{2}\sum_{i=1}^{N_p}(\mathbf{v}^{\mathbf{i}})^{\mathrm{T}}\mathrm{m}^{\mathbf{i}}\mathbf{v}^{\mathbf{i}} \tag{11.25}$$

where N_p is the number of particles in the domain and $\mathbf{v}^i$ and m^i the velocity vector and mass for particle i, respectively.

The internal energy term, W_{internal}, at time $t+\Delta t$, is given by

$$W_{\text{internal}}^{t+\Delta t} = W_{\text{strain}}^{t+\Delta t} - W_{\text{friction}}^{t+\Delta t} = \frac{1}{2}\sum_{i=1}^{N_c}\left(\frac{|F_n|^2}{K_n} + \frac{|F_s|^2}{K_s}\right) - \left[W_{\text{friction}}^{t} + \sum_{i=1}^{N_c}(F_s^i)(\Delta u_s^i)\right] \tag{11.26a}$$

where W_{strain} represents the strain energy (considering all particles), W_{friction} the energy dissipated by frictional sliding (only at contacts), N_c the number of contacts in the domain, F_n^i the normal force at contact i, F_s^i the shear force at contact i, vector u_s^i the incremental tangential displacement at the contact and K_n and K_s the normal and tangential spring stiffness, respectively.

The external energy term is given by

$$W_{\text{external}} = W_{\text{BF}}^{t+\Delta t} + W_{\text{AF}}^{t+\Delta t} + W_{\text{WF}}^{t+\Delta t} \tag{11.26b}$$

$$W_{\text{BF}}^{t+\Delta t} = W_{\text{BF}}^{t} + \sum_{i=1}^{N_p} m^i(\mathbf{b}^i)(\Delta\mathbf{u}^i) \tag{11.26c}$$

$$W_{\text{AF}}^{t+\Delta t} = W_{\text{AF}}^{t} + \sum_{i=1}^{N_p}(\mathbf{F}_{\text{Applied}}^i)(\Delta\mathbf{u}^i) \tag{11.26d}$$

$$W_{\text{WF}}^{t+\Delta t} = W_{\text{WF}}^{t} + \sum_{i=1}^{N_p}(\mathbf{F}_{\text{contact}}^i)(\Delta\mathbf{u}_c^i) \tag{11.26e}$$

where W_{BF} is the energy associated with the body forces, such as gravity, acting on the particles, W_{AF} the energy associated with any externally applied loads (such as boundary stresses/forces), W_{WF} the work done on the system by the rigid boundaries (with numbers of contacts N_c), b^i the vector of body forces acting on particle i, vector $\Delta\mathbf{u}^i$ the incremental displacement of particle i, vector $\mathbf{F}^i_{\text{Aplied}}$ the applied external force vectors acting on particle i, vector $\mathbf{F}^i_{\text{contact}}$ the contact force at contact i and vector $\Delta\mathbf{u}^i_c$ the incremental displacement at contact i. The incremental displacements are the displacements over the current time increment, $t \rightarrow t + \Delta t$.

11.5 Cosserat Continuum Equivalence to Particle Systems

11.5.1 Fundamental Concepts

One of the primary objectives of the particle models is the establishment of the relations between microscopic and macroscopic variables/parameters of the particle systems, mainly through micro-mechanical constitutive relations at contacts. Compared with a continuum, particles have an additional degree of freedom of rotation which enables them to transmit couple stresses, besides forces through their translational degrees of freedom. The equivalent continua of the particle systems with this additional rotational degree of freedom are called Cosserat continua or micro-polar continua (Cosserat and Cosserat, 1909). Description of a Cosserat continuum requires evaluation of couple stress and rotation gradient tensors from forces, torques and translational and rotational displacements at contacts, besides the classical Cauchy stress and strain tensors. The theoretical foundation is well presented in Eringen (1999), and a recent comprehensive review is given in Kruyt (2003).

The Cosserat micro-polar continuum concept has also been introduced into rock and soil mechanics and engineering fields, such as the work by Papamichos et al. (1990) for layered elastic media with concerns for rock burst and exfoliation, Dawson and Cundall (1995) with Cosserat plasticity models for layered rocks, Dai et al. (1996) for fractured rock mechanics, Cerrolaza et al. (1999) for blocky structures, Sulem and Cerrolaza (2000) for rock slopes, Morris et al. (2004) for underground excavations and Durand et al. (2006) for rock and soil stability, to mention just a few.

When characterizing particle systems as equivalent Cosserat continua, two approaches are taken to derive the corresponding macro-microscopic properties. One uses the continuum mechanics principles directly, such as that used in Chang and Ma (1991, 1992) and Oda and Iwashita (2000). Another uses the micro-mechanics approach based on characterization of particle contacts, such as Kruyt (2003). Although the macroscopic continuum mechanics analogue approach is more often applied in rock mechanics applications using the Cosserat medium concept, the micro-mechanics approach has closer ties with the DEM for particle systems. Therefore, we chose this approach for the more detailed and systematic summary description below, based mainly on Kruyt (2003).

The Cosserat continuum and DEM particle (or block) models have been applied hand-in-hand due to the fact that the DEM can explicitly and most naturally simulate particle contacts and rotations, and therefore contact forces, torques and frictional moments, so that the effects of couple stress and rotation gradient tensors can be evaluated discretely and explicitly. In this section, we provide a brief summary of the Cosserat micro-polar mechanics theory and its relation to DEM, so that a preliminary understanding of the fundamentals of this approach can be reached before going further into more systematic and in-depth studies on the subject leading to applications for rock engineering.

One of the basic concepts in Cosserat micro-mechanics theory is homogenization. Since detailed knowledge of the precise stress field at the microscopic scale (particle scale) is generally not required in studies of the macroscopic behavior of granular materials, it is natural to represent the assembly of

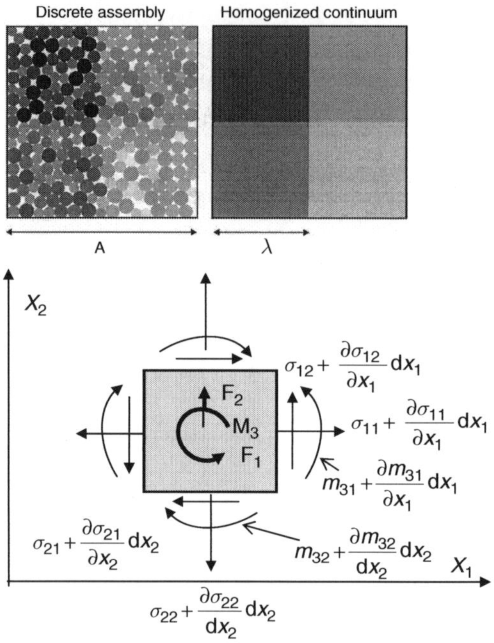

(a) Discrete assembly and homogenized continuum; gray scale indicates level of stress, with $R \ll \lambda < \Lambda$ (Kruyt, 2003).

(b) Stress equilibrium at a differential Cosserat continuum point with differential side lengths dx_1 and dx_2 in x_1- and x_2-directions, respectively, in two dimensions.

Fig. 11.23 (a) Homogenized Cosserat continuum from particle systems and (b) equilibrium stress state for two-dimensional problems.

particles by an equivalent homogenized continuum. Homogenization is needed since both geometrical (such as non-uniform distribution in size and locations) and mechanical (such as non-uniform stress distribution in individual particles, see Timoshenko and Goodier, 1970) heterogeneities of particle assemblies need to be treated by averaging over certain support domains or volumes for deriving effective or equivalent macroscopic properties using Cosserat continuum principles. It is clear that the length scale associated with the variation of the stress inside the particles is much smaller than the particle radius R. For this homogenization process to be meaningful, the representative length scale λ of the variation of the homogenized stress σ_{ij}, therefore, must be significantly larger than the particle radius R, but smaller than the domain (volume) size λ under study, very much like a REV scale as introduced before (Fig. 11.23a).

The equilibrium equation for an arbitrary point in a Cosserat continuum is then written (Fig. 11.23b),

$$\begin{bmatrix} \frac{\partial}{\partial x} & 0 & \frac{\partial}{\partial y} & 0 & 0 & 0 \\ 0 & \frac{\partial}{\partial x} & 0 & \frac{\partial}{\partial y} & 0 & 0 \\ 0 & 1 & -1 & 0 & \frac{\partial}{\partial x} & \frac{\partial}{\partial y} \end{bmatrix} \begin{Bmatrix} \sigma_{11} \\ \sigma_{12} \\ \sigma_{21} \\ \sigma_{22} \\ m_{31} \\ m_{32} \end{Bmatrix} + \begin{Bmatrix} \mathbf{F}_1 \\ \mathbf{F}_2 \\ \mathbf{M}_3 \end{Bmatrix} = \begin{Bmatrix} 0 \\ 0 \\ 0 \end{Bmatrix} \tag{11.27}$$

where $\mathbf{F_1}$, $\mathbf{F_2}$ and $\mathbf{M_3}$ are the resultant forces and moment of the differential area in Fig. 11.23b. Note that $\sigma_{12} \neq \sigma_{21}$, i.e., the stress tensor is not symmetric due to the micro-structure and the couple stress effects included in the formulation, which may have a significant impact on material failure mechanisms. The corresponding Cosserat continuum deformation is characterized by

$$\begin{bmatrix} \frac{\partial}{\partial x} & 0 & 0 \\ 0 & \frac{\partial}{\partial x} & -1 \\ \frac{\partial}{\partial y} & 0 & 1 \\ 0 & \frac{\partial}{\partial y} & 0 \\ 0 & 0 & \frac{\partial}{\partial x} \\ 0 & 0 & \frac{\partial}{\partial y} \end{bmatrix} \begin{Bmatrix} \dot{u} \\ \dot{v} \\ \dot{\omega} \end{Bmatrix} = \begin{Bmatrix} \dot{\gamma}_{11} \\ \dot{\gamma}_{12} \\ \dot{\gamma}_{21} \\ \dot{\gamma}_{22} \\ \dot{\kappa}_{31} \\ \dot{\kappa}_{32} \end{Bmatrix} \tag{11.28}$$

where $\dot{u}$, $\dot{v}$ and $\dot{\omega}$ are the deformation rate (velocities) in the x- and y-directions and rotation velocity, respectively, $\dot{\kappa}_{31}$ and $\dot{\kappa}_{32}$ are the rates (velocities) of rotational gradient in the x- and y-directions, respectively, indicating that the averaged Cosserat medium can sustain bending caused by the torques. The deformation rate vector is not written in similar form to that of the strain vector in continuum mechanics due to this difference. Equations (11.27) and (11.28) represent a direct analogue to a continuum mechanics interpretation.

11.5.2 Basic Concepts in the Micro-Mechanics Approach for Homogenization of Particle Systems

When Cosserat continua are chosen as the macroscopic representation for particle assemblies, the basic kinematic quantities are therefore the displacement field $U_i(\mathbf{x})$ and rotation field $\omega_i(\mathbf{x})$, and the basic force quantities are the homogenized Cauchy stress tensor σ_{ij} and the couple stress tensor μ_{ij}. Due to the additional couple stress and rotation gradients, the local mean values of the particle stresses and the local averages of the homogenized Cauchy stresses do not need to be equal, but the mechanical behavior of the discrete assembly of particles and the homogenized continuum has to be equivalent. Thus, the force traction vector over a boundary B of the homogenized stress tensor σ_{ij} must be equal to the resultant of the discrete forces that act on the boundary of the assembly (Kruyt, 2003)

$$\int_B n_i \sigma_{ij} \, \mathrm{d}B = \sum_{\beta \in B} f_i^{\beta} \tag{11.29a}$$

where n_i is the outward unit normal vector of boundary B, the sum is over the contacts $\beta \in B$ and $f_{i\beta}$ is the force acting at contact β.

Similarly, the couple traction vector (over a boundary B) of the homogenized couple stress tensor μ_{ij} must be equal to the resultant of the discrete couples acting on the boundary of the assembly (Kruyt, 2003)

$$\int_B n_i \mu_{ij} \, \mathrm{d}B = \sum_{\beta \in B} \kappa_i^{\beta} \tag{11.29b}$$

where $\kappa_{i\beta}$ is the couple acting at contact $\beta \in B$, without involving contact forces.

The mathematical foundation for homogenization of particle assemblies into continua is based on the statements below.

Since $R << \lambda$, a position-dependent density of contacts, $m_V(\mathbf{x})$, the number of contacts per unit volume can be defined. More generally, a position-dependent average (over contacts) quantity, $\langle \psi(\mathbf{x}) \rangle$, and a general contact quantity, ψ^{c}, such as components of contact forces or displacements, may also be defined.

Then the contact density and average general contact quantity will satisfy the following relations (Kruyt, 2003)

$$\sum_{c\in C} 1 = \int_V m_V(\mathbf{x})\mathrm{d}V \tag{11.30a}$$

$$\sum_{c\in C} \psi^{\mathrm{c}} = \int_V m_V(\mathbf{x})\langle\psi(\mathbf{x})\rangle\mathrm{d}V \tag{11.30b}$$

where C represents the complete set of contacts in the domain of interest with a volume V, $\mathbf{x}$ the position vector of particles and the left-hand side of (11.30a) is simply the total number of contacts. Relation (11.30b) is a natural extension of (11.30a) since $\langle\psi(\mathbf{x})\rangle$ is a constant for a specific problem.

In homogenizing particle system properties, a position-dependent quantity $\phi(\mathbf{x})$ that generally varies slowly across the domain of interest is frequently encountered. Since $\phi(\mathbf{x})$ is assumed to vary slowly across the domain we may write the averaged values of the product of the two variables in the form

$$\langle\phi\psi(\mathbf{x})\rangle = \phi(\mathbf{x})\langle\psi(\mathbf{x})\rangle \tag{11.31a}$$

Assuming that the sums are volume additive, a sum over contacts may be replaced by an equivalent volume integral (Kruyt, 2003)

$$\sum_{c\in C} \phi(\mathbf{C}^{\mathrm{c}})\psi^{\mathrm{c}} = \int_{\mathbf{V}} m_V(\mathbf{x})\phi(\mathbf{x})\langle\psi(\mathbf{x})\rangle\mathrm{d}V \tag{11.31b}$$

where $\mathbf{C}^{\mathrm{c}}$ is the position vector of the point of contact c.

The relation (11.31b) is the basis for deriving linking relations between equivalent continuum and a discrete particle assembly.

11.5.3 Micro-Mechanics – Kinematic Quantities

11.5.3.1 Contact quantities

Let f_i^{pq} and κ_i^{pq} be the vectors of contact force and couple at the contact between particles p and q (exerted by particle q on particle p). The relative displacement Δ_i^{pq} and the relative rotation Ω_i^{pq} (the kinematic variables at contact) are defined by (Kruyt, 2003)

$$\begin{cases} \Delta_i^{pq} = [U_i^q + e_{ijk}\omega_j^q r_k^{qp}] - [U_i^p + e_{ijk}\omega_j^p r_k^{pq}] \\ \Omega_i^{pq} = \omega_i^q - \Omega_i^p \end{cases} \tag{11.32}$$

where U_i^p and ω_i^p are the incremental displacement and rotation of particle p and e_{ijk} is the three-dimensional permutation symbol. The vectors $r_i^{pq} = C_i^{pq} - X_i^p$ and $r_i^{qp} = C_i^{pq} - X_i^q$ denote the distances from the centers of particles p and q, X_i^p and X_i^q, to the contact point C_i^{pq}, respectively (Fig. 11.24).

For a contact between particle p and the boundary at point $\beta \in B$, the relative displacement and the relative rotation are defined by (Kruyt, 2003)

$$\begin{cases} \Delta_i^{p\beta} = [U_i^\beta] - [U_i^p + e_{ijk}\omega_j^p r_k^{p\beta}] \\ \Omega_i^{p\beta} = \omega_i^\beta - \Omega_i^p \end{cases} \tag{11.33}$$

where U_i^β and ω_i^β are the displacement and rotation at the boundary. Slip is allowed in the formulation.

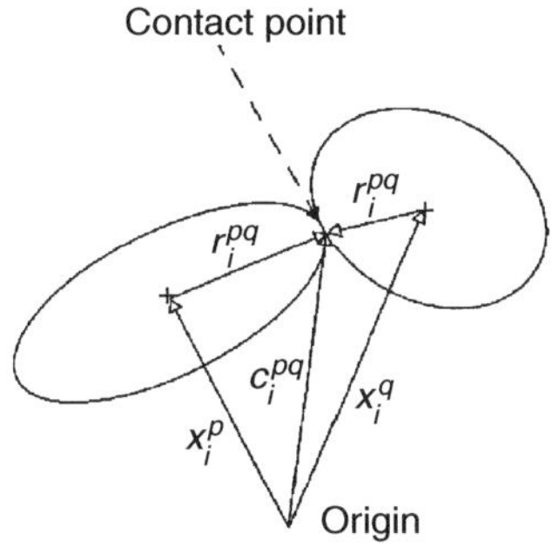

Fig. 11.24 Definitions of geometrical vectors (Kruyt, 2003).

The relative displacement Δ_i^{pq} is the sum of two parts, δ_i^{pq} due to translational displacements of particle centers and ρ_i^{pq} due to particle rotations (Kruyt, 2003)

$$\begin{cases} \delta_i^{pq} = U_i^q - U_i^p \\ \rho_i^{pq} = e_{ijk}\omega_j^q r_k^{qp} - e_{ijk}\omega_j^p r_k^{pq} \end{cases} \tag{11.34}$$

During rigid body motions of the whole assembly, there will be no relative displacement vectors between pairs of particles p and q, $\Delta_i^{pq} = \Omega_i^{pq} = 0$, but not δ_i^{pq}. The pair $(\Delta_i^{pq},\ \Omega_i^{pq})$ is better suited to formulate the micro-mechanical expressions for the strain and rotation gradient tensors.

The forces f_i^{pq} and couples κ_i^{pq} at contacts satisfy Newton's third law of action and reaction forces, $f_i^{qp} = -f_i^{pq}$, $\kappa_i^{qp} = -\kappa_i^{pq}$, and their relative displacements and relative rotations at contacts satisfy similar relations $\Delta_i^{qp} = -\Delta_i^{pq}$, $\quad \Omega_i^{qp} = -\Omega_i^{pq}$. The exact relation between static quantities f_i^{pq} and κ_i^{pq} and kinematic quantities Δ_i^{pq} and Ω_i^{pq} is described by the contact constitutive relations that are problem specific.

11.5.3.2 Discrete equilibrium equations

In the absence of body forces, for the quasi-static force and moment the discrete equilibrium equations of the particle assembly are (Kruyt, 2003)

$$\begin{cases} \sum_q f_j^{pq} + f_j^{p\beta} = 0 \\ \sum_q \left(\kappa_j^{pq} + e_{jkl}C_k^{pq} f_l^{pq}\right) + \left(\kappa_j^{p\beta} + e_{jkl}C_k^{p\beta} f_l^{p\beta}\right) = 0 \end{cases} \tag{11.35}$$

where the sums are over the particles q that are in contact with particle p, f_i^{pq} and κ_i^{pq} are the force and the couple exerted by the boundary on particle p (if present), C_i^{pq} and $C_i^{p\beta}$ the position vectors of the contact points between particles p and q and between particle p and the boundary. Note that multiple contacts between two particles are not allowed in the current formulation, i.e., only the convex particles are considered.

11.5.3.3 Graphical representation of particle contacts

Graphical or digraph representation of the topological patterns of contacts of particles is often used in formulating compatibility equations for the relative displacements and relative rotations at contacts in two-dimensions (Satake, 1992). In this graphical representation the internal nodes (or vertices) represent the centers of the particles and the boundary nodes represent the contact points at the boundary (Fig. 11.25).

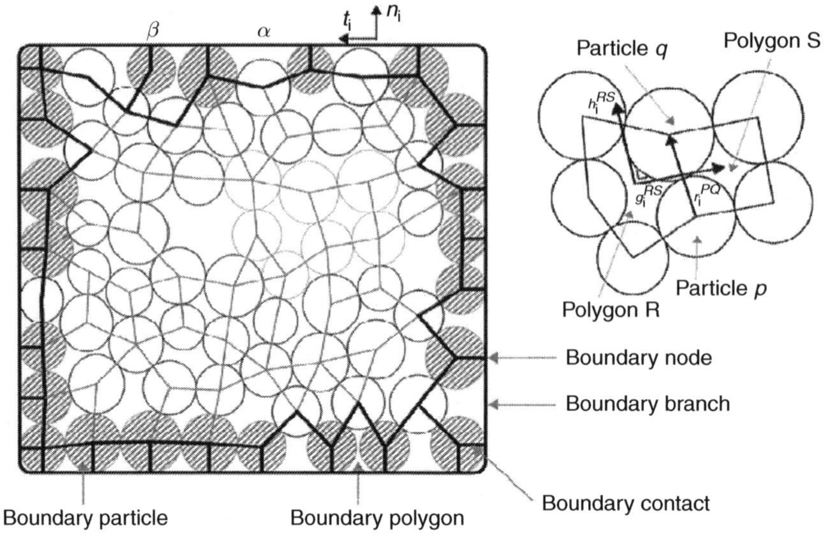

Fig. 11.25 Graphical representation of a particle assembly: polygons, edge (branch) vector, polygon vector and rotated polygon vector (Kruyt, 2003).

Edges between nodes represent the contacts, between either particle pairs or between particles and a boundary. Additional boundary edges between adjacent contact points at the boundary are added to ensure that the graph is a complete digraph satisfying the Euler–Poincáre relation (cf. Chapter 6).

Let the number of particles be N_p and the number of contact points at the boundary be N_c^B, then the number of nodes N_v in the graph is $N_v = N_\mathrm{p}+N_\mathrm{c}^\mathrm{B}$. The set of all contacts C consists of the set C^I of all internal contacts (between two particles) and the set C^B of all boundary contacts (between a particle and the boundary). The numbers of internal contacts in C^I, boundary contacts and the total number of contacts are N_c^I, N_c^B and N_c, with $N_\mathrm{c} = N_\mathrm{c}^\mathrm{I} + N_\mathrm{c}^\mathrm{B}$. The number of edges N_e is $N_\mathrm{c} + N_\mathrm{c}^\mathrm{B}$, where N_c^B is the number of boundary edges (branches).

Edge (contact) vectors l_i^{pq} are defined as the vectors that connect the centres of particles p and q that are in contact. They form closed loops, or polygons, as depicted in Fig. 11.25. For two adjacent polygons R and S, a vector, g_i^RS, connecting their two geometric centers is defined. Rotating g_i^{RS} counter-clockwise for 90° and another polygon vector h_j^{RS} is defined (Kruyt and Rothenbeug, 1996), see Fig. 11.25. These polygon vectors form a right-handed system and are needed in deriving the compatibility equations of the relative displacements and relative rotation of the discrete particle systems. Contacts can be identified by either the particles involved or by the polygons involved. The former is indicated in equations with particle labels in lowercase superscripts, *pq*; while the latter is indicated with polygon labels in uppercase superscripts, RS.

The polygons shown in Fig. 11.25 include 'boundary' polygons, i.e., those polygons that share an edge with the boundary (thick lines in Fig. 11.25). The polygon vector h_j^RS for a boundary contact between two polygons R and S is defined from the midpoint of the boundary edge of polygon R to the midpoint of the boundary edge of polygon S. The number of these boundary polygons equals the number of boundary contacts N_c^B. The number of internal polygons is N_l^I and the total number of polygons is $N_\mathrm{l} = N_\mathrm{l}^\mathrm{I} + N_\mathrm{c}^\mathrm{B}$.

The edge vector for a contact of particle p with the boundary, $l_i^{p\beta}$, points from the center of particle p to the contact point $C_i^{p\beta}$, i.e., $l_i^{p\beta} = C_i^{p\beta} - X_i^p$.

Using the above definitions, the Euler–Poincáre relation for a connected graph is

$$N_v - N_e + N_l = 1 \tag{11.36}$$

where the one face (polygon) formed by all the boundary edges with its normal pointing backwards from the graph plane is not counted. In Eqn (11.36), N_v is the number of nodes (vertices), N_l the number of polygons (loops) and N_e the number of edges of the graph. Using the relations $N_v = N_p + N_c^B$ and $N_e = N_c + N_c^B$, it follows that

$$N_p - N_c + N_l = 1 \tag{11.37}$$

A geometrical relation between edge vectors l_i^c and polygon (face) vectors h_j^c is (Kruyt and Rothenburg, 1996)

$$I_{ij} = \frac{1}{A} \sum_{c \in C} l_i^c h_j^c \tag{11.38}$$

where I_{ij} is the two-dimensional identity tensor and A is the area of the region of interest. This relation is based on the fact that the polygons tessellate the area. Note that Eqn (11.38) is valid for any particle shape, not only for disks.

11.5.4 Micro–Macro Equivalence Expressions For the Static and Kinematic Quantities Between Cosserat Continua and Particle Systems

After establishing the contact quantities, equilibrium equations and graphical representation of contacts, both between pairs of particles (using indices p and q) and between contact groups (polygons, using indices R and S), the remaining tasks are derivations of the expressions for compatibility equations, stress and strain tensors, stress couple and rotation gradient tensors and virtual works and complementary virtual work of the systems, so that macroscopic properties and quantities can be readily determined from particle contact configurations and constitutive relations for models of particle systems.

The theoretical framework is summarized in Table 11.5. Note that the two-dimensional case is considered in order to emphasize the dualities present and that the boundary terms have been omitted for clarity in the equilibrium and compatibility equations. In this table, index c is used for the set of particle contacts and index S is used for set of polygon contacts for simplicity. Details of derivation of the expressions are given in the Appendix at the end of this book.

Kruyt and Rothenburg (2002) defined uniform field expressions for the kinematic and static quantities at contacts in the two-dimensional case. The relative displacements and relative rotations at contacts that correspond to uniform strain and uniform rotation gradient are given by

$$\Delta_i^c = l_k^c \varepsilon_{ki}, \quad \Omega^c = l_k^c \frac{\partial \omega}{\partial x_k} \tag{11.39}$$

Similarly, the forces and couples at contacts that correspond to uniform stress and uniform couple stress are given by

$$f_i^c = h_k^c \sigma_{ki}, \quad \kappa^c = h_k^c \mu_k \tag{11.40}$$

These uniform field expressions are used to derive rigorous bounds for the effective elastic moduli of two-dimensional assemblies with bonded contacts (Kruyt and Rothenburg, 2002). The details of derivation of micro-mechanics expressions for stress and stress couple tensors of particle systems as equivalent Cosserat continua are given in the Appendix as demonstrative examples. Derivations for other micro-mechanics quantities as listed in Table 11.5 can be seen in Kruyt (2003).

Table 11.5 Summary of the theoretical framework of the discrete micro-mechanics approach of the Cosserat continuum using particle systems (Kruyt, 2003)

Statics		Kinematics	
Quantities	Expression	Quantities	Expression
Contact edge vector	l_i^c	Contact polygon vector	h_i^c
Couple	κ_i^c	Relative rotation	Ω^c
Equilibrium equations of the particles	$\sum_q f_j^{pq} = 0$ $\sum_q (\kappa^{pq} + e_{ij} C_i^{pq} f_j^{pq}) = 0$	Compatibility equations for contact polygons	$\sum_S (\Delta_i^{RS} + e_{ij} C_j^{RS} \Omega^{RS}) = 0$ $\sum_S \Omega^{RS} = 0$
Stress	$\langle \sigma_{ij} \rangle = (1/A) \sum_{c\in C}, l_i^c f_j^c$	Strain	$\langle \varepsilon_{ij} \rangle = (1/A) \sum_{c\in C} h_i^c \Delta_j^c$
Couple stress	$\langle \mu_i \rangle = (1/A) \sum_{c\in C} l_i^c \kappa^c$	Rotation gradient	$\langle \frac{\partial \omega}{\partial x_j} \rangle = \frac{1}{A} \sum_{c\in C} h_j^c \Omega^c$
Virtual work	$\sum_{c\in C} f_i^c \Delta_i^{c*} + \sum_{c\in C} \kappa_i^c \Omega_i^{c*}$ $= \sum_{\beta\in B} f_i^\beta U_i^{\beta*} + \sum_{\beta\in B} \kappa^\beta \omega^{\beta*}$	Complementary virtual work	$\sum_{c\in C} \Delta_i^c f_i^{c*} + \sum_{c\in C} \Omega_i^c \kappa_i^{c*}$ $= \sum_{\beta\in B} U_i^\beta f_i^{\beta*} + \sum_{\beta\in B} \omega^\beta \kappa^{\beta*}$
Geometry		$I_{ij} = (1/A) \sum_{c\in C} l_i^c h_j^c$	

Different formulations have been used by Chang and Ma (1991, 1992) and Oda and Iwashita (2000) for couple stress tensors. However, the contact polygon approach in Kruyt (2003) has been shown to be consistent with the macroscopic displacement gradient tensor derived from boundary displacements (Cambou et al., 2000; Bagi and Bojtár, 2001) and provides useful contact information for topological treatment of contact configurations of the particle systems.

One important assumption, by default but not explicitly stated in the above-presented micro-mechanics approach, is that the averaging process is performed over the region that has adequate size not less than a REV scale that may vary with size and spatial distributions of the particles and the constitutive relations at contacts. The validity of this assumption needs to be verified for practical applications.

11.6 Summary Remarks

The purpose of this chapter is to provide a brief introduction to the DEM approach for granular materials; therefore, the presentation focuses on its main features and assumptions and its relations to micro-mechanics research since it is one of its main areas of interest and advantage. The presentation on particle motion algorithm is kept brief due to the fact that excellent books about the particle mechanics approach using DEM have already been published, such as Oda and Iwashita (1999) and Mohammadi (2003), and there is therefore no need to repeat the details here. Some of the attractive applications using the DEM approach for granular materials are provided in Chapter 12 as test cases and the readers can learn more from such test cases in terms of the flexibility, limitation and merits of the approach. In summary, some of the outstanding issues in the DEM approach for granular materials are outlined below:

- In reality, the grain sizes of rock materials or soils/sands are very small indeed and are most often at micro-meter scales. How to represent the rock materials and soils/sands at this scale with more realistic particle shapes, instead of regular shapes of spheres (circles) and ellipsoids (ellipses), is a challenging difficulty that comes from not only the extremely high demand on computational power and resources, but also characterizing the contacts between grains and soil particles at micrometer scale.
- The effects of fluid flow in particle flow codes or other DEM models for granular materials are not commonly considered, due to the complexity in numerical treatment of fluid–solid interactions. Considering fluid effects demands intrinsic coupling, at pore scale, between the pore fluid and skeletal structure represented as solid particles in the particle models with challenging requirements for detailed characterizations of structure and connectivity of the pores and deformability of the skeletons.
- The DEM models for granular materials have three intrinsic difficulties: initial states of particle packing (configuration) and porosity distribution, the deformability of the particles themselves and the requirements of parameter calibration for particle models against experiments. The initial particle packing state and porosity distribution are practically unknown in many cases of rock mechanics problems. The effects of particle deformation, even breaking, can be simulated in the DEM approaches, but the computational cost is also increased considerably since numbers of particles for practical problems are almost always very large. The most striking difficulty is the third, i.e., the necessity of calibration of particle parameters, such as size or size distribution, contact stiffness or bonding parameters (friction, cohesion, bonding strength, etc.). Since these parameters are not conventional geotechnical properties and cannot be measured readily at the required scale at laboratory, the usual way of getting around this difficulty is by using measured macroscopic behavior of rocks, such as the complete stress–strain curves obtained from laboratory tests using standard-sized rock samples, and the particle models using DEM are then iteratively modified to provide a best fitting for the measured curves and calculated properties. The parameter set best fitting the measured behavior is said to represent the 'Real' parameters characterizing the particle system representing the target rock. This calibration is problematic since the uniqueness of the solution is not guaranteed, the selection of the parameters may not be physically optimal and/or unique and some of the model parameters may or may not be realistically determined (e.g., grain shapes and contact behavior). A recent study on the micro–macro properties of particle systems is presented in Yang et al. (2006).

Despite the outstanding issues as highlighted above, the particle mechanics approach for modeling mechanical behavior of the rock matrix is gaining momentum rapidly in rock engineering and rock mechanics/rock physics fields. This is because the approach is still, so far, the only tool that helps the researchers and practising engineers to improve their understanding of the behavior of rocks near the microscopic scales – so that some important issues in rock mechanics and rock physics, such as EDZ, fracture initiation and growth, micro-seismic events (such as acoustic emission) and damage evolution in the close near-field, which cannot be properly simulated using equivalent continuum methods or DEM approaches using block systems, can be approximated by the particle mechanics approach. The important thing is to remain aware of the qualitative nature of the approach when evaluating the quantitative data it yields regarding the objectives of the research or application and the confidence level that is required for such numerical modeling work.

References

Amadei, B. (ed.), Proc. of the 3rd Int. Conf. of Discontinuous Deformation (ICADD-III), Vail, Colorado, 1999.

Anandarajah, A., Discrete element method for simulating behavior of cohesive soils. Journal of Geotechnical Engineering, ASCE, 1994;120(9):1993.

Bagi, K. and Bojtár, I., Different microstructural strain tensors for granular assemblies. In: Bicanic, N. (ed.), Proc. ICADD-4 Discontinuous Deformation Analysis, pp. 261–270, 2001.

Bangash, T. and Munjiza, A., Experimental validation of a computationally efficient beam for combined finite–discrete element modeling of structures in distress. Computational Mechanics, 2003;30 (5/6):366–373.

Bathurst, R. J. and Rothenburg, L., Micromechanical aspects of isotropic granular assemblies with linear contact interactions. Transaction of the ASME, Journal of Applied Mechanics, 1988a;55(1):17–23.

Bathurst, R. J. and Rothenburg, L., Note on a random isotropic granular material with negative Poisson's ratio. International Journal of Engineering Science, 1988b;26(4):373–383.

Bathurst, R. J. and Rothenburg, L., Observations on stress–force–fabric relationships in idealized granular materials. Mechanics of Materials, 1990;9(1):65–80.

Bathurst, R. J. and Rothenburg, L., Investigation of micromechanical features of idealized granular assemblies using DEM. Engineering Computations, 1992;9:199–210.

Bathurst, R. J. and Rothenburg, L., Investigation of plane elliptical particle assemblies under stress rotations,. In: Siriwardane, H.J. and Zaman, M.M. (eds), Computer methods and advances in geomechanics, pp. 1325–1330. Balkema, Rotterdam, 1994.

Belytschko, T., An overview of semidiscretization and time integration procedures. In: Belytschko, T. and Hughes, T. J. R. (eds), Computational methods for transient analysis, Computational Methods in Mechanics Series, Vol. 1. pp. 1–65. North Holland, New York, NY, 1983.

Belytschko, T., Liu, W. K. and Moran, B., Nonlinear finite elements for continua and structures. Wiley, New York, NY, 2000.

Biarez, J. and Gourvés, R. (eds), Proc. of the Int. Conf. Micromechanics of Granular Media, Clermont-Ferrand, December 4–8, 1989. Balkema, Rotterdam, 1989.

Bićanić, N. (ed.), Proc. of the 4th Int. Conf. on Analysis of Discontinuous Deformation (ICADD-4), June 6–8, 2001, University of Glasgow, Scotland, UK, 2001. ISBN 0-85261-735-6.

Blair, S. C. and Cook, N. G. W., Statistical model of rock fracture. In: Tillerson and Wawersik (eds), Rock Mechanics, pp. 729–735. Balkema, Rotterdam, 1992.

Cambou, B., Chaze, M. and Dedecker, F., Change of scale in granular materials. European Journal of Mechanics A – Solids, 2000;19:999–1014.

Campbell, C. S. and Brennen, C. E. O., Computer simulation of shear flows of granular material. In: Jenkins, J. T. and Satake, M. (eds), Mechanics of granular materials: New models and constitutive relations, pp. 313–326, 1983.

Cerrolaza, M., Sulem, J. and El Bied, A., A Cosserat non-linear finite element analysis software for blocky structures. International Journal of Advances in Engineering Software, 1999;30:69–83.

Chang, C. S. and Ma, L., A micromechanical-based micropolar theory for deformation of granular solids. International Journal of Solids and Structures, 1991;28(1):67–86.

Chang, C. S. and Ma, L., Elastic material constants for isotropic granular solids with particle rotation. International Journal of Solids and Structures, 1992;29(8):1001–1018.

Cosserat, E. and Cosserat, F., Théorie des Corps Deformable. Hermann, Paris, France, 1909.

Cundall, P. A., BALL – a program to model granular media using the distinct element method. Dames and Moore Advanced Technology Group, London, 1978.

Cundall, P. A. and Hart, R. D., Numerical modeling of discontinua. Engineering Computation, 1992;9(2):101–113.

Cundall, P. A. and Strack, O. D. L., A discrete numerical model for granular assemblies. Géotechnique, 1979a;29(1):47–65.

Cundall, P. A. and Strack, O. D. L., The distinct element method as a tool for research. Report to NSF concerning grant ENG76-20711, Department of Civil and Mineral Engineering, University of Minnesota, Minneapolis, MN, 1979b.

Cundall, P. A. and Strack, O. D. L., The development of constitutive laws for soil using the distinct element method. In: Wittke, W. (ed.), Numerical methods in geomechanics, pp. 289–298. Balkema, Rotterdam, 1979c.

Cundall, P. A. and Strack, O. D. L., The distinct element method as a tool for research in Granular media. Report to NSF concerning grant ENG 76-20711, Part II, Department of Civil Engineering, University of Minnesota, Minneapolis, MN, 1979d.

Cundall, P. A. and Strack, O. D. L., Modeling of microscopic mechanisms in granular material. Proc. US–Japan Sem. New Models Const. Rel. Mech. Gran. Mat., Ithaca, 1982. Elsevier, Amsterdam, pp. 137–149.

Cundall, P. A. and Strack, O. D. L., Modeling of microscopic mechanisms in granular material. In: Jenkins, J. T. and Satake, M. (eds), Mechanics of granular materials: New models and constitutive relations, pp. 137–149, 1983.

Cundall, P. A., Konietzky, H. and Potyondy, D. O., PFC – ein neues Werzeug für numerische Modellierungen. Bautechnik, 1996b;73(8):492–498.

Cundall, P. A., Potyondy, D. O. and Lee, C. A., Micromechanic-based models for fracture and breakout around the mine-by tunnel. Proc. of Excavation Disturbed Zone Workshop on Designing the Excavation Disturbed Zone for a Nuclear Repository in Hard Rock, Winnipeg, pp. 113–122, 1996a.

Dai, C., Mohlhaus, Meek, H. J., and M. Duncan Fama, Modelling of blocky rock masses using the Cosserat method. International Journal of Rock Mechanics and Mining Sciences and Geomechanics Abstracts, 1996;33(4):425–432.

Dawson, P. E. M. and Cundall, P. A., Cosserat plasticity for modeling layered rock. In: Myer, L. R., Cook, N. W. G., Goodman, R. E. and Tsang, C.-F. (eds), Fractured and jointed rock masses, pp. 267–274. Balkema, Rotterdam, 1995.

Donzé, F. V., Bouchez, J. and Magnier, S. A., Modeling fractures in rock blasting. International Journal of Rock Mechanics and Mining Sciences, 1997;34(8):1153–1163.

Durand, A. F., Vargas, E. A. and Vaz, L. E., Application of numerical limit analysis (NLA) to stability problems of rock and soil masses. International Journal of Rock Mechanics and Mining Sciences, 2006;43(3):408–425.

Eringen, A. C., Microcontinuum field theories: Foundations and solids. Springer-Verlag, New York, NY, 1999.

Fox, P. J., Edil, T. B. and Malkus, D. S., Discrete element model for compression of peat. In: Siriwardane, H. J. and Zaman, M. M. (eds), Proc. 8th Int. Conf. on Computer Methods and Advances in Geomechanics, pp. 815–819. Balkema, Rotterdam, 1994.

Ghaboussi, J. and Barbosa, R., Three-dimensional discrete element method for granular materials. International Journal for Numerical and Analytical Methods in Geomechanics, 1990;14(7):451–472.

Golub, G. H. and Van Loan, C. F., Matrix-computations, North Oxford Academic, Oxford, 1983.

Hatzor, Y. H. (ed.), Stability of rock structures. Proc. of the 5th Int. Conf. on Analysis of Discontinuous Deformation (ICADD-5), Israel, October 6–10, 2002. Balkema, Rotterdam, 2002.

Hawkins, G. W., Simulation of granular flow. In: Jenkins, J. T. and Satake, M. (eds), Mechanics of granular materials: New models and constitutive relations, pp. 305–312, 1983.

Huang, J. I. and Kim, K., Fracture process zone development during hydraulic fracturing. International Journal of Rock Mechanics and Mining Sciences and Geomechanics Abstracts, 1993;30(7):1295–1298.

Issa, J. A. and Nelson, R. B., Numerical analysis of micromechanical behaviour of granular materials. Engineering Computations, 1992;9(2):211–223.

Itasca Consulting Group, Inc., PFC^{2D} (particle flow code in 2 dimensions), Version 1.1, Minneapolis, MN, 1995a.

Itasca Consulting Group, Inc., PFC^{3D} (particle flow code in 3 dimensions), Version 1.0, Minneapolis, MN, 1995b.

Itasca Consulting Group Inc., PFC^{2D} 2.00 particle flow code in two dimensions, Minneapolis, MN, 1998.

Iwashita, K., Tarumi, Y., Casaverde, L., Uemura, D., Meguro, K. and Hakuno, M., Granular assembly simulation for ground collapse. In: J. T. Jenkins and Satake, M. (eds), Micromechanics of Granular Materials, pp. 125–132. Elsevier Science, Amsterdam, 1988.

Jenkins, J. T. and Satake, M., Mechanics of granular materials: New models and constitutive relations. Elsevier Science, Amsterdam, 1983.

Jensen, R. P., Edil, T. B., Bosscher, P. J., Plesha, M. E. and Kahla, N. B., Effect of particle shape on interface behavior of DEM-simulated granular materials. International Journal of Geomechanics, 2001a;1(1):1–19.

Jensen, R. P., Plesha, M. E., Edil, T. B., Bosscher, P. J., and Kahla, N. B., DEM simulation of particle damage in granular media – structure interfaces. International Journal of Geomechanics, 2001b;1(1):21–39.

Kawai, T. and Takeuchi, N., New discrete models and their application to mechanics of granular materials, In: Jenkins, J. T. and Satake, M. (eds), Mechanics of granular materials: New models and constitutive relations, pp. 151–172, 1983.

Ke, T.-C. and Bray, J. D., Modeling of particulate media using discontinuous deformation analysis. Journal of Engineering Mechanics, ASCE, 1995;121(11):1234–1243.

Kim, K. and Yao, C., The influence of constitutive behaviour on the fracture process zone and stress field evolution during hydraulic fracturing. In: Nelson, P. P. and Laubach, S. E. (eds), Rock mechanics: Models and measurements – challenges from industry. pp. 193–200. Balkema, Rotterdam, 1994.

Kishino, Y. and Thornton, C., Discrete element approaches. In: Oda, M. and Iwashita, K. (eds), Introduction to mechanics of granular materials, pp. 147–221. Balkema, Rotterdam, 1999.

Kiyama, H., Fujimura, H., Nishimura, T. and Tanimoto, C., Distinct element analysis of the Fenner–Pacher type characteristic vurve for tunneling. In: Wittke, W. (ed.), Proc. of the 7th Cong. of ISRM, Aachen, Germany, Vol. 1, pp. 769–772, 1991.

Komodromos, P. I. and Williams, J. R., On the simulation of deformable bodies using combined discrete and finite element methods. Geotechnical Special Publication, 2002a; No. 117: 138–144.

Komodromos, P. I. and Williams, J. R., Utilization of Java and database technology in the development of a combined discrete and finite element multibody dynamics simulator. Geotechnical Special Publication, 2002b; No. 117: 118–124.

Koyama, T. and Jing, L., PFC modelling for EDZ effects in rocks for the DECOVALEX-THMC project, presentation at project workshop in Ottawa, Canada, October 3–4, 2005.

Kruyt, N. P., Statics and kinematics of discrete Cosserat-type granular materials. International Journal of Solids and Structures, 2003;40(3):511–534.

Kruyt, N. P. and Rothenburg, L., Micromechanical definition of the strain tensor for granular materials. Journal of Applied Mechanics (Transactions of the ASME), 1996;63(3):706–711.

Kruyt, N. P. and Rothenburg, L., Micromechanical bounds for the elastic moduli of granular materials. International Journal of Solids and Structures, 2002;39(2):311–324.

Kuraoka, S. and Bosscher, P. J., Parallelization of the distinct element method. In: Yan, J.-X. (ed.), Proc. 9th Int. Conf. on Computer Methods and Advances in Geomechanics, Wuhan, China, pp. 501–506. Balkema, Rotterdam, 1997.

Lee, K. W., Ryu, C. H., Synn, J. H. and Park, C., Rock fragmentation with plasma blasting method. In: Lee, H. K., Yang, H. S. and Chung, S. K. (eds), Environmental and safety concerns in underground construction, pp. 147–152. Balkema, Rotterdam, 1997.

Li, C., Wang, C. Y. and Sheng, J. (eds), Proc. of the First Int. Conf. on Analysis of Discontinuous Deformation (ICADD-1), National Central University, Chungli, Taiwan, 1995.

Lin, X. and Ng, T., Numerical modelling of granular soil using random arrays of three-dimensional elastic ellipsoids. In: Siriwardane, H. J. and Zaman, M. M. (eds), Proc. 8th Int. Conf. on Computer Methods and Advances in Geomechanics, pp. 605–610. Balkema, Rotterdam, 1994.

Lin, X. and Ng, T.-T., A three-dimensional discrete element model using arrays of ellipsoids. Geotechnique, 1997;47(2):319–329.

Lorig, L., Board, M. P., Potyondy, D. O. and Coetzee, M. J., Numerical modeling of caving using continuum and micro-mechanical models. Proc. 3rd Canadian Conf. on Computer Application in the Mineral Industry, Montreal, pp. 416–425, 1995a.

Lorig, L., Gibson, W., Alvial, J. and Cuevas, J., Gravity flow simulation with the particle flow code (PFC). ISRM News Journal, 1995b;3(1):18–24.

Maeda, Y., Maruoka, Y., Makino, H. and Nomura, H., Squeeze molding simulation using distinct element method considering green sand properties. Journal of Material Processing Technology, 2003;135(2–3):172–178.

Martin, C. L. and Bouvard, D., Study of the cold compaction of composite powders by the discrete element method. Acta Materialia, 2003;51(2):373–386.

Meegoda, J. N., Micro-mechanics and microscopic modeling in geotechnical engineering. Current status and future. In: Yan, J.-X. (ed.), Proc. 9th Int. Conf. on Computer Methods and Advances in Geomechanics, Wuhan, China, pp. 197–206. Balkema, Rotterdam, 1997.

Mindlin, R. D. and Deresiewicz, H., Elastic spheres in contact under varying oblique forces. Journal of Applied Mechanics, 1953;20(3):327–344.

Mirghasemi, A. A., Rothenburg, L. and Matyas, E. L., Effect of confining pressure on angle of internal friction of simulated granular materials. In: Siriwardane, H. J., and Zaman, M. M. (eds), Proc. 8th Int. Conf. on Computer Methods and Advances in Geomechanics, pp. 629–633. Balkema, Rotterdam, 1994.

Miyata, M., Nakagawa, M. and Mustoe, G. G. W., Design considerations of rubble rock foundations based on a discrete superquadric particle simulation method. In: Topping, B.H.V. (ed.), Proc. of 5th Int. Conf. on Engineering Computational Technology, pp. 213–218. Civil-Comp Press, Edinburgh, 2000.

Mohammadi, S., Discontinuum mechanics: Using finite and discrete elements. WIT Press, Southampton, 2003.

Mohammadi, S., Owen, D. R. J. and Peric, D., Combined finite/discrete element algorithm for delamination analysis of composite. Finite Elements in Analysis and Design, 1998;28(4):321–336.

Morris, J. P., Rubin, M. B., Blair, S. C., Glenn, L. A. and Heuze, F. E., Simulations of underground structures subjected to dynamic loading using the distinct element method. Engineering Computations, 2004;21(2/3/4):384–408.

Munjiza, A. and Andrews, K. R. F., Penalty function method for combined finite–discrete element systems comprising large number of separate bodies. International Journal for Numerical Methods in Engineering, 2000;49(11):1377–1396.

Munjiza, A., Owen, D. R. J. and Bicanic, N., Combined finite–discrete element method in transient dynamics of fracturing solids. Engineering Computations, 1995;12(2):145–174.

Munjiza, A., Andrews, K. R. F. and White, J. K. Combined single and smeared crack model in combined finite-discrete element analysis. International Journal for Numerical Methods in Engineering, 1996a;44(1):41–57.

Munjiza, A., Latham, J. P. and Andrews, K. R. F., Challenges of a coupled combined finite–discrete element approach to explosive induced rock fragmentation, FRAGBLAST. International Journal for Blasting and Fragmentation, 1996b;3(3):237–250.

Munjiza, A., Latham, J. P. and John, N. W. M., 3D dynamics of discrete element systems comprising irregular discrete elements – integration solution for finite rotations in 3D. International Journal for Numerical Methods in Engineering, 2003; 56(1):35–55.

Munjiza, A., Bangash, T. and John, N. W. M., The combined finite–discrete element method for structural failure and collapse. Engineering Fracture Mechanics, 2004;71(4/6):469–483.

Mustoe, G. G. W. and DePoorter, G., A numerical model for the mechanical behavior of particulate media containing non-circular shaped particles. Powders and Grains, 93(1993):421–427.

Mustoe, G. G. W. and Miyata, M., Material flow analysis of noncircular-shaped granular media using discrete element methods. Journal of Engineering Mechanics, 2001;127(10):1017–26.

Mustoe, G. G. W., Henriksen, M. and Huttelmaier, H. P. (eds), Proc. of the 1st US Conference on Discrete Element Methods. Colorado School of Mines Press (CSM Press), Golden, CO, 1989.

Mustoe, G. G. W., Miyata, M. and Nakagawa, M., Discrete element methods for mechanical analysis of systems of general shaped bodies. In: Topping, B. H. V. (ed.), Proc. of 5th Int. Conf. on Computational Structure Technology, pp. 219–224. Civil-Comp Press, Edinburgh, 2000.

Ng, T. Numerical simulation of granular soil under monotonic and cyclic loading: A perticulate mechanics approach. Ph.D. Thesis, Department of Civil Engineering, Rensselaer Polytechnic Institute, Troy, New York, 1989.

Ng, T-T., Numerical simulations of granular soil using elliptical particles. Computers and Geotechnics, 1994;16(2):153–69.

Ng, T. and Lin, X., Numerical simulations of naturally deposited granular soil with ellipsoidal elements. In: Williams, J. R. and Mustoe, G. G. W. (eds), Proc. 2nd Int. Conf. on Discrete Element Methods (DEM), MIT, March 18–19, AFORS, NSF (Geomechanical, Geotechnical and Environmental Engineering Division) and Department of Civil and Environmental Engineering, MIT, Cambridge, MA, USA, pp. 557–567, 1993.

O'Connor, R. M., Torczynski, J. R., Preece, D. S., Klosek, J. T. and Williams, J. R., Discrete element modeling of sand production. International Journal of Rock Mechanics and Mining Sciences, 1997;34(3/4):373.

Oda, M. and Iwashita, K. (eds), Mechanics of granular materials – an introduction. Balkema, Rotterdam, 1999.

Oda, M. and Iwashita, K., Study on couple stress and shear band development in granular media based on numerical simulation analyses. International Journal of Engineering Science, 2000;38(15):1713–1740.

Oda, K., Shigematsu, T. and Onishi, N., A new numerical method for analyzing liquid–solid flows – it's application to analyzing behavior of solid particles dumped into water. In: Williams, J. R. and Mustoe, G. G. W. (eds), Proc. 2nd Int. Conf. on Discrete Element Methods (DEM), MIT, March 18–19, AFORS, NSF (Geomechanical, Geotechnical and Environmental Engineering Division) and Department of Civil and Environmental Engineering, MIT, Cambridge, MA, USA, pp. 177–187, 1993.

Oelfke, S. M., Mustoe, G. G. W. and Kripakov, N. P., Application of the discrete element method toward simulation of ground control problems in underground mines. Proc. of the 8th Int. Conf. on Computer Methods and Advances in Geomechanics, Vol. 3, p. 1865, 1995.

Ohnishi, Y. (ed.), Proc. of the 2nd Int. Conf. of Discontinuous Deformation (ICADD-II), Kyoto, 1997.

Oner, M., Analysis of fabric changes during cyclic loading of granular soils. Proc. 8th World Conf. on Earthquake Engr., pp. 55–62, 1984.

O'Sullivan, C. and Bray, J. D., A comparative evaluation of two approaches to discrete element modeling of particulate media. In: Bicanic, N. (ed.), Proc. of the Fourth Int. Conf. on Discontinuous Deformation, University of Glasgow, Scotland, UK, pp. 97–110, 2001.

O'Sullivan, C. and Bray, J. D., Selecting a suitable time step for discrete element simulations that use the central difference time integration scheme. Engineering Computations, 2004;21(2/3/4):278–303.

Owen, D. R. J. and Feng, Y. T., Parallelised finite/discrete element simulation of multi-fracturing solids and discrete systems. Engineering Computations, 2001;18(3/4):557–576.

Owen, D. R. J., Feng, Y. T., Cottrel, M. G. and Yu, J., Discrete/finite element modeling of industrial applications with multi-fracturing and particulate phenomena. Geotechnical Special Publication, 2002;No. 117: 11–16.

Papamichos, E., Vardoulakis, I. and Muhlhaus, H. B., Buckling of layered elastic media: A Cosserat-continuum approach and its validation. International Journal for Numerical and Analytical Methods in Geomechanics, 1990;14(7):473–498.

Potyondy, D. O. and Cundall, P. A., Modeling notch-formation mechanisms in the URL mine-by test tunnel using bonded assemblies of circular particles. Proc. 3rd North American Rock Mechanics Symp. NARM'98, 1998.

Potyondy, D. O. and Cundall, P. A., A bonded-particle model for rock. International Journal of Rock Mechanics and Mining Sciences, 2004;41(8):1329–1364.

Potyondy, D. O. and Fairhurst, C. E., The value of numerical modeling in understanding the complete load/deformation behavior of cohesive-frictional materials. Non-destructive and automated testing for soil and rock Properties, ASTM STP 1350, American Society for Testing and Materials, 1998.

Potyondy, D. O., Cundall, P. A. and Lee, C. A., Modeling rock using bonded assemblies of circular particles. Proc. 2nd North American Rock Mechanics Symposium, Montreal, pp. 1937–1944. Balkema, Rotterdam, 1996.

Preece, D. S., Rock motion simulation and prediction of porosity distribution for a two-level retort. Oil Shale Symposium Proceedings, Colorado School of Mines Press, Golden, USA, pp. 62–67, 1990.

Preece, D. S., A numerical study of bench blast row delay timing and its influence on percent-cast. In: Siriwardane, H. J. and Zaman, M. M. (eds), Computer methods and advances in geomechanics, pp. 863–870. Balkema, Rotterdam, 1994.

Preece, D. S. and Knudsen, S. D., Computer modeling of gas flow and gas loading of rock in a bench blasting environment. In: Tilerson, J. R. and Wawersik, W. R. (eds), Rock mechanics, pp. 295–303. Balkema, Rotterdam, 1992.

Preece, D. S. and Scovira, D. S., Environmentally motivated tracking of geologic layer movement during bench blasting using discrete element method. In: Nelson, P. P. and Laubach, S. E. (eds), Rock mechanics, pp. 615–622. Balkema, Rotterdam, 1994.

Preece, D. S., Burchell, S. L. and Scovira, D. S., Coupled explosive gas flow and rock motion modeling with comparison to bench blast field data. In: Rossmanith, H.-P. (ed.), Rock Fragmentation by Blasting, pp. 239–245. Balkema, Rotterdam, 1993.

Preece, D. S., Jensen, R. A. and Chung, S. H., Development and application of a 3-D rock blast computer modeling capability using discrete elements – DMCBLAST__3D. Proceedings of the Annual Conference on Explosives and Blasting Technique, Vol. 1, pp. 11–18, 2001.

Ratnaweera, P. and Meegoda, N. J., Microscopic modeling of drained shear strength and stress-strain behavior of saturated granular soils, In: Williams, J. R. and Mustoe, G. G. W. (eds), Proc. 2nd Int. Conf. on Discrete Element Methods (DEM), MIT, March 18–19, AFORS, NSF (Geomechanical, Geotechnical and Environmental Engineering Division) and Department of Civil and Environmental Engineering, MIT, Cambridge, MA, USA, pp. 535–546, 1993.

Rothenburg, L., Micromechanics of idealised granular materials. Ph.D. Thesis, Department of Civil Engineering, Carleton University, Ottawa, Ontario, Canada, 1980.

Rothenburg, L. and Bathurst, R. J., Analytical study of induced anisotropy in idealized granular materials. Géotechnique, 1989;39(4):601–614.

Rothenburg, L. and Bathurst, R. J., Numerical simulation of idealize granular assemblies with plane elliptical particles. Computers and Geotechnics, 1991;11(4):315–329.

Rothenburg, L. and Bathurst, R. J., Micromechanical features of granular assemblies with planar elliptical particles. Géotechnique, 1992;42(1):79–95.

Sack, R. L., Matrix structural analysis. Waveland Press, Illinois, 1989.

Sadd, M. H., Tai, Q. and Shukla, A., Contact law effects on wave propagation in particulate materials using distinct element modeling. International Journal of Non-Linear Mechanics, 1993;28:251–265.

Salami, M. R. and Banks, D. (eds), Discontinuous deformation analysis (DDA) and simulations of discontinuous media. TSI Press, Albuquerque, NM, 1996.

Saltzer, S. D., Numerical modeling of crustal scale faulting using the distinct element method. In: Williams, J. R. and Mustoe, G. G. W. (eds), Proc. 2nd Int. Conf. on Discrete Element Methods (DEM), MIT, March 18–19, AFORS, NSF (Geomechanical, Geotechnical and Environmental Engineering Division) and Department of Civil and Environmental Engineering, MIT, Cambridge, MA, pp. 511–522, 1993.

Satake, M., A discrete mechanical approach to granular materials. International Journal of Engineering Science, 1992;30(10):1525–1533.

Satake, M. and Jenkins, J. T. (eds), Micromechanics of granular materials. Elsevier Science Amsterdam, 1988.

Sawada, S. and Pradhan, T. B. S., Analysis of anisotropy and particle shape by distinct element method. In: Siriwardane, H. J. and Zaman, M. M. (eds) Proc. 8th Int. Conf. on Computer Methods and Advances in Geomechanics. pp. 665–670. Balkema, Rotterdam, 1994.

Shi, G.-H., Discontinuous deformation analysis, a new numerical model for the statics and dynamics of block systems. Ph.D. Thesis, University of California, Berkeley, CA, 1988.

Shimizu, Y. Three-dimensional simulation using fixed coarse-grid thermal-fluid scheme and conduction heat transfer scheme in distinct element method. Powder Technology, 165(2006):140–152.

Shimizu, Y., Hart, R. D. and Cundall, P. A. (eds), Numerical modelling in via particle methods – 2004. Proc. of the 2nd Int. PFC Symp., Kyoto, Japan October 28–29, 2004. Balkema, Rotterdam, 2004.

Siriwardane, H. J. and Zaman, M. M. (eds), Proc. of the 8th Int. Conf. on Computer Methods and Advances in Geomechanics, West Virginia, USA, May 22–28, 1994. Balkema, Rotterdam, 1994.

Strack, O. D. L. and Cundall, P. A., The distinct element method as a tool for research in granular media. Research Report, Part I. Department of Civil Engineering, University of Minnesota, 1978.

Sulem, J. and Cerrolaza, M., Slope stability analysis in blocky rock. Proc. of the Int. Conf. on Geotechnical & Geological Engineering, Melbourne, Australia, 2000.

Taylor, L. M. and Preece, S. D., DMC – a rigid body motion code for determining the interaction of multiple spherical particles. Research Report SND-88-3482. Sandia National Laboratory, USA, 1989.

Taylor, L. M. and Preece, S. D., Simulation of blasting induced rock motion using spherical element models. Engineering Computations, 1992;9(2):243–252.

Thallak, S., Rothenburg, L. and Dusseault, M., Simulation of multiple hydraulic fracture in a discrete element system. In: Roegiers, J.-C., (ed.), Rock mechanics as a multidisciplinary science, pp. 271–280. Balkema, Rotterdam, 1991.

Thornton, C. (ed.), Proc. of the 2nd Int. Conf. Micromechanics of Granular Media, Birmingham, UK, July 12–16, 1993. Balkema, Rotterdam, 1993.

Thornton, C. and Barnes, D. J., Computer simulated deformation of compact granular assemblies. Acta Mechanica, 1986;64(1–2):45–61.

Thornton, C. and Zhang, L., Numerical simulations of the direct shear test. Chemical Engineering and Technology, 2003;26(2):153–156.

Thornton, C., Lian, G. and Adams, M. J., Modelling of liquid bridges between particles in DEM simulations of particle systems. In: Williams, J. R. and Mustoe, G. G. W. (eds), Proc. 2nd Int. Conf. on Discrete Element Methods (DEM), MIT, March 18–19, AFORS, NSF (Geomechanical,

Geotechnical and Environmental Engineering Division) and Department of Civil and Environmental Engineering, MIT, Cambridge, MA, USA, pp. 177–187, 1993.

Timoshenko, S. P. and Goodier, J. N., Theory of elasticity. McGraw-Hill, New York, 1970.

Ting, J. M. and Corkum, B. T., Strength behaviour of granular materials using discrete numerical modeling. Proc. 6th Int. Conf. Numerical Models in Geomechanics, Innsbruck, Austria, pp. 305–310, 1988.

Ting, J. M., Corkum, B. T. and Greco, C., Application of distinct element method in geotechnical engineering. Proc. 2nd Int. Symp. on Numerical Models in Geomech., Ghent, Belgium, 1986.

Ting, J. M., Maechum, L. and Rowell, J., Effect of particle shape on the strength and deformation mechanisms of ellipse-shaped granular assemblages. Engineering Computations, 1995;12(2):99–108.

Trent, B. C. and Margolin, L. G., A numerical laboratory for granular solids. Engineering Computations, 1992;9(2):191–197.

Walton, O. R. Explicit particle dynamics model for granular materials. Proc. 4th Int Conf. on Numerical Methods in Geomech., Edmonton, Canada, pp. 1261–1268, 1982.

Walton, O. R. Particle-dynamics calculations of shear flow. In: Jenkins, J. T. and Satake, M. (eds), Mechanics of granular materials: New models and constitutive relations, pp. 327–328, 1983.

Williams, J. R. and Mustoe, G. G. W. (eds), Proceedings 2nd Int. Conf. on Discrete Element Methods (DEM), MIT, March 18–19, 1993, AFORS, NSF, MIT, Cambridge, MA, USA, 1993.

Williams, J. R. and O'Connor, R., Linear complexity intersection algorithm for discrete element simulation of arbitrary geometries. Engineering Computations, 1995;2(2):185–201.

Williams, J. R. and Pentland, A. P., Superquadrics and modal dynamics for discrete elements in interactive design. Engineering Computations, 1992;9(2):115–127.

Williams, J. R., Rege, N., O'Connor, R. and Amaratunga, K., Dynamic wave propagation in particulate materials with different particle shapes using a discrete element method. Proceedings of Engineering Mechanics, 1995, Vol. 1, pp. 493–496. ASCE, New York, USA.

Yamane, K., Nakagawa, M., Altobelli, S. A., Tanaka, T. and Tsuji, Y., Steady particulate flows in a horizontal rotating cylinder. Physics of Fluids, 1998;10(6):1419–27.

Yang, B., Jiao, Y. and Lei, S., A study on the effects of microparameters on macroproperties for specimens created by bonded particles. International Journal for Computer-Aided Engineering and Software, 2006;23(6):607–631.

Zhai, E. D., Miyajima, M. and Kitaura, M., DEM simulation of rise of excess pore water pressure of saturated sands under vertical ground motion. In: Yuan (ed.), Computer methods and advances in geomechanics, Vol. 1, pp. 535–539. Balkema, Rotterdam, 1997.

Zhang, R. and Li, J., Simulation on mechanical behavior of cohesive soil by distinct element method. Journal of Terramechanics, 2006;43(3):303–316.

Zhang, R., Mustoe, G. G. W. and Nelson, K. R., Simulation of hydraulic phenomena using the discrete element method. In: Williams, J. R. and Mustoe, G. G. W. (eds), Proc. of the 2nd Int. Conf. on Discrete Element Methods (DEM), MIT, AFORS, NSF, March 18–19, pp. 189–200, 1993.

Zienkiewicz, O. and Taylor, R., The finite element method, volume 1, the basis, 5th Edition. Butterworth-Heinemann, Oxford, 2000.

Part Four
Application Studies

12 CASE STUDIES OF DISCRETE ELEMENT METHOD APPLICATIONS IN GEOLOGY, GEOPHYSICS AND ROCK ENGINEERING

12.1 Introduction

Today, the advances in discrete modeling make it tempting to attempt to construct models that are more and more complicated by including a large number of geological structures and rock types with different properties. This leads to a dilemma if the model result is to have practical use in rock engineering (Hart, 1990): on one hand, the simulation of a large number of fractures and rock types with different properties satisfies the conceptualization of the rock mass structures and characteristics; on the other hand, the problem might become too complicated so that the engineering understanding of the model results becomes less effective. In addition, for engineering projects as more details are added, the computing power requirements may quickly exceed what is available.

Sometimes the lack of geological and rock mechanics data is the major stumbling block to acceptance of modeling in rock engineering. Starfield and Cundall (1988) discussed modeling guidelines in the context of ecological modeling and converted them into a preliminary set of guidelines for modeling rock mechanics problems. These guidelines, as applied to rock engineering problems, are:

- Make it clear why a model is built and what question it is trying to resolve;
- Use a model at an early stage of a project to generate both data and understanding;
- Identify the important mechanics of the problem;
- Make simple numerical experiments at the start to bracket the true case;
- Run more complex models to explore neglected aspects of geology and mechanics;
- Perform a sensitivity analysis for the most important parameters; and
- Perform design runs using more complex models and derive simple equations based on the mechanisms revealed by the models.

Starfield and Cundall (1988) pointed out that modeling should be used in rock mechanics with the same care and curiosity that applies to good laboratory work. Visualizing and anticipating solutions before running a model is an important discipline and great care should be taken at the beginning to gain the right concept for the model geometry, the boundary conditions and material models and properties. In the following sections we present first a number of case studies in geology, geophysics and rock engineering to illustrate the essence of the above guidelines. The case studies are selected to demonstrate the large variability in applications of DEM to rock engineering problems such as:

- large rock structures (natural rock slopes, faulting and seismic instability);
- stability of underground openings (tunnels, rock chambers, rock reinforcement);
- mine structures (open pits, underground mines, pillars, subsidence, blast loading);
- radioactive waste disposal (block tests, shaft sinking, far and near-field stabilities, THM coupling); and
- groundwater flow (geothermal energy, dam foundation, pump testing).

At the end of this chapter, a comprehensive generic study on deriving hydro-mechanical properties using the UDEC code is presented to demonstrate the power of DEM approach in fundamental research on coupled hydro-mechanical behaviors of fractured rocks.

12.2 Geologic Structures and Processes

12.2.1 Crustal Deformation

Numerical modeling using the distinct element method (DEM) and discontinuous deformation analysis (DDA) have been performed to give additional insights into the mechanisms of crustal deformation in various geological and tectonic terrains. The purpose has been to obtain a better understanding of how different geological formations and the imposed loading and boundary conditions can influence rock structures. The results of such modeling are often used in combination with data from geological mapping and geophysical investigations to interpret geological structures and measured stresses and displacements. DEM methods are useful for studying these types of problems as they allow large deformations of complex structures with linear or non-linear material properties.

The DEM methods can be applied to the analysis of physical experiments of a geological process and, at the same time, be applied to study real problems at the natural scales (Cundall and Hart, 1992a,b). They have been applied to faulting and crustal deformation by Saltzer (1993), Dupin et al. (1993), Schelle and Grünthal (1994), Holmberg et al. (1997, 2004), Morgan (1999), Saltzer and Pollard (1999), Burbidge and Braun (2002), Pascal and Gabrielsen (2001) and Pascal (2002). Rosengren and Stephansson (1993) used the UDEC code to simulate the effect of glaciation and deglaciation on the states of stress and displacements of large fracture zones for a 2.5 km long and 2.0 km deep crustal section at Finnsjön, Central Sweden. The DEM has also been used to simulate tectonic processes, such as folding, faulting, thrusting and fracturing. One particular category of fault-related folds, so-called forced folds, have been studied by Finch et al. (2003). This study is used as a case study example below to illustrate the applicability of DEM to crust tectonics and related issues.

12.2.1.1 Case study – folding above rigid faulting

Fault-propagation folds are common structures where sedimentary rocks overlay basement rocks in sedimentary basins and other areas. They are important structures as they can form oil and gas traps and have recently been recognized for their importance in generating potential seismic hazards. Many of the fault-propagation folds form as upward widening zones of distributed deformation (monoclines) above discrete faults at depth (Finch et al., 2003). Figure 12.1a illustrates some natural examples of basement

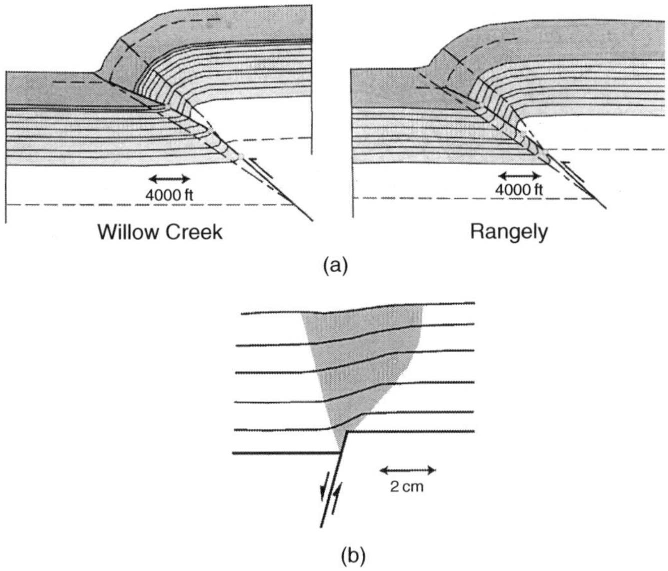

Fig. 12.1 Folding above a rigid basement fault block (Finch et al., 2003): (a) natural examples of anticline structures from Willow Creek and Rangely in the USA; (b) physical model in clay of extensional folding above block faulting in the basement.

faulting and related folding mechanisms in a sedimentary overburden and Fig. 12.1b shows a physical clay model of a soft overburden formed by extensional rigid basement fault displacement.

In an effort to better understand the development of fault-propagation folding above rigid basement fault blocks, Finch et al. (2003) developed a 2D discrete element model of sedimentary cover deformation in response to basement thrust faulting. The model was used to study the influence of the dip of the basement fault and the strength of the sedimentary overburden on the geometry of the folds generated by block movements in the basement and the rate of fault propagation. The discrete element model used circular particles connected by breakable elastic springs. Particles are bounded until the separation between them reaches a defined breaking strain and the bond breaks. Finch et al. (2003) applied their discrete element model to the problem of fault-propagation folding in a setting where a low-strength sedimentary sequence of rocks is deformed in response to faulting movement of a rigid basement block. It was assumed that in the basement there is a discrete, pre-existing fault of 45° of dip angle that reactivates and displaces the basement into two blocks. The displacement of the upward moving (hanging wall) block was given a fixed displacement of 0.000025 units at each iteration step. The scale of 40 units, together with the initial configuration of the model with the moving fault in the basement, is presented in Fig. 12.2.

The basement is assumed to be rigid and the particles lying on the basement are fixed, as are particles at the vertical boundaries of the model. The overburden is 40 units thick and contains 10 marker layers of particles, all with the same properties, subjected to gravity. Each model was run for a total of more than one million time steps on a supercomputer, which took about 90 h of CPU time to complete. As slip occurs on the basement fault the overlaying sequence of strata is deformed.

Figure 12.3 shows the sequential development of a model with the moving fault after 300 000, 660 000, 858 000 and 1 056 000 time steps. The corresponding units' displacements and the successive development of the fold structure are shown in the figure. A upward-widening monocline structure forms above the fault in the basement. As deformation proceeds, the limbs of the fold steepen and the fold

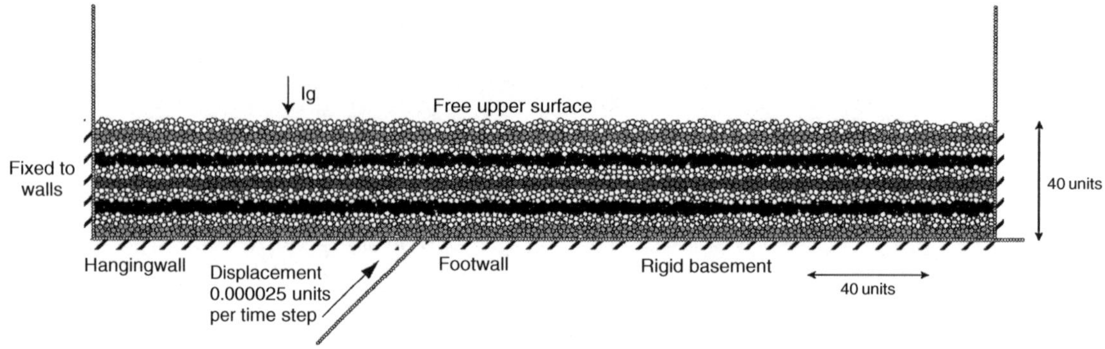

Fig. 12.2 Discrete element model for simulation of fault-propagation folding above a rigid basement with a pre-existing basement fault dipping 45° (Finch et al., 2003).

becomes tighter and the individual beds show thinning in the anticline regions and thickening in the syncline regions.

By increasing or decreasing the dip angle of the fault, the zone of deformation remains broadly triangular. The effect of increasing the dip of the fault up to 80° is illustrated in Fig. 12.3, which increases the amount of extensional faulting within the hinge zone of the monocline. Also notice the thickening of the strata in the synclinal hinge zone of the fold. This thickening is even more pronounced as the overburden strata are made softer.

One of the advantages of applying DEM to study tectonic processes and mechanisms is its ability to record the development of structures with large deformations, e.g., to observe the difference in model behavior between strong and weak sedimentary covers.

Finch et al. (2003) made a comparison of velocity vectors resulting from different cover strengths for identical horizontal unit displacements. They used the total displacement between 132 000 and 165 000 time steps corresponding to 0.825 units, which corresponds to about 3% total horizontal displacement, to calculate the different velocity fields above the basement fold for models with very weak overburden (Fig. 12.4a) and very strong overburden (Fig. 12.4b). For both models, there is a rigid body translation above the hanging wall of the fault and almost zero displacement above the footwall block. Between the zone of translation and the zone of zero displacement, there is a zone of transition that dips antithetically at about 25° to the main fault. All these features are reported to be similar to the velocity vectors found in kinematic models. Figure 12.4c depicts the velocity vectors for the model with the strong overburden between 660 000 and 693 000 time steps. The zone of rigid translation at the hanging wall and the zero velocity above the footwall zone remain, while the transition zone has almost disappeared. Notice also the disturbance in the velocity field close to the tip of the fault.

In summary, the DEM model proved to be of great help in studying tectonic processes and related geological structures. Its capacity to simulate large deformations and non-linear material properties are of particular importance. In this case study, the DEM model has been applied to study a problem containing both brittle (faulting) and ductile (folding) deformation mechanisms. The model of fault-propagation folding above the rigid basement presented reproduces many of the features observed in analogue physical modeling and reported from field studies.

12.2.2 Earthquakes and Seismic Hazards

The discrete element method provides a useful tool for understanding a wide spectrum of dynamic problems in fractured rock masses, including potential sources of seismic excitation, e.g.,

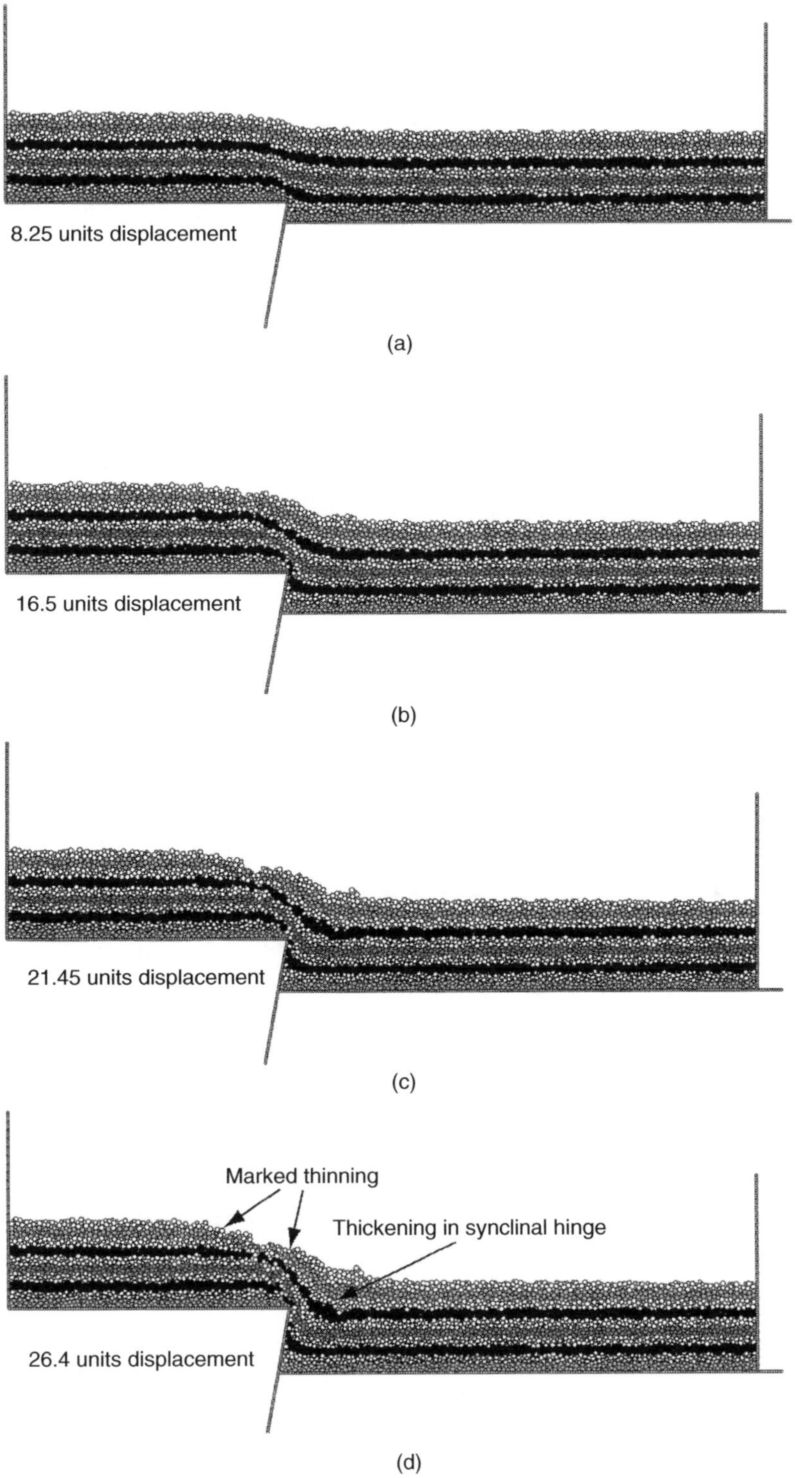

Fig. 12.3 Sequential evolution of fault-propagation folding in a sequence of soft layers overlaying a stiff basement with a 80° dipping thrust fault. The unit displacements are shown after (a) 3 330 000, (b) 660 000, (c) 858 000 and (d) 1 056 000 time steps and the total displacement is 26.4 units (Finch et al., 2003).

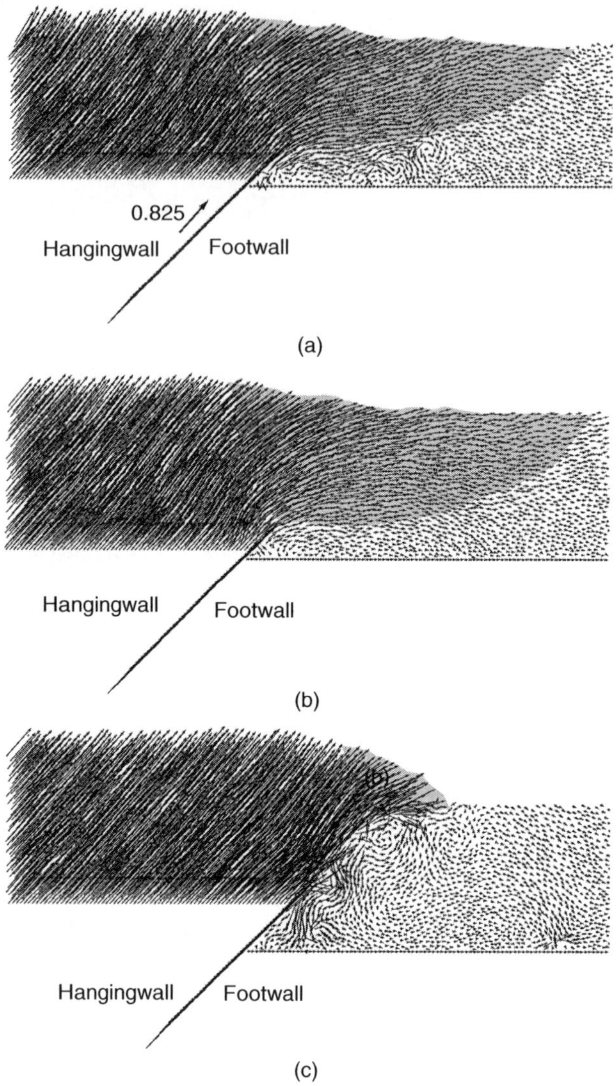

Fig. 12.4 Velocity vectors for the central part of fault-propagation folding model with a fault dip of 45°. (a) Model with very weak sedimentary overburden; (b) model with strong overburden with velocity vectors calculated between time steps of 132 000 and 165 000; (c) model with strong overburden with velocity vectors calculated between time steps of 660 000 and 693 000. Velocity vectors are lengthened by a factor of 10 and the transition zone between hanging wall and footwall velocity vectors are indicated in gray in (a) and (b) (Finch et al., 2003).

earthquakes and rock bursts, as well as the effects of dynamic loading as in blasting. The DEM method has the capability to simulate and visualize displacements and stresses with real time, model large displacements and post-peak behavior of faults and block systems and stress wave propagation. Consequently, large displacements and rotations,general non-linear constitutive behavior for both the rock mass and the faults and joints, and time domain calculations can be accommodated in a straightforward manner. The pioneering work in the dynamic analysis of discontinuous rock masses using DEM took place in the mid-to-late-1980s by Bardet and Scott (1985), Lemos (1987), Lemos et al.

(1985, 1987), Cundall and Lemos (1988) and in the 1990s and later by Lorig and Hobbs (1990) and Lemos and Cundall (1999).

Numerical modeling with DEM offers the possibility to study fault failure and earthquake mechanism of complex geometries, including the interactive effects of intersecting faults. Lemos et al. (1985) applied the UDEC code to study the effects of fault geometry on the likelihood of progressive failure of two intersecting *en echelon* arrays of faults subjected to a 2D stress field where the major principal compressive stress acts at an angle of 12 degrees to one of the fault traces. Fault segments were allowed to first shear elastically and then progressively fail. As a result, tensile failure occurs on the short joints that belong to one of the faults and stress reorientation after failure is prominent. The results illustrate the possibility of progressive failure of an *en echelon* fault structure, even when the far-field major principal stress is oriented nearly parallel with the fault trace.

Numerical simulations of fault instabilities with the continuously yielding model of fractures were presented by Lemos (1987), Cundall and Lemos (1988) and Lemos and Cundall (1999). Lemos (1987) simulated dynamic fault slip in a numerical experiment where a block of elastic medium containing a planar fault was subjected to a given history of far-field stresses. Absorbing boundaries were used to simulate the effect of the surrounding rock mass. Unstable slip was observed in the model when the applied rock stiffness was less than the descending slope of the given stress–displacement curve of the fault. At the center of the fault, a sudden jump in slip and simultaneous stress drop were observed. When applying higher rock stiffness, slip occurred in a stable manner throughout the descending part of the stress–displacement curve for the fault. The change in the elastic strain energy in the rock blocks gives the total energy available in the system. Part of this energy is dissipated by friction along the fault. The rate of energy release was recorded for the case where the fault was given a small initial displacement. Cundall and Lemos (1988) demonstrated how there exists time periods during loading when more energy is being dissipated on the fault than when released from the elastic rock mass. Although the energy estimates obtained with the two-dimensional model are not meaningful in absolute terms of physics, they gave conceptual understanding for comparisons between different model geometries and material properties. The numerical representation of dynamic systems in geomechanics also requires the use of boundary conditions that permit adequate energy radiation. Lemos and Cundall (1999) used absorbing boundaries to meet this requirement in their earthquake analysis of concrete gravity dams.

Lorig and Hobbs (1990) demonstrated the ability to model frictional sliding and stick–slip behavior of faults with the DEM for problems where the coefficients of friction of the faults depend on the instantaneous velocity of sliding, as well as on other phenomenological state variables. An extensive verification study was conducted by comparing numerical results for a system of loaded rock masses with analytical results using a number of different constitutive laws. The results of the study show the importance of stiffness of the surrounding rock mass in understanding slip instabilities of single faults. To resolve this problem in modeling work, Lorig and Hobbs (1990) recommend the coupling of a discrete element formulation to represent effects of the elasticity of the near field with a boundary element formulation. Stephansson and Shen (1991) and Ma and Brady (1992) have presented UDEC analyses of the dynamic performance of an underground excavation in jointed rock under single and repeated seismic loading, respectively.

To illustrate the capacity of modeling earthquake phenomena, a numerical example of simulating the dynamic process of the famous Tangshan earthquake in China, on 28 July 1976, is presented below. In this case study, Lagrangian discontinuous deformation analysis (LDDA) method is applied. The method is based on DDA with a Lagrangian description of motion and a domain decomposition algorithm (Cai et al., 2000). In LDDA, the contact detection algorithm in DDA is used to determine the contact between the elastic blocks, and a domain decomposition method is used to seek the solution of the system. The method can deal with dynamic problems with complex geological structures and material properties, and one does not need to define slide lines or surfaces in advance, as required in the finite element methods.

12.2.2.1 *Case study – simulation of the M = 7.8 Tangshan earthquake of 28 July 1976*

The Tangshan earthquake in July 1976 was one of the most destructive seismic hazards of the 20th century. The earthquake caused tremendous life losses and infrastructure damages to the city. The 250 km × 240 km geometrical model used to simulate the Tangshan earthquake was divided into 25 elastic sub-regions, see Fig. 12.5.

The model consists of 10 909 nodal points and the 60 km long earthquake fault was simulated with 25 nodal points, with the epicenter located at point B in Fig. 12.5. The fault is also divided into three segments by dashed lines with the aim of simulating the different frictional strengths of the model. A plane stress and homogeneous elastic model is adopted with a Young's modulus of 97.2 GPa and a Poisson's ratio of 0.25. The faults (thick lines in Fig. 12.5) in the area are simulated by the contact interface with Coulomb's friction law with an frictional coefficient of 0.6 when it is in a 'sticking' state. A frictional coefficient of 0.5 is used for the other faults and this will drop to 0.2 once the shear stress on a fault is greater than the frictional resistances of the fault, which means that an earthquake occurs (Cai et al., 2000). The applied stresses on the boundaries of the model are 100 MPa on the left and right boundaries and 50 MPa on the upper and lower boundaries, respectively. The initial time step is set to 0.5 s and a large damping value is adopted to generate a quasi-static initial stress field in the model. A reduction of the frictional coefficient from 0.6 to 0.2 at the middle segment of the Tangshan earthquake fault causes the system state change from static to dynamic. When sliding is detected in the model, the time steps will be adjusted automatically to simulate the unstable process. The total dynamic solution of the LDDA modeling lasted 30 s.

Figure 12.6a–d shows the calculated displacement contours generated by the LDDA model of the Tangshan earthquake at different times. From the sequence of plots of the displacement contours, one can follow how the rupture started at the middle section of the Tangshan fault (Fig. 12.6a) and moved to the southwest segment first (Fig. 12.6b) and later to the northeast section (Fig. 12.6c). The simulated displacements vary between 0.5 and 1.0 m in magnitude.

Figure 12.7a shows the displacement vector field simulated with LDDA after the Tangshan earthquake. The displacement vectors are largest close to the Tangshan earthquake fault and diminish away from the fault. Further, the vectors on the concave side of the fault are greater than those on the convex side, which might be due to the slightly curved shape of the fault. Also notice how the displacement

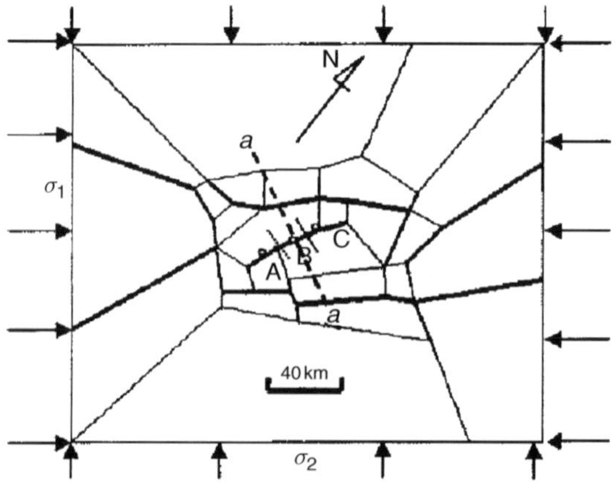

Fig. 12.5 Model geometry and boundary conditions for simulating the Tangshan earthquake. Solid lines are faults and the letter B indicates the location of the epicenter (Cai et al., 2000).

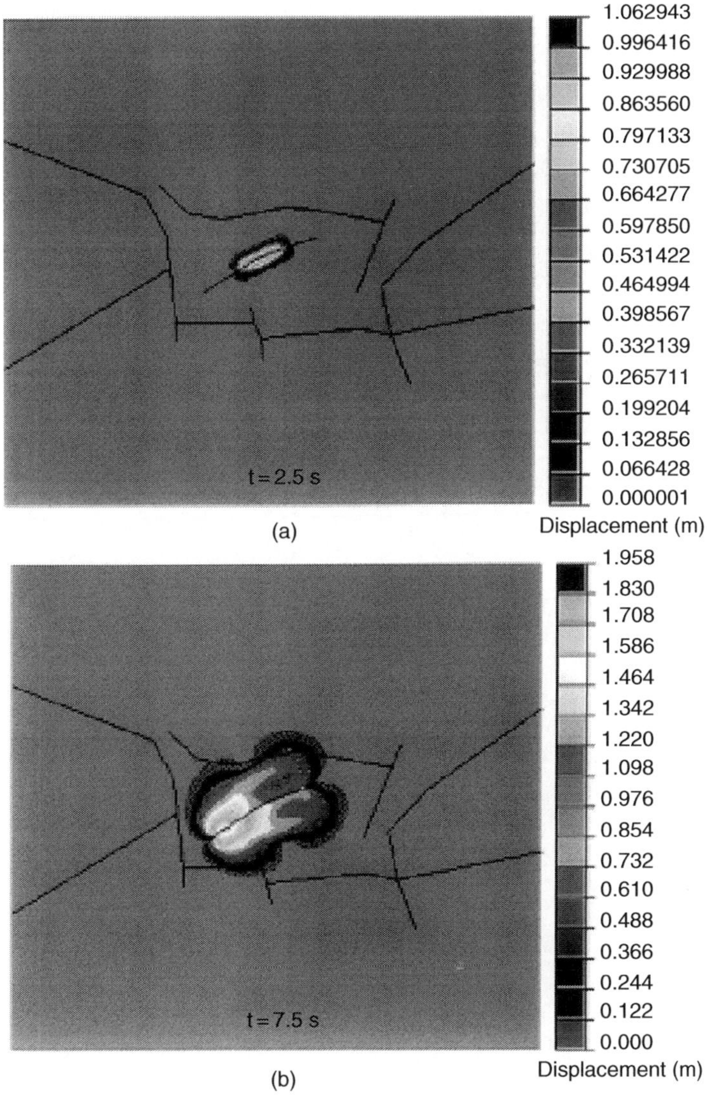

Fig. 12.6 LDDA simulation of displacements in the duration of the Tangshan earthquake in 1976 (a)–(b) (Cai et al., 2000).

vectors are greatest at the center of the fault and attenuate towards the ends of the Tangshan earthquake fault. The pattern of displacement vector field agrees fairly well with observed data presented by the Earthquake Editorial Group of the 1976 Tangshan Earthquake, State Seismological Bureau (SSB), China, as shown in Fig. 12.7b. The values in the legend of Fig. 12.7b denote: 1, seismic fault from crustal observations; 2, seismic fault inverted from geodesy data; 3, horizontal displacement vector; and 4, displacement vector scale, respectively.

The results presented in this case study demonstrate that the discrete numerical approaches, LDDA method in particular, can be applied for modeling the process of earthquakes. The displacement, seismic source time function, shear stress drop and rupture velocity of an earthquake fault can be readily captured. The methods can handle inhomogeneous material distributions, complex geometry and seismic sources and boundary conditions. According to Cai et al. (2000), the method can also be used to solve contact and impact problems, besides earthquakes.

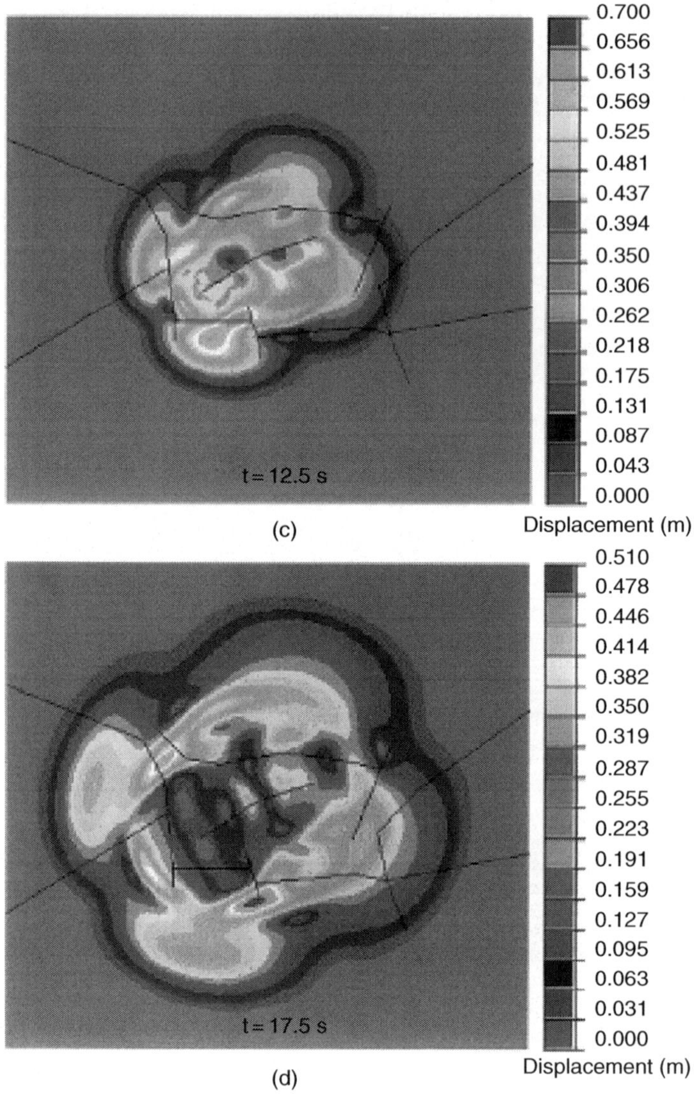

Fig. 12.6 LDDA simulation of displacements in the duration of the Tangshan earthquake in 1976 (c)–(d) (Cai et al., 2000).

12.2.3 Rock Stresses

Natural stresses in rock masses are of considerable importance for the stability analysis of almost all underground rock structures. Stress in rock is usually described within the context of continuum mechanics. Because of its definition, as a second-order Cartesian tensor acting at a point, rock stress is an enigmatic and fictitious quantity creating challenges in its characterization, measurement and application to rock engineering. Stresses in rock cannot be measured directly and can only be inferred by disturbing the rock mass, e.g., by drilling a hole or making a slot and recording the rock mass response to the imposed disturbance.

The available methods to measure rock stresses and how the results are analyzed and applied to rock engineering problems are presented by Amadei and Stephansson (1997). In general, rock stresses cannot be determined accurately due to the complex nature of the rock. If the rock mass quality is from good to

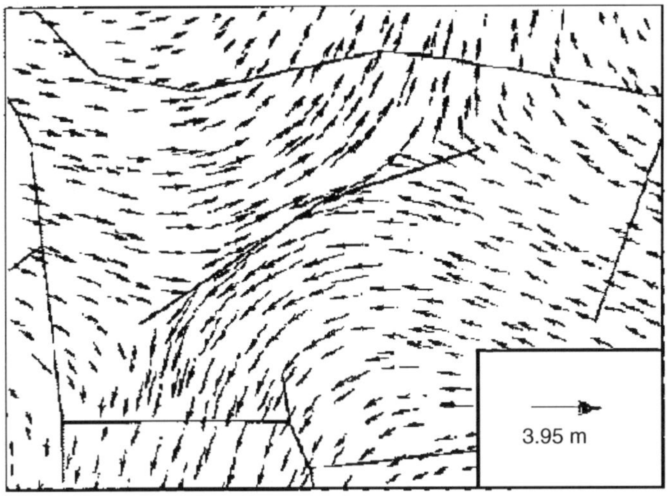

(a)

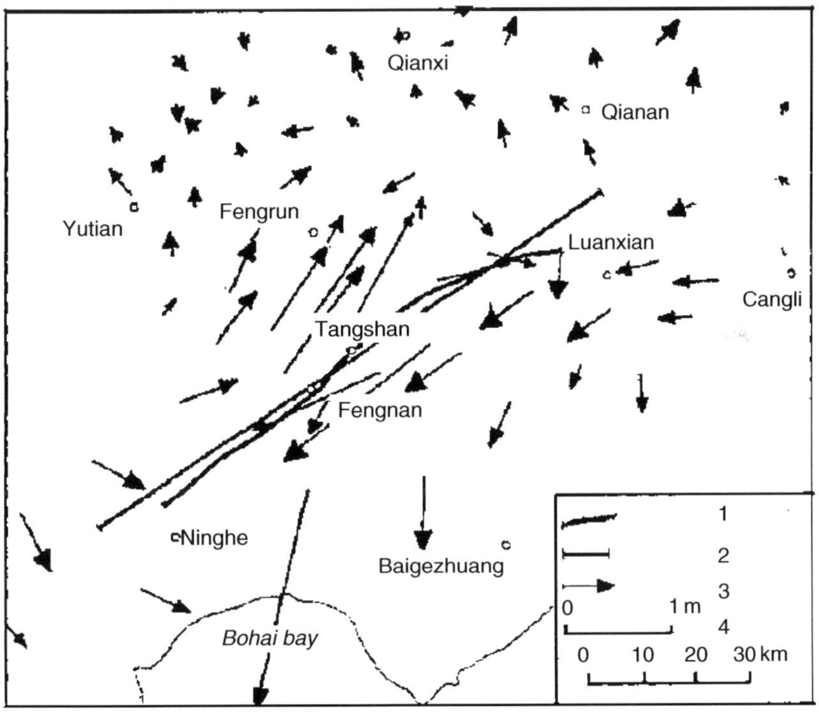

(b)

Fig. 12.7 Displacement vector fields after the Tangshan earthquake 1976. (a) Simulated displacement vectors by LDDA analysis. (b) Observed displacement vector field after the earthquake according to the State Seismological Bureau of China (Cai et al., 2000).

very good and of CHILE-type, i.e., continuous, homogeneous, isotropic and linearly elastic, and located between well-defined geological boundaries, the stresses can be determined with an error of ±10–20% for magnitudes and ±10–20 degrees for orientations. In poor-quality rocks, the measurement of rock stresses is difficult and the rate of success is low. With respect to the complex nature of rock materials, the results of numerical modeling of the state of stress for natural geological structures, and for discontinuous structures in particular, are helpful in the interpretation of magnitude and orientation of stresses.

In a recent paper, Hart (2003) presents several applications of numerical analysis to evaluate the influence of different factors, e.g., topography, excavation, loading history and geological structures, on the state of stress in rock. The method selected for generating the proper stress state in the model should attempt to approximate the type of geological processes that are believed to have occurred in the field. This procedure can involve several simulations to calibrate the numerical model with the measured stresses in the field as demonstrated by, e.g., McKinnon and Lorig (1999) and McKinnon (2001) for the El Teniente Mine in Chile.

In an attempt to study the irrecoverable movement that may be induced on existing fracture surfaces during cycles of regional tectonic activity, Brady et al. (1986) used DEM to analyze an assembly of rock blocks defined in a circular domain of radius 25 m and embedded in an infinite elastic continuum. After an initial hydrostatic and isotropic loading, the model is subjected to a stress ratio varying from one to four. Stress concentrations are introduced by the simulated tectonic activity wherever slipping fractures terminate at other intersecting fractures. These stresses are termed 'locked in' since they persist after the loading or tectonic activities have ceased. The results of the numerical analysis show that rock masses containing sets of non-persistent fractures may indeed be subject to locally varying field stresses. The contours of normal principal stresses plotted after a cycle of loading, unloading and reloading again clearly depict the complicated stress patterns at fracture intersections. The stresses that are locked in after applying the boundary loadings are consequences of the stress concentrations that develop at the fracture intersections in the model. The applied DEM models were able to show these phenomena for a fractured rock mass.

In the Provence coal basin in Southern France, high stresses and large stress ratios have been measured. To interpret these measurements, Homand et al. (1997) performed a large-scale DEM modeling using the UDEC code. Three different models were generated with a size of 27 km 22 km in 2D and depict the geological situation at a depth of 1100 m. The models have different boundary conditions and simulate orientations of major fault zones reflecting historical and current tectonics in the area. Results of calculations lead to a better explanation of the orientation of the principal stresses in the area, which are in accordance with the observation of a NE–SW stress orientation and existing geological information. *In situ* stress measurements were carried out in the area and the results of the DEM analyses are in global agreement with those measured.

Stephansson et al. (1991a) studied the displacements and state of stress of a faulted rock mass consisting of two intersecting faults and three blocks subjected to a far-field stress field, and Su and Stephansson (1999) explored the effect of a single fault on the orientation and magnitude of *in situ* stresses under the influence of a regional stress field using the two-dimensional distinct element method code UDEC. Hakami et al. (2006) applied 3DEC to depict a plausible *in situ* stress distribution for the two-candidate sites for radioactive waste disposal in Sweden. These applications demonstrate that with proper conceptualization and system/material characterizations, the DEM models can be applied to study evolutions of *in situ* stress fields under complex loading mechanisms and conditions.

12.2.3.1 Case study – state of stress at URL in Canada

The Underground Research Laboratory (URL) of the Atomic Energy of Canada (AECL) has been selected as a case study for rock stresses – as it represents one of the most comprehensive and best-documented sites for rock stress measurement and its interpretation. The work conducted at URL has the purpose of examining the concept of deep geological disposal of nuclear fuel and radioactive waste in plutonic rocks at depths ranging between 500 and 1000 m. A comprehensive summary of the

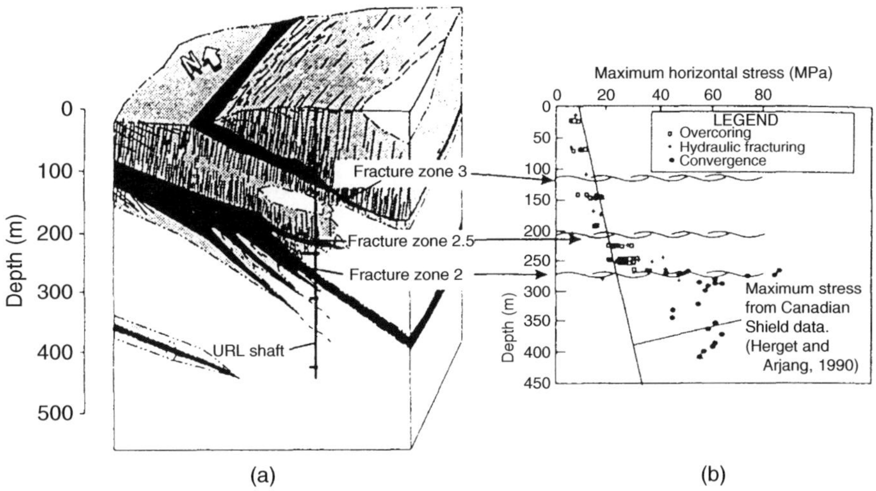

Fig. 12.8 Geological setting at the Underground Research Laboratory, Canada; (a) location of fracture zones and URL Shaft; (b) maximum horizontal stress versus depth recorded with different measuring techniques and stress anomalies at the location of fracture zones (Chandler and Martin, 1994).

geomechanics work from URL is presented by Martin and Simmons (1993). The recorded stresses in particular are presented in Amadei and Stephansson (1997). In an attempt to study the influence of reverse-dip displacement along low-dipping major faults at URL, Chandler and Martin (1994) constructed numerical models using both the finite difference method with the code FLAC and the discrete element method with the use of the UDEC code. Some of the results are presented in the following section.

The URL is located within the Lac du Bonnet granite batholith on the western edge of the Canadian Shield in the province of Manitoba, Canada. The batholith is 75 km × 25 km in surface area and is thought to reach a depth of about 10 km. Within the volume of rock immediately surrounding the URL, there exist three fracture zones of which fracture zone 2 is the most predominant one (Fig. 12.8a).

These fracture zones are typical thrust faults, indicating that one of the horizontal stresses was the major stress component and the vertical stress was the minor stress component at the time of the faulting event. Reverse-dip displacement of about 7 m has been observed along fracture zone 2. The maximum horizontal stress measured at URL can be grouped into three well-defined domains, as can be observed in Fig. 12.8b. Above fracture zone 2.5, horizontal stresses are similar to the mean horizontal stress measured elsewhere in the Canadian Shield at the same depth. At depths below fracture zone 2, the horizontal stress measured at URL is anomalously high compared to the mean state of stress in the Canadian Shield and other shield areas of the world. In the domain between fracture zones 2.5 and 2 the horizontal stress is intermediate in magnitude. Below fracture zone 2.5 the rock mass consists of an essentially massive, grey granite with few fractures except for the fracture zone 2 where the granite is more fractured. Also, the directions of the principal stresses vary from one stress domain to another, e.g., the direction of the maximum horizontal stress rotates by 90°, from the strike of the fault to its dip direction below fracture zone 2.

To determine if these observations could be replicated, Chandler and Martin (1994) simulated a 2-km long by 1.5-km deep section of the rock mass surrounding the URL and performed modeling with the finite difference code FLAC and the discrete element model UDEC assuming plane strain conditions. Fracture zones 2 and 3 were simulated by simple linear structures dipping at 25° and separated vertically by 180 m. The geometry of the model and the boundary conditions are shown in Fig. 12.9.

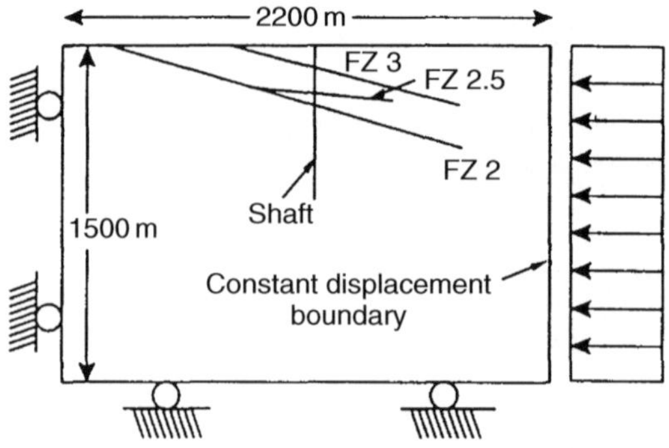

Fig. 12.9 Geometry and boundary conditions imposed in the UDEC and FLAC numerical model of the URL (Chandler and Martin, 1994).

To satisfy the condition of continuity with the surrounding rock mass, a constant displacement boundary is applied at the right-hand edge of the model. Then the horizontal distance between the vertical boundaries was decreased until stresses at the 420-m depth were in agreement with the average measured value of 55 MPa in the dip direction and 48 MPa in the strike direction. In the absence of the fracture zones, this assumption will cause a horizontal stress, which varies linearly with depth and is in agreement with the Poisson's effect.

Figure 12.10 compares the dip direction stress measured at URL with the horizontal stress calculated with UDEC. Also shown in the figure is the out-of-plane plane stress determined with UDEC. The dip direction stresses are the horizontal stresses calculated by the model and the strike direction stresses are

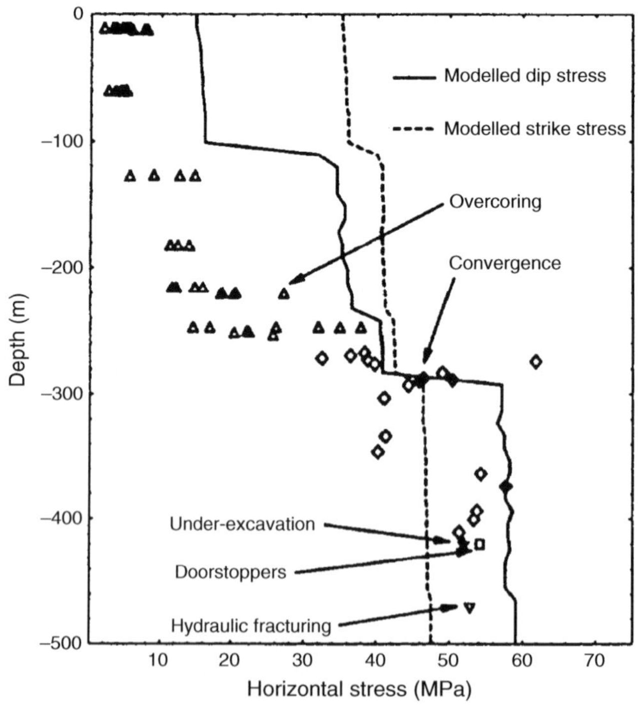

Fig. 12.10 Measured and calculated horizontal stresses using UDEC versus depth in the dip direction of fracture zone 2 of URL (Chandler and Martin, 1994).

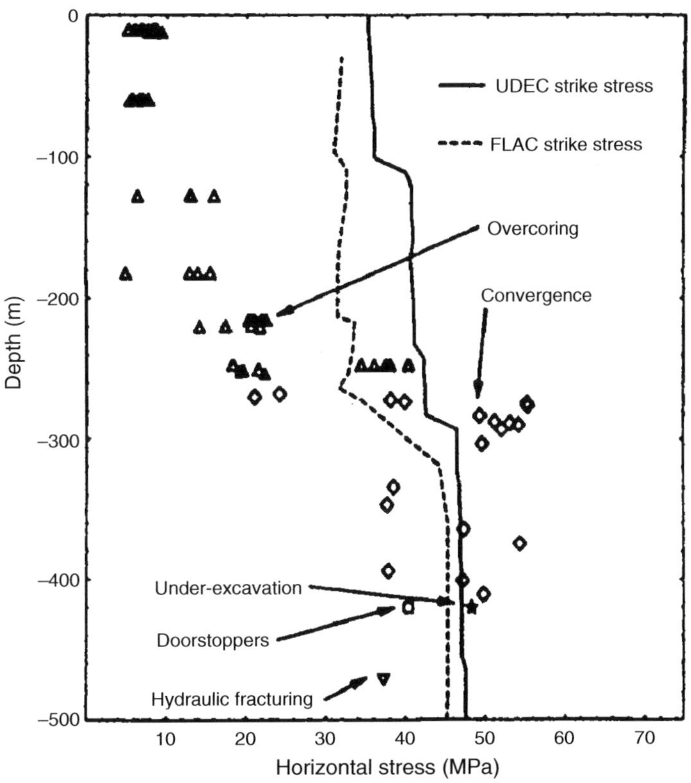

Fig. 12.11 Measured and calculated horizontal stress using both UDEC and FLAC versus depth in the strike direction of fracture zone 2 (out-of-plane stress) (Chandler and Martin, 1994).

the out-of-plane stresses that satisfy the condition of plane strain. Slips along the faults cause stepwise increases of stresses with depth and the direction of the maximum horizontal stress rotates from the dip direction below fracture zone 2 to the strike direction above the fracture zone. This is in full agreement with the results of the stress measurements in the field. Further, the modeling results show that the stresses between fracture zones remain essentially constant with depth.

A FLAC modeling gives similar results to the stresses calculated by the UDEC model except that the stress drop above fracture zone 2 is greater in the FLAC model. Both model approaches overestimate the horizontal stresses in the strike direction at shallower depth (Fig.12.11).

The fracture zones at URL are known to be disk shaped, while the models simulate them as planar features. The curvature of the fractures should lead to higher stresses occurring in the strike direction of the fracture zones at shallower depth. This feature cannot be simulated in the two-dimensional models.

The models presented by Chandler and Martin (1994) are geometrically very simple, but are able to demonstrate stress relief caused by fault slip and indicate the tendency of almost constant stress magnitudes in the domains between fracture zones. It also shows that the horizontal stresses below the deepest fracture zone do not increase significantly with depth.

The magnitude of *in situ* stresses at a potential site for deep disposal of radioactive wastes is one of the factors that determine the selection of the location and the depth of the repository tunnels and deposition vaults. The study by Chandler and Martin (1994) demonstrates that simple numerical simulations, of the type mentioned in the introduction to this chapter, provide insight into the variation of near-surface stresses with depth and also how stresses remain almost constant between fracture zones.

Hence, numerical modeling may also provide valuable indications about the states of stresses at greater depth where direct stress measurements are not available.

12.2.4 Instability of Natural Rock Slopes

The instability of natural rock slopes is often due to sliding. Slides can be classified according to the geometry of the sliding mass, e.g., wedge, block or slab. In terms of the type of motion and failure, they can displace as planar, translational, rotational and toppling structures. Sliding mechanisms can also be classified according to movement and rate of movement as creep, slip and stick–slip. Slides in hard rocks are governed almost entirely by movement along pre-existing fracture surfaces. Whether or not the slope will remain stable depends on the orientation of the sliding surfaces and the slope face, persistence of fractures and their roughness, strength and the water pressure acting within them. Slides in soft rocks can develop by generation of new sliding surfaces in intact rocks or by failure planes along pre-existing fractures. A special failure called toppling occurs when near-vertical slabs and columns of rock are undermined and the center of gravity moves outside the area of basal support (Call, 1992). The slabs can rotate either independently or under the action of gravity forces transmitted by blocks situated higher up the slope.

Most stability analysis of natural rock slopes is conducted by limiting equilibrium analyses (Hoek and Bray, 1977). However, the blocky structure of most natural rock slopes makes the DEM an attractive method for stability and rock slope remediation analysis (Vengeon et al., 1996). The application of DEM to toppling failure of natural slopes has been presented by Ishida et al. (1987) and Adachi et al. (1991) for natural slopes in Japan, by Esaki et al. (1999a,b) for the Belden Siphon rock slope remediation project in California and by Hsu and Nelson (1995) for rock slopes located in the Eagle Ford Shale, Texas, USA. The analyses are two-dimensional and were performed with the UDEC code. The same code has been applied to block modeling of undermined chalk cliffs along the Loire Valley in France by Homand-Etienne et al. (1990). Allison and Kimber (1998) used the UDEC code to examine rates and mechanisms of change in rock slopes and cliffs in the limestone formations of the Isle of Purbeck, central southern England. Rachez and Durville (1996) conducted numerical modeling of a bridge foundation on a fractured rock slope with the UDEC code. A fully dynamic, two-dimensional, stability analysis of the highly discontinuous rock slopes of Mount Masada in Israel using DDA is demonstrated by Hatzor et al. (2002) and Hatzor (2003). Ohnishi et al. (2006) discuss the effects of different types of damping in DDA applications to slope stability and rock fall problems in Japan.

12.2.4.1 Case study – Corvara Cliff in Italy

The stability condition of the calcareous cliff of Corvara in the Abruzzo Region of Italy is used to illustrate a two-dimensional DEM analysis of a natural rock slope (Lanaro et al., 1996). The cliff at Corvara extends for almost 4 km in the N–S direction, is about 500 m in width and is cut by a sub-vertical fault system that determines a stepped-line shape of the slope, with steep dipping faces and rock battlements. The rock mass consists of calcarenitic, micritic and bioclastic limestones with sub-horizontal bedding and breccias, overthrusted on top of a sandy clay formation with chalk layers. In addition to the fault structures, five sets of fractures with average spacing of 0.1–1.7 m intersect the cliff. Since the early 1900s, the village, located on the northern corner of the massif, was declared unstable and moved. From that time on, some limited toppling and rock falls have occurred in the vicinity of provincial and municipal roads 60 m below the old town. The predominant instability phenomena of the Corvara cliff are due to the presence of wedge-shaped blocks and tabular prisms defined by major fault zones and a network of fractures dipping against the slope face. Moreover, the clay formation at the toe of the cliff is subject to creep. Extensive field investigations and laboratory testing have been conducted.

For the numerical modeling of the Corvara cliff, both a discrete element model with deformable blocks using the UDEC code and a continuum model using the FLAC code were performed (Lanaro et al., 1996).

A vertical section, oriented E–W, that intersects the top of the cliff and the toe of the high face of the slope and encounters the boreholes drilled for field investigations was chosen for the model constructions (Fig. 12.12). The section comprises faults and traces of the most important sets of fractures that affect the rock mass stability. A plane strain analysis was performed where vertical sides of the model have zero horizontal displacement, the bottom has zero vertical displacement and the top surface is free. To represent *in situ* conditions, the model was subjected to specified boundary velocity constrains and gravity acceleration. For that purpose the model was given low deformability (linear elastic) and high strength (Mohr–Coulomb) for both rock materials and fractures. For the FLAC model a ubiquitous material model was used to represent the rock mass with a series of embedded weak zones. On one hand, the presence of well-defined, weak major faults justifies the distinct block assumption and use of the DEM and the UDEC code. On the other hand, the large number of fracture sets of small spacing justifies application of an equivalent continuum approach with a ubiquitous material model. In addition, the results of the discrete modeling are highly conditioned by the blocky structure, so that large blocks may become loose and produce rockslide, raveling and rock fall.

Figure 12.13a shows the block movements by shear along fractures and plastic yielding indicators from the UDEC model. The layer situated deeper down in the cliff displaces more slowly, mainly because of retention by the clay formation at the toe of the cliff. The vector of displacements and the contour of the horizontal displacement in the clay formation by the FLAC model are illustrated in Fig. 12.13b. At the contact between the stiff limestone in the cliff and the soft clay formation at the toe, high compressive stresses are generated in the clay formation. The vertical extension of the yielding zones (Fig. 12.14) and the contour of horizontal displacements in the clay formation (Fig. 12.13b) by the FLAC model indicate the formation of circular failure modes. Sliding along the sub-vertical faults is limited and the surface of the cliff is deformed because of differential displacements along the fractures. These observations are in agreement with the observed damage of the old houses in the town of Corvara and at the toe of the cliff.

The computed displacements from the continuous FLAC model are concentrated at the top of the cliff and almost linearly diminish with depth. Relatively large shear displacements of 4.0 and 4.6 cm were obtained along the fault at the top of the cliff and at the toe, respectively. Slip and tensile failure along ubiquitous fractures and extension of the plastic yield zone in the clay formation are illustrated in Fig. 12.14. The absence of a final stable state in the model is governed by the presence of unbalanced forces still acting on the system at the end of the computation.

It may be concluded that both the discrete UDEC model and the continuous FLAC model gave reasonable and compatible results and satisfactorily captured the failure modes of the Corvara cliff. Both computations indicate that the significant part, which controls the stability of the cliff, is the strength and deformability of the clay formation at the toe of the cliff.

The sensitivity of the modeling results for slight variations in material properties suggests that the slope is in a limiting equilibrium condition and that minor alteration of the geometry, material properties and groundwater condition can have significant effects on the stability. Sliding phenomena are localized at the top of the cliff and extend behind it for about 50 m. Housing located in this zone is invariably subjected to differential movements at the foundation level with severe damage.

One of the characteristics of natural rock slopes is their ability to remain intact at a stage of limited equilibrium. Minor changes in geometry, strength of intact rocks and fractures due to rainfall, snow, ice, frost heaving, groundwater fluctuation and weathering can cause instability and result in failure. Consequently, the modeling exercises have to be performed at the very limit of stability and therefore should include sensitivity analyses on the geometry and governing parameters. In addition, the modeler has to solve the problem of selecting relevant material properties of intact rocks and fractures for large rock masses, including the scale effects. For certain natural rock slopes in mountainous terrain, the loading from the virgin stress field has to be included as well. For the majority of cases considering gravity loading and consolidation may be sufficient.

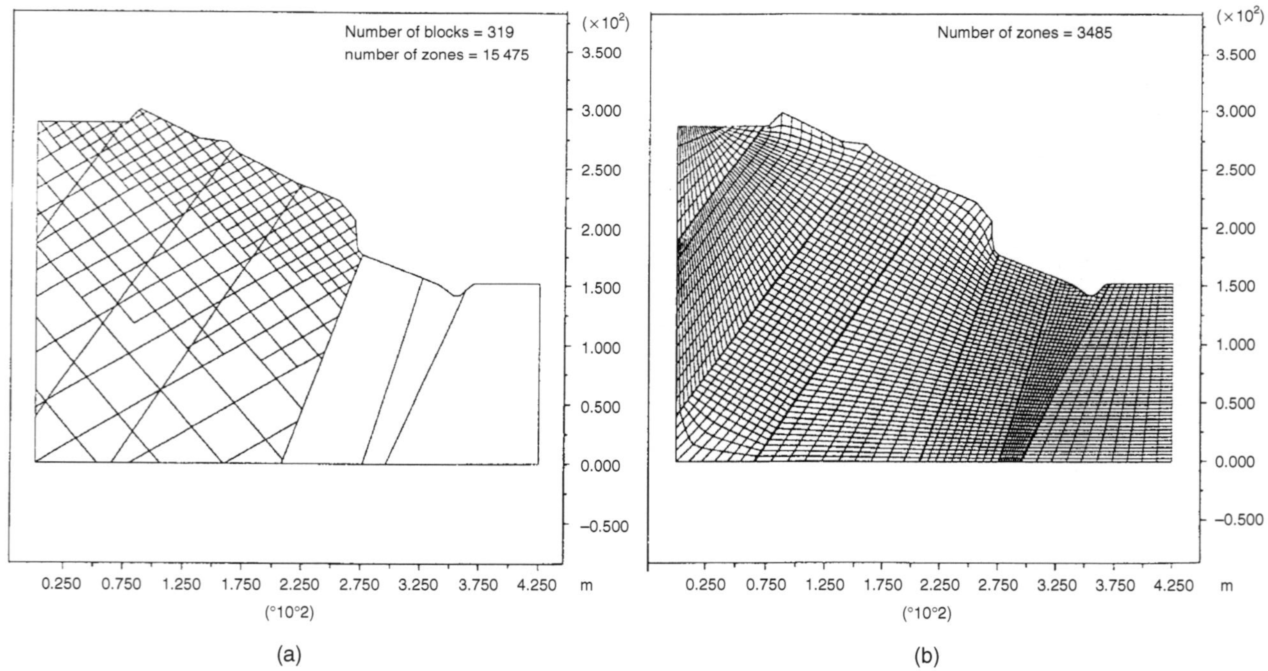

Fig. 12.12 Numerical model of Corvara Cliff, Italy; (a) discrete element model for UDEC analysis; (b) finite difference model for FLAC analysis (Lanaro et al., 1996).

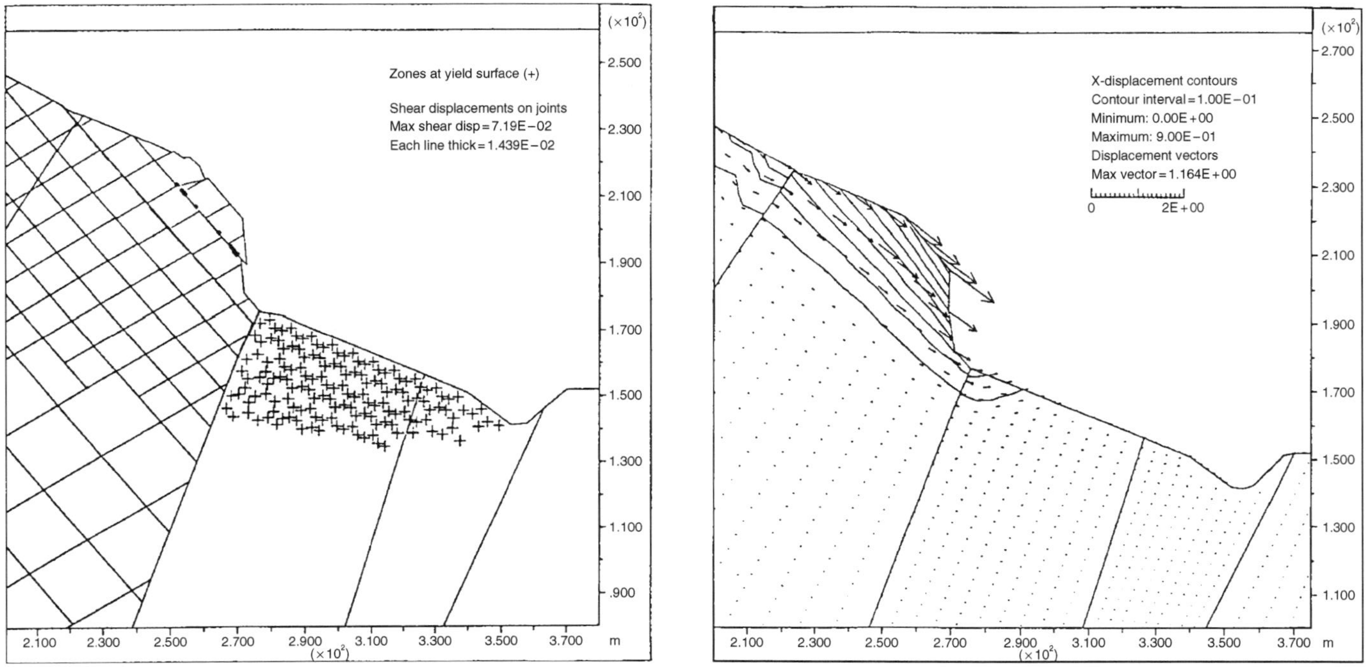

(a) (b)

Fig. 12.13 Displacement results of Corvara cliff, Italy: (a) block movement along fractures by UDEC model; (b) displacement vectors by FLAC model (Lanaro et al., 1996).

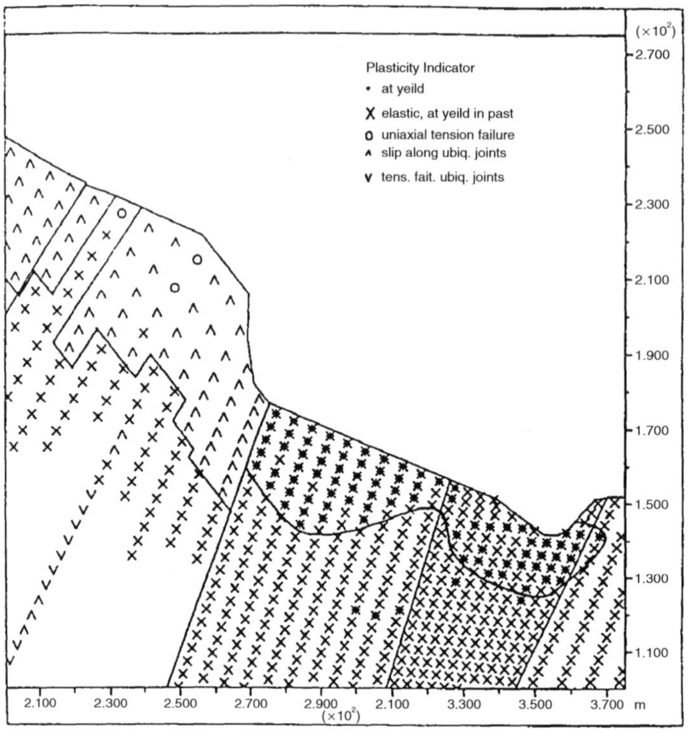

Fig. 12.14 Plasticity indicators in modeling the stability of Corvara cliff, Italy; (a) shear displacement of joints and yielding zones for the UDEC model; (b) plasticity indicator at yield for the continuous FLAC model (Lanarao et al., 1996).

12.3 Underground Civil Structures

Demands for underground spaces follow the growth of infrastructures of urban communities, particularly large cities in the world. The need for land for expansion, the environmental aspects and the price of surface properties in modern cities make underground spaces an attractive alternative to surface constructions. The design of underground civil structures involves determination of stresses and displacements of the rock mass surrounding the openings and for that purpose the DEM is well suited. Any design analysis of such excavations requires also that the effect of certain types of reinforcement be defined in a quantitative manner. The interactive nature of reinforcement mechanics requires that both the rock mass and supports or reinforcements should be represented adequately in the models. In this section some applications of the DEM for such problems are presented, for tunnels, rock chambers for civic works and for design of rock reinforcement.

12.3.1 Tunnels

Despite the large number of tunnels constructed for various civic applications, there is not so much published literature concerning DEM applications to tunnel design and performance so far. Karaca (1995) studied the effect of fractures on the stability of shallow tunnels in weak rocks and Kosugi et al. (1995) made a comparison of results from UDEC modeling and displacement measurements in boreholes surrounding tunnel excavations in Japan. Monsen et al. (1992) and Shen and Barton (1997) studied the excavation disturbed zone (EDZ) around tunnels in fractured rock masses with the UDEC-BB code, which is a special version of the UDEC code containing the Barton–Bandis model for fractures. The contribution by Monsen et al. (1992) deals with EDZ phenomenon observed in the site characterization and validation drift of the Stripa Mine, Sweden. In a series of papers, Souley and Homand (1993, 1996) and Souley et al. (1997) studied the influence of fracture constitutive laws on the stability of an experimental gallery excavated in slate in France. Mitarashi et al. (2006) applied a 3D distinct element method to study the effect of long face bolting for shallow tunnels.

12.3.1.1 Case study – centrifuge models and DEM simulations of tunnel face reinforcement by bolting

Tunnel face stability is of utmost importance for the safe excavation of tunnels. For tunneling in poor rock conditions, auxiliary excavation methods, such as face bolting and forepoling, can be used to stabilize the tunnel face and prevent raveling and collapse of the rock mass at the tunnel face. Face bolting implies that rock anchors or steel bars are installed parallel with the tunnel axis and at the face of the tunnel, while forepoling implies that bolts or anchors are installed parallel with the tunnel axis but only in a ca. 120° wide sector at the crown of the tunnel periphery.

In order to clarify the effect of the different reinforcement methods on the tunnel face stability, Kamata and Mashimo (2003) carried out centrifugal model tests with various arrangements and lengths of bolts, and the results were compared with results from UDEC modeling. The centrifuge was used to simulate the effect of gravity only.

Sand with a unit weight of 15.1 kN/m^3, water content of 6.5%, a cohesion of 4.6 kPa and a friction angle of 34.5° was used as the model material. It was poured into a container and compacted in layers with a fixed H/D ratio of 1.0 in all tests (Fig. 12.15). Once the tunnel was made the container was taken to the centrifuge and the centrifuge acceleration was applied. When the acceleration of the centrifuge reached the predetermined value an aluminum plate previously installed in the model was moved out to release the stress at the tunnel face and the relative stability of the face and its surroundings were observed and documented.

Figure 12.16 shows sketches of the failure states of two models with no reinforcement from the centrifuge tests at accelerations of 25g and 30g, respectively, where g stands for the Earth's gravitation acceleration. For both cases, a slip surface originated at the tunnel invert and progressed upwards towards

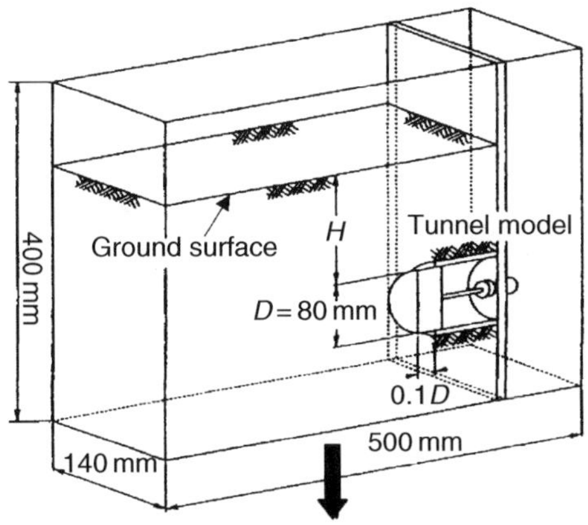

Fig. 12.15 Container for centrifuge models of tunnel stability. The container is inserted in a centrifuge to give the predetermined acceleration. The tunnel model is made of compacted sand and an 80 mm diameter tunnel is made with or without reinforcement. At the predetermined acceleration, an aluminum plate is pulled out to release the stress at the face (Kamata and Mashimo, 2003).

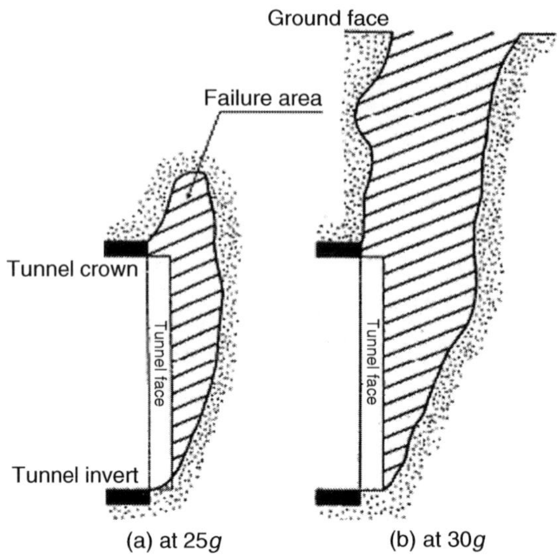

Fig. 12.16 Failure state of centrifuged tunnel models of sand: (a) with a centrifugal acceleration of 25*g*; (b) with a centrifugal acceleration of 30*g* (Kamata and Mashimo, 2003).

the crown of the tunnel and beyond was observed. At the acceleration of 25*g*, the failure area above the crown of the tunnel was about 0.4*D* (diameter of the tunnel). At the acceleration of 30*g*, the failure area reached all the way to the simulated ground surface.

Bolting for reinforcement of the rock at the face was simulated with bronze bolts of 1.0–1.2 mm in diameter and coated with sand for friction. Face bolting, vertical pre-reinforcement bolting and forepoling were simulated with different bolt patterns and lengths. Figure 12.17 shows the failure states of centrifuged models at 25*g* acceleration and reinforced with face bolting of the full tunnel section, lower half section and upper half section, respectively. The results indicate that installation of face bolting in the upper half of the tunnel face is more effective and safer than bolting in the lower half section. The fact that bolts longer than half of the tunnel diameter has a favorable effect on the face stability is another conclusion from the experiments.

Kamata and Mashimo (2003) performed DEM simulations of the centrifuged models with application of the two-dimensional UDEC code, version 3.0, which has the option to simulate rock bolts and cables

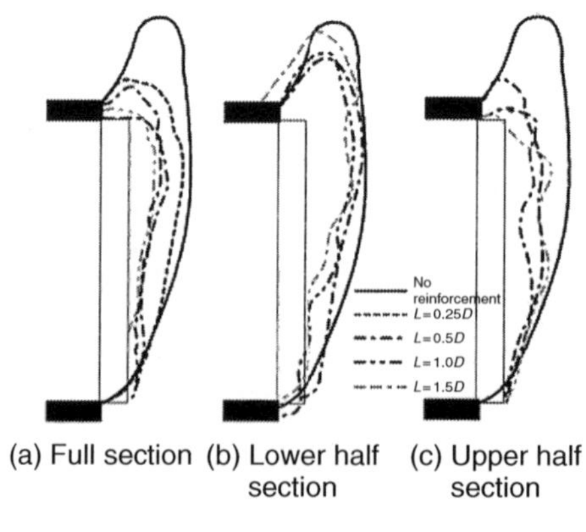

Fig. 12.17 Failure states of centrifuged models with different types of face bolting and bolt lengths (Kamata and Mashimo, 2003).

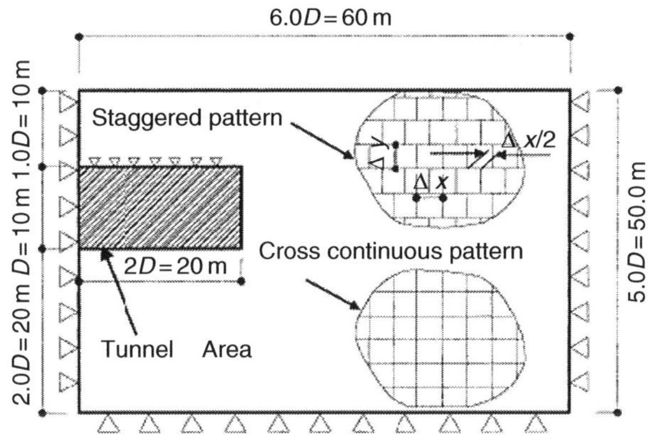

Fig. 12.18 Geometry and boundary condition of DEM model for simulation of centrifuged experiments of tunnel bolting (Kamata and Mashimo, 2003).

and their influence on the excavation stability. Since the stability of the tunnel face depends on the block size and the fracture properties, the authors decided to first study the influence of these parameters by simulating the case of no reinforcement from the centrifuged models. Once the DEM models without reinforcement were showing similar results, under the same conditions, the effects of bolting on the face stability could be modeled. The numerical model, with the dimensions, tunnel area and the two different fracture patterns, staggered and cross-continuous, is shown in Fig. 12.18.

For simulation of the displacement of the tunnel face for the case of no reinforcement, the block elasticity (E) and the normal and shear stiffness (k_n, k_s) of interfaces between the blocks were kept constant, while the cohesion (C), friction angle (φ) and block size (0.5×0.5 m and 0.25×0.25 m) were varied. Simulations were performed at an acceleration of $1g$. Figure 12.19a shows the block geometry for the case of a model with staggered fracture pattern and 0.5×0.5 m block size. Failure occurred and a slip surface was formed from the tunnel invert to the top of the ground surface. The failure area was defined by the generation of block separation within the model. According to Kamata and Mashimo (2003), these results are in good agreement with the experimental results and give confidence in selecting parameters

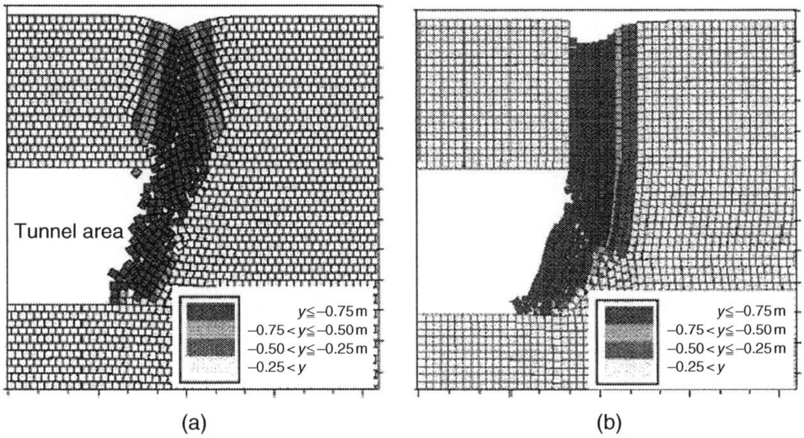

Fig. 12.19 DEM simulation of displacements with blocks 0.5×0.5 m in size without reinforcement at the tunnel face: (a) staggered fracture pattern; (b) cross-continuous fracture pattern (Kamata and Mashimo, 2003).

for simulation of bolting in the next step of modeling. Figure 12.19b shows the case for the cross-continuous fracture pattern and block size 0.5 × 0.5 m. At the tunnel face, the rock mass fails along pre-existing fracture surfaces and displaces as a plug, so-called chimney failure. This pattern was very different from that obtained in the centrifuge experiments, so therefore the staggered fracture pattern was used in simulating the influence of bolting.

The UDEC simulation of face bolting was performed for a staggered fracture pattern and 0.5 × 0.5 m block size and the parameter values listed in Table 12.1. The cable elements of the UDEC code generating axial forces along cables were applied and the result of inserting five bolts, of half the tunnel diameter in length, as face bolting is shown in Fig. 12.20. The figure illustrates the tensile axial forces along the length of

Table 12.1 Calculation parameters for UDEC simulation of face bolting (after Kamata and Mashimo, 2003)

	Young's modulus, E (MPa)	20
Sand (block)	Poisson's ratio	0.35
	Unit weight, γ (kN/m^3)	16
Fractures (between blocks)	Normal stiffness, k_n (kN/m)	2.2×10^5
	Shear stiffness, k_s (kN/m)	7.0×10^4
	Cohesion, C (kPa)	1.0
	Friction angle, φ	0.0
Bolts (cable elements)	Young's modulus, E (MPa)	2.06×10^5
	Area (m^2)	5.1×10^{-4}
	Tensile and compressive strength (kN)	123
	K bond (MN/m, per m)	12
	S bond (kN/m)	96

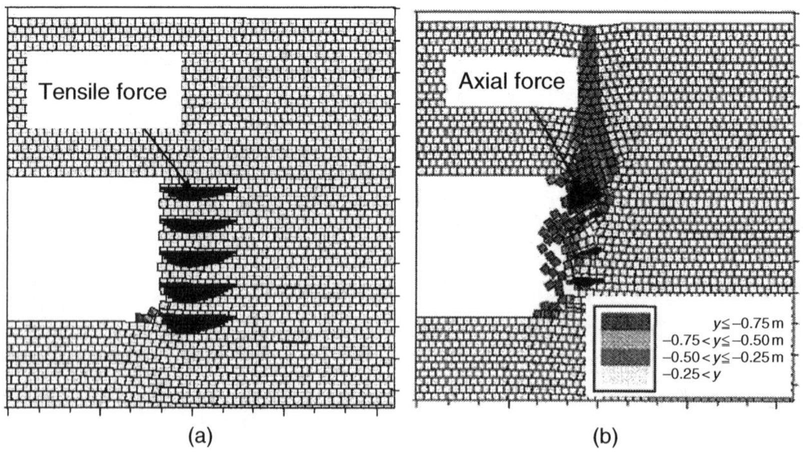

Fig. 12.20 Forces in bolts of a face-bolted tunnel in 0.5 × 0.5 m blocks. (a) Tensile force along the five bolts half the length of the tunnel diameter and a few loose blocks at the tunnel invert; (b) axial force along the five bolts of one-quarter length of the tunnel diameter and block raveling at the tunnel face (Kamata and Mashimo, 2003).

the five bolts where the maximum tensile forces appear at the center of the bolt (Fig. 12.20a). Failure in the model appeared only at the tunnel invert. In the case of using bolts of the same length as the tunnel diameter, the face state remains stable. However, for the case of a bolt length less than or equal to one quarter of the tunnel diameter, the calculations never reached equilibrium state and, at the final stage, the failure was as depicted in Fig. 12.20b. Large displacements reach deep into the rock mass behind the tunnel face.

The results of the DEM modeling of tunnel face reinforcement are consistent with the centrifugal experimental results. This demonstrates that results from the centrifuged models of non-reinforced models can be used for calibration of the model parameters for the DEM models. Both centrifuged models and DEM models show that, in order for face bolting to provide a proper reinforcement of the tunnel face, it is important to install bolts of a sufficient length. The results presented in this case study show that face bolting longer than half the tunnel diameter has a favorable effect on face stability in soft grounds.

12.3.2 Rock Caverns

Rock caverns accommodate an increasing variety of storage services and provide several important advantages when compared with facilities at the surface. Some of the main uses and benefits of underground spaces are presented by Franklin and Dusseault (1991). The basic goals of cavern design and construction are to minimize the disturbance to the surrounding rock, to conserve or enhance natural stability of the cavern by an optimal geometry and orientation and to install a minimum of rock reinforcement to satisfy the long-term functionalities of the caverns. A designer of caverns needs to assess the feasibility of the project based on data from site investigations and apply numerical analysis to help in decisions on the cavern location, orientation with respect to virgin stresses and geological structures, construction procedure and support and monitoring strategies. The ideal site for an underground storage cavern is the one that minimizes intersection with major fractures or fracture zones, in particular those that are weak, highly stressed or de-stressed and water bearing. As rocks are usually fractured and major fractures and fracture zones almost always exist in large rock masses, the discrete element method is found to be a useful tool in analysis for helping design, construction and monitoring of underground rock caverns (Brady, 1987).

The two-dimensional UDEC code has been extensively applied to cavern design of power plants, energy storage and underground sport facilities, e.g., by Akky et al. (1994), Barton et al. (1991, 1994), Baroudi et al. (1991), Bhasin et al. (1996), Chryssanthakis and Barton (1995) and Larsson et al. (1989). Bhasin and Høeg (1998) applied the UDEC code with a non-linear rock fracture model, called UDEC-BB, for a parametric study of large hydropower stations in the Himalayas. The three-dimensional 3DEC code was applied to examine the effect of major shear zones on the behavior of the wall of a powerhouse for a hydropower project in India by Dasgupta et al. (1995). The Viikinmäki underground sewage treatment plant outside Helsinki in Finland is one of the largest underground facilities in Europe, for which Johansson and Kuula (1995) conducted a series of back analyses using the 3DEC code. The results of calculations were compared with field measurements to verify the model and to examine the stability of the caverns concerned. Jiang et al. (2006) studied deformations, stability and rock reinforcement for a large underground pump storage cavern in Japan with the 3DEC code. The DDA method has been used for stability analysis of underground caverns of various sizes.

For this section, the Gjövik underground Olympic Ice Hockey Cavern in Norway is presented as a case study for large rock caverns using a DEM approach.

12.3.2.1 Case study – the underground ice hockey cavern in Norway

The 62 m span Olympic Ice Hockey Cavern for 5300 spectators in Gjövik, Norway is a valuable case record for assessing the viability of applying discrete element modeling for stability analyses of large caverns (Barton et al., 1991,1994; Chryssanthakis and Barton, 1995). The excavation of the ice hockey

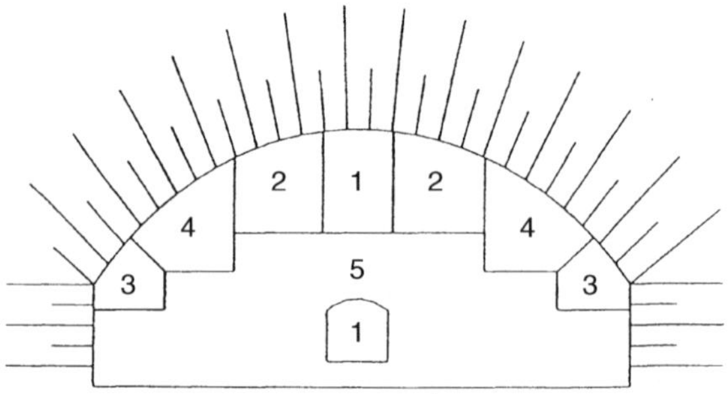

Fig. 12.21 Excavation sequence and bolting pattern for the Olympic Ice Hockey Cavern with a 62 m span at Gjövik, Norway (Chryssanthakis and Barton, 1995).

cavern was completed in 1992 to be ready for the 1994 Winter Olympic games in Lillehammer. The cavern is located in a small hillside adjacent to an existing underground swimming pool, as well as being a station for telecommunication and a civil defense facility. The ice hockey cavern has a finished span of 62 m, a length of 91 m and a height of 24 m, and is, at the time of writing, the largest cavern for public works in the world. The cavern layout and cross-section with rock bolt pattern and sequence of excavation are shown in Fig.12.21.

The Precambrian gneiss at the Gjövik site is well fractured with a mean RQD of about 70%, Barton et al. (1991). The fractures are generally rough and well interlocked and the foliation is weak. The rock mass quality varies from typical best quality, $Q = 30$, to typical poorest quality $Q = 1.1$ with a weighted mean of $Q = 12$. Based on a series of rock stress measurements with overcoring and hydraulic fracturing, the major horizontal stress was estimated to be 4.4 MPa at the level of the arch of the cavern. A bilinear variation of horizontal stress versus depth is applied to the boundary of the UDEC model as shown in Fig. 12.22. The vertical stress is generated by the weight of the overburden. Large-scale waviness of fractures was measured in the existing rock caverns and the roughness of fractures was determined from drill cores. Input parameters for the Barton–Bandis non-linear joint model used in the UDEC-BB code and applied to the Gjövik ice hockey cavern by Chryssanthakis and° Barton (1995) are as follows:

$$\mathrm{JRC}_0 = 7.5 \qquad \sigma_c = 100\,\mathrm{MPa} \qquad L_n = 0.5\,\mathrm{m}$$
$$\mathrm{JCS}_0 = 75\,\mathrm{MPa} \qquad \varphi_r = 27^\circ \qquad i = 6^\circ$$

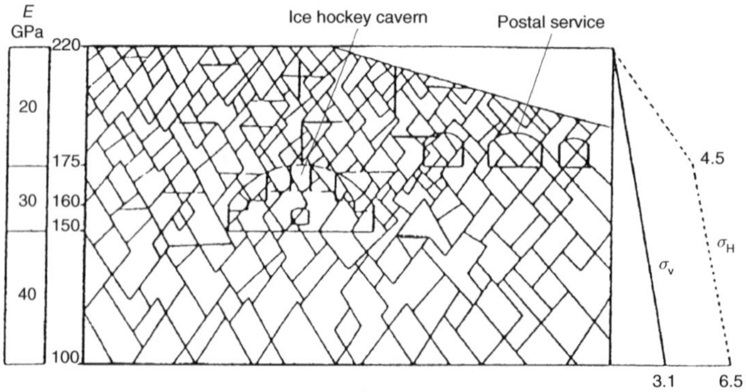

Fig. 12.22 UDEC-BB model of the Olympic Ice Hockey Cavern at Gjövik showing joint geometry, excavation geometry, stress boundary condition and variation of Young's modulus with depth (Chryssanthakis and Barton, 1995).

Table 12.2 Summary of the calculations with the UDEC-BB code for the Olympic Ice Hockey Cavern in Gjövik, Norway (Chryssanthakis and Barton, 1995)

Parameter	Step					Cavern excavation			Comments
	1	2	3	4	5	1st	2nd	3rd	
Maximum principal stress (MPa)	9.29	11.49	9.91	8.39	8.37	8.56	8.71	8.83	
Maximum displacement (mm)									Total deformation along horizontal joints
Total	1.85	1.80	2.63	6.99	8.16	8.28	8.43	8.65	
Wall	–	–	–	1.33	3.78	3.88	3.92	3.97	
Crown (vertical component)	0.50	1.08	2.62	4.05	4.33	4.39	4.87	7.01	
Maximum shear displacement (mm)									
Along horizontal joint	1.11	1.54	2.49	3.51	4.67	5.67	5.54	5.56	
Crown	1.11	1.54	2.49	3.51	3.70	3.70	4.10	6.85	
Maximum hydraulic aperture (mm)-crown	0.69	1.01	1.62	2.64	2.86	3.68	3.72	4.13	
Maximum axial forces on bolts (tnf)	7.0	25	25	25	25	25	25	25	

An increase of Young's modulus with depth is assumed and the effect of the nearby postal service caverns on the stability of the cavern was also evaluated in the numerical analysis (Fig. 12.22). The results of UDEC-BB modeling for each of the excavated steps are presented in Table 12.2 and the redistribution of stresses that occurs between the 4th and 5th excavation steps is shown in Fig. 12.23.

After the second excavation step, there is a progressive increase in displacement and the maximum deformation occurs along the sub-horizontal fractures on the right- and left-hand sides of the cavern. In the crown of the cavern, the maximum calculated deformation calculated after the 5th excavation sequence is 4.3 mm (Table 12.2). This value continues to increase after each postal cavern is excavated and the final deformation vectors after the completion of excavation are shown in Fig. 12.24. Notice the tendency for horizontal displacements of the blocks on each side of the crown of the cavern. The maximum shear displacement along the horizontal fractures on the crown after the 5th excavation sequence is about 4 mm and increases to almost 7 mm after the excavation of the third postal service cavern. Calculated maximum hydraulic aperture and axial force on rock bolts are also presented in Table 12.2.

An extensive monitoring program of rock deformation was conducted for the cavern at Gjövik. The near-surface location of the cavern meant that extensometers could be placed in boreholes from the surface to a distance of 1.5 and 2.0 m above the arch of the cavern. Surface precision leveling were conducted at the head of the extensometers and the total deformation from adding the extensometer readings and measured subsidence gave a total deformation of 8.4 mm at the southern end and 8.7 mm at

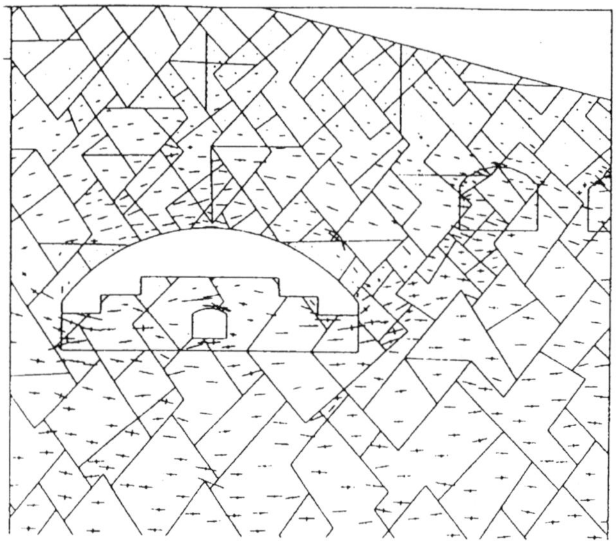

Fig. 12.23 Redistribution of principal stresses between the 4th and 5th excavation steps of the Olympic Ice Hockey Cavern in Gjövik, Norway (Chryssanthakis and Barton, 1995).

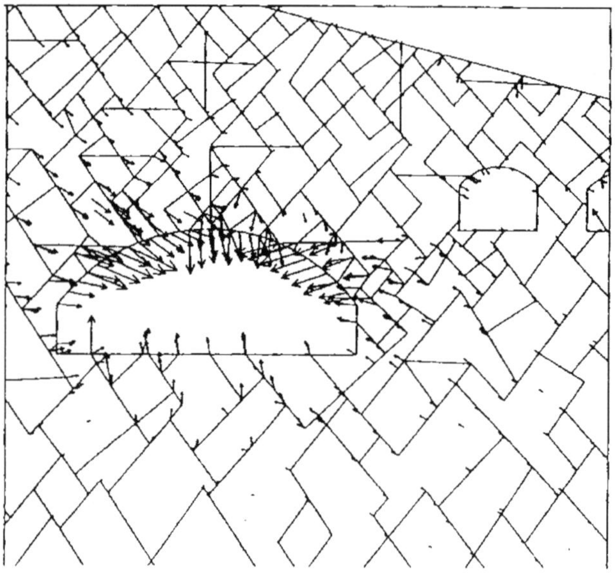

Fig. 12.24 Displacement between the 4th and 5th excavation steps of the Olympic Ice Hockey Cavern in Gjövik, Norway (Chryssanthakis and Barton, 1995).

the northern end of the cavern. The corresponding numerical results showed 4–6 mm after the 5th sequence of excavation. Deformations effectively ceased after about 1 month of completion of the excavation of the cavern.

In conclusion, the rock mechanics investigations of the large rock cavern at Gjövik indicate that the entire roof is a self-supporting structure and that the small displacements are the result of the relatively high horizontal virgin stresses and the irregular pattern of fractures with rough and undulating surfaces. The study demonstrates that two-dimensional DEM modeling provides useful information on details of behavior of the roof stability, particularly those caused by the local shearing along the fractures. Application of the UDEC-BB model shows satisfactory agreements with global measurements of deformation. In a recent study by Scheldt et al. (2002), with the aim of comparing the results obtained from discrete element and finite element modeling, in principle the same results were obtained. The good agreement between predictions and measurements can be considered as a validation of the UDEC-BB code subject to the assumptions made in the 2D DEM model (Barton et al., 1994) and its ability to model excavation effects in fractured rock masses.

12.4 Mine Structures

The essence of mining is to make excavations as entries from the ground surface to the mineral deposits and then to extract ores. Whether the openings lie at the surface or are placed underground determines the location of the mine. Different methods of extraction are selected depending on whether the ore body is thin or thick, tabular or massive, horizontal or steeply dipping. This, together with the fracturing and strength properties of the ore body and the host rocks, will determine the mining method to be applied. In principle, a mine goes through five stages in its lifetime: prospecting, exploration, development, exploitation and remediation. During the last four stages, rock mechanics principles and numerical modeling are of utmost importance. At the stage of exploration, different mining methods will be evaluated based, among others, on modeling results of mine stability, excavation sequences and rock reinforcement. During development of the mine, detailed modeling is needed for selection of geometry, orientation and location of the different mine structures, e.g., slopes, benches and ramps for open pit mines and shafts, ramps, raises, haulages passes, stopes (the openings created by mining), pillars and ore passes for underground mines. A primary concern in mining exploitation and operation, either on the surface or underground, is the ground control, i.e., control of the deformations and displacements of host rocks surrounding the various excavations generated by the mining activities. During different stages of mining, numerical modeling can be conducted on a frequent basis to support daily decisions for mine planning and ground control. In recent years, remediation of mines has become an important environmental issue, and numerical modeling becomes an important tool to examine the safties of old mines and the infrastructures located above them and their impacts on the local or regional environments.

The underground mine layouts are designed so that mining can be conducted without the risk of sudden mechanical instability and unacceptable displacements of rock masses, which may or may not have negative impacts on local or regional groundwater regimes. Failure of the rock masses is often governed by reaching the strength of existing faults and major fracture zones in the vicinity of the mine openings. The intensity and extent of mining-induced fault slip and related rock bursts are a function of mining depth, geology, rock strength and mining method. The application of the DEM to excavation-induced faulting and fault-slip-induced rock bursting in mines are presented by Hart et al. (1988), Hart (1993) and Tinucci and Hanson (1990) for the Strathcona Mine in Canada, by Lightfoot et al. (1994) for the deep-seated Western Deep Levels South Mine in South Africa and by Board (1996) for the Lucky Friday Mine, Idaho, USA and deep mines in Canada.

Mining-induced seismic events is an important subject, particularly for mining at great depth, where numerical modeling is often required for understanding the main mechanisms, major influence variables and parameters and optimizations of mining layouts and sequences. For such analyses, after defining a geological model of the mining area, a numerical analysis is often conducted to estimate the mining-induced stress state for a given mining sequence. Then the stress state is superimposed on a fracture model and the excess shear stress (ESS) approach suggested by Ryder (1987) is used to determine the resulting seismic source terms. The ESS at any point along a fault is the magnitude of the shearing stress in excess of the shear dynamic frictional strength. Positive values of the ESS indicate that slip is possible under dynamic conditions. The calculated source terms are used in comparison to the recorded seismic data as a means of calibrating the models.

Board (1996) applied a mining-induced seismicity simulation method to examine the seismicity phenomenon associated with possible slip on soft inter-beds of argillite in hard quartzite of the Lucky Friday Mine as a longwall mining face was advanced. Slip potential on the terminating fault structures was examined by explicit modeling of the slip using the 3DEC code.

Discrete element modeling has been widely applied in analyses of mine structures and their stability (e.g., Hardy et al., 1999). In this section, some applications of the DEM for mining-related issues are presented for modeling slope behavior for open pits and quarries, the stability of underground mines structures, such as stope and pillar operations, and mine subsidence.

12.4.1 Open Pits and Quarries

In the design of open pits and quarries, increasing the slope angle decreases the stripping of waste rocks and increases the percentage of recoverable ore, which gives a higher benefit margin. However, increasing the slope angle affects the stability of the slope and the risk of increasing the number, size and movement rate of slope failures. At steeper slope angles, the cost of cleaning up failed materials, lost production and ores losses from slope instability may increase more rapidly than the benefit of having the steeper slope. Minor rock falls and bench failures are often taken as an indication that the geometrical configuration is not too conservative. Attempts are not often made to stabilize temporary slopes with anchoring in open pits and quarries and therefore the design of slopes primarily is based on the basis of the relations between the geometry of the geological structures and that of the slopes.

To make a quantitative estimate of the stability of a slope, analytical or numerical models can be used. The requirements of these models are the failure geometry, material properties, virgin stresses, water conditions and boundary conditions. The sliding block failure mode is the most common type of failure and refers to a situation in which displacement occurs along one or more geological structures, and the potential moving rock mass is considered to be a or a number of rigid blocks. Shear planes, slabbing, stepping paths and wedge failures are the most common sliding block geometries. Together with toppling failure, with blocks of large height-to thickness ratios, they are the failure geometries that are most suitable for DEM modeling because of its capacity for large displacement simulations.

Hencher et al. (1996) presented results from a UDEC modeling of the complex failure of the foot wall slope in the large Aznalcóllar open pit mine in southern Spain. The model was composed of blocks containing the main set of parallel cleavage fractures, which were assumed to be the main control feature of the slope behavior, together with a set of steeply dipping release fractures. Horizontal stresses were assumed to be equal to the vertical stress from the weight of the overburden. Intact rocks and fractures were modeled as elastic–plastic materials with shear strength defined by the Mohr–Coulomb criterion. Modeling results show that the zones of rock that underwent significant displacements in the UDEC model agree well with the observed extent and depth of failure in the field. Local displacement vectors also corresponded well to field observations and gave confidence that the dominating mechanisms of instabilities were dealt with correctly and that the model with its complex fracture system geometry can be applied to changing conditions of future mining, slope geometry and drainage.

Similar modeling works were performed by Zhu et al. (1996) and Coulthard et al. (1997) for the slope stability problems of the Daye open pit iron mine in China, using both UDEC and 3DEC codes, and by Sjöberg (1999) for the high rock slopes of the Aitik open pit in Northern Sweden, by Hutchison et al. (2000) for flexural toppling failure in an Australian mine and Xu et al. (2000) for stability and risk assessment of pit walls in an Australian iron mine.

Deangeli et al. (1996) and Coggan and Pine (1996) performed rock mechanics studies and the 2D DEM modeling to improve slope stability conditions in quarries. A complex failure mechanism involving distinct planar surfaces caused a major failure of the western face at Delabole slate quarry, Cornwall, England in 1967. Also, deep-seated flexural toppling may have contributed to the instability of the slope. Coggan and Pine (1996) studied this slope failure with the UDEC code to improve the understanding of possible failure mechanisms, rather than provide a detailed analysis. Throughout the modeling, sensitivity analyses were performed via critical input parameters and the simulations demonstrated the importance of the pattern, shear strength and stiffness of fractures for the model behavior. Groundwater modeling confirmed the detrimental influence of a raised water table on the stability of the slope and also indicated that the observed cyclic opening and closing of tension cracks could be related to seasonal variations of the groundwater level.

12.4.2 Underground Mines

A computer model for underground mining excavations should accommodate variability in rock types and properties and mining sequences in space and time. This includes non-linear constitutive behavior and the ability to simulate large deformations. The model should also be able to simulate multiple, intersecting fractures and rock deformation or motion along these features under different loading and stress conditions. A general discussion on the prerequisite of discrete analysis for deep-seated excavations and mines is presented in Hart (1990), Cundall and Hart (1992a), and Oelfke et al. (1994).

Cundall and Hart (1992b) and Hart (1990) reviewed the attributes of the various discrete element and limit equilibrium methods and found that the distinct element method is best suited to model the mechanisms identified as important for deep underground excavations and mines. The distinct element codes UDEC (for two-dimensional analysis) and 3DEC (for three-dimensional analysis) have been applied extensively for mining applications. Below are some of the examples:

(a) Application examples of the UDEC code (Itasca, 1995a):

- mine excavation design (Coulthard et al., 1992; Gomes et al. 1993; Eve and Squelch, 1994; Kullman et al., 1994; Squelch et al., 1994; Ravi and Dasgupta, 1995; Shen and Duncan Fama 1997, 2000; Duncan Fama et al. 1999; MacLaughlin et al., 2001; Mac Laughlin and Clapp, 2002);
- design and stability analyses of mine pillars (Sjöberg, 1992; Esterhuizen, 1994; Nordlund et al., 1995; Sansone and Ayres da Silva, 1998; Hardy et al., 1999);
- slip-induced rockbursting (Lightfoot, 1993; Lightfoot et al., 1994);
- cavability for ore extraction (Hassan et al., 1993); and
- determination of longwall shield capacity in longwall coal mining (Gilbride et al., 1998).

(b) Application examples of the 3DEC code (Itasca, 1995b):

- long-hole stoping mining at Brusada , central Italian Alps (Wojtkowiak et al., 1995);
- large blasthole sublevel stooping in Kiirunavaara Mine, Sweden (Jing and Stephansson, 1991);
- mine cavability for ore extraction (Lorig et al.,1989; Hart, 1990);
- fault-slip rock burst phenomena in Strathcona mine, Canada, and elsewhere (Hart et al., 1988; Hart, 1990; Tinucci and Hanson, 1990; Bigarre et al., 1993; Lightfoot, 1993; Board, 1996); and
- stability of mine pillars and hanging walls (Antikainen et al. 1993; Nordlund et al., 1995; Board et al., 2000; Ferrero et al., 1995).

12.4.2.1 Case study – three-dimensional DEM modeling of sub-level stoping at Kiirunvaara Mine, Sweden

The use of distinct element codes to study mechanisms associated with excavation in fractured rocks is demonstrated by a 3DEC analysis of the large-scale sub-level stoping test mining in the LKAB Kiirunavaara Mine, Kiruna, Northern Sweden (Jing and Stephansson, 1991). This modeling is one of the very first mining applications of the 3DEC code. The traditional mining method in the Kiirunavaara Mine is sub-level caving. To increase the productivity and reduce the ore losses and waste dilution, a large-scale sub-level stoping technique with large diameter blastholes was tested. Large diameter blastholes are drilled from a drilling drift within the ore body and the blasting runs for each stope are proceeded upwards (Fig. 12.25). The design allows for separation of high phosphorus content ore (type B) and low phosphorus content ore (type D) The blasted ore is loaded at draw points at loading cross-cuts and transported to nearby ore passes.

Four large open stopes and four temporary pillars between the stopes and four crown pillars were included in the test, located in the OSCAR area, defined by sections Y29 and Y30 between levels –654 and –586 m of the mine (Fig.12.26). The orebody at the OSCAR site was divided into four large open

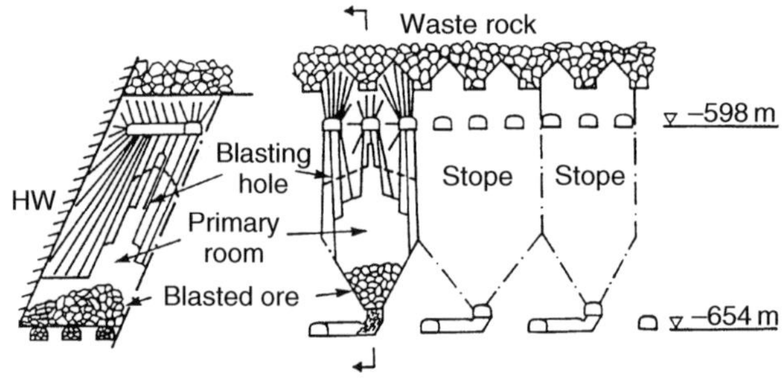

Fig. 12.25 Test mining concept of large-scale sub-level stoping in Kiirunavaara Mine, Sweden (Jing and Stephansson, 1991).

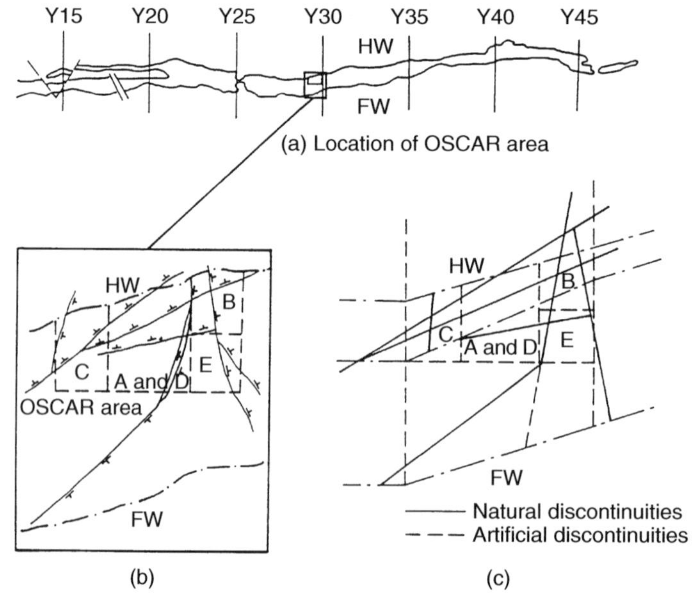

Fig. 12.26 Location and geological structures of OSCAR test area and stope configuration of test mining (Jing and Stephansson, 1991).

Table 12.3 Material properties for rocks and faults for 3DEC modeling of the OSCAR test case (Jing and Stephansson, 1991)

Rock type	Normal stiffness (GPa/m)	Shear stiffness (GPa/m)	Friction angle (degrees)	Cohesion (MPa)	Bulk modulus (GPa)	Shear modulus (GPa)	Density (kg/m^3)
Ore (D type)	2.67	0.21	40	1	32	19	4500
Ore (B type)	3.33	0.27	40	1	40	24	4500
Hanging wall	3.33	0.27	40	1	40	24	2700
Foot wall	3.33	0.27	40	1	40	24	2700

stopes: stopes B, C, AD and E. Because the ore drawing is performed separately from drilling, charging and blasting at different levels, continuous production can be achieved. When the primary room reached a height of 44 m from the loading level, the temporary roof pillar of the room is removed by mass blasting and the blasted ore is drawn. There were primarily nine major faults and three sets of fractures in the test area. Only the large faults were considered in the computational model with the 3DEC code (Fig. 12.26).

The material properties used in the modeling are listed in Table 12.3. The rock blocks are treated as linear elastic materials and faults were modeled by a Mohr–Coloumb friction law with constant stiffness. For the faults, or parts of the faults, located within the test stopes next to the stopes of undergoing mining, the friction angles are reduced to 30° and cohesion reduced to 0.1 MPa. This is to take account of the degradation of the cohesion and friction due to block movements caused by the mining. Rock mass response and deformation were monitored with stress monitoring cells and extensometers to validate the model.

Simulation of test mining was performed on a regional computational model whose boundary stresses were generated from a larger global model. The sub-structuring of the global model containing the regional model and the geometry of the large mining stopes in the OSCAR area is shown in Fig. 12.27. The regional model consists of 256 blocks and 9746 finite difference elements and the disturbed stresses

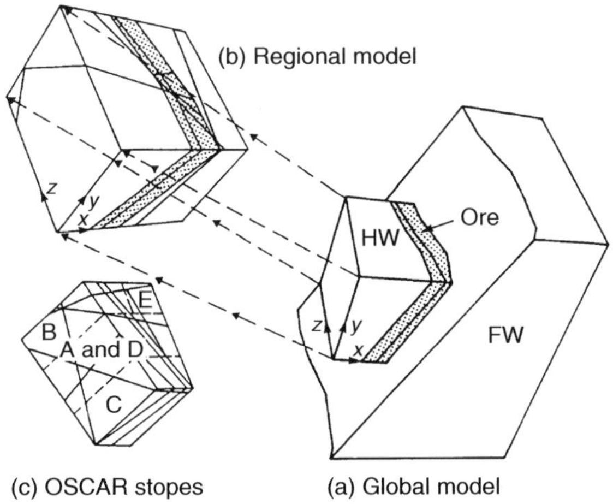

Fig. 12.27 (a) Sub-structuring of global model of the Kiirunavaara Mine, (b) regional model of the OSCAR test site and (c) stopes (Jing and Stephansson, 1991).

from the sub-level caving above the OSCAR area were simulated. Displacements were monitored at eight points in the regional model corresponding to the anchor points of extensometer No. 6 in stope AD and three stress components at one point from the stress monitoring point at the loading level. Also monitored were displacements and velocities at points in the center of the stopes to examine the state of equilibrium of the regional model.

The simulation of test mining was performed with the mining advances stope-by-stope. The stability condition of the temporary pillars was predicted by the numerical modeling and the calculated results were checked against the measurements from extensometers and stress monitoring cells. The modeling was divided into three stages:

(1) the simulation of mining stopes B and C and validation of the numerical model;

(2) simulation of mining stopes AD and E with alternative mining sequences; and

(3) design and simulation of completely new stope configuration.

For the simulation of mining stopes B and C, no major failure was predicted for the temporary pillars and the test mining of the two stopes was successful. The numerical model was validated at this stage of modeling against extensometer measurements and stress monitoring. The parameters listed in Table 12.3, especially the friction angle of faults, were determined from this analysis and later used throughout the modeling. The good agreement between measured and calculated relative axial displacement versus relative distance of anchors for extensometer No. 6 after mining stope B and stopes B and C is shown in Fig. 12.28. Good agreement was also obtained for two of the stress components measured at the loading level.

To avoid large ore losses and total collapse of stope AD, which was the largest stope in the test area, two alternative mining sequences were analyzed by using the regional model, namely mining stope E prior to stope AD as originally planned or mining stope AD prior to stope E. Figure 12.29 compares the zones of tensile stresses (shadowed area) caused by the two alternative mining sequences. If stope E were mined first, the zone of tensile stress would be increased significantly. The same would occur to deformation of rock masses as proven by the increase of relative axial displacement of the extensometers. Therefore, it was suggested that the original mining sequence should be changed with mining stope AD

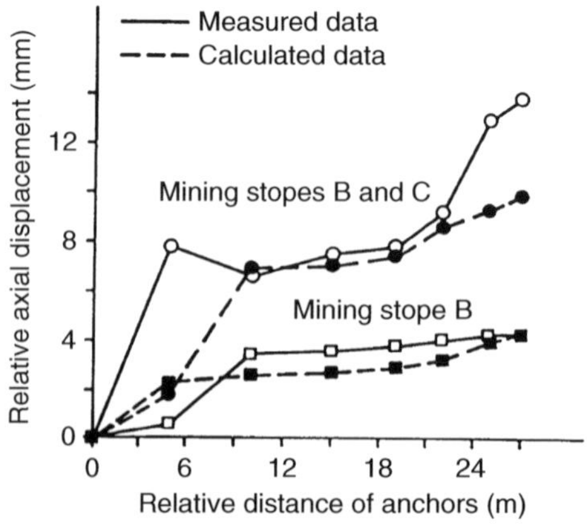

Fig. 12.28 Measured and calculated relative axial displacements of eight measuring points along extensometer no. 6 of the OSCAR test case after mining stope B and stopes B and C (Jing and Stephansson, 1991).

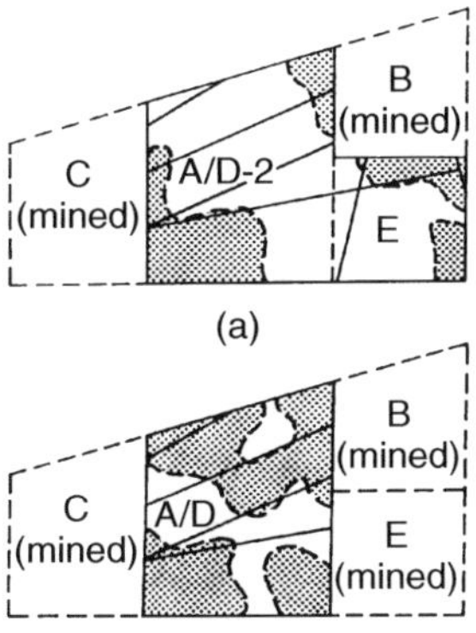

Fig. 12.29 Development of zones of tensile stresses in remaining ore as a function of the mining sequences of the OSCAR test case; (a) after mining stope B and primary room A/D-2 and prior to mining stope E; (b) after mining stope E and prior to mining stope A/D (Jing and Stephansson, 1991).

prior to mining stope E and the dimension of the primary open room of stope AD should be reduced. This recommendation was accepted for the test case and no major failure occurred during mining stope A.

In the third stage of the 3DEC modeling, the validated regional model was used to study alternative stope configurations with a different mining sequence for the test mining case. For this new design, the test site was divided into five stopes with the mining sequence C1 → C2 → A1 → A2 → D1 → D2 → B1 → B2 → E1 → E2 (see Fig. 12.30). A 'retreat' type of mining sequence was used in this case.

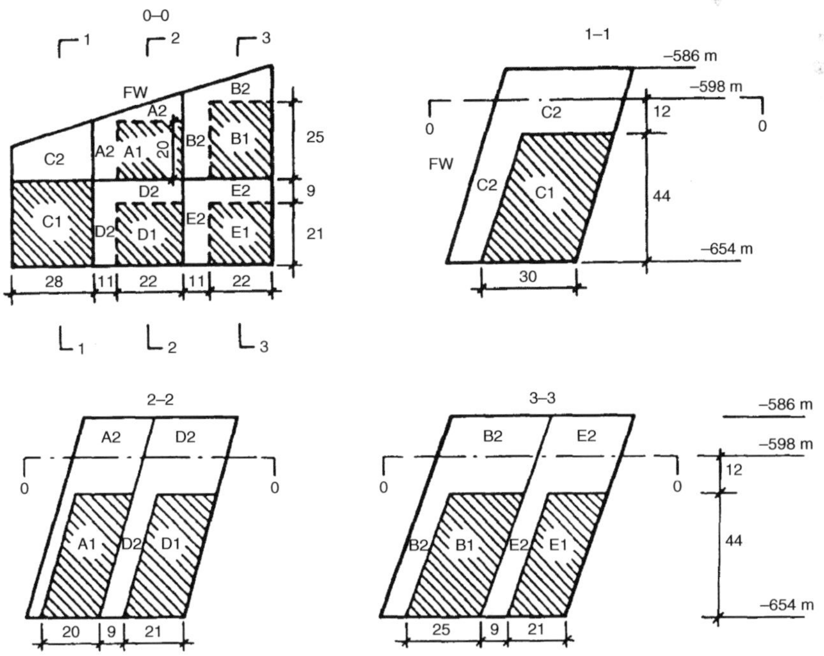

Fig. 12.30 Suggested stope design for test mining at the OSCAR site, Kiirunavaara Mine (Jing and Stephansson, 1991).

The modeling results for the same material properties and boundary stresses show that stopes C, A and B could be safely mined. The overall volume ratio of primary rooms over the whole stopes reached 31.9%, which is slightly above the theoretical threshold used for the test mining at OSCAR.

Jing and Stephansson (1991) draw the following general conclusions from the modeling results:

- the 3D underground mine structure of large rooms and pillars in the Kiirunavaara Mine cannot be adequately represented by two-dimensional computational models;
- the presence of major faults and other geological structures requires a distinct representation of these structures, both geometrically and mechanically;
- the success of modeling using the DEM depends to a great extent on the knowledge of the geometrical distribution, size and properties of the dominating faults;
- conventional numerical methods based on continuum analysis are not suited for modeling the type of problems presented in the test case of OSCAR;
- field measurements of displacements and stresses are prerequisites for successful application and verification/validation of DEM modeling or the application of any other numerical methods.

In conclusion, the results of the distinct element modeling of large blasthole sub-level stoping in the OSCAR area of the Kiirunavaara Mine demonstrated that, once a numerical model has been validated against field measurements, the model can be used in forward simulations of alternative stope configurations and mining sequences to support mine design.

12.4.3 Mine Subsidence

Mining-induced subsidence is the settlement of the ground surface following underground extraction of ores. According to Brady and Brown (1985), subsidence can be regarded as being of two types – continuous and discontinuous:

- Continuous subsidence causes a smooth surface subsidence profile that is free of steep changes and is usually associated with the mining of flat-dipping orebodies with overburdens of weak and ductile rocks, often of sedimentary origin. Room and pillar mining and longwall mining of flat-laying coal seams are activities that render continuous subsidence and related problems with possible settlement of buildings and infrastructures at the ground surface.
- Discontinuous subsidence is characterized by large surface displacements over limited surface areas where the displacements cause major steps in the subsidence profiles. Chimney caving, sinkholes, piping and funneling are examples of such failure mechanisms, caused by progressive migration of unsupported mining cavities through the overlaying strata to the surface, are typical discontinuous subsidence types which can have disastrous consequences.

Continuous subsidence and mine roof stability have been studied with empirical prediction methods (UK National Coal Board, 1975) and numerical modeling with DEM (Coulthard and Dutton, 1988; Choi and Coulthard, 1990; O'Connor and Dowding, 1990, 1992a, 1992b; Ahola, 1992; Siekmeier and O'Connor, 1994; Coulthard, 1995). In order to simulate the significant modes of rock mass behavior associated with continuous subsidence, O'Connor and Dowding (1992a) developed a hybrid numerical code by combining the Northwestern Rigid Block Model (NURBM) with the RBM code developed by Cundall (1974).

12.4.3.1 Case study – subsidence in the coalfields in New South Wales, Australia

Surface and sub-surface deformations were measured over two longwall panels, which were mined in a virgin area at the Angus Place Colliery in the coalfields west of Sydney in Australia. Numerical predictions of subsidence at various stages of the mining were performed prior to the release of the field data (Kay et al., 1991). The results of a series of UDEC analyses performed by Coulthard (1995) for the Angus Place case study are reported in this case study example. The rock strata overlying the 2.5 m thick coal seam consist of about 250–300 m inter-bedded sandstones, claystones, shales and coal. The longwall panels form the basis for the study where each are 211 m wide and 1600 m long and separated by a 35.2-m-wide chain pillar.

Subsidence monitoring grid lines were established prior to mining – two cross lines, one near the start of panel 11 (grid line 'A'), one across the center of panels 11 and 12 (grid line 'X') and a longitudinal one along the center of panel 11, see Fig. 12.31.

A mechanical sub-surface monitoring system was installed in a borehole at the intersection of the longitudinal gridline and gridline 'X'. Little information was available about the fracturing of the rock mass, except for the sub-horizontal bedding planes.

The UDEC analyses were performed on a vertical section along grid line 'X' and the mesh extended from the surface to 25 or 50 m below the coal seam. Bedding planes in the uppermost strata were spaced at 12.5 m, but the spacing became smaller within the 100 m above the coal seam. Sub-vertical fractures were included in the strata between the bedding planes and were given variable lateral spacing in the UDEC models. A total of 2 300 blocks were used for the analysis of the section across panels 11 and 12. The Mohr–Coulomb elasto-plastic constitutive model was applied to the internal blocks. Bedding planes were modeled as standard UDEC fractures and vertical fractures were treated as intact rock – becoming functional fractures once the rock shear or tensile strength was reached. The material properties used in the simulation are presented in Table 12.4. The vertical and horizontal virgin stresses were specified as: $\sigma_v = 0.02y$ and $\sigma_H = 2/3\sigma_v$ where y is in meters and the stresses are in MPas.

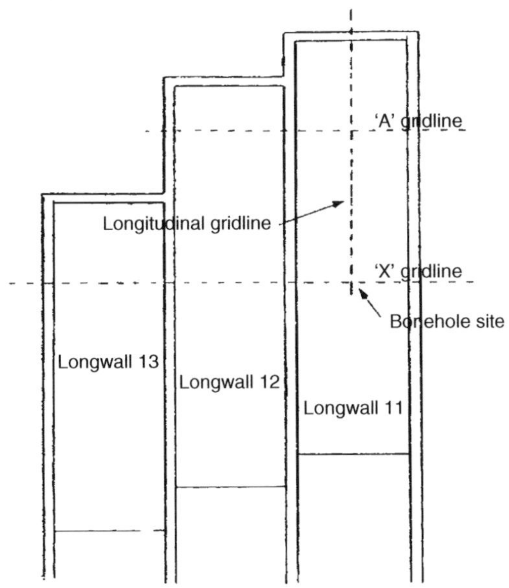

Fig. 12.31 Longwall panels and subsidence gridlines for Angus Place Colliery case study, NSW, Australia (Coulthard, 1995).

Table 12.4 Material properties used in Angus Place case study (Coulthard, 1995)

Material	Parameter	Unit	Experimental range	Mean	Value used in UDEC
Rock strata	Young's modulus	GPa	2.6–20.2	10	10
	Poisson's ratio	–	0.19–0.46	0.3	0.25
	Density	Mg/m^3	1.83–2.73	2.3	2.0
	UCS	MPa	6–126	50	–
	Cohesion	MPa	11–28	21	12
	Friction angle	Degree	11–41	24	35
	Dilation angle	Degree	–	–	25
	Tensile strength	MPa	0.2–10.8	4	5
Coal	Young's modulus	GPa	2.7–7.3	4.5	5.0
	Poisson's ratio	–	0.20–0.47	0.3	0.25
	Density	Mg/m^3	1.29–1.42	1.36	1.3
	UCS	MPa	14–54	30	–
	Cohesion	MPa	–	–	6
	Friction angle	Degree	–	–	20
	Tensile strength	MPa	–	–	1
Bedding plane	Cohesion	MPa	0–0.45	0.2	0
	Friction angle	Degree	19–21	20	20
	Dilation angle	Degree	–	–	0
	Tensile strength	MPa	–	–	0
	Normal stiffness	GPa/m	–	–	100
	Shear stiffness	GPa/m	–	–	10

Results of the UDEC analyses are presented for the twin panel analyses along gridline 'X' and the seven variations in modeling approaches (2A–3A) as presented in Table 12.5. All the analyses predicted that the maximum subsidence over the panel longwall 12 would become less than that over the panel longwall 11, see Fig. 12.32. When a finer discretization was used in the upper strata (approach 2E in Table 12.5), the displacements came closer to the measured values. Finer zoning in the floor strata underneath the coal seam (approach 2G in Table 12.5) increased the calculated subsidence at each stage. Reducing the dilation angle from 25 to 10° has a minimal effect on deformation for this problem as the mechanisms of deformation is dominated by separation of bedding planes and less by shearing on the sub-vertical fractures.

Details of the deformation pattern of the roof strata as calculated by the UDEC code are shown in Fig. 12.33. The model depicts the collapse of the lowermost strata at the roof of the panel, the failure of the sub-vertical fractures above the abutments and in the center of the model, where deflection and bending stresses are the maximum, together with the detachment of the strata along bedding planes.

As pointed out by Coulthard (1995), the overall picture of roof failure and deformation matches measurements very well without any artificial tuning of material properties or other parameters after the field data became available.

The calculated asymmetry in the subsidence profile as shown in Fig. 12.32 could not be obtained from a linear or non-linear continuum analysis. The most important limitations of the presented UDEC

Table 12.5 UDEC analysis and calculated subsidence from mining panels 11 and 12 for Angus Place case study (Coulthard, 1995)

Analysis No.	Details	First panel max (m)	Both panels (m)		
			Max 1	Min	Max 2
2A	'Standard analysis', UDEC 1.3	0.24	0.34	0.18	0.31
2D	'Standard analysis', UDEC 1.7	0.26	0.38	0.17	0.31
2E	2A + finer zoning in upper strata	0.70	0.87	0.19	0.72
2F	2E + finer zoning and lower strength in coal	0.56	0.85	0.23	0.66
2G	2F + finer zoning in floor	0.84	1.00	0.16	0.63
2I	2E + rock dilation angle reduced to 10°	0.70	0.87	0.19	0.71
3A	2D, but reversed extraction order	0.27	0.42	0.15	0.37

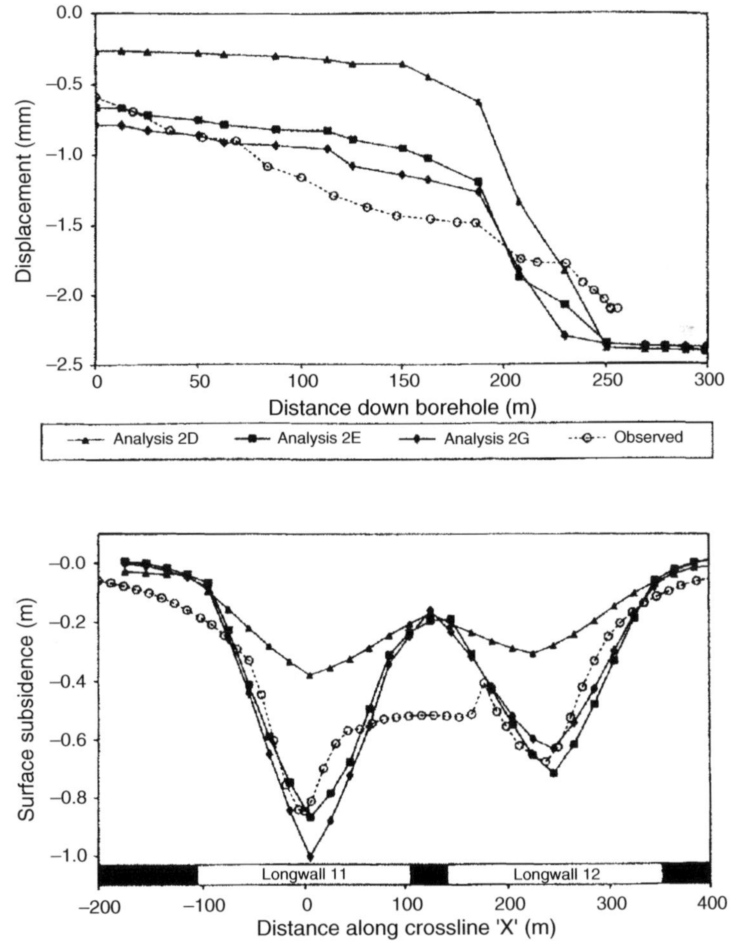

Fig. 12.32 Comparison of measured and calculated deformations for Angus Place case study. Top: vertical movement down the borehole over the center of panel 11 after mining that panel; (bottom) subsidence profiles on grid line 'X' after excavation of panels 11 and 12 (Coulthard, 1995).

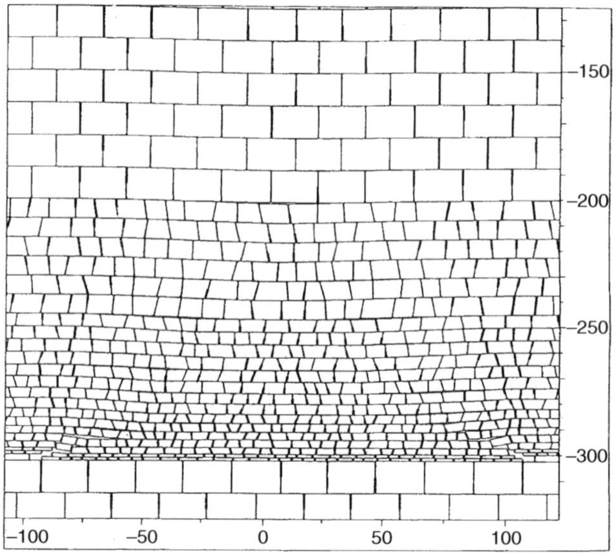

Fig. 12.33 A UDEC analysis for approach 2E over panel 11 of the Angus Place case study showing roof collapse, separation of bedding planes and sub-vertical fractures (heavy lines) (Coulthard, 1995).

model are its failure to reproduce the complexity of block rotation and bulking which occurs when the immediate roof collapses, thus forming the bulking of broken material (goaf), the relative smaller predicted amount of opening of bedding planes in the upper strata as compared with measured displacements in the boreholes and the considerable underestimates of subsidence over the chain pillar.

12.5 Radioactive Waste Disposal

In many countries, geological disposal with underground repositories is one of the several possible ways under study for permanent disposal of radioactive wastes and spent nuclear fuels. To evaluate the various geological media considered for such disposal, it is necessary to obtain adequate understanding and data on the characteristics of different host rocks and to consider different site locations and repository designs. It is also of utmost importance to evaluate the safety of the disposal system in rocks, especially the impacts of rock fractures of various sizes and possibilities of newly created fractures by repository engineering. An important part of the safety analysis is an assessment of the coupling of rock mass deformation and stability, tectonic stresses, groundwater flow and the thermal loading from the waste decay.

To demonstrate the isolation effectiveness of the nuclear waste repositories from the biosphere, it is necessary to have the capability of numerical modeling for predicting the effects of main mechanisms and coupled processes on performance and safety of repositories over long periods of time unprecedented in rock engineering. The important consequence of the existence of liquids and gases in the rock mass and the heating effect of the waste decay points to the significance of coupling between the mechanical, hydrological, thermal and chemical processes in engineered barrier (buffer materials such as bentonite) and natural (geological) burrier, the fractured rocks. The term 'coupled processes' implies that one process affects the initiation and progress of another. Important to note that the fractures of different sizes in the rock masses play more important roles than the intact rock matrix in terms of coupled mechanical, thermal and hydraulic behaviors that can differ by orders of magnitudes.

Mathematical models and numerical methods are applied to study the responses of nuclear waste disposal because closed-form solutions to the problems concerned generally do not exist, and full-scale experimental studies are impractical because of the large volumes and long time involved. The states of the arts of the numerical modeling for the coupled THM processes in fractured rocks and buffer material for nuclear waste repositories are presented in Stephansson et al. (1996, 2004). Jing et al. (1996a,b, 2002) presented reviews of numerical modeling of coupled thermo-hydro-mechanical processes of fractured rocks with continuum and discrete approaches as applied to nuclear waste repositories.

The first experimental tests that were performed with the aim to verify and validate numerical codes for radioactive waste disposal were designed as large block tests, e.g., the block test of the Basalt Waste Isolation Project at Hanford, Washington, USA (Hart et al., 1985; Cundall and Fairhurst, 1986; Donovan et al., 1987). A large block of closely fractured basalt was released and later subjected to triaxial loading by flat jacks and a system of pre-reinforced tendon, and the response was recorded and later compared to the modeling results. Hart et al. (1985) and Cundall and Fairhurst (1986) concluded that, in order to understand and model basalt deformation behavior, rotation and slip of the individual basalt columns in the rock mass must be incorporated.

In other block tests, fluid and heat load were added to the mechanical loading, and temperature and flow rate were measured, in addition to displacements and strains. The Colorado School of Mines block test, Colorado, USA consists of a 2-m cube of Precambrian gneiss intersected by three large, sub-vertical fractures (Chryssanthakis et al., 1991a). The block was subjected to biaxial and uniaxial loading at ambient and elevated temperatures, using flat jacks and a line of borehole heaters for measuring the hydro-thermo-mechanical response of the rock. An extensive monitoring program was installed and a series of supporting tests to the modeling program was run, e.g., stress monitoring with different instrumentation systems during flat jack loading of the block. Two versions of the UDEC code have been tested against the measured results from the CSM block test: the UDEC Linear code with the fracture stiffness varying linearly with applied stress and the UDEC-BB code with the fracture stiffness varying non-linearly with stress in accordance with the Barton–Bandis fracture model.

A large block test of welded tuff with the dimensions $3 \times 3 \times 4.5$ m was tested as a part of the field-scale thermal testing program of the Yucca Mountain Project in the USA. Two-dimensional forward and backwards analyses using the DDA method was performed and reported by Yeung and Blair (1999).

The French Nuclear Inspectorate (IRSN) has conducted a large-scale *in situ* heating test in granite at an experimental site in the abandoned uranium mine at Fanay-Augér, France. The test site is located at a depth of about 100 m below the ground surface. At the floor of the relatively dry test chamber and at a depth of 3 m, five 1.5 m long cylindrical heaters were installed. Six major fractures intersected the test block and an extensive monitoring program was installed to record the rock mass response. Simulating the heater test at Fanay-Augér mine by a discrete approach with five major fractures, the results show that three-dimensional finite element and discrete element methods are both adequate for modeling the thermo-mechanical behavior of a set of hard rock blocks bounded by fractures with specific mechanical properties (Rejeb, 1996).

Coupled thermo-mechanical impact on the fractured rock mass surrounding a nuclear waste repository is an important aspect of the performance and safety assessment of the repository (Jing et al., 2002). Considerations must be given to the ultimate thermal load from the waste and the temperature distribution over time due to the temperature decay of the heat source. Changes of strain and stress fields, repository stability and system response to future loading mechanisms, such as glaciation, glacial rebound and sea level changes, are processes to be considered in the performance assessment of a waste repository. These loadings and processes belong to the so-called far-field problems and have to be studied by means of models on the kilometer-scale. Some modeling work using DEM approaches are reported in Johansson et al. (1991), Tolppanen et al. (1996), Kwon et al. (2000) and Blair et al. (2001).

The effect of excavation of drifts and deposition holes for the waste canisters, the heat release from the radioactive wastes or spent nuclear fuels and the swelling pressure of the bentonite backfills are usually studied in the so-called near-field models. Barton et al. (1995) used a UDEC-BB model to analyze the rock mass response to boring of access tunnels and excavation of caverns for low level wastes at the Sellafield site in the UK. Shen and Stephansson (1991), Stephansson and Shen (1991), Stephansson et al. (1991b), Hansson et al. (1995) and Jing et al. (1997) applied two- and three-dimensional DEM and BEM modeling to the rock mass response of a KBS-3 deposition concept in hard rocks suggested by the Swedish Nuclear Fuel and Waste Management Company (SKB, 1992). To explore the differences between finite element continuum representation and discrete fracture representation and their responses to thermo-mechanical loading, Jung and Brown (1995) and Chen et al. (2000) performed drift responses in the fractured rocks at the Yucca Mountain site in Nevada, USA. Two-dimensional DEM modeling of the effect of glaciation, ice flow and deglaciation on large faulted rock masses for a radioactive waste repository were performed by Jing and Stephansson (1987) and Cryssanthakis et al. (1991a,b). Some other representative contributions by the authors of this book can be seen in Stephansson et al. (1994, 2001, 2004), Jing et al. (1995, 1996a,b) and Tsang et al. (2004).

12.5.1 Case study example 1 for radioactive waste disposal – 3D DEM prediction of repository performance under thermal and glacial loadings

To demonstrate modeling of coupled thermo-mechanical response of a potential repository under loading of *in situ* stresses, heating from spent nuclear fuel and a glaciation–deglaciation cycle, the contribution by Hansson et al. (1995) and Jing et al. (1997), are presented as a typical application example. The primary purpose of the studies was to investigate: (i) temperature distribution and thermal gradients in the vicinity of the repository; (ii) stress distribution and fracture deformation due to thermal loading and glaciation; and (iii) stability conditions of both the global rock formation and the near field around the tunnel and deposition holes. To achieve these aims, four DEM computational models were used: one large-scale far-field model without excavation of tunnels and deposition holes; and three small-scale near-field models established from three different fracture network realization models and with consideration of excavation of tunnels and deposition holes and swelling pressure from the bentonite backfill. Data concerning geology, fracture zones, fracture networks, virgin state of stress and material properties were taken from the Äspö area and the Äspö Hard Rock Laboratory, Southern Sweden. The hypothetical repository is located 500 m below the ground surface of the Äspö area. Figure 12.34 illustrates the geological structures (mainly fracture zones) and the $4 \times 4 \times 4$ km computational 3DEC model containing 23 major fracture zones, 1024 blocks and 84 574 finite difference elements.

An inner region model of $1.5 \times 1.5 \times 1.5$ km in size contains the repository and was specified with a finer mesh. The potential repository is assumed to have 4000 canisters in total and they are placed in the floors of 10 parallel tunnels with a spacing of 40 m between the tunnels. Along each tunnel, 40 deposition holes are arranged with an average spacing of 6 m. The orientation of the model is such that the axes of the two principal horizontal *in situ* stress components are normal to the vertical boundaries and parallel to the assigned z- and x-axes of the coordinate system. The bottom surface was fixed and the top surface was free, representing the ground surface. The initial temperature field in the model was specified with a gradient of 16°C/km and 6°C at the top surface. Adiabatic thermal boundary conditions were used for all thermal calculations. The boundary stresses on the vertical boundaries of the far-field model vary with depth and are extracted from rock stress measurements.

The near-field model is based on a tunnel-deposition hole system adopted from the KBS-3 concept (SKB, 1998). Three different DEM models were generated with 139, 202 and 126 fractures,

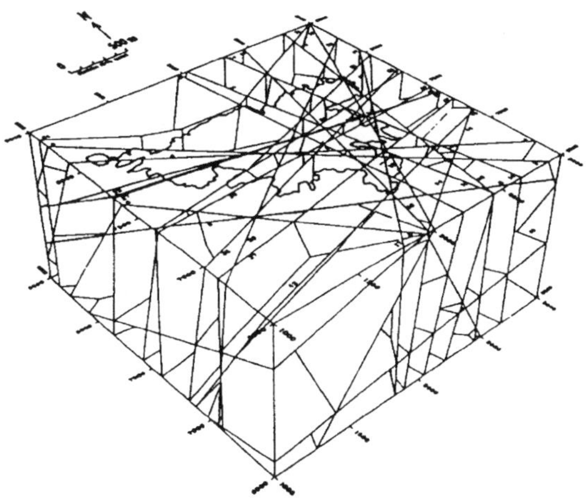

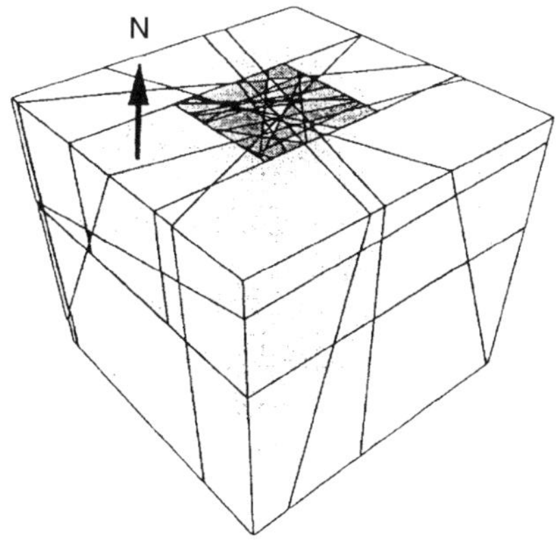

Fig. 12.34 Structural model of the fracture zones in the Äspö area and 3D far-field model (Jing et al., 1997).

respectively. The dimensions of the models were 25 × 25 × 18 m, containing an inner region model of 7.5 × 7.5 × 7.5 m in size. The fracture pattern of one of the fracture system had realizations with 139 fractures and the corresponding DEM model after simplifications is shown in Fig. 12.35. Boundary stresses were governed by the state of stress at 500 m depth. The rock mass were assumed to be a linearly elastic material for the far-field model and to be a Mohr–Coulomb plastic material for the near-field models. A Coulomb friction law with constant shear and normal stiffness were assumed for the fractures. For far-field models, the stiffness values are directly proportional to the rock elasticity and inversely proportional to the width of the fracture zones. Two different heat source functions were used for the time-dependent heat release of the canisters. Detailed information about the temperature-release functions of the wastes and material properties used in the modeling can be seen in Hansson et al. (1995) and Jing et al. (1997).

The impacts of five major events were analyzed: (i) virgin stresses; (ii) thermal loading due to heat generated by emplaced wastes; (iii) weight of a future ice sheet of 2 200 m in thickness due to glaciation and later ice removal by melting; (iv) excavation of the tunnel and deposition hole; and (v) the swelling pressure of 10 MPa on the wall of the deposition hole due to expansion of the compacted bentonite backfill. Only impacts from events (i), (ii) and (iii) were considered for far-field modeling.

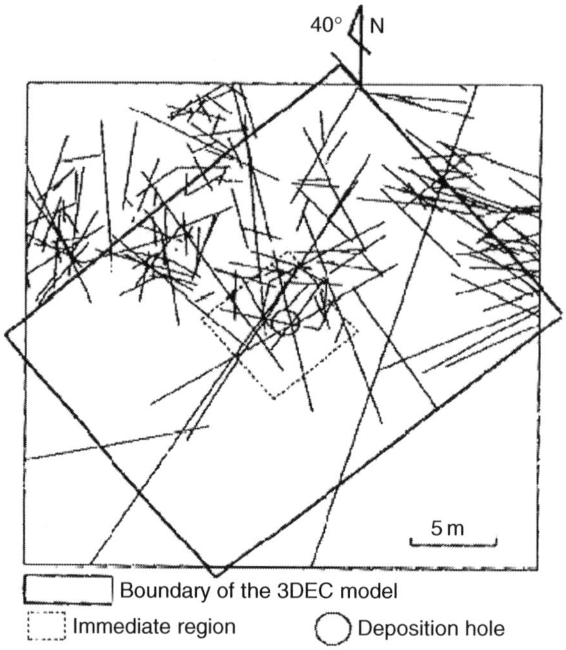

(a)

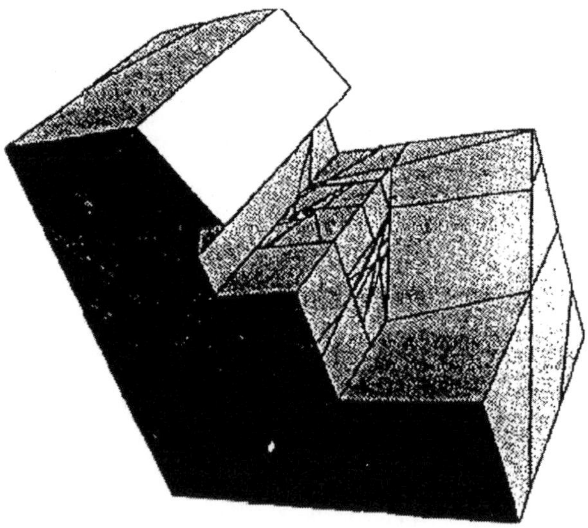

(b)

Fig. 12.35 (a) Fracture network model of Äspö rock mass and (b) 3D near-field model after simplification (Jing et al., 1997).

Results of the far-field 3DEC model gave a peak temperature of about 48°C at the center of the repository 200 years after emplacement of the canisters. The peak temperature will then remain for about another 200 years, followed by a gradual decrease until returning to the original temperature field at 60 000 years (Fig. 12.36). Deformation of the fracture zones is in general larger during ice loading than during heating. At repository level, the mode of fracture deformation is in general closure during heating,

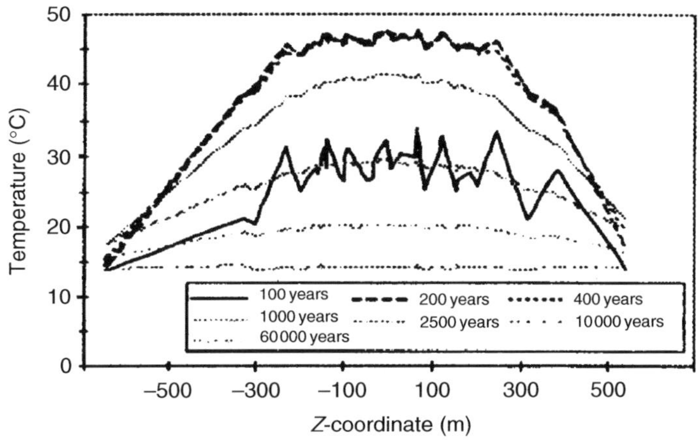

Fig. 12.36 Temperature distribution along a central profile across the far-field model of a potential repository during heating (Hansson et al., 1995).

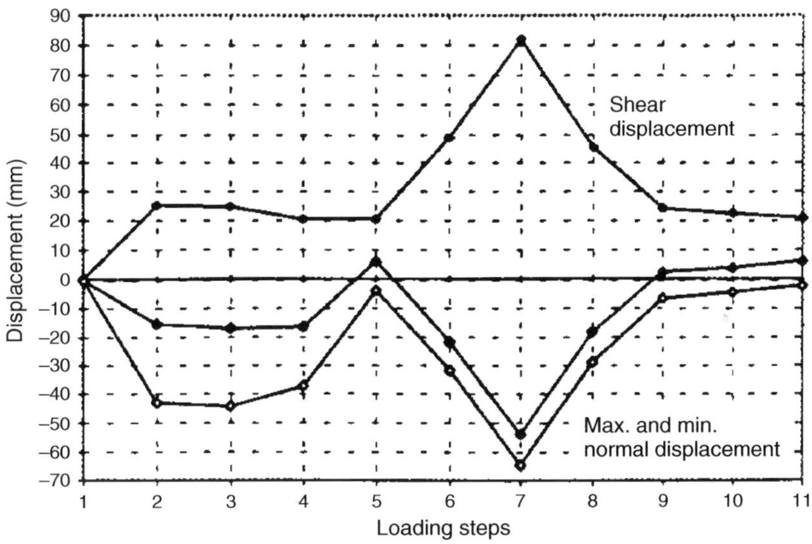

Fig. 12.37 Shear and normal displacements for fracture zone No. 43 in the 3DEC model during loading. Opening of fracture zone is positive; (1) present conditions; (2) 200 years heating; (3) 400 years heating; (4) 1000 years heating; (5) 60 000 years heating; (6) 1000 m ice thickness; (7) 2200 m ice thickness; (8) 1000 m ice thickness; (9) 200 m ice cliff; (10) 100 m water; (11) end of glaciation (Hansson et al., 1995).

opening during cooling and shearing due to the ice loading. The variation of maximum shear and normal displacement of one of the fracture zones (No. 43) intersecting the potential repository area is presented in Fig. 12.37. The maximum shear and normal (closure) displacements of this fracture zone are 82 and 65 mm, respectively. The fracture opening in the normal direction (positive value) is only 6 mm and appears after 60 000 years of heating and at the end of the glaciation cycle.

The measured stresses in borehole KAS03 on Äspö Island and the calculated stresses at the virgin state in the model are shown in Fig. 12.38. The measured stresses from a nearby borehole KAS02 were used to define the horizontal boundary stresses in the 3DEC model. The maximum net increase of stress magnitude due to heating is 10 MPa in the horizontal direction and 3 MPa in the vertical direction, and

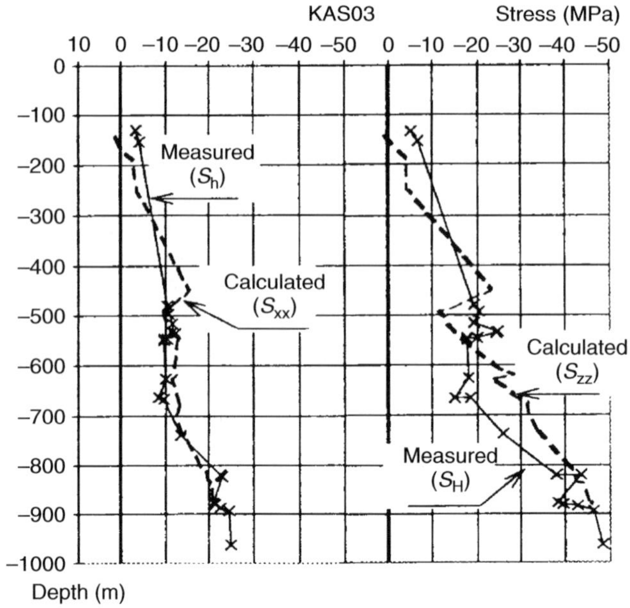

Fig. 12.38 Comparison of measured and calculated horizontal stress components along borehole KAS03 at Äspö (Hansson et al., 1995).

10–22 MPa in the horizontal direction and 10–30 in the vertical direction, due to glaciation. Large variation of stress occurs near the intersections of fracture zones, with either stress concentrations or relief, depending on the relative movement of blocks. This has been confirmed in a recent study of the stress field at Äspö Hard Rock Laboratory located on the island of Äspö (Ask, 2004). Stress variation of the order of ±10 MPa is likely to occur at 500 m depth due to the existence of fracture zones.

Results from the near-field 3DEC models show a major change of stress state and fracture displacement in the vicinity of the tunnel and deposition hole due to excavation, swelling pressure, thermal loading and glaciation, Fig. 12.39. Thermal loading after 200 years from waste emplacement has the most severe effects on both the near-field stress state and fracture deformation. The swelling pressure from the compacted bentonite can reduce the degree of stress concentration around the deposition holes by ca. 12% and is, therefore, a positive supporting measure in terms of the stability of the wall of the deposition hole. Plastic deformation or yielding appeared in some local spots near the wall of the tunnel and deposition hole due to excavation. This yielding becomes more widely spread in the model with heating and loading from glaciation.

In conclusion, the cyclic processes of heating–cooling and glaciation–deglaciation cycles were found to have significant impacts on the stability and safety of the repository. The modeling results show that it is possible for these events to cause instability for both far-field and near-field rock masses, cause fracture initiation, propagation and coalescence at some locations near major faults or fracture zones, and thereby may change pathways for groundwater flow. The swelling pressure from the compacted bentonite in the deposition holes is found to have minor but positive effects on the rock stability and fracture deformation. Closure is the major mode of fracture deformation during heating, which reduces the water inflow into the tunnel and deposition hole, and therefore may increase the resaturation time, but may also increase the stress concentration in the intact rock around the tunnel and deposition hole, which may, in turn, have a negative impact on the stability. Cooling of the repository and deglaciation cause a stress relief and reopening of the fractures, thereby changing the pathways for the groundwater flow,

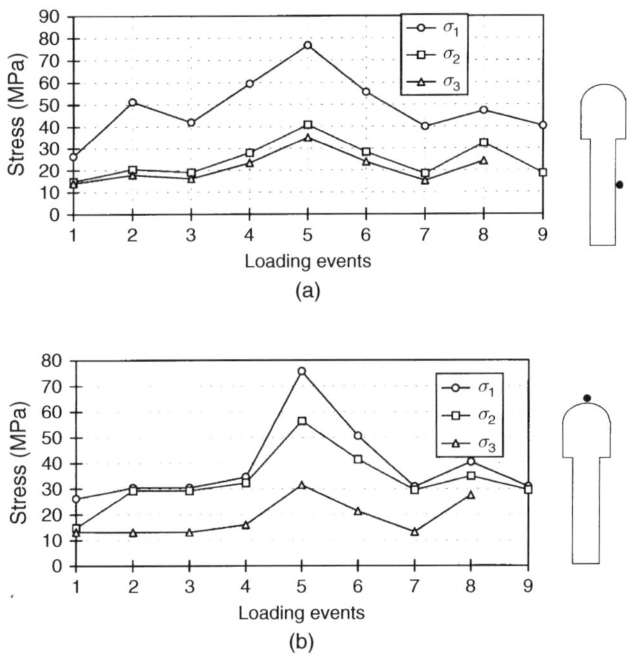

Fig. 12.39 3DEC modeling of the near-field response at the wall of the deposition hole as a function of different loading events; (a) principal stresses; (b) normal and shear displacements of fractures, where fracture closure is taken as positive. Loading events: (1) virgin stress; (2) excavation; (3) swelling pressure from compacted bentonite; (4) heating for 1 year; (5) heating for 200 years; (6) heating for 1000 years; (7) heating for 10 000 years; (8) glaciation; (9) deglaciation (Jing et al., 1997).

reducing the shear strength of the fractures and also reducing the stress concentration of the intact rock. A summary of the mechanical responses due to different loading mechanisms obtained from the three-dimensional DEM modeling is presented in Table 12.6.

Results from the modeling work presented by Hansson et al. (1995) and Jing et al. (1997) have been used to evaluate the impacts on safety and performance assessment measures of deep geological disposal in a hard rock repository by different loading mechanisms. Only the thermo-mechanical mechanisms and their couplings have been simulated in the presented work. The analysis considering fully coupled thermo-hydro-mechanical-chemical processes in fractured rocks is needed for more comprehensive analysis of performance and safety assessments of nuclear waste repositories considering not only mechanical stability but also the transport processes of the nuclides. The challenge for the reliability of such sophisticated analysis is not only on the computing power and numerical solution techniques, but more on reliable characterization of fracture systems and uncertainty estimations, representations and treatment.

12.5.2 Case study example 2 for radioactive waste disposal – 3D DEM study of water inflow into a deposition hole

The groundwater inflow into emplacement holes for nuclear waste disposal is an important factor to be considered for performance and safety assessments, and is influenced by a number of processes such as stress–permeability coupling, temperature, degassing of water, etc. These processes may depend on the host rock mass characteristics such as the arrangement of fractures in the near field, the mechanical

Table 12.6 Summary of mechanical responses from 3DEC modeling of a hypothetical repository due to different loading mechanisms (Jing et al., 1997)

Response	Excavation (near-field only)	Heating	Glaciation
Maximum Temperature		25–31°C[a] 48°C[b]	
Hole convergence	0.8 mm		
Maximum fracture shear displacement	0.4 mm	1.0 mm[a] 25 mm[b]	0.85 mm[a] 82 mm[b]
Maximum fracture normal displacement	0.14 mm (opening)	0.4 mm closure[a] 42 mm closure [b] 6.0 mm opening[b]	0.5 mm[a] 65 mm[b]
Material failure (plastic yielding)	Close to the tunnel and deposition hole	Widespread around fracture intersections	Widespread around fracture intersections

[a] From near-field model.
[b] From far-field model.

and hydraulic properties of the fractures and the initial stress field. Large-scale field tests were performed in Stripa and Äspö Hard Rock Laboratory, Sweden, where prediction of inflow into tunnels or deposition holes has been studied. The results from these tests show that the inflow into the larger holes is often less than predicted. Mas Ivars et al. (2004) have performed a generic numerical study with the 3DEC code containing an option of hydro-mechanical coupling for fluid flow and deformation in fractures, with the aim to improve understanding of the influence of the coupled hydro-mechanical processes and the different rock mass characteristics involved in the prediction of inflow into deposition holes. Realistic data from Äspö Hard Rock Laboratory, Sweden have been used whenever possible.

The model consists of a rock mass block of 20 × 20 × 20 m in size with a single sub-vertical fracture intersecting the block. A vertical deposition hole of 8 m in length and 2 m in diameter was excavated in the center of the rock cube so that the fracture intersected the hole (Fig. 12.40).

The simulated fracture has a strike of 127° and a dip of 84.2°. The maximum principal stress at Äspö has an approximate trend of 150° and a plunge of 0°. As the maximum principal stress in the model

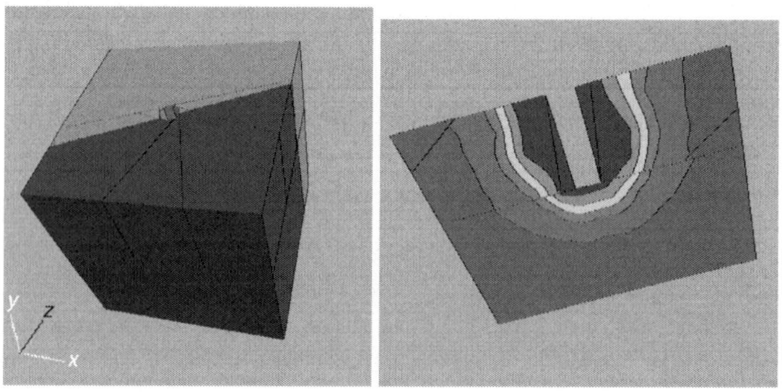

(a) (b)

Fig. 12.40 (a) Model geometry (20 × 20 × 20 m) with fracture, (b) fracture plane with aperture contours (*Mas Ivans et al., 2004*).

Table 12.7 Fracture properties used in the simulation of inflow to a deposition hole with 3DEC

[k_n[a] (GPa/m), k_s[b] (GPa/m)]	[20, 12], [61.5, 35.5], [360, 210]
Friction angle (°)	25, 30, 40
Dilation angle (°)	0, 5
Cohesion (MPa)	0
Initial hydraulic aperture (μm)	30
Residual hydraulic aperture (μm)	5
Maximum hydraulic aperture (μm)	60

[a] k_n, normal stiffness.
[b] k_s, shear stiffness.

corresponds to σ_{xx}, the fracture dip direction was set to 337°, and the dip angle to 84.2° in order to keep the same relative orientation between orientation of the single fracture and the *in situ* stress at Äspö. The model has roller boundary conditions on all sides. The boundary water pressures at all six boundaries were set according to the hydrostatic pressure gradient in the center of the model at 500 m depth, with a water pressure of 4.9 MPa at the top surface and 5.1 MPa at the bottom, respectively.

The initial stress field was set according to the depth-dependent linear relation

$$\begin{aligned}\sigma_{xx} &= 0.0373(-z) + 4.3 \ \ [\text{MPa}]\\ \sigma_{yy} &= 0.027(-z) \ \ [\text{MPa}]\\ \sigma_{zz} &= 0.0174(-z) + 3.3 \ \ [\text{MPa}]\end{aligned}$$

where z is the depth in meters. Then, the principal stress magnitudes in the center of the model become $\sigma_{xx} = 22.95$ MPa, $\sigma_{yy} = 13.5$ MPa and $\sigma_{zz} = 12$ MPa, respectively.

The intact rock was modeled as an isotropic, homogeneous and linearly elastic material with a density of 2700 kg/m^3, a Young's modulus of 40 GPa and a Poisson's ratio of 0.22. An elasto-perfectly plastic Mohr–Coulomb model was chosen for the fracture behavior. The uncertainty in the rock mass response was considered by studying different sets of values and combinations of fracture properties. Table 12.7 shows the list of fracture properties used in the simulations.

In Fig. 12.41 the aperture contours of the fracture intersecting the emplacement hole are shown for different normal and shear stiffness, Kn and Ks. There is a clear increase of water inflow to the deposition hole as the fracture stiffness increases. Figure 12.42 shows the measured inflow for simulations with Ks = 3 and 210 GPa/m, kn = 360 GPa/m and 5° dilation angle of the fracture. The influence of the variation of dip of the fracture zone on the water inflow was also investigated. Keeping the dimension of the model 20 × 20 × 20 m and all the fracture parameters as defined in Table 12.7 and the rock mass initial characteristics and boundary conditions constant, the dip angle was changed from 84.2° to 90° and its effect on the inflow studied. In general, the inflow is higher for the case of the dip angle of 90° due to the lower effective normal stress acting across the fracture and the difference in the maximum shear displacement is found to be more than double for the cases with the dip angle of 84.2° than that with the dip angle of 90°.

In total, 560 3DEC simulations were run with different parameter values for model size, fracture stiffness, orientation of the fracture zone, *in situ* stress magnitude, fluid bulk modulus and pore pressure, initial and residual hydraulic aperture and fracture friction angle and dilation angle with the uncertainty ranges listed in Table 12.8 and the fracture stiffness listed in Table 12.7. The aim of the systematic simulations was to improve the understanding concerning which of the listed parameters have the most significant influence on the inflow to the deposition hole. A Pearson correlation analysis

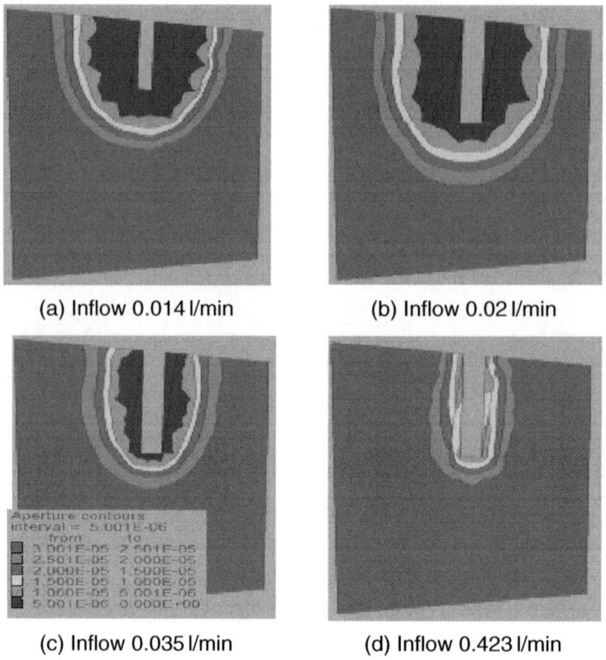

Fig. 12.41 Fracture aperture contours around the excavation and inflow measured with friction angle 30°, dilation 0° and different k_n (GPa/m), k_s (GPa/m): (a) for 30 × 30 × 30 m model; (b)– (c) and (d) 20 × 20 × 20 m model.

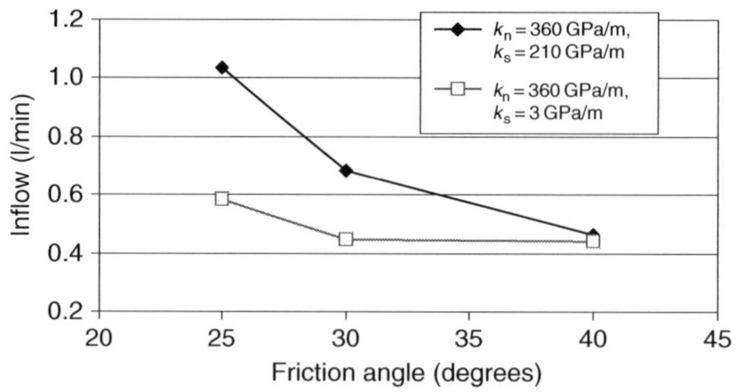

Fig. 12.42 Measured inflow versus friction angle of the fracture for a constant normal stiffness (360 GPa), a low shear stiffness (3 GPa/m) and a high shear stiffness (210 GPa/m) with a dilation angle of 5°.

Table 12.8 Uncertainty range considered in the correlation analysis

Fracture dip angle (°)	84.2 and 90
Fracture dip direction (°)	315, 337 and 360
Magnitude of *in situ* principal stresses	± 20%
Fracture pore pressure	± 20%
Fluid bulk modulus (GPa)	0.2 and 2
Model size (m)	20 × 20 × 20 and 30 × 30 × 30
Grid size (m)	1 and 2

Fracture properties according to data in Table 12.7.

Table 12.9 Correlation analysis of inflow, maximum and minimum shear and normal displacements from totally 560 3DEC simulations (Mas Ivars et al., 2004)

Parameters	Inflow	Max shear displacement	Max normal displacement
Normal stiffness, k_n	0.329[a]	0.053	−0.620[a]
Shear stiffness, k_s	0.327[a]	0.040	−0.600[a]
Friction angle, degrees	−0.098[b]	−0.272[a]	0.008
Dilation angle, degrees	0.123[a]	−0.190[a]	−0.052
Principal normal stress, σ_{xx}	0.046	0.256[a]	0.056
Principal normal stress, σ_{yy}	0.006	0.032	0.007
Principal normal stress, σ_{zz}	−0.161[a]	−0.307[a]	0.016
Pore pressure, P	−0.002	0.039	0.025
Dip angle, degrees	−0.033	−0.183[a]	−0.082
Dip direction, degrees	−0.021	−0.318[a]	− 0.172[a]
Fluid bulk modulus, K_f	−0.010	0.015	−0.001
Fracture length	−0.035	0.241[a]	0.032
Model size	−0.014	0.006	−0.009
Grid size	0.015	0.050	0.008
Maximum	18.072 l/min	4.9×10^{-3} m	9.1×10^{-4} m
Minimum	0.017 l/min	1.1×10^{-5} m	

[a] Correlation significant at the 0.01 level (two-tailed).
[b] Correlation significant at the 0.05 level (two-tailed).

was then conducted for all the factors related to the inflow, maximum shear and normal displacements (without considering the residual aperture limit). The initial and residual hydraulic apertures were not considered in the correlation analysis. The two most relevant parameters in the correlation analysis for the inflow to the deposition hole were found to be $k_n = 0.329$ and $k_s = 0.327$ (Table 12.9). Water pressure, fluid bulk modulus, model size and grid size of the models have minor correlations with respect to inflow to the deposition hole. Friction angle, state of stress in two of the normal directions, dip direction and fracture length are relevant parameters in the correlation analysis for the maximum shear displacement, while the normal and shear stiffness are the overall dominant parameters for the maximum normal displacement.

On the other hand, a low correlation does not mean that the inflow, maximum shear displacement or maximum normal mechanical displacements were not influenced by the parameters in question. Rather, it means that the uncertainty ranges of the low correlated parameters are less significant than the uncertainty ranges of the highly correlated ones in this study.

12.6 Rock Reinforcement

The aim of ground reinforcement of rock masses is to maintain stable conditions above and around excavations, to provide safe working conditions during construction and to provide a structure with long-term performance characteristics. Rock bolts, steel sets, wire mesh and shotcrete are the most common types used in the stiffer rocks. Rock bolts are used alone when the rock consists of large durable blocks, but are better used in combination with wire mesh and shotcrete, with or without steel fibers, when the rock mass is weak. Rock bolts are of two types: point anchored with or without pre-stressing; and cement or resin-grouted rebars. Rock bolts are used to obtain local reinforcement of the rock mass. Cables or tendons are used to reinforce large underground structures in rock masses.

Observations of excavations in highly fractured rock masses show that when large displacements occur, relaxation-type of numerical methods, such as the DEM, appear to be suitable to model the main responses. Rock reinforcement is often designed to interact with the rock mass and therefore was implemented in the DEM codes for analysis of rock strength and deformability.

The behavior of point-anchored reinforcement was assumed to act as a 1D bar element by Lorig (1984). Later, Lorig (1985) included the effect of shear stiffness and strength from the so-called dowel action of fully bonded reinforcement to simulate fully cement grouted un-tensioned rebar. Large shear displacements are accounted for in the simulation. The numerical model consists of two springs located at the discontinuity interface and oriented parallel and perpendicular to the reinforcement axis. The loads mobilized in the model by local deformation are related to the displacements by the axial and shear stiffness of the bolt.

In a rock mass subjected to a mixed mode of deformation, the spatially extensive reinforcement is mechanically more appropriate than the local deformation model described above. Because local resistance to shear is considered insignificant for the spatially extensive reinforcement, a one-dimensional constitutive model is adequate for describing the axial performance. Brady and Lorig (1988) suggested that the finite difference representation of the axial extension of a cable or tendon could be simulated by dividing the cable into separate segments where each segment is simulated by a spring of stiffness equivalent to the shear stiffness and limited by a plastic yield criterion. Interaction between the cable and the rock is modeled by a spring-slider unit, with the spring representing the stiffness of the grout and with the limiting shear resistance of the slider representing the ultimate shear load capacity of the grout or any of the contacts between the rock/grout or grout/cable. The reinforcement models suggested by Lorig (1985) and Brady and Lorig (1988) have been implemented in the UDEC, 3DEC and FLAC codes and applied to large stope hanging wall reinforcement (Brady and Lorig, 1988), tunnel support (Lorig, 1987; Makurat et al., 1990a; Wong et al., 1993) and supports of mines and mining structures (Brady and Brown, 1985; Brady and Lorig, 1988; Rosengren et al., 1992; Ng et al., 1993).

Shotcrete and, more recently, fiber reinforce shotcrete have been widely used as temporary and/or permanent support systems for underground excavations and tunnels. Chryssanthakis and Barton (1995) have established an algorithm for simulation of the behavior of fiber-reinforced shotcrete in multilayers in underground structures and have implemented the routine in the UDEC and UDEC-BB codes. The area of application of the shotcrete to the rock surface of an excavation is specified and UDEC automatically creates the elements necessary to represent a uniformly applied layer. Moment–thrust diagrams are used to illustrate the maximum force that can be applied to a typical section for various eccentricities of the excavation. The ultimate strength of fiber-reinforced and non-reinforced shotcrete are the same, while the reinforced shotcrete has a residual strength that remains after failure at the ultimate load. The outputs of UDEC modeling of rock reinforcement are axial forces on rock bolts, forces, moments and failure mode on the shotcrete elements. A shallow test tunnel excavated in fractured biotite gneiss in Korea has been used for demonstrating the capacity of the fiber-reinforced shotcrete sub-routine of UDEC. Stability analyses of a tunnel supported by a system of shotcrete and rock bolts were published by Hwang et al. (2001).

12.7 Groundwater Flow and Geothermal Energy Extraction

The fluid behavior of the fractured rock masses is of considerable interest in such areas as hydrocarbon recovery, nuclear waste disposal, geothermal energy extraction, underground storage and transport of fluids and contaminant transport. Fractures affect both the mechanical and hydraulic behavior of a rock mass and the behavior is coupled – in the sense that fracture conductivity is dependent on fracture deformation and, conversely, the fluid pressure in the fracture affects the mechanical response. This two-way coupling is of importance in conducting hydro-mechanical simulations with discrete element modeling.

Early formulation of the UDEC code was strictly limited to steady-state confined flow. Nevertheless, essential features of the hydro-mechanical behavior of engineering problems were captured. An analysis of flow in fractured rock beneath concrete gravity dams is reported in Lemos (1987); Fairhurst and Lemos (1988) used the early version of the UDEC code to study the influence of fractures on water losses in pressure tunnels for a hydro-electrical power plant and the validity of the hydraulic fracturing test as an indicator for determination of the need for installation of lining to such tunnels. Lemos and Lorig (1990) developed an efficient algorithm for fully coupled mechanical-hydraulic analysis in which water pressures in fractures were taken into account in the mechanical computations. Both steady-state and transient fluid flow problems can be analyzed for confined flow and flow with a free surface. Ferrero et al. (1993) used the UDEC code to study the hydro-mechanical behavior and stability of the long shore Pellestrina breakwaters of the Venice lagoon dam in Italy. The breakwaters are made up of large stone blocks fixed by silt and pozzolanic cement and are subjected to sea-storm action and sea erosion.

Gutierrez and Barton (1994) used the UDEC code with the hydro-mechanical coupling functions (Itasca, 1991) to study the hydraulic and mechanical properties of a single fracture in a 100 mm long and 20 mm thick fractured block. The fracture is simulated by connecting a series of straight segments, which are obtained from digitized profiles of the fracture surface. Each segment is allocated the appropriate hydraulic and mechanical properties, where the hydraulic aperture is governed by the sum of the initial hydraulic aperture and the normal deformation of the fracture. Kim and Lee (1995) applied the fully coupled hydro-mechanical capacity of the UDEC code (version 1.83) to study the performance and stability of six oil storage caverns in hard rocks in Korea with water curtain tunnels above them. The authors conclude that the redistribution of stresses from excavation compressed fractures with an orientation perpendicular to the maximum stress direction, resulting in an increase of water pressure in the fractures. However, the effect of changing groundwater pressure on convergence measures and stability was small.

12.7.1 Case study – flow rate injection test for the Hot Dry Rock (HDR) project in Cornwall, England

Modeling of rock mass hydraulics by conventional techniques for continuous porous media gives no insight into the fluid–rock interaction process, particular at hydraulic pressures near or above minimum virgin rock stress levels where fractures might open and propagate. For this reason, Pine and Cundall (1985) developed the Fluid–Rock Interaction Program (FRIP) that is a special version of the early distinct element program that incorporates fluid pressure and flow in an assembly of deformable but impermeable blocks. The FRIP model consists of rectangular blocks separated by fractures in which laminar fluid flow occurs. The fracture aperture changes with fluid pressures with fracture compliance and, under certain conditions, fracture dilation by shearing can occur. The solution technique in FRIP is similar to that of the UDEC code and the successive development steps and the main features of the FRIP code are presented in Pine and Cundall (1985).

The FRIP code was used for the modeling of fluid flow of high pressures in fractured rocks, in conjunction with field observations at the Camborne School of Mines Geothermal Project, with the results from modeling two different rocks and flow rate injection tests relating to the deep HDR system in Cornwall. The first model simulated a medium flow rate injection into a 2 km deep well with a flow rate of 24 l/s for about 9 min. Introduction of elastic boundaries permitted the model to be run on a 30×15 block grid where the injection hole is located at the center of the model. The measured and modeled results agreed well in terms of pressures (versus time), see Fig. 12.43. The kink in the modeled curve is

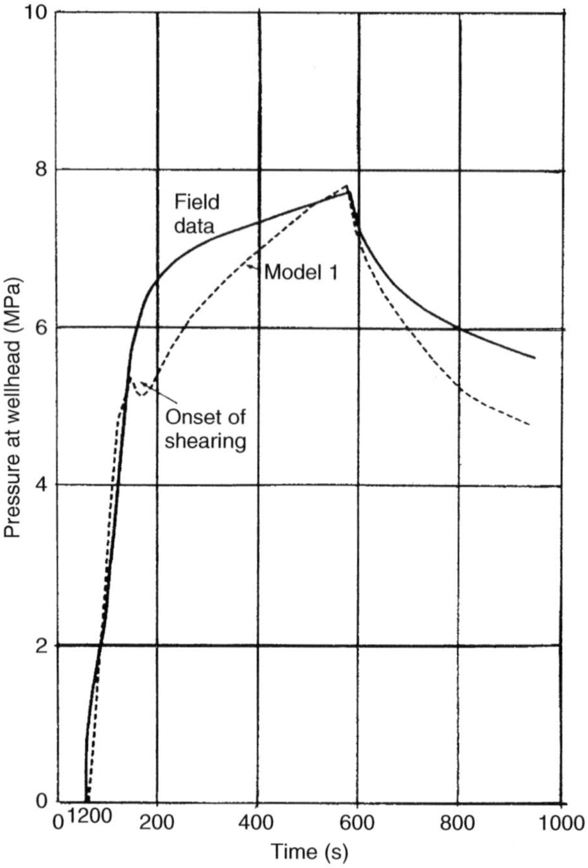

Fig. 12.43 Medium flow rate injection at the Camborne School of Mines HDR project in Cornwall, England; FRIP modeling and field data of wellhead pressure versus injection time (Pine and Cundall, 1985).

the onset of the shearing of joints near the well and the pressure drop is caused by dilation of the fractures. Shearing of the fractures close to the well resulted in stress redistribution and the appearance of zones of dilation and pore pressure reduction at locations remote from the well.

In a subsequent FRIP modeling study, stimulation in rocks with two sets of persistent fractures intersecting at right angles was simulated for a 2:1 horizontal effective stress ratio, as shown in Fig. 12.44. Since the horizontal stresses are not aligned with the fracture direction, shear stresses were present in the fractures. Fluid injection in the well induced fracture slip when the effective stress was less than or equal to $\tau/\tan\phi$ where τ is the shear stress along the fracture and ϕ is the friction angle. The slip causes dilation of the fractures in an *en echelon* pattern and the direction of the induced propagation lies between the orientation of the major principal horizontal stress and the strike of one of the fracture sets, as depicted in Fig. 12.44. This trend of fracture dilation and propagation was consistent with the observation of micro-seismic locations of the main hydraulic injection test. The fracture dilation was mostly due to strike–slip shearing and the thousands of detected micro-seismic emissions also indicated strike–slip shearing consistent with the FRIP model (Pine and Cundall, 1985; Pine and Nicol, 1988). The results of the Carnmenellis granite field test, part of the Camborne School of Mines HDR project and the associated FRIP modeling, led to several key points in the understanding of fluid–rock interaction and the hydro-mechanics of fractured rock masses, e.g., the nature and location of the micro-seismic events, *in situ*

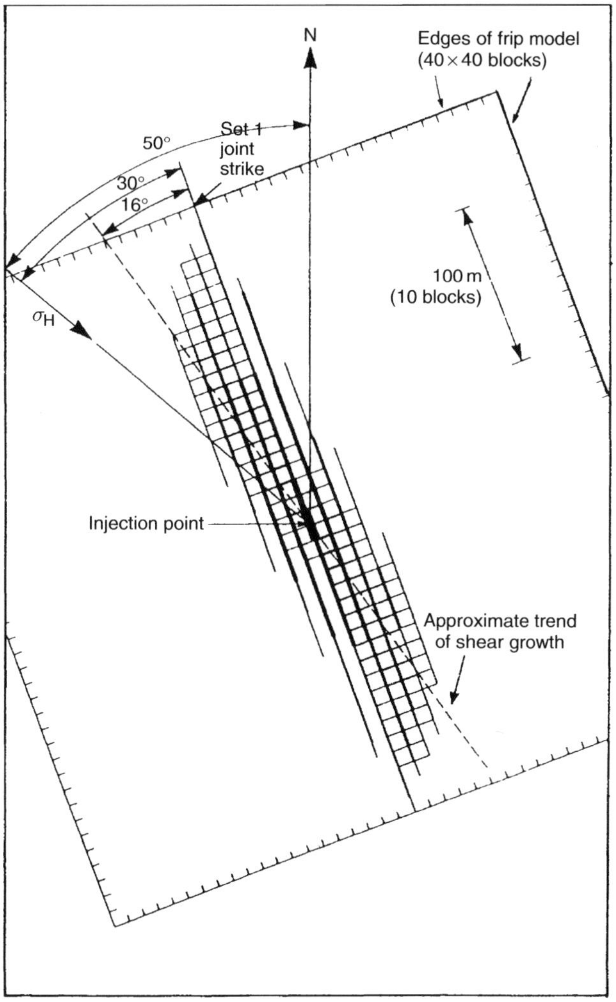

Fig. 12.44 Fracture dilation (black lines) pattern from FRIP modeling of a large hydraulic test for the HDR project in Cornwall (Pine and Cundall, 1985).

fracture and stress measurements and the pressure/flow/time records for the injection tests. The main conclusions are that high-pressure injections must take account of the pre-existing stress and fracture regimes, which create pressure-dependent anisotropies during active stimulation. In areas where large stress anisotropies are anticipated, shear-dominated stimulation processes are likely to result, with the dilation and shear displacement directions governed by stress gradients and the relative fracture and stress orientations.

Although the FRIP code proved to be a valuable tool for studying the water injection into a fractured geothermal reservoir, some important aspects/capabilities were missing, e.g., the block geometry is restricted to be rectangular, no material non-linearity can be treated and large displacements cannot be accommodated. To overcome these limitations, Last and Harper (1990) made improvements to the flow modeling. Fluid flow through the block system is modelled as the diffusion of a single, saturated compressible fluid through the connected network of porous blocks and fractures. Individual, fluid-filled elements (blocks, fractures and voids) are assumed to form uniformly pressured reservoirs between which the fluid transfer occurs according to simple one-dimensional flow laws. The dominant path for fluid flow is along the fractures, but flow through blocks is also an option.

The modified code by Last and Harper (1990) has been used to analyze the injection of a fluid into a jointed rock mass under conditions that might be representative of a hydrocarbon reservoir at about 3 km depth (Harper and Last, 1990a,b). Fluid leak-off through fracture walls and fracture dilation during shear are modeled for different combinations of fracture geometry and applied stresses.

In summary, the DEM models and codes have been widely and extensively applied in simulating many aspects of mechanical, hydraulic and thermal behavior of fractured rocks for different engineering projects or geological and seismic hazards. The success of the DEM approach is beyond doubt and will continue to be applied and further developed in future. However, geometric and physical characterization of the fracture system and proper simplification to capture the dominating issues of influence regarding the objectives of engineering projects and geophysical processes remain to be the basis of successful applications, as pointed out by the guidelines at the start of this chapter.

12.8 Derivation of Equivalent Hydro-mechanical Properties of Fractured Rocks

12.8.1 Fundamental Concepts in the Continuum Approximation of Fractured Rocks

Continuum models with equivalent properties are necessary for large-scale applications when explicit representation of a large number of fractures of different sizes becomes computationally intractable. The equivalent continuum approach depends on that: (i) the existence of a representative elementary volume (REV) of the fracture systems can be established; and (ii) the derived equivalent physical properties at the REV scale can be approximately expressed as tensors. An example is the equivalent hydraulic permeability tensor established with the criterion that the reciprocal of the square root of directional permeability in the direction of hydraulic gradient can be approximated by an ellipse in a polar diagram – for 2D (Bear, 1972; Long et al., 1982; Khaleel, 1989).

In theory, a REV is defined by the minimum volume of sampling domains beyond which the characteristic properties of the material in the domain remain essentially constant (Bear, 1972). This can be achieved only when the fracture system in the region of interest is homogeneous in terms of the distribution of its geometrical parameters such as density, size, orientation and connectivity. In reality, the fracture systems are generally heterogeneous and the geometric properties of the fracture systems often exhibit scale dependence. Therefore, generally speaking, there is no guarantee that a REV always exists for a given fractured rock mass (Neuman 1987; Panda and Kulatilake 1996). In practice, when fracture mapping data sets have sufficient quantity and quality so that the region of interest can be divided into a number of sub-regions with each of them having approximate homogeneity in terms of fracture system geometry, the equivalent continuum properties may be established for such sub-regions. Another technique is to treat the large-scale features such as fracture zones, large-scale faults and fault zones as deterministic objects that can be separately added to the numerical models, and the REV and equivalent properties are determined by considering only the fracture population of lesser, i.e., moderate or small, sizes (also see Section 5.4 and Fig. 5.16).

The existence of a REV is a necessary condition for the applicability of the equivalent continuum approach, but alone it is not sufficient. The sufficiency depends also on the existence of the property tensors that can adequately approximate the continuum behavior of the system. This implies that, even though REVs may be established for a fracture system of interest, the equivalent hydraulic properties determined may or may not have tensor qualities and hence may or may not be suitable for continuum analysis. Therefore, careful examinations as to the appropriateness of the equivalent continuum representation of fractured rock masses are needed, and the problem is basically site specific.

Since it is impossible to establish all the geometric information concerning the sub-surface fractures, the discrete fracture network (DFN) approach relies largely on stochastic realizations of fracture systems using the Monte Carlo simulation technique. Each realization should only be regarded as one possible partial representation of a real sub-surface fracture system in a statistical sense. Therefore, the fracture network realizations generated from the statistical properties of mapped fractures will have different geometric patterns, thus leading to different hydraulic behavior from realization to realization, although they may follow the same statistical distributions of the geometric parameters of the fracture population. This is the manifestation of the effects of the stochastic description of the, in reality, deterministic sub-surface structure systems, the so-called 'stochastic REV approach' (Min and Jing, 2003; Min et al. 2004a).

In the sections below, we present a hybrid DFN/DEM approach to derive the REV sizes and equivalent permeability and elastic compliance tensors of a fractured rock; this will be done as a generic 2D study but using realistic data from field investigations and a study of the stress effect on permeability, based on the established REV model. The aim is to establish a rigorous mathematical basis for solving such problems in theory and to identify the shortcomings of the 2D approximations, i.e., not aiming to provide representative properties of the site where the data were obtained.

12.8.2 REV and Derivation of the Equivalent Continuum Permeability of a Fractured Rock Mass Using a DEM Approach

The fracture system for this example study is based on part of the results from a site characterization at the Sellafield area, Cumbria, England (Nirex, 1995, 1997). The results are based on formations in the Borrowdale Volcanic Group, a thick sequence of Ordovician volcanoclastic rocks. It should be pointed out that a limited part of the characterization results were taken for this example study as part of a benchmark test of an international cooperative research project DECOVALEX (Andersson and Knight, 2000), and therefore the result presented here does not necessarily represent the complete results and conclusions from the entire site characterization at the Sellafield area.

12.8.2.1 Fracture system characterization

The benchmark problem is defined to be a two-dimensional generic problem. The basics of the fracture system characterization are described in Section 5.4. In addition, the minimum and maximum cut-offs of trace lengths are set to be 0.5 m and 250 m, respectively, corresponding to the measured fracture distributions. The cumulative probability density function of trace length (L) is then derived using the fractal dimension (D) (cf. Fig. 5.12)

$$L = (\mathrm{cut}_{\min}^{-D} - F(\mathrm{cut}_{\min}^{-D} - \mathrm{cut}_{\max}^{-D}))^{-\frac{1}{D}} \tag{12.1}$$

where $\mathrm{cut}_{\min}$ and $\mathrm{cut}_{\max}$ denote the minimum and maximum cut-offs of the trace lengths, F the random probability of a uniform distribution in the range $0 \leq F \leq 1$ and L the trace length of the fractures.

Orientations of the fractures are assumed to follow the Fisher distributions with the cumulative probability density function deviation angle (θ) from the mean orientation angle being derived as a function of the Fisher constant K, given by (Priest, 1993)

$$\theta = \cos^{-1}\left(\frac{\ln\left(e^K - F(e^K - e^{-K})\right)}{K}\right) \tag{12.2}$$

As the deviation angle is a one-dimensional expression measured from the mean normal direction of a fracture set, this must be converted to a three-dimensional form by rotating the generated unit normal vectors about the mean normal of the fracture set through a random angle taken from a uniform distribution in the range of 0 to 2π (Priest, 1993), using spherical coordinates and transformation of axis (Dershowitz et al., 1998). The generated 3D orientations are then converted to a 2D apparent orientation for this study.

The Monte Carlo Method is then used to generate the trace lengths and the orientations of fractures based on Eqns (12.1) and (12.2), after generating the locations of the fractures (cf. Fig. 5.13) following a Poisson process according to the following recursive equation

$$R_{i+1} = 27.0R_i - \mathrm{INT}(27.0R_i) \tag{12.3}$$

where R_i is a random number of uniform distribution in the range $0 \leq R_i \leq 1$, INT ( ) the integer part of the number inside ( ) and an initial value of R_0 is generated from the multiplicative congruential algorithm.

The fluid flow through individual fractures is assumed to follow the Cubic Law and the hydraulic aperture values are indirectly calculated from the laboratory shear flow tests using the following equation (Nirex, 1995)

$$e = \sqrt[3]{\frac{12Q\nu}{gwi}} \tag{12.4}$$

where e the hydraulic aperture (m), w the width of flow path (m), ν the kinematic viscosity (m^2/s), Q the flow rate (m^3/s), i the hydraulic gradient and g the acceleration of gravity (m/s^2). The calculated aperture from four samples ranged from about 30 to 100 μm, with a mean value of 65 μm.

The algorithm of the generation of stochastic DFN models (cf. Fig. 5.14) follows basically the approach suggested in Priest (1993) and Jang et al. (1996) and with consideration of minimization of the boundary effects (Samaniego and Priest, 1984).

12.8.2.2 Stochastic generation of DFN models

To avoid the boundary effect, 10 sufficiently large 'parent' DFN models of 300 m by 300 m in size were first generated based on the characterization parameters of the *in situ* fracture system. From each of the 10 large parent network models, 12 smaller DFN analysis models were extracted with varying sizes from 0.25 m by 0.25 m to 10 m by 10 m (Fig. 12.45). The side lengths of 0.25 m and 0.5 m were chosen to establish the

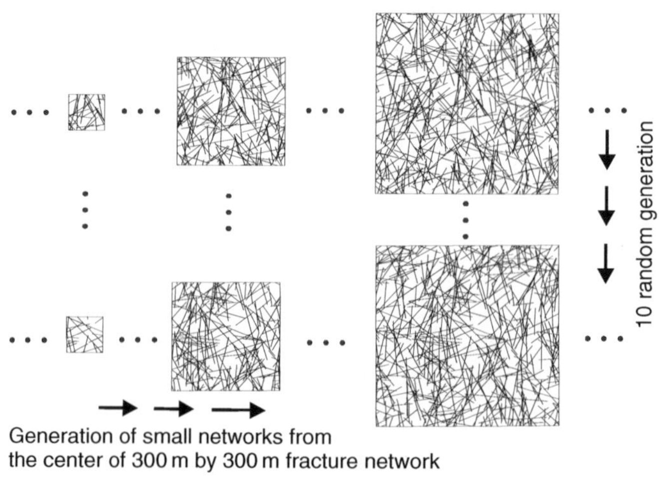

Fig. 12.45 Schematic view of the stochastic procedure for DFN model generation. The sizes of the models shown in the figure are 1 m × 1 m, 3 m × 3 m and 5 m × 5 m (Min and Jing, 2003).

influence of small domains since the minimum fracture length in this study was 0.5 m. The resultant DFN models, 120 in total (with 10 examples of $5\,\mathrm{m} \times 5\,\mathrm{m}$ in size shown in Fig. 5.14), were then used as the geometrical DFN models for the calculation of equivalent permeability values, using the two-dimensional distinct element code UDEC (Itasca, 2000). In order to check whether permeability has a tensor characteristic, one series of the DFN models was rotated at an interval of 30° in the clockwise direction for the calculation of the directional permeability values of the DFN models using the same generic boundary conditions.

12.8.2.3 Calculations of the REV size and the permeability tensor

The basic assumptions for this analysis are that the rock matrix is impermeable and the fluid flow occurs only through the fractures and obeys the Cubic Law, without considering the effect of roughness.

The fractured rock represented by the DFN model is then assumed to follow a generalized Darcy's law for anisotropic and homogeneous porous media (Bear, 1972) and is given by

$$Q_i = A \frac{k_{ij}}{\mu} \frac{\partial P}{\partial x_j} \tag{12.5}$$

where Q_i is the flow rate vector, A the cross-section area of the DFN model, k_{ij} the permeability tensor, μ the dynamic viscosity and P the hydraulic pressure applied. The elevation head is neglected.

Figure 12.46 shows the two sets of linearly independent boundary conditions for the calculation of permeability tensors. The flow rates in the x- and y-directions were calculated with a constant hydraulic pressure gradient in the x- and y-directions. Complete components of the 2D permeability tensors k_{ij} of the DFN models can be obtained for each DFN numerical experiment (Long et al., 1982). The numerical experiments were repeated for the rotated DFN models for calculations of directional permeability. If the directional permeability of the rotated DFN models can be approximated by an ellipse equation with two principal permeability values, k_a and k_b, defined (Bear, 1972) by equation

$$\frac{x^2}{\left(\frac{1}{k_\mathrm{a}}\right)} + \frac{y^2}{\left(\frac{1}{k_\mathrm{b}}\right)} = 1 \tag{12.6}$$

we say that the permeability can be approximated by as a tensor and continuum analysis can be applied.

The rotational transformation of a permeability tensor in a rotated Cartesian system in two dimensions is needed for the comparisons and establishment of the permeability tensor, and it is performed by the rotation mapping operations given as

$$k'_{pq} = k_{ij}\alpha_{pi}\alpha_{qj} \tag{12.7}$$

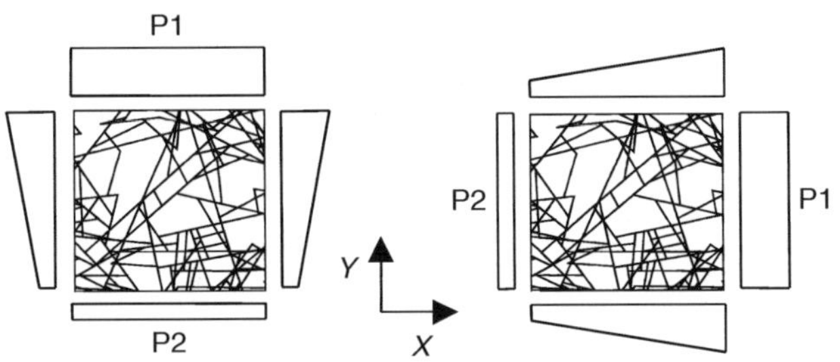

Fig. 12.46 Generic boundary conditions for calculation of the permeability tensor. P1 and P2 indicate the hydraulic pressure (Min et al., 2004a).

where k_{ij} and k'_{pq} are the permeability tensors in the original and rotated axes, respectively, and α_{pi} and α_{qi} are the direction cosines. The reciprocals of square roots of the directional permeability from the rotated models were plotted in a polar diagram to see whether they could be fitted to approximate an ellipse. In order to make such comparisons, an average permeability tensor, $\bar{k}_{ij}$, was first calculated at a given side length scale by averaging over the permeability values from all rotated models and transforming the average tensor to the pertinent rotation angles. The average tensor was calculated as

$$\bar{k}_{ij} = \frac{1}{N} \sum_{r=1}^{N} k^r_{pq} \alpha_{ip} \alpha_{jq} \tag{12.8}$$

where N denotes the number of rotations and k^r_{pq} is the calculated permeability in each rotated model.

Figure 12.47 shows the results of values of directional permeability components in a polar diagram at different model sizes. It is shown that at sizes of 5 and 10 m, acceptable approximations to ellipses of directional permeability can be achieved.

The results illustrate the fact that determination of a REV is a function of the resolution requirement for a given purpose. In order to assist in choosing an acceptable REV size with given resolution needs, two measures, the coefficient of variation and the prediction error, are suggested for the determination of a REV. The prediction error is defined to evaluate the error, or goodness of fitting, involved in deriving the permeability tensor and represents the relative errors of diagonal components in the permeability tensor, given by

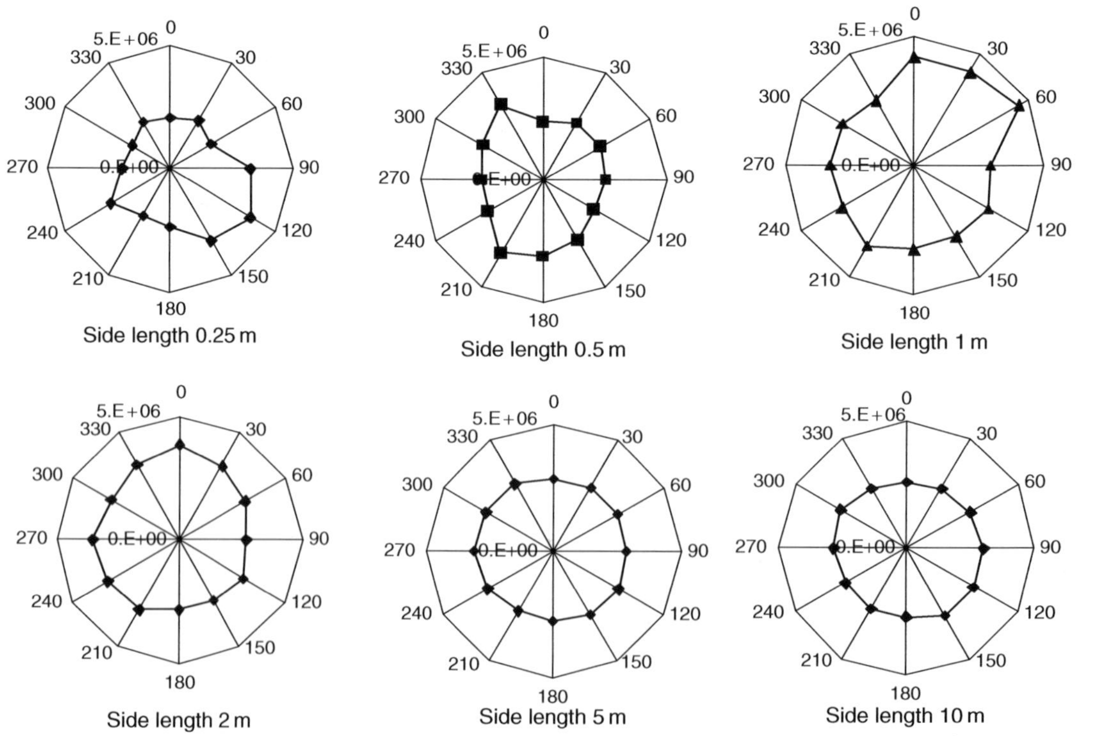

Fig. 12.47 Approximation of the equivalent permeability tensor with increasing model sizes (expressed in $K^{-1/2}(\theta)$) (Min et al., 2004a).

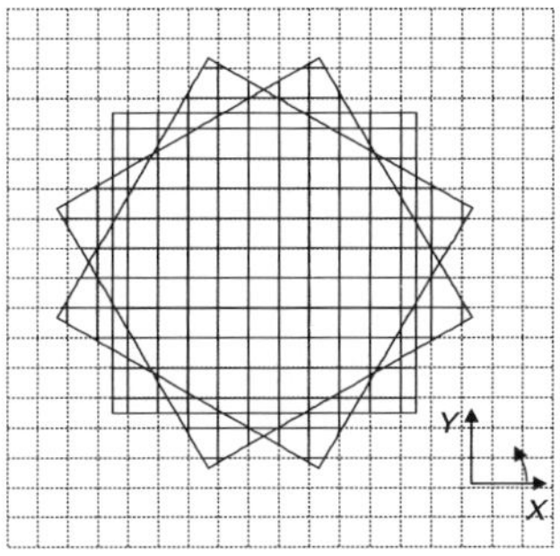

Fig. 12.48 Rotated computational models with two perpendicular sets of persistent fractures. The models in the figures are examples of 0°, 30° and 60° rotations.

$$\mathrm{EP}p_1 = \frac{1}{N}\frac{\sum_{r=1}^{N}|k_{11}^r - \bar{k}_{11}|}{\bar{k}_{11}}, \qquad \mathrm{EP}p_2 = \frac{1}{N}\frac{\sum_{r=1}^{N}|k_{22}^r - \bar{k}_{22}|}{\bar{k}_{22}} \tag{12.9}$$

where $\mathrm{EP}p_i$ is the prediction error of permeability tensor in the i-direction ($i = x, y$), $\bar{k}_{11}$ and $\bar{k}_{22}$ the diagonal components in the average permeability tensor and k_{11}^r and k_{22}^r the diagonal components in the permeability tensors from numerical experiments at the rth rotated state and N the number of rotations (six rotations for this study with 0°, 30°, 60°, 90°, 120° and 150°, respectively, see Fig. 12.48).

Note that, in evaluating the prediction error, the influence of non-diagonal components of permeability tensor was omitted for simplicity. The mean values of the prediction error ($\mu_{\mathrm{EP}p}$) were then defined as

$$\mu_{\mathrm{EP}p} = \frac{1}{2}\sum_{i=1}^{2}\mathrm{EP}p_i \tag{12.10}$$

At both the 5 and 10 m side lengths, the permeability ellipses could be approximated with 5% of mean prediction errors for this study. As a result, the final REV size can be established as 5 m (with 20% coefficient of variation and 5% prediction error) or 10 m (with 10% coefficient of variation and 5% prediction error). Figures 12.49 and 12.50 show the variations of k_{xx} and k_{yy} values at different values of the resolution measures.

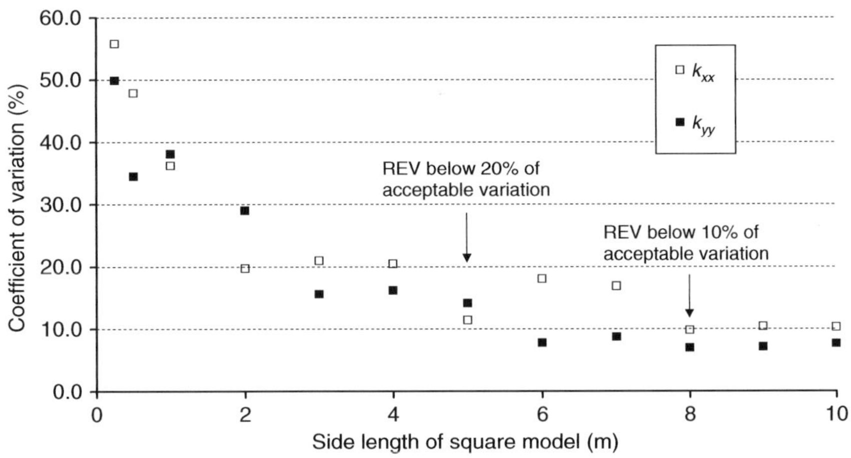

Fig. 12.49 The coefficients of variation of k_{xx} and k_{yy} with increasing side lengths of the DFN models (Min et al., 2004a).

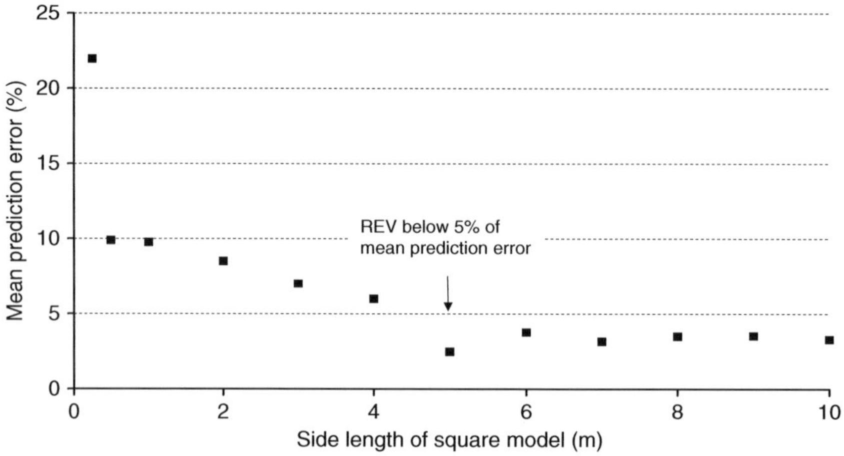

Fig. 12.50 Mean prediction error of the permeability tensors with increasing side lengths of the DFN models (Min et al., 2004a).

To undertake stochastic analysis of results, at the side lengths of 0.25, 0.5, 1, 5 and 10 m, the number of multiple realizations was extended to 50 in order to derive the statistical range of permeability values. Figure 12.51 shows the frequency of k_{yy} for the 50 stochastic DFN models with a side length of 5 m, which approximately follow a normal distribution.

Similar results were obtained at different sizes (Fig. 12.52). The determined permeability tensor at the 8 m scale is given by

$$k_{ij} = \begin{pmatrix} 1.00 \pm 0.099 & 0.0348 \pm 0.0745 \\ 0.0348 \pm 0.0745 & 1.29 \pm 0.09 \end{pmatrix} \times 10^{-13}\,(\mathrm{m}^2) \tag{12.11}$$

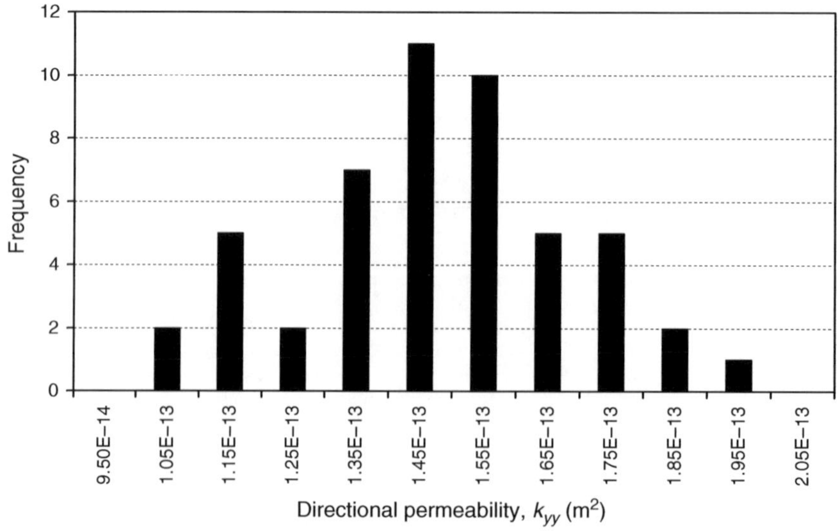

Fig. 12.51 Frequency of k_{yy} for DFN models of side length 5 m from 50 realizations (Min et al., 2004a).

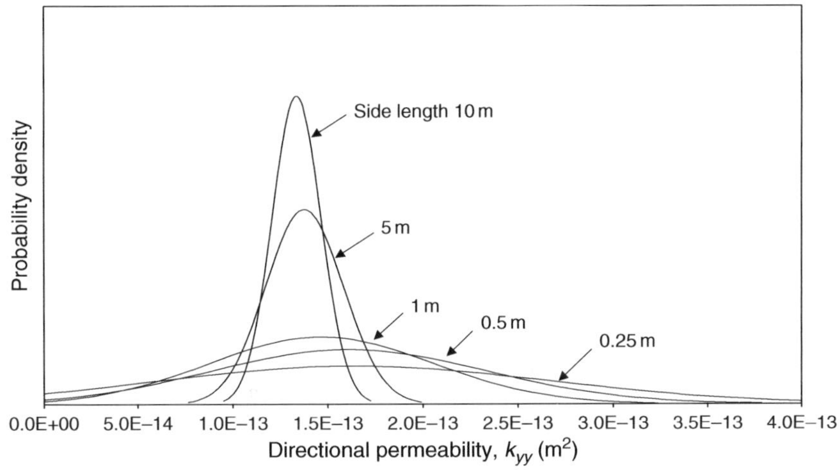

Fig. 12.52 The probability density functions of k_{yy} at different DFN model scales (Min et al., 2004a).

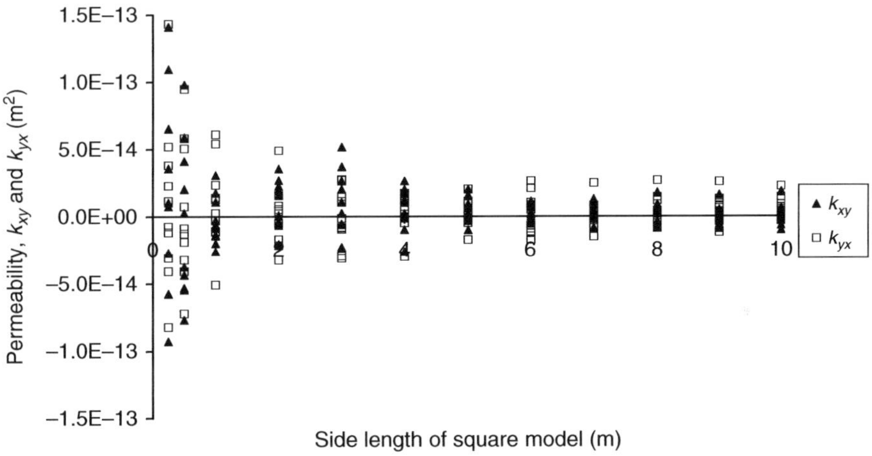

Fig. 12.53 Variations in the off-diagonal permeability components k_{xy} and k_{yx} with model size (Min et al., 2004a).

in the original x–y coordinate system with a 95% confidence level. Note that, from the numerical modeling results, k_{xy} and k_{yx} are close but not equal (Fig. 12.53), i.e., strictly speaking, the resultant matrix of the permeability is not symmetric, although it is close to that. The symmetric values of the off-diagonal terms in Eqn (12.11) are the results of averaging, so that a symmetric permeability tensor can be presented to represent the approximate permeability tensor of the fractured rocks as an equivalent continuum.

12.8.3 REV and the Elastic Compliance Tensor of a Fractured Rock

12.8.3.1 Constitutive equation of anisotropic elastic solids and the compliance tensor

The constitutive relation for general linear elasticity can be expressed as (Ting, 1996)

$$\varepsilon_{ij} = S_{ijkl}\sigma_{kl} \tag{12.12}$$

in tensor form or in a contracted matrix form as

$$\begin{Bmatrix} \varepsilon_x \\ \varepsilon_y \\ \varepsilon_z \\ \gamma_{yz} \\ \gamma_{xz} \\ \gamma_{xy} \end{Bmatrix} = \begin{bmatrix} s_{11} & s_{12} & s_{13} & s_{14} & s_{15} & s_{16} \\ s_{21} & s_{22} & s_{23} & s_{24} & s_{25} & s_{26} \\ s_{31} & s_{32} & s_{33} & s_{34} & s_{35} & s_{36} \\ s_{41} & s_{42} & s_{43} & s_{44} & s_{45} & s_{46} \\ s_{51} & s_{52} & s_{53} & s_{54} & s_{55} & s_{56} \\ s_{61} & s_{62} & s_{63} & s_{64} & s_{65} & s_{66} \end{bmatrix} \begin{Bmatrix} \sigma_x \\ \sigma_y \\ \sigma_z \\ \tau_{yz} \\ \tau_{xz} \\ \tau_{xy} \end{Bmatrix} \quad \text{or} \quad \{\varepsilon_i\} = [s_{ij}]\{\sigma_j\} \tag{12.13}$$

where ε_{ij} and σ_{kl} are stress and strain tensors of a second-order rank and S_{ijkl} is the compliance tensor of a fourth-order rank, involving 21 independent entries due to elastic energy symmetry considerations. The matrix $[S_{ij}]$ is called the compliance matrix. The symbols of ε_{i} and γ_{ij} $(i, j=x, y, z)$ denote the normal and shear strains, and the symbols σ_i and τ_{ij} $(i, j=x, y, z)$ denote the normal and shear stresses, respectively. Each entry of the compliance matrix has a physical meaning in terms of elastic moduli, Poisson ratios, shear moduli and other constants of the solids (Lekhnitskii, 1963). Note that

$$\begin{bmatrix} s_{11} & s_{12} & s_{13} & s_{14} & s_{15} & s_{16} \\ s_{21} & s_{22} & s_{23} & s_{24} & s_{25} & s_{26} \\ s_{31} & s_{32} & s_{33} & s_{34} & s_{35} & s_{36} \\ s_{41} & s_{42} & s_{43} & s_{44} & s_{45} & s_{46} \\ s_{51} & s_{52} & s_{53} & s_{54} & s_{55} & s_{56} \\ s_{61} & s_{62} & s_{63} & s_{64} & s_{65} & s_{66} \end{bmatrix} = \begin{bmatrix} \frac{1}{E_x} & -\frac{\nu_{yx}}{E_y} & -\frac{\nu_{zx}}{E_z} & \frac{\eta_{x,yz}}{G_{yz}} & \frac{\eta_{x,xy}}{G_{xz}} & \frac{\eta_{x,xy}}{G_{xy}} \\ -\frac{\nu_{xy}}{E_x} & \frac{1}{E_y} & -\frac{\nu_{yz}}{E_z} & \frac{\eta_{y,yz}}{G_{yz}} & \frac{\eta_{y,xz}}{G_{xz}} & \frac{\eta_{y,xy}}{G_{xy}} \\ -\frac{\nu_{xz}}{E_x} & -\frac{\nu_{yz}}{E_y} & \frac{1}{E_z} & \frac{\eta_{z,yz}}{G_{yz}} & \frac{\eta_{z,xz}}{G_{xz}} & \frac{\eta_{z,xy}}{G_{xy}} \\ \frac{\eta_{yz,x}}{E_x} & \frac{\eta_{yz,y}}{E_y} & \frac{\eta_{yz,z}}{E_z} & \frac{1}{G_{yz}} & \frac{\mu_{yz,xz}}{G_{xz}} & \frac{\mu_{yz,xy}}{G_{xy}} \\ \frac{\eta_{xz,x}}{E_x} & \frac{\eta_{xz,y}}{E_y} & \frac{\eta_{xz,z}}{E_z} & \frac{\mu_{xz,yz}}{G_{yz}} & \frac{1}{G_{xz}} & \frac{\mu_{xz,xy}}{G_{xy}} \\ \frac{\eta_{xy,x}}{E_x} & \frac{\eta_{xy,y}}{E_y} & \frac{\mu_{xy,z}}{E_z} & \frac{\mu_{xy,yz}}{G_{yz}} & \frac{\mu_{xy,xz}}{G_{xz}} & \frac{1}{G_{xy}} \end{bmatrix} \tag{12.14}$$

where E_x, E_y, E_z are the equivalent elastic moduli in the x, y and z-directions; G_{yz}, G_{xz}, G_{xy} are the equivalent shear moduli in the yz, xz and xy planes; ν_{ij}, $(i, j=x, y, z)$ are the equivalent Poisson's ratios as the ratio of strain in the j-direction due to the strain in the i-direction caused by a stress acting in the i-direction. The symbols $\mu_{ij,kl}$ $(i, j, k, l=x, y, z)$ are the coefficients of Chentsov characterizing the shear strain in the plane parallel to the one defined by indices ij that induces the tangential stress in the plane parallel to the one defined by indices kl; symbols $\eta_{k,ij}$ $(i, j, k=x, y, z)$ are the coefficients of mutual influence of the first kind characterizing the stretching in the k-direction induced by the shear stress acting within a plane parallel to the one defined by indices ij; symbols $\eta_{ij,k}$ $(i, j, k=x, y, z)$ are the coefficients of mutual influence of the second kind, characterizing a shear in the plane defined by indices ij under the influence of a normal stress acting in the k-direction.

From the symmetry of the compliance matrix, the following relation is valid:

$$\frac{\nu_{ij}}{E_i}=\frac{\nu_{ji}}{E_j}, \qquad \frac{\mu_{ik,jk}}{G_{jk}}=\frac{\mu_{jk,ik}}{G_{ik}}, \qquad \frac{\mu_{ij,k}}{E_k}=\frac{\eta_{k,ij}}{G_{ij}} \tag{12.15}$$

Since S_{ijkl} is a fourth-order tensor, its rotational transformation can be also defined by the following mapping operations (Ting, 1996):

$$S'_{ijkl}=\beta_{im}\beta_{jn}\beta_{kp}\beta_{lp}S_{mnpq} \tag{12.16}$$

where S'_{ijkl} and S_{mnpq} are the compliance tensors in the transformed and the original axes, respectively, and β_{im} are direction cosines representing rotational operations. Equation (12.16) is mathematically elegant, but not convenient for practical calculations because it involves fourth-order tensor operations. The following mapping operation with a 6×6 matrix for the transformation of compliance matrix is introduced to simplify the operations (Lekhnitskii, 1963)

$$S'_{ij} = S_{mn} q_{mi} q_{nj} \tag{12.17}$$

where S'_{ij} is the compliance matrix in the transformed axes and S_{mn} is the one in the original axes. Defining the directional cosines between the corresponding axes of the original (x–y–z) and the rotated (x'–y'–z') systems as $\alpha_1 = \cos(x, x')$, $\beta_1 = \cos(y, x')$, $\gamma_1 = \cos(z, x')$, $\alpha_2 = \cos(x, y')$, $\beta_2 = \cos(y, y')$, $\gamma_2 = \cos(z, y')$, $\alpha_3 = \cos(x, z')$, $\beta_3 = \cos(y, z')$, $\gamma_3 = \cos(z, z')$, respectively, the matrix q_{ij} is written as

$$q_{ij} = \begin{bmatrix} \alpha_1^2 & \alpha_2^2 & \alpha_3^2 & 2\alpha_2\alpha_3 & 2\alpha_3\alpha_1 & 2\alpha_1\alpha_2 \\ \beta_1^2 & \beta_2^2 & \beta_3^2 & 2\beta_2\beta_3 & 2\beta_3\beta_1 & 2\beta_1\beta_2 \\ \gamma_1^2 & \gamma_2^2 & \gamma_3^2 & 2\gamma_2\gamma_3 & 2\gamma_3\gamma_1 & 2\gamma_1\gamma_2 \\ \beta_1\gamma_1 & \beta_2\gamma_2 & \beta_3\gamma_3 & \beta_2\gamma_3 + \beta_3\gamma_2 & \beta_1\gamma_3 + \beta_3\gamma_1 & \beta_1\gamma_2 + \beta_2\gamma_1 \\ \gamma_1\alpha_1 & \gamma_2\alpha_2 & \gamma_3\alpha_3 & \gamma_2\alpha_3 + \gamma_3\alpha_2 & \gamma_1\alpha_3 + \gamma_3\alpha_1 & \gamma_1\alpha_2 + \gamma_2\alpha_1 \\ \alpha_1\beta_1 & \alpha_2\beta_2 & \alpha_3\beta_3 & \alpha_2\beta_3 + \alpha_3\beta_2 & \alpha_1\beta_3 + \alpha_3\beta_1 & \alpha_1\beta_2 + \alpha_2\beta_1 \end{bmatrix} \tag{12.18}$$

When rotation about only one axis is involved, the transformation form becomes simple. If the angle of anti-clockwise rotation about the z-axis is φ, the matrix of the direction cosines is then given by

$$\beta_{ij} = \begin{bmatrix} \cos\varphi & \sin\varphi & 0 \\ -\sin\varphi & \cos\varphi & 0 \\ 0 & 0 & 1 \end{bmatrix} \tag{12.19}$$

and the final matrix form for q_{ij} is:

$$q_{ij} = \begin{bmatrix} \cos^2\varphi & \sin^2\varphi & 0 & 0 & 0 & -2\sin\varphi\cos\varphi \\ \sin^2\varphi & \cos^2\varphi & 0 & 0 & 0 & 2\sin\varphi\cos\varphi \\ 0 & 0 & 1 & 0 & 0 & 0 \\ 0 & 0 & 0 & \cos\varphi & \sin\varphi & 0 \\ 0 & 0 & 0 & -\sin\varphi & \cos\varphi & 0 \\ \sin\varphi\cos\varphi & -\sin\varphi\cos\varphi & 0 & 0 & 0 & \cos^2\varphi - \sin^2\varphi \end{bmatrix} = Q \tag{12.20}$$

when a 2D simplification with rotation about the out-of-plane axis (z-axis) is presented, such as in the case under consideration in this section.

Using Eqns (12.18–12.20), it is possible to express the elastic constants, i.e., elastic modulus or Poisson's ratio, in the rotated axes in terms of direction cosines and components with reference to the original axes. The above rotational mapping relations are important for deriving elastic compliance tensor components for different DFN models in the latter part of this section.

Since the shear strain and shear stress components associated with the z-direction can be removed for two-dimensional plane strain analyses, Eqn (12.12) can be reduced to the following compacted form, after removing the components in the 4th and 5th columns/rows in the matrix of Eqn (12.14)

$$\begin{Bmatrix} \varepsilon_x \\ \varepsilon_y \\ \varepsilon_z \\ \gamma_{xy} \end{Bmatrix} = \begin{bmatrix} S_{11} & S_{12} & S_{13} & S_{16} \\ S_{21} & S_{22} & S_{23} & S_{26} \\ S_{31} & S_{32} & S_{33} & S_{36} \\ S_{61} & S_{62} & S_{63} & S_{66} \end{bmatrix} \begin{Bmatrix} \sigma_x \\ \sigma_y \\ \sigma_z \\ \tau_{xy} \end{Bmatrix} \tag{12.21}$$

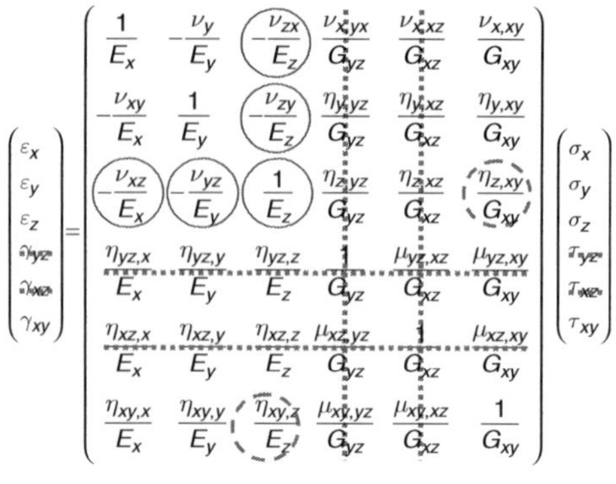

Fig. 12.54 Matrix consideration for the determination of compliance matrix (Min and Jing, 2003).

For the plane strain assumption adopted here, the components related to normal stress in the z-direction need to remain since the generated σ_z due to the strain confinement in the z-direction affects the deformation in the x and y-directions. Considering the fact that there is no effect of out-of-plane fractures (which means that the fractures in the UDEC model conceptually have a strike in the direction of the out-of-plane z-direction) the elastic modulus in the z-direction (E_z) can be pre-determined as the elastic modulus of the intact rock. Furthermore, if we consider the symmetry conditions, $S_{13} = S_{31}$, $S_{23} = S_{32}$ and $S_{36} = S_{63}$, which are associated with the z-directional stress and strains, eventually only three components are independent in each row (or column) of the matrix in Eqn (12.20). Therefore, three linearly independent stress boundary conditions are sufficient to determine all components in the matrix shown in Eqn (12.21). In fact, all the components related to σ_z can be pre-determined. As shown in Fig. 12.54, the values of elements S_{13} and S_{23} can be determined considering that the ν_{zx} and ν_{zy} are the same as the Poisson's ratio of intact rock (Min and Jing, 2003). Also, S_{36} will be zero since the shear stress τ_{xy} does not affect the z-directional deformation.

The compliance tensors of a rock mass with random fracture systems do not exist in closed form, except for some simplified cases with regular fracture systems, such as the case with three sets of orthogonal persistent fractures in three-dimensions as developed by Amadei and Goodman (1981) and as presented below.

$$\begin{bmatrix} \frac{1}{E_x}+\frac{1}{K_{nx}S_x} & -\frac{\nu_{xy}}{E_y} & -\frac{\nu_{xz}}{E_z} & 0 & 0 & 0 \\ -\frac{\nu_{yx}}{E_y} & \frac{1}{E_y}+\frac{1}{K_{ny}S_y} & -\frac{\nu_{zy}}{E_z} & 0 & 0 & 0 \\ -\frac{\nu_{zx}}{E_z} & -\frac{\nu_{zy}}{E_z} & \frac{1}{E_z}+\frac{1}{K_{nz}S_z} & 0 & 0 & 0 \\ 0 & 0 & 0 & \frac{1}{G_{yz}}+\frac{1}{K_{sy}S_y}+\frac{1}{K_{sz}S_z} & 0 & 0 \\ 0 & 0 & 0 & 0 & \frac{1}{G_{xz}}+\frac{1}{K_{sx}S_x}+\frac{1}{K_{sz}S_z} & 0 \\ 0 & 0 & 0 & 0 & 0 & \frac{1}{G_{xy}}+\frac{1}{K_{sx}S_x}+\frac{1}{K_{sy}S_y} \end{bmatrix} \tag{12.22}$$

where E_x, E_y, E_z are the elastic moduli of the intact rock in the x, y and z-directions, G^i_{yz}, G^i_{xz}, G^i_{xy} the shear moduli of the intact rock in the yz, xz and xy planes and ν_{yx}, ν_{zx}, ν_{zy} the Poisson ratios of the intact rock. The symbol ν_{ij} is the Poisson's ratio of normal strain in the j-direction to that in the i-direction when stress is applied in the i-direction. K_{nx}, K_{ny}, K_{nz} are the normal stiffness of fractures in the x, y and z-directions, K_{sx}, K_{sy}, K_{sz} the shear stiffness of fractures in the x, y and z-directions and S_x, S_y, S_z the spacing of fracture sets measured in the x, y and z-directions. Note that the fractures do not have to be evenly spaced. The compliance matrix can also be expressed in the transformed axes using Eqn (12.17).

12.8.3.2 Numerical techniques for DEM modeling and verification

The generated DFN models as described in Section 12.8.2.2 for hydraulic analysis are used again for the mechanical analysis. They are discretized with triangular zones of constant strains for distinct element analysis using the UDEC code, with the linear size of the triangular zones varying from 0.002 to 0.1 m. The maximum number of internal zones to be analyzed for this study was about 50 000, with 5000 blocks for a DEM model of 8 m × 8 m in size. Figure 12.55 shows the blocks and internal finite difference zones (elements) for a model of 3 m × 3 m in size.

Figure 12.56 illustrates the three linearly independent boundary conditions (BC1–BC3) used to produce the two-dimensional compliance matrix in Eqn (12.21). BC1 consists of biaxial normal stresses

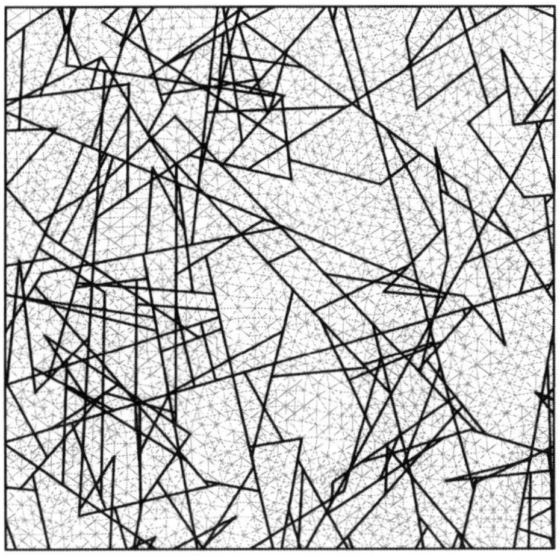

Fig. 12.55 Example of a DEM model for a 3 m × 3 m sample with deformable blocks (bounded by thick lines) internally discretized with 6234 triangular finite difference zones (Min and Jing, 2003).

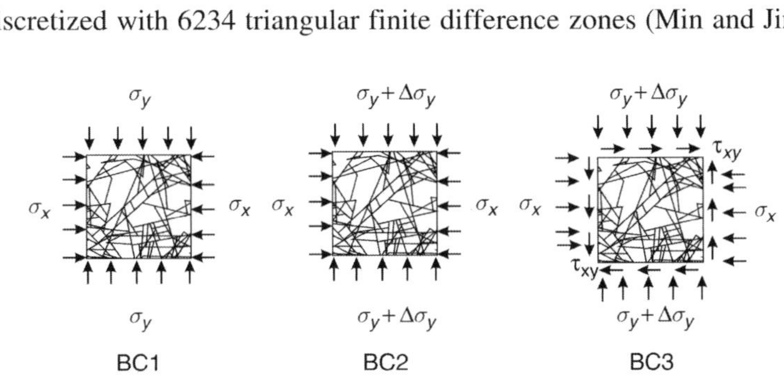

Fig. 12.56 Three linearly independent boundary conditions for the numerical experiments (Min and Jing, 2003).

and BC2 is created by sequentially increasing the normal stress in the y-direction. Even though one additional normal stress increment in the y-direction suffices for the purpose of BC2, two to four sequential y-directional normal stress increments were applied to check the linearity of the mechanical behavior. Similarly, BC3 is created by sequentially superimposing shear stress increments over the stress conditions of the final BC2. Three to five sequential shear stress increments were applied to consider the linearity of the model response. If linearity is observed, the final stress and strain states of the DFN models for each boundary condition were used for calculation of the compliance tensor.

In order to verify the methodology proposed above, a model with two orthogonal fracture sets was used for the comparison between the elastic properties produced by the numerical experiments and the closed-form solutions using rotated models, with an equation derived from the simplification of Eqn (12.22). The computational models were rotated in intervals of 10° to evaluate the variations of elastic moduli in the rotated directions in a fashion similar to Fig. 12.48. Note also that the model boundary is located at the center of the fracture spacing to avoid the need to adjust the equivalent spacing. The mechanical properties used for this verification example are: elastic modulus and Poisson's ratio of the intact rock of 84.6 GPa and 0.24, respectively. The initial normal and shear stiffness of the fractures were 434 and 43.4 GPa/m and the spacing of the fracture sets was 0.5 m. The ratio of the normal stiffness of fracture set 1 (vertical) over that of fracture set 2 (horizontal), K, varied between $K=0.1$ and 0.5 in order to generate the necessary numerical points in a polar diagram so that they can be compared with closed-form solutions obtained by Wei (1988). The three boundary conditions described above were applied to obtain the compliance matrix. Since the model is symmetric about the x and y-axes when the properties of the two fracture sets are the same, only values from 0° to 90° are compared. As shown in Fig. 12.57, the two sets of results show an almost perfect agreement. Therefore, it is reasonable to conclude that the methodology adopted for the numerical approach using the UDEC code can be used for more complex cases of irregular fracture system geometry.

Figure 12.58 shows the contours of the y-directional displacement of the fractured rock mass under biaxial loading (BC2) from one of the UDEC models for the purpose of a demonstration. Note that zero displacement lies in the center of the model since stresses were applied on both sides. It shows that, although the displacement distribution is much influenced by the deformation of fractures, a general trend of a roughly uniform deformation gradient can still be observed due to the collective contributions of large numbers of fractures. The averaging process is therefore applicable for deriving mean values of displacement gradients and strains. The dead-ends of fractures are not incorporated in the existing UDEC code and their effect is assumed not to be significant.

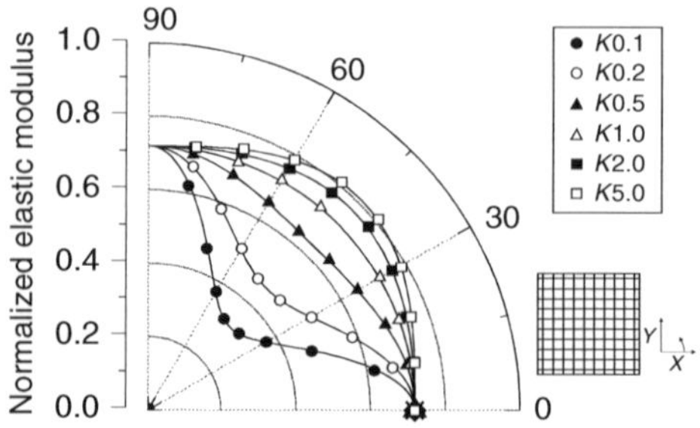

Fig. 12.57 Verification of UDEC modeling results against the analytical solution for a fractured rock mass with two orthogonal sets of fractures. The symbol points correspond to the numerical results with different K ratio values and the lines are the analytical solutions (Min and Jing, 2003).

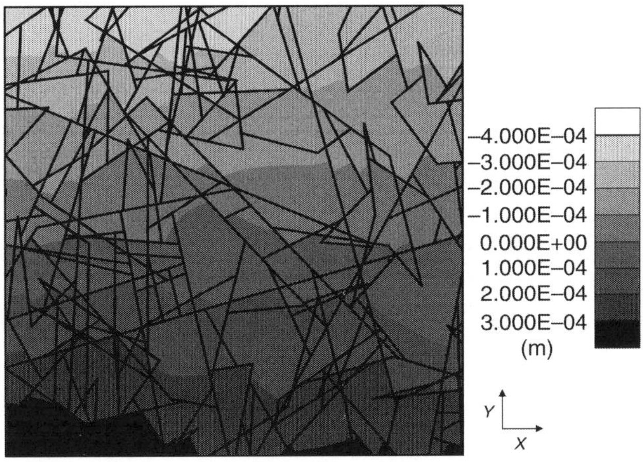

Fig. 12.58 The contours of the *Y*-displacements under biaxial compressive loading ($\sigma_x = -5$ MPa, $\sigma_y = -10$ MPa, size 3 m × 3 m, DFN3) (Min and Jing, 2003).

12.8.3.3 Results of scale dependency and tensor characteristics of the mechanical properties for fractured rock masses

In order to evaluate the directional properties of the fractured rock mass under study, the DFN models are rotated in six directions with 30° intervals (0°, 30°, 60°, 90°, 120° and 150°) to perform similar UDEC simulations in order to calculate the compliance matrices of the six rotated DFN models. In order to examine whether the calculated elastic properties can be represented by an elastic compliance tensor, an average compliance matrix is calculated by averaging six compliance matrices from the six rotated DFN models obtained from the numerical experiments, at different scales. It is important to have a reference axis when such matrices are to be compared or averaged (Hudson and Harrison, 1997) and the original x–y axes were used for this purpose. The average compliance matrix $\bar{S}_{ij}$ is then given by

$$\bar{S}_{ij} = \frac{1}{N} \sum_{r=1}^{N} S^r_{pq} q_{pi} q_{qj} \tag{12.23}$$

where S^r_{pq} are the calculated compliance matrixes for the rotated p–q axes and q_{pi} and q_{qj} the matrices of the direction cosines of the rotation as defined in Eqns (12.18–12.20).

If the average compliance matrix has an acceptably low difference from all six compliance matrices at a certain scale that scale will determine the size of the REV. Figure 12.59 presents the averaged and calculated trace curves of the elastic moduli in the x-direction from all the rotated models at different scales for the DFN1 series. The solid curve in the figure for the averaged values represents the elastic moduli in the x-direction with rotations and the numerical results obtained from the six rotated DFN models at their corresponding rotation directions are plotted as separated symbols.

Figure 12.59 shows that the numerical results from the small-scale models do not match well with the averaged values – until the model side length reaches 5 m, indicating that a REV size of 5 m can be established and the properties at this size and above can be approximated by a fourth-order tensor represented by an elastic compliance matrix in a compact form with a certain degree of accuracy through the 'homogenization' (averaging) process as presented above.

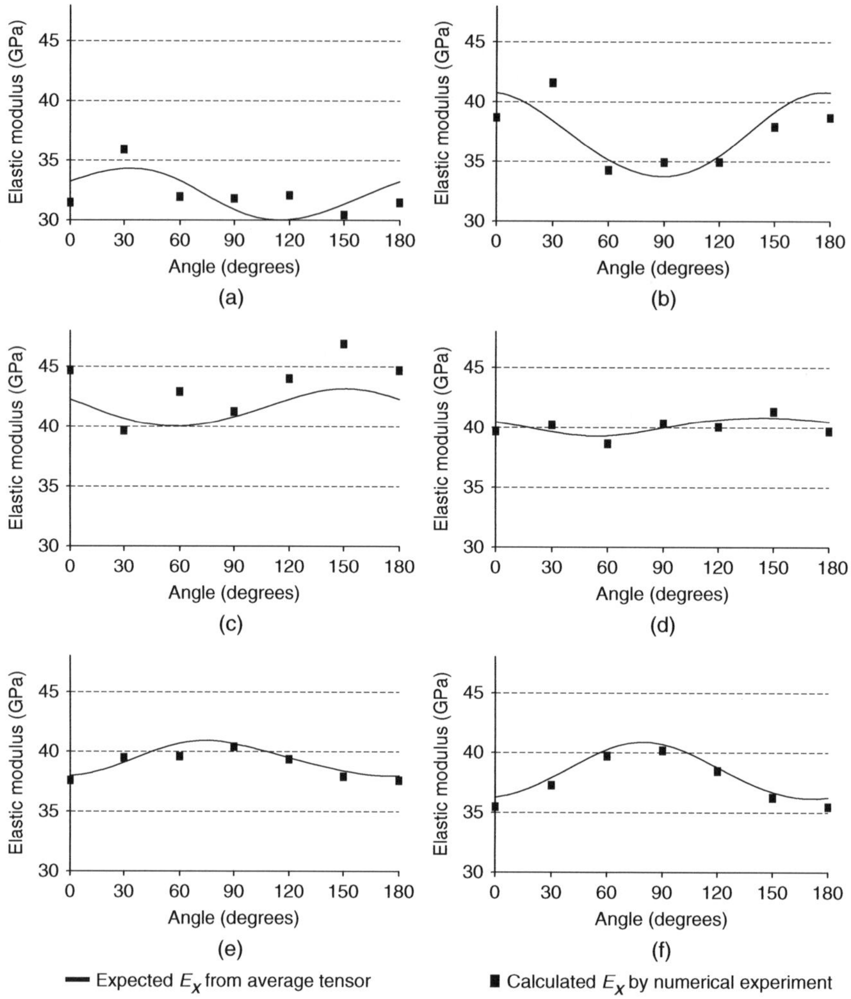

Fig. 12.59 Comparison between expected and calculated elastic moduli (E_x) in rotated axes. The expected values are calculated using Eqn (12.26). The side lengths of the DFN models are: (a) 0.25 m, (b) 0.5 m, (c) 1 m, (d) 3 m, (e) 5 m, (f) 7 m (Min and Jing, 2003).

Figure 12.60 shows the results for the Poisson's ratio as functions of the model side length and two cases of the ratio of shear and normal stiffness of fractures $k_s/k_n = 1$ and 0.2.

Similar to the permeability tensor case as presented in Section 12.7.2, a 'coefficient of variation' is defined as the ratio of standard deviation over the mean value of a certain scalar property (say, elastic modulus or Poisson's ratio) obtained from all random DFN models at a given scale, in order to determine the REV size with a given acceptable variation, on the condition that the average value remains constant with increasing scale. To evaluate discrepancies between tensors or matrices, a measure of 'prediction error' of the elastic compliance matrices is defined to represent the discrepancies in the compliance matrices, considering only the effects of the normal strains for simplicity since they are the dominating components. For two-dimensional cases, this implies that only the first two elements of the compliance matrix need to be considered, and the prediction errors are therefore defined as

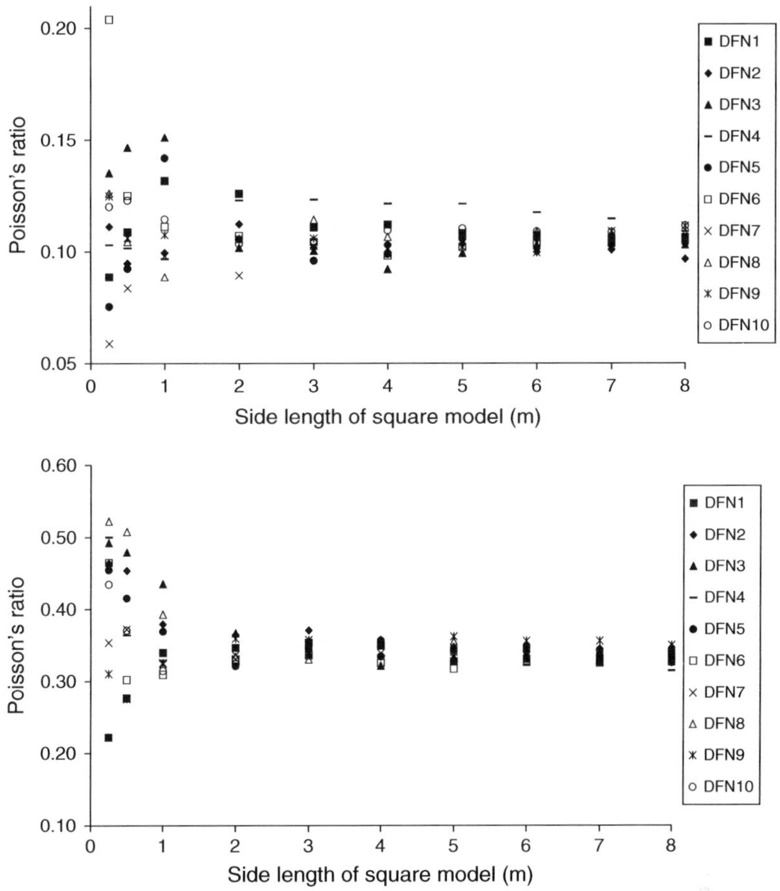

Fig. 12.60 Variation of Poisson's ratio (ν_{xy}) with the increase of side lengths of square models: (above) k_s/k_n 1.0, (below) $k_s/k_n = 0.2$ (Min and Jing, 2003).

$$\mathrm{EP}c_i = \frac{1}{N} \frac{\sum_{r=1}^{N} \sum_{j=1}^{2} |s_{ij}^r - \bar{s}_{ij}|}{\sum_{j=1}^{2} |\bar{s}_{ij}|} \tag{12.24}$$

where $\mathrm{EP}c_i$ is the prediction error of the compliance matrix in the i-direction ($i = x,y$), $\bar{S}_{ij}$ the average compliance matrix as defined above, $s^r{}_{ij}$ the compliance matrix from numerical experiments at rotated DFN models and N the number of cases of DFN model rotations ($N = 6$ for this study). Note also that comparison is conducted for the same reference axes. A mean prediction error (μ_{EPc}) can then be defined as the mean values of the two $\mathrm{EP}c_i$, given by

$$\mu_{\mathrm{EPc}} = \frac{1}{2} \sum_{i=1}^{2} \mathrm{EP}c_i \tag{12.25}$$

and this was used as the collective measure of prediction error for comparisons at different scales. The size at which the compliance matrix can be approximated by a fourth-order compliance tensor depends on a chosen 'acceptable variation' of mean prediction error.

For the case studied in this section, an average compliance matrix obtained from 10 realizations at the 6 m × 6 m scale is presented below for different fracture stiffness ratios (K), (Min and Jing, 2003).

(a) For $k_s/k_n = 1.0$,

$$\begin{pmatrix} S_{11} & S_{12} & S_{13} & S_{16} \\ S_{21} & S_{22} & S_{23} & S_{26} \\ S_{31} & S_{32} & S_{33} & S_{36} \\ S_{61} & S_{62} & S_{63} & S_{66} \end{pmatrix} = \begin{pmatrix} 2.7709 & -0.2933 & -0.2800 & -0.0445 \\ -0.2933 & 2.3934 & -0.2814 & -0.0405 \\ -0.2800 & -0.2814 & 1.1820 & -0.0051 \\ -0.0591 & -0.0653 & -0.0051 & 5.5066 \end{pmatrix} \times 10^{-11}\ (1/\mathrm{Pa}) \tag{12.26}$$

(b) For $k_s/k_n = 0.2$,

$$\begin{pmatrix} S_{11} & S_{12} & S_{13} & S_{16} \\ S_{21} & S_{22} & S_{23} & S_{26} \\ S_{31} & S_{32} & S_{33} & S_{36} \\ S_{61} & S_{62} & S_{63} & S_{66} \end{pmatrix} = \begin{pmatrix} 3.8389 & -1.3073 & -0.2808 & -0.2039 \\ -1.3097 & 3.4556 & -0.2812 & 0.1151 \\ -0.2808 & -0.2812 & 1.1820 & -0.0012 \\ -0.2336 & 0.1018 & 0.0012 & 10.683 \end{pmatrix} \times 10^{-11}\ (1/\mathrm{Pa}) \tag{12.27}$$

Note that the compliance matrices in Eqns (12.26) and (12.27) are almost symmetric, diagonally dominant and with almost zero contributions from the shear stresses to the out-of-plane normal strain ε_z and the out-of-plane normal stress σ_z to the in-plane shear strain γ_{xy}, represented by very small values of elements S_{36} and S_{63}, respectively. Because the symmetry assumption of $S_{31} = S_{13}$ and $S_{32} = S_{23}$ was initially used for determination of the components, the symmetry appearing in the matrices in Eqns (12.26) and (12.27) is not related to the accuracy of numerical experiments. However, the near symmetrical feature of the other components, namely S_{12} and S_{21}, S_{16} and S_{61} and S_{26} and S_{62}, are determined by the numerical solutions. This is, in part, due to the averaging process from the 10 realizations where the non-symmetry contributions from different realizations are compensated to a certain extent. However, the symmetry deviation in each individual realization was also below 5% for the diagonal components. The accuracy of the calculated results can be further evaluated by comparing S_{13}, S_{23} and S_{36} using Fig. 12.54. S_{13} and S_{23} are calculated as 0.28370×10^{-11} (/Pa) with a Poisson's ratio of 0.24 and elastic modulus of 84.6 GPa for the intact rock. The difference between the numerically calculated value (0.28×10^{-11}) and the theoretical solution is small (less than 2%). Furthermore, S_{36} is close to zero, which is also similar to the assumed behavior shown in Fig. 12.54. This result supports the numerical experiment approach presented.

With the determined nearly symmetric compliance matrix as shown above, it is easy to derive a symmetric compliance tensor $[\mathbf{S}_{ij}]$ by a simple averaging operation as

$$(\mathbf{S_{ij}}) = \frac{1}{2}\left[(S_{ij}) + (S_{ij})^{\mathrm{T}}\right] \tag{12.28}$$

where $(S_{ij})^{\mathrm{T}}$ is the transpose of the compliance matrix (S_{ij}) that is supported by a REV of 6 m in linear scale and can be used as the elastic property for continuum analysis.

12.8.4 Stress Effects on Fluid Flow Fields: Stress-Induced Flow Channeling

12.8.4.1 Basic requirements for simulating coupled stress-flow process using DEM

Laboratory investigations on single rock fractures show that normal closure and shear dilation can significantly change fracture transmissivity (Makurat et al., 1990b; Olsson and Barton, 2001). For fractured hard rocks, the changes in the transmissivities of critically oriented and connected fractures (compared with orientations of shear stresses) could have significant changes in the local or regional permeability fields due to shear of fractures (Barton, N. et al., 1995). These observations are supported by applications in geothermal reservoir engineering for detecting thermal fluid flow pathways (Ito and Hayashi, 2003).

Analytical models of stress-dependent permeability of fractured rock masses based on orthogonal and persistent fracture sets are available (Bai and Elsworth, 1994; Chen and Bai, 1998; Bai et al., 1999), which consider the normal closures of fractures and constant shear dilation in both fractured and fractured-porous media. Oda's permeability tensor approach considers stress dependency of permeability in complex fracture networks (Oda, 1986). However, fracture connectivity and complex fracture interactions, which are important factors affecting the overall hydro-mechanical behavior of the fractured rock masses, cannot be considered in this approach. In addition, when the shear failures and dilation of fractures are to be considered, analytical solutions do not in general exist. Therefore, numerical modeling has to be applied to deepen the understanding of the critical mechanisms and the stress contributions to the overall permeability from both normal and shear stresses.

The influence of stresses on the permeability of fractured rock masses was extensively investigated considering various geometries of fracture systems in Zhang and Sanderson (1996, 1997, 2002) and Zhang et al. (1996), using a DEM approach with the UDEC code. Further studies in this direction by Min et al. (2004b) go beyond the above works by requiring that the following three conditions must be satisfied for investigating problems involving coupled stress-flow processes of rock fractures.

- *REV size of the numerical model*: the model size for the numerical experiments should be properly established so that the initial mechanical and hydraulic behavior of the model is statistically representative when the equivalent behavior of fractured rock masses are considered. This means that the mechanical and hydraulic REVs of the numerical model must be reached before modeling hydraulic behavior with stress effects.
- *Non-linear normal stress–aperture relation*: the normal stress–displacement relation of fractures should be realistically represented because of the significant role of fracture aperture changes on the overall permeability of the rock mass concerned. Laboratory tests showed that the normal stress–closure behavior of fractures is highly non-linear (Barton, 1982; Bandis et al., 1983) and this feature has to be considered for accurate prediction of aperture change due to stress effect. A linear relation between normal stress and fracture closure (therefore constant normal stiffness) will not be able to serve the purpose since it generally underestimates the change of fracture transmissivity at lower normal stress magnitudes.
- *Shear dilation*: shear dilation caused by the roughness of fracture surfaces should be modeled since the effect of fracture dilation on permeability could be significant, especially for rough fractures (Bawden et al., 1980). The assumption of zero dilation (therefore assuming smooth fracture surfaces) in the fracture models will not only lead to underestimated fracture tranmissivity changes in general, but will also lead to loss of the capability for simulating the clustered fluid flow phenomenon induced by critical stress states in fractured rock masses (Barton et al., 1995a).

12.8.4.2 A DEM approach for stress effects on the permeability of fractured rocks

A DEM approach for the problem of the stress-dependent permeability of fractured rock masses based on the above three conditions is presented in this section (Min et al., 2004b), using the DEM code UDEC. The DEM model (Fig. 12.61) used is 5 m × 5 m at a REV size, as established for the calculations

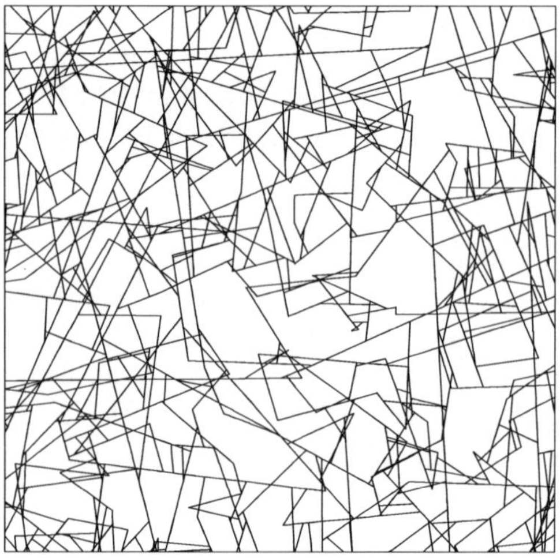

Fig. 12.61 Geometry of fracture system in the DFN model (Min et al., 2004b).

of the equivalent mechanical and hydraulic properties – as presented in Sections 12.8.2 and 12.8.3. A stepwise normal stress–normal closure model of fracture, established according to laboratory experiments (Fig. 12.62a) and an elasto-perfectly plastic constitutive model for the fracture shear with friction and dilation is adopted (Fig. 12.62a) and implemented with a friction angle of 24.9°, a maximum dilation of 50 μm, an initial aperture of 30 μm and a residual aperture of 5 μm, respectively. The fractures were characterized with a low JRC (joint roughness coefficient) value of 3.85. All three conditions mentioned above are satisfied.

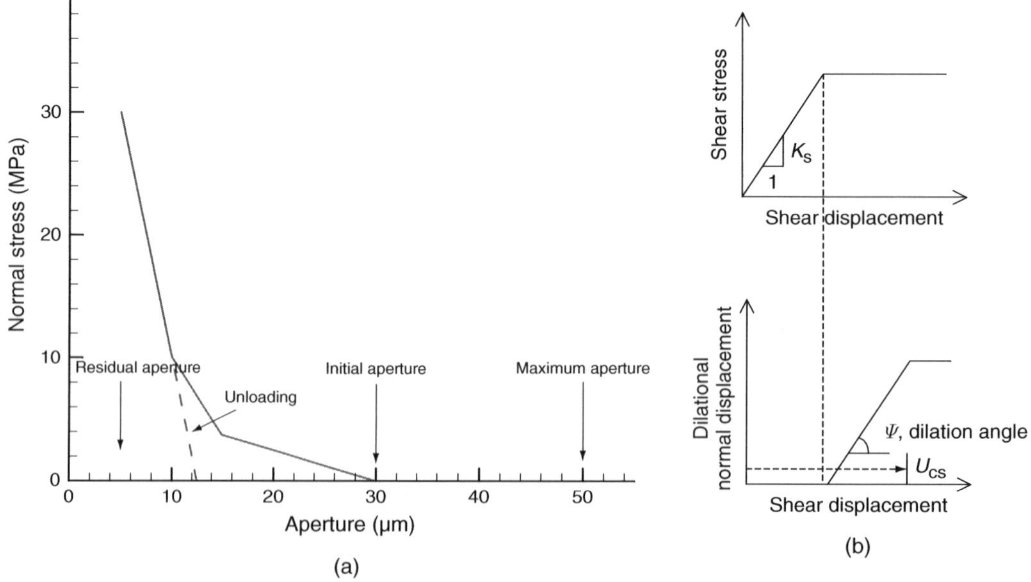

Fig. 12.62 Constitutive model of the fractures: (a) a stepwise normal stress–aperture relation as experimentally determined; and (b) simplified shear and dilational behavior of fracture adapted from UDEC code manual (Itasca, 2002; Min et al., 2004b).

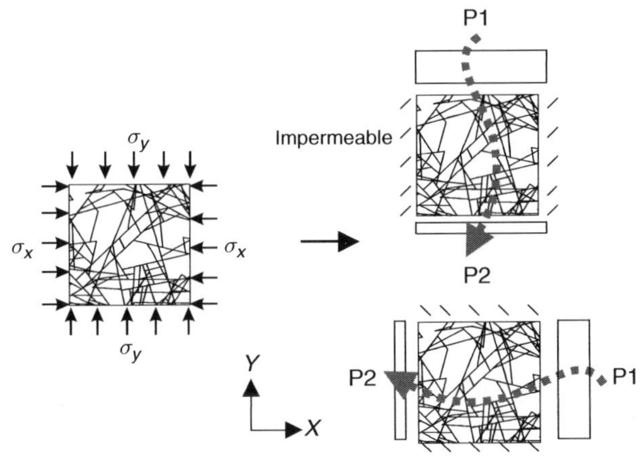

Fig. 12.63 Stress (σ_x and σ_y) and hydraulic (P1 and P2, with P1 > P2) boundary conditions for calculation of equivalent permeability in the *x*- and *y*-directions (Min et al., 2004b).

The boundary conditions for this study were selected in two ways (Fig. 12.63). Firstly, horizontal and vertical compressive stresses were applied and increased incrementally, with a fixed ratio of horizontal to vertical boundary stresses, $k = 1.3$, based on the data provided from the Sellafield site (Andersson and Knight, 2000). Secondly, the horizontal normal boundary stress was increased incrementally, while keeping the vertical normal stress constant and thus producing increased shear stresses in the fractures. The exercise is taken as a demonstration of the stress effect on flow behavior of fractured rocks with realistic fracture system geometry and has no relation to the *in situ* conditions at Sellafield at all.

Figure 12.64 shows the calculated equivalent permeability change with increasing magnitudes of boundary stresses while keeping a fixed ratio of horizontal/vertical stress components equal to 1.3.

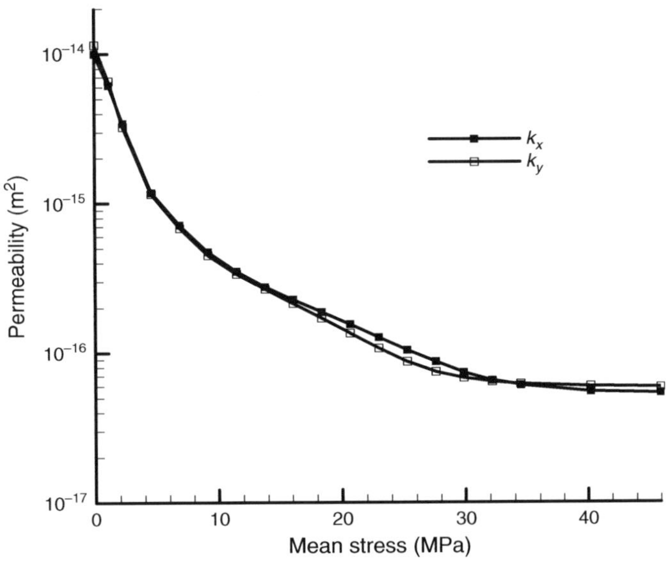

Fig. 12.64 Permeability (k_x and k_y) versus mean stress change with a fixed ratio of horizontal to vertical stresses = 1.3 (Min et al., 2004b).

Because of the dominating fracture closure with the increasing normal stresses without significant shear failure, the equivalent permeability of the model decreases hyperbolically with stress increase, with higher gradients at lower stress levels. The reduction of permeability is more than two orders of magnitude, reaching a sill at 30 MPa when residual aperture is reached in most of the fractures. The anisotropy of permeability is not significant. This result is comparable to the permeability variation with depth as reported in Wei et al. (1995) with the depth corresponding to the stress changes. When constant normal stiffness of fractures is applied (Zhang and Sanderson, 1996; Zhang et al., 1996), the hyperbolic variation of equivalent permeability will not be readily produced with increase of stresses.

In contrast to Fig. 12.64, with shear failure in fractures with the stress ratio higher than 2.7, significant changes in the magnitude and anisotropy of equivalent permeability can be observed in Fig. 12.65, caused by the shear dilation of critically oriented (about 33°) and well-connected fractures. When dilation starts, anisotropy becomes much more significant and the maximum anisotropy ratio reaches approximately 8. The fracture network used in this example shows a near-random orientation pattern with slightly more vertical or near-vertical fractures. The change of anisotropic permeability during the increase in differential stresses could be higher if the model had more fractures of critical or near-critical orientations for shear failure.

Figure 12.66 shows the change in flow patterns with increasing stress ratio. A notable channeling flow effect caused by stress-induced fracture dilation is observed. As larger shear dilations are concentrated in a smaller part of the fracture population with near-critical orientations, good connectivity and longer trace lengths, the rest of the fracture population, especially the sub-vertical ones, still undergo normal closures without any shear dilation. This situation causes a much clustered flow field where a few fractures with more increased apertures become the major pathways of fluid flow – the so-called 'flow localization'.

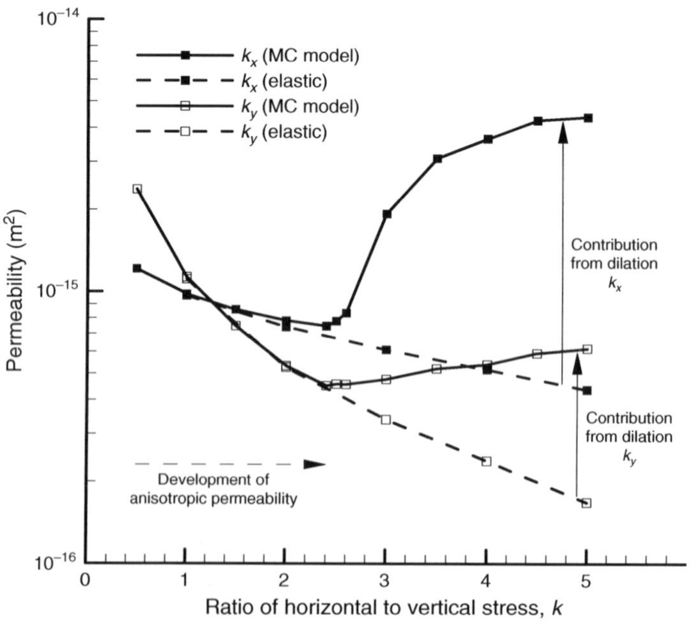

Fig. 12.65 Equivalent permeability (k_x and k_y) change with stress ratio. The dashed curves were obtained when the elastic model (no shear failure) of fractures was enforced (Min et al., 2004b).

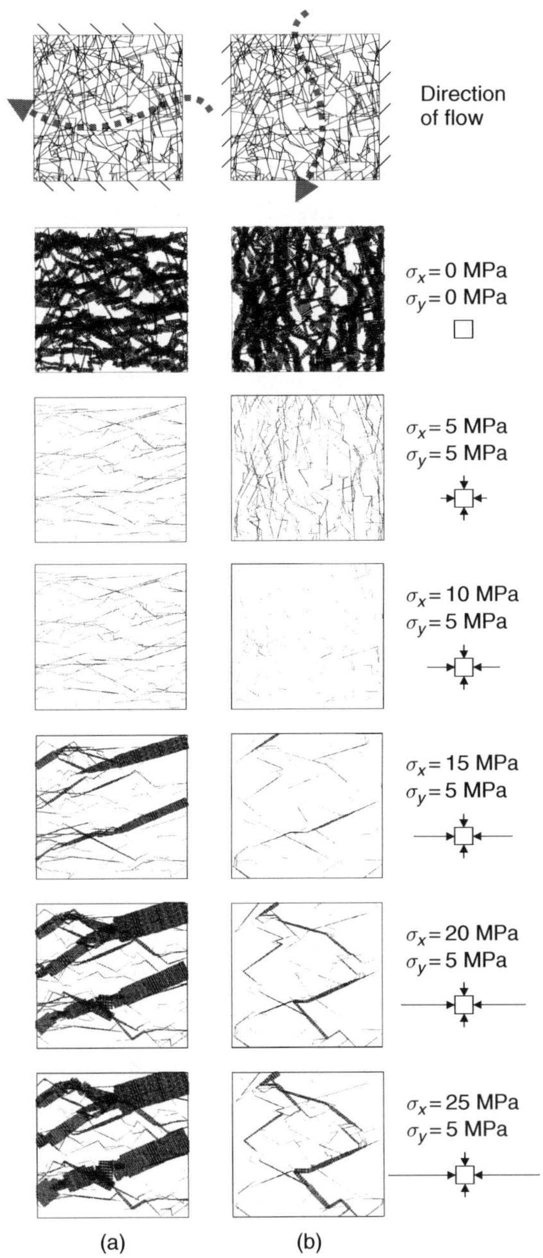

Fig. 12.66 Fluid pathways during stress applications with the direction of hydraulic pressure gradient (a) from right to left, (b) from top to bottom. Thickness of the line represents the magnitude of flow rates. A thin line indicates the flow rate of 10^{-9} m^3/; flow rates smaller than this value are not indicated (Min et al., 2004b).

The above demonstrative simulation of stress-induced flow localization is based on much simplified (and well-connected) 2D fracture system geometry, constitutive models and boundary conditions. In 3D situations, stress-induced fluid flow anisotropy in individual fractures and fracture system connectivity in the third direction will play significant roles. These 3D factors may lead to different behaviors as shown in Fig. 12.66, even under similar stress conditions. More comprehensive discussion is given in the next section.

12.8.5 Discussion on Outstanding Issues

The results presented in this section serve as an illustration of a DEM numerical approach to homogenize and upscale the hydro-mechanical properties of fractured rocks using discrete element models. The results should be interpreted with an awareness of the necessary simplifications and assumptions concerning problem dimensions and constitutive models of the rock matrix and fractures, especially the limitations introduced by the uncertainty of the geometric data of the fracture system and the hydraulic and mechanical properties of individual fractures. Since the aperture value plays a major role in the hydraulic behavior of fractured rock masses, especially the initial and residual aperture values, a reliable and realistic stress–aperture relation, or normal stiffness, such as the Barton–Bandis model, is essential for capturing the coupled stress-flow behavior of the fractured rocks. The approach depends on realistic characterization and modeling of both normal closure and shear dilation of fractures. Missing any one of these will render the approach incapable of simulating the change of permeability induced by stress.

The above-presented clustered (localized) fluid flow in fractures induced by stress changes indicates that previous histories of stress state (such as tectonic stresses), which may cause irreversible fracture shear failure at certain orientations, may be one of the reasons for highly clustered (channeled) flow patterns observed in fractured hard rocks. Such stress-induced flow channeling is of special significance to many rock engineering problems where groundwater flow is the major concern, such as nuclear waste repositories or geothermal reservoir engineering, although the approach presented is constrained by a number of limitations, especially the 2D simplification and ignorance of the fracture size–aperture correlation.

The two-dimensional analysis, as presented above, is useful for conceptual understanding of the hydro-mechanical behavior of fractured rock masses. However, for site-specific engineering applications, the limitation of the two-dimensional models is often not acceptable and certainly needs to be checked. Firstly, in two-dimensional analysis, the orientations of fractures (the apparent dip) depend on the orientation of the reference plane containing the computational model and the geometry, especially the connectivity, in the two-dimensional model varies with the locations of the reference plane cutting the parent fracture system in three-dimensional space, with the strikes of the fractures assumed to be normal to the model plane. Secondly, three-dimensional effects of fractures, especially stress-induced flow anisotropy such as reported in Yeo et al. (1998) and Koyama et al. (2004, 2006), cannot be considered in two-dimensional DEM models.

In essence, the problems of fractured rocks, especially when fluid flow in fracture system is dominant, are three-dimensional and should be solved in three dimensions. However, the 3D nature of the flow in fractures, especially when stress-induced flow anisotropy is concerned, requires much increased computational efforts. Three-dimensional modeling for coupled hydro-mechanical behavior of fractured rocks with explicit representation of large number of fractures may easily become computationally too intensive for practical applications at present, therefore equivalent continuum approach is still much preferred numerical approach for large-scale problems. However, direct 3D approach with hybrid DEM–DFN models for coupled stress-flow problems of large field scales may become feasible in the near future in line with the rapid increase of computer technology. The reliability of the solutions, however, depends always on the characterization of the fracture systems.

References

Adachi, T., Onishi, Y. and Arai, K., Investigation of toppling slope failure at Route 305 in Japan. Proc. 7th Int. Congr. of Rock Mechanics, Aachen, pp. 843–846. Balkema, Rotterdam, 1991.

Ahola, M. P., Application of the discrete element method toward roof stability problems in underground coal mines. Engineering Computations, 1992;2:242–247.

Akky, M. R., Rosidi, D., Madianos, M. N. and Kaneshiro, J.Y., Dynamic analysis of large underground caverns using discrete-element codes – verification and reliability. In: Salem, A. (ed.), Proc. of Int. Conf. Tunneling and Ground Conditions, pp. 477–484. Balkema, Rotterdam, 1994.

Allison, R. J. and Kimber, O. G., Modelling failure mechanisms to explain rock slope change along the Isle of Purbeck coast, UK. Earth Surf. Process. Landforms, 1998;23:731–750.

Amadei, B. and Goodman, R. E., A 3-D constitutive relation for fractured rock masses. In: Selvadurai, A. P. S. (ed.), Proc. of the Int. Symp. on the Mechanical Behavior of Structured Media, Ottawa, Part B, pp. 249–268, 1981.

Amadei, B. and Stephansson, O., Rock stress and its measurement. Chapman and Hall, London, 1997.

Andersson, J. and Knight, L., Understanding the impact of upscaling THM processes on performance assessment, DECOVALEXIII project, TASK 3, BMT2 protocol, version 6.0. Unpublished project report, 2000.

Antikainen, J., Simonen, A. and Makinen, I., 3D modelling of the central pillar in the Pyhasalmi Mine. Proc. of the Int. Congr. on Mine Design, Kingston, Ontario, Canada, pp. 631–640. Balkema, Rotterdam, 1993.

Ask, D., New developments of the integrated stress determination method and application to the Äspö Hard Rock Laboratory, Sweden. PhD Thesis, Royal Institute of Technology, Stockholm, 2004.

Bai, M. and Elsworth, D., Modeling of subsidence and stress-dependent hydraulic conductivity for intact and fractured porous media. Rock Mechanics and Rock Engineering, 1994;27(4):209–234.

Bai, M., Meng, F., Elsworth, D. and Roegiers, J. C., Analysis of stress-dependent permeability in nonorthogonal flow and deformation fields. Rock Mechanics and Rock Engineering, 1999; 32(3):195–219.

Bandis, S. C., Lumsden, A. C. and Barton, N., Fundamentals of rock joint deformation. International Journal of Rock Mechanics and Mining Sciences and Geomechanics Abstracts, 1983;20(6):249–268.

Bardet, J.-P. and Scott, R., Seismic stability of fractured rock masses with the distinct element method. Proc. 26th US Symp. on Rock Mechanics, Rapid City, pp. 139–149, 1985.

Baroudi, H., Piguet, J. P., Josien, J. P., Arif, I. and Lebon, P., Modelling of underground cavity storage and consideration of rock mass discontinuity. Proc. 7th Int. Cong. on Rock Mechanics, Aachen, Vol. 1, pp. 675–678. Balkema, Rotterdam, 1991.

Barton, C. A., Zoback, M. D. and Moos, D., Fluid flow along potentially active faults in crystalline rock. Geology 1995;23(8):683–686.

Barton, N., Modelling rock joint behaviour from in situ block tests: Implications for nuclear waste repository design. Office of Nuclear isolation, Columbus, OH, ONWI-308, 1982.

Barton, N., Tunbridge, L., Løset, F., Westerdahl, H., Kristianssen, J., Vik, G. and Chryssanthakis, P., Norwegian Olympic Ice Hockey Cavern of 60 m Span. Proc. 7th Int. Congr. on Rock Mechanics, Aachen, Vol. 2, pp. 1073–1081. Balkema, Rotterdam, 1991.

Barton, N., Yufin, S. A. and Swoboda, G., Engineering decisions in rock mechanics based on numerical modeling. Proc. 8th Int. Conf. on Computer Methods and Advances in Geomechanics, Morgantown, West Virginia, USA, pp. 2221–2228. Balkema, Rotterdam, 1994.

Barton, N., Løset, F., Vik, G., Rawlings, C., Chryssanthakis, P., Hansteen, H., Smallwood, A. and Ireland, T., Radioactive waste repository design using Q and UDEC-BB. Proc. of the Conf. on Fractured and Jointed Rock Masses, Lake Tahoe, CA, June 1992, pp. 709–716. Balkema, Rotterdam, 1995.

Bawden, W. F., Curran, J. H. and Roegiers, J.-C., Influence of fracture deformation on secondary permeability – a numerical approach. International Journal of Rock Mechanics and Mining Sciences and Geomechanics Abstracts, 1980;17(5):265–279.

Bear, J., Dynamics of fluids in porous media. Elsevier, New York, 1972.

Bhasin, R. and Høeg, K., Parametric study for a large cavern in jointed rock using a distinct element model (UDEC-BB). International Journal of Rock Mechanics and Mining Sciences, 1998;35:17–29.

Bhasin, R. K., Barton, N., Grimstad, E., Chryssanthakis, P. and Shende, F. P., Comparison of predicted and measured performance of a large cavern in the Himalayas. International Journal of Rock Mechanics and Mining Sciences and Geomechanics Abstracts, 1996;33:607–626.

Bigarre, P., Tinucci, J., Ben Slimane, K., Besson, J. C. and Piquet, J. P., 3-dimensional modelling of fault-slip rockbursting. Proc. of the Int. Symp. on Rock Bursts and Seismicity in Mines 93, pp. 315–319. Balkema, Rotterdam, 1993.

Blair, S. C., Carlson, S. R. and Wagoner, J. L., Distinct element modelling of the Drift Scale Test. Proc. 38th US Rock Mechanics Symp., Washington, DC, Vol. 1, pp. 527–531. Swetz & Zeitlinger, Lisse, The Netherlands, 2001.

Board, M., Numerical examination of mining-induced seismicity. In: Barla, G. (ed.), Special Lectures, Eurock'96, Prediction and Performance in Rock Mechanics and Rock Engineering, Torino, pp. 89–111, 1996.

Board, M., Seldon, S., Brummer, R. and Palkanis, R., Analysis of the failure of a large hangingwall wedge: Kidd Mine Division, Falconbridge, Ltd. Bulletin of the Canadian Institute of Mining, 2000;93:89–97.

Brady, B. H. G., Boundary element and linked methods for underground excavation design. In: Brown, E. T. (ed.), Analytical and computational methods in engineering rock mechanics, pp. 164–204. Allen and Unwin, London, 1987.

Brady, B. H. G. and Brown, E. T., Rock mechanics for underground mining, 527 pages. Allen and Unwin, London, 1985.

Brady, B. H. G. and Lorig, L., Analysis of rock reinforcement using finite difference methods. Computers and Geotechnics, 1988;5:123–149.

Brady, B. H. G., Lemos, J. V. and Cundall, P. A., Stress measurement schemes for jointed and fractured rock. In: Stephansson, O. (ed.), Proc. Int. Symp. on Rock Stress and Rock Stress Measurements, Stockholm, pp. 167–176. CENTEC, Lulea, 1986.

Burbidge, D. R. and Braun, J., Numerical models of the evolution of accretionary wedges and folds-and-thrust belts using distinct-element method, 2002;148:542–561.

Cai, Y., He, T. and Wang, R., Numerical simulation of dynamic process of the Tangshan earthquake by a new method – LDDA. Pure appl. Geophys., 2000;157:2083–2104.

Call, R. D., Slope stability. In: Hartman, H. L. (Sen. ed.), SME Mining Engineering Handbook, 2nd Edition, pp. 881–896. Society for Mining, Metallurgy, and Exploration, Littleton, CO, 1992.

Chandler, N. and Martin, D., The influence of near surface faults on in situ stresses in the Canadian Shield. In: Nelson, P. P. and Laubach, S. E. (eds), Proc. 1st North American Rock Mechanics Symp., University of Texas at Austin, pp. 369–376. Balkema, Rotterdam, 1994.

Chen, M. and Bai, M., Modeling stress-dependent permeability for anisotropic fractured porous rocks. International Journal of Rock Mechanics and Mining Sciences, 1998;35(8):1113–1119.

Chen, R., Ofoegbu, G. I. and Hsiung, S., Modeling drift stability in fractured rock mass at Yucca Mountain, Nevada – Discontinuum approach. Proc. 4th North American Rock Mechanics Symp., Seattle, pp. 945–952, 2000.

Choi, S. K. and Coulthard, M. A., Modelling of jointed rock masses using the distinct element method. Proc. of the Int. Conf. on Mechanics of Jointed and Faulted Rock, Vienna, pp. 471–477. Balkema, Rotterdam, 1990.

Chryssanthakis, P. and Barton, N., Predicting performance of the 62 m span ice hockey cavern in Gjövik, Norway. Proc. of the Conf. on Fractured and Jointed Rock Masses, Lake Tahoe, CA, Vol. 3, pp. 655–662. Balkema, Rotterdam, 1995.

Chryssanthakis, P., Monsen, K. and Barton, N., Validation of UDEC-BB against the CSM block test and large scale application to glacier loading of jointed rock masses. Proc. of 7th Int. Congr. on Rock Mechanics, Aachen, Vol. 1, pp. 693–698. Balkema, Rotterdam, 1991a.

Chryssanthakis, P., Barton, N., Jing, L. and Stephansson, O., Modelling the effect of glaciation, ice flow and deglaciation on large faulted rock masses. Proc. of NEA/SKI Symp. on Validation of Geosphere Flow and Transport Models (GEOVAL), May 1990, OECD, Paris, pp. 530–541, 1991b.

Coggan, J. S. and Pine, R. J., Application of distinct-element modelling to assess slope stability at Delabole slate quarry, Cornwall, England. Trns. Insttn. Min. Metall. (Sect. A: Min. industry), 1996;105:A22–A30.

Coulthard, M. A., Distinct element modelling of mining-induced subsidence – a case study. Proc. of the Conf. on Fractured and Jointed Rock Masses, Lake Tahoe, CA, June 1992, pp. 725–732. Balkema, Rotterdam, 1995.

Coulthard, M. A. and Dutton, A. J., Numerical modelling of subsidence induced by underground coal mining. In: Cundall, P. A., Sterling, R. L. and Starfield, A. M. (eds), Proc. of 29th U.S. Rock Mechanics Symp., Minneapolis, pp. 529–536. Balkema, Rotterdam, 1988.

Coulthard, M. A., Journet, N. J. and Swindells, C. F., Integration of stress analysis into mine excavation design. In: Tillerson, J. R. and Wawersik, W. R. (eds), Rock Mechanics (Proc. of the 33rd US Rock Mechanics Symp., Santa Fe, June 3–5, 1992), pp. 451–460. Balkema, Rotterdam, 1992.

Coulthard, M. A., Niu, J. K., Stewart, D. P., Sun, Z. N., Swindells, C. E. and Xiong, C. Z., Monitoring and numerical modelling of pit slope stability at Daye Iron Ore Mine. In: Yuan, J. (ed.), Proc. of the 9th Int. Conf. on Computer Methods and Advances in Geomechanics, Wuhan, China, pp. 1675–1680. Balkema, Rotterdam, 1997.

Cundall, P. A., Rational design of tunnel supports: A computer model for rock mass behaviour using interactive graphics for the input and output of geomaterial data. Technical Report MRD-2-74, Missuri River Division, US Army Corps of Engineers, NTIS Report No. AD/A-001 602, 1974.

Cundall, P. A. and Fairhurst, C., Correlation of discontinuum models with physical observations – an approach to the estimation of rock mass behaviour. Felsbau, 1986;4:197–202.

Cundall, P. A. and Hart, R. D., Numerical modelling of discontinua. Engineering Computations, 1992a;9:101–113.

Cundall, P. A. and Hart, R. D., Numerical modelling of discontinua. Keynote address. Engineering Computations, 1992b;2:132–152.

Cundall, P. and Lemos, J., Numerical simulation of fault instabilities with the continuously-yielding joint model. Proc. 2nd Int. Conf. on Rockburst in Mines, University of Minnesota, pp. 73–96. Balkema, Rotterdam, 1988.

Dasgupta, B., Dham, R. and Lorig, L., Three dimensional discontinuum analysis of the underground power house for Sardar Sarovar Project, India. Proc. 8th Int. Congr. on Rock Mechanics, Tokyo, Vol. 2, pp. 551–554. Balkema, Rotterdam, 1995.

Deangeli, C., Del Greco, O., Ferrero, A. M., Pancotti, G. and Giani, G. P., Rock mechanics studies to improve intact rock block exploitation and slope stability conditions in a quarry basin. Proc. Eurock'96, Predictions and Performance in Rock Mechanics and Rock Engineering, Torino, pp. 561–567. Balkema, Rotterdam, 1996.

Dershowitz, W. S., Lee, G., Geier, J., Foxford, T., La Pointe, P. and Thomas, A., FRACMAN User Documentation v.2.6. Golder Associates, Washington, 1998.

Donovan, K. S., Lehman, G. A., Pearce, R. S. and Kim, K., Rationale for development of constitutive models of a basalt rock mass for design of a nuclear waste repository at the Hanford site. Proc. of the 28th US Rock Mechanics Symp., Tucson, AZ, pp. 715–724, 1987.

Duncan Fama, M. E., Shen, B. T., Craig, M. S., Kelly, M., Follington, I. L. and Leisemann, B. E., Layout design and case study for highwall mining of coal. Proc. of the 9th ISRM Congr. on Rock Mechanics, Paris, Vol. 1, pp. 265–268. Balkema, Rotterdam, 1999.

Dupin, J.-M., Sassi, W. and Angelier, J., Homogeneous stress hypotheses and actual fault slip: A distinct element analysis. Journal of Structural Geology, 1993;15:1033–1043.

Esaki, T., Jiang, Y., Bhattarai, T. N., Nozaki, A. and Mizokami, T., Stability analysis and reinforcement system design in a progressively failed steep rock slope by the distinct element method. International Journal of Rock Mechanics and Mining Sciences, 1999a;35(4–5):Paper No. 31.

Esaki, T., Jiang, Y., Bhattarai, T. N., Maeda, T., Nozaki, A. and Mizokami, T., Modelling jointed rock masses and prediction of slope stabilities by DEM. Proc. 37th US Rock Mech. Symp., Vail, CO, June 1999, Vol. 1, pp. 83–90. Balkema, Rotterdam, 1999b.

Esterhuizen, G. S., The application of numerical models to assess the effect of discontinuities on coal pillar strength. Proc. of the Conf. on the Application of Numerical Modelling in Geotechnical Engineering, pp. 97–99. South African National Group of ISRM, Pretoria, South Africa, 1994.

Eve, R. A. and Squelch, A. P., Modeling of slope and gully support systems. Proc. of the Conf. on the Application of Numerical Modelling in Geotechnical Engineering, pp. 127–134. South African National Group of ISRM, Pretoria, South Africa, 1994.

Fairhurst, C. and Lemos, J. V., Influence of in-situ stresses on fluid penetration in jointed rock from unlined pressure tunnels. Proc. Int. Symp. on Rock Mechanics and Power Plants, pp. 231–264. Balkema, Rotterdam, 1988.

Ferrero, A. M., Giani, G. P., Verga, F. and Lemos, J. V., Back analysis of a rock block barrier subject to dynamic load using distinct element method and physical models. Proc. ISRM Int. Symp. on Safety and Environmental Issues in Rock Engineering, Eurock'93, Lisboa, pp. 107–113, 1993.

Ferrero, A. M., Giani, G. P., Kapenis, A. and Harrison, J. P., Theoretical and experimental study on geometric instability of pillars in discontinuous rock. Proc. of 8th Int. Congr. on Rock Mechanics, Tokyo, Vol. 2, pp. 555–558. Balkema, Rotterdam, 1995.

Finch, E., Hardy, S. and Gawthorpe, R., Discrete element modelling of contractional fault-propagation folding above rigid basement fault blocks. Journal of Structural Geology, 2003;25:515–528.

Franklin, J. A. and Dusseault, M. B., Rock engineering applications, 431 pages. McGraw-Hill, New York, 1991.

Gilbride, L. J., Richardson, A. M. and Agapito, J. F. T., Use of block models for longwall shield capacity determinations. International Journal of Rock Mechanics and Mining Sciences, 1998;35(4–5):Paper 44.

Gomes, M., Galiza, A. and Toscano, D., Long hole sublevel stoping at the Mina do Moinho. Proc. ISRM Int. Symp. on Safety and Environmental Issues in Rock Engineering, Eurock'93, Lisboa, pp. 551–559, 1993.

Guitierrez, M. and Barton, N., Numerical modelling of the hydro-mechanical behavior of single fractures. Proc. 1st North American Rock Mechanics Symp., University of Texas at Austin, pp. 165–172. Balkema, Rotterdam, 1994.

Hakami, E., Hakami, H. and Christiansson, R., Depicting a plausible *in situ* stress distribution by numerical analysis – examples from two candidate sites in Sweden. In: Lu, M., Li, C. C., Kjörholt, H. and Dahle, H. (eds), In-situ rock stress. Measurement, interpretation and application, pp. 473–481. Taylor & Francis, London, 2006.

Hansson, H., Jing, L. and Stephansson, O., 3-D DEM modelling of coupled thermo-mechanical response for a hypothetical nuclear waste repository. Proc. of the 5th Int. Symp. on Numerical Models in Geomechanics – NUMOG V, Davos, Switzerland, pp. 257–262. Balkema, Rotterdam, 1995.

Hardy, M. P., Lin, M. and Black, K., Hydrologic stability study of the crown pillar, Cardon Deposit, in support of mine permit application. Proc. of 37th US Rock Mechanics Symp., Vail, CO, Vol. 1, pp. 473–480. Balkema, Rotterdam, 1999.

Harper, T. R. and Last, N. C. J., Response of fractured rock subject to fluid injection. Part II. Characteristic behaviour. Tectonophysics, 1990a;172:33–51.

Harper, T. R. and Last, N. C. J., Response of fractured rock subject to fluid injection. Part III. Practical application. Tectonophysics, 1990b;172:53–65.

Hart, R. D., Lecture: discontinuum analysis for deep excavations in jointed rock. In: Maury, V. and Fourmaintraux, D. (eds), Rock at Great Depth, Vol. 3, pp. 1123–1130. Balkema, Rotterdam, 1990.

Hart, R. D., An introduction to distinct element modeling for rock engineering. In: Hudson, J. A. (ed.), Comprehensive rock engineering, Vol. 2, pp. 245–260. Pergamon Press, Oxford, 1993.

Hart, R. D., Enhancing rock stress understanding through numerical analysis. International Journal of Rock Mechanics and Mining Sciences, 2003;40:1089–1097.

Hart, R. D., Cundall, P. A. and Cramer, M. L., Analysis of a loading test on a large basalt block Proc. of the 26th US Rock Mechanics Symp., Rapid City, pp. 759–768. Balkema, Rotterdam, 1985.

Hart, R. D., Board, M., Brady, B., O'Hearn, B. and Allen, G., Examination of fault-slip induced rockbursting at the Strathcona Mine. In: Cundall, P. A., Sterling, R. L., and Starfield, A. M. (eds), Proc. 29th US Rock Mechanics Symp., Minneapolis, pp. 369–379. Balkema, Rotterdam, 1988.

Hassan, F. H., Spinnler, L. and Fine, J., A new approach for rock mass cavability modeling. International Journal of Rock Mechanics and Mining Sciences and Geomechanics Abstracts, 1993;30:1379–1385.

Hatzor, Y. H., Fully dynamic stability analysis of jointed rock slopes. 10th ISRM Cong. South Africa – Technology Road Map for Rock Mechanics South African Institute of Mining and Metallurgy, pp. 503–513, 2003.

Hatzor, Y. H., Arzi, A. A. and Tsesarsky, M., Realistic dynamic analysis of jointed rock slopes using DDA. In: Hatzor, Y. H. (ed.), Stability of rock structures. Balkema, Lisse, pp. 47–56. 2002.

Hencher, S. R., Liao, Q.-H. and Monaghan, B. G. Modelling slope behaviour for open-pits. Trns. Insttn. Min. Metall. (Sect. A: Min. industry), 1996;105:A37–A47.

Herget, G. and Arjang, B., Update on ground stressess in the Canadian Shield, in *Proc. Conf. on Stresses in Underground Structures*, Ottawa, CANMET, pp. 33–47, 1990.

Hoek, E. and Bray, J. W., Rock slope engineering, 402 pages. Institution of Mining and Metallurgy, London, 1977.

Holmberg, C., Hu, J. C., Angelier, J., Bergerat, F. and Lacombe, O., Characterization of stress perturbations near major fault zones: Insights from 2-D distinct-element numerical modelling and field studies (Jura mountains). Journal of Structural Geology, 1997;19:703–718.

Holmberg, C., Angelier, J., Bergerat, F. and Lacombe, O., Using stress deflections to identify slip events in fault systems. Earth and Planetary Science Letters, 2004;217:409–424.

Homand-Etienne, F., Rode, N. and Schwartzmann, R., Block modelling of jointed cliffs. In: Rossmanith, H.-P. (ed.), Proc. of Int. Conf. on Mechanics of Jointed and Faulted Rock, Vienna, pp. 819–825. Balkema, Rotterdam, 1990.

Homand, F., Souley, M., Gaviglio, P. and Mamane, I., Modelling natural stresses in the Arc syncline and comparison with *in situ* measurements. International Journal of Rock Mechanics and Mining Sciences, 1997;34:1091–1107.

Hsu, S. C. and Nelson, P. P., Analyses of slopes in jointed weak rock masses using distinct element method. In: Rossmanith, H.-P. (ed.), Proc. of the Second Int. Conf. on Mechanics of Jointed and Faulted Rock – MJFR-2, Vienna, pp. 589–594. Balkema, Rotterdam, 1995.

Hudson, J. A. and Harrison, J. P., Engineering rock mechanics, 444 pages. Pergamon, Amsterdam, 1997.

Hutchison, B., Dugan, K. and Coulthard, M., Analysis of flexural toppling at Australian Bulk Minerals Savage River Mine. Proc. of the Int. Conf. on Geotechnical & Geological Engineering, Melbourne, Australia, Paper SNES 1250, 2000.

Hwang, H., Jung, H., Ra, S., Kim, D. and Baek, H., Stability analysis of a tunnel supported by shotcrete-rockbolt system in hard bedrock. Proc. 38th US Rock Mechanics Symp., Washington, DC, Vol. 2, pp. 1029–1034, 2001.

Ishida, T., Chigira, M. and Hibino, S., Application of the distinct element method for analysis of toppling observed on a fissured rock slope. Rock Mechanics and Rock Engineering, 1987;20:177–283.

Itasca Consulting Group, Inc., UDEC – Universal Distinct Element Code, ver. ICG 1.7; User manual. Itasca Consulting Group, Inc., Minneapolis, MN, 1991.

Itasca Consulting Group, Inc., UDEC – Universal Distinct Element Code. Itasca Consulting Group, Inc., Minneapolis, MN, 1995a.

Itasca Consulting Group, Inc., 3DEC – Three-dimensional Distinct Element Code. Itasca Consulting Group, Inc., Minneapolis, MN, 1995b.

Itasca Consulting Group, Inc., UDEC user's guide, Minnesota, USA, 2000.

Itasca Consulting Group, Inc., UDEC user's guide, Minnesota, USA, 2002.

Ito, T. and Hayashi, K., Role of stress-controlled flow pathways in HDR geothermal reservoirs. Pure Appl. Geophys., 2003;160:1103–1124.

Jang, H. I., Chang, K. M. and Lee, C. I., Groundwater flow analysis of discontinuous rock mass with probabilistic approach. Journal of Korean Society for Rock Mechanics, 1996;6:30–38 (in Korean).

Jiang, Y. J., Li, B., Yamashita, Y., Etou, Y., Tanabashi, Y. and Zhao, X. D., Behaviour study of large-scale underground opening in discontinuous rock masses by using distinct element method. In: Leung, C. F. and Zhou, Y. X. (eds), Rock mechanics in underground construction. 4th Asian Rock Mechanics Symp., Singapore, p. 178. World Scientific, Singapore, 2006.

Jing, L., A review of techniques, advances and outstanding issues in numerical modelling of rock mechanics and rock engineering. International Journal of Rock Mechanics and Mining Sciences, 2003;40:283–353.

Jing, L. and Stephansson, O., Distinct element modelling of the influence of glaciation and deglaciation on the state of stress in faulted rock masses. SKB Technical Report No. 88–23, SKB, Stockholm, 1987.

Jing, L. and Stephansson, O., Numerical modeling of large-scale sublevel stoping by three-dimensional distinct element method. In: Wittke, N. (ed.), Proc. 7th Int. Cong. ISRM, Aachen, Germany, Vol. 1, pp. 741–746. Balkema, Rotterdam, 1991.

Jing, L., Tsang, C.-F. and Stephansson, O., DECOVALEX – an international co-operative research project on mathematical models of coupled THM processes for safety analysis of radioactive waste repositories. International Journal of Rock Mechanics and Mining Sciences and Geomechanics Abstracts, 1995;32:389–398.

Jing, L., Tsang, C.-F., Stephansson, O. and Kautsky, F., Validation of mathematical models against experiments for radioactive waste repositories – DECOVALEX experience. In: Stephansson, O., Jing, L. and Tsang, C.-F. (eds), Coupled thermo-hydro-mechanical processes of fractured media, mathematical and experimental studies. Developments in Geotechnical Engineering, 79, pp. 25–56. Elsevier Science, Amsterdam, 1996a.

Jing, L., Stephansson, O., Tsang, C.-F. and Kautsky, F., DECOVALEX – mathematical models of coupled T-H-M processes for nuclear waste repositories. Executive summary for phases I, II and III. SKI Report 96:58, Swedish Nuclear Power Inspectorate, Stockholm, Sweden, 1996b.

Jing, L., Hansson, H., Stephansson, O. and Shen, B., 3D DEM study of thermo-mechanical responses of a nuclear waste repository in fractured rocks – far- and near-field problems. In: Yuan, J. (ed.), Proc. of the 9th Int. Conf. on Computer Methods and Advances in Geomechanics, Wuhan, China, pp. 1207–1214. Balkema, Rotterdam, 1997.

Jing, L., Min, K.-B. and Stephansson, O., Numerical models for coupled thermo-hydro-mechanical processes in fractured rocks – continuum and discrete approaches. In: Hatzor, Y. H. (ed.), Stability of rock structures, pp. 57–67. Balkema, Lisse, 2002.

Johansson, E. and Kuula, H., Three-dimensional back-analysis calculations of Viikinmäki underground sewage treatment plant in Helsinki. Proc. 8th Int. Cong. on Rock Mechanics, Tokyo, Vol. 2, pp. 597–600. Balkema, Rotterdam, 1995.

Johansson, E. M., Hakala, M., Peltonen, E., Salo, J.-P. and Lorig, L. J., Comparison of two-and three-dimensional thermomechanical response of jointed rock. Proc. 7th Int. Cong. on Rock Mechanics, Aachen, Vol. 1, pp. 115–119. Balkema, Rotterdam, 1991.

Jung, J. and Brown, S. R., A study of discrete and continuum joint modeling techniques. In: Myer, L. R., Cook, N. G. W., Goodman, R. E. and Tsang, C.-F. (eds), Fractured and jointed rock masses. Proc. of the Conf. on Fractured and Jointed Rock Masses, Lake Tahoe, CA, June 1992, pp. 671–678. Balkema, Rotterdam, 1995.

Kamata, H. and Mashimo, H., Centrifuge model test of tunnel face reinforcement by bolting. Tunneling and Underground Space Technology, 2003;18:205–212.

Karaca, M., The effects of discontinuities on the stability of shallow tunnels in weak rock. In: Myer, L. R., Cook, N. G. W., Goodman, R. E. and Tsang, C.-F. (eds), Fractured and jointed rock masses. Proc. of the Conf. on Fractured and Jointed Rock Masses, Lake Tahoe, CA, June 1992, pp. 733–737. Balkema, Rotterdam, 1995.

Kay, D. R., McNabb, K. E. and Carter, J. P., Numerical modelling of mine subsidence at Angus Place Colliery. Proc. of the 7th Int. Conf. on Computer Methods and Advances in Geomechanics, Cairns, Queensland, pp. 999–1004. Balkema, Rotterdam, 1991.

Khaleel, R., Scale dependence of continuum models for fractured basalts. Water Resour Research, 1989;25(8):1847–1855.

Kim, T. K. and Lee, H. K., Numerical analysis for coupling of groundwater flow with the behavior of jointed rocks around a cavern. Proc. 8th Int. Cong. on Rock Mechanics, Tokyo, Vol. 2, pp. 783–785. Balkema, Rotterdam, 1995.

Kosugi, M., Ishihara, H. and Nakagawa, M., Tunneling method coupled with joint Monitoring and DEM analysis. In: Myer, L. R., Cook, N. G. W., Goodman, R. E. and Tsang, C.-F. (eds), Fractured and jointed rock masses. Proc. of the Conf. on Fractured and Jointed Rock Masses, Lake Tahoe, CA, June 1992, pp. 739–744. Balkema, Rotterdam, 1995.

Koyama T. Fardin, N. and Jing, L., Shear induced anisotropy and heterogeneity of fluid flow in a single rock fracture with translational and rotary shear displacements – a numerical study, International Journal of Rock Mechanics and Mining Sciences, 2004;41(3):426 (Paper No. 2A08).

Koyama, T., Fardin, N., Jing, L. and Stephansson, O., Numerical simulation of shear induced flow anisotropy and scale dependent aperture and transmissivity evolutions of fracture replicas. International Journal of Rock Mechanics and Mining Sciences, 2006;43(1):89–106.

Kullman, D. H., Stewart, R. D. and Lightfoot, N., Verification of a discontinuum model used to investigate rock mass behaviour around a deep level stope. Proc. of the Conf. on the Application of Numerical Modelling in Geotechnical Engineering, pp. 63–67. South African National Group of ISRM, Pretoria, South Africa, 1994.

Kwon, S., Park, B. Y., Kang, C. H. and Chang, K., Structural stability analysis for a high-level underground nuclear waste repository in granite. Proc. of the 4th North American Rock Mechanics Symp., Seattle, pp. 1279–1285, 2000.

Lanaro, F., Barla, G., Jing, L. and Stephansson, O., Continuous and discontinuous modelling of the Corva cliff. Proc. Eurock'96, Predictions and Performance in Rock Mechanics and Rock Engineering, Torino, pp. 583–588. Balkema, Rotterdam, 1996.

Larsson, H., Glamheden, R. and Ahrling, G., Storage of natural gas at high pressure in lined rock caverns – rock mechanics analysis. Proc. Int. Conf. on Storage of Gases in Rock Caverns, Trondheim, Norway, pp. 177–184. Balkema, Rotterdam, 1989.

Last, N. C. and Harper, T. R., Response of fractured rock subject to fluid injection. Part I. Development of a numerical model. Tectonophysics, 1990;172:1–31.

Lekhnitskii, S. G., Theory of elasticity of an anisotropic elastic body, 404 pages. Holden Day Inc., San Francisco, 1963.

Lemos, J., A distinct element model for dynamic analysis of jointed rock with application to dam foundation and fault motion. Ph.D. Thesis, University of Minnesota, 1987.

Lemos, J. V. and Cundall, P. A., Earthquake analysis of concrete gravity dams on jointed rock foundation. In: Sharma, V. M., Saxena, K. R. and Wood, R. D. (eds), Distinct element modelling in geomechanics, pp. 117–143. Oxford & IBH Publishing, New Delhi, 1999.

Lemos, J. V. and Lorig, L. J., Hydromechanical modelling of jointed rock masses using the Distinct Element Code. Proc. of the Int. Conf. on Mechanics of Jointed and Faulted Rock, Vienna, pp. 605–612. Balkema, Rotterdam, 1990.

Lemos, J., Hart, R. D. and Cundall, P. A., A generalized distinct element program for modelling jointed rock mass: A keynote lecture. In: Stephansson, O. (ed.), Proc. of the Int. Symp. on Fundamentals of Rock Joints, Björkliden, Sweden, pp. 335–343. Centek Publisher, Lulea, 1985.

Lemos, J., Hart, R. and Lorig, L., Dynamic analysis of discontinua using the distinct element method. Proc. 6th Int. Cong. on Rock Mechanics, Montréal, Vol. 2, pp. 1079–1084, 1987.

Lightfoot, N., The use of numerical modeling in rockburst control. Proc. of the Int. Cong. on Mine Design, Kingston, Ontario, Canada, pp. 355–360. Balkema, Rotterdam, 1993.

Lightfoot, N., Kullman, D. H. and Leach, A. R., A conceptual model of a hard rock, deep level, tabular ore body that incorporates the potential for face bursting as a natural product of mining. Proc. 1st North American Rock Mechanics Symp., Austin, TX, pp. 903–910. Balkema, Rotterdam, 1994.

Long, J. C. S., Remer, J. S., Wilson, C. R. and Witherspoon, P. A., Porous media equivalents for networks of discontinuous fractures. Water Resour Research, 1982;18(3):645–658.

Lorig, L., A hybrid computational model for excavation and support design in jointed media. Ph.D. Thesis, University of Minnesota, 1984.

Lorig, L., A simple numerical representation of fully bonded passive rock reinforcement for hard rocks. Computers and Geotechnics, 1985;1:79–97.

Lorig, L., Distinct element – structural element analysis of support systems in jointed rock masses. Proc. 6th Australian Tunneling Conf., Melbourne, Australasian Institute of Mining and Metallurgy, pp. 173–182, 1987

Lorig, L. J. and Hobbs, B. E., Numerical modelling of slip instability using the distinct element method with state variable friction laws. International Journal of Rock Mechanics and Mining Sciences and Geomechanics Abstracts, 1990;27:525–534.

Lorig, L. J., Hart, R. D., Board, M. P. and Swan, G., Influence of discontinuity orientation and strength on cavability in a confined environment. In: Khair, A. W. (ed.), Rock mechanics as a guide for efficient utilization of natural resources, pp. 167–174. Balkema, Rotterdam, 1989.

Ma, M. and Brady, B. H., Analysis of the dynamic performance of an underground excavation in jointed rock under repeated seismic loading. Geotechnical and Geological Engineering, 1992;17:1–20.

MacLaughlin, M. M. and Clapp, K. K., Discrete element analysis of an underground opening in blocky rock: An investigation of the difference between UDEC and DDA results. Proc. 3rd Int. Conf. on Discrete Element Methods, Santa Fe, Mew Mexico, pp. 329–334, 2002.

MacLaughlin, M. M., Langston, R. B. and Brady, T. M., Numerical geomechanical modelling for Stillwater Mine, Nye, Montana. Proc. 38th US Rock Mechanics Symp., Washington, DC, Vol. 1, pp. 423–432. Swets & Zeitlinger, The Netherlands, 2001.

Makurat, A., Barton, N., Vik, G. and Bandis, S., Investigation of disturbed zone effects and support strategies for the Fjellinjen road tunnels under Oslo. In: Lo, K. Y. (ed.), Proc. of the Int. Cong. on Progress and Innovation in Tunneling, Toronto, pp. 125–134. TAC/NRC/ITA, Toronto, 1990a.

Makurat A., Barton, N., Rad, N. S. and Bandis, S., Joint conductivity variation due to normal and shear deformation. In: Barton, N. and Stephansson, O. (eds), Proc. Int. Symp. Rock Joints, pp. 535–540. Balkema, Rotterdam, 1990b.

Martin, C. D. and Simmons, G. R., The Atomic Energy of Canada Limited Underground Research Laboratory: An overview of geomechanics characterization. In: Hudson, J. A. (ed.), Comprehensive rock engineering, Vol. 3, pp. 915–950. Pergamon Press, Oxford, 1993.

Mas Ivars, D., Hakami, E. and Stephansson, O., Influence of rock mass characteristics on inflow into deposition holes for nuclear waste disposal – a coupled hydro-mechanical analysis using 3DEC. In: Konietzky, H. (ed.), Proc. of 1st Int. UDEC/3DEC Symp., Bochum, Germany, pp. 95–103, 2004.

McKinnnon, S. D., Analysis of stress measurements using a numerical model methododology. International Journal of Rock Mechanics and Mining Sciences, 2001;38:699–709.

McKinnon, S. D. and Lorig, L., Considerations for three-dimensional modelling in analysis of underground excavations. In: Sharma, V. M., Saxena, K. R. and Wood, R. D. (eds), Distinct element modelling in geomechanics, pp. 145–166. Oxford & IBH Publishing, New Delhi, 1999.

Min, K.-B. and Jing, L., Numerical determination of the equivalent elastic compliance tensor for fractured rock masses using the distinct element method. International Journal of Rock Mechanics and Mining Sciences, 2003;40(6):795–816.

Min, K.-B., Jing, L. and Stephansson, O., Fracture system characterization and evaluation of the equivalent permeability tensor of fractured rock masses using a stochastic REV approach. Hydrogeol. Journal, 2004a;12(5):497–510.

Min, K.-B., Rutqvist, J., Tsang, C.-F. and Jing, L., Stress-dependent permeability of fractured rock masses: A numerical study. International Journal of Rock Mechanics and Mining Sciences, 2004b;41:1191–1210.

Mitarashi, Y., Tezuka, H., Okabe, T., Morimoto, S. and Jiang, Y., The evaluation of the effect of long face bolting by 3D distinct element method. In: Leung, C. F. and Zhou, Y. X. (eds), Rock mechanics in underground construction. 4th Asian Rock Mechanics Symp., Singapore, p. 256. World Scientific, Singapore, 2006.

Monsen, K., Makurat, A. and Barton, N., Numerical modelling of disturbed zone phenomena at Stripa. In: Hudson, J. A. (ed.) Proc. ISRM Symp.: Eurock'92, Chester, pp. 354–359. British Geotechnical Society, London, 1992.

Morgan, J. K., Numerical simulation of granular shear zones using the distinct element method 2. Effects of particle size distribution and interparticle friction on mechanical behaviour. Journal of Geophysical Research, 1999;104:2703–2719.

National Coal Board, Subsidence Engineers Handbook, 2nd Edition. National Coal Board Mining Department, London, 1975.

Neuman, S. P., Stochastic continuum representation of fractured rock permeability as an alternative to the REV and fracture network. 28th US Symp. on Rock Mechanics, Tucson, pp. 533–561, 1987.

Ng, L. K. W., Swan, G. and Board, M., The application of energy approach in fault models for support design. Proc. of the Int. Symp. on Rockbursts and Seismicity in Mines 93, Kingston, Canada, pp. 387–391, 1993.

Nirex (Nirex UK Ltd), Geotechnical studies at Sellafield, Executive summary of NGI/WSA work from 1990–1994. Nirex Report 801, 1995.

Nirex (Nirex UK Ltd), Evaluation of heterogeneity and scaling of fractures in the Borrowdale Volcanic Group in the Sellafield Area. Nirex Report SA/97/028, 1997.

Nordlund, E., Radberg, G. and Jing, L., Determination of failure modes in jointed pillars by numerical modelling. In: Myer, L. R., Cook, N. G. W., Goodman, R. E. and Tsang, C.-F. (eds), Fractured and jointed rock masses. Proc. of the Conf. on Fractured and Jointed Rock Masses, Lake Tahoe, CA, June 1992, pp. 345–350. Balkema, Rotterdam, 1995.

O'Connor, K. M. and Dowding, C. H., Monitoring and simulation of mining-induced subsidence. In: Rossmanith, H.-P. (ed.), Proc. of the Int. Conf. on Mechanics of Jointed and Faulted Rock, Vienna, pp. 781–787. Balkema, Rotterdam, 1990.

O'Connor, K. M. and Dowding, C. H., Hybride discrete element code for simulation of mining-induced strata movements. Engineering Computations, 1992a;2:235–242.

O'Connor, K. M. and Dowding, C. H., Distinct element modeling and analysis of mining-induced subsidence. Rock Mechanics and Rock Engineering, 1992b;25:1–24.

Oda, M., An equivalent continuum model for coupled stress and fluid flow analysis in jointed rock masses, Water Resource Research, 1986;22(13):1845–1856.

Oelfke, S. M., Mustoe, G. G. W. and Kripakov, N. P., Application of the discrete element method toward simulation of ground control problems in underground mines. Proc. of the 8th Int. Conf. on Computer Methods and Advances in Geomechanics, Morgantown, West Virginia, pp. 1865–1870. Balkema, Rotterdam, 1994.

Ohnishi, Y., Nishiyama, S. and Sasaki, T., Development and application of discontinuous deformation analysis. In: Leung, C. F. and Zhou, Y. X. (eds), Rock mechanics in underground construction. 4th Asian Rock Mechanics Symp., Singapore, pp. 59–70. World Scientific, Singapore, 2006.

Olsson, R. and Barton, N., An improved model for hydromechanical coupling during shearing of rock joints. International Journal of Rock Mechanics and Mining Sciences, 2001;38(3):317–329.

Panda, B. B. and Kulatilake, P. H. S. W., Effect of block size on the hydraulic properties of jointed rock through numerical simulation. Rock Mechanics: Tools and techniques. Proc. of 2nd North American Rock Mechanics Symp., NARMS 96, Montreal, Canada, pp. 1969–1976, 1996.

Pascal, C., Interaction of faults and perturbation of slip: Influence of anisotropic stress state in the presence of fault friction and comparison between Wallace-Bott and 3D distinct element models. Tectonophysics, 2002;356:307–322.

Pascal, C. and Gabrielsen, R. H., Numerical modelling of Cenozoic stress patterns in the Mid Norwegian Margin and the northern North Sea. Tectonics, 2001;20:585–599.

Pine, R. J. and Cundall, P. A., Application of the fluid–rock interaction program (FRIP) to the modelling of hot dry rock geothermal energy systems. In: Stephansson, O. (ed.), Proc. of the Int. Symp. on Fundamentals of Rock Joints, Björkliden, pp. 293–302. Centek Publishing, Luleå, 1985.

Pine, R. J. and Nicol, D. A., Conceptual and numerical models of high pressure fluid–rock interaction. Proc. 29th US Rock Mechanics Symp., Minneapolis, pp. 495–502. Balkema, Rotterdam, 1988.

Priest, S. D., Discontinuity analysis for rock engineering. Chapman and Hall, London, 1993.

Rachez, X. and J.-L. Durville, Numerical modelling of a bridge foundation on a jointed rock slope with the distinct element method. Proc. Eurock'96, Predictions and Performance in Rock Mechanics and Rock Engineering, Torino, pp. 535–541. Balkema, Rotterdam, 1996.

Ravi, G. and Dasgupta, B., Influence of joints on the stability of shallow seated underground structures at Zawarmala mine. Proc. of the 2nd Int. Conf. on the Mechanics of Jointed and Faulted rock – MJFR-2, Vienna, pp. 645–650. Balkema, Rotterdam, 1995.

Rejeb, A., Mathematical simulations of coupled THM processes of Fanay-Augères field test by distinct element and discrete finite element methods. In: Stephansson, O., Jing, L. and Tsang, C.-F. (eds), Coupled thermo-hydro-mechanical processes of fractured media, mathematical and experimental studies. Developments in Geotechnical Engineering, 79, pp. 341–368. Elsevier Science, Amsterdam, 1996.

Renshaw, C. E. and Park, J. C., Effect of mechanical interactions on the scaling of fracture length and aperture. Nature, 1997;386:482–484.

Rosengren, L. and Stephansson, O., Modelling of rock mass response to glaciation at Finnsjön, Central Sweden. Tunneling and Underground Space Technology, 1993;8:75–82.

Rosengren, L., Board, M., Krauland, N. and Sandström, S., Numerical analysis of the effectiveness of reinforcement methods at the Kristineberg Mine in Sweden. Proc. of the Int. Symp. on Rock Support, Sudbury, Canada, pp. 507–514. Balkema, Rotterdam, 1992.

Ryder, J. A., Excess shear stress (ESS): An engineering criterion for assessing unstable slip and associated rockburst hazards. Proc. 6th International Congress of Rock Mechanics, Montreál, pp. 1211–1214. Balkema, Rotterdam, 1987.

Saltzer, S., Numerical modelling of crustal scale faulting using the distinct element method Proc. of the 2nd Int. Conf. on Discrete Element Methods (DEM), pp. 511–522. IESL Publications, 1993.

Saltzer, S. D. and Pollard, D., Distinct element modelling of structures formed in sedimentary overburden by extensional reactivation of basement normal faults. Tectonics, 1999;11:165–174.

Samaniego, J. A. and Priest, S. D., The prediction of water flows through discontinuity networks into underground excavation. Proc. Symp. on the Design and Performance of Underground Excavations, Cambridge, pp. 157–164. International Society for Rock Mechanics, 1984.

Sansone, E. C. and Ayres da Silva, L. A., Numerical modelling of the pressure arch in underground mines. International Journal of Rock Mechanics and Mining Sciences, 1998;35:Paper 111.

Scheldt, T., Lu, M. and Myrvang, A., Numerical analysis of Gjovik olympic cavern: A comparison of continuous and discontinuous results by using Phase^2 and DDA. In: Hatzor, Y. H. (ed.), Stability of rock structures, pp. 125–131. Balkema, Lisse, 2002.

Schelle, H. and Grünthal, G., Modelling of crustal block rotations in southern California – conclusions for geotechnics. In: Savidis, S. A. (ed.), 2nd Conf. on Earthquake Resistant Construction and Design, DGEB Publication 8, Vol. 1, pp. 11–18, 1994.

Shen, B. and Barton, N., The disturbed zone around tunnels in jointed rock masses. International Journal of Rock Mechanics and Mining Sciences, 34(1997):117–125.

Shen, B. and Duncan Fama, M. E., A laminated span failure model for highwall mining span stability assessment. In: Yuan, J. (ed.), Proc. of the 9th Int. Conf. on Computer Methods and Advances in Geomechanics, Wuhan, China, pp. 1585–1591. Balkema, Rotterdam, 1997.

Shen, B. T. and Duncan Fama, M. E., Review of highwall mining experience in Australia and case study. Proc. of the Int. Conf. on Geotechnical and Geological Engineering, Melbourne, Australia, Paper UW 1083, 2000.

Shen, B. and Stephansson, O., Rock mass response to glaciation and thermal loading from nuclear waste. Symp. on Validation of Geosphere Flow and Transport Models, GEOVAL-1990, pp. 550–558. OECD, Paris, 1991.

Siekmeier, J. A. and O'Connor, K. M., Explicit modeling of rock mass discontinuities and mining subsidence simulation. Proc. of the 8th Int. Conf. on Computer Methods and Advances in Geomechanics, Morgantown, West Virginia, pp. 1895–1900. Balkema, Rotterdam, 1994.

Sjöberg, J., Failure modes and pillar behaviour in the Zinkgruvan mine. Proc. of the 33rd US Rock Mechanics Symp., Santa Fe, pp. 491–500. Balkema, Rotterdam, 1992.

Sjöberg, J., Analysis of large scale rock slopes. Ph.D. Thesis, Luleå Technical University, 788 pages, 1999.

SKB (Swedish Nuclear Waste and Fuel Management Co.), Project on alternative systems study (PASS). SKB Technical Report 93–04, SKB, Stockholm, Sweden, 1992.

SKB (Swedish Nuclear Waste and Fuel Management Co.), RD&D-Programme 98: Treatment and final disposal of nuclear waste. Programme for research, development and demonstration of encapsulation and geological disposal, Stockholm, Sweden, 1998.

Souley, M. and Homand, F., Influence of joint constitutive laws on the stability of jointed rock masses. Proc. ISRM Int. Symp. on Safety and Environmental Issues in Rock Engineering, Eurock'93, Lisboa, pp. 203–208, 1993.

Souley, M. and Homand, F., Stability of jointed rock masses evaluated by UDEC with an extended Saeb-Amadei constitutive law. International Journal of Rock Mechanics and Mining Sciences and Geomechanics Abstracts, 1996;33:233–244.

Souley, M., Homand, F. and Thoraval, A., The effect of joint constitutive laws on the modelling of an underground excavation and comparison with *in situ* measurements. International Journal of Rock Mechanics and Mining Sciences, 1997;34:97–115.

Squelch, A. P., Roberts, M. K. C. and Taggart, P. N., The development of a rationale for the design of stope and gully support in tabular South African mines. Proc. 1st North American Rock Mechanics Symp., Austin, TX, pp. 919–926. Balkema, Rotterdam, 1994.

Starfield, A. M. and Cundall, P. A., Towards a methodology for rock mechanics modeling. International Journal of Rock Mechanics and Mining Sciences and Geomechanics Abstracts, 1988;25:99–106.

Stephansson, O. and Shen, B., Modelling of rock masses for site location of a nuclear waste repository. Proc. 7th Int. Cong. on Rock Mechanics, Aachen, Vol. 1, pp. 157–162. Balkema, Rotterdam, 1991.

Stephansson, O., Ljunggren, C. and Jing, L., Stress measurements and tectonic implications for Fennoscandia. Tectonophysics, 1991a;189:317–322.

Stephansson, O., Shen, B. and Lemos, J., Modelling of excavation, thermal loading and bentonite swelling pressure for a waste repository. Proc. of 2nd Annual Int. Conf. on High-Level Waste Management, Las Vegas, NV, pp. 1375–1381. American Nuclear Society/ASCE, New York, 1991b.

Stephansson, O., Tsang, C.-F. and Kautsky, F., Development of coupled models and their validation against experiments – DECOVALEX project. Proc. of GEOVAL'94 Validation Through Model Testing, pp. 349–360. Nuclear Energy Agency/OECD, Paris, 1994.

Stephansson, O., Jing, L. and Tsang, C.-F. (eds), Coupled thermo-hydro-mechanical processes of fractured media, mathematical and experimental studies. Recent development of DECOVALEX Project for radioactive waste repositories. Developments in geotechnical engineering, 79, 575 pages. Elsevier Science, Amsterdam, 1996.

Stephansson, O., Tsang, C.-F. and Kautsky, F., Foreword (second phase of DECOVALEX). International Journal of Rock Mechanics and Mining Sciences, 2001;38:1–4.

Stephansson, O., Jing, L. and Hudson, J. A. (eds), Coupled T-H-M-C processes in geosystems: Fundamentals, modeling, experiments and applications. Elsevier, Oxford, 2004.

Su, S. and Stephansson, O., Effect of fault on *in situ* stress studied by distinct element method. International Journal of Rock Mechanics and Mining Sciences, 1999;36:1501–1506.

Ting, T. C. T., Anisotropic elasticity – theory, applications, 570 pages. Oxford University Press, Oxford, 1996.

Tinucci, J. P. and Hanson, D. S. G., Assessment of seismic fault-slip potential at the Strathcona Mine. Proc. Rock mechanics Contribution and Challenges, pp. 753–760. Balkema, Rotterdam, 1990.

Tolppanen, P. J., Johansson, E. J. W. and Salo, J. P., Rock mechanical analyses of in-situ stress/strength ratio at the Posiva Oy investigation sites, Kivetty, Olkiluoto and Romuvaara, in Finland. Proc. Eurock'96, Predictions and Performance in Rock Mechanics and Rock Engineering, Torino. pp. 435–442. Balkema, Rotterdam, 1996.

Tsang, C.-F., Stephansson, O., Kautsky, F. and Jing, L., Coupled THM processes in geological systems and the DECOVALEX Project. In: Stephansson, O. Hudson, J. A. and Jing, L. (eds), Coupled thermo-hydro-mechanical-chemical processes in geo-systems: Fundamentals, modelling, experiments & applications, pp. 3–16. Elsevier, Oxford, 2004.

Vengeon, J. M., Hantz, D., Giraud, A. and Ract, D., Numerical modelling of rock slope deformation. Proc. Eurock'96, Predictions and Performance in Rock Mechanics and Rock Engineering, Torino. pp. 659–666. Balkema, Rotterdam, 1996.

Wei, Z. Q., A fundamental study of the deformability of rock masses. Ph.D thesis, Imperial College, University of London, 1988.

Wei, Z. Q., Egger, P. and Descoeudres, F., Permeability predictions for jointed rock masses. International Journal of Rock Mechanics and Mining Sciences and Geomechanics Abstracts, 1995;32(3):251–261.

Wojtkowiak, F., Soukatchoff, V., Peila, D., Ferraro, A. M. and Manfroi, I., Monitoring and numerical modelling of soft rock mass behaviour in a long-hole stoping mine at great depth. Proc. Int. Cong. on Rock Mechanics, Tokyo, Japan, Vol. 2, pp. 683–687. Balkema, Rotterdam, 1995.

Wong, F., Tennant, D., Vaughan, D., Isenberg, J., Lorig, L. and Dawson, E., Modeling of rock blocks and tunnel liner interaction, a comparative study. In: Noor, A. K. (Chair.), Abstracts of the 2nd US National Cong. on Computational Mechanics, Washington, DC, p. 90. NASA, Washington, DC, 1993.

Xu, D., Lilly, P. and Walker, P., Stability and risk assessment of pit walls and BHP Iron Ore's Mt Whaleback Mine. Proc. of the Int. Conf. on Geotechnical & Geological Engineering, Melbourne, Australia, Paper SNES 0067, 2000.

Yeo, I. W., De Freitas, M. H. and Zimmerman, R. W., Effect of shear displacement on the aperture and permeability of a rock fracture. International Journal of Rock Mechanics and Mining Sciences, 1998;35(8):1051–1070.

Yeung, M. R. and Blair, S. C., Analysis of large block test data using the DDA method. In: Amadei, B. (ed.), Third Int. Conf. on Analysis of Discontinuous Deformation – From Theory to Practice, June 3–4, Vail, CO, ARMA, pp. 141–150. Balkema, Rotterdam, 1999.

Zhang, X. and Sanderson, D. J., Effects of stress on the 2-D permeability tensor of natural fracture networks. Geophys J Int. 1996;125:912–924.

Zhang, X. and Sanderson, D. J., Effects of loading direction on localized flow in fractured rocks. In: Yuan, J. (ed.), Computer methods and advances in geomechanics, pp. 1027–1032. Balkema, Rotterdam, 1997.

Zhang, X. and Sanderson, D. J., Numerical modelling and analysis of fluid flow and deformation of fractured rock masses, 288 pages. Pergamon, Oxford, 2002.

Zhang, X., Sanderson, D. J., Harkness, R. M. and Last, N. C., Evaluation of the 2-D permeability tensor for fractured rock masses. International Journal of Rock Mechanics and Mining Sciences and Geomechanics Abstracts, 1996;33(1):17–37.

Zhu, F., Stephansson, O. and Wang, Y., Stability investigation and reinforcement for slope at Daye Open Pit Mine, China. Proc. Eurock'96, Predictions and Performance in Rock Mechanics and Rock Engineering, Torino, pp. 621–625. Balkema, Rotterdam, 1996.

APPENDIX: DERIVATION OF EXPRESSIONS FOR STRESS AND STRESS COUPLE TENSORS OF PARTICLE SYSTEMS AS EQUIVALENT COSSERAT CONTINUA

This appendix presents detailed derivations of expressions for stress and stress couple tensors of particle systems as examples for deriving equivalent macroscopic quantities of the homogenized Cosserat continua, following strictly the principles, logic and term definitions in Kruyt (2003). The detailed derivations for expressions for displacement compatibility equation, couple stress and rotation gradient vectors, and virtual and complementary virtual works can be obtained using similar techniques. The conventions for variables, indices and definitions are the same as those used in Section 11.5.

In order to establish equivalent Cosserat continuum models for discrete particle systems, moments are needed to obtain the discrete equilibrium equations (i.e., the equations of motion under static conditions). The main principle is that, by multiplying these equations by 1 and summing over all particles, the continuum equilibrium equations can be retrieved. By multiplying the equilibrium equations by position vectors and summing over all particles, the micro-mechanics expression for the averaged Cauchy stress tensor and the averaged couple stress tensor can be obtained. An alternative approach to obtain these expressions is to use the continuum mechanical definition of these tensors, but this approach will not be included.

A.1 Continuum Equilibrium Equations

Continuum equivalents of the discrete equilibrium equations of the particle assembly, Eqn (11.35), are obtained by multiplying these equations by 1 and summing over all particles. From the force equilibrium equation in Eqn (11.35), the summation leads to

$$\sum_p \sum_q f_j^{pq} + \sum_p f_j^{p\beta} = 0 \tag{A1}$$

The first double summation in Eqn (A1) consists of terms $f_j^{pq} + f_j^{qp}$ as action and reaction contact forces between particle p and q, which equals zero since $f_j^{pq} = -f_j^{qp}$. The second single summation has a continuum equivalent of $\int_B n_k \sigma_{kj}\, \mathrm{d}B$ (see Eqn (11.29a)), which can be expressed as $\int_V \partial\sigma_{kj}/\partial x_k\, \mathrm{dV}$, using the divergence theorem.

The continuum equivalent of the discrete force equilibrium equations is given as

$$\frac{\partial \sigma_{kj}}{\partial x_k} = 0 \tag{A2}$$

since the result holds for any sub-volume. It is identical to the classical continuum force equilibrium equations for quasi-static deformations without body forces.

Similarly from the discrete moment equilibrium equation in Eqn (11.35), one can obtain

$$\sum_p \sum_q (\kappa_j^{pq} + e_{jkl} C_k^{pq} f_l^{pq}) + \sum_p (\kappa_j^{p\beta} + e_{jki} C_k^{p\beta} f_i^{p\beta}) = 0 \tag{A3}$$

The first double summation consists of action and reaction rotational torques

$$(\kappa_j^{pq} + e_{jkl} C_k^{pq} f_l^{pq}) + (\kappa_j^{qp} + e_{jkl} C_k^{qp} f_l^{qp}) = 0 \tag{A4}$$

since $\kappa_j^{pq} = -\kappa_j^{qp}$, $C_j^{pq} = -C_j^{qp}$ and $f_j^{pq} = -f_j^{qp}$. Equation (A3) then is reduced to

$$\sum_p (\kappa_j^{p\beta} + e_{jkl} C_k^{p\beta} f_l^{p\beta}) = 0 \tag{A5}$$

which has a continuum equivalent of (cf. Eqn (11.29a))

$$\int_B n_m (\mu_{mj} + e_{jkl} x_k \sigma_{ml}) \mathrm{d}B = 0 \tag{A6}$$

Using the divergence theorem again on this expression gives

$$\int_V \frac{\partial(\mu_{mj} + e_{jkl} x_k \sigma_{ml})}{\partial x_m} \mathrm{d}V = 0 \tag{A7}$$

Hence the continuum equivalent of the discrete moment equilibrium equations is

$$\frac{\partial(\mu_{mj} + e_{jkl} x_k \sigma_{ml})}{\partial x_m} = 0 \tag{A8}$$

since the result holds for any sub-volume. Expanding Eqn (A8) and using the continuum force equilibrium equation (A2), one finally obtains

$$\frac{\partial \mu_{mj}}{\partial x_m} + e_{jkl} \sigma_{kl} = 0 \tag{A9}$$

which is identical to the continuum moment equilibrium equations for quasi-static deformations. It should be noted that it is not possible to obtain this result without using Eqn (11.29b), with the couple traction vector independent of the translational contact forces whose effects on moments are represented in the second term in Eqn (A9)).

A.2 Moments of Equilibrium Equations

By multiplying the force equilibrium equation (the first Eqn in (11.35)) by X_i^p and summing over all particles, one obtains

$$\sum_p \sum_q X_j^p f_j^{pq} + \sum_p X_i^p f_j^{p\beta} = 0 \tag{A10}$$

The first double summation contains components

$$X_j^p f_j^{pq} + X_j^q f_j^{qp} = -(X_i^q - X_i^p) f_j^{pq} = -l_i^{pq} f_j^{pq} \tag{A11}$$

since $f_j^{pq} = -f_j^{qp}$. Since $l_i^{pq} f_j^{pq} = l_i^{qp} f_j^{qp}$ is a proper contact property, then the first term of Eqn (A10) can be written as

$$\sum_p \sum_q X_j^p f_j^{pq} = -\sum_{c \in C^I} l_i^c f_j^c \tag{A12}$$

Using the definition of the edge vector for a boundary contact, $l_i^{p\beta} = C_i^{p\beta} - X_i^p$ as defined in Section 11.5, we obtain a new version of Eqn (A10)

$$-\sum_{c \in C^I \cup C^B} l_i^c f_j^c + \sum_{\beta \in B} C_i^\beta f_j^{p\beta} = 0 \tag{A13}$$

Recall Eqn (11.29a), the second term in Eqn (A13) has a continuum equivalent of

$$\sum_{\beta \in B} C_i^\beta f_j^{p\beta} = \int_B x_i n_k \sigma_{kj} \, \mathrm{d}B = \int_V \frac{\partial (x_i \sigma_{kj})}{\partial x_k} \mathrm{d}V \tag{A14}$$

after applying the divergence theorem. Using the continuum force equilibrium Eqn (A2) once more, the micro-mechanics expression for the average homogenized stress tensor $\langle \sigma_{ij} \rangle$ becomes

$$\langle \sigma_{ij} \rangle = \frac{1}{V} \int_V \sigma_{ij} \, \mathrm{d}V = \frac{1}{V} \sum_{c \in C} l_i^c f_j^c \tag{A15}$$

Multiplying the moment equilibrium equation in Eqn (A2) by X_i^p and summing over all particles, an expression for the average homogenized couple stress tensor is obtained by

$$\sum_p \sum_q X_i^p (\kappa_j^{pq} + e_{jkl} C_k^{pq} f_l^{pq}) + \sum_p X_i^p (\kappa_j^{p\beta} + e_{jkl} C_k^{p\beta} f_l^{p\beta}) = 0 \tag{A16}$$

Using the same technique with $X_i^p \kappa_j^{pq} + X_i^q \kappa_j^{qp} = -(X_i^q - X_i^p)\kappa_j^{pq} = -l_i^{pq} \kappa_j^{pq}$ (since $\kappa_j^{pq} = -\kappa_j^{qp}$) and $l_i^{pq} \kappa_j^{pq} = l_i^{qp} \kappa_j^{qp}$, the first term and second term in the double summation can be written as $-\sum_{c \in C^I} l_i^c \kappa_j^c$ and $-e_{jkl} \sum_{c \in C^I} l_i^c C_k^c f_l^c$. Using the same definition of the edge vector for a boundary contact, $l_i^{p\beta} = C_i^{p\beta} - X_i^p$, then leads to

$$-\sum_{c \in C^I \cup C^B} l_i^c \kappa_j^c - e_{jkl} \sum_{c \in C^I \cup C^B} l_i^c C_k^c f_l^c + \sum_{\beta \in B} C_i^\beta (\kappa_j^\beta + e_{jkl} C_k^\beta f_l^\beta) = 0 \tag{A17}$$

The continuum equivalent of the third term in Eqn (A17) is

$$\int_B x_i n_m (\mu_{mj} + e_{jkl} x_k \sigma_{ml}) \mathrm{d}B = \int_V \frac{\partial [x_i (\mu_{mj} + e_{jkl} x_k \sigma_{ml})]}{\partial x_m} \mathrm{d}V \tag{A18}$$

after using the divergence theorem. Recalling the continuum force and moment equilibrium equations (A2) and (A10–A12), the averaged couple stress tensor $\langle \mu_{ij} \rangle$ is given by

$$\langle \mu_{ij} \rangle = \frac{1}{V} \int_V \mu_{ij} \, \mathrm{d}V = \frac{1}{V} \left[\sum_{c \in C} l_i^c \kappa_j^c + e_{jkl} \left(\sum_{c \in V} C_k^c l_i^c f_l^c - \int_V x_k \sigma_{il} \, \mathrm{d}V \right) \right] \tag{A19}$$

Using relation (11.31) for the equivalence of a sum over contacts with a volume integral, expression (A15) for the averaged stress tensor can be written as

$$\int_V \sigma_{ij}\,\mathrm{d}V = \int_V m_V(\mathbf{x})\langle l_i f_j(\mathbf{x})\rangle \mathrm{d}V \tag{A20}$$

Since this relation holds for any volume V, it follows that

$$\sigma_{ij}(\mathbf{x}) = m_V(x)\langle l_i f_j(\mathbf{x})\rangle \tag{A21}$$

Similarly one obtains

$$\sum_{c\in V} C_k^c l_i^c f_l^c = \int_V m_V(x) x_k \langle l_i f_j(x)\rangle \mathrm{d}V = \int_V x_k \sigma_{ij}(x)\mathrm{d}V \tag{A22}$$

by using Eqns (A21) and (11.31). Hence the last two terms in Eqn (A19) cancel each other, the micromechanics expression for the averaged couple stress tensor becomes

$$\langle \mu_{ij}\rangle = \frac{1}{V}\int_V \mu_{ij}\,\mathrm{d}V = \frac{1}{V}\sum_{c\in C} l_i^c \kappa_j^c = \frac{1}{V}\int_V m_V(x)\langle l_i \kappa_j(x)\rangle \mathrm{d}V \tag{A23}$$

Since this relation holds for any volume V, it follows that

$$\mu_{ij} = m_V(x)\langle l_i \kappa_j(x)\rangle \tag{A24}$$

Recalling the moment equilibrium equation (11.35), an alternative form for the second equation in (11.35), by taking the moments with respect to the particle centers, may be written as

$$\sum_q \kappa_j^{pq} + \kappa_j^{p\beta} + e_{jkl}\sum_q r_k^{pq} f_l^{pq} + e_{jkl} r_k^{p\beta} f_l^{p\beta} = 0 \tag{A25}$$

which uses the relative co-ordinates. Multiplication of this equation by 1 and then summing it over all particles, one obtains

$$\sum_p \sum_q (\kappa_j^{pq} + e_{jkl} r_k^{pq} f_l^{pq}) + \sum_p (\kappa_j^{p\beta} + e_{jkl} r_k^{p\beta} f_l^{p\beta}) = 0 \tag{A26}$$

Using the same arguments as before, the first term of the double summation equals zero since each contact between particles p and q will contribute $\kappa_j^{pq} + \kappa_j^{qp} = \kappa_j^{pq} - \kappa_j^{qp} = 0$. Since $l_i^{pq} = r_i^{pq} - r_i^{qp}$ and $f_j^{qp} = -f_j^{pq}$, the contribution of each contact in the second term in the double summation becomes

$$e_{jkl}(r_k^{pq} f_l^{pq} + r_k^{qp} f_l^{qp}) = e_{jkl}(r_k^{pq} - r_k^{qp}) f_l^{pq} = e_{jkl} l_i^{pq} f_l^{pq} \tag{A27}$$

and the total sum can be expressed simply as $e_{jkl}\sum_{c\in C^I} l_k^c f_l^c$. Using the definition of the boundary edge (or branch) vector l_i^β (cf. Section 11.5.3), the contribution of the contact forces in the single summation in Eqn (A26) is $e_{jkl}\sum_{c\in C^B} l_k^c f_l^c$. The moment equilibrium equation (A26) is simplified as

$$\sum_{\beta=B} \kappa_j^\beta + e_{jkl}\sum_{c\in C^I\cup C^B} l_k^c f_l^c = 0 \tag{A28}$$

SUBJECT INDEX